COMPUTER GRAPHICS
THROUGH OPENGL®
From Theory to Experiments

Chapman & Hall/CRC Computer Graphics, Geometric Modeling, and Animation Series

Series Editor

Brian A. Barsky

Professor of Computer Science and Vision Science
Affiliate Professor of Optometry
University of California, Berkeley

Aims and Scope

The computer graphics, geometric modeling, and animation series aims to cover a wide range of topics from underlying concepts and applications to the latest research advances. The series will comprise a broad spectrum of textbooks, reference books, and handbooks, from gentle introductions for newcomers to the field to advanced books for specialists that present the latest research results. The advanced books will describe the latest discoveries, including new mathematical methods and computational techniques, to facilitate researchers with their quest to advance the field. The scope of the series includes titles in the areas of 2D and 3D computer graphics, computer animation, rendering, geometric modeling, visualization, image processing, computational photography, video games, and more.

Published Titles

An Integrated Introduction to Computer Graphics and Geometric Modeling
Ronald Goldman

Computer Graphics Through OpenGL®: From Theory to Experiments
Sumanta Guha

Chapman & Hall/CRC Computer Graphics, Geometric Modeling, and Animation Series

COMPUTER GRAPHICS
THROUGH OPENGL®
From Theory to Experiments

Sumanta Guha

CRC Press
Taylor & Francis Group
Boca Raton London New York

CRC Press is an imprint of the
Taylor & Francis Group an **informa** business

A CHAPMAN & HALL BOOK

Cover Designed by Somying Pongpimol.

Chapman & Hall/CRC
Taylor & Francis Group
6000 Broken Sound Parkway NW, Suite 300
Boca Raton, FL 33487-2742

© 2011 by Taylor and Francis Group, LLC
Chapman & Hall/CRC is an imprint of Taylor & Francis Group, an Informa business

No claim to original U.S. Government works

Printed in India by Replika Press Pvt. Ltd.
10 9 8 7 6 5 4 3 2

International Standard Book Number: 978-1-4398-4620-9 (Hardback)

Library of Congress Cataloging-in-Publication Data

Guha, Sumanta.
　Computer graphics through OpenGL : from theory to experiments / author, Sumanta Guha.
　　p. cm. -- (Chapman & Hall/CRC computer graphics, geometric modeling, and animation series)
　Includes bibliographical references and index.
　ISBN 978-1-4398-4620-9 (hardcover : alk. paper)
　1. Computer graphics. 2. OpenGL. I. Title. II. Series.

　T385.G85 2011
　006.6'6--dc22   2010030217

Visit the Taylor & Francis Web site at
http://www.taylorandfrancis.com

and the CRC Press Web site at
http://www.crcpress.com

To my late parents Santa and Utpal Chandra
and to Kamaladi

Contents

VI   Lights, Camera, Equation   403

11   COLOR AND LIGHT   405

Foreword

Over the last five years, I have witnessed the development of this book as a labor of love for Sumanta Guha. I am pleased to invite him to publish his manuscript in my new *Computer Graphics, Geometric Modeling, and Animation Series* that I am editing for Chapman & Hall/CRC Press.

Sumanta has crafted *Computer Graphics Through OpenGL: From Theory to Experiments* based on his notes for a course he has taught, first at the University of Wisconsin-Milwaukee and then at the Asian Institute of Technology, into the 800-page textbook you are reading today.

What distinguishes this book from other computer graphics texts is the experimental approach of explaining theory through programs, providing a balanced treatment of computer graphics theory and programming. The book enables students to "learn by doing" by running code themselves through a hands-on approach with theory and practice intertwined.

Computer Graphics Through OpenGL: From Theory to Experiments provides 140 programs with 200 experiments based on them, and over 600 exercises, 100 worked examples, and 600 color illustrations, and there is a comprehensive website. The book offers an engaging and fun approach to the study of computer graphics.

Brian A. Barsky
Professor of Computer Science and Vision Science
University of California, Berkeley

July 2010

Preface

When in 1996 the chair of our computer science department at the University of Wisconsin-Milwaukee was looking for someone to modernize the undergrad computer graphics course and base it on a new lab stocked with SGI O2s, he didn't have a faculty member specialized in the area to whom to turn. However, he figured that computational geometry, my research interest, was close enough. I was it, though I had never taken a computer graphics course in my life (you don't say no to the chair, especially so close to tenure).

So began my self-study of CG. It turned out to be probably the most fun new thing that I ever learned. Since finishing my Ph.D. in parallel algorithms for geometric problems, I had been primarily a theoretician. CG opened my eyes to the importance of implementation. It wasn't enough just to understand concepts on paper; they had to be translated into code and *seen* to be true. And, the OpenGL API was a joy to this end.

Worried about messing up, I spent hours writing out detailed notes, slides and demo programs for each lecture, not just for the class, but for myself to make sure that I knew what I was talking about. In a few years, these grew to the point of being complete enough that I could dispense with textbooks: the only required material for my CG course was a set of my notes and the red book (the OpenGL programming guide).

After moving to the Asian Institute of Technology in Thailand in 2002, I began the project of turning the notes into a book in my spare time. This turned out to be a much longer undertaking than I thought it would be at first. Class notes can lean on the instructor, but the book had to be a self-contained introduction to CG. Around 300 pages grew to more than 800 over the next few years and thousands of hours of research, writing, revising and coding. It was arduous, but I actually had a blast, the latter probably for the one reason that it was CG.

CG is so stimulating because theory and practice play off one another constantly. It's hard to do one without the other. You really can't claim to have understood, say, how transformations are composed or the part material reflectances play in the lighting model until you are capable of demonstrating these on the screen. Conversely, you are not going to be particularly efficient in coding animation or lighting a scene until you understand their underlying principles.

In fact, that is where the title of the book came from, as well as its spirit. It's all about going from theory to experiments – that's code – and back.

A few final words before we get to specifics: CG is fun! Boy, is it fun! It's lobster-for-lunch, evening-on-the-beach, dancing-till-dawn fun. Yes, of course, there's hard work too. And I make no attempt to hide this. But, from my own experience self-studying, and from teaching intro CG for over a decade, I know very well where the going gets rough for newcomers, and do my best to help them through those stretches. Still, at the end of the day, after all the sweat, there should be joy. You should see this as you read this book.

About the Book

This is an introductory textbook on computer graphics with an equal emphasis on theory and practice. After the first fourteen chapters – the undergraduate core of the book – the reader should have a good grasp of the concepts underlying 3D computer graphics, as well as an ability to code fairly sophisticated 3D scenes and animation, including games and movies. The remaining six chapters, though advanced, but still mainstream, could be selected topics for an undergraduate course or part of a second course.

The programming language used is C++, with OpenGL as the graphics API, which means calls are made to the OpenGL library from C++ programs. This book serves as an introduction to OpenGL as well (including version 2.0 and the GLSL).

The book has been written to be used both as a textbook for a first college course, as well as for self-study.

Specs

This book comes with approximately 140 programs, 200 experiments based on these programs, 600 exercises, including programming exercises and projects, 100 worked examples, and 600 color illustrations. The book was typeset using LaTeX and figures drawn in Adobe Illustrator.

Pedagogical Style

Code and theory have been intertwined as far as possible in what may be called a discuss-experiment-repeat loop: often, following a theoretical discussion, the reader is asked to perform validating experiments (run code, that is); sometimes, too, the other way around, an experiment is followed by an explanation of what is observed.

Needless to say, I am not a believer in an API-free approach to teaching CG, where the focus is on principles only, with no programming practice. Undergrads, typically, love to code and make things happen. And, OpenGL is so well-designed, and the learning curve short enough, that there is little justification to denying the new student the pleasure of creating scenes, movies and games, not to mention the pride of achievement. Moreover, OpenGL is supported on every OS platform and by drivers for almost every graphics card on the market, so it's at hand for anyone who owns a computer.

It seems wasteful, as well, not to leverage the way code and theory reinforce one another in CG when teaching the subject.

Note to student: Our pedagogical style means that for most parts of the book you want a computer handy to run experiments. So, unfortunately, curling up with it at night might not be a good option (unless you sleep with a laptop).

Note to instructor: Lectures on most topics – both of the theory and programming practice – are best based around the book's experiments, as well as those you develop yourself. The *Experimenter* teaching resource makes this convenient. Slides, otherwise, are rarely necessary.

Target Audience

(1) Students in a first university CG course, typically offered by a CS department at a junior/senior level (though, often, graduate students can take it for credit). Such a course could be based on the first fourteen chapters – the undergrad core of the book – with selections from the remaining six.

(2) Students in a second or advanced CG course, who may use the book as preparation or reference, depending on the goals. For example, the book would be a useful reference for a study of 3D design – particularly, Bézier, B-spline and NURBS theory – and of projective transformations and their applications to CG.

(3) Students in a non-traditional setting, e.g., studying alone or in a short course or an on-line program. The author has tried to be especially considerate of the reader on her own.

(4) Professional programmers, to use the book as a reference.

Prerequisites

Zero knowledge of computer graphics is presumed. However, the student is expected to know the following:

(1) Basic C++ programming. There is no need to be an expert programmer. The C++ program serves mainly as an environment for

the OpenGL calls, so there's rarely need for fancy footwork in the C++ part itself.

(2) Basic math. This includes coordinate geometry, trigonometry and linear algebra, all at college first-course level (or, even strong high school in some cases). For intended readers of the book who may be unsure of their math preparation, we have a self-test in Appendix C, with solutions in Appendix D. The test should tell exactly how ready you are and where the weaknesses are.

Resources

The following are available through the book's website `www.sumantaguha.com`:

(1) Program source code – which runs on Windows, Mac OS and Linux platforms.

(2) Guide to installing OpenGL and running the programs.

(3) Multiplatform *Experimenter* software to help run the experiments – which is a pdf file containing all the experiments, each being clickable to bring up the related program and, in a Windows environment, the workspace as well. (This is only an aid and not mandatory – the program for each experiment is stand-alone and runs independently.)

(4) Book figures in pdf format.

(5) Instructor's manual with solutions to 100 problems from Chapters 1-14 (only for instructors who have adopted this textbook).

(6) Contributory resource bank with homework and examination questions, experiments and other teaching and learning aids.

(7) Discussion forum for interaction between users of the book, including the author.

(8) Other resources as they are developed.

Organization and Course Development

Capsule Chapter Descriptions

Part I: Hello World

Chapter 1: An Invitation to Computer Graphics
A non-technical introduction to the field of computer graphics.

Chapter 2: On to OpenGL and 3D Computer Graphics

Begins the technical part of the book. It introduces OpenGL and the fundamental principles of 3D CG.

Part II: Tricks of the Trade

Chapter 3: An OpenGL Toolbox

Describes a collection of OpenGL programming devices, including mouse and key interaction, creation of pop-up menus, text drawing and the programming of multiple viewports, among others.

Part III: Movers and Shapers

Chapter 4: Transformation, Animation and Viewing

Describes the fundamentals of animation and the use of the virtual camera, both theory and programming. It also describes how to code user interactivity through object selection. This chapter lays the foundations for game and movie programming.

Chapter 5: Inside Animation: The Theory of Transformations

Focuses on the theory underlying animation, particularly linear and affine transformations in 3D.

Chapter 6: Advanced Animation Techniques

Describes frustum culling, as well as orienting animation using both Euler angles and quaternions, which are techniques essential to programming games and busy scenes.

Part IV: Geometry for the Home Office

Chapter 7: Convexity and Interpolation

Explains the theory of convexity and the role it plays in interpolation, which is the procedure of spreading material properties from the vertices of a primitive, such as a line segment or triangle, to its interior.

Chapter 8: Triangulation

Describes how and why complex objects should be split into triangles for efficient rendering.

Chapter 9: Orientation

Describes how the orientation of a primitive is used to determine the side of it that the camera sees, and the importance of consistently orienting a collection of primitives which make up a single object.

Part V: Making Things Up

Chapter 10: Modeling in 3D Space

Systematizes the principles of modeling both curves and surfaces, including

Bézier and fractal. This chapter is foundational for scene design.

Part VI: Lights, Camera, Equation

Chapter 11: Color and Light

This foundational chapter establishes the theory of light and material color, the interaction between the two, and describes how to program light and color into 3D scenes.

Chapter 12: Textures

Explains the theory of texturing and how to apply textures to objects.

Chapter 13: Special Visual Techniques

Describes a set of special techniques to enhance the visual quality of a scene, including, among others, blending, billboarding, environment mapping and stencil buffer methods.

Part VII: Pixels, Pixels, Everywhere

Chapter 14: Raster Algorithms

Describes low-level rendering algorithms to determine the set of pixels on the screen corresponding to a line segment or a polygon.

Part VIII: Anatomy of Curves and Surfaces

Chapter 15: Bézier

Describes the theory and programming of Bézier primitives, including curves and surfaces.

Chapter 16: B-Spline

Describes the theory and programming of (polynomial) B-spline primitives, including curves and surfaces.

Chapter 17: Hermite

Introduces the basics of Hermite curves and surfaces.

Part IX: The Projective Advantage

Chapter 18: Applications of Projective Spaces

Uses the theory of projective spaces to deduce the projection transformation in the graphics pipeline, as well as introduce rational Bézier and B-spline, particularly NURBS, theory and practice.

Part X: The Time is Pipe

Chapter 19: Fixed-Functionality Pipelines

Gives a detailed view of the synthetic-camera and ray tracing graphics pipelines and introduces radiosity. Also explains the use of the BSP (binary space partitioning) tree in the pipeline.

Chapter 20: Programmable Pipelines

Introduces the programmable graphics pipeline, particularly with use of the OpenGL GLSL.

Appendix A: Projective Spaces and Transformations
A CG-oriented introduction to the mathematics of projective spaces and transformations. It provides a complete theoretical background for Chapter 18 on applications of projective spaces.

Appendix B: Installing OpenGL and Running Code
How to install OpenGL on Windows, Linux and Mac OS platforms and run the book's code.

Appendix C: Math Self-Test
A self-test to assess math readiness for intended readers.

Appendix D: Math Self-Test Solutions
Solutions for the preceding self-test.

Suggested Course Outlines

(1) Undergraduate first CG course:

This course should be based on Chapters 1-14, though full coverage might be ambitious for one semester. Instructors should pick topics to emphasize or skip, depending on their goals for the course and the chapter dependence chart below.

For example, for more practice and less theory, a possible sequence would be $1 \rightarrow 2 \rightarrow 3 \rightarrow 4 \rightarrow 6$ (only frustum culling) $\rightarrow 7 \rightarrow 8 \rightarrow 9 \rightarrow 10$ (skip curve/surface theory) $\rightarrow 11 \rightarrow 12 \rightarrow 13 \rightarrow 14$.

Section 19.1 is a capstone section piecing together a synthetic-camera pipeline from end to end, based almost entirely on material from Chapters 2, 4, 5, 10, and 14; however, it assumes, as well, the projection transformation to convert a viewing frustum to the canonical box, which is deduced in Chapter 18, but might be taken for granted at a UG level.

Note to instructor: The most effective teaching method with this book is to use a "discuss-experiment-repeat" style. In other words, base discussion around experiments – both from the book and those that surely you will develop yourself. Teach in a lab setting or, at least, where students can use their notebooks, so they can be involved in the experiments. Minimize use of slides except, possibly, to show book figures.

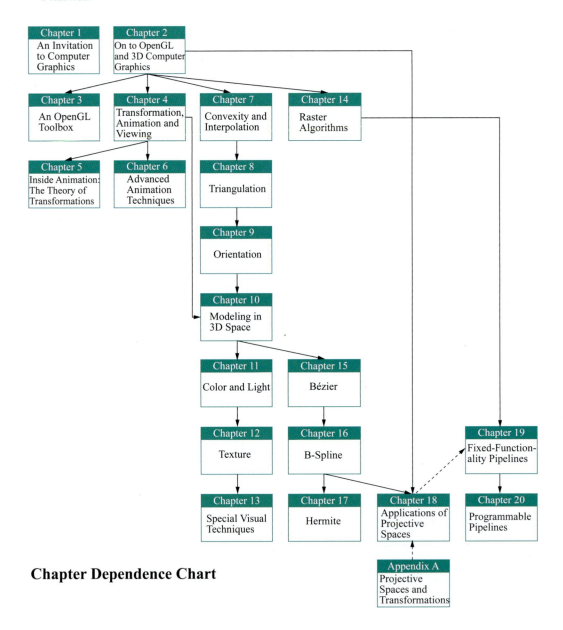

Chapter Dependence Chart

(2) Advanced CG courses:

This book could serve as a reference for a study of 3D design – particularly, Bézier (Chapter 15), B-spline (Chapter 16) and rational Bézier and NURBS theory (Chapter 18) – and of projective transformations and their applications (Appendix A and Chapter 18). From a practical point of view, Chapter 20 introduces the second generation of OpenGL and the GLSL.

(3) Self-study:

A recommended first pass for an introduction to CG would be 1 → 2 → 4 → 7 → 8 → 9 → 10 (skip curve/surface theory) → 11 → 12.

A second pass could pick up the missing pieces from Chapters 1-14 according to taste and need. The material in Chapters 15-20 is for those who want to dig deeper.

Note to the aspiring professional: To program B-splines you will need Chapter 16 and, for NURBS, Chapter 18; the second generations of OpenGL and the GLSL are introduced in Chapter 20.

Code

All the book's programs are in C++ with OpenGL and can be run on Windows, Mac OS and Linux platforms. They can be downloaded from www.sumantaguha.com, where there is also a guide to installing OpenGL and running the programs on the three different platforms.

Acknowledgments

I owe a lot to many people, most of all students whom I have had the privilege of teaching in my CG classes over the years at UW-Milwaukee and the Asian Institute of Technology.

I thank Tarun Mukherjee at Jadavpur University for being a constant source of inspiration, not to mention help with various technical questions.

I thank KV, Ichiro Suzuki, Glenn Wardius, Mahesh Kumar, Le Phu Binh, Maria Sell and, especially, Paul McNally, for their support at UWM, where I began to teach CG and learn OpenGL.

I am grateful to my colleagues and the staff and students at AIT for such a pleasant environment, which allowed me to combine my teaching and research commitments with the writing of a book.

Particular thanks at AIT to Vu Dinh Van, Nguyen Duc Cong Song, Ahmed Waliullah Kazi, Hameedullah Kazi, Long Hoang, Songphon Klabwong, Robin Chanda, Sutipong Kiatpanichgij, Samitha Kumara, Somchok Sakjiraphong, Pyae Phyo Myint Soe, K. Siriporn and K. Tong.

I am grateful to Kumpee Teeravech, Kanit Tangkathach and Thanapoom Veeranitinun, students in my CG course at AIT, for allowing me to use programs that they wrote. I thank Somying Pongpimol for many of the Illustrator drawings.

I am especially grateful to Brian Barsky for encouraging me to persevere after seeing an early and awkward draft a few years ago, and subsequently inviting the book to his series. I want to acknowledge the production team at Taylor & Francis who went out of their way for this book. Particularly, I want to thank my editor Randi Cohen who has been simply tremendous to work with. I really appreciate her making so stress-free the "business" of publishing.

I am grateful to the numerous reviewers whose comments helped immeasurably to improve the manuscript. I acknowledge the many persons and businesses who were kind enough to allow me to include images to which they own copyrights.

On a personal note, I express my deep gratitude to Dr. Anupam De for keeping Kamaladi healthy enough the last few years that I could concentrate on the book.

By far my biggest debt of gratitude is to my student and friend Chansophea Chuon. Chansophea developed the LaTeX style sheet, supervised the drawings while doing many of the Illustrator figures himself, laid out the manuscript, developed the multi-platform program template, designed the *Experimenter* software to help run the book experiments, all the while putting out countless fires as they happened. Chansophea's layout and illustrations, in particular, transformed a rather dowdy set of notes into a handsome four-color textbook. There is no doubt that, without Chansophea working shoulder to shoulder with me over the past year, this book would not have been completed.

Finally, I must say that had I not had the opportunity to study computer science in the US and teach there, I would never have reached a position where I could even contemplate writing a textbook. It's true, too, that had I not moved to Thailand, this book would never have begun to be written. This is an enchanting country with a strangely liberating and lightening effect – to which thousands of expats can attest – that encourages one to express oneself.

Website and Contact Information

The book's website is at `www.sumantaguha.com`. In addition to various other resources, users of the book will find there a forum to interact with each other and the author. The author welcomes feedback, corrections and suggestions for improvement. These can be posted to the forum or sent to the author at `sg@sumantaguha.com`.

About the Author

Sumanta Guha obtained a Ph.D. in mathematics from the Indian Statistical Institute, Kolkata, in 1987. From 1984 to 1987 he taught mathematics at Jadavpur University in Kolkata. He left in 1987 to study computer science at the University of Michigan in Ann Arbor, where he obtained a Ph.D. in 1991. On graduating from Michigan he joined the computer science department of the University of Wisconsin-Milwaukee where he taught from 1991 to 2002. In 2002 he moved to the information management and computer science program of the Asian Institute of Technology in Thailand, where he is currently an associate professor. His research interests include computational geometry, computer graphics, computational topology, robotics and data mining.

Part I

Hello World

CHAPTER 1

An Invitation to Computer Graphics

C omputer graphics, or CG as it is often simply called, is the use of computers to generate images. This is as opposed to, say, the mechanical capture of images of real-world objects, which would be photography, or, say, the work of an artist with pencil and paper.

To not see some kind of computer-generated imagery (CGI) through your day, you would have to be on a deserted island. Images on the screen of the cell phone you probably check first thing on waking are digitally synthesized by a processor. Almost every frame on the TV showing the morning news has CGI in some part. If you commute, then the vehicle in which you travel to school or work likely communicates with its operator through multiple computer-managed console panels, displaying information ranging from fuel level to geographical location.

Figure 1.1: A cell phone, news opening graphics, car dashboard.

At work, if at all you use a computer, then, of course, there you are sitting right at a fountainhead of computer graphics. And, CGI probably plays an even more important role in your recreational life. Even the most casual video games that amuse commuters heading home nowadays have sophisticated interactive 3D graphics. The web on which we spend so many hours a day is increasingly becoming a multimedia experience synthesizing animation, movie clips, CGI and sound.

Figure 1.2: A computer at work, handheld game player, AIT home page (used with permission of the Asian Institute of Technology).

When you watch a movie you are seeing a product from an industry, which together with the gaming industry, has the biggest relationship with CG of any other, not only as a consumer of the latest and greatest in technique, but also as a promoter, with hundreds of millions of dollars in investment, of cutting-edge research. A little blue elephant that grows into a mighty warrior, an eccentric mouse with a ribald sense of humor, not to mention a massive dinosaur looking so hungrily for food that you would think its species had never really become extinct more than fifty million years ago – to contemplate such achievements is to be in awe of the human imagination, as well as the ingenuity of the engineers and programmers who materialize these fantastical conceptions as palpable and believable digital presences.

Figure 1.3: Khan Kluay, the first 3D animated Thai movie (courtesy Kantana Animation), an anthropomorphic mouse, a massive (fortunately herbivorous) dinosaur.

Then there's the quiet CG that impacts our lives some would say even more profoundly than its more flamboyant manifestations. Doctors and surgeons practice their craft in simulated environments detailed to the tiniest capillaries. Commercial pilots put in hundreds of hours in a flight simulator before entering a real cockpit. (Flight simulators are a sentimental favorite in CG because they were the first killer app, drawing attention and investment dollars to the then nascent field in the sixties.) Automobiles, airplanes and almost any fairly complex manufactured object that we see around us are designed, fabricated and even put through regulatory tests as virtual entities – which exist entirely as a collection of bits perceptible only as an image on a monitor – gestating often for years before the first real prototype is ever built. Supercomputers implement extremely complex mathematical models of the weather, but their predictions have to be visualized in order to be meaningful to a human.

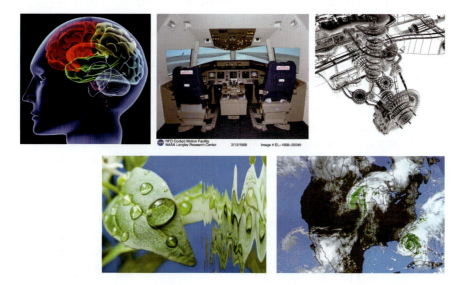

Figure 1.4: Clockwise from top left: Image of the human brain, flight simulator cockpit (from NASA), engine design, hurricane over Florida, water drop on a leaf.

Because its business is the creation of pictures, computer graphics has an immediate allure. But, it is a science as well, with intellectual challenges ranging from the routine to about as deep and hard as you please. Think of modeling a drop of water rolling off a leaf. There would be a fair amount of physics and, probably, a differential equation or two to solve on the way to getting just the mechanics of the rolling drop right, not to mention texturing the leaf, creating a translucent (and changing) shape for the drop, and determining illumination.

The field of computer graphics brings particular pleasure to students and practitioners alike because it's always about making something – just like sculpting or painting. Part by part you watch your creation come together and then even alive if it is animated. Aside from the aesthetic, there are more tangible rewards to be had, too. You would be hard pressed to name a sphere of social or scientific or industrial activity where computer-generated pictures don't have a role. Wherever it is that ultimately you want to be, medicine or fashion, rocket science or banking, weapons development or teaching yoga, sales and marketing or environmental modeling, your CG skills not only can make a difference, but make you a career.

1.1 Brief History of Computer Graphics

Although the term "computer graphics" itself was coined in 1960 by William Fetter, a designer at Boeing, to describe his own job, the field could be said to have first arrived with the publication in 1963 of Ivan Sutherland's Sketchpad program, as part of his Ph.D. thesis at MIT.

Sketchpad, as its name suggests, was a drawing program. Beyond the interactive drawing of primitives such as lines and circles and their manipulation – in particular, copying, moving and constraining – with use of the then recently invented light pen, Sketchpad had the first fully-functional graphical user interface (GUI) and the first algorithms for geometric operations such as clip and zoom. Interesting, as well, is that Sketchpad's innovation of an object-instance model to store data for geometric primitives foretold object-oriented programming. And, on the hardware side, the year 1963 saw also the invention by Douglas Engelbart at the Stanford Research Institute of the mouse, the humble device that even today carries so much of GUI on its skinny shoulders.

Figure 1.5: Ivan Sutherland operating Sketchpad on a TX-2 (courtesy of Ivan Sutherland), Douglas Engelbart's original mouse (courtesy of John Chuang).

Although Sketchpad ran on a clunky Lincoln TX-2 computer with only 64KB in memory and a bulky monochrome CRT monitor as its front-end, nevertheless, it thrust CG to the attention of the early computer researchers by showing what was possible. Subsequent advances through the sixties came thick and fast: raster algorithms, the implementation of parametric surfaces, hidden-surface algorithms and the representation of points by homogeneous coordinates, the latter crucially presaging the foundational role of projective geometry in 3D graphics, to name a few. Flight simulators were the killer app of the day and companies such as General Electric and Evans & Sutherland, co-founded by Douglas Evans and Ivan Sutherland, wrote simulators with real-time graphics.

Interestingly, the advent of flight simulators actually predated that of CG – at least Sutherland and his Sketchpad – by nearly two decades, when the US Navy began the funding of Project Whirlwind at MIT during the second World War for the purpose of creating simulators to train bomber crews. Those early devices had little graphics and consisted essentially of a simulated instrument panel reacting in real-time to control input from the pilots, but Project Whirlwind helped fund the talent and research environment at MIT which enabled Sutherland to create Sketchpad, launch computer graphics and, finally, complete the circle by establishing a company to make flight simulators.

The next decade, the seventies, brought the z-buffer for hidden surface removal, texture mapping, Phong's lighting model – all crucial components of the OpenGL API (Application Programming Interface) which we'll be using soon – as well as keyframe-based animation. Photorealistic rendering of animated movie keyframes almost invariably deploys ray tracers, which were born in the seventies too. Emblematic of the advances in 3D design was Martin Newell's 1975 Utah teapot, composed entirely of bicubic Bézier patches. The latter half of the decade also saw the Apple I and II personal computers make their debut, bringing CG for the first time to the mass market.

Figure 1.6: Utah teapot (from Wikimedia), Apple II Plus (courtesy of Steven Stengel), SIGGRAPH 2006 expo floor in Boston (courtesy of Jason Della Rocca).

From the academic point of view, particularly important were the establishment in 1969 of SIGGRAPH (Special Interest Group in Graphics) by the ACM (Association for Computing Machinery, the premier academic society for computers and computing) and, subsequently, the first annual SIGGRAPH conference in 1973. These two developments signaled the emergence of computer graphics as a major subdiscipline of computer science. The SIGGRAPH conference since then has become probably the foremost annual event in the CG world. In addition to being the most prestigious forum for research papers, it hosts a giant exhibition which attracts hundreds of companies, from software developers to book publishers, who set up booths to promote their wares and recruit talent.

Since the early eighties, CG, both software and hardware, began rapidly to assume the form we see today. The IBM PC, the Mac and the x86 chipsets all arrived, sparking off the race to become faster (processor), smaller (size), bigger (memory) and cheaper. As computers became consumer goods, the market for software spilled over from academia to individuals and businesses and exploded. Nintendo released Donkey Kong in 1981, its first successful arcade video game, and soon after Wavefront Technologies released its Preview software, used then to create opening graphics for television programs. Now, of course, Nintendo dominates the video games industry with the Wii console and Wavefront has morphed into Alias (owned by Autodesk) whose 3D graphics modeling package Maya is ubiquitous in the design world.

Figure 1.7: Donkey Kong arcade game (from Wikimedia), Maya screenshot of Scary Boris (courtesy of Sateesh Malla at www.sateeshmalla.com), 2D characters on the left versus 3D on the right (© Mediafreaks Cartoon Pte. Ltd., 2006. All rights reserved.).

3D graphics, in particular, began to displace its plainer 2D sister through the nineties as hardware increasingly became capable of supporting the real-time rendering needs of 3D models. The only difference as such between 2D and 3D graphics is that models in the latter are created in a (virtual) 3D world, geometrically the same as the real world, and then projected onto the viewing screen, while all drawings in 2D graphics are on a flat plane. Models drawn in 3D are more realistic, but they are more complex as well; moreover, the projection step, which is non-existent for 2D graphics, is computation-intensive too. Graphics cards, manufactured by companies such as ATI and Nvidia, which not only manage the image output to the display unit, but have additional hardware support for rendering of 3D primitives, are now inexpensive enough that desktops and even notebooks can run high-end 3D applications. How well they run 3D games is often, in fact, used to benchmark personal computers.

Through the nineties, as well, the use of 3D effects in movies became pervasive. The Terminator and Star Wars series, and Jurassic Park, were among the early movies to set the standard for CGI. Toy Story from Pixar, released in 1995, has special importance in the history of 3D CGI as the first movie to be entirely computer-generated – no scene was ever pondered by the director through a glass lens, nor any recorded on a photographic reel!

Figure 1.8: Toy Story movie (© 1995 Disney Enterprises, Inc.), Quake 1 game (courtesy of Quake ® © 1996 id Software LLC, a ZeniMax Media Company, All Rights Reserved).

Quake, released in 1996, and the first of the hugely popular Quake series of games, was the first fully 3D game.

Another landmark from the nineties of particular relevance to us was the release in 1992 of OpenGL, the open-standard cross-platform and cross-language 3D graphics API, by Silicon Graphics. OpenGL is actually a library of calls to perform 3D tasks, which can be accessed from programs written in various languages and running over various operating systems. That OpenGL was high-level (in that it frees the applications programmer from having to care about such low-level tasks as representing primitives like lines and triangles in the raster, or rendering them to the window) and easy to use (much more so than its predecessor 3D graphics API, PHIGS, standing for Programmer's Hierarchical Interactive Graphics System) first brought 3D graphics programming to the "masses". What till then had been the realm of a specialist was now open to a casual programmer following a fairly short learning curve. Since its release OpenGL has been rapidly adopted throughout academia and industry. It's only among game developers that Microsoft's proprietary 3D API, Direct3D, which came soon after OpenGL and is optimized for the Windows platform, is more popular.

Figure 1.9: OpenGL and OpenGL ES logos (used with permission of Khronos).

The story of the past decade has been one of steady progress, rather than spectacular innovations in CG. Hardware continues to get faster, better, smaller and cheaper, continually pushing erstwhile high-end software downmarket, and raising the bar for new products. The almost complete displacement of CRT monitors by LCD and the emergence of high-definition television are familiar consequences of hardware evolution. Of likely even greater economic impact is the migration of sophisticated software applications – ranging from web browsers to 3D games – to handheld devices like smartphones, on the back of small yet powerful processors. CG has now been untethered from immobile devices and placed into the hands and pockets of consumers. In fact, a lightweight subset of OpenGL called OpenGL ES – ES for Embedded Systems – released by the Khronos Group in 2003, is now the most popular API for programming 3D graphics on small devices.

1.2 Overview of a Graphics System

The operation of a typical graphics systems can be split into a three-part sequence:

$$\text{Input} \longrightarrow \text{Processing} \longrightarrow \text{Output}$$

The simplest example of this is when you click on a thumbnail image in, say, YouTube, and a video clip pops up and begins to play. Your click is the input. Your computer then reacts to this input by processing, which involves downloading the movie file and running it through the Adobe Flash Player, which in turn outputs video frames to your monitor.

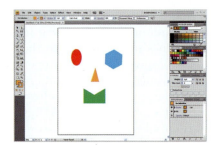

Figure 1.10: YouTube and Adobe Illustrator screenshots.

Graphics systems can be of two types, non-interactive and interactive. The playing of a YouTube clip is an example of a non-interactive one: beyond the first click to get the movie started you have little further say over the output process, other than maybe to stop it or manipulate the window. On the other hand, if, say, you are using a package like Adobe Illustrator, then the output – what you have drawn – changes in real-time in response to the input that you provide by pressing keys and moving and clicking the mouse; e.g., you can move objects, color them, create shapes and so on. In an interactive system, output continuously reacts to input via the processor.

Input/output devices (or I/O devices, or peripheral devices, as they are also called) are of particular importance in interactive systems because they determine the scope of the interaction. For example, an input device that functions like a steering wheel would be essential to a video game to race cars; simulating flight through a virtual 3D environment, on the other hand, needs something like a joystick that is used to maneuver an aircraft.

Because it is, in fact, interactive computer graphics – theory and programming – which we'll be studying the next nineteen chapters, let's quickly survey first the most common I/O devices found in graphics systems nowadays. As for the processors that may come between the 'I' and the 'O', from the point of view of CG, essentially, they are just boxes to be coded in order to obtain the desired input-to-output mapping. For the sake of completeness, though, here's a list of the important ones, all somewhat different one from the other in the context of CG (Figure 1.11 pictures them):

Computer: As far as we are concerned, this category includes PC's, workstations, servers and the like.

Portable computer: This, of course, is simply a small and light computer with a built-in display, keyboard and pointing device. Because of the size constraint, limited power supply and also the lack of space for a large cooling fan, CPU's and graphics cards in portable computers tend to underperform their desktop counterparts. Software writers need to take this into account, especially for graphics-intensive applications.

Figure 1.11: Processing devices clockwise from top: laptop, smartphone, game console (used with permission from Microsoft), computer box.

Handheld device: The size-weight constraint on this class of devices – of which the mobile phone is the most visible example – is even more severe than for portable computers. Handhelds are expected to travel in handbags and pockets. Low-end handhelds often have no peripheral other than a limited keypad, while higher-end ones may come equipped with a full QWERTY keypad or even a touchscreen. In addition to the small RAM and anemic CPU, another consideration to keep in mind for graphics developers for handhelds is the limited real estate of the display: busy scenes tend to become "chaotic" on a handheld.

Game consoles: All stops are off for programming these devices. Running graphics-intensive applications at blinding speeds is what these machines were born to do.

1.2.1 Input Devices

The following is by no means a complete list of input devices, but it does cover the ones we are most likely to encounter in everyday use. The devices are all pictured in Figure 1.12, ringing the processing devices in the middle, and our list goes clockwise starting from the top rightmost.

Keyboard: This device is a mandatory peripheral for any computer. Its alphanumeric keys, evidently derived from the traditional typewriter, are used to enter text strings, while additional keys, such as the arrow and function keys, perform special actions.

Mouse: This is an example of a *pointing device* which inputs spatial data to the computer. As the mouse is moved by the user's hand on a flat surface, a mechanical ball or optical sensor at its base signals the amount of movement to the computer, which correspondingly moves a cursor on the screen. Effectively, then, the user determines the location of the cursor in a 2D space. Strictly speaking, a mouse is more than just a pointing device if it has buttons, as most do, which can be clicked to give binary input.

Touchpad: Another 2D pointing device, particularly common on portable computers, the touchpad is a small rectangular area embedded with electronic sensors to determine the position of a touching finger or stylus. Movement of the finger or stylus is echoed by movement of the cursor.

Pointing stick: Yet another 2D pointing device common on portable computers, the pointing stick is, typically, a rubber peg located between the 'G', 'H' and 'B' keys, which moves the cursor in response to pressure applied with a finger.

Trackball: This is essentially an upside-down mouse, with a socket containing a ball which the user rotates using her hand, resulting in cursor motion.

Spaceball: This is a pointing device with six degrees of freedom versus the two of an ordinary mouse. It is used in special applications such as manipulating a camera in a 3D scene: not only is the camera moved, but

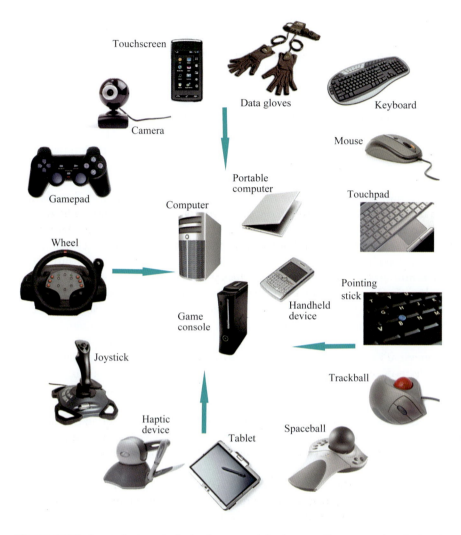

Figure 1.12: Input devices clockwise from top right (surrounding processing devices in the middle): keyboard, mouse, touchpad, pointing stick (courtesy of Long Zheng), trackball, spaceball (courtesy of Logitech), tablet, haptic device (© SensAble Technologies, Inc.), joystick, wheel, gamepad, webcam, touchscreen, data gloves (courtesy of www.5dt.com).

also rotated, affording it multiple degrees of freedom, each of which the user controls. The spaceball itself consists of a pressure-sensitive ball which can distinguish different kinds of forces, including forward/backward, lateral and twist, and responds by moving and orienting the selected object.

Tablet: This is a *digitizing device* which has a surface embedded with sensors to pick up the successive coordinates of a stylus head as it travels over the surface (in effect converting the motion of the stylus into digital data). The user can write or draw on a tablet with the stylus, just as on paper with a pen, the output being displayed on the monitor. The monitor is usually separate, though, on devices like a tablet PC, the display and the sensing surface are the same.

Haptic device: This is a pointing device which gives physical feedback to the user based on the location of the cursor or, possibly, that of an object being moved along with the cursor. The easiest way to understand the functioning of a haptic device, if you have never used one, is to imagine a mouse with a mechanical ball which is (somehow) programmed to lock and stop rolling when the cursor reaches the side of the screen. The reaction the user then has is of that of the cursor running into a physical obstacle at the edge of the screen, though evidently it is moving in virtual space. The device depicted in Figure 1.12 is not a haptic mouse, of course, but is the one most commonly seen in HCI (human-computer interaction) labs. The three-link arm swiveling on a ball gives it six degrees of freedom.

Haptics has numerous applications, a couple of noteworthy ones being the teleoperation of robots (where the operator is given haptic feedback as she manipulates a robot in either a virtual or a remote real environment) and simulated surgery training in medicine (which is similar to training pilots on a flight simulator, except that surgery has the added component of tactile feedback, mostly absent in flying).

Joystick: This is an input device popular in video games and applications such as flight simulators. It originated from its namesake found in real aircraft cockpits. A joystick pivots around a fixed base, gaining thus two degrees of freedom, and usually has buttons which can be depressed to provide additional input. In a game or simulator setting a joystick is typically used to control an object traveling through space. Nowadays, high-end joysticks have embedded motors to provide haptic feedback to user motion, e.g., resistance as a plane is banked.

Wheel: This again is a specialized input device for games and simulators, obviously derived from the car steering wheel, and provides rotational input in an exactly similar manner, most often to a virtual automobile. Again, haptic feedback to give the user a sense of the vehicle's response, and even of the terrain over which it is traveling, is becoming increasingly popular.

Gamepad: This device is the standard controller for most modern game consoles (except the Wii). Standard features include action buttons operated usually with the right thumb and a cross-shaped directional controller with the left.

Camera: Although this input device needs no introduction, it's worth noting the increasingly sophisticated uses a peripheral camera is being put to with the help, e.g., of software to recognize faces, gestures and expressions.

Touchscreen: Increasingly popular as the interface of handheld devices such as smartphones, a touchscreen is a display which can accept input via touch. It is similar to touchpads and tablets in that it senses the location of a finger or stylus – one or the other is usually preferable based on the particular technology used to make the screen – on the display area. A common application of touchscreens is to eliminate the need for a physical keyboard by displaying a virtual one which responds to taps on the display.

Touchscreens often respond not only to the location of the touch, but also the motion of the touching object. For example, a flicking motion with a finger may cause a window to scroll. Multi-touch enhancements, now increasingly common, enable the device to respond to gestures with more than one finger, e.g., pinching and spreading with two fingers.

Data gloves: This device is used particularly in virtual reality environments which are programmed to react to the position of the gloves, the direction in which fingers are pointing, as well as to hand motion and gestures. The gloves themselves are wired to transmit not only their location, but also their configuration and orientation to the processor, so that the latter can display the environment accordingly. For example, an index finger pointing at a particular atom in a virtual-reality display of a molecule may cause that atom to zoom up to the viewer.

1.2.2 Output Devices

Again, the following list is not meant to be comprehensive, but, rather, representative of the most common output devices. We go clockwise around the outer ring of devices pictured in Figure 1.13 beginning with the rightmost.

CRT (cathode-ray tube) monitor: A CRT monitor has phosphors of the three primary colors – R(ed), G(reen) and B(lue) – located at each one of a rectangular array of pixels, called the raster. Additionally, it has three electron guns that each fires a beam at phosphors of one color. A mechanism to aim and control their intensities causes the beams to travel together, striking one pixel after another, row after row, exciting the RGB phosphors at each pixel to the values specified for it in the color buffer. Figure 1.14(a) shows the electron beams striking one pixel on a dog.

From the point of view of OpenGL and, indeed, most CG theory, what matters is that the pixels in a monitor are, in fact, arranged in a rectangular raster (as depicted in Figure 1.14(b)). For, this layout is the basis of the lowest-level CG algorithms, the so-called raster algorithms, which actually select and color the pixels to represent user-specified shapes such as lines and triangles on the monitor. Figure 1.14(b), for example, shows the rasterization

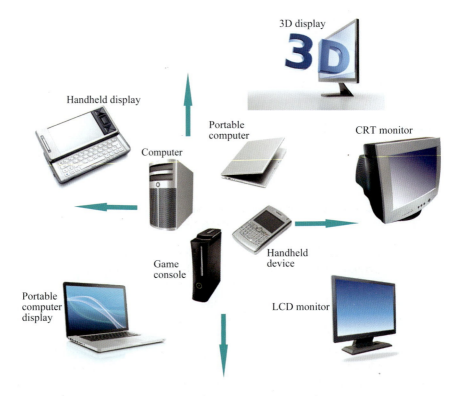

Figure 1.13: Output devices clockwise from the rightmost (surrounding processing devices in middle): CRT monitor, LCD monitor, notebook, mobile phone, 3D LCD monitor.

of a right-angled triangle (with terrible jaggies because of the low resolution).

The number of rows and columns of pixels in the raster determines the monitor's resolution. Typical for a CRT monitor is a resolution in the range of 1024×768 (which means 1024 columns and 768 rows). High-definition monitors (as needed, say, for high-definition TV, or HDTV as it's acronymed) have higher resolution, e.g., 1920×1080 is common.

Moreover, a memory location called the color buffer, either in the CPU or graphics card, contains, typically, 32 bits of data per raster pixel – 8 bits for each of RGB, and 8 for the alpha value (used in blending). It is the RGB values in the color buffer which determine the corresponding raster pixel's color intensities. The values in the color buffer are read by the raster – in other words, the raster is refreshed – at a rate called the monitor's refresh rate. Beyond this, the technology underlying the particular display device, no matter how primitive or how fancy, really matters little to the CG programmer.

LCD (liquid crystal display) monitor: Pixels in an LCD monitor each consist

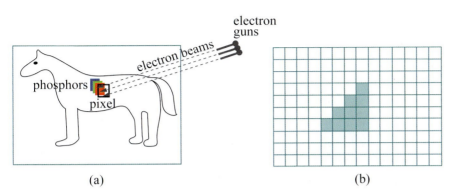

(a)   (b)

Figure 1.14: (a) Color CRT monitor with electron beams aimed at a pixel with phosphors of the 3 primaries (b) A raster of pixels showing a rasterized triangle.

of three subpixels made of liquid crystal molecules, which separately filter lights of the primary colors. The amount of light emerging through a subpixel is controlled by an electric charge whose intensity is determined by subpixel's corresponding value in the color buffer. The absence of electron guns allows LCD monitors to be made flat and thin – unlike CRT monitors – so they are one of the class of flat panel displays.

Technologies other than LCD, e.g., plasma and OLED (organic light emitting diode), are used as well in flat panel displays, though LCD is by far the most common one found with computers.

Again, for all practical purposes, the view to keep in mind of the LCD monitor, as of other flat panel displays, is as a rectangular raster of pixels whose RGB intensities are individually set by values in the computer's color buffer.

Portable computer display: This display is again a raster of pixels whose RGB values are read from a color buffer. The technology employed, typically, is TFT-LCD, a variant of LCD which uses thin film transistors to improve image quality.

Handheld display: Handheld displays, such as those on devices like mobile phones, commonly use the same TFT-LCD technology as portable computers. The resolution, though, is necessarily smaller, e.g., 320×480 would be in the ballpark of most.

3D display: Almost all 3D displays are based on the principle of stereoscopy, in which an illusion of depth is created by showing either eye images of the scene captured by one of two cameras slightly offset from one another. If the cameras have been positioned, respectively, at the left and right eye of a hypothetical viewer, then the illusion is authentic (see Figure 1.15). Once the scene has been recorded with two cameras, it is in ensuring that each eye sees frames only from one of them, called stereoscopic viewing, that there are primarily two competing technologies.

Figure 1.15: Dual cameras filming a motorbike for subsequent 3D viewing.

In the first, frames alternately from either camera are displayed on the monitor, a process called alternate frame sequencing. Simultaneously, the viewer wears LCD shutter glasses embedded with a polarizing filter which can be darkened with an electrical signal. The glasses are synchronized with the monitor's refresh rate via a link such as Bluetooth, either lens being alternately darkened with successive frames. Consequently, each eye sees images from only one of the two cameras, resulting in a stereoscopic effect. Typically, the frame rate is increased to 48 per second as well, so that both eyes experience a smooth-seeming 24 frames each second. The great advantage of LCD shutter glasses is that they can be used with any computer which has a monitor with a refresh rate fast enough to support alternate frame sequencing, as well as a graphics card with enough buffer space for two video streams. So with these glasses even a high-end home system would qualify to play 3D movies and games.

Polarized 3D glasses, on the other hand, are used to view two images, from either camera, projected simultaneously on the same screen through orthogonal polarizing filters. The lenses too contain orthogonal polarizing filters, each allowing through only light of like polarization. Consequently, either lens sees images from only one or other camera, engendering a stereoscopic view. Polarized 3D glasses are significantly less expensive than LCD shutter glasses and, moreover, require no synchronization with the monitor. However, the projection system is complicated and expensive and primarily used to equip theaters for 3D viewing.

OpenGL, the API we'll be using, is well-suited to making scenes and movies for 3D viewing because it allows one or more (virtual) cameras to be positioned arbitrarily.

1.3   Quick Preview of the Adventures Ahead

To round out this invitation to CG we want to show you three programs written by students in their first college 3D CG course, taught using a draft of this book. They were written in C++ with OpenGL.

So what exactly is OpenGL? You may have been wondering this. We said earlier in the section on CG history that OpenGL is a cross-platform 3D graphics API. It consists of a library of nearly 300 calls to perform 3D tasks, which can be accessed from programs written in various languages. Here's something concrete – an example snippet from a C++ environment to draw 10 red points:

```
glColor3f(1.0, 0.0, 0.0);
glBegin(GL_POINTS);
   for(int i = 0; i < 10; i++)
   {
      glVertex3i(i, 2*i, 0);
   }
glEnd();
```

The first function call declares the red drawing color, while the loop bracketed between the `glBegin(GL_POINTS)` and `glEnd()` calls draws a point at $(i, 2i, 0)$ in each of ten iterations. There are other calls in the OpenGL library to draw straight lines, triangles, create light sources, apply textures, move and rotate objects, maneuver the camera and do many other things – in fact, pretty much all one would need to create and animate a detailed and realistic 3D scene. To learn more you'll really have to start reading from the next chapter!

Getting back to the student programs, the code itself is not of importance at this time and would actually be a distraction. Instead, just running the programs and viewing the output will give an idea of what can be accomplished after a fairly short time (ranging from 3 weeks to 3 months for the different programs) by persons coming to CG with little more than a good grasp of C++ and some basic math. We'll get a feel as well for what goes into making 3D scenes.

Experiment 1.1. Open the folder `Invitation/Ellipsoid` in the `Code` directory and, hopefully, you'll then be able to run at least one of the two executables there for the `Ellipsoid` program – one for Windows and one for the Mac. The program draws an ellipsoid (an egg shape). The left of Figure 1.16 shows the initial screen. There's plenty of interactivity to try as well. Press any of the four arrow keys, as well as the page up and down keys, to change the shape of the ellipsoid, and 'x', 'X', 'y', 'Y', 'z' and 'Z' to turn it.

It's a simple object, but the three-dimensionality of it comes across rather nicely does it not? As with almost all surfaces that we'll be drawing ourselves, the ellipsoid is made up of triangles. To see these press the space bar to enter wireframe mode. Pressing space again restores the filled mode. The wireframe reveals the ellipsoid to be made of a mesh of triangles decorated with large points. A color gradient has apparently been applied toward the poles as well.

That's it. There's really not much more to this program: no lighting or blending or other effects you may have heard of as possible using OpenGL (the program was written just a few weeks into the semester). It's just a bunch of colored triangles and points laid out in 3D space. The magic is in those last two words: *3D space*. 3D modeling is all about making things in 3D space – not just on a flat plane – to create an illusion of depth, even when they are viewed on a flat screen. **End**

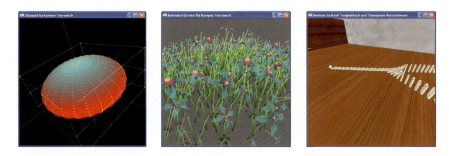

Figure 1.16: Screenshots of `Ellipsoid`, `AnimatedGarden` and `Dominos`.

Experiment 1.2. Our next program is animated. It creates a garden that grows and grows and grows. You will find executables in the folder `Invitation/AnimatedGarden` in the `Code` directory. Press enter to start the animation; enter again to stop it. The delete key restarts the animation, while the period (full stop) key toggles between the camera rotating and not. Again, the space key toggles between wireframe and filled. The middle of Figure 1.16 is a screenshot a few seconds into the animation.

As you can see from the wireframe, there's again a lot of triangles (in fact, the flowers might remind you of the ellipsoid from the previous program). The plant stems are thick lines and, if you look carefully, you'll spot points as well. The one special effect this program has that `Ellipsoid` did not is blending. **End**

Experiment 1.3. Our final program is a movie which shows a Rube Goldberg domino effect with "real" dominos. The executables are in the folder `Invitation/Dominos` in the `Code` directory. Simply press enter to start and stop the movie. The screenshot on the right of Figure 1.16 is from part way through.

This program has a bit of everything – textures, lighting, camera movement and, of course, a nicely choreographed animation sequence, among others. **End**

Acknowledgments: Kumpee Teeravech wrote the `Ellipsoid` and `Animated-Garden` programs. Kanit Tangkathach and Thanapoom Veeranitinunt wrote the `Dominos` program.

CHAPTER 2

On to OpenGL and 3D Computer Graphics

The primary goal for this chapter is to get acquainted with OpenGL and begin an exploration of computer graphics using this API (Application Programming Interface) as our vehicle of choice. We shall apply an experiment-discuss-repeat approach where we run code and ask questions of what is seen, acquiring thereby an understanding not only of the way the API functions, but of the underlying CG concepts as well. Particularly, we want to gain insight into the following two:

(a) The synthetic-camera model to record 3D scenes, which OpenGL implements.

(b) The approach of approximating curved objects, such as circles and spheres, with the help of straight and flat geometric primitives, such as straight line segments and triangles, which is fundamental to object design in computer graphics.

We begin in Section 2.1 with our first OpenGL program to draw a square, the computer graphics equivalent of "Hello World". Simple though it is, with a few careful experiments and their analysis, `square.cpp` yields a surprising amount of information through Sections 2.1-2.3 about orthographic projection, the fixed world coordinate system OpenGL sets up and how the so-called viewing box in which the programmer draws is specified in this system. We gain insight as well into the 3D-to-2D rendering process.

Adding code to `square.cpp` we see in Section 2.4 how parts of objects outside the viewing box are clipped off. Section 2.5 discusses OpenGL as a state machine. We have in this section as well our first glimpse of property values, such as color, initially specified at the vertices of a primitive, being interpolated throughout its interior.

Next is the very important Section 2.6 where all the drawing primitives of OpenGL are introduced. These are the parts at the application programmer's disposal with which to assemble objects from thumbtacks to spacecrafts.

The first use of straight primitives to approximate a curved object comes in Section 2.7: a curve (a circle) is drawn using straight line segments. To create more interesting and complex objects use must be made of OpenGL's famous three-dimensionality. This involves learning first in Section 2.8 about perspective projection and also hidden surface removal using the depth buffer.

After a bunch of drawing exercises in Section 2.9 for the reader to practice her newly-acquired skills, the topic of approximating curved objects is broached again in Section 2.10, this time to approximate a surface with triangles, rather than a curve with straight segments as in Section 2.7. We conclude with a summary, brief notes and suggestions for further reading in Section 2.11.

2.1 First Program

Experiment 2.1. Run `square.cpp`.

Note: See Appendix B for how to install OpenGL and run our programs on Windows, Linux and Mac OS platforms.

In the OpenGL window appears a black square over a white background, as shown in Figure 2.1 (where blue stands in for white to distinguish it from the paper). We are going to understand next how the square is drawn, and gain some insight as well into the workings behind the scene. **End**

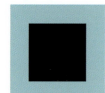

Figure 2.1: OpenGL window of `square.cpp` (blue pretending to be white).

Figure 2.2: The coordinate axes on the OpenGL window of `square.cpp`? *No.*

The following six statements in `square.cpp` create the square:

```
glBegin(GL_POLYGON);
    glVertex3f(20.0, 20.0, 0.0);
    glVertex3f(80.0, 20.0, 0.0);
    glVertex3f(80.0, 80.0, 0.0);
    glVertex3f(20.0, 80.0, 0.0);
glEnd();
```

The corners of the square are specified by the four vertex declaration statements between `glBegin(GL_POLYGON)` and `glEnd()`. Let's determine how the `glVertex3f()` statements correspond to corners of the square.

If, suppose, the vertices are specified in some coordinate system that is embedded in the OpenGL window – which certainly is plausible – and if we knew the axes of this system, the matter would be simple. For example, *if* the x-axis increased horizontally rightwards and the y-axis vertically downwards, as in Figure 2.2, then `glVertex3f(20.0, 20.0, 0.0)` would correspond to the upper-left corner of the square, `glVertex3f(80.0, 20.0, 0.0)` to the upper-right corner and so on.

However, even assuming that there do exist these invisible axes attached to the OpenGL window, how do we find out where they are or how they are oriented? One way is to "wiggle" the corners of the square! For example, change the first vertex declaration from `glVertex3f(20.0, 20.0, 0.0)` to `glVertex3f(30.0, 20.0, 0.0)` and observe which corner moves. Having determined in this way the correspondence of the corners with the vertex statements, we ask the reader to deduce the orientation of the hypothetical coordinate axes. Decide where the origin is located too.

Well, it seems then that `square.cpp` sets up coordinates in the OpenGL window so that the increasing direction of the x-axis is horizontally rightwards, that of the y-axis vertically upwards and, moreover, the origin seems to correspond to the lower-left corner of the window, as in Figure 2.3. We're making progress but there's more to the story, so read on!

The last of the three parameters of a `glVertex3f(*, *, *)` declaration is evidently the z coordinate. Vertices are specified in *3-dimensional* space (simply called 3-space or, mathematically, $\mathbb{R}^3$). Indeed, OpenGL allows us to draw in 3-space and create truly 3D scenes, which is its major claim to fame. However, we *perceive* the 3-dimensional scene as a picture *rendered* to a 2-dimensional part of the computer's screen, the rectangular OpenGL window. Shortly we'll see how OpenGL converts a 3D scene to its 2D rendering.

2.2 Orthographic Projection, Viewing Box and World Coordinates

What exactly do the vertex coordinate values mean? For example, is the vertex at (20.0, 20.0, 0.0) of `square.cpp` 20 mm., 20 cm. or 20 pixels away from the origin along both the x-axis and y-axis, or is there some other absolute unit of distance native to OpenGL?

Experiment 2.2. Change the `glutInitWindowSize()` parameter values of `square.cpp`* – first to `glutInitWindowSize(300, 300)` and then `glutInitWindowSize(500, 250)`. The square changes in size, and even shape, with the OpenGL window. Therefore, coordinate values appear not to specify any kind of absolute units on the screen. **End**

Remark 2.1. Of course, you could have reshaped the OpenGL window directly by dragging one of its corners with the mouse, rather then resetting `glutInitWindowSize()` in the program.

---

*When we refer to `square.cpp`, or any `program.cpp`, it's always to the original version in the `Code` directory, so if you've modified the code for an earlier experiment you'll need to copy back the original.

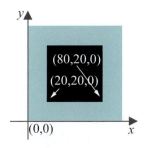

Figure 2.3: The coordinate axes on the OpenGL window of `square.cpp`? *Almost there...*

Understanding what the coordinates actually represent involves understanding first OpenGL's rendering mechanism, which itself begins with the program's *projection statement*. In the case of `square.cpp` the projection statement is

```
glOrtho(0.0, 100.0, 0.0, 100.0, -1.0, 1.0)
```

in the `resize()` routine, which determines an imaginary *viewing box* inside which the programmer draws. Generally,

```
glOrtho(left, right, bottom, top, near, far)
```

sets up a viewing box, as in Figure 2.4, with corners at the 8 points:

(left, bottom, −near), *(right, bottom, −near)*, *(left, top, −near)*, *(right, top, −near)*, *(left, bottom, −far)*, *(right, bottom, −far)*, *(left, top, −far)*, *(right, top, −far)*

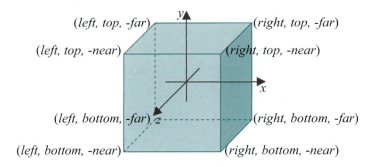

Figure 2.4: Viewing box of `glOrtho(`*left, right, bottom, top, near, far*`)`.

It's a box with sides aligned along the axes, whose span along the x-axis is from *left* to *right*, along the y-axis from *bottom* to *top*, and along the z-axis from *−far* to *−near*. Note the little quirk of OpenGL that the *near* and *far* values are flipped in sign.

The viewing box corresponding to the projection statement `glOrtho(0.0, 100.0, 0.0, 100.0, -1.0, 1.0)` of `square.cpp` is shown in Figure 2.5(a). The reader may wonder at this time how the initial coordinate axes are *themselves* calibrated – e.g., is a unit along an axis one inch, one centimeter or something else – as the size of the viewing box and that of the objects drawn inside it depend on this. The answer will be evident once the rendering process is explained momentarily.

For the drawing now, though, a vertex declared by `glVertex3f(x, y, z)` corresponds to the point (x, y, z). For example, the corner of the square declared by `glVertex3f(20.0, 20.0, 0.0)` is at the point $(20.0, 20.0, 0.0)$. The entire square of `square.cpp`, then, is as depicted in Figure 2.5(b).

Once the programmer has drawn the entire scene, if the projection statement is `glOrtho()` as in `square.cpp`, then the rendering process is two-step:

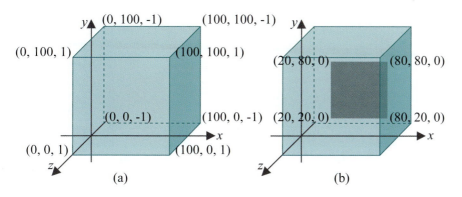

Figure 2.5: (a) Viewing box of square.cpp (b) With the square drawn inside.

1. *Shoot*: First, objects are *projected perpendicularly* onto the front face of the viewing box, i.e., the face on the $z = -near$ plane. For example, the square in Figure 2.6(a) (same as Figure 2.5(b)) is projected as in Figure 2.6(b). The front face of the viewing box is called the *viewing face* and the plane on which it lies the *viewing plane*.

   This step is like shooting the scene on film. In fact, one can think of the viewing box as one of those archaic *box cameras* where the photographer ducks behind the film – the viewing face – and covers her head with a black cloth. Mind that with this analogy that there's *no* lens, only the film!

2. *Print*: Next, the viewing face is *proportionately scaled* to fit the rectangular OpenGL window. This step is like printing the film on paper. In the case of square.cpp, printing takes us from Figure 2.6(b) to (c).

   If, say, the window size of square.cpp were changed to one of *aspect ratio* (= width/height) of 2, by replacing glutInitWindowSize(500, 500) with glutInitWindowSize(500, 250), printing would take us from Figure 2.6(b) to (d) (which actually distorts the square into a rectangle).

   The answer to the earlier question of how to calibrate the coordinate axes of the space in which the viewing box is created should be clear now: the 2D rendering finally displayed is the same *no matter* how they are calibrated, because of the proportionate scaling of the viewing face of the box to fit the OpenGL window. So it does not matter what unit we use, be it an inch, millimeter, mile, ...! Here's a partly-solved exercise to drive home the point.

Exercise 2.1.

(a) Suppose the viewing box of square.cpp is set up in a coordinate system where one unit along each axis is 1 cm. Assuming pixels to

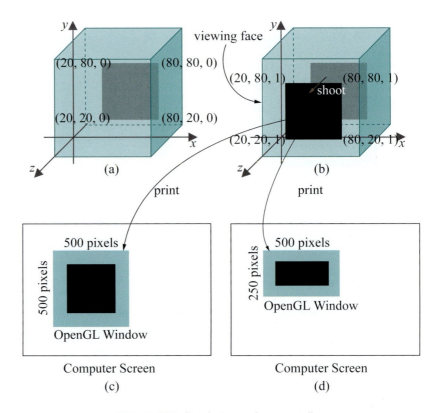

Figure 2.6: Rendering with glOrtho().

be 0.2 mm. × 0.2 mm. squares, compute the size and location of the square rendered by shoot-and-print to a 500 pixel × 500 pixel OpenGL window.

(b) Suppose next that the coordinate system is re-calibrated so that a unit along each axis is 1 meter instead of 1cm., everything else remaining same. What then are the size and location of the rendered square in the OpenGL window?

(c) What is rendered if, additionally, the size of the OpenGL window is changed to 500 pixel × 250 pixel?

Part answer:

(a) Figure 2.7 on the left shows the square projected to the viewing face, which is 100 cm. square. The viewing face is then scaled to the OpenGL window on the right, which is a square of sides 500 pixels = 500 × 0.2 mm. = 100 mm. The scaling from face to the window, therefore, is a factor of 1/10 in both dimensions. It follows that the rendered square

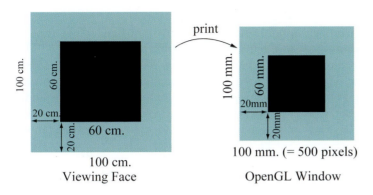

100 cm.
Viewing Face

100 mm. (= 500 pixels)
OpenGL Window

Figure 2.7: The viewing face for `square.cpp`, given that one unit along each coordinate axis is 1 cm., scaled to a 500 pixel × 500 pixel OpenGL window.

is 60 mm. × 60 mm., with its lower-left corner located both 20 mm. above and to the right of the lower-left corner of the window.

(b) Exactly the same as in part (a) because, while the viewing box and viewing face are now 10 times larger, the scaling from face to window is now a factor of 1/100, rather than 1/10.

We conclude that the size and location of the rendering in each coordinate direction are independent of how the axes are calibrated, but determined rather by the *ratio* of the original object's size to that of the viewing box in that direction.

Although the calibration of the axes doesn't matter, nevertheless, we'll make the sensible assumption that all three are calibrated *identically*, i.e., one unit along each axis is of equal length (yes, oddly enough, we could make them different and still the rendering would not change, which you can verify yourself by re-doing Exercise 2.1(a), after assuming that one unit along the x-axis is 1 cm. and along the other two 1 meter). The only other assumptions about the initial coordinate system that we make are conventional ones:

(a) It is *rectangular*, i.e., the three axes are mutually perpendicular.

(b) The x-, y- and z-axes in that order form a *right-handed* system in the following sense: the rotation of the x-axis 90° about the origin so that its positive direction matches with that of the y-axis appears *counter-clockwise* to a viewer located on the positive side of the z-axis (Figure 2.8).

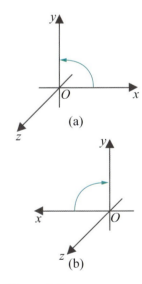

Figure 2.8: (a) The x-, y- and z-axes are rectangular and form (a) a right-handed system (b) a left-handed system.

Fixed World System

To summarize, set up an initial rectangular right-handed coordinate system located wherever you like in space, but with axes all calibrated identically. Call a unit along each axis just "a unit" as we know it doesn't matter what the unit is. Then leave it *fixed forever* – imagine it nailed to the top of your desk! See Figure 2.9. This system coordinatizes *world space* and, in fact, we shall refer to it as the *world coordinate system*. All subsequent objects, including the viewing box and those that we create ourselves, inhabit world space and are specified in world coordinates. These are all virtual objects, of course!

Figure 2.9: A dedicated 3D graphics programmer in a world all her own.

Remark 2.2. Because it's occupied by user-defined objects, world space is sometimes called *object space*.

Experiment 2.3. Change only the viewing box of `square.cpp` by replacing `glOrtho(0.0, 100.0, 0.0, 100.0, -1.0, 1.0)` with `glOrtho(-100, 100.0, -100.0, 100.0, -1.0, 1.0)`. The location of the square in the new viewing box is different and, so as well, the result of shoot-and-print. Figure 2.10 explains how.   **End**

Exercise 2.2. (Programming) Change the viewing box of `square.cpp` by replacing `glOrtho(0.0, 100.0, 0.0, 100.0, -1.0, 1.0)` successively with the following, in each case trying to predict the output before running:

(a) `glOrtho(0.0, 200.0, 0.0, 200.0, -1.0, 1.0)`

(b) `glOrtho(20.0, 80.0, 20.0, 80.0, -1.0, 1.0)`

(c) `glOrtho(0.0, 100.0, 0.0, 100.0, -2.0, 5.0)`

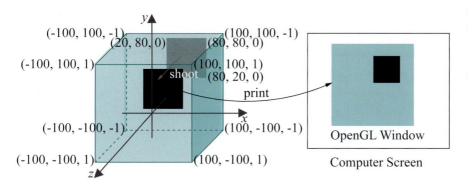

Figure 2.10: The viewing box of `square.cpp` defined by `glOrtho(-100, 100.0, -100.0, 100.0, -1.0, 1.0)`.

Exercise 2.3. (Programming) We saw earlier that, as a result of the print step, replacing `glutInitWindowSize(500, 500)` with `glutInitWindowSize(500, 250)` in `square.cpp` causes the square to be distorted into a rectangle. By changing *only one* numerical parameter elsewhere in the program, eliminate the distortion to make it appear square again.

Incidentally, it's clear now that our working hypothesis after the first experiment in Section 2.1, that the OpenGL window comes with axes fixed to it, though not unreasonable, was not accurate either. The OpenGL window it turns out is simply an empty target rectangle on which the front face of the viewing box is printed. This rectangle is called *screen space*.

So there are two spaces we'll be interacting with: world and screen. The former is a virtual 3D space in which we create our scenes, while the latter is a real 2D space where images concocted from our scenes by shoot-and-print are rendered for viewing.

Exercise 2.4. (Programming) Alter the z coordinates of each vertex of the "square" – we should really call it a polygon if we do this – of `square.cpp` as follows (Block 1*)::

```
glBegin(GL_POLYGON);
    glVertex3f(20.0, 20.0, 0.5);
    glVertex3f(80.0, 20.0, -0.5);
    glVertex3f(80.0, 80.0, 0.1);
    glVertex3f(20.0, 80.0, 0.2);
glEnd();
```

The rendering does not change. Why?

Remark 2.3. Always set the parameters of `glOrtho(`*left, right, bottom, top, near, far*`)` so that *left < right, bottom < top,* and *near < far*.

---

*To cut-and-paste you can find the block in text format in the file `chap2codeModifications.txt` in the directory `Code/CodeModifications`.

Remark 2.4. The aspect ratio (= width/height) of the viewing box should be set same as that of the OpenGL window or the scene will be distorted by the print step.

Remark 2.5. The perpendicular projection onto the viewing plane corresponding to a `glOrtho()` call is also called *orthographic projection* or *orthogonal projection* (hence the name of the call). Yet another term is *parallel projection* as the lines of projection from points in the viewing box to the viewing plane are all parallel.

2.3 The OpenGL Window and Screen Coordinates

We've already had occasion to use the `glutInitWindowSize(`w`, `h`)` command which sets the size of the OpenGL window to width w and height h measured in pixels. A companion command is `glutInitWindowPosition(`x`,` y`)` to specify the location (x, y) of the upper-left corner of the OpenGL window on the computer screen.

Experiment 2.4. Change the parameters of `glutInitWindowPosition(`x`,` y`)` in `square.cpp` from the current (100, 100) to a few different values to determine the location of the origin (0, 0) of the computer screen, as well as the orientation of the screen's own x-axis and y-axis. **End**

The origin $(0, 0)$ of the screen it turns out is at its upper-left corner, while the increasing direction of its x-axis is horizontally rightwards and that of its y-axis vertically downwards; moreover, one unit along either axis is *absolute* and represents a pixel. See Figure 2.11, which shows as well the coordinates of the corners of the OpenGL window initialized by `square.cpp`.

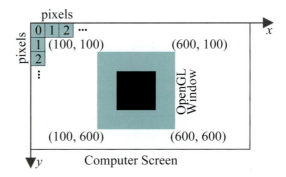

Figure 2.11: The screen's coordinate system: a unit along either axis is the pitch of a pixel.

Note the inconsistency between the orientation of the screen's y-axis and the y-axis of the world coordinate system, the latter being directed *up* the

OpenGL window (after being projected there). One needs to be careful about this, especially when coding programs where data is read from one system and used in the other.

2.4 Clipping

A question may have come to the reader's mind about objects which happen to be drawn outside the viewing box. Here are a few experiments to clarify how they are processed.

Experiment 2.5. Add another square by inserting the following right after the code for the original square in **square.cpp** (Block 2):

```
glBegin(GL_POLYGON);
    glVertex3f(120.0, 120.0, 0.0);
    glVertex3f(180.0, 120.0, 0.0);
    glVertex3f(180.0, 180.0, 0.0);
    glVertex3f(120.0, 180.0, 0.0);
glEnd();
```

From the value of its vertex coordinates the second square evidently lies entirely outside the viewing box.

If you run now there's no sign of the second square in the OpenGL window! This is because OpenGL *clips* the scene to within the viewing box before rendering, so that objects or parts of objects drawn outside are not rendered. Clipping is a stage in the graphics pipeline. We'll not worry about its implementation at this time, only the effect it has. **End**

Exercise 2.5. (Programming) In the preceding experiment can you redefine the viewing box by changing the parameters of the **glOrtho()** statement so that both squares are visible?

Experiment 2.6. For a more dramatic illustration of clipping, first replace the square in the original **square.cpp** with a triangle; in particular, replace the polygon code with the following (Block 3):

```
glBegin(GL_POLYGON);
    glVertex3f(20.0, 20.0, 0.0);
    glVertex3f(80.0, 20.0, 0.0);
    glVertex3f(80.0, 80.0, 0.0);
glEnd();
```

Figure 2.12: Screenshot of a triangle clipped to a quadrilateral.

Next, lift the first vertex up the z-axis by changing it to **glVertex3f(20.0, 20.0, 0.5)**; lift it further by changing its z-value to 1.5 (Figure 2.12 is a screenshot), then 2.5 and, finally, 10.0. Make sure you believe that what you see in the last three cases is indeed a triangle clipped to within the viewing box – Figure 2.13 may be helpful. **End**

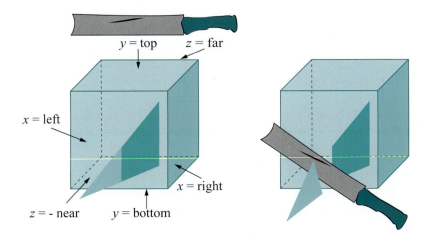

Figure 2.13: Six clipping planes of the `glOrtho`(*left*, *right*, *bottom*, *top*, *near*, *far*) viewing box. The lightly shaded part of the triangle sticking out of the box is clipped by a "clipping knife".

Exercise **2.6.** (**P**rogramming) A triangle was clipped to a quadrilateral in the viewing box in the preceding experiment. What is the maximum number of sides of a figure to which you can clip a triangle in the box (quadrilateral, pentagon, hexagon, ...)? Code and show.

Exercise **2.7.** Use pencil and paper to guess the output if the polygon declaration part of `square.cpp` is replaced with the following (Block 4):

```
glBegin(GL_POLYGON);
    glVertex3f(-20.0, -20.0, 0.0);
    glVertex3f(80.0, 20.0, 0.0);
    glVertex3f(120.0, 120.0, 0.0);
    glVertex3f(20.0, 80.0, 0.0);
glEnd();
```

The viewing box has six faces that lie on different planes and, effectively, OpenGL clips *off* the scene on one side of each of these six planes, accordingly called *clipping planes*. Imagine a knife slicing down each plane as in Figure 2.13. Specifically, in the case of the viewing box set up by `glOrtho`(*left*, *right*, *bottom*, *top*, *near*, *far*), clipped off is to the left of the plane *x* = *left*, to the right of the plane *x* = *right* and so on.

Remark 2.6. As we shall see in Chapter 3, the programmer can define clipping planes in addition to the six that bound the viewing box.

2.5 Color, OpenGL State Machine and Interpolation

Experiment 2.7. The color of the square in `square.cpp` is specified by the three parameters of the `glColor3f(0.0, 0.0, 0.0)` statement in the `drawScene()` routine, each of which gives the value of one of the three primary components, *blue*, *green* and *red*.

Determine which of the three parameters of `glColor3f()` specifies the blue, green and red components by setting in turn each to 1.0 and the others to 0.0. In fact, verify the following table:

Call	Color
`glColor3f(0.0, 0.0, 0.0)`	Black
`glColor3f(1.0, 0.0, 0.0)`	Red
`glColor3f(0.0, 1.0, 0.0)`	Green
`glColor3f(0.0, 0.0, 1.0)`	Blue
`glColor3f(1.0, 1.0, 0.0)`	Yellow
`glColor3f(1.0, 0.0, 1.0)`	Magenta
`glColor3f(0.0, 1.0, 1.0)`	Cyan
`glColor3f(1.0, 1.0, 1.0)`	White

End

Generally, the `glColor3f(`*red, green, blue*`)` call specifies the *foreground color*, or *drawing color*, which is the color applied to objects being drawn. The value of each color component, which ought to be a number between 0.0 and 1.0, determines its *intensity*. For example, `glColor3f(1.0, 1.0, 0.0)` is the brightest yellow while `glColor3f(0.5, 0.5, 0.0)` is a weaker yellow

Remark 2.7. The color values are each *clamped* to the range $[0, 1]$. This means that, if a value happens to be set greater than 1, then it's taken to be 1; if less than 0, it's taken to be 0.

Exercise 2.8. (Programming) Both `glColor3f(0.2, 0.2, 0.2)` and `glColor3f(0.8, 0.8, 0.8)` should be grays, having equal red, green and blue intensities. Guess which is the darker of the two. Verify by changing the foreground color of `square.cpp`.

The call `glClearColor(1.0, 1.0, 1.0, 0.0)` in the `setup()` routine specifies the *background color*, or *clearing color*. Ignore for now the fourth parameter, which is the *alpha* value. The statement `glClear(GL_COLOR_BUFFER_BIT)` in `drawScene()` actually clears the window to the specified background color, which means that every pixel in the color buffer is set to that color.

Experiment 2.8. Add the additional color declaration statement `glColor3f(1.0, 0.0, 0.0)` just after the existing one `glColor3f(0.0, 0.0,`

0.0) in the drawing routine of `square.cpp` so that the foreground color block becomes

```
// Set foreground (or drawing) color.
glColor3f(0.0, 0.0, 0.0);
glColor3f(1.0, 0.0, 0.0);
```

The square is drawn red because the *current* value of the foreground color is red when each of its vertices is specified.   **End**

Foreground color is one of a collection of variables, called *state variables*, which determine the state of OpenGL. Among other state variables are point size, line width, line stipple, material properties, etc. We'll meet several as we go along or you can refer to the red book* for a full list. OpenGL remains and functions in its current state until a declaration is made changing a state variable. For this reason, OpenGL is often called a *state machine*. The following simple experiment illustrates a couple of important points about how state variables control rendering.

Experiment 2.9. Replace the polygon declaration part of `square.cpp` with the following to draw two squares (Block 5):

```
glColor3f(1.0, 0.0, 0.0);
glBegin(GL_POLYGON);
    glVertex3f(20.0, 20.0, 0.0);
    glVertex3f(80.0, 20.0, 0.0);
    glVertex3f(80.0, 80.0, 0.0);
    glVertex3f(20.0, 80.0, 0.0);
glEnd();

glColor3f(0.0, 1.0, 0.0);
glBegin(GL_POLYGON);
    glVertex3f(40.0, 40.0, 0.0);
    glVertex3f(60.0, 40.0, 0.0);
    glVertex3f(60.0, 60.0, 0.0);
    glVertex3f(40.0, 60.0, 0.0);
glEnd();
```

Figure 2.14: Screenshot of a green square drawn in the code after a red square.

A small green square appears inside a larger red one (Figure 2.14). Obviously, this is because the foreground color is red for the first square, but green for the second. One says that the color red *binds* to the first square – or, more precisely, to each of its four specified vertices – and green to the second square. These bound values specify the color *attribute* of either square. Generally, the values of those state variables which determine how it is rendered collectively form a primitive's attribute set.

---

*The *OpenGL Programming Guide* [100] and its companion volume, the *OpenGL Reference Manual* [101], are the canonical references for the OpenGL API and affectionately referred to as the red book and blue book, respectively, by the CG community.

Flip the order in which the two squares appear in the code by cutting the seven statements that specify the red square and pasting them after those to do with the green one. The green square is overwritten by the red one and no longer visible because OpenGL draws in *code order*: primitives are rendered to the screen as they are specified in the code. This is called *immediate mode* graphics. One could also call it *memory-less* graphics, as primitives are not stored in the rendering pipeline, but drawn (and forgotten). **End**

Remark 2.8. Immediate mode is OpenGL's default and most commonly used. However, OpenGL has *retained mode* graphics as well, which allows the user to store drawing commands into a so-called display list to be invoked later in the code. We'll learn about display lists in Chapter 3.

Experiment 2.10. Replace the polygon declaration part of `square.cpp` with (Block 6):

```
glBegin(GL_POLYGON);
   glColor3f(1.0, 0.0, 0.0);
   glVertex3f(20.0, 20.0, 0.0);
   glColor3f(0.0, 1.0, 0.0);
   glVertex3f(80.0, 20.0, 0.0);
   glColor3f(0.0, 0.0, 1.0);
   glVertex3f(80.0, 80.0, 0.0);
   glColor3f(1.0, 1.0, 0.0);
   glVertex3f(20.0, 80.0, 0.0);
glEnd();
```

The different color values bound to the four vertices of the square are evidently *interpolated* over the rest of the square as you can see in Figure 2.15. In fact, this is most often the case with OpenGL: numerical attribute values specified at the vertices of a primitive are interpolated throughout its interior. In a later chapter we'll see exactly what it means to interpolate and how OpenGL goes about the task. **End**

Figure 2.15: Screenshot of a square with differently colored vertices.

2.6 OpenGL Geometric Primitives

The geometric primitives – also called drawing primitives or, simply, primitives – of OpenGL are the parts that programmers use in Lego-like manner to create mundane objects like balls and boxes, as well as elaborate spacecrafts, the worlds to which they travel, and pretty much everything in between. The only one we've seen so far is the polygon. It's time to get acquainted with the whole family.

Experiment 2.11. Replace `glBegin(GL_POLYGON)` with `glBegin(GL_-POINTS)` in `square.cpp` and make the point size bigger with a call to `glPointSize(5.0)`, so that the part drawing the polygon is now

```
glPointSize(5.0); // Set point size.
glBegin(GL_POINTS);
    glVertex3f(20.0, 20.0, 0.0);
    glVertex3f(80.0, 20.0, 0.0);
    glVertex3f(80.0, 80.0, 0.0);
    glVertex3f(20.0, 80.0, 0.0);
glEnd();
```

End

Experiment 2.12. Continue, replacing GL_POINTS with GL_LINES, GL_-LINE_STRIP and, finally, GL_LINE_LOOP. **End**

In the explanation that follows of how OpenGL draws, assume that the n vertices declared in the code between glBegin(*primitive*) and glEnd() are $v_0, v_1, \ldots, v_{n-1}$ in that order, i.e., the declaration of the primitive is of the form:

```
glBegin(primitive);
    glVertex3f(*, *, *); // v0
    glVertex3f(*, *, *); // v1
    ...
    glVertex3f(*, *, *); // vn-1
glEnd();
```

Refer to Figure 2.16 as you read (note that the primitives drawn there are general, their vertex positions not necessarily corresponding to the modified square.cpp).

GL_POINTS draws a point at each vertex

$$v_0, v_1, \ldots, v_{n-1}$$

GL_LINES draws a *disconnected* sequence of straight line segments (henceforth, we'll simply use the term "segment") between the vertices, taken two at a time. In particular, it draws the segments

$$v_0 v_1, \ v_2 v_3, \ \ldots, \ v_{n-2} v_{n-1}$$

if n is even. If n is not even then the last vertex v_{n-1} is simply ignored.

GL_LINE_STRIP draws the *connected* sequence of segments

$$v_0 v_1, \ v_1 v_2, \ \ldots, \ v_{n-2} v_{n-1}$$

Such a sequence is called a *polygonal line* or *polyline*.

GL_LINE_LOOP is the same as GL_LINE_STRIP, *except* that an additional segment $v_{n-1} v_0$ is drawn to complete a loop:

$$v_0 v_1, \ v_1 v_2, \ \ldots, \ v_{n-2} v_{n-1}, \ v_{n-1} v_0$$

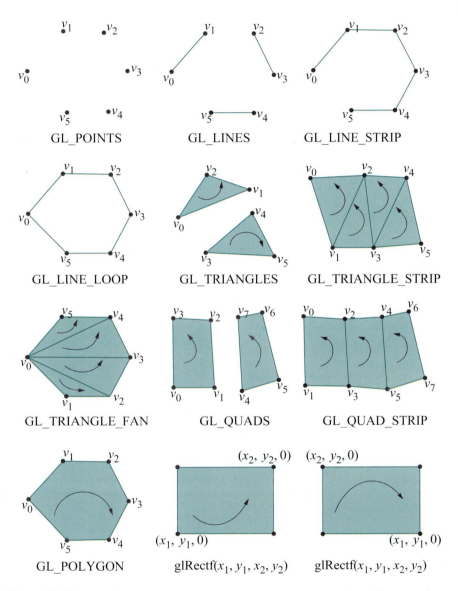

Figure 2.16: OpenGL's geometric primitives. Vertex orders are indicated by curved arrows. All except `glRectf()` have vertices specified within a `glBegin(`*primitive*`)` and `glEnd()` pair.

Such a segment sequence is called a *polygonal line loop*.

The thickness of lines can be set by a `glLineWidth(`*width*`)` call.

Remark 2.9. In world space points have zero dimension and lines zero width; values specified by `glPointSize()` and `glLineWidth()` are used only for rendering. Otherwise, it would be rather hard to see a point actually of zero dimension or a line of zero width!

Why does OpenGL provide separate primitives to draw polygonal lines and line loops when both can be viewed as a collection of segments and drawn using GL_LINES? For example,

```
glBegin(GL_LINE_STRIP);
    v0;
    v1;
    v2;
    ...
glEnd();
```

is equivalent to

```
glBegin(GL_LINES);
    v0;
    v1
    v1;
    v2
    v2;
    ...
glEnd();
```

The answer is first to avoid redundancy in vertex data. Second, possible rendering error is avoided as well because OpenGL does not know, say, that the two `v1`s in the GL_LINES specification above are supposed to represent the same vertex and may render the two at slightly different locations because of differences in floating point round-offs.

Exercise 2.9. (Programming) This relates to the brief discussion on interpolation at the end of Section 2.5. Replace the polygon declaration part of `square.cpp` with (Block 7):

```
glLineWidth(5.0);
glBegin(GL_LINES);
    glColor3f(1.0, 0.0, 0.0);
    glVertex3f(20.0, 20.0, 0.0);
    glColor3f(0.0, 1.0, 0.0);
    glVertex3f(80.0, 20.0, 0.0);
glEnd();
```

Can you say what the color values should be at the midpoint $(50.0, 20.0, 0.0)$ of the segment drawn? Check your answer by drawing a point with those color values just above the midpoint, say at $(50.0, 22.0, 0.0)$.

Experiment 2.13. Replace the polygon declaration part of `square.cpp` with (Block 8):

```
glBegin(GL_TRIANGLES);
    glVertex3f(10.0, 90.0, 0.0);
    glVertex3f(10.0, 10.0, 0.0);
    glVertex3f(35.0, 75.0, 0.0);
    glVertex3f(30.0, 20.0, 0.0);
    glVertex3f(90.0, 90.0, 0.0);
    glVertex3f(80.0, 40.0, 0.0);
glEnd();
```

End

GL_TRIANGLES draws a sequence of triangles using the vertices three at a time. In particular, the triangles are

$$v_0 v_1 v_2, \ v_3 v_4 v_5, \ \ldots, \ v_{n-3} v_{n-2} v_{n-1}$$

if n is a multiple of 3; if it isn't, the last one, or two, vertices are ignored.

The given order of the vertices for each triangle, in particular, v_0, v_1, v_2 for the first, v_3, v_4, v_5 for the second and so on, determines its *orientation* as perceived by a viewer. Figure 2.16 indicates vertex orders with curved arrows.

Orientation is important because it enables OpenGL to decide which side of a primitive, front or back, the viewer sees. We'll deal with this important topic separately in Chapter 9. Till then disregard orientation when drawing, listing the vertices of a primitive in any order you like.

GL_TRIANGLES is a *2-dimensional* primitive and, by default, triangles are drawn filled. However, one can choose a different drawing mode by applying the `glPolygonMode(face, mode)` command where *face* may be one of GL_FRONT, GL_BACK or GL_FRONT_AND_BACK, and *mode* one of GL_FILL, GL_LINE or GL_POINT. Whether a primitive is front-facing or back-facing depends, as said above, on its orientation. To keep matters simple for now, though, we'll use only GL_FRONT_AND_BACK in a `glPolygonMode()` call, which applies the given drawing mode to a primitive regardless of which face is visible. The GL_FILL option is, of course, the default filled option for 2D primitives, while GL_LINE draws the primitive in *outline* (or *wireframe* as it's also called), and GL_POINT only the vertices.

Experiment 2.14. In fact, it's often easier to decipher a 2D primitive by viewing it in outline. Accordingly, continue the preceding experiment by inserting the call `glPolygonMode(GL_FRONT_AND_BACK, GL_LINE)` in the drawing routine and, further, replacing GL_TRIANGLES with GL_TRIANGLE_-STRIP. The relevant part of the display routine then is as below:

```
// Set polygon mode.
glPolygonMode(GL_FRONT_AND_BACK, GL_LINE);
```

```
// Draw a triangle strip.
glBegin(GL_TRIANGLE_STRIP);
    glVertex3f(10.0, 90.0, 0.0);
    glVertex3f(10.0, 10.0, 0.0);
    glVertex3f(35.0, 75.0, 0.0);
    glVertex3f(30.0, 20.0, 0.0);
    glVertex3f(90.0, 90.0, 0.0);
    glVertex3f(80.0, 40.0, 0.0);
glEnd();
```

End

GL_TRIANGLE_STRIP draws a sequence of triangles – called a *triangle strip* – as follows: the first triangle is $v_0 v_1 v_2$, the second $v_1 v_3 v_2$ (v_0 is dropped and v_3 brought in), the third $v_2 v_3 v_4$ (v_1 dropped and v_4 brought in) and so on. Figure 2.16 should make clear the scheme to specify successive triangles. Formally, the triangles in the strip are

$$v_0 v_1 v_2, \ v_1 v_3 v_2, \ v_2 v_3 v_4, \ \ldots, \ v_{n-3} v_{n-2} v_{n-1} \qquad \text{(if } n \text{ is odd)}$$

or

$$v_0 v_1 v_2, \ v_1 v_3 v_2, \ v_2 v_3 v_4, \ \ldots, \ v_{n-3} v_{n-1} v_{n-2} \qquad \text{(if } n \text{ is even)}$$

Again, this is a 2-dimensional primitive and the given order of the vertices of each triangle determines its orientation.

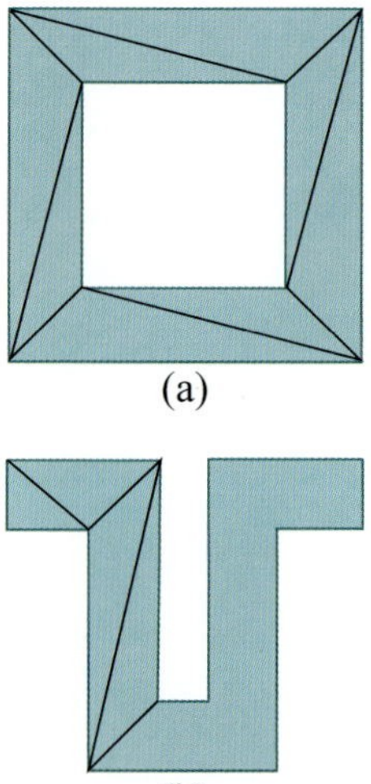

Exercise 2.10. (Programming) Create a square annulus as in Figure 2.17(a) using a *single* triangle strip. You may first want to sketch the annulus on graph paper to determine the coordinates of its eight corners. The figure depicts one possible *triangulation* – division into triangles – of the annulus.
Hint: A solution is available in `squareAnnulus1.cpp`.

Exercise 2.11. (Programming) Create the shape of Figure 2.17(b) using a single triangle strip. A partial triangulation is indicated.

Experiment 2.15. Replace the polygon declaration part of `square.cpp` with (Block 9):

```
glBegin(GL_TRIANGLE_FAN);
    glVertex3f(10.0, 10.0, 0.0);
    glVertex3f(15.0, 90.0, 0.0);
    glVertex3f(55.0, 75.0, 0.0);
    glVertex3f(80.0, 30.0, 0.0);
    glVertex3f(90.0, 10.0, 0.0);
glEnd();
```

(a)

(b)

Figure 2.17: (a) Square annulus – the region between two bounding squares – and a possible triangulation (b) A partially triangulated shape.

Apply both the filled and outlined drawing modes.

End

GL_TRIANGLE_FAN draws a sequence of triangles – called a *triangle fan* – around the first vertex as follows: the first triangle is $v_0 v_1 v_2$, the second $v_0 v_2 v_3$ and so on. The full sequence is

$$v_0 v_1 v_2, \ v_0 v_2 v_3, \ \ldots, \ v_0 v_{n-2} v_{n-1}$$

Exercise 2.12. (Programming) Create a square annulus using *two* triangle fans. First sketch a triangulation different from that in Figure 2.17(a).

Experiment 2.16. Replace the polygon declaration part of `square.cpp` with (Block 10):

```
glBegin(GL_QUADS);
    glVertex3f(10.0, 90.0, 0.0);
    glVertex3f(10.0, 10.0, 0.0);
    glVertex3f(40.0, 20.0, 0.0);
    glVertex3f(35.0, 75.0, 0.0);
    glVertex3f(15.0, 80.0, 0.0);
    glVertex3f(20.0, 10.0, 0.0);
    glVertex3f(90.0, 20.0, 0.0);
    glVertex3f(90.0, 75.0, 0.0);
glEnd();
```

Apply both the filled and outlined drawing modes. **End**

GL_QUADS draws a sequence of quadrilaterals using the vertices, taken four at a time. In particular, the first quadrilateral is $v_0 v_1 v_2 v_3$, the second $v_4 v_5 v_6 v_7$ and so on. If n is not a multiple of 4 then the last one, two or three vertices is dropped.

Experiment 2.17. Replace the polygon declaration part of `square.cpp` with (Block 11):

```
glBegin(GL_QUAD_STRIP);
    glVertex3f(10.0, 90.0, 0.0);
    glVertex3f(10.0, 10.0, 0.0);
    glVertex3f(30.0, 80.0, 0.0);
    glVertex3f(40.0, 15.0, 0.0);
    glVertex3f(60.0, 75.0, 0.0);
    glVertex3f(60.0, 25.0, 0.0);
    glVertex3f(90.0, 90.0, 0.0);
    glVertex3f(85.0, 20.0, 0.0);
glEnd();
```

Apply both the filled and outlined drawing modes. **End**

GL_QUAD_STRIP draws a sequence of quadrilaterals as follows: the first is $v_0 v_1 v_3 v_2$, the second $v_2 v_3 v_5 v_4$, the third $v_4 v_5 v_7 v_6$ and so on. Note the somewhat quirky sequence of vertices in the member quadrilaterals.

Exercise 2.13. List formally the full sequence of quadrilaterals composing a GL_QUAD_STRIP primitive specified by the n vertices $v_0, v_1, \ldots, v_{n-1}$, in the same manner that we earlier listed the triangles composing a GL_TRIANGLE_STRIP.

Needless to say, both GL_QUADS and GL_QUAD_STRIP are 2D primitives and the given order of their vertices determines the orientation of each component quadrilateral. We've already used the 2D primitive GL_POLYGON to draw squares and triangles. Generally:

GL_POLYGON draws a polygon with the vertex sequence

$$v_0 \, v_1 \, \ldots \, v_{n-1}$$

(n must be at least 3 for anything to be drawn).

Finally:

glRectf($x1$, $y1$, $x2$, $y2$) draws a rectangle lying on the $z = 0$ plane with sides parallel to the x- and y-axes. In particular, the rectangle has diagonally opposite corners at ($x1$, $y1$, 0) and ($x2$, $y2$, 0). The full list of four vertices is ($x1$, $y1$, 0), ($x2$, $y1$, 0), ($x2$, $y2$, 0) and ($x1$, $y2$, 0). The rectangle created is 2-dimensional and its vertex order depends on the situation of the two vertices ($x1$, $y1$, 0) and ($x2$, $y2$, 0) with respect to each other, as indicated by the two drawings at the lower right of Figure 2.16.

Obviously, a glRectf() call can be replaced by a suitable GL_POLYGON or GL_QUADS call. OpenGL offers it as a useful macro because rectangles are drawn so often.

Extremely important: The programmer should ensure when using GL_QUADS, GL_QUAD_STRIP or GL_POLYGON that each individual quadrilateral, or the polygon, is a *plane convex* figure, i.e., it lies on one plane and has no "bays" or "inlets" (see Figure 2.18). Otherwise, rendering is unpredictable. We'll see the reason for this in Chapter 8 when we discuss triangulation but, in the meantime, here are a couple of experiments, the second one being rather curious.

Experiment 2.18. Replace the polygon declaration of **square.cpp** with (Block 12):

```
glBegin(GL_POLYGON);
    glVertex3f(20.0, 20.0, 0.0);
    glVertex3f(50.0, 20.0, 0.0);
    glVertex3f(80.0, 50.0, 0.0);
    glVertex3f(80.0, 80.0, 0.0);
    glVertex3f(20.0, 80.0, 0.0);
glEnd();
```

Not planar, not convex   Planar, not convex   Planar and convex

Figure 2.18: OpenGL polygons should be planar and convex.

You see a convex 5-sided polygon (Figure 2.19(a)).   **End**

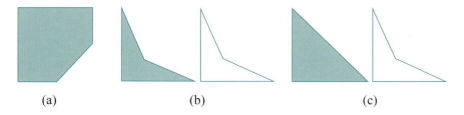

(a)   (b)   (c)

Figure 2.19: Experimental outputs.

Experiment 2.19. Replace the polygon declaration of `square.cpp` with (Block 13):

```
glBegin(GL_POLYGON);
    glVertex3f(20.0, 20.0, 0.0);
    glVertex3f(80.0, 20.0, 0.0);
    glVertex3f(40.0, 40.0, 0.0);
    glVertex3f(20.0, 80.0, 0.0);
glEnd();
```

Display it *both* filled and outlined using appropriate `glPolygonMode()` calls. A non-convex quadrilateral is drawn in either case (Figure 2.19(b)). Next, keeping the *same* cycle of vertices as above, list them starting with `glVertex3f(80.0, 20.0, 0.0)` instead (Block 14):

```
glBegin(GL_POLYGON);
    glVertex3f(80.0, 20.0, 0.0);
    glVertex3f(40.0, 40.0, 0.0);
    glVertex3f(20.0, 80.0, 0.0);
    glVertex3f(20.0, 20.0, 0.0);
glEnd();
```

Make sure to display it both filled and outlined. When filled it's a triangle, while outlined it's a non-convex quadrilateral identical to the one output earlier (Figure 2.19(c))! Because the cyclic order of the vertices is unchanged, shouldn't it be as in Figure 2.19(b) both filled and outlined?   **End**

45

We'll leave the apparent anomaly* of this experiment as a mystery to be resolved in Chapter 8. However, if you are impatient to settle it right now, then here's a tip: there's little between here and Chapter 8 that you need for that later chapter, which itself is a fairly easy read. Don't forget to come back though!

Exercise 2.14. (Programming) Verify, by cycling the vertices, that no such anomaly arises in the case of the convex polygon of Experiment 2.18.

Exercise 2.15. (Programming) Draw the double annulus (a figure '8') shown in Figure 2.20 in two ways: (i) using as few triangle strips as possible, and (ii) using as few quad strips as possible. Introduce additional vertices on the three boundary components if you need to (in addition to the original twelve).

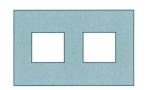

Figure 2.20: Double annulus.

Note: Such additional vertices are called *Steiner vertices*.

Remark 2.10. Here's an interesting semi-philosophical question. OpenGL claims to be a 3D drawing API. Yet, why does it not have a single 3D drawing primitive, e.g., cube, tetrahedron or such? All its primitives are 0-dimensional (`GL_POINTS`), 1-dimensional (`GL_LINE*`) or 2-dimensional (`GL_TRIANGLE*`, `GL_QUAD*`, `GL_POLYGON`, `glRectf()`).

The answer lies in how we humans (the regular ones that is and not supers with X-ray vision) perceive 3D objects such as cubes, tetrahedrons, chairs and spacecraft: *we see only the surface, which is two-dimensional*. It makes sense for a 3D API, therefore, to be able to draw only as much as can be seen.

2.7 Approximating Curved Objects

Looking back at Figure 2.16 we see that the OpenGL geometric primitives are composed of points, straight segments and flat pieces – triangles, quads and polygons – bounded by straight edges. How, then, to draw curved objects such as discs, ellipses, spirals, beer cans and flying saucers? The answer is to *approximate* them with straight and flat OpenGL primitives well enough that the viewer cannot tell the difference. As a wag once put it, "Sincerity is a very important human quality. If you don't have it, you *gotta* fake it!" In the next experiment we fake a circle.

Experiment 2.20. Run `circle.cpp`. Increase the number of vertices in the line loop

```
glBegin(GL_LINE_LOOP);
    for(i = 0; i < numVertices; ++i)
```

*The rendering depends on the particular OpenGL implementation. However, all implementations that we are aware of show identical behavior.

```
{
    glColor3ub(rand()%256, rand()%256, rand()%256);
    glVertex3f(X + R * cos(t), Y + R * sin(t), 0.0);
    t += 2 * PI / numVertices;
}
glEnd();
```

by pressing '+' till it "becomes" a circle, as in the screenshot of Figure 2.21. Press '-' to decrease the number of vertices. The `glColor3ub()` statement is for eye candy.   End

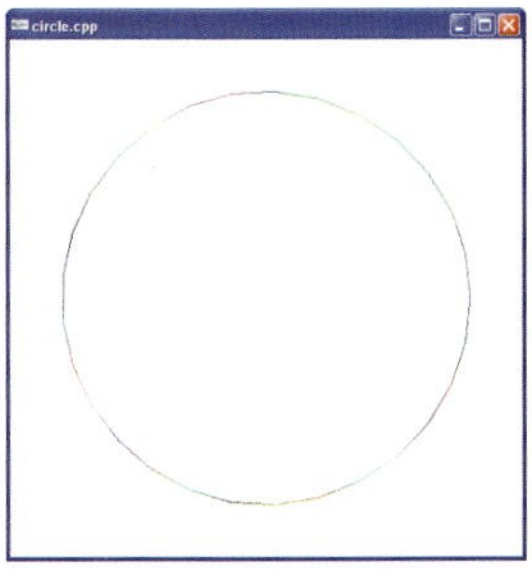

The vertices of the loop of `circle.cpp`, which lie evenly spaced on the circle, are collectively called a *sample of points* or, simply, *sample* from the circle. See Figure 2.22(a). The denser the sample evidently the better the approximation.

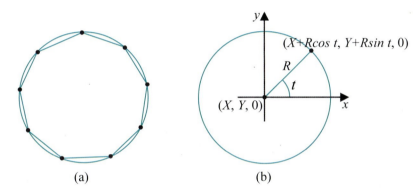

Figure 2.21: Screenshot of `circle.cpp`.

(a)   (b)

Figure 2.22: (a) A line loop joining a sample of points from a circle (b) Parametric equations for a circle.

The parametric equations of the circle implemented are

$$x = X + R\cos t, \ y = Y + R\sin t, \ z = 0, \quad 0 \le t \le 2\pi \qquad (2.1)$$

where $(X, Y, 0)$ is the center and R the radius of the circle. See Figure 2.22(b). A `numVertices` number of sample points equally spaced apart is generated by starting with the angle $t = 0$ and then incrementing it successively by $2\pi/\texttt{numVertices}$.

Observe that the vertex specifications occur within a loop construct, which is pretty much mandatory if there is a large number of vertices.

Incidentally, the program `circle.cpp` also demonstrates output to the command window, as well as non-trivial user interaction via the keyboard. The routine `keyInput()` is registered as the key handling routine in `main()` by the `glutKeyboardFunc(keyInput)` statement. Note the calls to `glutPostRedisplay()` in `keyInput()` asking the display to be redrawn after each update of `numVertices`.

Follow these conventions when writing OpenGL code:

1. Program the "Esc" key to exit the program.

2. Describe user interaction at two places:

   (a) The command window using `cout()`.

   (b) Comments at the *top* of the source code.

Exercise 2.16. (Programming) Draw a disc (a filled circle) by way of (a) a polygon and (b) a triangle fan.

Here's a parabola.

Experiment 2.21. Run `parabola.cpp`. Press '+/-' to increase/decrease the number of vertices of the approximating line strip. Figure 2.23 is a screenshot with enough vertices to make a smooth-looking parabola.

The vertices are equally spaced along the x-direction. The parametric equations implemented are

$$x = 50 + 50t, \ y = 100t^2, \ z = 0, \quad -1 \leq t \leq 1$$

the constants being chosen so that the parabola is centered in the OpenGL window. **End**

Figure 2.23: Screenshot of `parabola.cpp`.

We'll be returning shortly to the topic of approximating curved objects, but it's on to 3D next.

2.8 Three Dimensions, the Depth Buffer and Perspective Projection

The reader by now may be getting impatient to move on from the plane (pun intended) and simple to full 3D. Okay then, let's get off to an easy start in 3-space by making use of the third dimension to fake a circular annulus. Don't worry, we'll be doing fancier stuff soon enough!

Experiment 2.22. Run `circularAnnuluses.cpp`. Three identical-looking red circular annuluses (see Figure 2.24) are drawn in three *different* ways:

   i) Upper-left: There is not a real hole. The white disc *overwrites* the red disc as it appears later in the code.

```
glColor3f(1.0, 0.0, 0.0);
drawDisc(20.0, 25.0, 75.0, 0.0);
glColor3f(1.0, 1.0, 1.0);
drawDisc(10.0, 25.0, 75.0, 0.0);
```

Figure 2.24: Screenshot of `circularAnnuluses.cpp`.

Note: The first parameter of `drawDisc()` is the radius and the remaining three the coordinates of the center.

ii) Upper-right: There is not a real hole either. A white disc is *drawn closer* to the viewer than the red disc thus blocking it out.

```
glEnable(GL_DEPTH_TEST);
glColor3f(1.0, 0.0, 0.0);
drawDisc(20.0, 75.0, 75.0, 0.0);
glColor3f(1.0, 1.0, 1.0);
drawDisc(10.0, 75.0, 75.0, 0.5);
glDisable(GL_DEPTH_TEST);
```

Observe that the z-value of the white disc's center is greater than the red disc's. We'll discuss the mechanics of one primitive blocking out another momentarily.

iii) Lower: A true circular annulus with a real hole.

```
if (isWire) glPolygonMode(GL_FRONT, GL_LINE);
else glPolygonMode(GL_FRONT, GL_FILL);
glColor3f(1.0, 0.0, 0.0);
glBegin(GL_TRIANGLE_STRIP);
  ...
glEnd();
```

Press the space bar to see the wireframe of a triangle strip. **End**

How one chooses to draw the annulus depends on the application. If all that the viewer must be shown is a front view of what *appears* to be a red disc with a hole in the middle, then either of the first two methods may suffice. If the viewer wants to look *through* a true hole at some object behind, then one must use the third method to create an authentic annulus.

Exercise 2.17. (Programming) Interchange in `circularAnnuluses.cpp` the drawing orders of the red and white discs – i.e., the order in which they appear in the code – in either of the top two annuluses. Which one is affected? (*Only the first!*) Why?

Remark 2.11. Note the use of a text-drawing routine in `circular-Annuluses.cpp`. OpenGL offers only rudimentary text-drawing capability but it often comes in handy, especially for annotation. We'll discuss text-drawing in fair detail in Chapter 3.

By far the most important aspect of `circularAnnuluses.cpp` is its use of the *depth buffer* to draw the upper-right annulus. Following is an introduction to this critical utility which enables realistic rendering of 3D scenes.

2.8.1   A Vital 3D Utility: The Depth Buffer

Enabling the depth buffer, also called the *z-buffer*, causes OpenGL to eliminate, prior to rendering, parts of objects that are *obscured* (or, *occluded*) by others. Precisely, a point of an object is not drawn if its projection – think of a ray from that point – toward the viewing face is obstructed by another object. See Figure 2.25(a) for the making of the upper-right annulus of `circularAnnuluses.cpp`. This process is called *hidden surface removal* or *depth testing* or *visibility determination*.

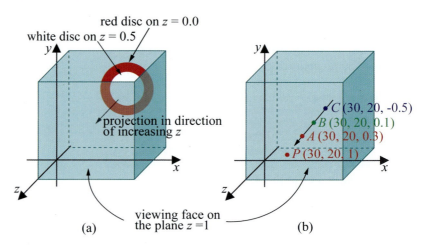

Figure 2.25: (a) The front white disc obscures part of the red one (b) The point A with largest z-value is projected onto the viewing plane so P is red.

Stated mathematically, the result of hidden surface removal in case of orthographic projection is as follows. Suppose that the set of points belonging to drawn objects in the viewing box, with their first two coordinate values particularly equal to X and Y, respectively, is $S = \{(X, Y, z)\}$, where z varies. Then only the point (X, Y, Z) of S, with the *largest* z-value, say, Z, lends its color attributes to their shared projection $(X, Y, -near)$ on the viewing face. The implication is that only (X, Y, Z) is drawn of the points in S, the rest obscured.

For example, in Figure 2.25(b), the three points A, B and C, colored red, green and blue, respectively, share the same x and y values, and all project to the point P on the viewing face. As A has the largest z coordinate of the three, it obscures the other two and P, therefore, is drawn red.

The z-buffer itself is a block of memory containing z-values, one per pixel. If depth testing is enabled, then, as a primitive is processed for rendering, the z-value of each of its points – or, more accurately, of each of the pixels comprising it – is compared with that of the one with the same (x, y)-values currently resident in the z-buffer. If an incoming pixel's z-value is greater, then its RGB attributes and z-value replace those of the current one; if not, the incoming pixel's data is discarded. For example, if the order in which the points of Figure 2.25(b) happen to appear in the code is C, A and B, here's how the color and z-buffer values at the pixel corresponding to P change:

```
draw C; // Pixel corresponding to P gets color blue
        // and z-value -0.5.
draw A; // Pixel corresponding to P gets color red
        // and z-value 0.3: A's values overwrite C's.
draw B; // Pixel corresponding to P retains color red
        // and z-value 0.3: B is discarded.
```

The preceding description of hidden surface removal with help from the z-buffer, though somewhat simplified, is adequate for almost all OpenGL applications. Particularly, note in `circularAnnuluses.cpp` the enabling syntax of hidden surface removal so that you can begin to use it:

1. The `GL_DEPTH_BUFFER_BIT` parameter of `glClear(GL_COLOR_BUFFER_BIT | GL_DEPTH_BUFFER_BIT)` in the `drawScene()` routine causes the depth buffer to be cleared.

2. The command `glEnable(GL_DEPTH_TEST)` in the `drawScene()` routine turns hidden surface removal on. The complementary command is `glDisable(GL_DEPTH_TEST)`.

3. The `GLUT_DEPTH` parameter of `glutInitDisplayMode(GLUT_SINGLE | GLUT_RGB | GLUT_DEPTH)` in `main()` causes the depth buffer to be initialized.

2.8.2 A Helix and Perspective Projection

We get more seriously 3D next by drawing a spiral or, more scientifically, helix. A helix, though itself 1-dimensional – drawn as a line strip actually – can be made authentically only in 3-space.

Open `helix.cpp` but don't run it as yet! The parametric equations implemented are

$$x = R\cos t, \ y = R\sin t, \ z = t - 60.0, \quad -10\pi \le t \le 10\pi \qquad (2.2)$$

See Figure 2.26. Compare these with Equation (2.1) for a circle centered at $(0, 0, 0)$, putting $X = 0$ and $Y = 0$ in that earlier equation. The difference is that the helix *climbs up* the z-axis, in addition to rotating circularly, with increasing t. Typically, one writes simply $z = t$ for the last coordinate; however, we tack on "-60.0" to push the helix far enough down the z-axis so that it's contained entirely in the viewing box.

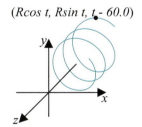

(Rcos t, Rsin t, t - 60.0)

Figure 2.26: Parametric equations for a helix.

Exercise 2.18. Even before viewing the helix, can you say from Equation (2.2) how many times it is supposed to coil around the z-axis, i.e., how many full turns it is supposed to make?

Hint: One full turn corresponds to an interval of 2π along t.

Experiment 2.23. Okay, run `helix.cpp` now. All we see is a circle as in Figure 2.27(a)! There's no sign of any coiling up or down. The reason, of course, is that the orthographic projection onto the viewing face flattens the helix. Let's see if it makes a difference to turn the helix upright, in particular, so that it coils around the y-axis. Accordingly, replace the statement

```
glVertex3f(R * cos(t), R * sin(t), t - 60.0);
```

in the drawing routine with

```
glVertex3f(R * cos(t), t, R * sin(t) - 60.0);
```

Hmm, not a lot better (Figure 2.27(b))! **End**

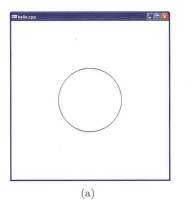
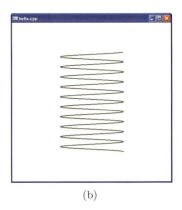

(a)   (b)

Figure 2.27: Screenshots of `helix.cpp` using orthographic projection with the helix coiling around the: (a) z-axis (b) y-axis.

Because it squashes a dimension, typically, orthographic projection is not suitable for 3D scenes. OpenGL provides, in fact, another kind of projection, called *perspective projection*, more appropriate for most 3D applications. Perspective projection is implemented with a `glFrustum()` call.

Instead of a viewing box, a `glFrustum(`*left, right, bottom, top, near, far*`)` call sets up a *viewing frustum* – a frustum is a *truncated pyramid* whose top has been cut off by a plane parallel to its base – in the following manner (see Figure 2.28):

The apex of the pyramid is at the origin. The front face, called the *viewing face*, of the frustum is the rectangle, lying on the plane $z = -near$, whose corners are (*left, bottom, $-near$*), (*right, bottom, $-near$*), (*left, top, $-near$*), and (*right, top, $-near$*). The plane $z = -near$ is called the *viewing plane*. The four edges of the pyramid emanating from the apex pass through the four corners of the viewing face. The base of the frustum, which is also the base of the pyramid, is the rectangle whose vertices are precisely where the pyramid's four edges intersect the $z = -far$ plane. By proportionality with the front vertices, the coordinates of the base verticcs are:

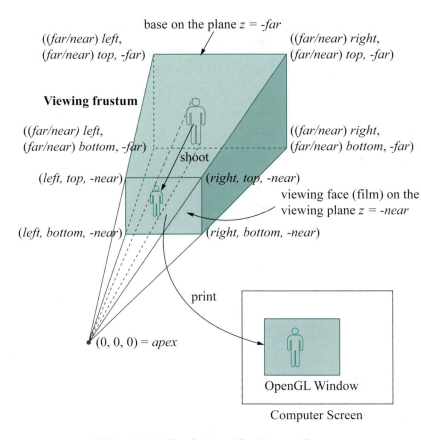

Figure 2.28: Rendering with glFrustum().

$((far/near) \; left, (far/near) \; bottom, \; -far),$
$((far/near) \; right, (far/near) \; bottom, \; -far),$
$((far/near) \; left, \; (far/near) \; top, \; -far),$
$((far/near) \; right, \; (far/near) \; top, \; -far)$

Values of the glFrustum() parameters are typically set so that the frustum lies symmetrically about the z-axis; in particular, *right* and *top* are chosen to be positive, and *left* and *bottom* their respective negatives. The parameters *near* and *far* should both be positive and *near* < *far*.

Example 2.1. Determine the corners of the viewing frustum created by the call glFrustum(-15.0, 15.0, -10.0, 10.0, 5.0, 50.0).

Answer: By definition, the corners of the front face are $(-15.0, -10.0, -5.0)$, $(15.0, -10.0, -5.0)$, $(-15.0, 10.0, -5.0)$ and $(15.0, 10.0, -5.0)$. The x and y values of the vertices of the base (or back face) are scaled from those on the front by a factor of 10 (= *far/near* = 50/5). These vertices are, therefore, $(-150.0, -100.0, -50.0)$, $(150.0, -100.0, -50.0)$, $(-150.0, 100.0, -50.0)$ and $(150.0, 100.0, -50.0)$.

53

Exercise 2.19. Determine the corners of the viewing frustum created by the call `glFrustum(-5.0, 5.0, -5.0, 5.0, 5.0, 100.0)`.

The rendering sequence in the case of perspective projection is a two-step shoot-and-print, similarly as for orthographic projection. The shooting step again consists of projecting objects within the viewing frustum onto the viewing face, *except that the projection is no longer perpendicular.* Instead, each point is projected along the line joining it to the apex, as depicted by the black dashed lines from the bottom and top of the man in Figure 2.28. Perspective projection causes *foreshortening* because objects farther away from the apex appear smaller (a phenomenon also called *perspective transformation*). For example, see Figure 2.29 where A and B are of the same height, but the projection pA is shorter than the projection pB.

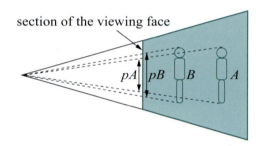

Figure 2.29: Section of the viewing frustum showing foreshortening.

Time now to see perspective projection turn on its magic!

Experiment 2.24. Fire up the original `helix.cpp` program. Replace orthographic projection with perspective projection; in particular, replace the projection statement

```
glOrtho(-50.0, 50.0, -50.0, 50.0, 0.0, 100.0);
```

with

```
glFrustum(-5.0, 5.0, -5.0, 5.0, 5.0, 100.0);
```

You can see a real spiral now (Figure 2.30(a)). View the upright version as well (Figure 2.30(b)), replacing

```
glVertex3f(R * cos(t), R * sin(t), t - 60.0);
```

with

```
glVertex3f(R * cos(t), t, R * sin(t) - 60.0);
```

A lot better than the orthographic version is it not?!   End

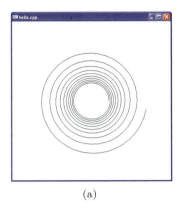
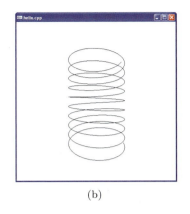

(a)   (b)

Figure 2.30: Screenshots of `helix.cpp` using perspective projection with the helix spiraling up the (a) z-axis (b) y-axis.

Perspective projection is more realistic than orthographic projection as it mimics how images are formed on the retina of the eye by light rays traveling toward a fixed point. And, in fact, it's precisely foreshortening that cues us humans to the distance of an object.

Remark 2.12. One can think of the apex of the frustum as the location of a *point camera* and the viewing face as its film.

The second rendering step where the viewing face is proportionately scaled to fit onto the OpenGL window is exactly as for orthographic projection. Similarly as in orthographic projection as well, the scene is clipped to within the viewing frustum by the 6 planes that bound the latter.

Remark 2.13. One might think of orthographic and perspective projections *both* as being along lines of projection to a single point, the center of projection (COP). In the case of orthographic projection, however, the COP is a "point at infinity" – i.e., infinitely far away – so that lines toward it are parallel.

Remark 2.14. There do exist 3D applications, e.g., in architectural design, where foreshortening amounts to distortion, so, in fact, orthographic projection is preferred.

Remark 2.15. It's because it captures the image of an object by intersecting rays projected from the object – either orthographically or perspectively – with a plane, which is similar to how a real camera works, that OpenGL is said to implement the *synthetic-camera* model.

Exercise 2.20. (**Programming**) Continuing from where we were at the end of the preceding experiment, successively replace the `glFrustum()` call as follows, trying in each case to predict the change in the display before running the code:

(a) `glFrustum(-5.0, 5.0, -5.0, 5.0, 5.0, 120.0)`

(b) glFrustum(-5.0, 5.0, -5.0, 5.0, 10.0, 100.0)

(c) glFrustum(-5.0, 5.0, -5.0, 5.0, 2.5, 100.0)

(d) glFrustum(-10.0, 10.0, -10.0, 10.0, 5.0, 100.0)

Parts (b) and (c) show, particularly, how moving the film forward and back causes the camera to "zoom" in and out, respectively.

Exercise 2.21. Formulate mathematically how hidden surface removal should work in the case of perspective projection, as we did in Section 2.8.1 for orthographic projection.

Experiment 2.25. Run moveSphere.cpp, which simply draws a movable sphere in the OpenGL window. Press the left, right, up and down arrow keys to move the sphere, the space bar to rotate it and 'r' to reset.

The sphere appears distorted as it nears the periphery of the window, as you can see from the screenshot in Figure 2.31. Can you guess why? Ignore the code, especially unfamiliar commands such as glTranslatef() and glRotatef(), except for the fact that the projection is perspective.

This kind of *peripheral distortion* of a 3D object is unavoidable in any viewing system which implements the synthetic-camera model. It happens with a real camera as well, but we don't notice it as much because the field of view when snapping pictures is usually quite large and objects of interest tend to be centered.   End

Figure 2.31: Screenshot of moveSphere.cpp.

2.9   Drawing Projects

Here are a few exercises to stretch your drawing muscles. The objects may look rather different from what we have drawn so far, but as programming projects aren't really. In fact, you can probably cannibalize a fair amount of code from earlier programs.

Exercise 2.22. (Programming) Draw a sine curve between $x = -\pi$ and $x = \pi$ (Figure 2.32(a)). Follow the strategy of circle.cpp to draw a polyline through a sample from the sine curve.

Exercise 2.23. (Programming) Draw an ellipse. Recall the parametric equations for an ellipse on the xy-plane, centered at (X, Y), with semi-major axis of length A and semi-minor axis of length B (Figure 2.32(b)):

$$x = X + A\cos t, \; y = Y + B\sin t, \; z = 0, \quad 0 \le t \le 2\pi$$

Again, circle.cpp is the template to use.

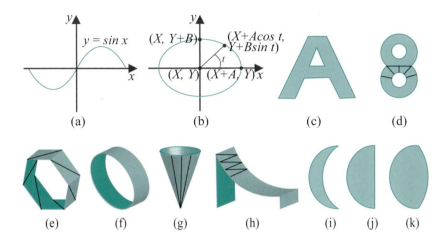

Figure 2.32: Draw these!

Exercise 2.24. (Programming) Draw the letter 'A' as a *two-dimensional* figure like the shaded region in Figure 2.32(c). It might be helpful to triangulate the figure first on graph paper.

Allow the user to toggle between filled and wireframe *a la* the bottom annulus of `circularAnnuluses.cpp`.

Exercise 2.25. (Programming) Draw the number '8' as the 2D object in Figure 2.32(d). Do this in two different ways: (i) drawing 4 discs and using the z-buffer and (ii) as a true triangulation, allowing the user to toggle between filled and wireframe. For (ii), a method of dividing the '8' into two triangle strips is suggested in Figure 2.32(d).

Exercise 2.26. (Programming) Draw a ring with cross-section a regular (equal-sided) polygon as in Figure 2.32(e), where a scheme to triangulate the ring in one triangle strip is indicated. Allow the user to change the number of sides of the cross-section. Increasing the number of sides sufficiently should make the ring appear cylindrical as in Figure 2.32(f). Use perspective projection and draw in wireframe.

Exercise 2.27. (Programming) Draw a cone as in Figure 2.32(g) where a possible triangulation is indicated. Draw in wireframe and use perspective projection.

Exercise 2.28. (Programming) Draw a children's slide as in Figure 2.32(h). Choose an appropriate equation for the cross-section of the curved surface – part of a parabola, maybe – and then "extrude" it as a triangle strip. (If you did Exercise 2.26 then you've already extruded a polygon.) Draw in wireframe and use perspective projection.

$Remark$ 2.16. Your output from Exercises 2.26-2.28 may look a bit "funny", especially viewed from certain angles. For example, the ring viewed head-on down its axis may appear as two concentric circles on a single plane. This problem can be alleviated by drawing the object with a different alignment or, equivalently, changing the viewpoint. In Experiment 2.26, coming up shortly, we'll learn code for the user to be able to change her viewpoint.

$Exercise$ 2.29. ($Programming$) Draw in a single scene a crescent moon, a half-moon and a three-quarter moon (Figures 2.32(i)-(k)). Each should be a true triangulation. Label each as well using text-drawing.

2.10 Approximating Curved Objects Once More

Our next 3-space drawing project is a bit more challenging: a hemisphere, which is a 2-dimensional object. We'll have an opportunity to get in place certain design principles which will be expanded in Chapter 10, which is dedicated to drawing (no harm starting early).

$Remark$ 2.17. A hemisphere is a 2-dimensional object because it is a surface. Recall that a helix is 1-dimensional because it's line-like. Now, both hemisphere and helix need 3-space to "sit in"; they cannot do with less. For example, you could sketch either on a piece of paper (2-space) but it would not be the real thing. On the other hand, a circle – another 1D object – does sit happily in 2-space.

Consider a hemisphere of radius R, centered at the origin O, with its circular base lying on the xz-plane. Suppose the spherical coordinates of a point P on this hemisphere are a longitude of θ (measured clockwise from the x-axis when looking from the plus side of the y-axis) and a latitude of ϕ (measured from the xz-plane toward the plus side of the y-axis). See Figure 2.33(a). The Cartesian coordinates of P are by elementary trigonometry

$$(R\cos\phi\cos\theta, \ R\sin\phi, \ R\cos\phi\sin\theta)$$

The range of θ is $0 \le \theta \le 2\pi$ and of ϕ is $0 \le \phi \le \pi/2$.

$Exercise$ 2.30. Verify that the Cartesian coordinates of P are as claimed. *Suggested approach*: From the right-angled triangle OPP' one has $|PP'| = R\sin\phi$ and $|OP'| = R\cos\phi$. $|PP'|$ is the y-value of P. Next, from right-angled triangle $OP'P''$ find $|OP''|$ and $|P'P''|$, the x and z values of P, respectively, in terms of $|OP'|$ and θ.

Sample the hemisphere at a mesh of $(p+1)(q+1)$ points P_{ij}, $0 \le i \le p$, $0 \le j \le q$, where the longitude of P_{ij} is $(i/p)*2\pi$ and its latitude $(j/q)*\pi/2$. In other words, $p+1$ longitudinally equally-spaced points are chosen along each of $q+1$ equally-spaced latitudes. See Figure 2.33(b), where $p = 10$ and

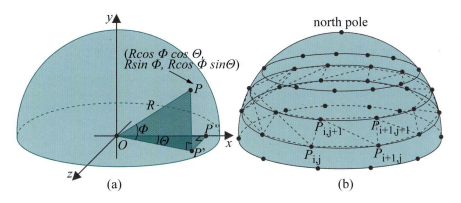

Figure 2.33: (a) Spherical and Cartesian coordinates on a hemisphere (b) Approximating a hemisphere with latitudinal triangle strips.

$q = 4$. The points P_{ij} are not all distinct. In fact, $P_{0j} = P_{pj}$, for all j, as the same point has longitude both 0 and 2π; and, the point P_{iq}, for all i, is identical to the north pole, which has latitude $\pi/2$ and arbitrary longitude.

The plan now is to draw one triangle strip with vertices at

$$P_{0,j+1}, \ P_{0j}, \ P_{1,j+1}, \ P_{1j}, \ldots, \ P_{p,j+1}, \ P_{pj}$$

for each j, $0 \leq j \leq q-1$, for a total of q triangle strips. In other words, each triangle strip takes its vertices alternately from a pair of adjacent latitudes and, therefore, approximates the circular band between them. Figure 2.33(b) shows one such strip. The stack of all q triangle strips approximates the hemisphere itself.

Experiment 2.26. Run `hemisphere.cpp`, which implements exactly the strategy just described. You can verify this from the snippet that draws the hemisphere:

```
for(j = 0; j < q; j++)
{
// One latitudinal triangle strip.
glBegin(GL_TRIANGLE_STRIP);
for(i = 0; i <= p; i++)
{
   glVertex3f(R * cos((float)(j+1)/q * PI/2.0) *
                      cos(2.0 * (float)i/p * PI),
            R * sin((float)(j+1)/q * PI/2.0),
            R * cos((float)(j+1)/q * PI/2.0) *
                      sin(2.0 * (float)i/p * PI));
   glVertex3f(R * cos((float)j/q * PI/2.0) *
                      cos(2.0 * (float)i/p * PI),
            R * sin((float)j/q * PI/2.0),
            R * cos((float)j/q * PI/2.0) *
                      sin(2.0 *(float)i/p * PI));
```

```
}
  glEnd();
}
```

Increase/decrease the number of longitudinal slices by pressing 'P/p'. Increase/decrease the number of latitudinal slices by pressing 'Q/q'. Turn the hemisphere about the axes by pressing 'x', 'X', 'y', 'Y', 'z' and 'Z'. See Figure 2.34 for a screenshot.   **End**

Figure 2.34: Screenshot of `hemisphere.cpp`.

Experiment **2.27.** Playing around a bit with the code will help clarify the construction of the hemisphere:

(a) Change the range of the hemisphere's outer loop from

```
for(j = 0; j < q; j++)
```

to

```
for(j = 0; j < 1; j++)
```

Only the bottom strip is drawn. The keys 'P/p' and 'Q/q' still work.

(b) Change it again to

```
for(j = 0; j < 2; j++)
```

Now, the bottom two strips are drawn.

(c) Reduce the range of both loops:

```
for(j = 0; j < 1; j++)
    ...
        for(i = 0; i <= 1; i++)
        ...
```

The first two triangles of the bottom strip are drawn.

(d) Increase the range of the inner loop by 1:

```
for(j = 0; j < 1; j++)
    ...
        for(i = 0; i <= 2; i++)
        ...
```

The first four triangles of the bottom strip are drawn.   **End**

There's syntax in `hemisphere.cpp` – none to do with the actual making of the hemisphere – which you may be seeing for the first time. The command `glTranslatef(0.0, 0.0, -10.0)` is used to move the hemisphere, drawn initially centered at the origin, into the viewing frustum, while the `glRotatef()` commands turn it. We'll explain these so-called *modeling transformations* in Chapter 4 but you are encouraged to experiment with them even now as the syntax is fairly intuitive. The set of three `glRotatef()`s, particularly, comes in handy to re-align a scene.

Exercise 2.31. (Programming) Modify `hemisphere.cpp` to draw:

(a) the bottom half of a hemisphere (Figure 2.35(a)).

(b) a 30° slice of a hemisphere (Figure 2.35(b)).

Make sure the 'P/p/Q/q' keys still work.

Exercise 2.32. (Programming) Just to get you thinking about animation, which we'll be studying in depth soon enough, guess the effect of replacing `glTranslatef(0.0, 0.0, -10.0)` with `glTranslatef(0.0, 0.0, -20.0)` in `hemisphere.cpp`. Verify.

And, here are some more things to draw.

Exercise 2.33. (Programming) Draw the objects shown in Figure 2.36. Give the user an option to toggle between filled and wireframe renderings.

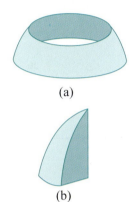

(a)

(b)

Figure 2.35: (a) Half a hemisphere (b) Slice of a hemisphere.

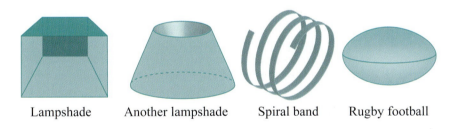

| Lampshade | Another lampshade | Spiral band | Rugby football |

Figure 2.36: More things to draw.

A suggestion for the football, or ellipsoid, is to modify `hemisphere.cpp` to make half of an ellipsoid (a hemi-ellipsoid?). Two hemi-ellipsoids back to back would then give a whole ellipsoid.

Remark 2.18. Filled renderings of 3D scenes, even with color, rarely look pleasant in the absence of lighting. See for yourself by applying color to 3D objects you have drawn so far (remember, you may have to invoke a `glPolygonMode(*, GL_FILL)` call). For this reason, we'll draw mostly wireframe till Chapter 11, which is all about lighting. You'll have to bear with this. Wireframe, however, fully exposes the geometry of an object, which is not a bad thing when one is learning object design.

2.11   Summary, Notes and More Reading

In this chapter we began the study of 3D CG, looking at it through the "eyes" of OpenGL. OpenGL itself was presented to the extent that the reader acquires functional literacy in this particular API. The drawing primitives were probably the most important part of the API's vernacular.

We discovered as well how OpenGL functions as a state machine, attributes such as color defining the current state. Moreover, we learned that quantifiable attribute values, e.g., RGB color, are typically interpolated from the vertices of a primitive throughout its interior. We saw that OpenGL clips whatever the programmer draws to within a viewing volume, either a box or frustum.

Beyond acquaintance with the language, we were introduced as well to the synthetic-camera model of 3D graphics, which OpenGL implements via two kinds of projection: orthographic and perspective. This included insights into the world coordinate system, the viewing box or frustum in which all drawings are made, the shoot-and-print rendering process to map a 3D scene to a 2D window, as well as hidden surface removal. We made a first acquaintance as well with another cornerstone of 3D graphics: the technique of simulating curved objects using straight and flat primitives like line segments and triangles.

Historically, OpenGL evolved from SGI's IRIS GL API, which popularized the approach to creating 3D scenes by drawing objects in actual 3-space and then rendering them to a 2D window by means of a synthetic camera. IRIS GL's efficiently pipelined architecture enabled high-speed rendering of animated 3D graphics and, consequently, made possible as well real-time interactive 3D. The ensuing demand from application developers for an open and portable (therefore, platform-independent) version of their API spurred SGI to create the first OpenGL specification in 1992, as well as a sample implementation. Soon after, the OpenGL ARB (Architecture Review Board), a consortium composed of a few leading companies in the graphics industry, was established to oversee the development of the API. Control of the OpenGL specification passed in 2006 to the Khronos Group, a member-funded industry consortium dedicated to the creation of open-standard royalty-free API's. (That no *one* owns OpenGL is a good thing.) The canonical, and very useful, source for information about OpenGL is its own home page [99].

Microsoft has a non-open Windows-specific 3D API – Direct3D [86, 135] – which is popular among game programmers as it allows optimization for the pervasive Windows platform. However, outside of the games industry, where it nonetheless competes with Direct3D, and leaving aside particular application domains with such high-quality rendering requirements that ray tracers are preferred, by far the dominant graphics API is OpenGL. It's safe to say that OpenGL is the de facto standard 3D graphics API. A primary reason for this, other than the extremely well thought out design which it had from inception – initial versions of Direct3D in contrast were notoriously buggy and hard to use – is OpenGL's portability. With their recent versions, though, OpenGL and Direct3D seem to be converging, at least in functionality (read an interesting comparison in Wikipedia [26]). It's worth knowing as well that, despite its intended portability, OpenGL can take advantage of platform-specific and card-specific capabilities via so-called extensions, at the cost of clumsier code.

An unofficial clone of OpenGL, Mesa 3D [89], which uses the same syntax, was originally developed by Brian Paul for the Unix/X11 platform, but there are ports now to other platforms as well.

Perhaps the best reason for OpenGL to be *the* API of choice for students of 3D computer graphics is – and this is a consequence of its almost universal adoption by the academic, engineering and scientific communities – the sheer volume of learning resources available. Not least among these is the number of textbooks that teach computer graphics with the help of OpenGL. Search amazon.com with the keywords "computer graphics opengl" and you'll see what we mean. Angel [2], Buss [21], Govil-Pai [56], Hearn & Baker [65], Hill & Kelley [68] and McReynolds & Blythe [88] are some introductions to computer graphics via OpenGL that the author has learned much from.

In case the reader prefers not to be distracted by code, here are a few API-independent introductions: Akenine-Möller, Haines & Hoffman [1], Foley et al. [44, 45] (the latter being an abridgment of the former), Shirley & Marschner [124], Watt [142] and Xiang & Plastock [148]. Keeping different books handy in the beginning is a good idea as, often, when you happen to be confused by one author's presentation of a topic, simply turning to another for help on just that may clear the way.

With regard to the code which comes with this book, we don't make much use of OpenGL-defined data types, which are prefixed with GL, e.g., GLsizei, GLint, etc., though the red book advocates doing so in order to avoid type mismatch issues when porting. Fortunately, we have not yet encountered a problem in any implementation of OpenGL that we've tried.

In addition to the code that comes with this book, the reader should try to acquire OpenGL programs from as many other sources as possible, as an efficient way to learn the language – any language as a matter of fact – is by modifying live code. Among the numerous sources on the web – there are pointers to several coding tutorials at the OpenGL site [99] – special mention must be made of Jeff Molofee's excellent tutorials at NeHe Productions [98], covering a broad spectrum of OpenGL topics, each illustrated with a well-documented program. The book by Wright, Lipchak & Haemel [146] is specifically about programming OpenGL and has numerous example programs. The red book comes with example code as well. Incidentally, in addition to the somewhat bulky red and blue books, a handy reference manual for OpenGL is Angel's succinct primer [3].

Hard-earned wisdom: Write experiments of your own to clarify ideas. Even if you are sure in your mind that you understand something, do write a few lines of code in verification. As the author has repeatedly been, you too might be surprised!

Part II

Tricks of the Trade

CHAPTER 3

An OpenGL Toolbox

Before getting to animation and other fun stuff in the next chapter, here are a few practical skills worth acquiring first. Our goal this chapter is to learn the following frequently-used OpenGL programming devices:

1. Vertex arrays: devices to store geometric data in a single location.

2. Display lists: "macros" to store frequently-invoked pieces of code.

3. Drawing of text.

4. Programming mouse buttons – for both clicks and motion.

5. Programming non-ASCII keys.

6. Programming pop-up menus.

7. Line stipples: applying patterns to lines.

8. GLUT objects: ready-made library objects.

9. Clipping planes: planes to clip a scene in addition to the automatic six that bound the viewing box or frustum.

10. `gluPerspective()`: a simpler version of the `glFrustum()` projection statement with fewer parameters.

11. Viewports: specifiable parts of the OpenGL window to which a drawing is rendered.

12. Multiple windows: multiple top-level OpenGL windows.

None is particularly challenging or deep, but an OpenGL programmer should at least be familiar with all items on this list. The next twelve sections follow the order of the list.

3.1   Vertex Arrays

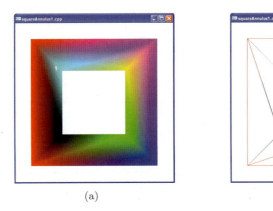

(a)   (b)

Figure 3.1: Screenshots of `squareAnnulus1.cpp`.

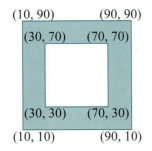

Figure 3.2: Square ann-
ulus (z coordinates all 0).

Experiment 3.1. Run `squareAnnulus1.cpp`. A screenshot is seen in Figure 3.1(a). Press the space bar to see the wireframe in Figure 3.1(b).

It is a plain-vanilla program which draws the square annulus diagrammed in Figure 3.2 using a single triangle strip (and multiply-colored vertices).

End

Experiment 3.2. Run `squareAnnulus2.cpp`.

It draws the same annulus as `squareAnnulus1.cpp`, except that the vertex coordinates and color data are now stored in two-dimensional global arrays, `vertices` and `colors`, respectively. A vector of coordinate values is retrieved by the *pointer form* (also called *vector* form) of vertex declaration, namely, `glVertex3fv(*pointer)`. Similarly, a vector of color values is retrieved with the pointer form `glColor3fv(*pointer)`.

Compared with `squareAnnulus1.cpp`, the obvious efficiency gained is in placing vertex and color data at one place in the code and then simply pointing to them from elsewhere.

End

It's always good programming practice to collect and place data for a program at a single location separate from the procedures which access the data. Redundancy and consequent errors tend to be eliminated, memory usage is more efficient and it's easier to modularize and debug those procedures which access the data.

OpenGL offers specific devices – the *vertex array* data structures – which make it easy and efficient for the user to centralize and share data. Let's learn them from live code.

Experiment 3.3. Run `squareAnnulus3.cpp`.

It again draws the same colorful annulus as before. The coordinates and color data of the vertices are stored in one-dimensional global vertex arrays, `vertices` and `colors`, respectively, as in `squareAnnulus2.cpp`, except, now, the arrays are flat and not 2D. The *i*th vector of values from both arrays is retrieved *simultaneously* with a single `glArrayElement(i)` call.

Note the initialization steps:

1. Two vertex arrays are enabled with calls to `glEnableClientState-`(*array*), where *array* is, successively, `GL_VERTEX_ARRAY` and `GL_COLOR_-ARRAY`. There are other possible values for the parameter *array* to store different kinds of data (we'll be storing normal and texture coordinates later).

2. The data for the two vertex arrays is specified with a call to `gl-VertexPointer`(*size, type, stride, *pointer*) and a call to `glColor-Pointer`(*size, type, stride, *pointer*). The parameter *pointer* is the address of the start of the data array, *type* declares the data type, *size* is the number of values per vertex (both coordinate and color arrays store 3 values for each vertex) and *stride* is the byte offset between the start of the values for successive vertices (0 indicates that values for successive vertices are not separated). **End**

Experiment 3.4. Run `squareAnnulus4.cpp`.

The code is even more concise with the application of a single call of the form `glDrawElements`(*primitive, count, type, *indices*) to draw the triangle strip. Parameter *primitive* is a geometric primitive, *indices* is the address of the start of an array of indices, *type* is the data type of the *indices* array and *count* is the number indices to use. The call itself is equivalent to the loop

```
glBegin(primitive);
    for(i = 0; i < count; i++) glArrayElement(indices[i]);
glEnd();
```

End

When there are multiple objects in a scene it's convenient to keep their data separately in different vertex arrays, as in the following program.

Experiment 3.5. Run `squareAnnulusAndTriangle.cpp`, which adds a triangle inside the annulus of the `squareAnnulus*.cpp` programs. See Figure 3.3 for a screenshot.

This program demonstrates the use of multiple vertex arrays. The vertex arrays `vertices1` and `colors1` contain the coordinate and color data, respectively, for the annulus, exactly as in `squareAnnulus3.cpp` and `squareAnnulus4.cpp`.

The single vertex array `vertices2AndColors2Intertwined` for the triangle, on the other hand, is *intertwined* in that it contains both coordinate

Figure 3.3: Screenshot of `squareAnnulusAnd-Triangle.cpp`.

and color data together. When pointing to data for the triangle, the *stride* parameter of both the `glVertexPointer()` and `glColorPointer()` calls is set to 6 times the number of bytes in a `float` data item, as there are 6 such items between the start of successive coordinate or color vectors in the intertwined array. **End**

Vertex arrays make for efficient OpenGL code. *Make a habit of using them!*

$\mathcal{R}em\alpha rk$ 3.1. Interestingly, OpenGL ES [75], an increasingly popular API for programming 3D graphics in small devices such as mobile phones, *requires* all vertex-related data to be stored in vertex arrays and provides no other option. OpenGL ES – ES stands for embedded systems – is a lightweight subset of OpenGL.

$\mathcal{R}em\alpha rk$ 3.2. Keep in mind that the display routine is called repeatedly if there is animation. It's particularly inefficient, therefore, and, unfortunately, a common beginner's mistake to store data in this routine, or perform computations there which actually can be done once initially and the results saved. Vertex arrays should be used to store data, while the initialization routine is the place for one-time computation.

3.2 Display Lists

A set of commands, e.g., to define an object such as a wheel or robot arm which is invoked repeatedly can be cached in a so-called *display list*. The display list is stored on the machine which runs the display unit and, often, pre-compiled and optimized. When the set of commands needs to be invoked, the program simply calls the display list rather than reissue them.

Note: Display lists implement *retained mode* graphics, versus OpenGL's default draw-and-forget immediate mode.

Display lists are particularly efficient in a client-server environment where the two communicate over a network and a goal is to minimize traffic. Once a display list has been saved by the server (the machine running the display unit), it can be invoked on a single command from the client (the machine running the program).

$\mathcal{E}xpe\mathcal{r}imen\mathbf{t}$ 3.6. Run `helixList.cpp`, which shows many copies of the same helix, variously transformed and colored. Figure 3.4 is a screenshot.

Here's the snippet from the initialization routine that makes a display list to draw the helix:

Figure 3.4: Screenshot of `helixList.cpp`.

```
aHelix = glGenLists(1);
glNewList(aHelix, GL_COMPILE);
glBegin(GL_LINE_STRIP);
for(t = -10 * PI; t <= 10 * PI; t += PI/20.0)
```

```
    glVertex3f(20 * cos(t), 20 * sin(t), t);
  glEnd();
  glEndList();
```

The call glGenLists(*range*) returns an integer which starts a block of size *range* of available display list indices. If a block of size *range* is not available, 0 is returned.

The set of commands to be cached in a display list – a helix-drawing routine in the case of helixList.cpp – is grouped between a glNewList(*listName*, *mode*) and a glEndList() statement. The parameter *listName* – aHelix in helixList.cpp – is the index which identifies the list. The parameter *mode* may be GL_COMPILE (only store, as in the program) or GL_COMPILE_AND_EXECUTE (store and execute immediately).

Finally, the drawing routine of helixList.cpp invokes glCallList(aHelix) six times to execute the display list. The glPushMatrix()-glPopMatrix() statement pairs, as also the modeling transformations (viz., glTranslatef(), glRotatef(), glScalef()) within these pairs, are used to position and scale copies of the helix. Ignore them if they don't make sense at present. End

Exercise 3.1. (**Programming**) Put the hemisphere-drawing routine of hemisphere.cpp into a display list and call the list twice to make a sphere – apply the scaling transformation glScalef(1.0, -1.0, 1.0) to one of the hemispheres to flip it over.

Exercise 3.2. (**Programming**) Make a ring of concentric circles of multiple colors on the xy-plane by repeatedly calling a display list containing a circle-drawing routine based on circle.cpp. Scale each invocation of the circle by a factor of u with a call to glScalef(u, u, 1.0).

There is a special mechanism in OpenGL to execute several display lists together.

Experiment 3.7. Run multipleLists.cpp. See Figure 3.5 for a screenshot. Three display lists are defined in the program: to draw a red triangle, a green rectangle and a blue pentagon, respectively.

The call glCallLists(n, *type*, **lists*) causes n display list executions (n is 6 in the program). The indices of the lists to be executed are obtained by adding the current display list base – this base is specified by glListBase(*base*) – to the successive offset values of type *type* in the array pointed by *lists*. End

Figure 3.5: Screenshot of multipleLists.cpp.

Exercise 3.3. (**Programming**) Modify multipleLists.cpp to draw a vertical black line between each object and the next. The line itself should be in a display list.

3.3   Drawing Text

Graphical text can be of two types: *bitmapped* (also called *raster*) and *stroke* (also called *vector*). Characters of bitmapped text are defined as a pattern of on and off bits in a rectangular block, while characters of stroke text are created using line primitives. For example, in Figure 3.6, the letter 'E' is represented as a bitmap consisting of 10 on bits and 5 off in a 3×5 raster, as well as in stroke form as a union of four straight segments.

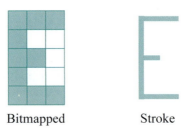

Bitmapped   Stroke

Figure 3.6: Bitmapped versus stroke text.

The GLUT library offers both bitmapped and stroke characters. The calls **glutBitmapCharacter**(**font*, *character*) and **glutStrokeCharacter**-(**font*, *character*) render *character* in the specified *font*.

Fonts available for bitmapped characters include:
GLUT_BITMAP_8_BY_13
GLUT_BITMAP_9_BY_15
GLUT_BITMAP_TIMES_ROMAN_10
GLUT_BITMAP_TIMES_ROMAN_24
GLUT_BITMAP_HELVETICA_10
GLUT_BITMAP_HELVETICA_12
GLUT_BITMAP_HELVETICA_18

Fonts available for stroke characters include:
GLUT_STROKE_ROMAN
GLUT_STROKE_MONO_ROMAN

Stroke characters offer an advantage over bitmapped ones in that they can be scaled in size and rotated, because line segments can be so transformed, whereas bitmapped characters, being fixed patterns, cannot.

Experiment 3.8. Run `fonts.cpp`. Displayed are the various fonts available through the GLUT library. See Figure 3.7.   **E**nd

Figure 3.7: Screenshot of fonts.cpp.

The canonical routine we use to draw bitmapped text is the following:

```
void writeBitmapString(void *font, char *string)
{
    char *c;
```

```
    for (c = string; *c != '\0'; c++) glutBitmapCharacter(font, *c);
}
```

Accordingly, a subsequent call block

```
glRasterPos3f(p, q, r);
writeBitmapString(font, string);
```

renders *string* in bitmapped *font* starting from position (p, q, r) in world coordinates. Keep in mind that these coordinates are transformed by prior modelview transformations, e.g., `glTranslatef()`, `glRotatef()` and such, which move objects around in world space.

Our canonical routine to draw stroke text is

```
void writeStrokeString(void *font, char *string)
{
    char *c;
    for (c = string; *c != '\0'; c++) glutStrokeCharacter(font, *c);
}
```

which renders the text starting from $(0, 0, 0)$ in world coordinates. Note that in addition to scaling and rotation, one can apply a `glLineWidth()` call to alter the thickness of stroke characters as well, as GLUT uses `GL_LINE*` primitives to draw them.

Exercise 3.4. (Programming) Locate the labels of `circularAnnuluses.cpp` in the white center of each annulus (you may have to split the labels into more than one line to fit them).

Exercise 3.5. (Programming) Modify `fonts.cpp` to be able to cycle through stroke fonts of different line widths by pressing the space bar.

3.4 Programming Mouse Buttons

Mouse buttons can be programmed to respond to both clicks and motion.

Clicks

Experiment 3.9. Run `mouse.cpp`. Click the left mouse button to draw points on the canvas and the right one to exit. Figure 3.8 is a screenshot of "OpenGL" scrawled in points. **End**

A mouse callback routine *mouse_callback_func()* is registered to handle mouse events by the GLUT statement `glutMouseFunc(`*mouse_callback_func*`)` in the main routine. In the case of `mouse.cpp`, the callback routine is `mouseControl()`:

Figure 3.8: Screenshot of `mouse.cpp`.

```
void mouseControl(int button, int state, int x, int y)
{
    if (button == GLUT_LEFT_BUTTON && state == GLUT_DOWN)
        points.push_back( Point(x, height - y) );
    if (button == GLUT_RIGHT_BUTTON && state == GLUT_DOWN) exit(0);
    glutPostRedisplay();
}
```

The callback routine itself has the form *mouse_callback_func*(*button*, *state*, *x*, *y*), where *button* is one of:

> GLUT_LEFT_BUTTON,   GLUT_RIGHT_BUTTON,   GLUT_MIDDLE_BUTTON

and *state* is one of:

> GLUT_UP,   GLUT_DOWN

The coordinates (x, y) return the location in the OpenGL window where the mouse event occurs. They are measured similarly as for screen coordinates – recall from Section 2.3 that screen coordinates are measured in pixels starting from the origin at the upper-left corner of the screen with the x-axis heading right and the y-axis down. The only difference in the case of a mouse click is that the origin is at the upper-left corner of the OpenGL window, rather than screen. Units are still pixels and the x-axis still heads right and the y-axis down. See Figure 3.9. This necessitates care when using the coordinates of a mouse event in the OpenGL program itself, because there is no *a priori* connection between the former and the world coordinates used by the latter.

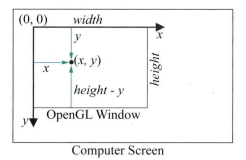

Figure 3.9: Mouse event coordinates (x, y).

In particular, note the following two steps in `mouse.cpp`:

1. The call

   > `glOrtho(0.0, (GLdouble)w, 0.0, (GLdouble)h, -1.0, 1.0);`

   in the reshape routine `resize(w, h)` ties screen coordinates to world coordinates by making the dimensions of the viewing face *equal* the actual physical dimensions of the OpenGL window, the latter being passed to the reshape routine via the parameters **w** and **h**. Because viewing face and OpenGL window are now the same size in their respective coordinate systems (world and screen), effectively, one unit along the viewing face along either the x or y axis is a pixel.

The only correction remaining to be made is owing to the y-axis being "upside down" going from one coordinate system to the other. This is done next.

2. The statement

```
points.push_back( Point(x, height - y) );
```

in the mouse callback routine to store points in the **points** vector when the mouse is clicked makes the final correction from the event's screen coordinates to world coordinates, as (x, y) on the screen corresponds to $(x, height - y, 0)$ in the world.

Remark 3.3. A point of note in **mouse.cpp** is the use of an STL **vector** to store **Point** objects. STL stands for the *Standard Template Library*, a C++ library of container classes, e.g., vector, list, set and so forth, together with routines to manipulate container objects. It is a part of the current ANSI C++ standard. The STL is extremely useful and saves a lot of repetitive programming. Readers not already familiar with the STL are well-advised to pick it up. There's no need to devote time separately for this purpose. Keeping a book like Schildt [119] handy while you code should be enough.

Exercise 3.6. (**Programming**) Write a program to draw a circle on a canvas after two left clicks of the mouse. The first click picks the center and the second a point on the circle.

Motion

Experiment 3.10. Run **mouseMotion.cpp**, which enhances **mouse.cpp** by allowing the user to drag the just-created point using the mouse with the left button still pressed. **End**

The additional capability of **mouseMotion.cpp** is obtained as follows. First, when the left mouse button is clicked, the mouse callback routine **mouseControl()** stores a point at the clicked position in the variable **currentPoint** of type **Point**. Only when the button is released is the new point added to the **points** vector by the same routine.

In the interim, between the press and release of the left mouse button, if the mouse moves, then its motion is tracked by the mouse motion callback routine **mouseMotion()**:

```
void mouseMotion(int x, int y)
{
    currentPoint.setCoords(x, height - y);
    glutPostRedisplay();
}
```

This routine simply keeps updating the coordinates of `currentPoint` with the current location of the mouse as the latter moves with the button pressed. The result is that this point, which is drawn separately in the `drawScene()` routine, travels with the mouse. Note that, just as the mouse callback routine is registered in main, so is the motion callback routine, the latter by `glutMotionFunc(mouseMotion)`.

One can also track so-called *passive* motion of the mouse – when it moves with no button pressed – via a passive motion callback function, which is registered in main with a `glutPassiveMotionFunc()` call.

Exercise 3.7. (Programming) Enhance the previous circle-drawing exercise by allowing the user to view the changing circle as the mouse is dragged with the second click.

3.5   Programming Non-ASCII Keys

In various programs to date, we've already interacted with the OpenGL window through keyboard entry by registering a handling function *keyboard_handling_func*() in the main routine via a call to `glutKeyboardFunc`(*keyboard_handling_func*). To interact with non-ASCII keys such as the arrow, F, and page up and down keys, one needs likewise to register a handling function *special_key_handling_func*() with a call to `glutSpecialFunc`(*special_key_handling_func*).

Experiment 3.11. Run `moveSphere.cpp`, a program we saw earlier in Experiment 2.25, where you can see a screenshot as well. Press the left, right, up and down arrow keys to move the sphere, the space bar to rotate it and 'r' to reset.

Note how the `specialKeyInput()` routine is written to enable the arrow keys to change the location of the sphere. Subsequently, this routine is registered in `main()` as the handling routine for non-ASCII entry.   **End**

Exercise 3.8. (Programming) Write a program to cycle through the GLUT fonts applied to the string "I am having so much fun with OpenGL it can't be legal!" by pressing the left and right arrow keys.

3.6   Menus

The GLUT library provides pop-up menus.

Experiment 3.12. Run `menus.cpp`. Press the right mouse button for menu options which allow you to change the color of the initially red square or exit. Figure 3.10 is a screenshot.

A `glutCreateMenu(`*menu_function*`)` declaration in the `makeMenu()` routine creates a menu, registers *menu_function*`()` as its callback function and returns a unique integer identifying the menu – to be used by any higher-level menu which may call the current one.

`glutAddMenuEntry(`*tag, returned_value*`)` creates a menu entry titled *tag* which, when clicked, returns *returned_value* to the callback function *menu_function*`()`. The latter, therefore, must be of the form *menu_function-*(*type_of_returned_value*).

`glutAddSubMenu(`*tag, sub_menu*`)` is similar to `glutAddMenuEntry()`, except that when *tag* is clicked a sub-menu pops up whose ID is *sub_menu*. Evidently, the statement creating a sub-menu must precede that for a higher-level menu which calls it, as the former's ID *sub_menu* is needed in order to create the latter.

`glutAttachMenu(`*button*`)` attaches the menu to a mouse button. **End**

Figure 3.10: Screenshot of `menus.cpp`.

Exercise 3.9. (Programming) Enhance `menus.cpp` to add two more items to the top-level menu:

(a) A "Mode" option allowing the rectangle to be shown either "Outlined" or "Filled".

(b) A "Size" option which leads to two sub-menu options "Width" and "Height", either of which has options "Small", "Medium" and "Large".

3.7 Line Stipples

One can create *stippled*, i.e., broken, lines in OpenGL by specifying and applying a stipple pattern.

Experiment 3.13. Run `lineStipple.cpp`. Press the space bar to cycle through stipples. A screenshot is shown in Figure 3.11. **End**

Stippling is enabled with a call to `glEnable(GL_LINE_STIPPLE)` and disabled by calling `glDisable(GL_LINE_STIPPLE)`. The stipple pattern itself is specified by the call `glLineStipple(`*factor, pattern*`)`.

Parameter *pattern* is a hex string of the form $0xX_3X_2X_1X_0$ where each X_i is a hexadecimal symbol (equivalent to 4 bits). Thus $X_3X_2X_1X_0$ represents a 16 bit string, say, $a_{15}a_{14}\ldots a_0$. Parameter *factor* is a positive integer.

The stipple pattern is applied as follows: if a_0 is 1, then the first *factor* pixels starting from the first vertex of the line primitive are set on; if a_0 is 0, the first *factor* pixels are off. If a_1 is 1, the next *factor* pixels of the line are on; if a_1 is 0, they are off. And so on... Note that the lower bits of the stipple pattern come first and that *factor* simply scales the pattern.

Figure 3.11: Screenshot of `lineStipple.cpp`.

For example, suppose the stipple is specified by `glLineStipple(1, 0x5555)`. Since 0x5555 = 0101010101010101, alternate pixels of the line are on and off with the first one being on. See Figure 3.12(a).

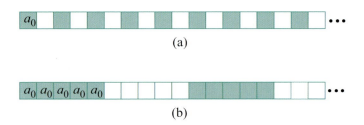

(a)

a_0 a_0 a_0 a_0 a_0

(b)

Figure 3.12: (a) Line stipple specified by `glLineStipple(1, 0x5555)` (b) Line stipple specified by `glLineStipple(5, 0x5555)`.

If the stipple is specified by `glLineStipple(5, 0x5555)` then alternate groups of five pixels on the line are on and off. See Figure 3.12(b).

Exercise 3.10. (Programming) Apply the different line stipples of `lineStipple.cpp` to the circle in `circle.cpp`.

GLUT stroke characters can be stippled to interesting effect as well, as `GL_LINE*` primitives are used to draw them.

Exercise 3.11. (Programming) Display the text "I am having so much fun with OpenGL it can't be legal!" using variously stippled stroke fonts.

Remark 3.4. One can enable polygon stippling as well with the call `glEnable(GL_POLYGON_STIPPLE)`. The call `glPolygonStipple(*mask)` specifies the stipple, a 32×32 bit pattern pointed by *mask*. The stipple is applied repeatedly over the polygon.

However, there are likely few applications where a programmer will prefer stipples to textures to apply to a 2D surface, so we'll not discuss them in any detail.

A 2D Drawing Program

Experiment 3.14. Run `canvas.cpp`, a simple program to draw on a flat canvas with menu and mouse functionality.

Left click on an icon to select it. Then left click on the drawing area to draw – click once to draw a point, twice to draw a line or rectangle. Right click for menu options. Figure 3.13 is a screenshot. **End**

Figure 3.13: Screenshot of `canvas.cpp`.

Exercise 3.12. (Programming) Enhance `canvas.cpp`:

(a) Add a polyline (multi-segment line) drawing capability. Create a suitable icon. Left clicking this icon picks the polyline option.

Subsequent left clicks pick successive segment endpoints until a middle click completes the polyline.

(b) Add a circle drawing capability. After left clicking the circle icon the next two left clicks pick the center and a point on the circle, respectively, following which the circle is drawn

(c) Add a regular (equal-sided) hexagon drawing capability. After left clicking the hexagon icon the next two left clicks pick the center and a vertex, respectively, following which the hexagon is drawn.

(d) Add text drawing capability.

(e) Give options for the grid size in the menu.

(f) Add color options through the menu.

(g) Add an outlined/filled option through the menu.

(h) Add stipple options through the menu.

Note: For the preceding three parts, the menu which pops up should depend on where the mouse is clicked. To keep it simple, for (f) color options could only be part of the menu which pops up when one of the icons is clicked, and not (objects in) the drawing area. A way to do this is by having the mouse callback routine, rather than the main routine, call the menu-making routine, so that mouse event coordinates may be passed to the latter.

(i) Use mouse motion tracking to allow the user to view in real-time a primitive changing as the mouse is moved, before it is saved with a final click.

3.8 GLUT Objects

The GLUT library offers a collection of standard objects. Each object is available in two flavors: solid and wireframe. The respective calls are shown in the following table.

Solid	Wireframe
glutSolidSphere(*radius*, *slices*, *stacks*)	glutWireSphere(*radius*, *slices*, *stacks*)
glutSolidCube(*size*)	glutWireCube(*size*)
glutSolidCone(*base*, *height*, *slices*, *stacks*)	glutWireCone(*base*, *height*, *slices*, *stacks*)
glutSolidTorus(*inRadius*, *outRadius*, *sides*, *rings*)	glutWireTorus(*inRadius*, *outRadius*, *sides*, *rings*)
glutSolidDodecahedron(*void*)	glutWireDodecahedron(*void*)
glutSolidOctahedron(*void*)	glutWireOctahedron(*void*)
glutSolidTetrahedron(*void*)	glutWireTetrahedron(*void*)
glutSolidIcosahedron(*void*)	glutWireIcosahedron(*void*)
glutSolidTeapot(*size*)	glutWireTeapot(*size*)

The objects are drawn centered at the origin. The parameters, if any, determine the object's size and the fineness of its triangulation. All the GLUT objects are depicted in wireframe in Figure 3.14.

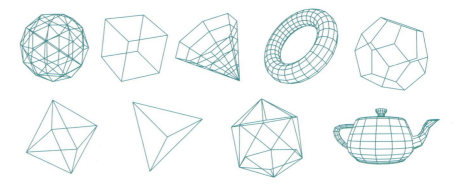

Figure 3.14: Wireframe GLUT objects.

Experiment **3.15.** Run `glutObjects.cpp`. Press the arrow keys to cycle through the various GLUT objects and 'x/X', 'y/Y' and 'z/Z' to turn them.

End

3.9 Clipping Planes

We saw in Section 2.4 that OpenGL clips a scene to within a viewing volume (box or frustum), a process which can be thought of as clipping the scene off on one side of each of the six planes which bound the volume. These six planes, called *clipping planes*, are automatically implied by the projection statement, such as `glOrtho()` or `glFrustum()`, which defines the box or frustum. However, the programmer can specify additional clipping planes. The call

glClipPlane(GL_CLIP_PLANEi, *equation*);

specifies an ith additional clipping plane, where *equation* points to an array giving the four coefficients of the equation

$$Ax + By + Cz + D = 0$$

of the new clipping plane. If this plane is enabled with the call glEnable(GL_CLIP_PLANEi), then the points (x, y, z) of objects which lie in the open half-space

$$Ax + By + Cz + D < 0$$

are clipped off; equivalently, only those points (x, y, z) of objects lying in the closed half-space

$$Ax + By + Cz + D \geq 0$$

are rendered. The ith additional clipping plane is disabled with a call to glDisable(GL_CLIP_PLANEi).

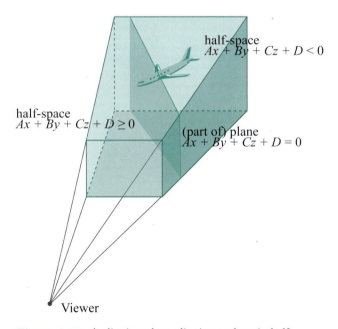

Figure 3.15: A clipping plane clipping a plane in half.

In Figure 3.15, for example, only the front part of the aircraft will be visible. Although we depict it in the figure, the clipping plane itself is not drawn by OpenGL.

Figure 3.16: Screenshot of `clippingPlanes.cpp`.

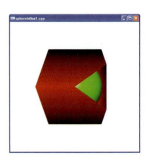

Figure 3.17: Screenshot of `sphereInBox1.cpp` with a corner clipped off.

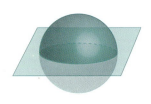

Figure 3.18: Clipping a sphere to make a hemisphere: the clipped half is computed *and* suppressed.

Experiment 3.16. Run `clippingPlanes.cpp`, which augments `circular-Annuluses.cpp` with two additional clipping planes which can be toggled on and off by pressing '0' and '1', respectively.

The first plane clips off the half-space $-z + 0.25 < 0$, i.e., $z > 0.25$, removing the floating white disc of the annulus on the upper-right. The second one clips off the half-space $x + 0.5y < 60.0$, which is the space below an angled plane parallel to the z-axis. Figure 3.16 is a screenshot of both clipping planes activated. **End**

Exercise 3.13. (Programming) Change the equations of the two clipping planes of `clippingPlanes.cpp` so that enabling both leaves only the red disc of the upper-right annulus visible.

Example 3.1. Replace the data

```
double eqn0[4] = 0.0, 0.0, -1.0, 0.25;
```

for the first clipping plane of `clippingPlanes.cpp` with

```
double eqn0[4] = 0.0, 0.0, 1.0, -0.25;
```

Apparently, we are replacing $-z + 0.25 = 0$ with $z - 0.25 = 0$, which are both equations of the same plane. Why, then, is the result of clipping different for the two?

Answer: The half-space clipped, given the equation $-z + 0.25 = 0$, is $-z + 0.25 < 0$, i.e., $z > 0.25$. On the other hand, the half-space clipped, given the equation $z - 0.25 = 0$, is $z - 0.25 < 0$, i.e., $z < 0.25$.

Exercise 3.14. (Programming) Add a clipping plane to `sphereIn-Box1.cpp` to clip off a corner of the box, revealing the sphere inside. Your output should look like Figure 3.17.
Hint: The plane $x + y + z - 3.0 = 0$ cuts off a corner at the front upper right.

Exercise 3.15. (Programming) Add a clipping plane to `moveSphere.cpp` to turn the movable sphere into a movable hemisphere.

Clipping planes cause OpenGL to not display parts of an object which are otherwise computed. For example, if one draws a hemisphere by clipping off half a sphere, then OpenGL first computes geometric data (vertices, etc.) for the entire sphere and then suppresses the part on one side of the clipping plane before rendering. See Figure 3.18. Clearly, this is doubly inefficient for the suppressed part, as OpenGL computes the location of each of its vertices, *and then* computes again to decide that they are actually on the invisible side of a clipping plane! It is always more efficient to draw an object as a whole. On the other hand, clipping planes are ideal for the purpose of displaying a *cut-away view* of an object, as in Figure 3.17.

Bottom line: Use clipping planes as a viewing and not a drawing device.

3.10 gluPerspective()

The statement

gluPerspective(*fovy, aspect, near, far*);

calls a utility library routine built on top of glFrustum(), the perspective projection command introduced in Section 2.8. It creates a viewing frustum as does glFrustum(). However, the frustum is specified differently:

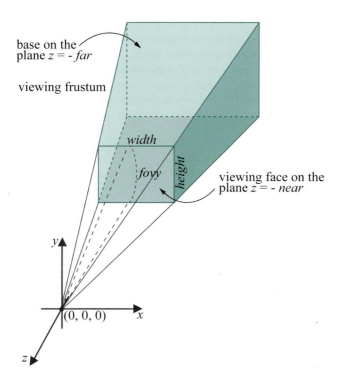

Figure 3.19: Viewing frustum created by gluPerspective(*fovy, aspect, near, far*).

The parameter *fovy*, called the *field of view angle*, is the angle along the *yz*-plane at the apex of the pyramid (of which the frustum is a truncation); *aspect* is the *aspect ratio = width/height* of the front face of the frustum; and *near* and *far* remain as for glFrustum(). See Figure 3.19. These four parameters it turns out are, in fact, enough for OpenGL to determine the eight vertices of a frustum which is *symmetric* about the *z*-axis, in other words, a frustum corresponding to a

glFrustum(*left, right, bottom, top, near, far*);

call where *left = −right* and *bottom = −top*. Such frustums are, in fact, most typical in applications and rarely does one have occasion to create one not symmetric about the *z*-axis.

Example **3.2.** The projection statement of `hemisphere.cpp` is the symmetric

```
glFrustum(-5.0, 5.0, -5.0, 5.0, 5.0, 100.0);
```

Determine the equivalent `gluPerspective()` call.

Answer: The aspect ratio of the front face of the frustum created by `glFrustum(-5.0, 5.0, -5.0, 5.0, 5.0, 100.0)` is 1 as both its width ($= right - left$) and height ($= top - bottom$) are 10.0. To determine $fovy$, see Figure 3.20, which shows the section of the viewing frustum by the yz-plane. By elementary trigonometry the half-angle at the apex is 45°, so that $fovy = 90.0$. Therefore, the equivalent call is

```
gluPerspective(90.0, 1.0, 5.0, 100.0);
```

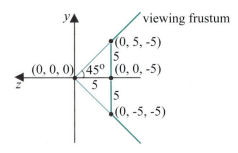

Figure 3.20: Section by the yz-plane (i.e., $x = 0$ plane) of the viewing frustum (bold) created by `glFrustum(-5.0, 5.0, -5.0, 5.0, 5.0, 100.0)`.

Check it out! Replace the current `glFrustum()` projection statement of `hemisphere.cpp` with the `gluPerspective()` statement just computed.

Exercise **3.16.** Change the *near* value of the projection statement of `hemisphere.cpp` as follows:

```
glFrustum(-5.0, 5.0, -5.0, 5.0, 10.0, 100.0);
```

Determine the equivalent `gluPerspective()` call.

Exercise **3.17.** Determine the equivalent `glFrustum()` call of the following projection statement:

```
gluPerspective(60.0, 2.0, 10.0, 100.0);
```

Hint: Use trigonometry in the yz-section to determine first the *top* and *bottom* values and then the aspect ratio to determine *left* and *right*.

Whether to define a perspective projection by a `glFrustum()` or a `gluPerspective()` call is a matter of personal preference. As we've seen, they are equivalent provided one is interested only in frustums symmetric about the z-axis.

However, a convenience of `gluPerspective()` in certain applications arises from the fact that the aspect ratio of the viewing face is an explicit parameter, making it easy to bind it to the aspect ratio of the OpenGL window itself. This comes in handy if you recall the final step of the rendering process when the viewing face is scaled to fit onto the OpenGL window – resulting in distortion if the aspect ratios of the two differ.

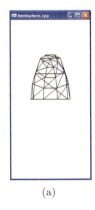
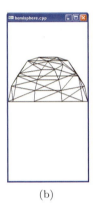

(a)   (b)

Figure 3.21: Screenshots of `hemisphere.cpp` with the window squished and the projection statement (a) `glFrustum(-5.0, 5.0, -5.0, 5.0, 5.0, 100.0)` and (b) gluPerspective(90.0, (float)w/(float)h, 5.0, 100.0).

Experiment 3.17. Run `hemisphere.cpp`.

The initial OpenGL window is a square 500×500 pixels. Drag a corner to change its shape, making it tall and thin. The hemisphere is distorted to become ellipsoidal (Figure 3.21(a)). Replace the perspective projection statement

```
glFrustum(-5.0, 5.0, -5.0, 5.0, 5.0, 100.0);
```

with

```
gluPerspective(90.0, 1.0, 5.0, 100.0);
```

As this is equivalent to the original `glFrustum()` call, there is still distortion if the window's shape is changed. Next, replace the projection statement with

```
gluPerspective(90.0, (float)w/(float)h, 5.0, 100.0);
```

which sets the aspect ratio of the viewing frustum equal to that of the OpenGL window. Resize the window – the hemisphere is no longer distorted (Figure 3.21(b))! **End**

GLU call gluOrtho2D()

There's a GLU projection statement, `gluOrtho2D()`, paired with `glOrtho()`. The call `gluOrtho2D(`*left, right, bottom, top*`)` is equivalent to `glOrtho(`*left, right, bottom, top,* −1, 1`)`. Additionally, z-coordinates of objects in the scene are assumed all to lie between −1 and 1, so that there is no need to clip along the z-axis.

In particular, if you're drawing a flat scene using 2D vertex commands, e.g., `glVertex2f()`, which implies z-coordinates are all zero, then it's more efficient to use `gluOrtho2D()`.

3.11   Viewports

The *viewport* of a scene is that region of the OpenGL window in which it is drawn. By default, it is the entire window. However, a `glViewPort()` call may be used to draw to a smaller rectangular subregion.

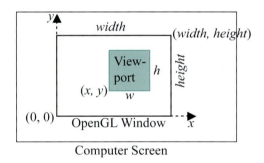

Figure 3.22: Viewport specified by `glViewPort(x, y, w, h)`.

The call `glViewPort(`x, y, w, h`)` specifies the viewport as the rectangular subregion of the OpenGL window which has its lower-left corner at the point (x, y), and is of width w and height h. Units are pixels and the coordinates in the OpenGL window are such that the origin is located at the lower-left corner, the increasing direction of the x-axis rightwards, and that of the y-axis upwards. See Figure 3.22.

Multiple viewports can be created in a single OpenGL window by invoking more than one `glViewPort()` call in the drawing routine. The contents of a particular viewport are defined by the statements between its defining `glViewPort()` call and the next one (if any).

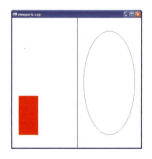

Figure 3.23: Screenshot of `viewports.cpp`.

Experiment 3.18. Run `viewports.cpp` where the screen is split into two viewports with contents a square and a circle, respectively. Figure 3.23 is a screenshot.

A vertical black line is drawn (in the program) at the left end of the second viewport to separate the two. As the aspect ratio of both viewports differs from that of the viewing face, the square and circle are squashed laterally. **End**

Viewports are particularly useful in games to show split-screen views of different scenes, or perhaps the same scene from different cameras. We'll see a nice application of this in the program `spaceTravel.cpp` coming up in the next chapter, where the user maneuvers a spacecraft through an asteroid field with a split-screen view in the OpenGL window – one from a global fixed camera and the other from the craft itself.

Exercise 3.18. (Programming) Change the orthographic projection statement of `viewports.cpp` so that the square and circle are no longer distorted.

Exercise 3.19. (Programming) Create a 2×2 grid of four equally-sized viewports with one of the words "This", "is", "so" and "easy" in each. Add lines to separate the viewports.

3.12 Multiple Windows

The `glutCreateWindow()` call of the GLUT library may be invoked more than once in the main routine to create multiple top-level OpenGL windows. Each call to `glutCreateWindow()` returns an integer *id* which is then passed to a `glutSetWindow(id)` call in the display routine to determine the current drawing window. Properties such as the display routine, resize routine, etc., of each top-level window may be specified independently.

Experiment 3.19. Run `windows.cpp`, which creates two top-level windows (Figure 3.24). **End**

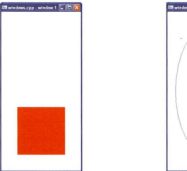 

Figure 3.24: Screenshot of `windows.cpp`.

Exercise 3.20. (Programming) Create three top-level windows with red, green and blue backgrounds, and containing the words "Red", "Green" and "Blue", respectively.

3.13 Summary, Notes and More Reading

In this chapter we learned a number of different coding utilities, none difficult, but all useful. The coverage of the syntax was by no means complete, nor was it intended to be. For OpenGL utilities the reader should refer to the red and blue books for a full description, while the canonical source for GLUT syntax is its own page [53] at the OpenGL site. Lighthouse 3D [83] has tutorials on the GLUT and display lists in addition to others. NeHe Productions [98] has, among many, tutorials on fonts and display lists.

Keep in mind that the GLUT was written (by Mark Kilgard) to provide a few easy-to-use window-system independent utilities to help build a simple GUI, not to provide a full-featured suite. GLUT is more than adequate for basic interactivity, and certainly for our purpose, as our interest is in the principles of CG more than in writing full-blown applications.

Programmers who do require sophisticated interfaces should employ utilities specific to their particular platform, e.g, the MFC Library for Windows. Readers may also find helpful Paul Rademacher's GLUI User Interface Library [52], which provides a collection of window-system independent GLUT-based utilities such as buttons and checkboxes. Trolltech's Qt [136] may be of interest to those planning commercial-grade GUI's.

Part III

Movers and Shapers

CHAPTER 4

Transformation, Animation and Viewing

The goal for this chapter is to understand how to move and manipulate objects, and maneuver the camera, skills essential for making movies and games. OpenGL will provide the laboratory for us to explore. The modeling transformations of OpenGL – including translation, scaling and rotation – control object motion, while the viewing transformation manages the camera. We'll examine the syntax of the transformation commands and how they are composed and applied to achieve animation. To efficiently and creatively animate it's essential to have some grasp of its implementation, so we'll examine parts of OpenGL's animation engine as well. An experiment-discuss-repeat approach is used throughout, each new idea introduced and illustrated with the help of live code.

When objects move, especially in an interactive and unscripted environment like that of a game, they can collide. We'll discuss collision detection, therefore, in the context of animation. Related to animation as well is the notion of the orientation of an object and we'll explain a way to measure orientation by so-called Euler angles.

Section 4.1 introduces the three modeling transformations – translation, rotation and scaling. Sections 4.2 and 4.3 discuss composing modeling transformations and how such compositions place multiple objects relative to one another. The modelview matrix stack facilitates the application of transformations to multiple objects, as we see in Section 4.4. In Section 4.5 we analyze a few instructional animation programs and conclude with a bunch of exercises.

The viewing transformation is introduced in Section 4.6. After assimilating its functioning we find that a viewing transformation is actually simulated by OpenGL with the help of modeling transformations. An understanding of the viewing transformation leads to a preliminary discussion in Section 4.6, as well, of orientation and how it is specified by Euler angles. We present as well an application of the viewing transformation to animate a camera, together with rudimentary collision detection, in a space-travel program.

More animation code, including programs to develop key-frame animation sequences for a man-like articulated figure, as well as simple shadow animation, is presented in Section 4.7.

Section 4.8 describes methods to enable a user to choose an object on the screen with a mouse-like device, a facility critical in interactive programs like games. Section 4.9 concludes the chapter with a summary, notes and suggestions for more reading.

This is a long chapter but it gets you well on the way to designing realistic 3D applications.

4.1 Modeling Transformations

Translation, scaling and rotation, the so-called *modeling transformations* of OpenGL, are applied to objects to change their location and shape.

4.1.1 Translation

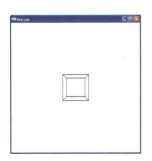

Eᴈperiment 4.1. Run box.cpp, which shows an axis-aligned – i.e., with sides parallel to the coordinate axes – GLUT wireframe box of dimensions $5 \times 5 \times 5$. Figure 4.1 is a screenshot. Note the foreshortening – the back of the box appears smaller than the front – because of perspective projection in the viewing frustum specified by the glFrustum() statement. Comment out the statement

```
glTranslatef(0.0, 0.0, -15.0);
```

What do you see now? *Nothing!* We'll explain why momentarily.   End

Figure 4.1: Screenshot of box.cpp.

The *translation* command glTranslatef(p, q, r) translates an object p units in the x-direction, q units in the y-direction and r units in the z-direction. Precisely, each point (x, y, z) of the object is mapped to the point $(x + p, y + q, z + r)$. See Figure 4.2, which also shows a whole box translated by glTranslatef(p, q, r).

Returning to box.cpp, the command glutWireCube(5.0) itself creates a box of side length 5 centered at the origin, with vertices, therefore, at $(\pm 2.5, \pm 2.5, \pm 2.5)$, each vertex corresponding to one of the eight possible combinations of signs. The box clearly lies entirely outside the viewing frustum specified by glFrustum(-5.0, 5.0, -5.0, 5.0, 5.0, 100.0) – in fact, entirely on the clipped side of the viewing plane $z = -5$. However, glTranslatef(0.0, 0.0, -15.0) pushes the box 15 units in the $-z$

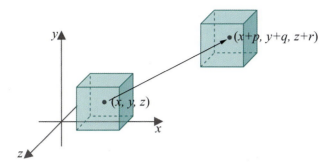

Figure 4.2: Translation: `glTranslatef(`*p*, *q*, *r*`)`.

direction, to place it inside the viewing frustum and make it visible (see Figure 4.3). That is why commenting out this statement results in a blank window.

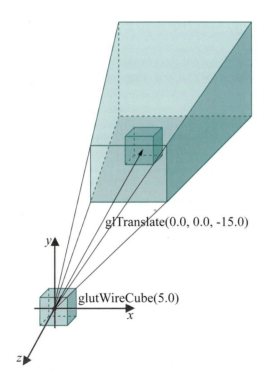

Figure 4.3: Translating into the viewing frustum.

Experiment 4.2. Successively replace the translation command of `box.cpp` with the following, making sure that what you see matches your understanding of where the command places the box. Keep in mind

93

foreshortening, as well as clipping to within the viewing frustum.

1. `glTranslatef(0.0, 0.0, -10.0)`

2. `glTranslatef(0.0, 0.0, -5.0)`

3. `glTranslatef(0.0, 0.0, -25.0)`

4. `glTranslatef(10.0, 10.0, -15.0)`

End

Exercise 4.1. To what point is $(-2.0, 3.0, 9.0)$ transformed by `glTranslatef(3.0, 1.0, -8.0)`?

Exercise 4.2. What is the OpenGL translation that takes $(3.0, -1.0, 2.0)$ to $(3.0, 5.0, 9.0)$?

Figure 4.4: Screenshot of Experiment 4.3.

4.1.2 Scaling

Experiment 4.3. Add a scaling command, in particular, replace the modeling transformation block of `box.cpp` with (Block 1*):

```
// Modeling transformations.
glTranslatef(0.0, 0.0, -15.0);
glScalef(2.0, 3.0, 1.0);
```

Figure 4.4 is a screenshot – compare with the unscaled box of Figure 4.1.

End

Remark 4.1. The `glTranslatef(0.0, 0.0, -15.0)` call is retained to "kick" the scaled box into the viewing frustum.

Precisely, the *scaling* command `glScalef(u, v, w)` maps each point (x, y, z) of an object to the point (ux, vy, wz). This has the effect of *stretching* objects by a factor of u in the x-direction, v in the y-direction, and w in the z-direction. See Figure 4.5.

As the box created by `glutWireCube(5.0)` is $5 \times 5 \times 5$ in size, `glScalef(2.0, 3.0, 1.0)` of the preceding experiment *apparently* scales it to the dimensions $2 * 5 \times 3 * 5 \times 1 * 5 = 10 \times 15 \times 5$. The way to actually verify this is by determining the vertices of the scaled box, whose coordinates are obtained from those of the original vertices by the transformation $(x, y, z) \mapsto (2x, 3y, 1z)$. For example, $(2.5, 2.5, 2.5) \mapsto (5.0, 7.5, 2.5)$, $(-2.5, 2.5, 2.5) \mapsto (-5.0, 7.5, 2.5)$ and so on. It's seen then that the new vertices are $(\pm 5.0, \pm 7.5, \pm 2.5)$, which, indeed, give a $10 \times 15 \times 5$ box.

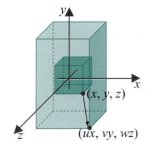

Figure 4.5: Scaling: glScalef(u, v, w).

---

*To cut-and-paste you can find the block in text format in the file `chap4codeModifications.txt` in the directory `Code/CodeModifications`.

Experiment 4.4. An object less symmetric than a box is more interesting to work with. How about a teapot? Accordingly, change the modeling transformation and object definition part of `box.cpp` to (Block 2):

```
// Modeling transformations.
glTranslatef(0.0, 0.0, -15.0);
glScalef(1.0, 1.0, 1.0);

glutWireTeapot(5.0); // Teapot.
```

Of course, `glScalef(1.0, 1.0, 1.0)` does nothing and we see the original unscaled teapot (Figure 4.6).

Next, successively change the scaling parameters by replacing the scaling command with the ones below. In each case make, sure your understanding of the command matches the change that you see in the shape of the teapot.

1. `glScalef(2.0, 1.0, 1.0)`

2. `glScalef(1.0, 2.0, 1.0)`

3. `glScalef(1.0, 1.0, 2.0)` $\qquad$ **End**

Exercise 4.3. (Programming) Continuing with the preceding experiment, try to guess first, for the scalings below, each of which has at least one negative parameter, the difference you will see from the initial configuration shown in Figure 4.6.
Hint: The transformation $(x, y, z) \mapsto (-x, y, z)$, for instance, is a mirror-like reflection about the yz-plane. See Figure 4.7.

4. `glScalef(-1.0, 1.0, 1.0)`

5. `glScalef(1.0, -1.0, 1.0)`

6. `glScalef(1.0, 1.0, -1.0)`

7. `glScalef(-1.0, -1.0, 1.0)`

Exercise 4.4. (Programming) Continue with the preceding exercise and replace the scaling command with the following, each of which has a zero parameter:

8. `glScalef(1.0, 1.0, 0.0)`

Hint: The transformation

$$(x, y, z) \mapsto (1x, 1y, 0z) = (x, y, 0)$$

"collapses" all z-values to 0.0.

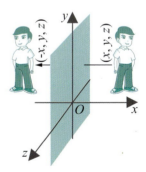

Figure 4.6: Screenshot of initial configuration of Experiment 4.4.

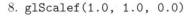

Figure 4.7: Reflection in the yz-plane.

10. `glScalef(1.0, 0.0, 1.0)`

11. `glScalef(0.0, 1.0, 1.0)`

Not very interesting the last two! A scaling transformation where one or more of the scaling factors is zero is said to be *degenerate*. Although not common, there is the occasional application where a degenerate scaling transformation comes in handy. We'll see one such in drawing a shadow later in this chapter in Experiment 4.34.

Exercise 4.5. To what point is $(-2.0, 3.0, 9.0)$ transformed by `glScalef-(3.0, 1.0, -8.0)`?

Exercise 4.6. What is the OpenGL scaling that transforms $(3.0, -1.0, 2.0)$ to $(3.0, 5.0, 9.0)$?

We have so far scaled only GLUT wire cubes and teapots, whose own axes are aligned with the coordinate axes, so that, effectively, they are only stretched and not skewed. Let's try one that's not so aligned.

Experiment 4.5. Replace the cube of `box.cpp` with a square whose sides are not parallel to the coordinate axes. In particular, replace the modeling transformation and object definition part of that program with (Block 3):

```
// Modeling transformations.
   glTranslatef(0.0, 0.0, -15.0);
// glScalef(1.0, 3.0, 1.0);

glBegin(GL_LINE_LOOP);
   glVertex3f(4.0, 0.0, 0.0);
   glVertex3f(0.0, 4.0, 0.0);
   glVertex3f(-4.0, 0.0, 0.0);
   glVertex3f(0.0, -4.0, 0.0);
glEnd();
```

See Figure 4.8(a). Verify by elementary geometry that the line loop forms a square with sides of length $4\sqrt{2}$ angled at $45°$ to the axes.

Uncomment the scaling. See Figure 4.8(b). The square now seems skewed to a non-rectangular parallelogram. Apply the transformation $(x, y, z) \mapsto (x, 3y, z)$ to each vertex of the original square to verify that the new shape is indeed a parallelogram. **End**

4.1.3 Rotation

Experiment 4.6. Add a rotation command by replacing the modeling transformation and object definition part – we prefer a teapot – of `box.cpp` with (Block 4):

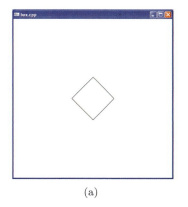 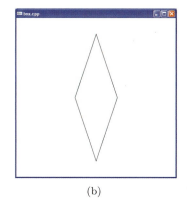

(a)   (b)

Figure 4.8: Screenshots of Experiment 4.5: (a) before scaling (b) after.

```
// Modeling transformations.
glTranslatef(0.0, 0.0, -15.0);
glRotatef(60.0, 0.0, 0.0, 1.0);

glutWireTeapot(5.0);
```

Figure 4.9 is a screenshot.

The *rotation* command `glRotatef(A, p, q, r)` rotates each point of an object about an axis along the line from the origin $O = (0, 0, 0)$ to the point (p, q, r). The amount of rotation is $A°$, measured counter-clockwise when looking *from* (p, q, r) to the origin. In this experiment, then, the rotation is 60° CCW (counter-clockwise) looking down the z-axis.   **End**

Let's make what happens on a rotation `glRotatef(A, p, q, r)` explicit as a physical process:

Assume that $(p, q, r) \neq O$ so that an axis l through (p, q, r) and O can be indeed drawn. Now, first, if a given point P lies on l itself then the situation is simple – the rotation does not move it. Suppose, then, that P does not lie on l. Here's how it's mapped by the rotation (see Figure 4.10):

Figure 4.9: Screenshot for Experiment 4.6

1. Drop the perpendicular from P to the point Q on l. Call the segment PQ as L. L lies on the plane h perpendicular to l through Q.

   Note that Figure 4.10 has Q and h on the same side of the origin as (p, q, r), but they could very well be on the other side, or even touching the origin, depending on where P is.

2. Locate a viewer at V far enough along l, on the side of (p, q, r), as to be able to see h when looking toward the origin.

3. Rotate the segment L about Q on h through an angle $A°$ counter-clockwise, as measured by the viewer.

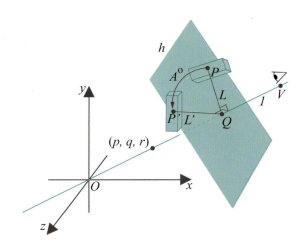

Figure 4.10: Rotation: `glRotatef`(A, p, q, r).

4. If L' is the new position of L after rotation, then P is mapped to the corresponding endpoint P' of L'.

Note: In Experiment 4.6, the axis of rotation, the z-axis, happens to intersect the object rotated, which is the teapot.

$\mathbf{E}$xperiment 4.7. Continuing with Experiment 4.6, successively replace the rotation command with the ones below, in each case trying to match what you see with your understanding of how the command should turn the teapot. (It can occasionally be a bit confusing because of the perspective projection.)

1. `glRotatef(60.0, 0.0, 0.0, -1.0)`

2. `glRotatef(-60.0, 0.0, 0.0, 1.0)`

3. `glRotatef(60.0, 1.0, 0.0, 0.0)`

4. `glRotatef(60.0, 0.0, 1.0, 0.0)`

5. `glRotatef(60.0, 1.0, 0.0, 1.0)` $\qquad$ $\mathbf{E}$nd

The alert reader probably noticed in the 4-step definition of rotation earlier, that the purpose of the point (p, q, r), apart from specifying the axis l joining it to the origin, is to specify the *side* of the origin on l that the viewer is located. If (p, q, r) were replaced by another point (p', q', r') on l on the same side of O as (p, q, r), then the rotation would be *exactly same*. This is illustrated in the next experiment.

$\mathbf{E}$xperiment 4.8. Appropriately modify `box.cpp` to compare the effects of each of the following pairs of rotation commands:

1. `glRotatef(60.0, 0.0, 0.0, 1.0)` and `glRotatef(60.0, 0.0, 0.0, 5.0)`

2. `glRotatef(60.0, 0.0, 2.0, 2.0)` and `glRotatef(60.0, 0.0, 3.5, 3.5)`

3. `glRotatef(60.0, 0.0, 0.0, -1.0)` and `glRotatef(60.0, 0.0, 0.0, -7.5)`

There is no difference in each case. One concludes that the rotation command `glRotatef`(A, p, q, r) is equivalent to `glRotatef`$(A, \alpha p, \alpha q, \alpha r)$, where α is any *positive* scalar. **End**

Exercise 4.7. Relate the three commands `glRotatef`(A, p, q, r), `glRotatef`$(-A, p, q, r)$ and `glRotatef`$(A, \beta p, \beta q, \beta r)$, where β is a *negative* scalar.

The description of rotation as a 4-step physical process was quite intuitive. Unfortunately, though, the general *formula* for how a point $P = (x, y, z)$ is mapped by the rotation `glRotatef(`A`, `p`, `q`, `r`)` is more complicated – in fact, significantly more so than the corresponding formulae in case of translation and scaling that we have already given – and we defer its deduction to the next chapter. However, we do ask the reader to derive the formula in three particularly simple cases – where the rotation is about a coordinate axis – in the following exercise.

Exercise 4.8. Deduce the formula for how $P = (x, y, z)$ is mapped by each of the rotations

(a) `glRotatef(`A`, 1.0, 0.0, 0.0)`

(b) `glRotatef(`A`, 0.0, 1.0, 0.0)`

(c) `glRotatef(`A`, 0.0, 0.0, 1.0)`

Part answer: See Figure 4.11 for (c). The axis of rotation is the z-axis. The point $P = (x, y, z)$ is mapped to $P' = (x', y', z')$. We'll find expression for x', y' and z' in terms x, y, z and the angle parameter A.

Draw $L = PQ$, the perpendicular from P to the z-axis. Further, draw the line k through Q parallel to the x-axis and the perpendicular PR from P to k. If $\angle PQR = \alpha$, then

$$\begin{aligned} x &= |RQ| &= |L| \cos \alpha \\ y &= |RP| &= |L| \sin \alpha \end{aligned}$$

Now, the rotated segment $L' = P'Q$ makes an angle of $\alpha + A$ with k, so that $\angle P'QR' = \alpha + A$, where R' is the foot of the perpendicular from P' to k. Therefore,

$$\begin{aligned} x' &= |L'| \cos(\alpha + A) &= |L| \cos(\alpha + A) \\ y' &= |L'| \sin(\alpha + A) &= |L| \sin(\alpha + A) \end{aligned}$$

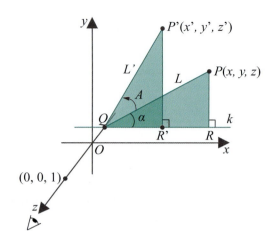

Figure 4.11: `glRotatef(A, 0.0, 0.0, 1.0)`.

because rotation does not change the length of L. Apply trigonometric formulae to expand the rightmost sides of the two equations above:

$$x' = |L| \cos \alpha \cos A - |L| \sin \alpha \sin A = x \cos A - y \sin A$$
$$y' = |L| \cos \alpha \sin A + |L| \sin \alpha \cos A = x \sin A + y \cos A$$

using the expressions for x and y derived earlier. And, of course,

$$z' = z$$

because rotation about the z-axis does not change the z-value.

Exercise 4.9. To what point is $(2.0, 3.0, 9.0)$ transformed by

(a) `glRotatef(90.0, 0.0, 0.0, 1.0)`?

(b) `glRotatef(90.0, 0.0, 0.0, 5.0)`?

(c) `glRotatef(90.0, 0.0, 0.0, -5.0)`?

(d) `glRotatef(60.0, 0.0, 0.0, 1.0)`?

(e) `glRotatef(180.0, 0.0, 1.0, 0.0)`?

(f) `glRotatef(45.0, 1.0, 0.0, 0.0)`?

4.2 Composing Modeling Transformations

In most of the previous experiments we successively applied more than one modeling transformation to an object, but never explained exactly how it is that OpenGL goes about *composing* multiple transformations. There is magic to this as we'll see, but first a couple of motivating experiments.

Experiment 4.9. Apply three modeling transformations by replacing the modeling transformations block of `box.cpp` with (Block 5):

```
// Modeling transformations.
glTranslatef(0.0, 0.0, -15.0);
glTranslatef(10.0, 0.0, 0.0);
glRotatef(45.0, 0.0, 0.0, 1.0);
```

It seems the box is *first* rotated 45° about the *z*-axis and *then* translated right 10 units. See Figure 4.12(a). The first translation `glTranslatef(0.0, 0.0, -15.0)`, of course, serves only to "kick" the box down the *z*-axis into the viewing frustum.

Next, interchange the last two transformations, namely, the rightward translation and the rotation, by replacing the modeling transformations block with (Block 6):

```
// Modeling transformations.
glTranslatef(0.0, 0.0, -15.0);
glRotatef(45.0, 0.0, 0.0, 1.0);
glTranslatef(10.0, 0.0, 0.0);
```

It seems that the box is now *first* translated right and *then* rotated about the *z*-axis causing it "rise". See Figure 4.12(b).   **End**

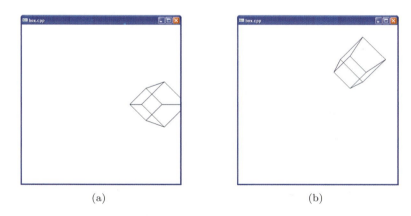

(a)   (b)

Figure 4.12: Screenshots from Experiment 4.9.

Exercise 4.10. (Programming) Again, apply three modeling transformations, this time by replacing the modeling transformations block of `box.cpp` with (Block 7):

```
// Modeling transformations.
glTranslatef(0.0, 0.0, -15.0);
glRotatef(45.0, 0.0, 0.0, 1.0);
glScalef(1.0, 3.0, 1.0);
```

Interchange the rotation and scaling by replacing the modeling transformation block with (Block 8):

```
// Modeling transformations.
glTranslatef(0.0, 0.0, -15.0);
glScalef(1.0, 3.0, 1.0);
glRotatef(45.0, 0.0, 0.0, 1.0);
```

Keeping the conclusions of the preceding experiment in mind, can you explain what you see?

Apparently transformations are applied to an object in *backward* order through the code from where the object is created! This is correct and, hopefully, once it's explained how the OpenGL engine composes transformations, will not seem as idiosyncratic as it might at first.

We need, though, a quick acquaintance first with a concept which will be discussed in depth in the next chapter – that transformations correspond to matrices. We'll present here just enough that the reader can follow along. Our goal is a conceptual understanding of how transformations are composed.

A vertex $V = (x, y, z)$ is represented in OpenGL as a 4×1 column matrix

$$\begin{bmatrix} x \\ y \\ z \\ 1 \end{bmatrix}$$

Take that extra 1 in the 4th row for granted for now – it's to do with so-called homogeneous coordinates. We'll use V to denote this column matrix as well.

A modeling transformation t is represented by a 4×4 matrix of the form

$$M = \begin{bmatrix} a_{11} & a_{12} & a_{13} & a_{14} \\ a_{21} & a_{22} & a_{23} & a_{24} \\ a_{31} & a_{32} & a_{33} & a_{24} \\ a_{41} & a_{42} & a_{43} & a_{44} \end{bmatrix}$$

Applying this transformation to the vertex V consists of multiplying V from the left by the transformation matrix. In particular, V is transformed by t to the vertex $t(V)$ where

$$t(V) = MV = \begin{bmatrix} a_{11} & a_{12} & a_{13} & a_{14} \\ a_{21} & a_{22} & a_{23} & a_{24} \\ a_{31} & a_{32} & a_{33} & a_{24} \\ a_{41} & a_{42} & a_{43} & a_{44} \end{bmatrix} \begin{bmatrix} x \\ y \\ z \\ 1 \end{bmatrix}$$

$$= \begin{bmatrix} a_{11}x + a_{12}y + a_{13}z + a_{14} \\ a_{21}x + a_{22}y + a_{23}z + a_{24} \\ a_{31}x + a_{32}y + a_{33}z + a_{24} \\ a_{41}x + a_{42}y + a_{43}z + a_{44} \end{bmatrix}$$

Here's an example.

Example 4.1. The transformation t_1 given by the translation command `glTranslatef(5.0, 0.0, 0.0)` corresponds to the matrix

$$M_1 = \begin{bmatrix} 1.0 & 0.0 & 0.0 & 5.0 \\ 0.0 & 1.0 & 0.0 & 0.0 \\ 0.0 & 0.0 & 1.0 & 0.0 \\ 0.0 & 0.0 & 0.0 & 1.0 \end{bmatrix}$$

This is verified by the multiplication

$$\begin{bmatrix} 1.0 & 0.0 & 0.0 & 5.0 \\ 0.0 & 1.0 & 0.0 & 0.0 \\ 0.0 & 0.0 & 1.0 & 0.0 \\ 0.0 & 0.0 & 0.0 & 1.0 \end{bmatrix} \begin{bmatrix} x \\ y \\ z \\ 1 \end{bmatrix} = \begin{bmatrix} x + 5.0 \\ y \\ z \\ 1 \end{bmatrix}$$

Similarly, verify that the transformation t_2 given by the translation command `glTranslatef(0.0, 10.0, 0.0)` corresponds to the matrix

$$M_2 = \begin{bmatrix} 1.0 & 0.0 & 0.0 & 0.0 \\ 0.0 & 1.0 & 0.0 & 10.0 \\ 0.0 & 0.0 & 1.0 & 0.0 \\ 0.0 & 0.0 & 0.0 & 1.0 \end{bmatrix}$$

Now, if one applies t_2 followed by t_1 to a vertex V, then V is mapped as follows:

$$V \mapsto t_1(t_2(V)) = M_1(M_2 V) = (M_1 M_2) V$$

(the associativity of matrix multiplication was applied in the second equality). The skeptical reader may multiply matrices as below to verify that

$$
\begin{aligned}
(M_1 M_2) V &= \left(\begin{bmatrix} 1.0 & 0.0 & 0.0 & 5.0 \\ 0.0 & 1.0 & 0.0 & 0.0 \\ 0.0 & 0.0 & 1.0 & 0.0 \\ 0.0 & 0.0 & 0.0 & 1.0 \end{bmatrix} \begin{bmatrix} 1.0 & 0.0 & 0.0 & 0.0 \\ 0.0 & 1.0 & 0.0 & 10.0 \\ 0.0 & 0.0 & 1.0 & 0.0 \\ 0.0 & 0.0 & 0.0 & 1.0 \end{bmatrix} \right) \begin{bmatrix} x \\ y \\ z \\ 1 \end{bmatrix} \\
&= \begin{bmatrix} 1.0 & 0.0 & 0.0 & 5.0 \\ 0.0 & 1.0 & 0.0 & 10.0 \\ 0.0 & 0.0 & 1.0 & 0.0 \\ 0.0 & 0.0 & 0.0 & 1.0 \end{bmatrix} \begin{bmatrix} x \\ y \\ z \\ 1 \end{bmatrix} \\
&= \begin{bmatrix} x + 5.0 \\ y + 10.0 \\ z \\ 1 \end{bmatrix}
\end{aligned}
$$

which indeed corresponds to the code sequence

```
glTranslatef(5.0, 0.0, 0.0);
glTranslatef(0.0, 10.0, 0.0);
```

Put simply, the matrix of the composition of two transformations is the product of their matrices. This generalizes. If one applies successively the transformations $t_n, t_{n-1}, \ldots, t_1$ (in that order, t_n being first) to a vertex V, then it is mapped to

$$t_1(t_2(\ldots t_n(V)\ldots)) = M_1(M_2(\ldots(M_nV)\ldots)) = (M_1M_2\ldots M_n)V \quad (4.1)$$

again with the help of associativity of matrix multiplication, where matrix M_i corresponds to transformation t_i, $1 \le i \le n$. One sees that the matrix of the composition is again the product of the matrices.

We now have enough to explain how OpenGL composes transformations. Consider the code sequence:

```
modelingTransformation 1;    // t₁
modelingTransformation 2;    // t₂
...
modelingTransformation n-1;  // tₙ₋₁
modelingTransformation n;    // tₙ
object;
```

where the transformation t_i corresponds to the statement `modelingTransformation i`.

Now, OpenGL maintains a 4×4 *modelview matrix*, call it M, which is initialized to the identity

$$I = \begin{bmatrix} 1.0 & 0.0 & 0.0 & 0.0 \\ 0.0 & 1.0 & 0.0 & 0.0 \\ 0.0 & 0.0 & 1.0 & 0.0 \\ 0.0 & 0.0 & 0.0 & 1.0 \end{bmatrix}$$

As the drawing routine is processed during runtime, the modelview matrix is multiplied on the *right* by the matrix corresponding to each successive modeling transformation encountered. For example, assuming the matrix of t_i is M_i and that there were no earlier transformations, the successive values of the modelview matrix M for the code sequence above are indicated in the comments below:

```
                             // M  =  I, initially
modelingTransformation 1;    // M  =  IM₁  =  M₁
modelingTransformation 2;    // M  =  M₁M₂
...
modelingTransformation n-1;  // M  =  M₁M₂... Mₙ₋₁
modelingTransformation n;    // M  =  M₁M₂...   Mₙ₋₁Mₙ
object;
```

An object drawing statement is processed by multiplying the object's vertices from the left by the current modelview matrix, e.g., for the code sequence above, each vertex V of `object` is transformed as follows:

$$V \mapsto MV = (M_1M_2\ldots M_{n-1}M_n)V$$

However, by associativity

$$(M_1 M_2 \ldots M_{n-1} M_n)V \;=\; M_1(M_2(\ldots M_{n-1}(M_n V)\ldots))$$
$$=\; t_1(t_2(\ldots t_{n-1}(t_n(V)\ldots)))$$

We see, from the last line of the preceding equation, that first transformation t_n is applied to V, then t_{n-1} and so on, until, finally, t_1, *indeed backward in code order*!

The conclusion, then, is that the backward order in which OpenGL applies transformations is simply a consequence of the particular way it processes their matrices. It does take a little getting used to, but (trust us) by the end of this chapter you will be quite comfortable applying multiple transformations.

Remark 4.2. Here's another more informal way to understand how multiple transformations are applied. Transformation t_n, given by the statement `modelingTransformation n` which is *closest* in the code to `object`, is applied first, then, t_{n-1}, given by the next closest statement `modelingTransformation n-1`, etc. Indeed, transformations are applied to the `object` as one works *away* from it, which is not unfamiliar if one recalls evaluating mathematical expressions such as $\cos(\exp(\sin x))$.

Remark 4.3. There is one other kind of transformation, in addition to the three modeling transformations, which can modify the modelview matrix by multiplication from the right – the *viewing transformation* `gluLookAt()`. We'll discuss viewing transformations in Section 4.6. Modelview matrices, in fact, get their name from these two kinds of transformations.

Exercise 4.11. For each of the following, give (x, y, z) coordinates of the point where the center of the sphere is transformed by the given piece of code in the display routine.

(a)
```
glTranslatef(0.0, 2.0, 2.0);
glTranslatef(4.0, 0.0, 2.0);
glutWireSphere(2.0, 10, 8);
```

(b)
```
glRotatef(90.0, 0.0, 0.0, 1.0);
glTranslatef(4.0, 0.0, 0.0);
glutWireSphere(2.0, 10, 8);
```

(c)
```
glRotatef(90.0, 1.0, 0.0, 0.0);
glTranslatef(4.0, 0.0, 0.0);
glRotatef(90.0, 0.0, 0.0, 1.0);
glutWireSphere(2.0, 10, 8);
```

(d)
```
glRotatef(90.0, 1.0, 0.0, 0.0);
glRotatef(90.0, 0.0, 1.0, 0.0);
glTranslatef(4.0, 0.0, 0.0);
glutWireSphere(2.0, 10, 8);
```

(e)
```
glScalef(1.0, 2.0, 3.0);
glRotatef(45.0, 1.0, 0.0, 0.0);
glRotatef(90.0, 0.0, 0.0, 1.0);
glTranslatef(4.0, 0.0, 0.0);
glutWireSphere(2.0, 10, 8);
```

(f)
```
glRotatef(90.0, 0.0, 0.0, 1.0);
glRotatef(45.0, 1.0, 0.0, 0.0);
glRotatef(90.0, 0.0, 1.0, 0.0);
glTranslatef(4.0, 0.0, 0.0);
glutWireSphere(2.0, 10, 8);
```

Example 4.2. Replace the object definition statement

```
glutWireCube(5.0); // Box.
```

of box.cpp with

```
glRectf(5.0, 5.0, 10.0, 10.0); // Square
```

to draw, instead of a box centered at the origin, an axis-aligned square some ways north-east of it, centered at $(7.5, 7.5, 0.0)$.

Now, add transformation(s) to rotate the square $45°$ counter-clockwise about its *own center*, as indicated in Figure 4.13(a).

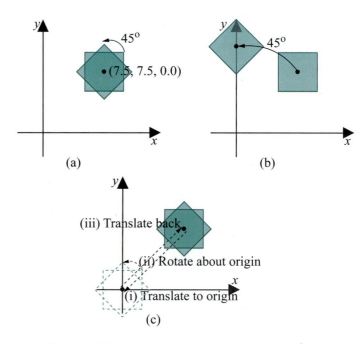

Figure 4.13: Rotating a square about its own center.

Answer: Inserting the command `glRotatef(45.0, 0.0, 0.0, 1.0)` just before `glRectf()` will not do as it rotates the square about the origin, and not its own center, as shown in Figure 4.13(b). What one must do instead (see Figure 4.13(c)) is first (i) translate the square so that its center is at the origin, then (ii) rotate it about the origin and, finally, (iii) translate it back. This is equivalent to the following modeling transformation block (Block 9):

```
// Modeling transformations.
glTranslatef(0.0, 0.0, -15.0);

glTranslatef(7.5, 7.5, 0.0); // Translate back.
glRotatef(45.0, 0.0, 0.0, 1.0); // Rotate about origin.
glTranslatef(-7.5, -7.5, 0.0); // Translate to origin.
```

Such a maneuver is unavoidable as OpenGL's own rotations, as we know from Section 4.1.3, are each about an axis through the origin, such being called a *radial* axis. Therefore, a non-radial axis needs to be translated to the origin and back again in order to be rotated about. This "tricky" maneuver and its variants come up so often that we'll give them a collective name: the Trick.

Exercise 4.12. (Programming) As in the preceding example, replace the object definition statement

```
glutWireCube(5.0); // Box.
```

of `box.cpp` with

```
glRectf(5.0, 5.0, 10.0, 10.0); // Square
```

Now, scale the square so that its center is unchanged, but its shape changes to a rectangle of aspect ratio 2. Use the Trick.

Exercise 4.13. Prove that a composition of multiple translations is a single translation and that a composition of multiple scalings is a single scaling.

Remark 4.4. A composition of multiple rotations is a single rotation as well, but this is much harder to prove generally and we'll leave it to Chapter 5.

Exercise 4.14. What is the *inverse* of a translation? Specifically, what modeling transformation composed with a translation `glTranslatef(`p`, `q`, `r`)` "undoes" its effect, so that all points remain stationary?

How about scalings and rotations? What are their inverses?

4.3   Placing Multiple Objects

We next consider the vital problem of applying modeling transformations to place multiple objects in a desired manner *relative* to one another.

Experiment 4.10. Replace the entire display routine of the original box.cpp with (Block 10):

```
void drawScene(void)
{
    glClear(GL_COLOR_BUFFER_BIT);
    glColor3f(0.0, 0.0, 0.0);
    glLoadIdentity();

    // Modeling transformations.
    glTranslatef(0.0, 0.0, -15.0);
    // glRotatef(45.0, 0.0, 0.0, 1.0);
    glTranslatef(5.0, 0.0, 0.0);

    glutWireCube(5.0); // Box.

    //More modeling transformations.
    glTranslatef (0.0, 10.0, 0.0);

    glutWireSphere (2.0, 10, 8); // Sphere.

    glFlush();
}
```

See Figure 4.14(a) for a screenshot. The objects are a box and a sphere.

End

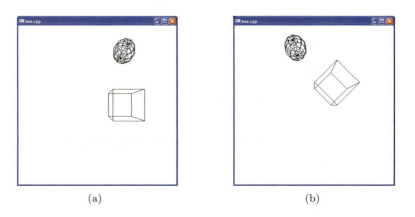

(a)   (b)

Figure 4.14: Screenshots from Experiments 4.10 and 4.11.

Let's understand the placement of the box and sphere in the preceding experiment individually first and then with respect to each other. It's fairly straightforward to understand the placements individually. For example, to place the sphere, work backwards from where it's created in the code, applying to it the successive modeling transformations (all translations in this case) encountered, and ignoring the one non-transformation statement `glutWireCube()`. The result is that the sphere is centered at $(5.0.10.0, -15.0)$. Likewise, the box is seen to be centered at $(5.0.0.0, -15.0)$.

The relative placement in this case is not difficult either. Clearly, the sphere is transformed by `glTranslatef(0.0, 10.0, 0.0)`, which is the transformation "between them", with respect to the box. The result is that the sphere's center is 10 units vertically above the box's.

Experiment 4.11. Continuing with the previous experiment, uncomment the `glRotatef()` statement. Figure 4.14(b) is a screenshot.

Again, the individual placements are fairly straightforward. Working backwards from where it is created we see that, after being translated to $(5.0, 10.0, 0.0)$, the sphere is rotated $45°$ counter-clockwise about the z-axis and, of course, finally pushed 15 units in the $-z$ direction. We'll not compute the exact final coordinates of its center. The individual placement of the box is simple to parse as well and left to the reader.

It's the relative placement which is particularly interesting in this case. The sphere is no longer vertically above the box, though the transformation between them is still `glTranslatef(0.0, 10.0, 0.0)`! Before trying to explain what's going on let's return to the basics for a moment. End

Consider the code sequence below which draws two objects:

```
modelingTransformation 1;    // t₁
modelingTransformation 2;    // t₂
...
modelingTransformation n-1;  // tₙ₋₁
modelingTransformation n;    // tₙ
object 1;
modelingTransformation n+1;  // tₙ₊₁
...
modelingTransformation m;    // tₘ
object 2;
```

Assuming that the transformation t_i specified by `modelingTransformation` i corresponds to the matrix M_i, for $1 \leq i \leq m$, the successive values of the modelview matrix M are indicated below:

```
                               // M = I, initially
modelingTransformation 1;      // M = IM₁ = M₁
modelingTransformation 2;      // M = M₁M₂
...
modelingTransformation n-1;    // M = M₁M₂... Mₙ₋₁
modelingTransformation n;      // M = M₁M₂... Mₙ₋₁Mₙ
object 1;                      // M does not change
modelingTransformation n+1;    // M = M₁M₂... Mₙ₋₁MₙMₙ₊₁
...
modelingTransformation m;      // M = M₁M₂... Mₙ₋₁MₙMₙ₊₁... Mₘ
```

object 2;

Accordingly, each vertex V of the final object 2 call is transformed according to:

$$V \mapsto (M_1 \ldots M_{m-1} M_m)V = t_1(\ldots t_{m-1}(t_m(V))\ldots) \qquad (4.2)$$

exactly as we would expect by working backwards in the code from object 2. Now, how about the placement of object 2 *with respect to* object 1?

Let's repeat the transformation for a vertex V of object 2 by stepping backward through the right side of Equation (4.2): first transform V by t_m to $t_m(V)$, then by t_{m-1} to $t_{m-1}(t_m(V))$, ..., then by t_{n+1} to $t_{n+1}(\ldots t_{m-1}(t_m(V))\ldots)$. *Stop!*

At this time object 1 is drawn. Imagine that a part of object 1 is a set of three directed line segments (drawn, say, using GL_LINES), aligned with the three world coordinate axes and calibrated identically. These lines are said to represent the *local coordinate system* of object 1. See Figure 4.15. At the time of its creation, the local coordinate system of object 1 coincides with the world coordinate system.

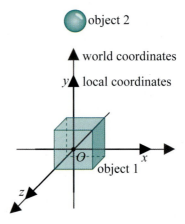

Figure 4.15: Local system (bold) coincides with the global initially. The global system is fixed.

Further, suppose at the time of object 1's creation that the so-far transformed V, i.e., $t_{n+1}(\ldots t_{m-1}(t_m(V))\ldots)$, is located at (a, b, c) in the local coordinates of object 1 (same as world coordinates, of course, at that moment).

Let's get back now to applying transformations backwards from where we had stopped. Next was t_n. Transformation t_n applies to *both* V and object 1. Three cases arise according to the type of t_n.

1. t_n is a translation specified by glTranslatef(p, q, r):

    This translation applies to V and object 1 and, so, to the local coordinate system of the latter as well. That is, they all "move together". Therefore, the location (a, b, c) of V with respect to the local coordinate system of `object 1` does not change. The location of V in world coordinates, of course, changes to $(a + p, b + q, c + r)$.

2. t_n is a rotation specified by `glRotatef(A, p, q, r)`:

   Same argument as for translation. Again, the location (a, b, c) of V with respect to the local coordinate system of `object 1` does not change.

3. t_n is a scaling specified by `glScalef(u, v, w)`:

   The location of V in world coordinates is changed by the scaling to (ua, vb, wc). However, as the same scaling applies to the axes of the local coordinate system of `object 1` – particularly the units calibrating them – the location (a, b, c) of V with respect to this system again does not change.

   Homely analogy: Prior to an experiment in a lab you measure yourself with a tape to be 6 feet tall. The experiment goes horribly wrong and radiation causes you to shrink by a factor of 12, leaving you at only 6 inches. However, if everything around you including the tape shrank by exactly the same factor, you would still believe yourself to be 6 feet.

Continue, applying transformations $t_{n-1}, t_{n-2}, \ldots t_1$, successively, and reason as above for each. The conclusion is that the location of V at the point (a, b, c) of the local coordinate system of `object 1` at the time of the latter's creation is not altered by any subsequent transformation, i.e., those in the code prior to `object 1`. Neither is it changed by transformations in the code after `object 2`, because their corresponding matrices multiply into the modelview matrix only after both `object 1` and `object 2` have already been drawn. We have, therefore, the following:

Proposition 4.1. *If* `object 1` *precedes* `object 2` *in the code, then the location of* `object 2` *in the local coordinate system of* `object 1` *is determined by the transformation statements between the two and nothing else.* □

What the proposition says is that, if `object 1` precedes `object 2` in the code, then the latter is *frozen* in the former's coordinate system at a position determined solely by the transformation statements between the two. Accordingly, moving `object 2` *with respect to* `object 1` requires changing transformations between them. The practical importance of this, as we'll see, cannot be over-emphasized.

Let's try and understand now the relative position of the sphere with respect to the box in Experiment 4.11 in light of the preceding proposition. We'll do this by the often useful technique of deconstructing code by incrementally adding back transformations.

Experiment 4.12. Repeat Experiment 4.11. The modeling transformation and object definition part are as below (Block 11):

```
// Modeling transformations.
glTranslatef(0.0, 0.0, -15.0);
glRotatef(45.0, 0.0, 0.0, 1.0);
glTranslatef(5.0, 0.0, 0.0);

glutWireCube(5.0); // Box.

//More modeling transformations.
glTranslatef (0.0, 10.0, 0.0);

glutWireSphere (2.0, 10, 8); // Sphere.
```

First, comment out the last two statements of the first modeling transformations block as below (the first translation is always needed to place the entire scene in the viewing frustum):

```
// Modeling transformations.
glTranslatef(0.0, 0.0, -15.0);
// glRotatef(45.0, 0.0, 0.0, 1.0);
// glTranslatef(5.0, 0.0, 0.0);

glutWireCube(5.0); // Box.

//More modeling transformations.
glTranslatef (0.0, 10.0, 0.0);

glutWireSphere (2.0, 10, 8); // Sphere.
```

The output is as depicted in Figure 4.16(a).

Next, uncomment glTranslatef(5.0, 0.0, 0.0) as below:

```
// Modeling transformations.
glTranslatef(0.0, 0.0, -15.0);
// glRotatef(45.0, 0.0, 0.0, 1.0);
glTranslatef(5.0, 0.0, 0.0);

glutWireCube(5.0); // Box.

//More modeling transformations.
glTranslatef (0.0, 10.0, 0.0);

glutWireSphere (2.0, 10, 8); // Sphere.
```

The output is as in Figure 4.16(b). Finally, uncomment glRotatef(45.0, 0.0, 0.0, 1.0) as follows:

```
// Modeling transformations.
glTranslatef(0.0, 0.0, -15.0);
glRotatef(45.0, 0.0, 0.0, 1.0);
glTranslatef(5.0, 0.0, 0.0);
```

```
glutWireCube(5.0); // Box.

//More modeling transformations.
glTranslatef (0.0, 10.0, 0.0);

glutWireSphere (2.0, 10, 8); // Sphere.

glFlush();
```

The result is seen in Figure 4.16(c). Figure 4.16 shows the box's local coordinate system as well after each transition. Observe that in this particular system the sphere is *always* 10 units vertically above the box, as one would expect from the glTranslatef (0.0, 10.0, 0.0) call between the two.

End

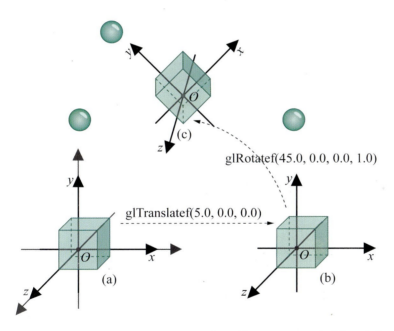

glRotatef(45.0, 0.0, 0.0, 1.0)

glTranslatef(5.0, 0.0, 0.0)

(a)

(b)

Figure 4.16: Transitions of the box, the box's local coordinates system (bold) and the sphere. The world coordinate system, which never changes, coincides with the box's initial local.

Experiment 4.13. Run composeTransformations.cpp. Pressing the up arrow key once causes the last statement, viz., **drawBlueMan**, of the following piece of code to be executed:

```
glScalef(1.5, 0.75, 1.0);
glRotatef(30.0, 0.0, 0.0, 1.0);
```

```
glTranslatef(10.0, 0.0, 0.0);
drawRedMan; // Also draw grid in his local coordinate system.
glRotatef(45.0, 0.0, 0.0, 1.0);
glTranslatef(20.0, 0.0, 0.0);
drawBlueMan;
```

With each press of the up arrow we go back a statement and successively execute that statement *and* the ones that follow it. The statements executed are written in black text, the rest white. Pressing the down arrow key goes forward a statement. Figure 4.17 is a screenshot after all transformations from the scaling on have been executed.

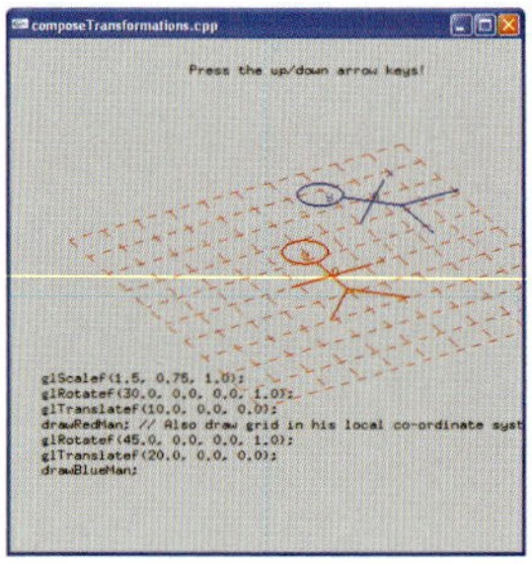

Figure 4.17: Screenshot of composeTrans-formations.cpp after all transformations from the scaling down have been executed.

The torso and arms of both men are aligned along their respective local coordinate axes. The world coordinate axes which never change are drawn in cyan. At the time of the red man's creation also drawn is a 10×10 grid of boxes in his local coordinate system, the sides of each box being 5 units long. With each transformation going back from the red man's creation, observe – focus on a point like the blue man's origin and trust your eyes – how the blue man stays static in the red man's local coordinate system. A simple calculation shows that the blue man's origin is actually at $(20/\sqrt{2},\ 20/\sqrt{2}) \simeq (14.14, 14.14)$ in the red man's system. Even when scaling skews the red man's system so that it's not rectangular any more, the blue man skews the same way as well, staying put in the red system.

End

Exercise 4.15. For the following two pieces of code in the drawing routine give (x, y, z) coordinates of the point to which the center of the sphere is transformed. Explain as well the relative positions of the sphere and box.

(a)
```
glRotatef(90.0, 1.0, 0.0, 0.0);
glutWireCube(1.0);
glRotatef(90.0, 0.0, 0.0, 1.0);
glTranslatef(4.0, 0.0, 0.0);
glutWireSphere(2.0, 10, 8);
```

(b)
```
glTranslatef(2.0, 0.0, 0.0);
glScalef(2.0, 2.0, 2.0);
glutWireCube(1.0);
glRotatef(90.0, 0.0, 1.0, 0.0);
glTranslatef(0.0, 0.0, 4.0);
glutWireSphere(2.0, 10, 8);
```

If you're impatient to get to animation hang on – there's one final piece to get in place!

4.4 Modelview Matrix Stack and Isolating Transformations

The modelview matrix, which we have described as being modified by modeling transformations by multiplication on the right, is actually the topmost one of a *modelview matrix stack*. This particular matrix is called the *current modelview matrix*. In fact, OpenGL maintains three different matrix stacks: modelview, projection and texture. A glMatrixMode(*mode*) command, where *mode* is GL_MODELVIEW, GL_PROJECTION or GL_TEXTURE, determines which stack is currently active.

Here's an experiment to motivate use of the modelview matrix stack:

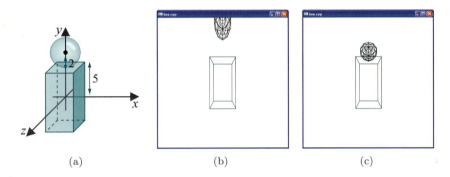

(a)　　　　　　(b)　　　　　　(c)

Figure 4.18: Planning a head on a torso: (a) The plan (b) Drawn without isolating the scaling (c) After isolating the scaling.

Expe$r$imen$t$ **4.14.** We want to create a human-like character. Our plan is to start by drawing the torso as an elongated cube and placing a round sphere as its head directly on top of the cube (no neck for now). To this end replace the drawing routine of **box.cpp** with (Block 12):

```
void drawScene(void)
{
   glClear(GL_COLOR_BUFFER_BIT);
   glColor3f(0.0, 0.0, 0.0);
   glLoadIdentity();

   glTranslatef(0.0, 0.0, -15.0);

   glScalef(1.0, 2.0, 1.0);
   glutWireCube(5.0); // Box torso.

   glTranslatef(0.0, 7.0, 0.0);
   glutWireSphere(2.0, 10, 8); // Spherical head.

   glFlush();
```

}

Our calculations are as follows: (a) the scaled box is $5 \times 10 \times 5$ and, being centered at the origin, is 5 units long in the $+y$ direction; (b) the sphere is of radius 2; (c) therefore, if the sphere is translated $5 + 2 = 7$ in the $+y$ direction, then it should sit exactly on top of the box (see Figure 4.18(a)).

It doesn't work: the sphere is no longer round and is, moreover, some ways above the box (Figure 4.18(b)). Of course, because the sphere is transformed by `glScalef(1.0, 2.0, 1.0)` as well! So, what to do? A solution is to *isolate* the scaling by placing it within a *push-pop pair* as below (Block 13):

```
void drawScene(void)
{
    glClear(GL_COLOR_BUFFER_BIT);
    glColor3f(0.0, 0.0, 0.0);
    glLoadIdentity();

    glTranslatef(0.0, 0.0, -15.0);

    glPushMatrix();
    glScalef(1.0, 2.0, 1.0);
    glutWireCube(5.0); // Box.
    glPopMatrix();

    glTranslatef(0.0, 7.0, 0.0);
    glutWireSphere(2.0, 10, 8); // Sphere.

    glFlush();
}
```

End

The resulting screenshot is Figure 4.18(c), which shows a round head on a neckless torso as desired.

What the `glPushMatrix()` command does is make a *copy* of the current (i.e., current topmost) matrix in the modelview matrix stack and place it on top of the stack; consequently, upon execution of a `glPushMatrix()`, the two top matrices of the stack are identical. The `glPopMatrix()` statement, on the other hand, deletes the topmost matrix of the modelview matrix stack so the one underneath it becomes the current one.

Let's follow the modelview matrix stack through the code above. Assume that the matrix corresponding to the translation `glTranslatef(0.0, 0.0, -15.0)` is M_1, to `glScalef(1.0, 2.0, 1.0)` is M_2, and to `glTranslatef(0.0, 7.0, 0.0)` is M_3. The transitions of the stack are shown in Figure 4.19, starting from the top.

As you see, the push-pop pair *stores* the current modelview matrix *prior* to the scaling transformation and then *restores* it once the cube has been drawn, effectively localizing the effect of the scaling to only the cube.

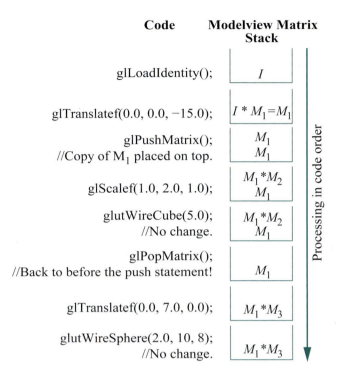

Code	Modelview Matrix Stack
glLoadIdentity();	I
glTranslatef(0.0, 0.0, −15.0);	$I * M_1 = M_1$
glPushMatrix(); //Copy of M_1 placed on top.	M_1 M_1
glScalef(1.0, 2.0, 1.0);	$M_1 * M_2$ M_1
glutWireCube(5.0); //No change.	$M_1 * M_2$ M_1
glPopMatrix(); //Back to before the push statement!	M_1
glTranslatef(0.0, 7.0, 0.0);	$M_1 * M_3$
glutWireSphere(2.0, 10, 8); //No change.	$M_1 * M_3$

Processing in code order

Figure 4.19: Transitions of the modelview matrix stack.

Exercise 4.16. Give (x, y, z) coordinates of the points where the centers of the four spheres in it are located by the drawing routine below, assuming no prior transformations.

```
glPushMatrix();

glTranslatef(2.0, 0.0, 0.0);
glutWireSphere(2.0, 10, 8); // Sphere A

glPushMatrix();
glScalef(2.0, 2.0, 2.0);
glutWireSphere(2.0, 10, 8); // Sphere B
glPopMatrix();

glPushMatrix();
glRotatef(90.0, 1.0, 0.0, 0.0);
glTranslatef(0.0, 0.0, 4.0);
glutWireSphere(2.0, 10, 8); // Sphere C
glPopMatrix();

glTranslatef(0.0, 4.0, 0.0);
glutWireSphere(2.0, 10, 8); // Sphere D
```

```
glPopMatrix();
```

Remark 4.5. It's generally good programming practice to enclose all the transformations in the drawing routine in one giant push-pop pair, as in the preceding exercise, so that at the end of the routine the modelview matrix stack is guaranteed to revert to its initial state of containing a single identity matrix.

4.5 Animation

We're there! Animation in computer graphics is achieved through a "transform-draw" loop: the scene is redrawn after transformations in the drawing routine change the location or shape, or both, of objects in the scene, redrawing being instigated either automatically or interactively.

4.5.1 Animation Technicals

Before analyzing animated programs we need first to explain a couple of animation-related technicalities.

Controlling Animation

OpenGL provides essentially three different methods to control animation:

1. Interactively, via keyboard or mouse input, with the help of their callback routines to invoke transformations.

   Experiment 4.15. Run `rotatingHelix1.cpp` where each press of space calls the `increaseAngle()` routine to turn the helix. Note the `glutPostRedisplay()` command in `increaseAngle()` which asks the screen to be redrawn. Keeping the space bar pressed turns the helix continuously. Figure 4.20 is a screenshot. **End**

Figure 4.20: Screenshot of `rotatingHelix1.cpp`.

2. Automatically, by specifying a function `idle_function`, called the *idle function*, with the statement `glutIdleFunc(idle_function)`. The idle function is called whenever no OpenGL event is otherwise pending.

   Experiment 4.16. Run `rotatingHelix2.cpp`, a slight modification of `rotatingHelix1.cpp`, where pressing space causes the routines `increaseAngle()` and NULL (do nothing) to be alternately specified as idle functions.

   The speed of animation is determined by the processor speed – in particular, the speed at which frames can be redrawn – and the user cannot influence it. **End**

3. Automatically, by specifying a routine *timer_function*, called the *timer function*, with a call to glutTimerFunc(*period, timer_function, value*). The timer function is called *period* milliseconds after the glutTimerFunc() statement is executed and with the parameter *value* being passed to it.

Experiment 4.17. Run rotatingHelix3.cpp, another modification of rotatingHelix1.cpp, where the particular timer function animate() is first called from the main routine 5 msecs. after the glutTimerFunc(5, animate, 1) command there. The parameter value 1 which is passed to animate() is not used in this program. The routine increaseAngle() called by animate() turns the helix as before.

Subsequent calls to animate() are made recursively from that routine itself after animationPeriod number of msecs., by means of its own glutTimerFunc(animationPeriod, animate, 1) call. The user can vary the speed of animation by changing the value of animationPeriod by pressing the up and down arrow keys. **End**

Remark 4.6. The speed of animation or, equivalently, *frame rate* – the rate at which the screen is redrawn – cannot be increased arbitrarily by lowering the value of animationPeriod because redrawing the scene takes some minimum amount of time, depending on its complexity and the speed of the processor (or graphics card).

Moreover, the frame rate can never exceed the monitor's installed refresh rate, e.g., if the latter is n Hz then the maximum achievable fps (frames per second) is n.

Exercise 4.17. (Programming) Enhance rotatingHelix3.cpp to display the frame rate on the screen. You will need to include an OS-dependent call in the display routine to read the system clock at each redraw. The time between two redraws is the current seconds per frame, the reciprocal of which is the current fps.

Double Buffering

The second technicality critical to smooth animation is *double buffering*.

   Space for two color buffers is provided in a double-buffered system in such a manner that one buffer, the *viewable buffer*, displays the current frame while the next frame is being drawn in the second buffer, the *drawable buffer*. When the drawing of the frame in the drawable buffer is complete, the contents of the buffers are swapped, so that the next frame now becomes viewable and, at the same time, the one following it begins to be drawn. This *draw-and-swap* loop repeats through the animation.

Terminology: The viewable buffer is often called the *front buffer* or *main buffer*, while the drawable buffer is called the *back buffer* or *swap buffer*. Either buffer is also called a *refresh buffer*.

Remark 4.7. There is a subtle difference between the "draw" in the transform-draw animation loop described earlier as how animation is implemented and the "draw" in the draw-and-swap loop just described as how double buffering operates. The first is a programmer-instigated operation – typically, with a `glutPostRedisplay()` call – in which the world space is projected and scaled (recall shoot-and-print from Chapter 2) *and* rasterized into the color buffer. The second actually draws the screen, in particular, the OpenGL window, with the contents of the color buffer.

Double buffering greatly improves the quality of animation by hiding transition between successive frames from the viewer. With single buffering, on the other hand, the viewer "sees" the next frame being drawn in the same buffer that contains the current one. The result can be unpleasant *ghosting*, so called because a prior image persists while the next is being created.

The double buffering display mode is enabled by calling `glutInit-DisplayMode()` in `main` with GLUT_DOUBLE as one of the arguments (instead of GLUT_SINGLE) *and* inserting a call to `glutSwapBuffers()` at the end of the drawing routine (*instead of* `glFlush()`).

Experiment 4.18. Disable double buffering in `rotatingHelix2.cpp` by replacing GLUT_DOUBLE with GLUT_SINGLE in the `glutInitDisplayMode()` call in `main`, and replacing `glutSwapBuffers()` in the drawing routine with `glFlush()`. Ghostly is it not?! **End**

4.5.2 Animation Code

Ball Flying About a Torus

Experiment 4.19. Run `ballAndTorus.cpp`. Press space to start the ball both flying around (longitudinal rotation) and in and out (latitudinal rotation) of the torus. Press the up and down arrow keys to change the speed of the animation. Press 'x/X', 'y/Y' and 'z/Z' to change the viewpoint. Figure 4.21 is a screenshot.

The animation of the ball is interesting and we'll deconstruct it. Comment out all the modeling transformations in the ball's block, except the last translation, as follows:

Figure 4.21: Screenshot of `ballAndTorus.cpp`.

```
// Begin revolving ball.
// glRotatef(longAngle, 0.0, 0.0, 1.0);

// glTranslatef(12.0, 0.0, 0.0);
// glRotatef(latAngle, 0.0, 1.0, 0.0);
// glTranslatef(-12.0, 0.0, 0.0);
```

```
glTranslatef(20.0, 0.0, 0.0);

glColor3f(0.0, 0.0, 1.0);
glutWireSphere(2.0, 10, 10);
// End revolving ball.
```

The ball is centered at $(20, 0, 0)$, its start position, by `glTranslatef(20.0, 0.0, 0.0)`. See Figure 4.22. There is no animation.

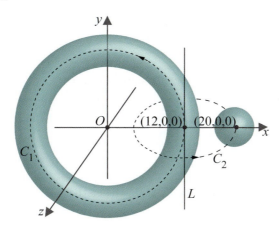

Figure 4.22: The ball's axis of latitudinal rotation from its start position is L.

The ball's intended latitudinal rotation is in and out of the circle C_1 through the middle of the torus. C_1's radius, called the *outer radius* of the torus, is 12.0, as specified by the second parameter of `glutWireTorus(2.0, 12.0, 20, 20)`. Moreover, C_1 is centered at the origin and lies on the xy-plane. Therefore, ignoring longitudinal motion for now, the latitudinal rotation of the ball *from its start position* is about the line L through $(12, 0, 0)$ parallel to the y-axis (L is tangent to C_1). This rotation will cause the ball's center to travel along the circle C_2 centered at $(12, 0, 0)$, lying on the xz-plane, of radius 8.

As `glRotatef()` always rotates about a radial axis, how does one obtain the desired rotation about L, a non-radial line? Employ the Trick (see Example 4.2, if you don't remember). First, translate left so that L is aligned along the y-axis, then rotate about the y-axis and, finally, reverse the first translation to bring L back to where it was. Accordingly, uncomment the corresponding three modeling transformations as below:

```
// Begin revolving ball.
// glRotatef(longAngle, 0.0, 0.0, 1.0);

glTranslatef(12.0, 0.0, 0.0);
glRotatef(latAngle, 0.0, 1.0, 0.0);
```

```
glTranslatef(-12.0, 0.0, 0.0);

glTranslatef(20.0, 0.0, 0.0);

glColor3f(0.0, 0.0, 1.0);
glutWireSphere(2.0, 10, 10);
// End revolving ball.
```

Press space to view only latitudinal rotation.

Note: The two consecutive translation statements could be combined into one, but then the code would be less easy to parse.

Finally, uncomment `glRotatef(longAngle, 0.0, 0.0, 1.0)` to implement longitudinal rotation about the *z*-axis. The angular speed of longitudinal rotation is set to be five times slower than that of latitudinal rotation – see the increments to `latAngle` and `longAngle` in the `animate()` routine. This means the ball winds in and out of the torus five times by the time it completes one trip around it. **End**

Exercise **4.18.** (**P**rogramming) It's instructive as well to uncomment the three modeling transformations used to apply the Trick in the preceding experiment one by one, rather than all together. So uncomment `glTranslatef(-12.0, 0.0, 0.0)` first, then `glRotatef(latAngle, 0.0, 1.0, 0.0)` and, last, `glTranslatef(12.0, 0.0, 0.0)`. Check if animation can be activated and explain the output at each step.

Exercise **4.19.** (**P**rogramming) Now here's something rather funny. Actually, what we'll show is not an uncommon accidental error. Cut the `glLoadIdentity()` call from the drawing routine of `ballAndTorus.cpp` and paste it as the last line of the window reshape routine (as, say, in `square.cpp`).

Oops! The ball and torus speed away together and are out of sight pretty quickly. Explain.

Hint: The current modelview matrix is not automatically cleared to identity between successive calls to the drawing routine.

Exercise **4.20.** (**P**rogramming) Add a red ball that starts a quarter of the way around the torus behind the blue ball and follows a similar rotate-revolve path.

Experimen**t** **4.20.** We want to add a satellite that tags along with the ball of `ballAndTorus.cpp`. The following piece of code added to the end of the drawing routine – just before `glutSwapBuffers()` – does the job (Block 14):

```
glTranslatef(4.0, 0.0, 0.0);

// Satellite
```

```
glColor3f(1.0, 0.0, 0.0);
glutWireSphere(0.5, 5, 5);
```

See Figure 4.23 for a screenshot. For a revolving satellite add the following instead (Block 15):

```
glRotatef(10*latAngle, 0.0, 1.0, 0.0);
glTranslatef(4.0, 0.0, 0.0);

// Satellite
glColor3f(1.0, 0.0, 0.0);
glutWireSphere(0.5, 5, 5);
```

Observe how Proposition 4.1 is being applied in both cases to determine the motion of the satellite *relative to the* ball by means of transformation statements between the two. **End**

Figure 4.23: Screenshot from Experiment 4.20.

Exercise 4.21. (**Programming**) Thinking that the Trick should be invoked to revolve the satellite about the ball, exactly as was done to obtain the latitudinal rotation of the ball itself, suppose we code the satellite as below (Block 16):

```
// Trick code block.
glTranslatef(4.0, 0.0, 0.0);
glRotatef(10*latAngle, 0.0, 1.0, 0.0);
glTranslatef(-4.0, 0.0, 0.0);

glTranslatef(4.0, 0.0, 0.0);

// Satellite.
glColor3f(1.0, 0.0, 0.0);
glutWireSphere(0.5, 5, 5);
```

The satellite still follows the ball, but does *not* revolve about it. Why? *Hint*: A good way to verify your answer is to stop the ball from moving by commenting out both `glRotatef()`'s in its definition block and observing only the satellite.

Exercise 4.22. (**Programming**) Continuing with Experiment 4.20, add a second satellite. Both should revolve around the ball, but in different orbits.

Throwing a Ball

Experiment 4.21. Run `throwBall.cpp`, which simulates the motion of a ball thrown with a specified initial velocity subject to the force of gravity. Figure 4.24 is a screenshot.

Press space to toggle between animation on and off. Press the right/left arrow keys to increase/decrease the horizontal component of the initial

Figure 4.24: Screenshot of `throwBall.cpp`.

123

velocity, up/down arrow keys to increase/decrease the vertical component of the initial velocity and the page up/down keys to increase/decrease gravitational acceleration. Press 'r' to reset. The values of the initial velocity components and of gravitational acceleration are displayed on the screen.

End

The equation determining the horizontal motion of the ball in `throwBall.cpp`, in terms of time t, is

$$x(t) = ht$$

where h is the horizontal component of the initial velocity; that determining vertical motion is

$$y(t) = vt - \frac{g}{2}t^2$$

where v is the vertical component of the initial velocity and g is gravitational acceleration.

Note: Most Physics 101 books should have a derivation of the equations of the motion of a rigid body under force, or you could even look up books on game physics like Bourg [19] and Eberly [38].

Motion is simulated by repeatedly redrawing the ball at the new location it's mapped to by `glTranslatef`($x(t)$, $y(t)$, 0), incrementing t by 1 each time.

Remark 4.8. The techniques to animate the spheres in `ballAndTorus.cpp` and `throwBall.cpp` are interesting to compare. One could say that the first is "physical" while the latter "equational".

Ball Facing Friction

Experiment 4.22. Run `ballAndTorusWithFriction.cpp`, which modifies `ballAndTorus.cpp` to simulate an invisible viscous medium through which the ball travels.

Press space to apply force to the ball. It has to be kept pressed in order to continue applying force. The ball comes to a gradual halt after the key is released. Increase or decrease the level of applied force by using the up and down arrow keys. Increase or decrease the viscosity of the medium using the page up and down keys. Press 'x/X', 'y/Y' and 'z/Z' to rotate the scene.

End

The equation of motion implemented takes the frictional drag (or, equivalently, deceleration) on the ball of `ballAndTorusWithFriction.cpp` to be proportional to its velocity, a valid assumption from physics [48]:

$$force_{drag} = -drag * velocity$$

where the *drag* (in real life) depends on the medium and shape of the moving object. This translates to the line of code

```
acceleration = applied_acceleration - drag*velocity;
```

for the total acceleration of the ball when space is pressed to apply an acceleration of `applied_acceleration` and to, simply,

```
acceleration = -drag*velocity;
```

in the absence of any external force. At every time step we find the change in velocity from the equation

$$\frac{\Delta(velocity)}{\Delta(time)} = acceleration$$

which is certainly true in the limit as $\Delta(time) \rightarrow 0$. However, we approximate change through a unit time step by setting $\Delta(time) = 1$ to get

$$\Delta(velocity) = acceleration$$

which is implemented by the program statement

```
velocity += acceleration;
```

Finally, change per time step in the `latAngle` and `longAngle` variables is taken proportional to the current value `velocity`.

Remark 4.9. The last two programs, `throwBall.cpp` and `ballAndTorus-WithFriction.cpp`, demonstrated simple applications of *physics in graphics*. This is a fascinating field – also known as *physically-based modeling* and *game physics* – of great importance in realistic animation. Plausible simulation of such phenomena as a wall of bricks crashing down, clothes and hair blowing in the wind, a drop of water rolling off a leaf, and smoke, fire and explosions, to mention a few, all require the programmer to take into account the real-world physics of the setting.

Special effects in a Hollywood production are almost always physics in graphics in action. There are two overarching and competing considerations in this discipline – realism versus computational efficiency.

A couple of books for the interested reader include Bourg [19] and Eberly [38]. A comprehensive list of pointers to ongoing research in the field is maintained by Simon Clavet [106].

Clown Head

Our next project is a program, which we'll develop incrementally, to draw a clown's head.

Experiment 4.23. We start with simply a blue sphere for the head. See `clown1.cpp`, which has the following drawing routine:

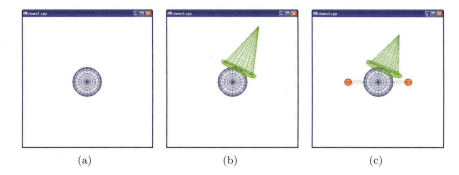

Figure 4.25: Screenshot of (a) `clown1.cpp` (b) `clown2.cpp` (c) `clown3.cpp`.

```
void drawScene(void)
{
   glClear(GL_COLOR_BUFFER_BIT);
   glLoadIdentity();

   // Place scene in frustum.
   glTranslatef(0.0, 0.0, -9.0);

   // Head.
   glColor3f(0.0, 0.0, 1.0);
   glutWireSphere(2.0, 20, 20);

   glutSwapBuffers();
}
```

Figure 4.25(a) is a screenshot.

Next, we want a green conical hat. The command `glutWireCone(`*base*, *height*, *slices*, *stacks*`)` draws a wireframe cone of base radius *base* and height *height*. The base of the cone lies on the xy-plane with its axis along the z-axis and its apex pointing in the positive direction of the z-axis. See Figure 4.26(a). The parameters *slices* and *stacks* determine the fineness of the mesh (not shown in the figure).

Accordingly, insert the lines

```
   // Hat.
   glColor3f(0.0, 1.0, 0.0);
   glutWireCone(2.0, 4.0, 20, 20);
```

in `clown1.cpp` after the call that draws the sphere, so that the drawing routine becomes (Block 17):

```
void drawScene(void)
{
   glClear(GL_COLOR_BUFFER_BIT);
   glLoadIdentity();
```

```
// Place scene in frustum.
glTranslatef(0.0, 0.0, -9.0);

// Head.
glColor3f(0.0, 0.0, 1.0);
glutWireSphere(2.0, 20, 20);

// Hat.
glColor3f(0.0, 1.0, 0.0);
glutWireCone(2.0, 5.0, 20, 20);

glutSwapBuffers();
}
```

Not good! Because of the way `glutWireCone()` aligns, the hat covers the clown's face. This is easily fixed. Translate the hat 2 units up the z-axis and rotate it $-90°$ about the x-axis to arrange it on top of the head. Finally, rotate it a rakish $30°$ about the z-axis! Here's the modified drawing routine of `clown1.cpp` at this point (Block 18):

```
void drawScene(void)
{
    glClear(GL_COLOR_BUFFER_BIT);
    glLoadIdentity();

    // Place scene in frustum.
    glTranslatef(0.0, 0.0, -9.0);

    // Head.
    glColor3f(0.0, 0.0, 1.0);
    glutWireSphere(2.0, 20, 20);

    // Transformations of the hat.
    glRotatef(30.0, 0.0, 0.0, 1.0);
    glRotatef(-90.0, 1.0, 0.0, 0.0);
    glTranslatef(0.0, 0.0, 2.0);

    // Hat.
    glColor3f(0.0, 1.0, 0.0);
    glutWireCone(2.0, 5.0, 20, 20);

    glutSwapBuffers();
}
```

Let's add a brim to the hat by attaching a torus to its base. The command `glutWireTorus(`*inRadius*, *outRadius*, *sides*, *rings*`)` draws a wireframe torus of inner radius *inRadius* (the radius of a circular section of the torus), and outer radius *outRadius* (the radius of the circle through the middle of the torus). The axis of the torus is along the z-axis and it is centered at the origin. See Figure 4.26(b). Insert the call `glutWireTorus(0.2, 2.2, 10, 25)` right after the call that draws the cone, so the drawing routine becomes (Block 19):

```
void drawScene(void)
{
    glClear(GL_COLOR_BUFFER_BIT);
    glLoadIdentity();

    // Place scene in frustum.
    glTranslatef(0.0, 0.0, -9.0);

    // Head.
    glColor3f(0.0, 0.0, 1.0);
    glutWireSphere(2.0, 20, 20);

    // Transformations of the hat and brim.
    glRotatef(30.0, 0.0, 0.0, 1.0);
    glRotatef(-90.0, 1.0, 0.0, 0.0);
    glTranslatef(0.0, 0.0, 2.0);

    // Hat.
    glColor3f(0.0, 1.0, 0.0);
    glutWireCone(2.0, 5.0, 20, 20);

    // Brim.
    glutWireTorus(0.2, 2.2, 10, 25);

    glutSwapBuffers();
}
```

Observe that the brim is drawn suitably at the bottom of the hat and stays there despite modeling transformations between head and hat – a consequence of Proposition 4.1.

To animate, let's spin the hat about the clown's head by rotating it around the y-axis. We rig the space bar to toggle between animation on and off and the up/down arrow keys to change speed. All updates so far are included in `clown2.cpp`. Figure 4.25(b) is a screenshot.

What's a clown without little red ears that pop in and out?! Spheres will do for ears. An easy way to bring about oscillatory motion is to make use of the function sin($angle$) which varies between -1 and 1. Begin by translating either ear a unit distance from the head, and then repeatedly translate each a distance of sin($angle$), incrementing $angle$ each time.

Note: A technicality one needs to be aware of in such applications is that angle is measured in *degrees* in OpenGL syntax, e.g., in `glRotatef`($angle$, p, q, r), while the C++ math library assumes angles to be given in *radians*. Multiplying by $\pi/180$ converts degrees to radians.

The ears and head are physically separate, though. Let's connect them with springs! Helixes are springs. We borrow code from `helix.cpp`, but modify it to make the length of the helix 1, its axis along the x-axis and its radius 0.25. As the ears move, either helix is scaled along the x-axis so that it spans the gap between the head and an ear. The completed program is `clown3.cpp`, of which a screenshot is seen in Figure 4.25(c).   **End**

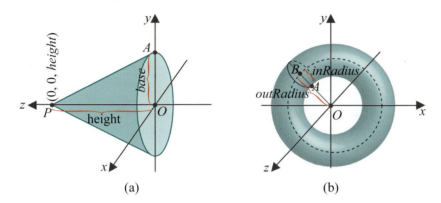

(a)   (b)

Figure 4.26: (a) Cone drawn by `glutWireCone`(*base*, *height*, *slices*, *stacks*) (b) Torus drawn by `glutWireTorus`(*inRadius*, *outRadius*, *sides*, *rings*). Note that the axes are depicted differently in each diagram.

Exercise **4.23. (P**rogramming**)** Comment out the push-pop pair isolating the hat and brim in `clown3.cpp`. Explain the new situation of the ears.

Exercise **4.24.** Proposition 4.1 came before our discussion of push-pop pairs, so the assumption there is that there are none. Do we have to revise the proposition to take into account possible push-pop pairs?

Blooming Flower

Experiment **4.24.** Run `floweringPlant.cpp`, an animation of a flower blooming. Press space to start and stop animation, delete to reset, and 'x/X', 'y/Y' and 'z/Z' to change the viewpoint. Figure 4.27 is a screenshot.   **End**

Figure 4.27: Screenshot of `floweringPlant.cpp` in mid-bloom.

The stem of the plant of `floweringPlant.cpp` consists of four straight segments, the sepal (base of the flower) is modeled as a hemisphere, while the six petals are circles. Both the hemisphere and circle are reshaped by scaling during the animation. The code to draw the two is modified from `circle.cpp` and `hemisphere.cpp`.

As calls to display lists cannot be parametrized, those defining a sepal and a petal have to be placed, unfortunately, in the drawing routine to allow them access to the changing global variable `t` via the variables – `angleFirstSegment`, ..., `petalOpenAngle` – at the top of `drawScene()`. Another option for modular code would have been to write those parts of the plant as C++ objects.

The parameters involved in configuring the stem, sepal and petal all change from a start value to an end one via linear interpolation using the animation parameter t. For example,

`hemisphereScaleFactor` $= (1 - t) * 0.1 + t * 0.75$

linearly changes `hemisphereScaleFactor` from 0.1 to 0.75 as t goes from 0 to 1.

4.5.3   Animation Projects

Exercise 4.25. (Programming) Starting from `clown3.cpp`, add to the clown's head a conical nose which changes in length and color, as well as eyes that rotate and change in size and color.

Exercise 4.26. (Programming) Animate a ball *rolling* down an inclined flat plane. See Figure 4.28(a). The ball should not slip or slide. Make the plane a wireframe mesh of triangles and the ball a wireframe sphere, as well, so that relative motion is apparent.

Next, add another plane at the bottom so that the ball rolls from the first onto the second. See Figure 4.28(b).

Exercise 4.27. (Programming) Roll a ball down the curved children's slide of Exercise 2.28 of Chapter 2, if you did that particular exercise.

Exercise 4.28. (Programming) Animate a ball bouncing up and down a box which itself moves in a straight line. See Figure 4.28(c).

First, code the straight-line motion of the box and, then, that of the ball relative to the box, which is straight, too. The resultant motion of the ball as viewed in the OpenGL window, which is, of course, as that seen by a stationary external observer, is parabolic.

Exercise 4.29. (Programming) Animate a ball traveling a helical path. See Figure 4.28(d). Make sure to do this physically *a la* `ballAndTorus.cpp`, and not equationally.

Exercise 4.30. (Programming) Animate four straight segments, which initially bound a square, smoothly opening into a straight line. See Figure 4.28(e), where the initial, final and two intermediate positions are depicted.

Exercise 4.31. (Programming) Animating a solar system is a canonical exercise for beginning 3D programmers. Animate a solitary planet, with two moons, in circular orbit around a stationary sun. See Figure 4.28(f). The planet rotates about its own axis as well, while its moons revolve about it at different speeds and on different orbital planes.

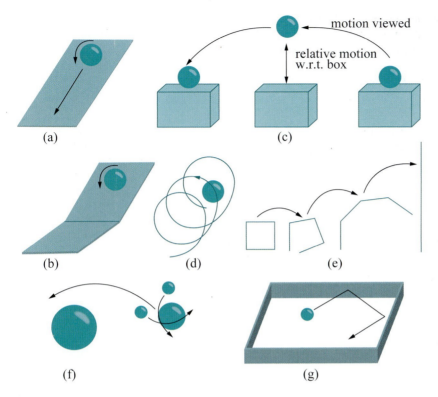

Figure 4.28: (a) Ball rolling down one plane (b) Ball rolling down two planes (c) Ball bouncing on a box (d) Ball traveling along a helix (e) Four segments opening from a square into a straight line (f) Solar system with a sun, one planet and two moons (g) Pool table with one ball.

Exercise 4.32. (Programming) Simulate a ball falling through air and then striking and continuing through a viscous medium such as water. Code in friction following `ballAndTorusWithFriction.cpp`.

There is no need to visually simulate splashing. To differentiate air and water use color, e.g., the upper half of your window can be white, while the lower blue.

Exercise 4.33. (Programming) Create an animated garden with the help of `floweringPlant.cpp`.

Exercise 4.34. (Programming) Animate a lone ball moving on a pool table. The pool table should simply be a rectangle enclosed by four low walls – no need to make pockets. See Figure 4.28(g)).

The ball should initially be stationary at a fixed position on the table. Then allow the user, with the help of a simple visual interface, to choose a direction and speed to get the ball moving.

Animate the subsequent motion of the ball as it *rolls* along the surface

131

of the table and *bounces* off its sides. You can either choose to not program in any deceleration, so that the ball keeps moving at uniform speed, or to incorporate frictional resistance to ultimately bring the ball to rest.

4.6 Viewing Transformation

We begin our discussion of the viewing transformation `gluLookAt()`, whose function is to arrange OpenGL's camera, by systematically deciphering its somewhat non-trivial syntax.

4.6.1 Understanding the Viewing Transformation

Think of the OpenGL camera as located at the origin with its lens pointing down the $-z$ direction (the *line of sight*) and with its top aligned along the $+y$ direction (the *up direction*). This, in fact, is the *default pose* of the OpenGL camera. See Figure 4.29(a).

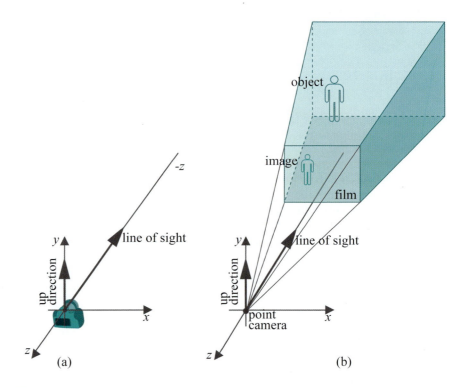

Figure 4.29: (a) The (conceptual) OpenGL camera's default pose (b) A (conceptual) point camera at the origin with film on the viewing plane of the frustum.

Keep in mind, though, that the OpenGL camera is merely a *conceptual* device! The rendering we see of objects drawn is determined solely, as described in Chapter 2, by the shape of the viewing box or frustum, which in turn is decided by the programmer-specified projection statement (e.g., `glOrtho()`, `glFrustum()`). Figure 4.29(b) reminds us of the process. There is *no* camera as such!

Nevertheless, it appeals to the intuition to imagine that what we're viewing is through a camera. In the case of a viewing frustum, particularly, one can imagine a point camera at the origin with the film lying in front of it on the viewing face, as indicated in Figure 4.29(b). It's intuitive as well to think of changing the view by moving and turning the camera. This is exactly where the viewing transformation `gluLookAt()` comes in.

The command `gluLookAt(`*eyex, eyey, eyez, centerx, centery, centerz, upx, upy, upz*`)` *simulates* – mark the word *simulates* – OpenGL's camera first being moved to the location *eye* = (*eyex, eyey, eyez*) and pointed at *center* = (*centerx, centery, centerz*). Next, it is rotated about its line of sight (*los*) – the line joining *eye* to *center* – so that its up direction is one determined from *up* = (*upx, upy, upz*). See Figure 4.30. We'll see shortly how the up direction is, in fact, determined from *up*.

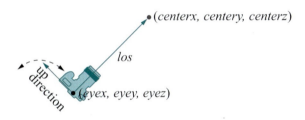

Figure 4.30: Camera pose determined by `gluLookAt(`*eyex, eyey, eyez, centerx, centery, centerz, upx, upy, upz*`)`.

Experiment 4.25. Replace the translation command `glTranslatef(0.0, 0.0, -15.0)` of `box.cpp` with the viewing command `gluLookAt(0.0, 0.0, 15.0, 0.0, 0.0, 0.0, 0.0, 1.0, 0.0)` so that the drawing routine is as below (Block 20):

```
void drawScene(void)
{
    glClear(GL_COLOR_BUFFER_BIT);
    glColor3f(0.0, 0.0, 0.0);
    glLoadIdentity();

    // Viewing transformation.
    gluLookAt(0.0, 0.0, 15.0, 0.0, 0.0, 0.0, 0.0, 1.0, 0.0);

    glutWireCube(5.0); // Box.
```

```
    glFlush();
}
```

There is no change in what is viewed. The commands `glTranslatef(0.0,`
`0.0, -15.0)` and `gluLookAt(0.0, 0.0, 15.0, 0.0, 0.0, 0.0, 0.0,`
`1.0, 0.0)` are exactly equivalent. To understand why, note that `gluLook-`
`At(0.0, 0.0, 15.0, 0.0, 0.0, 0.0, 0.0, 1.0, 0.0)` places the *eye* at
$(0, 0, 15)$ looking down the z-axis toward the *center* at $(0, 0, 0)$. Now, compare
Figures 4.31(a) and (b): should the box appear different to the viewer in
one from the other? *No*, because its position relative to the frustum is the
same in both.

The advantage of the command `gluLookAt()` over `glTranslatef()` is
that it allows one to write code according to one's conception of where the
camera is situated and how it's pointed at the scene. **End**

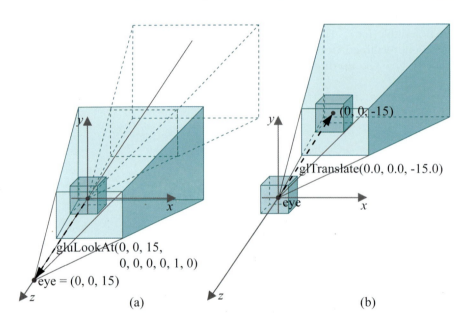

Figure 4.31: (a) `gluLookAt()`: the broken frustum is the original viewing frustum, the
unbroken one is where it's "moved" by the `gluLookAt()` call, the box doesn't move. (b)
`glTranslatef()`: the viewing frustum doesn't move, rather the box is translated by the
`glTranslatef()` call.

It's worth repeating that the OpenGL camera remains *always* at its
default pose at the origin or, more accurately, that the viewing volume
(frustum or box) remains *always* at the shape and location in world space
specified by the program's projection statement. Moreover, the `gluLookAt()`
command is *simulated* by OpenGL via modeling transformations, and we'll
find out soon enough how.

As box.cpp with gluLookAt() instead of glTranslatef(), as in the preceding experiment, is used often, the modified program is stored as boxWithLookAt.cpp.

Experiment 4.26. Continue the previous experiment, or run boxWith-LookAt.cpp, successively changing only the parameters *centerx*, *centery*, *centerz* – the middle three parameters – of the gluLookAt() call to the following:

1. $0.0, 0.0, 10.0$

2. $0.0, 0.0, -10.0$

3. $0.0, 0.0, 20.0$

4. $0.0, 0.0, 15.0$ End

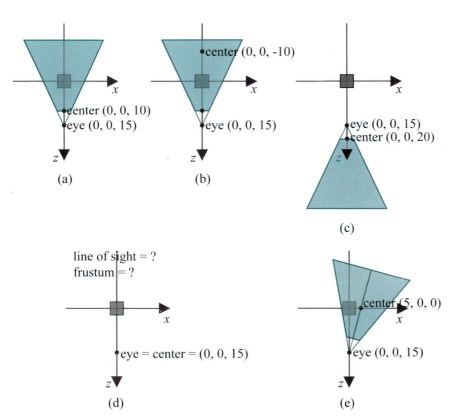

Figure 4.32: Sectional diagrams of the (simulated) configuration of the eye and frustum for various gluLookAt() calls in boxWithLookAt.cpp.

The view does not change with the first two parameter sets of the experiment as the viewer's line of sight from eye to center does not change.

Figures 4.32(a) and (b) show the respective simulated configurations. The third set (Figure 4.32(c)) produces a blank screen because the eye is looking the "wrong way". The last set (Figure 4.32(d)) confuses OpenGL because eye and center coincide, making it impossible to decide a line of sight. Again a blank screen appears. Note in all cases that the shape of the frustum is not changed by `gluLookAt()`, only its placement and alignment.

Here are a few more *center* sets for you to try.

Exercise 4.35. (Programming) Restore the original `boxWithLookAt.cpp` program with its `gluLookAt(0.0, 0.0, 15.0, 0.0, 0.0, 0.0, 0.0, 1.0, 0.0)`. Next, successively change only the parameters *centerx*, *centery*, *centerz* – the middle three parameters of `gluLookAt()` – to the following, drawing diagrams as in Figure 4.32 to explain what is seen in each case:

1. $5.0, 0.0, 0.0$ (*Answer*: See Figure 4.32(e))

2. $-5.0, 0.0, 0.0$

3. $0.0, 5.0, 0.0$

4. $0.0, -5.0, 0.0$

5. $5.0, 5.0, 0.0$

Let's change the *eye* next. It's still pretty much the same game as changing *center*.

Exercise 4.36. (Programming) Restore the original `boxWithLookAt.cpp` program with its `gluLookAt(0.0, 0.0, 15.0, 0.0, 0.0, 0.0, 0.0, 1.0, 0.0)` call. First, replace the box with a `glutWireTeapot(5.0)`, a non-symmetric object. Next, successively change only the parameters *eyex*, *eyey*, *eyez* – the first three parameters of `gluLookAt()` – to the following, drawing diagrams as in Figure 4.32 to explain what is seen in each case:

1. $0.0, 0.0, 10.0$

2. $0.0, 0.0, 25.0$

3. $0.0, 0.0, -15.0$

4. $15.0, 0.0, 15.0$

5. $15.0, 0.0, 0.0$

6. $15.0, 15.0, 15.0$

Let's get a feel now for how the up vector $up = (upx, upy, upz)$ works.

Experiment 4.27. Restore the original `boxWithLookAt.cpp` program with its `gluLookAt(0.0, 0.0, 15.0, 0.0, 0.0, 0.0, 0.0, 1.0, 0.0)` call and, again, first replace the box with a `glutWireTeapot(5.0)`. Run: a screenshot is shown in Figure 4.33(a). Next, successively change the parameters *upx, upy, upz* – the last three parameters of `gluLookAt()` – to the following:

1. $1.0, 0.0, 0.0$

2. $0.0, -1.0, 0.0$

3. $1.0, 1.0, 0.0$

Screenshots of the successive cases are shown in Figures 4.33(b)-(d). The camera indeed appears to rotate *about* its line of sight, the *z*-axis, so that its up direction points along the *up* vector (*upx, upy, upz*) each time. **End**

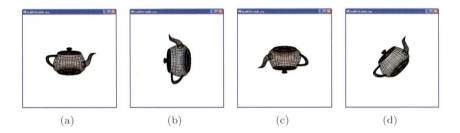

(a) (b) (c) (d)

Figure 4.33: Screenshots from Experiment 4.27.

Before we can state the rule for how the *up* vector determines the camera's up direction generally, here are some facts about the dot product of vectors which we'll need. Skip this part if you already have dot product basics.

Sidebar on Dot Products

The *dot product* (also called *scalar product*) of two vectors u and v in $\mathbb{R}^3$ is a scalar, denoted $u \cdot v$, defined as follows:

(a) if either of u or v is zero, then $u \cdot v$ is zero;

(b) if not, then the value of $u \cdot v$ is $|u||v|\cos\theta$, where θ is the angle between u and v.

See Figure 4.34.

It turns out that $u \cdot v$ is given by the following simple formula, where $u = (u_x, u_y, u_z)$ and $v = (v_x, v_y, v_z)$:

$$u \cdot v = u_x v_x + u_y v_y + u_z v_z \qquad (4.3)$$

This makes the dot product useful in calculating angles between pairs of vectors.

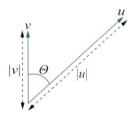

Figure 4.34: Taking the dot product:
$u \cdot v = |u||v|\cos\theta.$

Example 4.3. Determine the angle θ between the two vectors $u = (1, 0, 2)$ and $v = (-2, 3, 4)$.

Answer:

$$|u||v| \cos \theta = u \cdot v = u_x v_x + u_y v_y + u_z v_z = 1 * -2 + 0 * 3 + 2 * 4 = 6$$

Therefore,

$$\cos \theta = \frac{6}{|u||v|} = \frac{6}{\sqrt{1^2 + 0^2 + 2^2}\sqrt{(-2)^2 + 3^2 + 4^2}} = \frac{6}{\sqrt{5}\sqrt{29}} \simeq 0.49827$$

which gives $\theta \simeq 60.11439°$.

Exercise 4.37. Determine the angle between each pair, from the three vectors $(1, 0, 0)$, $(\frac{1}{\sqrt{2}}, \frac{1}{\sqrt{2}}, 0)$ and $(\frac{1}{\sqrt{3}}, \frac{1}{\sqrt{3}}, \frac{1}{\sqrt{3}})$.

Exercise 4.38. Prove the following about dot products where u, v and w are any three vectors and c any scalar:

(a) Assuming that they are both non-zero, u and v are perpendicular if and only if $u \cdot v = 0$ (*perpendicularity test*)

(b) $u \cdot u = |u|^2$

(c) $u \cdot v = v \cdot u$ (*dot product is commutative*)

(d) $(cu) \cdot v = u \cdot (cv) = c(u \cdot v)$

(e) $u \cdot (v + w) = u \cdot v + u \cdot w$ (*dot product distributes over a sum*)

(f) $|u \cdot v| \le |u||v|$

A particularly useful application of the dot product is when one wants to split a given vector v as $v = v_1 + v_2$, where the components v_1 and v_2 are, respectively, parallel and perpendicular to another given non-zero vector u. See Figure 4.35(a). An intuitive way to think of v_2 is as the shadow of v cast on a plane p perpendicular to u by a light shining from the direction of u, as depicted in Figure 4.35(b).

The component v_1 is the perpendicular projection of v onto the line of u, so its *signed length* is

$$|v| \cos \theta = \frac{|u||v| \cos \theta}{|u|} = \frac{u \cdot v}{|u|}$$

where θ is the angle between u and v. Multiplying the value of the signed length by the unit vector in the direction of u, which is $u/|u|$, one obtains the formula for v_1:

$$v_1 = \frac{u \cdot v}{|u|^2} u \tag{4.4}$$

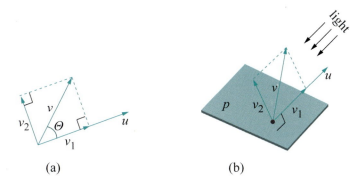

(a)   (b)

Figure 4.35: (a) Splitting v into components v_1 and v_2, parallel and perpendicular to u, respectively (b) v_2 as the "shadow" of v on a plane p perpendicular to u.

The formula for the component v_2 of v that is perpendicular to u follows, as the sum of v_1 and v_2 is v:

$$v_2 = v - v_1 = v - \frac{u \cdot v}{|u|^2} u \tag{4.5}$$

The preceding formulae have particularly simple forms if u is a unit vector, as we ask the reader to show next.

Exercise 4.39. If u is a unit vector and v arbitrary, prove the following:

(a) The component of v parallel to u is $v_1 = (u \cdot v)\, u$.

(b) The component of v perpendicular to u is $v_2 = v - (u \cdot v)\, u$.

Example 4.4. Split $v = (-2, 3, 4)$ into components parallel and perpendicular to $u = (1, 0, 2)$.

Answer: The component parallel to u is

$$v_1 = \frac{u \cdot v}{|u|^2} u = \frac{6}{5} (1, 0, 2) = \left(\frac{6}{5},\, 0,\, \frac{12}{5} \right)$$

and that perpendicular to u

$$v_2 = v - v_1 = (-2, 3, 4) - \left(\frac{6}{5},\, 0,\, \frac{12}{5} \right) = \left(-\frac{16}{5},\, 3,\, \frac{8}{5} \right)$$

The following worked example shows a neat matrix expression for the component of one vector parallel to another. The vectors themselves are written as column matrices.

Example 4.5. Show that if $u = [u_x \; u_y \; u_z]^T$ and $v = [v_x \; v_y \; v_z]^T$ are two vectors in $\mathbb{R}^3$, such that u is not zero, then the component v_1 of v parallel to u is given by

$$v_1 = \frac{1}{|u|^2} \begin{bmatrix} u_x^2 & u_x u_y & u_x u_z \\ u_x u_y & u_y^2 & u_y u_z \\ u_x u_z & u_y u_z & u_z^2 \end{bmatrix} v$$

Answer: We have that the component of v parallel to u is

$$
\begin{aligned}
v_1 &= \frac{u \cdot v}{|u|^2} u \\[2mm]
&= \frac{1}{|u|^2} (u^T v) u \qquad (u^T v \text{ denotes its lone scalar entry, viz., } u \cdot v, \\
&\qquad\qquad\qquad\qquad\quad \text{as a } 1 \times 1 \text{ matrix product)} \\[2mm]
&= \frac{1}{|u|^2} u(u^T v) \qquad (\text{as } (u^T v)u = u(u^T v), \text{ where } u^T v \text{ denotes a scalar} \\
&\qquad\qquad\qquad\qquad\quad \text{on the LHS and a } 1 \times 1 \text{ matrix on the RHS}) \\[2mm]
&= \frac{1}{|u|^2} (uu^T)v \qquad (\text{by associativity}) \\[2mm]
&= \frac{1}{|u|^2} \begin{bmatrix} u_x^2 & u_x u_y & u_x u_z \\ u_x u_y & u_y^2 & u_y u_z \\ u_x u_z & u_y u_z & u_z^2 \end{bmatrix} v \quad (\text{multiplying as matrices } u \text{ and } u^T)
\end{aligned}
$$

For a more thorough discussion of dot products refer to any book on linear algebra, e.g., Banchoff and Wermer [7].

It's simple now to explain how OpenGL uses the $up = (upx, upy, upz)$ vector to align the top of its camera – in other words, determine its up direction – upon the call `gluLookAt`($eyex, \; eyey, \; eyez, \; centerx, \; centery, \; centerz, \; upx, \; upy, \; upz$).

Denote the camera's line of sight vector $(centerx, centery, centerz) - (eyex, eyey, eyez)$ by *los*. What OpenGL does is split *up* into components up_1 and up_2 parallel and perpendicular, respectively, to *los*. *The up direction is then taken to be up_2.* In particular, think of the camera, which is located at $eye = (eyex, eyey, eyez)$ and pointing down *los*, as being rotated about *los* till its top points in the direction parallel to up_2.

For an example, see Figure 4.36. Imagine the camera lying with its back on this page (call it the plane p) facing up, so that the line of sight *los* emerges perpendicularly from p (toward the reader). The specified *up* vector is drawn in the figure starting from the camera, as also its components up_1 and up_2, the latter lying on the page. The camera, then, is rotated about *los* with its back always on the page till its top points along up_2.

The magnitude of *up* or of up_2 is of no consequence as long as it's not zero, because it's only the direction that matters in aligning the top; if either

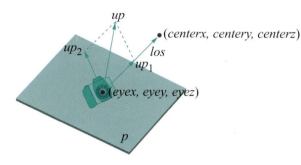

Figure 4.36: The camera is seen face-forward so that its back-plane p lies on the page. The line of sight *los* comes perpendicularly up from the page toward the reader. The components of the vector *up*, parallel and perpendicular to *los*, respectively, are up_1 and up_2 (the latter lying on the page).

is zero, then OpenGL is unable to determine the alignment and renders a blank screen.

Exercise 4.40. Of course, if *up* is zero then its component up_2 is zero. Can it happen that *up* is non-zero and yet up_2 is zero?

Experiment 4.28. Replace the wire cube of `boxWithLookAt.cpp` with a `glutWireTeapot(5.0)` and replace the `gluLookAt()` call with:

```
gluLookAt(0.0, 0.0, 15.0, 0.0, 0.0, 0.0, 1.0, 1.0, 1.0)
```

The vector $los = (0.0, 0.0, 0.0) - (15.0, 0.0, 0.0) = (-15.0, 0.0, 0.0)$, which is down the z-axis. The component of $up = (1.0, 1.0, 1.0)$, perpendicular to the z-axis, is $(1.0, 1.0, 0.0)$, which then is the up direction. Is what you see the same as Figure 4.33(d), which, in fact, is a screenshot for `gluLookAt(0.0, 0.0, 15.0, 0.0, 0.0, 0.0, 1.0, 1.0, 0.0)`? **End**

Exercise 4.41. (Programming) Change (upx, upy, upz) of `gluLookAt()` in `boxWithLookAt.cpp` to $(0.0, 0.0, 1.0)$. What do you see? *Nothing!* Why?

Exercise 4.42. Compute the direction of the top of the camera for each of the following viewing transformations:

(a) `gluLookAt(0.0, 0.0, 5.0, 5.0, 0.0, 0.0, 0.0, 1.0, 1.0)`

(b) `gluLookAt(0.0, 5.0, 5.0, 0.0, 0.0, 0.0, 1.0, 1.0, 1.0)`

(c) `gluLookAt(10.0, 5.0, 5.0, 0.0, 5.0, 0.0, 5.0, 1.0, 1.0)`

(d) `gluLookAt(10.0, 10.0, 5.0, 0.0, 5.0, 0.0, 1.0, 2.0, 3.0)`

Part answer:

(a) The line of sight vector

$$los = (centerx, centery, centerz) - (eyex, eyey, eyez)$$
$$= (5,0,0) - (0,0,5) = (5,0,-5)$$

The component of $up = (0,1,1)$ perpendicular to los is

$$up_2 = up - \frac{los \cdot up}{|los|^2} los$$
$$= (0,1,1) - \frac{(5,0,-5) \cdot (0,1,1)}{50} (5,0,-5)$$
$$= (0,1,1) + \frac{1}{10} (5,0,-5)$$
$$= (0.5,1,0.5)$$

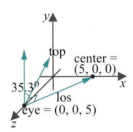

Figure 4.37: Solution to Exercise 4.42(a).

which, therefore, is the direction of the top of the camera. It is perpendicular, of course, to the line of sight and, as easily verified, tilted about $35.3°$ from the direction of the y-axis. See Figure 4.37.

$\mathcal{R}emark$ 4.10. Collectively, the modeling transformations glTranslatef(), glScalef() and glRotatef() and the viewing transformations gluLook-At() are called *modelview transformations*.

Exercise 4.43. (Programming) Program a camera flying at a height of 3 units over a sequence of balls arranged along the x-axis, looking ahead and down at the balls. See Figure 4.38.
Hint: Coordinates for the eye and center are suggested in the figure.

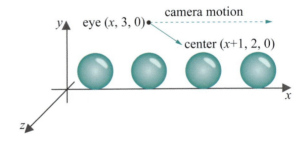

Figure 4.38: Camera flying over balls.

Exercise 4.44. (Programming) Place a wire teapot centered at the origin. Program a camera which can be moved by the user anywhere on an imaginary sphere enclosing the teapot, the direction of the camera being always toward the origin. Suitably program keys to move the camera. See Figure 4.39, where a couple of positions of the camera are indicated.

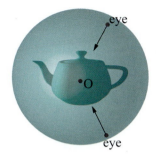

Figure 4.39: Camera rotated on an imaginary sphere enclosing a teapot.

The next two subsections are somewhat theoretical: about how a viewing transformation is implemented by modeling transformations, and about the use of so-called Euler angles to specify the orientation of a camera. If you are more interested in practice, then skip straight to Section 4.6.4 for a nice application of the viewing transformation in a space-travel program.

4.6.2 Simulating a Viewing Transformation with Modeling Transformations

You can skip this section on a first reading.
When we introduced `gluLookAt()` we said that it *simulates* OpenGL camera movement. This is exactly right. The OpenGL camera *never* leaves its default pose at the origin with its lens pointing down the $-z$ direction and with its top aligned along the $+y$ direction. The viewing transformation is simulated by replacing it with an equivalent sequence of modeling transformations. We actually saw a simple example of this earlier in Section 4.6.1 where the commands `glTranslatef(0.0, 0.0, -15.0)` and `gluLookAt(0.0, 0.0, 15.0, 0.0, 0.0, 0.0, 0.0, 1.0, 0.0)` were found to be equivalent.

Here's a simple motivating thought experiment for the general case:

You are out in an open field with a friend and a camera. She stands 10 meters in front of you, but looking through the viewfinder you think she should be, say, 3 meters closer. There are two options: (1) you, i.e., the camera, translate (walk) 3 meters toward her, or (2) she, i.e., the scene, translates 3 meters toward you. See the top left of Figure 4.40. The picture is the same in either case. Ignore the backdrop, as it's a homogeneous open field!

Here's another way to arrive at the equivalence of the two options. Say you had already applied (1) when the guy you had borrowed the camera from starts yelling that it's really expensive and would you mind not moving it around but just keep it where it was first set up. In other words, you have to manage by rearranging the scene instead. So, to undo the effect of (1) and bring the camera back to its original position, you apply the reverse of (1) to *both* camera and scene (so as to not alter the picture). The result,

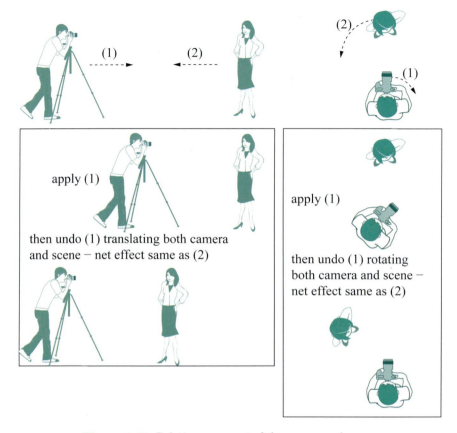

apply (1)

then undo (1) translating both camera
and scene − net effect same as (2)

apply (1)

then undo (1) rotating
both camera and scene −
net effect same as (2)

Figure 4.40: Relative movement of the camera and scene.

of course, is the same as applying just (2) in the first place. See the two diagrams in the big box on the left of Figure 4.40.

Looking through the viewfinder again, you feel it'll be a nicer composition if your friend stands not at the center of the frame but to a side. Again, (1) you can rotate the camera, say, 45° *clockwise*, or (2) your friend can sidle 45° *counter-clockwise* along a circle centered where you are, as in the top right of Figure 4.40. The picture is exactly the same in either case. And, again, one can imagine arriving at (2) by first applying (1), and then undoing it by applying the reverse of (1) to both camera and scene, as the two diagrams in the big box on the right of Figure 4.40 indicate.

It should now be fairly straightforward understanding the equivalence of a viewing transformation to a sequence of modeling transformations. The transformation

gluLookAt(*eyex, eyey, eyez, centerx, centery, centerz, upx, upy, upz*)

asks that the camera be (i) first translated to the position (*eyex, eyey, eyez*), then, (ii) rotated at that position till it's pointing at (*centerx, centery, centerz*) and, finally, (iii) rotated about its line of sight till its up direction is parallel to the vector up_2, the component of (upx, upy, upz) perpendicular to the line of sight.

Let's move the camera as asked. Figure 4.41(a) shows the resulting configuration. Next – it's the owner yelling again – we'll restore the camera to its default pose by incrementally undoing its movements, moving instead the scene as in the thought experiment. The sum total, then, of these reverse movements to bring the camera back to default will be equivalent to the viewing transform.

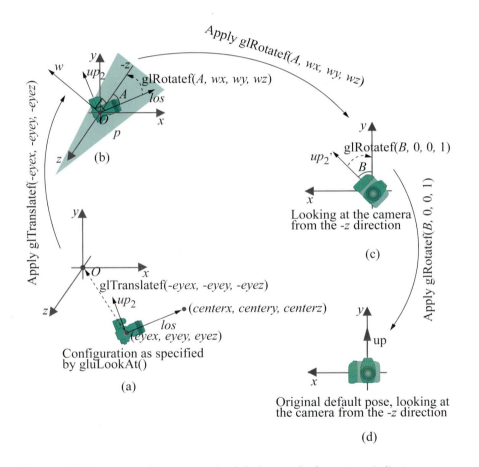

Figure 4.41: Restoring the camera to its default pose: broken arrows indicate movements which applied take the camera to the next configuration in the sequence (a)-(d).

The first translation is undone by applying `glTranslatef`($-eyex$, $-eyey$, $-eyez$). The camera is then at the origin, but still pointing parallel to the line of sight vector

$$los = (centerx, centery, centerz) - (eyex, eyey, eyez)$$

and with its top still parallel to up_2. See Figure 4.41(b).

Suppose that p is a plane that contains both los and the z-axis – shaded in the figure. If los does not lie along the z-axis then p is unique; if it does then p can be any plane that contains the common line. Choose a non-zero vector $w = (wx, wy, wz)$ perpendicular to p, i.e., w is perpendicular to both los and the z-axis. Let A be the angle from los to $-z$ on the plane p measured counterclockwise when looking down from w. Applying $\texttt{glRotatef}(A, wx, wy, wz)$ then rotates the camera till its line of sight matches the $-z$ direction. Moreover, its top then is parallel to the vector, call it up_2', which is the result of $\texttt{glRotatef}(A, wx, wy, wz)$ applied to up_2. Now, up_2', the new top direction, is perpendicular to the new line of sight down $-z$, because both were obtained by the same rotation $\texttt{glRotatef}(A, wx, wy, wz)$ applied to perpendicular vectors up_2 and los. Therefore, up_2' lies on the xy-plane. See Figure 4.41(c) where the camera is seen from the negative side of the z-axis.

Finally, all that remains to restore the camera to its default position is a rotation $\texttt{glRotatef}(B, 0.0, 0.0, 1.0)$, of angle B about the z-axis, to align its top along $+y$. See Figure 4.41(d).

We conclude that the viewing transformation

$\texttt{gluLookAt}(eyex, \ eyey, \ eyez, \ centerx, \ centery, \ centerz, \ upx, \ upy, \ upz)$

is, indeed, equivalent to a sequence of modeling transformations, in particular, a translation followed by two rotations:

```
glRotatef(B,  0.0,  0.0,  1.0);
glRotatef(A,  wx,  wy,  wz);
glTranslatef(-eyex,  -eyey,  -eyez);
```

Experiment 4.29. Replace the display routine of box.cpp with (Block 21):

```
void drawScene(void)
{
    glClear(GL_COLOR_BUFFER_BIT);
    glColor3f(0.0, 0.0, 0.0);

    glLoadIdentity();

    // Viewing transformation.
    gluLookAt(0.0, 0.0, 15.0, 15.0, 0.0, 0.0, 0.0, 1.0, 0.0);

    // Modeling transformation block equivalent
    // to the preceding viewing transformation.
    // glRotatef(45.0, 0.0, 1.0, 0.0);
    // glTranslatef(0.0, 0.0, -15.0);
```

```
    glutWireCube(5.0);

    glFlush();
}
```

Run.   Next, both comment out the viewing transformation and uncomment the modeling transformation block following it. Run again. The displayed output, shown in Figure 4.42, is the same in both cases. The reason, as Figures 4.43(a)-(c) explain, is that the viewing transformation is equivalent to the modeling transformation block. In particular, the former is undone by the latter. End

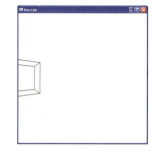

Figure 4.42: Screenshot from Experiment 4.29.

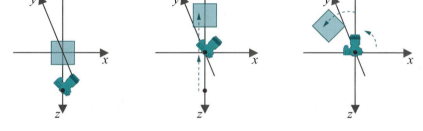

gluLookAt(0.0, 0.0, 15.0,
15.0, 0.0, 0.0, 0.0, 1.0, 0.0);   glTranslatef(0.0, 0.0, -15.0);
          (a)   (b)

glRotatef(45.0, 0.0, 1.0, 0.0);
glTranslatef(0.0, 0.0, -15.0);
          (c)

Figure 4.43: Viewing transformation equivalent to a sequence of modeling transformations.

Exercise 4.45. (Programming) Replace the display routine of box.cpp with (Block 22):

```
void drawScene(void)
{
    glClear(GL_COLOR_BUFFER_BIT);
    glColor3f(0.0, 0.0, 0.0);

    glLoadIdentity();

    // Viewing transformation.
    gluLookAt(-30.0, 0.0, 30.0, 0.0, 0.0, 0.0, 0.0, 1.0, 0.0);

    // Modeling transformation block equivalent
    // to the preceding viewing transformation.
    // glRotatef(45.0, 0.0, 1.0, 0.0);
    // glTranslatef(30.0, 0.0, -30.0);

    glutWireCone(3.0, 10.0, 20, 20);
```

```
    glFlush();
}
```

Draw diagrams as in Figure 4.43 to show the equivalence of the viewing transformation and the modeling transformation block following it.

Exercise 4.46. Show that the viewing transformation

```
gluLookAt(30.0, 0.0, 30.0, 0.0, 0.0, 0.0, 1.0, 0.0, -1.0);
```

is equivalent to the sequence

```
glRotatef(90, 0.0, 0.0, 1.0);
glRotatef(45, 0.0, -1.0, 0.0);
glTranslatef(-30.0, 0.0, -30.0);
```

of modeling transformations. Pay particular attention to the alignment of the top of the camera.

Exercise 4.47. The sequence of modeling transformations equivalent to a given viewing transformation is not unique. In fact, for the preceding exercise find a sequence of modeling transformations, different from the one given, yet equivalent to the viewing transformation there.

Exercise 4.48. What sequence of modeling transformations is equivalent to each of the following viewing transformations:

(a) `gluLookAt(0.0, 0.0, 0.0, -15.0, 0.0, 15.0, 0.0, 1.0, 0.0)`

(b) `gluLookAt(0.0, 0.0, 15.0, 0.0, 0.0, 0.0, 1.0, 0.0, 0.0)`

(c) `gluLookAt(0.0, 0.0, 15.0, -15.0, 0.0, 0.0, 1.0, 0.0, -1.0)`

(d) `gluLookAt(0.0, 0.0, 15.0, 0.0, 1.0, 14.0, 0.0, 1.0, 0.0)`

(e) `gluLookAt(0.0, 0.0, 15.0, 0.0, 1.0, 14.0, 1.0, 0.0, 0.0)`

Part answer:

(c) One solution:

```
glRotatef(90, 0.0, 0.0, 1.0);
glRotatef(-45, 0.0, 1.0, 0.0);
glTranslatef(0.0, 0.0, -15.0);
```

Exercise 4.49. What is the viewing transformation equivalent to the following sequence of modeling transformations:

```
glRotatef(45.0, 0.0, 1.0, 0.0);
glTranslatef(0.0, 0.0, -5.0);
```

Remark 4.11. It's almost good programming practice for there to be a single viewing transformation in a program, whose statement comes before those of all modeling transformations; in other words, the viewing transformation is applied last. This implies that objects are drawn first and placed as desired with respect to each other using modeling transformations and then a `gluLookAt()` is finally applied to transport the *entire* scene together, presumably into the viewable box or frustum.

Remark 4.12. We have been insistent that the viewing transformation `gluLookAt()`'s purported manipulation of the camera is simulated entirely by modeling transformations. Indeed, we showed in this section how this can be done. But, is this *really* what OpenGL does? For, it's plausible that OpenGL actually does move the viewing frustum, with the camera at its apex, as directed by a `gluLookAt()` call, rather than apply any modeling transformations. For example, is Figure 4.31(a) or (b) in Section 4.6.1 the "truth"?

Here's how to decide. Modeling transformations change the current modelview matrix at the top of the modelview matrix stack. The viewing frustum, on the other hand, is determined by the current projection matrix at the top of the projection matrix stack, which is altered, among others, by projection statements such as `glFrustum()`. So, a way to find out what really happens inside the OpenGL engine is to read both the current modelview and projection matrices, both before and after issuing a `gluLookAt()`, and to see which changes.

Answer: Only the current modelview matrix changes! The current projection matrix remains at the value it had prior to the `gluLookAt()` call. Take this in good faith now – you'll be able to verify the claim when we learn to access the modelview and projection matrix stacks in Chapter 5. In fact, we'll see then that the modelview matrix changes exactly as if multiplied on the right by the matrices corresponding to a sequence of modeling transformations equivalent to the given viewing transformation.

Remark 4.13. An interesting upshot of all this is that viewing transformations are not really needed, as any such transformation can always be manufactured from modeling transformations! As a matter of fact, this economizing approach has been adopted by OpenGL ES, the small version of OpenGL for embedded systems: OpenGL ES has no `gluLookAt()` call.

4.6.3 Orientation and Euler Angles

This section may be skipped on a first reading. You will need it, though, before Section 6.2 about animating orientation with the help of Euler angles. The viewing transformation leads nicely to a method of specifying the orientation of a camera. Recall the conclusion just before Experiment 4.29 that the viewing transformation

`gluLookAt(`*eyex, eyey, eyez, centerx, centery, centerz, upx, upy, upz*`)` is equivalent to a translation followed by two rotations:

```
glRotatef(B, 0.0, 0.0, 1.0);
glRotatef(A, wx, wy, wz);
glTranslatef(−eyex, −eyey, −eyez);
```

The axis of the particular rotation $\texttt{glRotatef}(A, wx, wy, wz)$ is variable and depends on the line of sight. It was chosen, in fact, perpendicular to both line of sight and the z-axis. It's possible, however, to find a translation followed by a sequence of rotations, each about a *fixed* axis, equivalent to the given viewing transformation. In particular, one can show that

```
gluLookAt(eyex, eyey, eyez, centerx, centery, centerz, upx, upy, upz)
```

is equivalent to:

```
glRotatef(−γ, 0.0, 0.0, 1.0);
glRotatef(−β, 0.0, 1.0, 0.0);
glRotatef(−α, 1.0, 0.0, 0.0);
glTranslatef(−eyex, −eyey, −eyez);
```

where rotations are each about a coordinate axis, for suitable angles α, β and γ (the minus signs are for simpler notation later on). Here's how. Figure 4.44 – an all-in-one version of Figure 4.41 – shows that the sequence of four transformations below restores the camera to its default pose from the one specified by

```
gluLookAt(eyex, eyey, eyez, centerx, centery, centerz, upx, upy, upz)
```

so, indeed, they are equivalent.

(*Note*: Figure 4.44 looks busy but it's not hard to read. The best way is to start with bold vector indicating the camera's initial configuration and follow the sequence (1)-(4) of transformations one by one.)

(1) $\texttt{glTranslatef}(−eyex, −eyey, −eyez)$, to bring the eye to the origin.

(2) $\texttt{glRotatef}(−\alpha, 1.0, 0.0, 0.0)$, $−\alpha$ chosen to rotate the *los* about the x-axis till it lies on the xz-plane.

(3) $\texttt{glRotatef}(−\beta, 0.0, 1.0, 0.0)$, $−\beta$ chosen to rotate the *los* about the y-axis till it points down the $−z$ direction.

(4) $\texttt{glRotatef}(−\gamma, 0.0, 0.0, 1.0)$, $−\gamma$ chosen to rotate the camera about its *los* (pointing down the z-axis) till its top is aligned in the $+y$ direction.

Remark 4.14. Evidently, we can reduce the number of parameters required to specify camera movement from the nine of $\texttt{gluLookAt()}$ to only six: α, β, γ, $eyex$, $eyey$ and $eyez$. This indicates redundancy in the construction of $\texttt{gluLookAt()}$, but it has the virtue of being intuitive to use.

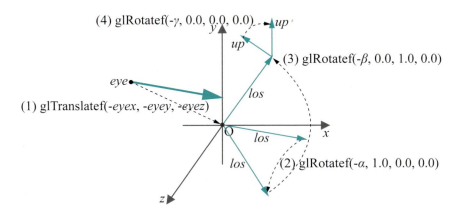

Figure 4.44: Applying a translation (1) and rotations (2)-(4) about the three coordinate axes to bring the camera back to its default pose. The original line of sight is bold. The up direction is shown only at the end.

Example 4.6. Express the viewing transformation

```
gluLookAt(0.0, 0.0, 0.0, 1.0, 1.0, 0.0, -1.0, 1.0, 0.0);
```

as a sequence of rotations about the coordinate axes (no translation is needed as the eye is already at the origin).

Answer:

```
glRotatef (90, 0.0, 0.0, 1.0);
glRotatef (135.0, 1.0, 0.0, 0.0);
glRotatef (90, 1.0, 0.0, 0.0);
```

See Figure 4.45 for how the camera is restored to its default pose by these three rotations.

Exercise 4.50. What sequence of rotations would have been found by the method of Section 4.6.2 as equivalent to the viewing transformation of the preceding example? Would they all have been about the coordinate axes?

If its first three parameters $(eyex, eyey, eyez) = (0, 0, 0)$, then a gluLookAt()'s translational component is zero, so that it alters only a camera's orientation, or pose. From the preceding discussion such a gluLookAt() call is equivalent to some sequence

```
glRotatef (−γ, 0.0, 0.0, 1.0);
glRotatef (−β, 0.0, 1.0, 0.0);
glRotatef (−α, 1.0, 0.0, 0.0);
```

because this sequence restores the camera *to* its default pose. In the opposite direction, therefore, the orientation of the camera resulting from this particular gluLookAt() call is obtained *from* its default pose by applying the inverse of the above sequence, viz.,

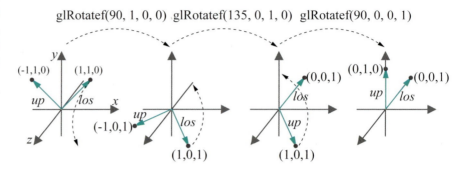

Figure 4.45: Solution to Example 4.6: the configuration of the camera given by `gluLookAt(0.0, 0.0, 0.0, 1.0, 1.0, 0.0, -1.0, 1.0, 0.0)` is at left; the line of sight and top vectors are indicated by blue arrows; rotations are both annotated at the top and indicated in the figures themselves by broken arrows, the result of each rotation being the next configuration.

```
glRotatef(α,  1.0,  0.0,  0.0);
glRotatef(β,  0.0,  1.0,  0.0);
glRotatef(γ,  0.0,  0.0,  1.0);
```

The angles α, β and γ are called the camera's *Euler angles*.

Euler angles specify the camera's orientation. Specifically, if they are α, β and γ, then the camera's orientation is obtained by applying first `glRotatef`$(\gamma, 0.0, 0.0, 1.0)$, then `glRotatef`$(\beta, 0.0, 1.0, 0.0)$ and, finally, `glRotatef`$(\alpha, 1.0, 0.0, 0.0)$ to its default pose.

Euler angles are not unique. For example, it's clear in Figure 4.44 that `glRotatef`$(-\alpha \pm 180°, 1.0, 0.0, 0.0)$ could have been applied in Step (2), instead of `glRotatef`$(-\alpha, 1.0, 0.0, 0.0)$, to still place the camera's *los* on the xz-plane. So the orientation given by Euler angles α, β and γ is the same as the ones given by $\alpha \pm 180°$, β' and γ', for some, possibly, new β' and γ'.

Exercise 4.51. What are the Euler angles of a camera

(a) at the origin pointing at $(1, 0, 0)$?

(b) at the origin pointing at $(1, 1, 1)$?

(c) at $(1, 0, 0)$ pointing at $(1, 1, 1)$?

Note: In each case assume the *up* vector to be $(0, 1, 0)$. To determine the Euler angles of a camera not at the origin simply translate it first to the origin.

Part answer:

(a) $0°$, $-90°$, $0°$ (one possible answer)

We'll see more of Euler angles when we discuss animating the orientation of rigid objects in Chapter 6.

4.6.4 Viewing Transformation and Collision Detection in Animation

Our next program makes use of viewing transformations to simulate a moving camera. It also has another aspect of interest, particularly to those programming interactive environments such as games, namely, collision detection.

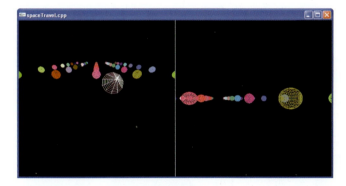

Figure 4.46: Screenshot of spaceTravel.cpp.

Experiment 4.30. Run spaceTravel.cpp. The left viewport shows a global view from a fixed camera of a conical spacecraft and 40 stationary spherical asteroids arranged in a 5 × 8 grid. The right viewport shows the view from a front-facing camera attached to the tip of the craft. See Figure 4.46 for a screenshot of the program.

Press the up and down arrow keys to move the craft forward and backward and the left and right arrow keys to turn it. Approximate collision detection is implemented to prevent the craft from crashing into an asteroid.

The asteroid grid can be changed in size by redefining ROWS and COLUMNS. The probability that a particular row-column slot is filled is specified as a percentage by FILL_PROBABILITY – a value less than 100 leads to a non-uniform distribution of asteroids. **End**

We'll discuss the two most interesting aspects of spaceTravel.cpp: (a) the viewing transformation that defines the scene in the right viewport and (b) collision detection.

Viewing Transformation

The shape of the craft is defined by the glutWireCone(5.0, 10.0, 10, 10) statement; precisely, it is a cone of base radius 5 and height 10. The configuration of the spacecraft is specified by the values of $xVal$, $zVal$ and $angle$, all three global variables of spaceTravel.cpp. In Figure 4.47(a) you see a generic configuration in section along the xz-plane. The coordinates of the center of the craft's base are $(xVal, 0, zVal)$, while the angle its axis makes with the negative z-direction is *angle*. The middle A of the craft's axis will be of use in collision detection.

The camera for the right viewport is situated at the tip of the craft pointing straight ahead. It's straightforward trigonometry, now, to calculate the coordinates of *eye*, where this camera is located, and of an imaginary point *center* to which it points, located 1 unit ahead of the tip along the cone's axis:

$$
\begin{aligned}
eye &= (xVal - 10\sin(angle),\ 0,\ zVal - 10\cos(angle)) \\
center &= (xVal - 11\sin(angle),\ 0,\ zVal - 11\cos(angle))
\end{aligned}
$$

These equations for *eye* and *center* explain the parameters of the `gluLook-At()` command for the right viewport.

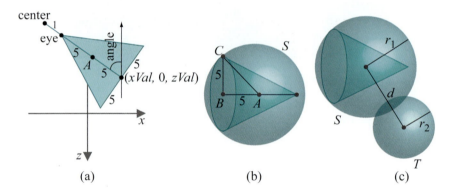

Figure 4.47: Spacecraft diagrams.

Collision Detection

Collision detection as implemented in `spaceTravel.cpp` is simple though approximate. The spacecraft is enclosed in an imaginary bounding sphere S centered at the middle A of the cone's axis, with radius equal to the distance $|AC|$ from A to a point C on the boundary of its base. See Figure 4.47(b).

If B is the center of the base, then it follows from the dimensions of the cone that $|AB| = |BC| = 5$; therefore,

$$|AC| = \sqrt{|AB|^2 + |BC|^2} = \sqrt{50} = 7.071\dots$$

Accordingly, we specify the radius of S to be 7.072 (slightly larger, in fact, than $|AC|$). The coordinates of the center A of S are obtained by trigonometry from Figure 4.47(a):

$$A = (xVal - 5\sin(angle),\ 0,\ zVal - 5\cos(angle))$$

To detect collision between the spacecraft and an asteroid T, we detect instead collision between the craft's bounding sphere S and T. It's easy to determine if there is a collision between the two spheres S and T – simply compare the distance d between their centers with the sum $r_1 + r_2$ of their radii. There is collision if $d \leq r_1 + r_2$ (e.g., as in Figure 4.47(c)), and not otherwise. This check is implemented in the routine `checkSphereCollision()`. This collision-detection test is approximate, in fact, conservative, as the craft's bounding sphere may intersect an asteroid even if the craft itself doesn't (as shown in Figure 4.47(c)).

The up and down arrow keys are programmed to move the craft a distance of 1 in either direction along its axis, and the left and right arrows to turn the craft an angle of 5°, *only if* there'll not be a collision with an asteroid in the new position (according to the conservative test above).

Exercise 4.52. (Programming) Modify `spaceTravel.cpp` as follows:

(a) Make the left viewport the view from the front of the spacecraft (currently, it is the right viewport).

(b) Make one of the asteroids the "big golden asteroid" by drawing it larger than the others and painting it suitably. Make it glow as well by oscillating the intensity of its color.

(c) Place a camera on the golden asteroid whose location is fixed but which rotates to track the spacecraft, i.e., its direction is pointed always toward the craft. Attach a tall antenna to the craft so that, even if it's obscured by other asteroids, at least the antenna will be visible from the big golden asteroid. Show the view from the golden asteroid's camera in the right viewport.

(d) When the spacecraft reaches the big golden asteroid, flash the text "You have found gold!".

Exercise 4.53. (Programming) Modify `spaceTravel.cpp` as follows:

(a) All the asteroids are currently colored spheres. Make them more interesting by using a few different GLUT objects, e.g., cube, tetrahedron, octahedron, etc. You can also combine more than one object, e.g., one sphere on top of another, or design your own.

(b) Currently, the spacecraft moves interactively. Change this to program an automated tour which takes a fixed but zig-zag path through the asteroids and returns to the start position. Plan a path so that the craft comes close to a few interesting asteroids, visible in the right viewport. Pressing space should start/stop the movement.

(c) Currently, the camera on the craft always points straight ahead. Program occasional rotation of the camera, e.g., when the craft passes a strange asteroid, pan the camera to keep it in view.

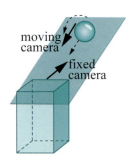

moving
camera

fixed
camera

Figure 4.48: Ball rolling toward a box.

Exercise 4.54. (Programming) Place a camera on top of the rolling ball of Exercise 4.26, pointing always down the plane. This camera does *not* rotate with the ball, but stays always at the top, so its motion is entirely linear.

Place a box just beyond the bottom of the plane so that the ball's camera sees an approaching object. Place an additional fixed camera on the box pointing at the plane to observe the ball. See Figure 4.48. Give a split-screen view as in `spaceTravel.cpp`.

The following experiment is to whet your appetite for the topic of *frustum culling*, critical to the efficient rendering of complex scenes.

Experiment 4.31. Run `spaceTravel.cpp` with ROWS and COLUMNS both increased to 100. The spacecraft begins to respond so slowly to key input that its movement seems clunky, unless, of course, you have a super-fast computer (in which case, increase the values of ROWS and COLUMNS even more). **End**

The reason for the degradation in the preceding experiment is that, every time an arrow key is pressed, OpenGL processes 10,000 asteroids, which is an enormous amount of computing. However, of these 10,000 only a few (about 100, or 1%) are ultimately rendered, as you can roughly count on the screen! The rest, of course, are found to lie outside the viewing frustum and clipped.

Unfortunately, by the time in the graphics pipeline the decision to clip is made, a large amount of computation has already been invested. Frustum culling is a technique to reduce this burden on OpenGL, whereby the programmer leverages her knowledge of the scene to pre-filter objects lying beyond the viewing frustum.

We'll discuss frustum culling in detail in Section 6.1. There's really not much more by way of prerequisites needed to read that particular section though, so if you're anxious to learn this technique, which is so important in coding busy games and movies, feel free to jump right there.

4.7 More Animation Code

4.7.1 Animating an Articulated Figure

Our next project is a "studio" to develop animation sequences for an articulated figure.

Figure 4.49: Screenshot of `animateMan1.cpp`.

Experiment 4.32. Run `animateMan1.cpp`. This is a fairly complex program to develop a sequence of key frames for a man-like figure, which can subsequently be animated. In addition to its spherical head, the figure consists of nine box-like body parts which can rotate about their joints. See Figure 4.49. All parts are wireframe. We'll explain the program next. **End**

It's advisable to learn to use the program before studying the code. There are two modes, develop and animate, and the program starts in the develop mode with the man facing you with his currently highlighted part, the torso, colored red. The rest of the body is black. Press the space bar to cycle through the man's movable parts, successively highlighting each. There are nine movable parts, all OpenGL wire cubes: the torso, the upper and lower arms on either side, and the upper and lower legs on either side.

Rotate the currently highlighted part by pressing the page-up and page-down keys. To move the man as a whole press the left/right and up/down arrow keys. The angles at which the 9 movable parts are currently rotated, as well as the vertical and horizontal translational components of the man as a whole, are shown as text data in the window in develop mode.

While arranging the man into a desired configuration, you can rotate your own viewpoint by pressing 'r/R', or zoom in and out pressing 'z/Z'.

Once the first configuration is completed to your satisfaction, press 'n'. This creates a new configuration which cannot be seen immediately as it's a copy of the previous one. Press, say, the right arrow key to separate the new configuration from the previous one. The (current) new configuration is bright, while the other(s) are ghosted. Again, use the space key to select a part, the page-up and page-down keys to rotate that part and the arrow keys to move the entire configuration until it is arranged suitably.

Press 'n' to create new configurations until the key frames sequence is complete. Figure 4.50 shows a screenshot part way through the develop mode. You can edit the sequence at any time as follows.

Press the tab key to cycle through the sequence of configurations – the currently selected configuration is bright, while the rest ghosted. Press backspace to reset the currently selected configuration, delete to remove it altogether or rearrange it using keys as already described.

When the key frames sequence is complete, pressing 'a' begins an animation which cycles through the programmer-created configurations. Pressing the up or down arrow keys speeds up or slows down the animation. Pressing 'a' again returns the program to develop mode.

Switching to animation mode also causes the program to write out to the file `animateManDataOut.txt` successive configurations of the animation sequence, stored currently in the vector `manVector`. Configuration are stored in successive lines of `animateManDataOut.txt`, each consisting of 11 floating point values – `partAngles[0]-[8]`, `upMove` and `forwardMove` – the same as are displayed on the screen in develop mode.

Now let's look at the code. From an OpenGL point of view, most interesting possibly is the drawing of a configuration by the function `Man::draw()`. The best way to understand it is to analyze the successive placement of parts. We'll do this our usual way of deconstructing a program by first commenting out most of it and then restoring code piece by piece.

Accordingly, first comment out all parts except the torso as below:

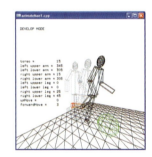

Figure 4.50: Screenshot of `animateMan1.cpp` in develop mode.

```
// Function to draw man.
void Man::draw()
{
    if (highlight||animateMode) glColor3fv(highlightColor);
    else glColor3fv(lowlightColor);

    glPushMatrix();

    // Up and forward translations.
    glTranslatef(0.0, upMove, forwardMove);

    // Torso begin.
    if (highlight && !animateMode) if (selectedPart == 0)
        glColor3fv(partSelectColor);

    glRotatef(partAngles[0], 1.0, 0.0, 0.0);

    glPushMatrix();
    glScalef(4.0, 16.0, 4.0);
    glutWireCube(1.0);
    glPopMatrix();
    if (highlight && !animateMode) glColor3fv(highlightColor);
    // Torso end.

    /*
    // Head begin.
    .

    .

    .

    // Right upper and lower leg with foot end.
    */

    glPopMatrix();
}
```

Next, uncomment the head:

```
// Function to draw man.
void Man::draw()
{
    if (highlight||animateMode) glColor3fv(highlightColor);
    else glColor3fv(lowlightColor);

    glPushMatrix();

    // Up and forward translations.
    glTranslatef(0.0, upMove, forwardMove);

    // Torso begin.
    if (highlight && !animateMode) if (selectedPart == 0)
```

```
      glColor3fv(partSelectColor);

    glRotatef(partAngles[0], 1.0, 0.0, 0.0);

    glPushMatrix();
    glScalef(4.0, 16.0, 4.0);
    glutWireCube(1.0);
    glPopMatrix();
    if (highlight && !animateMode) glColor3fv(highlightColor);
    // Torso end.

    // Head begin.
    glPushMatrix();

    glTranslatef(0.0, 11.5, 0.0);
    glPushMatrix();
    glScalef(2.0, 3.0, 2.0);
    glutWireSphere(1.0, 10, 8);
    glPopMatrix();

    glPopMatrix();
    // Head end.

    /*
    // Left upper and lower arm begin.
    .

    .

    .
    // Right upper and lower leg with foot end.
    */

    glPopMatrix();
}
```

Continue – as you successively uncomment each body part, it'll be clear how it's being placed with respect to existing ones.

The creation of the **camera** as an object of the **Camera** class may be of interest as well and we'll leave the reader to relate the parameter values of the **gluLookAt()** command to the member variables **viewDirection** and **zoomDistance** of the **Camera** class.

Much of the rest of the code consists simply of managing and using **manVector**, which itself stores the sequence of configurations.

Remark 4.15. Even though he himself is 3D, the man moves and rotates his parts always parallel to the yz-plane, so he's not really capable of 3D motion!

Experiment 4.33. Run **animateMan2.cpp**. This is simply a pared-down version of **animateMan1.cpp**, whose purpose is to animate the sequence of configurations listed in the file `animateManDataIn.txt`, typically generated from the develop mode of `animateMan1.cpp`. Press 'a' to toggle between animation on/off. As in `animateMan1.cpp`, pressing the up or down arrow key speeds up or slows down the animation. The camera functionalities via the keys 'r/R' and 'z/Z' remain as well.

The current contents of `animateManDataIn.txt` cause the man to do a handspring over the ball. Figure 4.51 is a screenshot.

Think of `animateMan1.cpp` as the studio and `animateMan2.cpp` as the movie.   End

Figure 4.51: Screenshot from Experiment 4.33.

Exercise 4.55. (Programming) Use `animateMan*.cpp` to animate a character kicking a football.

Exercise 4.56. (Programming) Enhance `animateMan*.cpp`:

(a) The character's body parts, except for the head, are currently all cubes. Make them more realistically rounded using cylinders.

(b) Add movement to the character's feet, which are currently fixed with respect to his lower legs. Give him movable hands as well.

(c) As remarked earlier, all the character's movements are currently parallel to a single plane. Enhance to true 3D.

Exercise 4.57. (Programming) Stick a camera 'to the front of the man's head and give a split-screen view of what he sees as he advances through an animation sequence, as well as from another fixed camera which follows the man.

Exercise 4.58. (Programming) By scaling individual body parts, create a second character who looks different from the first, though with identical functionality. Make a simple movie with the two.

It would be particularly effective in such a sequence to occasionally switch to a camera located in front of either one of their heads, to record how one sees the other.

Exercise 4.59. (Programming) Smoothly animating even a short movie requires several key frames (approximately 20 per second). However, the "important" ones are likely far fewer in number. For example, if a man kicks a ball, these are probably his wound-up pose ready to kick, the pose when his foot makes contact with the ball, a follow-through pose having kicked and, possibly, a few more in between to guide the sequence; certainly, far fewer than the 40 or so key frames needed for even a 2-second kicking sequence.

A movie-maker, therefore, saves a lot of tedious labor by simply drawing the important key frames, leaving an interpolating routine to fill in enough frames to make the animation smooth, a process called *tweening*.

Write a simple tweening routine based on `animateMan*.cpp`. In particular, use linear interpolation to fill configurations in between successive programmer-created ones.

4.7.2 Simple Shadow Animation

When the scaling transformation was introduced at the beginning of this chapter we said that degenerate scalings have the occasional application. Here's one to create and animate a simple shadow.

Experiment **4.34.** Run `ballAndTorusShadowed.cpp`, based on `ballAndTorus.cpp`, but with additional shadows drawn on a checkered floor. Press space to start the ball traveling around the torus and the up and down arrow keys to change its speed. Figure 4.52 is a screenshot. **End**

Figure 4.52: Screenshot of `ballAndTorus-Shadowed.cpp`.

There are parts of the program to make the picture look nice, e.g., lighting and material properties, which may not make sense currently, but neither are they relevant to drawing shadows. We focus now on the latter.

Note, first, that the routine `drawFlyingBallAndTorus()` repositions the ball and torus from `ballAndTorus.cpp` horizontally so that their shadow, thrown supposedly by a distant overhead light source, falls on the floor. That the (imaginary) light source is vertically far above is important, as it justifies drawing the shadows as if cast by rays parallel to the y-axis. The actual drawing itself is quite simple – the following few lines of code in the drawing routine do the trick:

```
glPushMatrix();
glScalef(1.0, 0.0, 1.0);
drawFlyingBallAndTorus(1);
glPopMatrix();
```

The argument value 1 to `drawFlyingBallAndTorus()` causes both ball and torus to be drawn black, while the degenerate scaling command `glScalef(1.0, 0.0, 1.0)` collapses the y-values of all their vertices to 0, creating a flat black object which is precisely their shadow on the xz-plane from light rays coming down parallel to the y-axis.

*R*emark *4.16.* Since `ballAndTorusShadowed.cpp` evidently contains code to light the scene, you might think that OpenGL can compute shadows automatically. This is not the case: OpenGL does not compute secondary consequences of lighting such as shadows and reflection. These have to be implemented separately by the programmer.

4.8 Selection and Picking

Strictly speaking, this section does not fit in a chapter about animation and viewing. However, countless interactive applications based on animation and viewing do ask the user to pick an object on the screen with a mouse or mouse-like device and, then, maybe even move it. Shoot-em-up games come to mind, as do drawing programs where the user chooses an object to reposition or modify its properties. We thought it important, therefore, to equip the reader at the end of this chapter with the means to incorporate this kind of interactivity in her applications.

Unfortunately, picking an object on the screen – which, effectively, means deciding to which object a picked pixel belongs – is not a simple operation given how the synthetic-camera pipeline functions. Objects enter the pipeline, are processed and emerge each as a set of fragments (fragment = pixel + color values), which are then rendered to the screen. Figure 4.53 is a conceptual diagram. The pipeline is not designed to be reversible, so there's no easy way to "climb back up" from screen space to world space. How then does one go about picking? Fortunately, OpenGL provides support for picking as well as a process it calls selection, which, in fact, enables picking. Let's begin with selection.

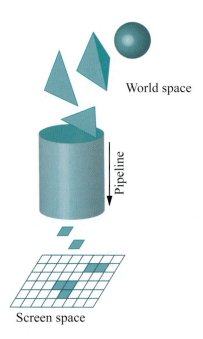

Figure 4.53: OpenGL's synthetic-camera pipeline (highly simplified!).

4.8.1 Selection

The idea underlying selection is simple. In a nutshell, it is to allow the user to specify a viewing volume and to then find the objects that intersect, or *hit*, this volume. To this end the user must first enter a rendering mode, called *selection mode*, by invoking `glRenderMode(GL_SELECT)`. In selection mode nothing is drawn to the frame buffer; rather, primitives are processed simply to determine their intersections with the specified viewing volume and generate so-called *hit records*.

To help determine from a hit record the primitive, or primitives, which produced it, OpenGL provides a so-called *name stack* which the user manipulates. The user can load names onto the name stack in a manner that establishes correspondence between primitives and names. A hit record contains the contents of the name stack at the time of its creation so, based upon the correspondence between primitives and names, the user can determine those involved in the hit. Let's get to specifics with the help of live code.

Experiment 4.35. Run `selection.cpp`, which is inspired by a similar program in the red book. It uses selection mode to determine the identity of rectangles, drawn with calls to `drawRectangle()`, which intersect the viewing volume created by the projection statement `glOrtho (-5.0, 5.0, -5.0, 5.0, -5.0, 5.0)`, this being a $10 \times 10 \times 10$ axis-aligned box centered at the origin. Figure 4.54 is a screenshot. Hit records are output to the command window. In the discussion following, we parse the program carefully.

End

Note: We'll call the viewing volume `glOrtho (-5.0, 5.0, -5.0, 5.0, -5.0, 5.0)`, used to "select" the rectangles intersecting it, the *selection volume*. It is different from the program's own viewing volume defined by the `glFrustum(-5.0, 5.0, -5.0, 5.0, 5.0, 100.0)` call in the `resize()` routine.

Displayed by the `drawConfiguration` routine is the outline of the selection volume and two rectangles, one red and one green, both inside it. If you don't trust the perspective view of the scene in Figure 4.54, as probably you shouldn't, verify from the parameters of the `drawRectangle()` call that the two rectangles indeed lie inside the selection volume.

The `selectHits()` routine, which comes next in the code, is where all the action is. Let's step through it carefully. The first statement

```
glSelectBuffer(1024, buffer);
```

specifies the array, called the *hit buffer*, to store hit records, as well as its size. The next statement

```
glRenderMode(GL_SELECT);
```

makes OpenGL enter selection mode. The next block of statements

```
glMatrixMode(GL_PROJECTION);
glPushMatrix();
```

Figure 4.54: Screenshot from `selection.cpp`.

```
glLoadIdentity();
glOrtho(-5.0, 5.0, -5.0, 5.0, -5.0, 5.0);
glMatrixMode(GL_MODELVIEW);
glLoadIdentity();
```

causes the matrix mode to change to projection, the current projection matrix (i.e., the one defined in the `resize()` routine) to be saved, that corresponding to the selection volume for hit testing to be placed on top of the projection matrix stack and, finally, modelview matrix mode to be re-entered and the current modelview matrix set to identity.

The statement pair next, viz.,

```
glInitNames();
glPushName(0);
```

initializes an empty name stack and pushes the name 0 on it (names are always non-negative integers). We'll not be using 0 to name any primitive, but push it on so that we have something to replace with "real" names when using `glLoadName()`. The initial configuration is depicted in Figure 4.55(a).

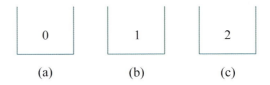

| 0 | 1 | 2 |
| (a) | (b) | (c) |

Figure 4.55: Name stack configurations: (a) Initial (b) When the red rectangle is drawn (c) When the green rectangle is drawn.

The following set of commands both manipulates the name stack and correspondingly "draws" primitives. Keep in mind that in selection mode nothing is *actually* drawn to the frame buffer, in other words, nothing is seen to happen.

```
glLoadName(1);
drawRectangle(0.0, 0.0, 3.0, 1.0, 0.0, 0.0); // Rect 1 (red).

glLoadName(2);
drawRectangle(0.0, 0.0, -3.0, 0.0, 1.0, 0.0); // Rect 2 (green).
```

Figures 4.55(b) and (c) depict the name stack as it is at the time of drawing of the first and second rectangles, respectively. The next statement

```
hits = glRenderMode(GL_RENDER);
```

takes OpenGL back to the default rendering mode where objects are indeed drawn to the frame buffer, at the same time returning the number of hit records in the hit buffer. Note that the return value of `glRenderMode()`

has meaning only when transiting out of selection mode or another mode called feedback, which we'll not use, and not when leaving rendering mode. Finally,

```
glMatrixMode(GL_PROJECTION);
glPopMatrix();
glMatrixMode(GL_MODELVIEW);
```

restore the projection matrix from the `resize()` routine and return OpenGL to modelview matrix mode.

As `selectHits()` was being executed, hit records were written into the hit buffer following rules we'll describe next. A hit record is written into the hit buffer when both of the following conditions hold:

(a) A name stack manipulation or `glRenderMode()` command is encountered, *and*

(b) a hit has occurred (i.e., a primitive drawn that intersects the selection volume) since the previous occurrence of such a command.

Each hit record contains four fields in the following order:

1. The number of names in the name stack at the time of writing the record.

2. The minimum z-value of vertices belonging to primitives which have hit the selection volume since the last hit record was written. This value is normalized by dividing by the depth of the selection volume to a number in the range $[0, 1]$, which is then multiplied by $2^{32} - 1$, rounded, and stored in the hit record as a 32-bit unsigned integer.

3. The maximum z-value of vertices belonging to primitives which have hit the selection volume since the last hit record was written, stored likewise.

4. The sequence of the names in the name stack at the time of writing the record with the bottom one first. (This sequence may be empty.)

It is the `processHitBuffer()` routine, called by `drawScene()`, which steps through the hit buffer, outputting its contents to the command window. Items 2 and 3 above, the minimum and maximum z-values of vertices, are normalized to between 0 and 1 by dividing by `maxint`.

There are two hit records, as you can see in the command window. The first one $(1, 0.2, 0.2, 1)$ was generated on processing the `glLoadName(2)` call because a hit (the red rectangle) occurred after the previous name stack manipulation command (`glLoadName(1)`). The contents of this record are easy to understand if one observes that the configuration of the name stack at the time of the record's creation was as in Figure 4.55(b); moreover, the depth of *all* vertices of the red rectangle from the front face of the viewing box is 2, which becomes $2/10 = 0.2$ when normalized by division by the box's depth.

The second hit record $(1, 0.8, 0.8, 2)$ is generated on processing the

```
hits = glRenderMode(GL_RENDER);
```

statement, and we leave the reader to parse its contents.

Exercise 4.60. (Programming) Add one more rectangle, but in two different ways. First, insert the statement

```
drawRectangle(0.0, 0.0, 0.0, 0.0, 0.0, 1.0);
```

in the selectHits() routine (a) *just before* the glPushName(0) call (and after glInitNames()), and (b) *just after* glPushName(0). What are the hit records generated in each case? When is each of these hit records generated?

Note: To *see* the new rectangle, make sure to add an identical drawing statement in the drawConfiguration() routine!

Part answer: In either case a new hit record comes before the two from the original program. When the statement is before glPushName(0), the new record is $(0, 0.5, 0.5, \quad)$ with an empty name list.

Exercise 4.61. (Programming) Restore the original selection.cpp program and insert the rectangle-drawing statement

```
drawRectangle(0.0, 0.0, 0.0, 0.0, 0.0, 1.0);
```

of the preceding exercise in the selectHits() routine, just after the statement that draws the first (red) rectangle. Explain the hit records.

Exercise 4.62. (Programming) Restore the original selection.cpp program and insert the pair of name stack manipulation commands

```
glLoadName(3);
glLoadName(4);
```

right after the statement that draws the second (green) rectangle. The output is the same as for the unmodified selection.cpp. Why?

Exercise 4.63. (Programming) Restore the original selection.cpp program and add a new name stack manipulation command

```
glPushName(3);
```

between the glLoadName(2) call and the statement that draws the green rectangle. Predict the output before running.

Remark 4.17. The previous exercise shows a way of tagging an object with multiple names (in this case the green rectangle with 2 and 3) which is particularly useful in a scene where there is a hierarchy of objects. For example, we may want to tag the tail fin of the fourth aircraft with the names 4 and 7 (if 7 is the part number of a tail fin).

Exercise 4.64. (Programming) The one remaining name stack manipulation command, which we have not used yet, is `glPopName()`, whose action the user can easily guess. Insert a `glPopName()` statement in the `selectHits()` routine of `selection.cpp` in such a manner that the second hit record generated is $(0, 0.8, 0.8,\ )$.

4.8.2 Picking

Now that we have an understanding of the selection process, let's move on to picking, which is really selection plus a little help from OpenGL in setting up a selection volume to track a user-specified point on the screen.

Figure 4.56 illustrates the idea. V is the viewing frustum defined by the projection statement of a program. Objects are, therefore, drawn to the OpenGL window following perspective projection to the viewing face of V (we'll identify V's viewing face with the OpenGL window without harm, because going from one to the other is a simple scaling). Accordingly, one can find the objects picked by choosing a point P in the OpenGL window by determining those that intersect a long thin frustum like V' surrounding P, because it's precisely these objects whose projections intersect P, with some error depending on the thickness of V'. Clearly, the thinner V' is the more accurate the picking. And, of course, it's in detecting intersections with V' that selection comes in.

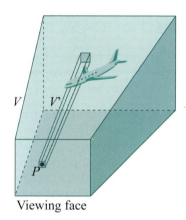

Viewing face

Figure 4.56: Clicking P "hits" the aircraft because the latter intersects V'.

In addition to the selection mechanism, there's even more help to be had from OpenGL: the GLU routine `gluPickMatrix()` defines a selection volume for use in picking, which is a frustum of user-specified size centered at a user-specified point. Here's how it works. The sequence of commands

```
glLoadIdentity();
gluPickMatrix(pickX, pickY, width, height, viewport[4]);
```

```
glFrustum(); or gluPerspective(); or glOrtho(); // Copied from the
                                                // reshape routine.
```

causes the top matrix of the projection matrix stack to be replaced by one corresponding to a selection volume whose front face is a *width* × *height* rectangle centered at the point of the OpenGL window with x and y world coordinate values equal to *pickX* and *pickY*, respectively. The *viewport[]* array supplies the current viewport boundaries and may be set by calling **glGetIntegerv(GL_VIEWPORT,** *viewport*). Functionally, the **gluPickMatrix()** command actually generates a matrix, called the *pick matrix*.

Let's get to work using the pick mechanism in a simple game-like application.

Experiment 4.36. Run **ballAndTorusPicking.cpp**, which preserves all the functionality of **ballAndTorus.cpp** upon which it is based and adds the capability of picking the ball or torus with a left click of the mouse. The picked object blushes. See Figure 4.57 for a screenshot. **End**

Figure 4.57: Screenshot of **ballAndTorus-Picking.cpp** moments after the ball has been clicked.

The **drawBallAndTorus()** routine of **ballAndTorusPicking.cpp** is pretty much the whole **drawScene()** routine of **ballAndTorus.cpp**, except with two main differences:

(a) If the value of the global **isSelecting** variable is 1 – it can be 0 or 1 – then **glLoadName()** is invoked to tag the torus with the name 1 and the ball with name 2.

(b) If one of the torus or ball is picked – the name being contained in the global **closestName** – it is painted red for as long as the global **highlightFrames** is greater than 0.

The mouse callback **pickFunction()** is written along the lines of **selectHits()** of **selection.cpp**. The important difference is that the selection volume for hits is specified with the help of a **gluPickMatrix()** call. And, of course, instead of drawing rectangles as in **selection.cpp**, **drawBallAndTorus()** is called with **isSelecting** set to 1.

The routine **findClosestHit()** called by **pickFunction()** is an interesting modification of the **processHitBuffer()** routine of **selection.cpp**. In case there is more than one hit record, implying that both ball and torus fell under the mouse click, **findClosestHit()** compares their min-z fields to determine the one closer to the viewer.

Note: Sometimes an object doesn't light up on what seems like a definite click or the farther object lights up when both fall under the same mouse click. That's because the click fell between mesh wires! Possible solutions include making the meshes finer or the picking less sensitive by increasing the *width* and *height* parameters of **gluPickMatrix()** from the current values of 3 for both.

Picking plus dragging with mouse motion (see Section 3.4 for the latter) make a potent duo. Give it a go next.

Exercise 4.65. (Programming) Enhance `canvas.cpp`, from the previous chapter, so that figures in the drawing area can be picked and moved.

Exercise 4.66. (Programming) Create a game, be it a shoot-em-up or drag-em-down or ... Use your imagination.

Picking by Color Coding

Yet another method to pick objects in OpenGL is by means of so-called *color coding*. We'll describe the idea briefly, but not go into detail, nor use it in a program. Picking by color coding requires use of the back buffer, so the program must run in double-buffered mode.

Here's how it works. When the user picks a pixel the entire scene is *redrawn* to the back buffer, but with objects of interest drawn in *different* colors. In other words, objects are color coded there. Next, data from the picked pixel is read from the back buffer with help of a `glReadPixels()` call and its color decoded to determine the picked object.

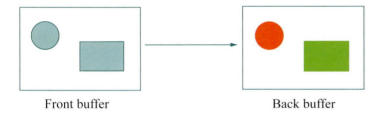

Front buffer   Back buffer

Figure 4.58: Color coding.

Figure 4.58 illustrates the idea – the disc and the rectangle are distinguishable in the back buffer by means of their color. So, e.g., if the picked pixel is red in the back buffer then the primitive picked is the disc.

4.9 Summary, Notes and More Reading

If animation is like a car, then we've just gotten our driver's license. In this chapter we learned the basics of the modeling and viewing transformations and how to use them to move objects and change their shape, as well as to manipulate the camera. We also peeked under the hood at OpenGL's engine, particularly in order to understand how transformations are composed and how they are used to place objects relative to one another. Collision detection, which is often crucial in interactive game programming, was discussed in the context of animation. We also began a discussion of orientation and Euler angles which will be continued in a more advanced chapter on animation. We learned as well the techniques of selection and picking, essential in interactive environments. And we saw plenty of live code along the way.

The topics covered in this chapter are at the very heart of computer graphics. Every introduction to the subject will have some coverage – see, for example, any of the introductory references, both OpenGL-based and API-independent, listed in Section 2.11 – differing perhaps in style and extent. It can only reinforce understanding to get more than one point of view, so the reader is encouraged to pick up other CG books which may be handy and turn to the relevant chapters. She will also find useful several of the on-line tutorials listed at the OpenGL site [99]. Particularly noteworthy is Nate Robins' finely-designed suite [96].

Collision detection is also to a greater or lesser extent covered in most introductory CG books, often in the context of ray tracing, which is a technique of rendering where light rays are followed from source to collision with an object (and possibly reflection again). For further reading about collision detection, however, the reader is well-advised to consult books on game programming, where it is especially important. See, e.g., Lengyel [81] and van Verth & Bishop [139]. Specialized books on collision detection include Ericson [40] and van den Bergen [138]. An extensive repository of resources on collision detection, including research papers and code, is at the UNC Gamma Research Group website [137].

This chapter is also a lead-in to the extremely important discipline of physics in graphics (popularly called game physics), which includes the study of multi-body kinematics and dynamics of rigid and deformable bodies, among other topics from real-world physics. Real-time game physics is particularly important in the creation of realistic interactive games. Introductory books for the interested reader include Bourg [19] and Eberly [38].

Undoubtedly, the best way for the reader to build on this chapter is to write lots and lots of animation code. In fact, this is a good time for her to begin coding, if she hasn't already done so, a significant project, e.g., a game or movie. She can get the essentials in place now and then embellish her project as we go along – with more complex objects, color, light and texture.

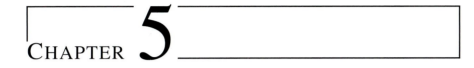

CHAPTER 5

Inside Animation: The Theory of Transformations

We studied transformations and their application to creating animation by moving and turning and scaling objects in Chapter 4. The goal for this chapter is to understand the underlying theory. We want to look under the hood of the graphics engine and understand exactly how transformations are implemented. What we'll encounter is the mathematics of geometric transformations.

We begin our discussion of geometric transformations in Section 5.1 in the simple surroundings of Flatland (2-dimensional space), the objective being to get concepts in place and prove results that will extend fairly easily to the real world. This program starts in Sections 5.1.1-5.1.4 with the expression of familiar geometric transformations, in particular, translations, scalings, rotations and reflections, by means of matrices.

Next, we briefly interrupt our pursuit of 2D geometric transforms to digress in Section 5.2 into linear algebra, particularly for an understanding of affine transformations. Affine transformations will provide a unifying perspective of all the geometric transformations that we encounter. In Sections 5.2.1-5.2.3 we define affine transformations as a generalization of linear transformations, prove that they are particularly pleasant in their geometric behavior, understand the central role they play in the design of a graphics API such as OpenGL and, finally, learn the use of homogeneous coordinates to facilitate the application of affine transformations.

We resume our study of 2D geometric transforms in Section 5.3 with our newfound knowledge of affine transformations. We begin in 5.3.1 by placing the transformations of Section 5.1 in context as affine geometric transformations. In 5.3.2 comes the notion of Euclidean transformations, and their subclass of rigid transformations, neither of which distorts the shape of an object. Consequently, these are the transformations to use to animate rigid objects. The exploration of 2D transformations concludes with a discussion of shears, a commonly occurring shape-distorting transformation, in 5.3.3.

Geometric transformations of the real world or 3-space – transformations that OpenGL actually implements – is the topic of Section 5.4. The development parallels that of the previous section on 2D transformations. Matrix expressions for translations, scalings and reflections generalize easily from their 2D counterparts in Sections 5.4.1-5.4.2 and 5.4.4. 3D rotations, however, require considerably more work in the longish 5.4.3. Observing in 5.4.5 that translations, scalings and rotations about radial axes are fundamental affine transformations, in the sense that they can be used to generate all other affine transformations, lends insight into the design of a CG animation engine. We realize that, however exciting the game or movie is that we happen to be enjoying, most of what is going on inside the machine is the distinctly unglamorous activity of multiplying 4×4 matrices – lots and lots of them and very very fast! We learn to access and manipulate the OpenGL modelview matrix stack in 5.4.6. Euclidean and rigid 3D transformations are discussed next in Section 5.4.7. The ability to access the modelview matrix stack comes in handy in 5.4.8 when we learn about 3D shears and how to manually code and insert their matrices into the stack.

We conclude in Section 5.5 with a summary, notes and suggestions for further reading.

Important: We assume for this chapter familiarity with basic linear algebra, as, say, would be found in a first college course or in a multitude of introductory texts, e.g., [7, 46, 69, 76, 80, 132] and others. Section 5.2 needs familiarity as well with the basics of convex sets, again from an introductory geometry text or course*.

You can safely defer this somewhat theoretical chapter to a second pass through the book.

5.1   Geometric Transformations in 2-Space

We begin discussion of geometric transformations in 2D space, rather than real-life 3D, in order to develop our intuition in a simpler setting. Much of what we say and prove, though, will generalize fairly easily to 3D.

---

*The needed material on convexity can also be found in our own Chapter 7. However, that chapter was not written to precede the current one and logically belongs to the group Chapters 7-9.

5.1.1 Translation

A translation is specified by a *displacement vector* $D = [d_x \; d_y]^T$, which is added to the location vector of each point. Precisely, the image of the point $P = [x \; y]^T$ by this translation is $P' = [x' \; y']^T$, where

$$\begin{bmatrix} x' \\ y' \end{bmatrix} = \begin{bmatrix} x \\ y \end{bmatrix} + \begin{bmatrix} d_x \\ d_y \end{bmatrix} = \begin{bmatrix} x + d_x \\ y + d_y \end{bmatrix} = \begin{bmatrix} 1 & 0 \\ 0 & 1 \end{bmatrix} \begin{bmatrix} x \\ y \end{bmatrix} + \begin{bmatrix} d_x \\ d_y \end{bmatrix}$$

See Figure 5.1. Concisely:

$$P' = \begin{bmatrix} 1 & 0 \\ 0 & 1 \end{bmatrix} P + D \qquad (5.1)$$

The matrix multiplication may seem redundant, but it serves to put the RHS in a form useful for when we come to affine transformations.

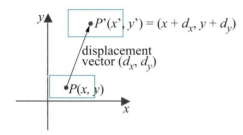

Figure 5.1: Translation.

Terminology: We'll use the coordinate notation (x, y) and the matrix notation $[x \; y]^T$ for a point interchangeably, particularly preferring the latter when we want to treat the point's location as a vector.

Exercise 5.1. Prove that the composition of translations is a translation and that the inverse of a translation is another translation.

Part answer: We'll prove that the composition of translations is again a translation the long way, using the matrix form of the translation equation, but it's good practice. So suppose the translations t_1 and t_2 are specified by the displacement vectors D_1 and D_2, respectively. Then

$$\begin{aligned} (t_1 \circ t_2)(P) &= t_1(t_2(P)) = t_1 \left(\begin{bmatrix} 1 & 0 \\ 0 & 1 \end{bmatrix} P + D_2 \right) \\ &= \begin{bmatrix} 1 & 0 \\ 0 & 1 \end{bmatrix} \left(\begin{bmatrix} 1 & 0 \\ 0 & 1 \end{bmatrix} P + D_2 \right) + D_1 \\ &= \left(\begin{bmatrix} 1 & 0 \\ 0 & 1 \end{bmatrix} P + D_2 \right) + D_1 = \begin{bmatrix} 1 & 0 \\ 0 & 1 \end{bmatrix} P + (D_2 + D_1) \end{aligned}$$

proving that $t_1 \circ t_2$ is indeed a translation, specified by the displacement $D_2 + D_1$.

5.1.2 Scaling

A scaling is specified by a scaling factor s_x along the x-axis and a scaling factor s_y along the y-axis. The image of a point P by this scaling is the one whose x coordinate value is s_x times that of P and y coordinate value s_y times that of P (see Figure 5.2). Precisely, the image of $P = [x\ y]^T$ is $P' = [x'\ y']^T$, where

$$\begin{bmatrix} x' \\ y' \end{bmatrix} = \begin{bmatrix} s_x x \\ s_y y \end{bmatrix} = \begin{bmatrix} s_x & 0 \\ 0 & s_y \end{bmatrix} \begin{bmatrix} x \\ y \end{bmatrix}$$

Concisely:

$$P' = \begin{bmatrix} s_x & 0 \\ 0 & s_y \end{bmatrix} P \tag{5.2}$$

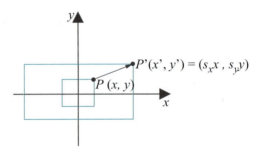

Figure 5.2: Scaling.

If either, or both, of the scaling factors s_x and s_y is zero, then the scaling is said to be *degenerate*; if neither is zero, it is *non-degenerate*. By a scaling we shall always mean a non-degenerate one, unless specifically stated otherwise.

Exercise 5.2. Show that the scaling given by Equation (5.2) is non-degenerate if and only if its matrix is non-singular.

Exercise 5.3. Use Equation (5.2) to prove that the composition of scalings is a scaling and that the inverse of a non-degenerate scaling is another non-degenerate scaling. Are degenerate scalings invertible?

5.1.3 Rotation

A rotation about the origin is specified by an angle θ measured counter-clockwise as seen by a viewer V located on the positive side of the z-axis (in a hypothetical right-handed 3D coordinate system made by adding a z-axis to the x and y axes of the given 2D plane). See Figure 5.3(a).

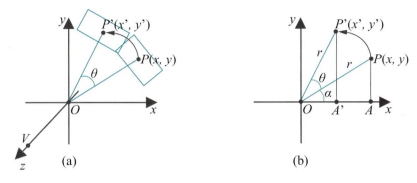

Figure 5.3: Rotation.

Note: We had to add the z-axis and place a viewer on a particular side of it because it's not enough to simply say that a rotation on the xy-plane is counter-clockwise: the same rotation appears counter-clockwise from one side and clockwise from the other.

In the future, to avoid tedious language, we'll always assume that a viewer is located at a point such as V, on the positive side of the z-axis.

We want to calculate the image $P' = [x' \; y']^T$ of the point $P = [x \; y]^T$ by this rotation. The method is exactly the same as the solution given for Exercise 4.8(c), in the case of a 3D rotation about the z-axis. The reader can refer again to that exercise or deduce herself the following equations from Figure 5.3(b):

$$
\begin{aligned}
x &= OA = r \cos \alpha \\
y &= PA = r \sin \alpha
\end{aligned}
$$

which are used in

$$
\begin{aligned}
x' &= OA' = r \cos(\alpha + \theta) = r \cos \alpha \cos \theta - r \sin \alpha \sin \theta \\
&= x \cos \theta - y \sin \theta \\
y' &= P'A' = r \sin(\alpha + \theta) = r \cos \alpha \sin \theta + r \sin \alpha \cos \theta \\
&= x \sin \theta + y \cos \theta
\end{aligned}
$$

Therefore, the image of $P = [x \; y]^T$ by a rotation of angle θ counter-clockwise about the origin is $P' = [x' \; y']^T$, where

$$
\begin{bmatrix} x' \\ y' \end{bmatrix} = \begin{bmatrix} x \cos \theta - y \sin \theta \\ x \sin \theta + y \cos \theta \end{bmatrix} = \begin{bmatrix} \cos \theta & -\sin \theta \\ \sin \theta & \cos \theta \end{bmatrix} \begin{bmatrix} x \\ y \end{bmatrix} \tag{5.3}
$$

or, concisely,

$$
P' = \begin{bmatrix} \cos \theta & -\sin \theta \\ \sin \theta & \cos \theta \end{bmatrix} P \tag{5.4}
$$

The matrix in the preceding equation is often called a *rotation matrix*.

Example 5.1. Write the matrix form as in (5.4) of a counter-clockwise rotation by an angle of $60°$ about the origin. To which point is $[1 \ -2]^T$ transformed by this particular rotation?

Answer: The given rotation will take $P = [x \ y]^T$ to $P' = [x' \ y']^T$, where

$$P' = \begin{bmatrix} \cos 60° & -\sin 60° \\ \sin 60° & \cos 60° \end{bmatrix} P = \begin{bmatrix} 1/2 & -\sqrt{3}/2 \\ \sqrt{3}/2 & 1/2 \end{bmatrix} P$$

Therefore, $[1 \ -2]^T$ is transformed to

$$\begin{bmatrix} 1/2 & -\sqrt{3}/2 \\ \sqrt{3}/2 & 1/2 \end{bmatrix} \begin{bmatrix} 1 \\ -2 \end{bmatrix} = \begin{bmatrix} 1/2 + \sqrt{3} \\ -1 + \sqrt{3}/2 \end{bmatrix} \simeq \begin{bmatrix} 2.23 \\ -0.13 \end{bmatrix}$$

Exercise 5.4. Is the matrix of a rotation about the origin always non-singular?

Exercise 5.5. Determine the matrix expression for a counter-clockwise rotation by an angle θ about an arbitrary point $O' = [a \ b]^T$, not necessarily the origin.

Suggested approach: Use the Trick of Example 4.2 to express this rotation as a composition of three successive transformations:

1. A translation by the displacement vector $[-a \ -b]^T$ taking O' to the origin.

2. A counter-clockwise rotation by θ about the origin.

3. A translation by the displacement vector $[a \ b]^T$ restoring O' to its original position.

Next, compose the expressions corresponding to these three transformations (make sure to do this in the right order). The result will be a transformation of the form $P \mapsto MP + D$.

Exercise 5.6. Determine the matrix expression for a rotation of $45°$ counter-clockwise about the point $[2 \ 3]^T$.

Exercise 5.7. Use Equation (5.4) to prove that the composition of rotations about the origin is another such and that so is the inverse of a rotation about the origin.

Exercise 5.8. How about the composition of rotations about some fixed point other than the origin? Is this again a rotation about that point?

Example 5.2. Is the composition of rotations about *different* points necessarily equivalent to a *single* rotation about some one point?

Answer: Consider rotations r_1 and r_2, both of $180°$, about the two points $O_1 = [0\ 0]^T$ and $O_2 = [1\ 0]^T$, respectively. We'll show that $r_2 \circ r_1$ is not a rotation about any point, answering the question asked in the negative.

It's easy to check that $(r_2 \circ r_1)(O_1) = r_2(r_1(O_1)) = r_2(O_1) = [2\ 0]^T$, while $(r_2 \circ r_1)(O_2) = r_2(r_1(O_2)) = r_2([-1\ 0]^T) = [3\ 0]^T$. See Figure 5.4(a).

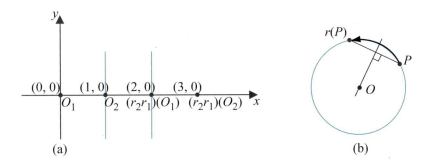

(a)   (b)

Figure 5.4: Illustrations for Example 5.2.

Next, observe that for any (non-identity) rotation r, and any point P which is not itself the center O of the rotation, O lies on the perpendicular bisector of the segment joining P and $r(P)$. Figure 5.4(b) indicates why.

Now, *if* $r_2 \circ r_1$ were indeed a rotation, its center, first, is not either O_1 or O_2, as both points are moved by $r_2 \circ r_1$. Therefore, its center must lie on the perpendicular bisector of the segment joining O_1 and $(r_2 \circ r_1)(O_1)$, as also on the perpendicular bisector of the segment joining O_2 and $(r_2 \circ r_1)(O_2)$. But this is not possible as the two bisectors are straight lines through $[1\ 0]^T$ and $[2\ 0]^T$, respectively, both parallel to the y-axis and, therefore, non-intersecting. One concludes that $r_2 \circ r_1$ is not a rotation about any point.

Exercise 5.9. The composition $r_2 \circ r_1$ of the preceding example, though not a rotation, is, nevertheless, a familiar kind of transformation. Can you identify it?
Hint: It's a translation!

Exercise 5.10. Can you give an example where the composition of two (non-trivial) rotations about different points *is* equivalent to a single rotation about some point?
Hint: Consider rotating $90°$ counter-clockwise around $[-1\ 1]^T$ and then around $[1\ 1]^T$ the same amount. Show that this composition is equivalent to a rotation of $180°$ around the origin.

Remark 5.1. From the preceding two exercises we see one case where the composition of two rotations is a translation and one where it is again a rotation. It turns out that these are the only two possibilities in general.

We'll prove this fact in Section 5.3.2, in particular, when we classify rigid transformations, and see an easy rule as well to decide the nature of a composition of rotations.

5.1.4   Reflection

The image of the point $P = [x \; y]^T$ by a reflection about a straight line l, called the *mirror*, is $P' = [x' \; y']^T$ such that:

(a) if P lies on l, then $P' = P$;

(b) if P does not lie on l, then P' is the point on the other side of l such that PP' is perpendicular to l, and P' is the same distance from l as P. See Figure 5.5.

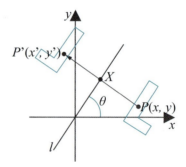

Figure 5.5: Reflection. $|XP| = |XP'|$.

A reflection is, therefore, specified by the mirror about which it occurs. Let's analyze first the reflection about a radial mirror l at an angle θ counterclockwise from the positive direction of the x-axis (as depicted in Figure 5.5). We claim that reflection about l maps the point P to P' where

$$P' = \begin{bmatrix} \cos 2\theta & \sin 2\theta \\ \sin 2\theta & -\cos 2\theta \end{bmatrix} P \qquad (5.5)$$

and leave the proof to the reader in the next exercise.

Note: A *radial* line or plane is one which passes through the origin.

Exercise 5.11. Verify Equation (5.5).

Suggested approach: Use the Trick to express this reflection as the composition of three successive transformations:

1. A rotation of $-\theta$ about the origin to align l along the x-axis.

2. A reflection about the x-axis. This is given by $[x\ y]^T \mapsto [x\ -y]^T$, which is simply scaling by a factor of 1 along the x-axis and -1 along the y-axis.

3. A rotation of θ about the origin to restore l to its original alignment.

Exercise 5.12. Write the matrix form, as in (5.5), of a reflection about the radial mirror at an angle of 30° to the positive x-axis. To which point is $[1\ 1]^T$ transformed by this reflection?

Exercise 5.13. What is the determinant of the matrix of a reflection about a radial mirror? Is the matrix always non-singular?

Exercise 5.14. Use the Trick to prove that a reflection about an arbitrary mirror, not necessarily radial, is a composition of two translations, two rotations about the origin and one scaling.

A consequence of the preceding exercise is that one of those highly-paid Flatland programmers developing a graphics API has only to implement translations, rotations about the origin and scalings to get reflections for free.

Exercise 5.15. What is the inverse of a reflection?

Exercise 5.16. Show that any non-identity translation can be obtained by composing reflections about two parallel mirrors. Show that any non-identity rotation can be obtained by composing reflections about two intersecting mirrors. (The identity transformation itself can be obtained obviously by composing reflections about the same mirror twice.)

Exercise 5.17. A reflection about a mirror l, followed by translation by a displacement vector (d_x, d_y) which is either zero or parallel to l, is called a *glide reflection*. See Figure 5.6.

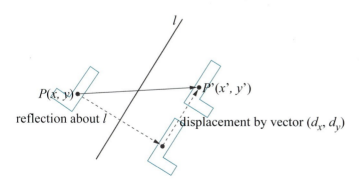

Figure 5.6: Glide reflection.

Determine the matrix expression for the glide reflection which uses the mirror l and displacement vector (d_x, d_y). Assume l goes through the point (a, b) and makes an angle θ measured counter-clockwise from the positive direction of the x-axis.

Exercise 5.18. Two transformations f_1 and f_2 are said to *commute* (or be *commutative*) if $f_1 \circ f_2 = f_2 \circ f_1$, in other words, if applying f_1 followed by f_2 is the same as applying f_2 followed by f_1.

(a) Do translations commute with each other?

(b) Do scalings commute with each other?

(c) Do rotations about the same point commute with each other?

(d) Does a rotation about one point commute with another about a different point?

Hint: A counter-example to show that, generally, rotations about different points don't commute can be obtained, in fact, from the configuration given in the answer to Example 5.2: consider if $(r_2 \circ r_1)(O_1)$ and $(r_1 \circ r_2)(O_1)$ are the same. If they are not, then, of course, $r_2 \circ r_1$ and $r_1 \circ r_2$ are not the same either.

(e) Do translations and rotations commute?

(f) Do reflections about two different mirrors *ever* commute?

Hint: Keep in mind the special case of perpendicular mirrors.

5.2   Affine Transformations

Before resuming our pursuit of geometric transformations in 2-space, we change pace a bit to learn about affine transformations because they will provide a unifying framework in which to locate the seemingly disparate geometric transformations that we have encountered (and will encounter).

5.2.1   Affine Transformations Defined

Affine transformations are a natural generalization of linear transformations, obtained by tacking on an additional translation to a non-singular linear transformation. We'll give the next couple of definitions in arbitrary dimensions – this generality costing nothing in difficulty. Down the road, of course, we can specialize to $\mathbb{R}^2$ or $\mathbb{R}^3$ depending on the setting.

For the record, here's the definition of a linear transformation, which is our starting point.

Definition 5.1. The *linear transformation* f^M of $\mathbb{R}^m$, with *defining matrix*

$$M = \begin{bmatrix} a_{11} & a_{12} & \cdots & a_{1m} \\ a_{21} & a_{22} & \cdots & a_{2m} \\ & \cdots & \cdots & \\ a_{m1} & a_{m2} & \cdots & a_{mm} \end{bmatrix}$$

is the transformation $f^M : \mathbb{R}^m \to \mathbb{R}^m$ specified by the equation

$$f^M(P) = MP, \qquad P \in \mathbb{R}^m \tag{5.6}$$

In other words, f^M maps $P = [x_1 \ x_2 \ \dots \ x_m]^T$ to $f^M(P) = [x_1' \ x_2' \ \dots \ x_m']^T$, where

$$
\begin{aligned}
x_1' &= a_{11}x_1 + a_{12}x_2 + \dots + a_{1m}x_m \\
x_2' &= a_{21}x_1 + a_{22}x_2 + \dots + a_{2m}x_m \\
&\cdots \ \cdots \\
x_m' &= a_{m1}x_1 + a_{m2}x_2 + \dots + a_{mm}x_m
\end{aligned} \tag{5.7}
$$

As promised, an additional translation next gives affine transformations.

Definition 5.2. An *affine transformation* of $\mathbb{R}^m$ is a transformation $g : \mathbb{R}^m \to \mathbb{R}^m$ specified by an equation of the form

$$g(P) = f^M(P) + D = MP + D \tag{5.8}$$

for $P \in \mathbb{R}^m$, where f^M is a non-singular linear transformation of $\mathbb{R}^m$ and D is an m-vector. The matrix M, which is non-singular as f^M is non-singular, is called the *defining matrix* of g. The vector D is called the *translational component* of g.

Accordingly, if

$$M = \begin{bmatrix} a_{11} & a_{12} & \cdots & a_{1m} \\ a_{21} & a_{22} & \cdots & a_{2m} \\ & \cdots & \cdots & \\ a_{m1} & a_{m2} & \cdots & a_{mm} \end{bmatrix} \text{ is non-singular and } D = \begin{bmatrix} d_1 \\ d_2 \\ \cdots \\ d_m \end{bmatrix} \text{ arbitrary,}$$

then the affine transformation g defined by $g(P) = MP + D$ maps $P = [x_1 \ x_2 \ \dots \ x_m]^T$ to $g(P) = [x_1' \ x_2' \ \dots \ x_m']^T$, where

$$
\begin{aligned}
x_1' &= a_{11}x_1 + a_{12}x_2 + \dots + a_{1m}x_m + d_1 \\
x_2' &= a_{21}x_1 + a_{22}x_2 + \dots + a_{2m}x_m + d_2 \\
&\cdots \ \cdots \\
x_m' &= a_{m1}x_1 + a_{m2}x_2 + \dots + a_{mm}x_m + d_m
\end{aligned} \tag{5.9}
$$

If its translational component is zero, then an affine transformation evidently reduces to a non-singular linear transformation.

Example 5.3. $g : \mathbb{R}^2 \to \mathbb{R}^2$ given by

$$g([x \; y]^T) = \begin{bmatrix} 2 & 1 \\ 0 & 4 \end{bmatrix} [x \; y]^T + [4 \; 6]^T$$

is affine. Writing out the formula for g we have

$$g([x \; y]^T) = \begin{bmatrix} 2x + y + 4 \\ 4y + 6 \end{bmatrix}$$

So, e.g.,

$$g([-1 \; 2]^T = [4 \; 14]^T \quad \text{and} \quad g([0 \; 3]^T = [7 \; 18]^T$$

Exercise 5.19. What are the images of the points $[0 \; 0 \; 0]^T$ and $[1 \; -1 \; 1]^T$ by the affine transformation $g : \mathbb{R}^3 \to \mathbb{R}^3$ given by

$$g([x \; y \; z]^T) = \begin{bmatrix} -1 & -2 & 3 \\ 4 & 0 & 2 \\ 0 & -3 & 1 \end{bmatrix} [x \; y \; z]^T + [-1 \; 6 \; 3]^T \; ?$$

The next example says that an affine transformation is respectful of convex combinations.

Example 5.4. Show that an affine transformation g of $\mathbb{R}^m$ preserves convex combinations and barycentric coordinates in that

$$g(c_1 P_1 + c_2 P_2 + \ldots + c_k P_k) = c_1 g(P_1) + c_2 g(P_2) + \ldots + c_k g(P_k)$$

for any m-vectors P_i and scalars c_i, $1 \le i \le k$, such that $0 \le c_i \le 1$ and $c_1 + c_2 + \ldots + c_k = 1$.

Answer: Suppose that $g(P) = MP + D$, where M is the defining matrix and D the translational component of g. Then

$$
\begin{aligned}
g(c_1 P_1 + c_2 P_2 + \ldots + c_k P_k) &= M(c_1 P_1 + c_2 P_2 + \ldots + c_k P_k) + D \\
&= M(c_1 P_1 + c_2 P_2 + \ldots + c_k P_k) + \\
&\quad (c_1 + c_2 + \ldots + c_k)D \\
&\quad (\text{because } c_1 + c_2 + \ldots + c_k = 1) \\
&= c_1 M P_1 + c_2 M P_2 + \ldots + c_k M P_k + \\
&\quad c_1 D + c_2 D + \ldots + c_k D \\
&= c_1 (M P_1 + D) + c_2 (M P_2 + D) + \ldots + \\
&\quad c_k (M P_k + D) \\
&= c_1 g(P_1) + c_2 g(P_2) + \ldots + c_k g(P_k)
\end{aligned}
$$

Exercise 5.20. Prove that an affine transformation which fixes the origin (i.e., maps the origin to itself) is a non-singular linear transformation.

Exercise 5.21. Prove that the composition of affine transformations is again an affine transformation.

Exercise 5.22. Determine the affine transformation $g_1 \circ g_2$, where

$$g_1(P) = \begin{bmatrix} 2 & 1 \\ 0 & 4 \end{bmatrix} P + [4\ 6]^T \quad \text{and} \quad g_2(P) = \begin{bmatrix} -1 & 3 \\ 1 & -2 \end{bmatrix} P + [-1\ 0]^T$$

Example 5.5. An affine transformation g is always invertible. In fact, if g is defined by $g(P) = MP + D$, then show that its inverse, also affine, is given by

$$g^{-1}(P) = M^{-1}P - M^{-1}D$$

Answer: For any $P \in \mathbb{R}^m$, we have

$$(g^{-1} \circ g)(P) = M^{-1}(MP + D) - M^{-1}D = P$$

proving that $g^{-1} \circ g$ is the identity on $\mathbb{R}^m$. Likewise, it can be seen that $g \circ g^{-1}$ is the identity, proving that g^{-1}, as defined above, indeed is the inverse of g.

Exercise 5.23. Determine the inverse of the affine transformation g of $\mathbb{R}^2$ given by

$$g([x\ y]^T) = \begin{bmatrix} 2 & 1 \\ 0 & 4 \end{bmatrix} [x\ y]^T + [4\ 6]^T$$

The following important proposition says that affine transformations are particularly well-behaved from a geometric point of view, in particular, that they preserve straightness, planarity, parallelism and convexity.

Proposition 5.1.

(a) An affine transformation g of $\mathbb{R}^2$ maps straight lines to straight lines. Moreover, it maps parallel straight lines to parallel straight lines and intersecting straight lines to intersecting straight lines.

(b) An affine transformation g of $\mathbb{R}^2$ maps convex sets to convex sets. Moreover, it maps the convex hull of $\{P_1, P_2, \ldots, P_k\}$ to the convex hull of $\{f^M(P_1), f^M(P_2), \ldots, f^M(P_k)\}$.

(c) An affine transformation g of $\mathbb{R}^3$ maps straight lines to straight lines and planes to planes. Moreover, it maps parallel straight lines to parallel straight lines, intersecting straight lines to intersecting straight lines, parallel planes to parallel planes and intersecting planes to intersecting planes.

(d) An affine transformation g of $\mathbb{R}^3$ maps a convex set lying on one plane of $\mathbb{R}^3$ to a convex set on another plane. Moreover, it transforms the convex hull of a set of points $\{P_1, P_2, \ldots, P_k\}$ on one plane to the convex hull of $\{g(P_1), g(P_2), \ldots, g(P_k)\}$ lying on another plane.

Figure 5.7 illustrates the actions of an affine transformation in $\mathbb{R}^3$.

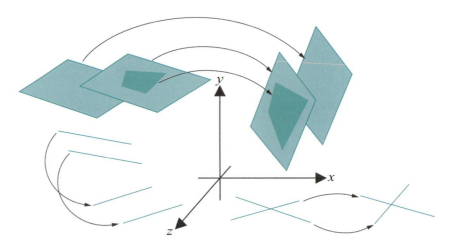

Figure 5.7: Affine transformation in $\mathbb{R}^3$.

Proof. (a) Say $g(P) = MP + D$, for $P \in \mathbb{R}^3$, where M is a non-singular 2×2 matrix and D a column 2-vector.

Suppose, first, that the equation of a given straight line l in $\mathbb{R}^2$ is $ax + by = c$, which can be written as $[a\ b][x\ y]^T = c$. If the point $[x\ y]^T$ is on l, let's see where its image $g([x\ y]^T) = M[x\ y]^T + D$ lies. Now

$$([a\ b]\,M^{-1})\,(M[x\ y]^T + D) = [a\ b][x\ y]^T + [a\ b]M^{-1}D = c + [a\ b]M^{-1}D$$

Writing $[a'\ b'] = [a\ b]M^{-1}$ and the scalar $c' = c + [a\ b]M^{-1}D$, the preceding equation gives

$$[a'\ b']\,(M[x\ y]^T + D) = c'$$

which shows that if $[x\ y]^T$ lies on l, then $g([x\ y]^T)$ lies on the straight line $[a'\ b'][x\ y]^T = c'$, proving that the image $g(l)$ of l is indeed a straight line.

Say next that l and l' are two parallel straight lines in $\mathbb{R}^2$, whose equations can then be written as $[a\ b][x\ y]^T = c$ and $[a\ b][x\ y]^T = d$, respectively, where $c \neq d$.

From the first part of the proof it's seen that $g(l)$ is the straight line whose equation is

$$[a'\ b'][x\ y]^T = c', \quad \text{where} \quad [a'\ b'] = [a\ b]M^{-1} \quad \text{and} \quad c' = c + [a\ b]M^{-1}D$$

Likewise, $g(l')$ is the straight line whose equation is

$$[a'\ b'][x\ y]^T = d', \quad \text{where} \quad [a'\ b'] \text{ is as before and } d' = d + [a\ b]M^{-1}D$$

Moreover, it follows from $c \neq d$, that $c' \neq d'$. We conclude that $g(l)$ and $g(l')$ are indeed parallel straight lines.

Finally, it's easy to see that two straight lines l and l' which intersect at P are mapped by g to straight lines which intersect at $g(P)$.

(b) Again, say, $g(P) = MP + D$, for $P \in \mathbb{R}^3$, where M is a non-singular 2×2 matrix and D a column 2-vector.

Suppose, first, that S is a convex subset of $\mathbb{R}^2$. To prove that $g(S)$ is convex as well, it is sufficient to show that $cP + (1 - c)Q \in g(S)$, given two points P and Q in $g(S)$ and c in $0 \le c \le 1$.

Since $P, Q \in g(S)$, there exist $P', Q' \in S$ such that $g(P') = P$ and $g(Q') = Q$. As S is convex $cP' + (1 - c)Q' \in S$. Applying g to both sides of the preceding inclusion we have that $g(cP' + (1 - c)Q') \in g(S)$, but

$$
\begin{aligned}
g(cP' + (1-c)Q') &= M(cP' + (1-c)Q') + D \\
&= M(cP' + (1-c)Q') + cD + (1-c)D \\
&= cMP' + cD + (1-c)MQ' + (1-c)D \\
&= c(MP' + D) + (1-c)(MQ' + D) \\
&= cg(P') + (1-c)g(Q') \\
&= cP + (1-c)Q
\end{aligned}
$$

proving that indeed $cP + (1 - c)Q \in g(S)$, so that the latter is a convex set.

We leave the proof of the second part of (b) as well as those of (c) and (d) to the reader. □

Exercise 5.24. Does an affine transformation of $\mathbb{R}^2$ or $\mathbb{R}^3$ necessarily map radial lines to radial lines or radial planes to radial planes? How about a linear transformation?

5.2.2 Affine Transformations and OpenGL

Proposition 5.1 says that affine transformations preserve straightness and flatness, among other properties, because they keep straight lines straight and planes plane. It's not hard to see, if one works through the proof, that this is a consequence of the fact that their defining Equations (5.9) (shown again below)

$$
\begin{aligned}
x_1' &= a_{11}x_1 + a_{12}x_2 + \ldots + a_{1m}x_m + d_1 \\
x_2' &= a_{21}x_1 + a_{22}x_2 + \ldots + a_{2m}x_m + d_2 \\
&\quad \ldots \quad \ldots \\
x_m' &= a_{m1}x_1 + a_{m2}x_2 + \ldots + a_{mm}x_m + d_m
\end{aligned}
$$

are of degree one, in particular, that the maximum degree of a variable x_i on the right side of each of these equations is one.

Here's an example of what happens if this were not the case, and if even we go to degree two.

Example 5.6. The quadratic transformation h of $\mathbb{R}^2$ defined by $h([x\ y]^T) = [x\ y^2]^T$ doesn't necessarily keep straight lines straight. In fact, we'll show that it maps at least one straight segment into an arc of a parabola.

Write the transformation as

$$\begin{aligned} x' &= x \\ y' &= y^2 \end{aligned}$$

Now consider how the straight line

$$y = x$$

is mapped. We have, using the preceding 3 equations

$$y' = y^2 \implies y' = x^2 \implies y' = x'^2$$

which is the equation of a parabola. It follows that h maps the straight segment between $(0,0)$ and $(1,1)$ to the arc of the parabola $y = x^2$ joining the same two points, as shown in Figure 5.8.

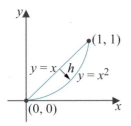

Figure 5.8: Quadratic transform h of $\mathbb{R}^2$ takes a straight segment to a parabolic arc.

One notices, further, that affine transformations are the *most general* class of transformations of degree one, because the right side of each one of the Equations (5.9) has its full complement of terms possible up to degree one – specifically, *every* x_i is present with degree one and there is, as well, the constant term d_i of degree zero.

One concludes that not only do affine transformations of $\mathbb{R}^2$ or $\mathbb{R}^3$ (or $\mathbb{R}^m$ in general) preserve straightness and flatness, they are the most general class of transformations to do so. Put another way, one cannot hope to go beyond affine transformations if straightness and flatness are not to be broken.

What has all this to do with OpenGL? Because they preserve straightness and flatness, affine transformations preserve as well the primitives of OpenGL, in particular, they map primitives of one type to another of the same type. The following exercise asks the reader to prove the specifics of this claim.

Exercise 5.25. Given an affine transformation g of $\mathbb{R}^2$ or $\mathbb{R}^3$, prove that

(a) g maps the straight segment joining two points P and Q to the straight segment joining $g(P)$ and $g(Q)$.

(b) g maps the triangle with vertices at P, Q and R to the triangle with vertices at $g(P)$, $g(Q)$ and $g(R)$.

(c) g maps the n-sided polygon with vertices at $P_1, P_2, \ldots, P_n$ to the n-sided polygon with vertices at $g(P_1), g(P_2), \ldots, g(P_n)$. If the original polygon is planar, then so is the transformed one. If the original polygon is convex, then so is the transformed one.

Hint: Use Proposition 5.1.

Non-affine transformations may not treat OpenGL primitives with quite as much respect, as the following exercise shows.

Exercise 5.26. We already saw in Example 5.6 the non-affine quadratic transformation h of $\mathbb{R}^2$, given by $h([x\ y]^T) = [x\ y^2]^T$, take a straight segment to a parabolic arc. How does h transform the triangle with corners at $(0,0)$, $(1,0)$ and $(1,1)$? Figure 5.9 is a gentle hint.

Figure 5.9: Hint for Exercise 5.26.

Now, it's desirable for a graphics API such as OpenGL to implement only modeling transformations which preserve its drawing primitives – mapping each one to another of the same type. Why? Consider OpenGL in particular. At the rendering end of its pipeline are evidently modules to render points, segments and triangles (mind that even a general polygon is triangulated prior to rendering). Suppose, then, that a particular scene is specified by the programmer as a list of n primitives:

$$primitive1,\ primitive2,\ \ldots,\ primitiveN$$

where each *primitiveI*, $1 \leq I \leq N$, is a point, segment or triangle. The scene is rendered essentially in a simple loop:

```
for (I = 1;  I ≤ N;  I ++) render primitiveI
```

where each iteration invokes the appropriate primitive rendering module.

Suppose, next, that a modeling transformation g is applied to the scene. The transformed scene is given by the list:

$$g(primitive1),\ g(primitive2),\ \ldots,\ g(primitiveN)$$

If g preserves primitives, then $g(primitiveI)$ is of the same class as *primitiveI*, for $1 \leq I \leq N$, and the transformed scene is rendered in the loop

```
for (I = 1;  I ≤ N;  I ++) render g(primitiveI)
```

invoking the same modules as before.

On the other hand, if g doesn't map primitives of one class to another of the same, e.g., if a triangle can change to something that is no longer one, as in Figure 5.9, then the situation becomes significantly more complicated. In this case, either there have to be modules to render all possible target objects of all the drawing primitives or modules to approximate them using existing primitives or, maybe, a combination of both. See Figure 5.10 for an illustration of both kinds of situations.

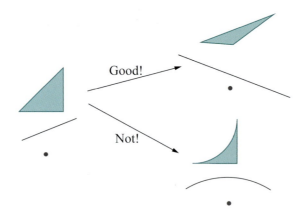

Figure 5.10: Transformations that are good from the API programmer's point of view, and not so good.

If the API designer is understandably reluctant to open this particular can of worms, then she should restrict herself to modeling transformations which do keep primitives within their class. In the case of OpenGL, this calls for transformations preserving straightness and flatness. However, even given this constraint, the designer would reasonably want as many as possible at her disposal. Transformations of degree one preserve straightness and flatness, so the designer would want them all if possible; in other words, *all affine transformations*.

And, in fact, we shall see that the designers of OpenGL have implemented, barring a few degenerate calls, exactly the class of affine transformations as their modeling transformations.

5.2.3 Affine Transformations and Homogeneous Coordinates

Despite their virtues listed in the previous two sections, there is potentially a serious computational problem with applying affine transformations rather than linear ones. The source lies in the difference in how the two are defined. A linear transformation is given by an equation of the form

$$f^M(P) = MP$$

while an affine transformation by one of the form

$$g(P) = MP + D$$

The former is expressed as a single matrix-vector multiplication, while the latter by a matrix-vector multiplication *followed* by a vector-vector sum. It is the additional sum step which cascades when composing affine transformations.

For example, if $f^{M_1}, f^{M_2}, f^{M_3}, \ldots$ are linear transformations of $\mathbb{R}^m$, then for an m-vector P,

$$
\begin{aligned}
f^{M_1}(P) &= M_1 P \\
(f^{M_2} \circ f^{M_1})(P) &= (M_2 M_1)P \\
(f^{M_3} \circ f^{M_2} \circ f^{M_1})(P) &= (M_3 M_2 M_1)P
\end{aligned}
\tag{5.10}
$$

and so on. On the other hand, if $g_1, g_2, g_3, \ldots$ are affine transformations of $\mathbb{R}^m$ given by $g_1(P) = M_1 P + D_1$, $g_2(P) = M_2 P + D_2$, $g_3(P) = M_3 P + D_3, \ldots$, respectively, then for an m-vector P,

$$
\begin{aligned}
g_1(P) &= M_1 P + D_1 \\
(g_2 \circ g_1)(P) &= (M_2 M_1)P + M_2 D_1 + D_2 \\
(g_3 \circ g_2 \circ g_1)(P) &= (M_3 M_2 M_1)P + M_3 M_2 D_1 + M_3 D_2 + D_3
\end{aligned}
\tag{5.11}
$$

It's not hard to see that the number of matrix operations grows quadratically with the number n of affine transformations $g_n \circ \ldots \circ g_2 \circ g_1$ being composed, versus linearly in the case $f^{M_n} \circ \ldots \circ f^{M_2} \circ f^{M_1}$ of linear transformations. Composing affine transformations, at least by means of equations as above, therefore, is highly inefficient. There is an elegant way, however, to rectify the problem. It is with the help of so-called homogeneous coordinates.

Definition 5.3. A point

$$P = [x_1 \ x_2 \ \ldots \ x_m]^T$$

belonging to $\mathbb{R}^m$ is represented in *homogeneous coordinates* by *any* $m+1$-tuple of the form

$$[cx_1 \ cx_2 \ \ldots \ cx_m \ c]^T$$

where c is a *non-zero* scalar. Homogeneous coordinates, therefore, are not unique.

Example 5.7. Possible homogeneous coordinates of the point $P = [3 \ 7]^T \in \mathbb{R}^2$ include $[3 \ 7 \ 1]^T$, $[16.5 \ 38.5 \ 5.5]^T$, $[-6 \ -14 \ -2]^T$, etc.

For our current purposes, though, it's good enough to fix the scalar c in Definition 5.3 to be 1. We'll have use for the general c later when studying projective spaces. So, for the present, assume that the point

$$P = [x_1 \ x_2 \ \ldots \ x_m]^T$$

is represented in homogeneous coordinates by

$$[x_1 \; x_2 \; \ldots \; x_m \; 1]^T$$

For example, $[3 \; 7]^T$ would be homogenized to $[3 \; 7 \; 1]^T$. To save space we'll often write $[x_1 \; x_2 \; \ldots \; x_m \; 1]^T$ as $[P \; 1]^T$.

Observe now that Equations (5.9)

$$
\begin{aligned}
x_1' &= a_{11}x_1 + a_{12}x_2 + \ldots + a_{1m}x_m + d_1 \\
x_2' &= a_{21}x_1 + a_{22}x_2 + \ldots + a_{2m}x_m + d_2 \\
&\ldots \; \ldots \\
x_m' &= a_{m1}x_1 + a_{m2}x_2 + \ldots + a_{mm}x_m + d_m
\end{aligned}
$$

defining an affine transformation g are equivalent to the *single* matrix equation

$$
\begin{bmatrix} x_1' \\ x_2' \\ \ldots \\ x_m' \\ 1 \end{bmatrix}
=
\begin{bmatrix}
a_{11} & a_{12} & \ldots & a_{1m} & d_1 \\
a_{21} & a_{22} & \ldots & a_{2m} & d_2 \\
& & \ldots & \ldots & \\
a_{m1} & a_{m2} & \ldots & a_{mm} & d_m \\
0 & 0 & 0 & 0 & 1
\end{bmatrix}
\begin{bmatrix} x_1 \\ x_2 \\ \ldots \\ x_m \\ 1 \end{bmatrix}
$$

(as is easily verified by multiplying the two matrices on the right). Concisely:

$$
\begin{bmatrix} g(P) \\ 1 \end{bmatrix}
=
\begin{bmatrix} P' \\ 1 \end{bmatrix}
=
\begin{bmatrix} M & D \\ 0 & 1 \end{bmatrix}
\begin{bmatrix} P \\ 1 \end{bmatrix}
\tag{5.12}
$$

Presto! Computation of the affine transformation g, which earlier required a matrix-vector multiplication followed by a vector-vector addition, has now become a *single* matrix-vector multiplication with the use of homogeneous coordinates and, albeit, a bigger matrix. The translational component has, evidently, been subsumed into the extra dimension of the larger matrix.

Example 5.8. The affine transformation $g : \mathbb{R}^2 \to \mathbb{R}^2$ given by

$$
g\left(\begin{bmatrix} x \\ y \end{bmatrix} \right)
=
\begin{bmatrix} x' \\ y' \end{bmatrix}
=
\begin{bmatrix} 2 & 1 \\ 0 & 4 \end{bmatrix}
\begin{bmatrix} x \\ y \end{bmatrix}
+
\begin{bmatrix} 4 \\ 6 \end{bmatrix}
$$

can be written using homogeneous coordinates as

$$
\begin{bmatrix} x' \\ y' \\ 1 \end{bmatrix}
=
\begin{bmatrix} 2 & 1 & 4 \\ 0 & 4 & 6 \\ 0 & 0 & 1 \end{bmatrix}
\begin{bmatrix} x \\ y \\ 1 \end{bmatrix}
$$

Let's give it a check for, say, the point $[1 \; 1]^T$. Now,

$$
g\left(\begin{bmatrix} 1 \\ 1 \end{bmatrix} \right)
=
\begin{bmatrix} 2 & 1 \\ 0 & 4 \end{bmatrix}
\begin{bmatrix} 1 \\ 1 \end{bmatrix}
+
\begin{bmatrix} 4 \\ 6 \end{bmatrix}
=
\begin{bmatrix} 7 \\ 10 \end{bmatrix}
$$

and with homogeneous coordinates

$$\begin{bmatrix} 2 & 1 & 4 \\ 0 & 4 & 6 \\ 0 & 0 & 1 \end{bmatrix} \begin{bmatrix} 1 \\ 1 \\ 1 \end{bmatrix} = \begin{bmatrix} 7 \\ 10 \\ 1 \end{bmatrix}$$

the RHS of the preceding equation indeed being the homogenization of the RHS of the one before it.

$E_{xercise}$ 5.27. Express the affine transformation g of $\mathbb{R}^2$ given by

$$g\left(\begin{bmatrix} x \\ y \end{bmatrix}\right) = \begin{bmatrix} -1 & 2 \\ -3 & 0 \end{bmatrix} \begin{bmatrix} x \\ y \end{bmatrix} + \begin{bmatrix} 2 \\ 1 \end{bmatrix}$$

as a single matrix vector multiplication using homogeneous coordinates.

The composition of affine transformations is drastically simplified with use of homogeneous coordinates. For example, the last equation of (5.11) becomes

$$\begin{bmatrix} (g_3 \circ g_2 \circ g_1)(P) \\ 1 \end{bmatrix} = \left(\begin{bmatrix} M_3 & D_3 \\ 0 & 1 \end{bmatrix} \begin{bmatrix} M_2 & D_2 \\ 0 & 1 \end{bmatrix} \begin{bmatrix} M_1 & D_1 \\ 0 & 1 \end{bmatrix}\right) \begin{bmatrix} P \\ 1 \end{bmatrix}$$

the number of matrix operations now growing linearly with the number of affine transformations being composed, instead of quadratically.

$E_{xercise}$ 5.28. If you did Exercise 5.22, then you have determined the affine transformation $f \circ g$, where

$$f(P) = \begin{bmatrix} 2 & 1 \\ 0 & 4 \end{bmatrix} P + [4\ 6]^T \quad \text{and} \quad g(P) = \begin{bmatrix} -1 & 3 \\ 1 & -2 \end{bmatrix} P + [-1\ 0]^T$$

Now, verify your answer by multiplying the 3×3 matrices corresponding to f and g, and checking if the result corresponds to the composed transformation already computed.

5.3 Geometric Transformations in 2-Space Continued

We resume our study of 2D geometric transformations, equipped now with a newfound grasp of affine transformations. Keep in mind that, as in the first section, by default we are in 2D space.

5.3.1 Affine Geometric Transformations

Are translations, scalings, rotations about the origin and reflections about radial mirrors, which we studied in the opening section, affine transformations? Of course, they all are! This follows easily from the non-singularity of the matrix on the RHS of each of the Equations (5.1), (5.2), (5.4) and (5.5).

Exercise 5.29. Prove that rotations about arbitrary points (not necessarily the origin) and reflections about arbitrary mirrors (not necessarily radial) are affine as well.

That translations, scalings, rotations and reflections are affine means they are geometrically well-behaved, preserving straightness, parallelism and convexity, as well. We record this fact as a proposition.

Proposition 5.2. *Let g be either a translation, a scaling, a rotation (about an arbitrary point) or a reflection (about an arbitrary mirror). Then:*

(a) *g maps straight lines to straight lines. Moreover, it maps parallel straight lines to parallel straight lines and intersecting straight lines to intersecting straight lines.*

(b) *g maps convex sets to convex sets. Moreover, it maps the convex hull of* $\{P_1, P_2, \ldots, P_k\}$ *to the convex hull of* $\{f^M(P_1), f^M(P_2), \ldots, f^M(P_k)\}$.

Proof. Follows from Proposition 5.1 for 2D affine transformations in general. $\square$

Geometric Transformation Equations Using Homogeneous Coordinates

In Section 5.2.3 we learned how to express an affine transformation as a single matrix-vector multiplication, after writing points in homogeneous coordinates. In the case of $\mathbb{R}^2$ this means writing $P = [x \ y]^T$ as $[x \ y \ 1]^T$ or $[P \ 1]^T$ for short. Let's rewrite Equations (5.1), (5.2), (5.4) and (5.5) using homogeneous coordinates:

Translation by displacement vector $[d_x \ d_y]^T$:

$$
\begin{bmatrix} P' \\ 1 \end{bmatrix} = \begin{bmatrix} 1 & 0 & d_x \\ 0 & 1 & d_y \\ 0 & 0 & 1 \end{bmatrix} \begin{bmatrix} P \\ 1 \end{bmatrix}
\tag{5.13}
$$

Scaling by a factor of s_x along the x-axis and s_y along the y-axis:

$$
\begin{bmatrix} P' \\ 1 \end{bmatrix} = \begin{bmatrix} s_x & 0 & 0 \\ 0 & s_y & 0 \\ 0 & 0 & 1 \end{bmatrix} \begin{bmatrix} P \\ 1 \end{bmatrix}
\tag{5.14}
$$

Rotation by an angle θ counter-clockwise about the origin:

$$
\begin{bmatrix} P' \\ 1 \end{bmatrix} = \begin{bmatrix} \cos\theta & -\sin\theta & 0 \\ \sin\theta & \cos\theta & 0 \\ 0 & 0 & 1 \end{bmatrix} \begin{bmatrix} P \\ 1 \end{bmatrix}
\tag{5.15}
$$

Reflection about a radial mirror l at an angle of θ counter-clockwise from the positive x-axis:

$$\begin{bmatrix} P' \\ 1 \end{bmatrix} = \begin{bmatrix} \cos 2\theta & \sin 2\theta & 0 \\ \sin 2\theta & -\cos 2\theta & 0 \\ 0 & 0 & 1 \end{bmatrix} \begin{bmatrix} P \\ 1 \end{bmatrix} \tag{5.16}$$

Exercise 5.30. Write the 3×3 matrix corresponding to each of the following affine transformations:

(a) Translation by the displacement vector $[-2\ 3]^T$.

(b) Scaling by a factor of 2 in the x-direction and 4 in the y.

(c) Counter-clockwise rotation by an angle of $-45°$ about the origin.

(d) Reflection about the radial mirror making an angle of $30°$ measured counter-clockwise from the positive direction of the x-axis.

Factoring Affine Transformations

We know then that affine transformations include translations, scalings and rotations. But are they more than just these three special kinds of transformations? It's extremely important that the answer is: *essentially no*. In fact, *any* affine transformation can be "made from" translations, scalings and rotations. Precisely, any affine transformation can be expressed as a composition of transformations of just these three kinds. Here is the formal statement:

Proposition 5.3. *Any affine transformation of $\mathbb{R}^2$ is the composition in some order of translations, scalings and rotations about the origin.*

In particular, any affine transformation $g : \mathbb{R}^2 \to \mathbb{R}^2$ can be factored into a composition $g = g_4 \circ g_3 \circ g_2 \circ g_1$, where g_1 is a rotation about the origin, g_2 a scaling, g_3 another rotation about the origin and g_4 a translation.

Proof. Let

$$g(P) = MP + D$$

where $M = \begin{bmatrix} a_{11} & a_{12} \\ a_{21} & a_{22} \end{bmatrix}$ is g's non-singular 2×2 defining matrix and the

2-vector $D = \begin{bmatrix} d_x \\ d_y \end{bmatrix}$ is its translational component.

We claim first that it is possible to find 2×2 matrices M_1, M_2 and M_3, corresponding, respectively, to a rotation about the origin, a scaling and another rotation about the origin, such that

$$M = M_3 M_2 M_1$$

Say M_1 corresponds to a rotation by angle θ, M_2 to scaling by a factor of s_x along the x-axis and s_y along the y-axis and M_3 to a rotation by angle ϕ. The preceding equation gives, therefore, that

$$\begin{bmatrix} a_{11} & a_{12} \\ a_{21} & a_{22} \end{bmatrix} = \begin{bmatrix} \cos\phi & -\sin\phi \\ \sin\phi & \cos\phi \end{bmatrix} \begin{bmatrix} s_x & 0 \\ 0 & s_y \end{bmatrix} \begin{bmatrix} \cos\theta & -\sin\theta \\ \sin\theta & \cos\theta \end{bmatrix} \quad (5.17)$$

which we'll show next can be solved to find ϕ, θ, s_x and s_y.

Multiply the three matrices on the RHS of the preceding equation and then equate terms of the resulting matrix with the corresponding ones on the LHS to see that:

$$\begin{aligned} a_{11} &= s_x \cos\phi \cos\theta - s_y \sin\phi \sin\theta \\ a_{12} &= -s_x \cos\phi \sin\theta - s_y \sin\phi \cos\theta \\ a_{21} &= s_x \sin\phi \cos\theta + s_y \cos\phi \sin\theta \\ a_{22} &= -s_x \sin\phi \sin\theta + s_y \cos\phi \cos\theta \end{aligned} \quad (5.18)$$

Four equations in four unknowns seems right. Check that:

$$\begin{aligned} a_{21} - a_{12} &= (s_x + s_y) \sin(\phi + \theta) \\ a_{11} + a_{22} &= (s_x + s_y) \cos(\phi + \theta) \\ a_{21} + a_{12} &= (s_x - s_y) \sin(\phi - \theta) \\ a_{11} - a_{22} &= (s_x - s_y) \cos(\phi - \theta) \end{aligned} \quad (5.19)$$

Assuming for the moment that neither $a_{11} + a_{22}$ nor $a_{11} - a_{22}$ is zero, divide the first equation above by the second and the third by the fourth to get:

$$\begin{aligned} \tan(\phi + \theta) &= \frac{a_{21} - a_{12}}{a_{11} + a_{22}} \\ \tan(\phi - \theta) &= \frac{a_{21} + a_{12}}{a_{11} - a_{22}} \end{aligned} \quad (5.20)$$

which implies:

$$\begin{aligned} \phi + \theta &= \tan^{-1}\left(\frac{a_{21} - a_{12}}{a_{11} + a_{22}}\right) \\ \phi - \theta &= \tan^{-1}\left(\frac{a_{21} + a_{12}}{a_{11} - a_{22}}\right) \end{aligned} \quad (5.21)$$

These two equations can be solved to determine ϕ and θ. Furthermore, the values of $\phi + \theta$ and $\phi - \theta$ can then be substituted back into equation set (5.19) to determine equations for $s_x + s_y$ and $s_x - s_y$, which can then be solved to find s_x and s_y.

The earlier claim that (5.17) can be solved to find ϕ, θ, s_x and s_y is proved and, therefore, $M = M_3 M_2 M_1$, in the manner claimed at the start of the proof as well – *except* when either or both of $a_{11} + a_{22}$ and $a_{11} - a_{22}$ is 0, in which case Exercise 5.32 completes the proof.

As $g(P) = MP + D = (M_3 M_2 M_1)P + D$ one concludes, finally, that indeed $g = g_4 \circ g_3 \circ g_2 \circ g_1$, where g_1 is the counter-clockwise rotation about the origin by an angle of θ, g_2 the scaling by a factor of s_x along the x-axis and s_y along the y-axis, g_3 the counter-clockwise rotation about the origin by an angle of ϕ and g_4 translation by the displacement vector D. □

Example 5.9. Factor the affine transformation

$$g(P) = \begin{bmatrix} 1 & \frac{\sqrt{3}}{2} \\ \frac{\sqrt{3}}{2} & 0 \end{bmatrix} P + \begin{bmatrix} 2 \\ 1 \end{bmatrix}$$

according to the proposition.

Answer: From equations (5.21) we have

$$\phi + \theta = \tan^{-1} 0 = 0° \quad \text{and} \quad \phi - \theta = \tan^{-1} \sqrt{3} = 60°$$

which solve to

$$\phi = 30° \quad \text{and} \quad \theta = -30°$$

Plugging the values of $\phi + \theta$ and $\phi - \theta$ into the second and fourth equations of (5.19) we have

$$1 = (s_x + s_y) \cos 0° = s_x + s_y \quad \text{and} \quad 1 = (s_x - s_y) \cos 60° = \frac{1}{2}(s_x - s_y)$$

which solve to

$$s_x = \frac{3}{2} \quad \text{and} \quad s_y = -\frac{1}{2}$$

(If the reader is wondering about the other two equations in (5.19) – the first and third – she may check that these are satisfied as well by the values found above for ϕ, θ, s_x and s_y.)

Therefore, $g = g_4 \circ g_3 \circ g_2 \circ g_1$, where g_1 is the clockwise rotation about the origin by an angle of $30°$, g_2 the scaling by a factor of $\frac{3}{2}$ along the x-axis and $-\frac{1}{2}$ along the y-axis, g_3 the counter-clockwise rotation about the origin by an angle of $30°$ and g_4 translation by the displacement vector $[2\ 1]^T$.

Exercise 5.31. Factor the affine transformation

$$g(P) = \begin{bmatrix} 1 & \frac{1}{2} \\ \frac{1}{2} & 0 \end{bmatrix} P + \begin{bmatrix} -1 \\ 1 \end{bmatrix}$$

according to the proposition.

Exercise 5.32. Fill in the gap in the proof of the preceding proposition, where it was assumed (just after equations (5.19)) that neither $a_{11} + a_{22}$ nor $a_{11} - a_{22}$ is zero. In particular, even if one or both of these quantities is zero, show how to proceed again from (5.19) to solve for ϕ, θ, s_x and s_y.

Exercise 5.33. Factor the affine transformation

$$g(P) = \begin{bmatrix} \frac{1}{2} & \sqrt{3} \\ 0 & \frac{1}{2} \end{bmatrix} P$$

according to the proposition.

Exercise 5.34. Give an example of an affine transformation which itself is not a translation, scaling or rotation about the origin, so one *has to* compose in order to obtain it.

Remark 5.2. The reader may have noticed that we never used the non-singularity of M in the proof of Proposition 5.3. As a matter of fact, even if M is singular, it can be written as $M = M_3 M_2 M_1$ as in the proposition, *except* that the scaling M_2 turns out to be degenerate.

Proposition 5.3 suggests that translations, scalings and rotations about the origin are fundamental in the sense that they can be used to generate *all* affine transformations, a particularly useful insight for anyone in Flatland trying to implement a graphics API. For, all such a programmer has to code is an implementation of each of those three special kinds of affine transformations, to get the rest automatically.

Since a non-singular linear transformation of $\mathbb{R}^2$ is simply an affine transformation with null translational component, we have also proved the following on the way to proving Proposition 5.3:

Proposition 5.4. *Any non-singular linear transformation of $\mathbb{R}^2$ is the composition successively of a rotation about the origin, a scaling and another rotation about the origin.* □

5.3.2   Euclidean and Rigid Transformations

Proposition 5.2 tells us that transformations such as translations, scalings, rotations and reflections are respectful of a bunch of geometric attributes, from straightness to convexity. How about that most important geometric attribute of all, though, namely, distance? We would say a transformation g *preserves distance* if it were true that, for any pair of points P and Q, the distance between $f(P)$ and $f(Q)$ is the same as that between P and Q.

It's clear, if one thinks of scalings, that distance is not preserved by transformations in general. However, there certainly are transformations that seem to preserve distance. Translations come to mind, as points are "carried together" by a translation, so neither pulled apart nor drawn closer together. Similar thoughts apply to rotations. We'll see soon that translations and rotations do indeed preserve distance.

Distance-preserving transformations are important in animation because they preserve *shape* as well. In fact, an object's shape is not changed precisely when the distance between *every* pair of points belonging to it is not changed.

See Figure 5.11. Comparing the pre-hit and post-hit heads, one observes that the distance between at least two pairs of points is different from those between the transformed pairs: the eyeballs, and P and Q. On the other hand, the distance between any pair of points of the book remains unchanged.

Figure 5.11: Square-headed student struck by a CG book: the shape of the head is distorted, but not that of the book.

Transformations preserving distances are the ones, therefore, to use when animating *rigid* objects such as balls, bats (not the flying kind) and houses. They are important enough, in fact, to have been honored with the name of the great ancient geometer Euclid. Here's a formal definition.

Definition 5.4. An *Euclidean transformation* (also called *isometry*) of $\mathbb{R}^2$ is one that preserves distance. Precisely, $f : \mathbb{R}^2 \to \mathbb{R}^2$ is Euclidean if $|f(P)f(Q)| = |PQ|$ for any two points $P, Q \in \mathbb{R}^2$.

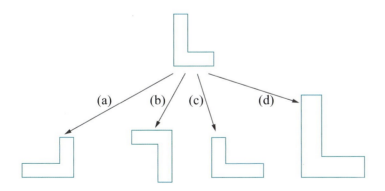

Figure 5.12: Transformations (a)-(c) are Euclidean, (d) is not.

See Figure 5.12 for three simple examples of Euclidean and one of non-Euclidean transformation. It may seem, as it cannot alter shape, that all an Euclidean transformation can do is "slide" an object around the plane, which, if true, would imply that it is merely a composition of translations and rotations. However, consider the Euclidean transformations in cases (a), (b) and (c) of Figure 5.12. The first two can certainly be obtained by sliding the top L around the page. However, it's not hard to convince oneself that (c) cannot and, therefore, is not a combination of translations and rotations.

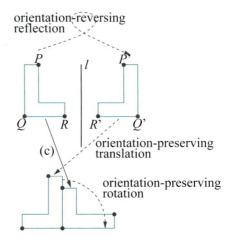

Figure 5.13: Executing (c) of Figure 5.12 by a reflection about the mirror l followed by translation and rotation.

Let's examine (c). As indicated in Figure 5.13, it can, in fact, be obtained by applying a reflection about a vertical mirror l, followed by translation and rotation. A reflection is required because (c) is a so-called orientation-reversing transformation. Here's the relevant definition:

Definition 5.5. An Euclidean transformation f of $\mathbb{R}^2$ is said to be *orientation-reversing* if there exist three non-collinear points P, Q and R in $\mathbb{R}^2$ such that, looking at $\mathbb{R}^2$ from a fixed side, one of the two sequences PQR and $f(P)f(Q)f(R)$ appears clockwise (CW) and the other counter-clockwise (CCW).

An Euclidean transformation that is not orientation-reversing is said to be *orientation-preserving*.

Orientation-preserving Euclidean transformations are also called *rigid transformations*.

Remark 5.3. The property of the transformation f described in the first paragraph of the preceding definition does not depend on the choice of the non-collinear points P, Q and R. In fact, as we'll see, if for *some* three non-collinear points P, Q and R it is true that PQR and $f(P)f(Q)f(R)$ appear oriented differently, then this is true for *any* three non-collinear points.

Rigid transformations are so called because they model the physical motion of a rigid object restricted always to a plane – such motion can never reverse orientation. Conceptually, reversing orientation requires the object to be "lifted off" the plane, "flipped" and "placed back" again.

The sequence PQR in Figure 5.13 appears CCW to the reader, while that of their images $P'Q'R'$ by reflection about l appears CW, proving that the reflection is indeed orientation-reversing and, therefore, not rigid.

Exercise 5.35. Show that an Euclidean transformation f preserves angles, i.e., $\angle ABC = \angle f(A)f(B)f(C)$, where A, B, C are any three points on the plane.

We'll see next how to determine algorithmically if PQR appears CW or CCW to a given viewer, which will in turn help decide if a transformation is orientation-preserving or not.

Lemma 5.1. *Let $P = [x_1 \ y_1]^T$, $Q = [x_2 \ y_2]^T$ and $R = [x_3 \ y_3]^T$ be three points on the plane. Define the scalar D by*

$$D = x_1 y_2 - x_2 y_1 + x_2 y_3 - x_3 y_2 + x_3 y_1 - x_1 y_3 = \begin{vmatrix} x_1 & x_2 & x_3 \\ y_1 & y_2 & y_3 \\ 1 & 1 & 1 \end{vmatrix}$$

the rightmost term being called the discriminant determinant.

Let V be a viewer on the positive side of the z-axis of a hypothetical right-handed system. We have then the following:

1. *If $D = 0$, then P, Q and R are collinear.*

2. *If $D < 0$, then V perceives the order PQR as CW.*

3. *If $D > 0$, then V perceives the order PQR as CCW.*

Note: The column vectors of the discriminant determinant are the coordinates of P, Q and R, respectively, homogenized; so, it can be written

$$D = \begin{vmatrix} P & Q & R \\ 1 & 1 & 1 \end{vmatrix}$$

Proof. We'll first prove the lemma assuming that $R = O$, the origin, in which case

$$D = \begin{vmatrix} x_1 & x_2 & 0 \\ y_1 & y_2 & 0 \\ 1 & 1 & 1 \end{vmatrix} = x_1 y_2 - x_2 y_1$$

If $P = O$ as well, then P, Q and R are trivially collinear, and it's easily seen that the determinant $D = 0$, too, which falls into case 1 of the lemma. Accordingly, suppose that $P \neq O$ as in Figure 5.14. The straight line l through P and R has the equation

$$x_1 y - y_1 x = 0$$

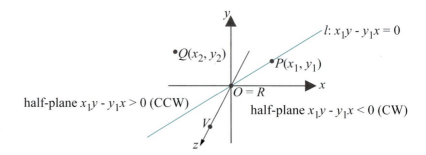

Figure 5.14: The orientation of PQR perceived by V depends on the half-plane of l containing Q (Q is depicted here in the half-plane $x_1y - y_1x > 0$).

If Q does not lie on l, then whether PQR appears CW or CCW to V depends on which half-plane of l contains Q. In particular, if Q lies in the half-plane $x_1y - y_1x > 0$ – the case depicted in the figure – then PQR appears CCW to V; if in the half-plane $x_1y - y_1x < 0$, then CW. Plugging in Q's coordinates means that PQR appears CCW to V if $x_1y_2 - y_1x_2 > 0$ and CW if $x_1y_2 - y_1x_2 < 0$. Of course, $x_1y_2 - y_1x_2 = 0$ if Q lies on l, in which case P, Q and R are collinear. Therefore, we've proved the lemma assuming $R = O$.

The case of arbitrary R can be reduced to that of $R = O$ by applying the translation $-R$ to all three points, because the relative dispositions of P, Q and R as they appear to V are the same as those of $P - R$, $Q - R$ and $R - R$ ($= O$). Now, $P - R = [x_1 - x_3 \ \ y_1 - y_3]^T$ and $Q - R = [x_2 - x_3 \ \ y_2 - y_3]^T$, and we leave it to the reader to use the case $R = O$ to finish up the proof. $\square$

Exercise **5.36.** Verify the lemma for the following triples by plotting the points on graph paper:

(a) $P = [1 \ 0]^T$, $Q = [0 \ 1]^T$, $R = [0 \ 0]^T$

(a) $P = [-1 \ \ -1]^T$, $Q = [-2 \ 1]^T$, $R = [3 \ 4]^T$

The following proposition is intuitively fairly clear but, nevertheless, has to be proved formally.

Proposition 5.5. *A translation or a rotation about an arbitrary point is a rigid transformation of 2-space. A reflection about an arbitrary mirror is an orientation-reversing Euclidean transformation of 2-space.*

Proof. We'll prove first that a translation t by the displacement vector $D = [d_x \ d_y]^T$ preserves both distance and orientation.

Let $P = [x_1 \ y_1]^T$ and $Q = [x_2 \ y_2]^T$ be two points in $\mathbb{R}^2$. The images of P and Q by t are, respectively, $P' = P + D = [x_1 + d_x \ \ y_1 + d_y]^T$ and $Q' = Q + D = [x_2 + d_x \ \ y_2 + d_y]^T$. Now,

$$|PQ| = \sqrt{(x_1 - x_2)^2 + (y_1 - y_2)^2}$$

and

$$\begin{aligned} |P'Q'| &= \sqrt{((x_1 + d_x) - (x_2 + d_x))^2 + ((y_1 + d_y) - (y_2 + d_y))^2} \\ &= \sqrt{(x_1 - x_2)^2 + (y_1 - y_2)^2} \end{aligned}$$

proving t indeed preserves distance.

Let $P = [x_1 \ y_1]^T$, $Q = [x_2 \ y_2]^T$ and $R = [x_3 \ y_3]^T$ be three points in $\mathbb{R}^2$, and V a viewer on the positive side of the z-axis. Lemma 5.1 says that PQR are collinear, appear CW to V, or CCW to V, according as the determinant

$$D = \begin{vmatrix} x_1 & x_2 & x_3 \\ y_1 & y_2 & y_3 \\ 1 & 1 & 1 \end{vmatrix}$$

is equal to, less than, or greater than 0.

The images of P, Q and R by t are $P' = P + D = [x_1 + d_x \ y_1 + d_y]^T$, $Q' = Q + D = [x_2 + d_x \ y_2 + d_y]^T$ and $R' = R + D = [x_3 + d_x \ y_3 + d_y]^T$, respectively. By another application of Lemma 5.1, $P'Q'R'$ are collinear, appear CW to V, or CCW to V, according as the determinant

$$D' = \begin{vmatrix} x_1 + d_x & x_2 + d_x & x_3 + d_x \\ y_1 + d_y & y_2 + d_y & y_3 + d_y \\ 1 & 1 & 1 \end{vmatrix}$$

is equal to, less than, or greater than 0.

However, subtracting d_x times the third row of D' from its first and d_y times the third row from its second, we see that, in fact, $D = D'$. It follows that the relative dispositions of PQR and of $P'Q'R'$ (either CCW or CW) with respect to V are identical, giving the conclusion that t indeed preserves orientation.

The proofs for rotations and reflections are left to the reader. $\square$

Exercise 5.37. Scalings in general are not Euclidean transformations, but for certain choices of scaling factors they are. List these choices and for each say if it preserves or reverses orientation.

Exercise 5.38. Show that the composition of two Euclidean transformations is Euclidean and that of two rigid transformations rigid.

Exercise 5.39. Show that the composition of two orientation-reversing Euclidean transformations is an orientation-preserving Euclidean transformation (in other words, rigid). Show that the composition of an orientation-preserving and an orientation-reversing Euclidean transformation is orientation-reversing.

We saw in Proposition 5.3 that an affine transformation can be factored as a composition of translations, scalings and rotations about the origin. The following proposition shows how Euclidean and rigid transformations can be factored. The first part verifies our intuition that a rigid transformation slides an object around the plane by translation and rotation, while the second says that an Euclidean transformation is at most one reflection away from being rigid.

Proposition 5.6. *A rigid transformation of $\mathbb{R}^2$ keeping the origin fixed is a rotation about the origin, while an arbitrary rigid transformation is a composition of a rotation about the origin followed by a translation.*

An Euclidean transformation of $\mathbb{R}^2$ is a composition of a rotation about the origin, followed by a translation, possibly followed again by a reflection.

Proof. Consider, first, a rigid transformation $f : \mathbb{R}^2 \to \mathbb{R}^2$ keeping the origin fixed, i.e., $f(O) = O$. Let $P \in \mathbb{R}^2$ be different from the origin. By the distance-preserving property

$$|OP| = |f(O)f(P)| = |Of(P)|$$

so both P and $f(P)$ lie on a circle c centered at O. See Figure 5.15, (a) or (b). Say the angle from OP to $Of(P)$ is θ measured counter-clockwise. We'll show for any point $Q \in \mathbb{R}^2$ that its image $f(Q)$ is obtained by rotating Q counter-clockwise by an angle of θ about the origin as well, proving the claim that f is a rotation about the origin.

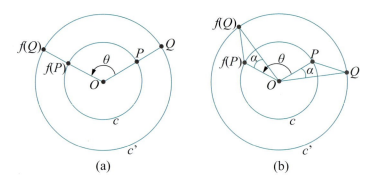

(a)   (b)

Figure 5.15: Illustrations for the proof of Proposition 5.6.

Let Q be an arbitrary point on the plane. Without loss of generality assume $Q \neq O$. Reasoning as before, then, both Q and $f(Q)$ lie on some circle c' centered at O. By the distance-preserving property

$$|PQ| = |f(P)f(Q)|$$

Consider first the case that Q lies on the straight line through O and P (Figure 5.15(a)). Because $|PQ| = |f(P)f(Q)|$ it's seen that $f(Q)$ lies at the intersection with c' of the straight line through O and $f(P)$, as all other points on c' are at a distance more than $|PQ|$ from $f(P)$. In which case, $f(Q)$ is indeed obtained by rotating Q counter-clockwise by an angle of θ about the origin.

Next, consider the case that Q does not lie on the straight line through O and P. First, suppose that the vertex order POQ of triangle POQ appears CCW to the viewer (Figure 5.15(b)). Let the angle POQ be α. The congruence of the triangles POQ and $f(P)Of(Q)$, a consequence of the distance-preserving property, implies that angle $f(P)Of(Q)$ is α as well. Furthermore, f being rigid preserves orientation, so $f(P)Of(Q)$ appears CCW to the viewer as well. It follows from simple angular arithmetic that $f(Q)$ is θ counter-clockwise about the origin from Q.

If the vertex order POQ appears CW instead, a similar conclusion can still be reached. This completes the proof that, if f is a rigid transformation keeping the origin fixed, then it is a rotation about the origin.

Suppose, next, that f is an arbitrary rigid transformation, not necessarily fixing the origin. Let $f(O) = O'$ and t be translation by the displacement vector $O'O$. Then the transformation $f' = t \circ f$ is a rigid transformation such that $f'(O) = O$, i.e., fixing the origin. Therefore, as proved earlier, f' is a rotation about the origin. Consequently, $f = t^{-1} \circ f'$ is a rotation about the origin followed by a translation, proving the statement of the proposition about arbitrary rigid transformations and completing the proof of the first paragraph.

Note: If f is itself a translation then, of course, the rotation f' about the origin is the identity, i.e., zero rotation.

For the second paragraph of the proposition, suppose that f is an orientation-reversing Euclidean transformation, because if f is orientation-preserving, then it is rigid, and there is nothing to prove after the first paragraph.

Let w be a reflection about any mirror, an orientation-reversing Euclidean transformation by Proposition 5.5. Then $f' = w \circ f$, being a composition of two orientation-reversing Euclidean transformations, is rigid by Exercise 5.39. By the first part of the proposition, f' is a rotation about the origin followed by a translation, implying that $f = w^{-1} \circ f'$ is the composition of a rotation about the origin, followed by a translation and, then, a reflection. This completes the proof of the second paragraph. $\qquad\square$

Exercise 5.40. Apply the proposition to show that a rigid transformation which keeps

(a) no point fixed is a translation.

(b) exactly one point fixed is a rotation (about the fixed point as center).

(c) more than one point fixed is the identity (which, therefore, keeps every point fixed).

Exercise 5.41. Use Exercises 5.16 and 5.40 to prove that any Euclidean transformation can be obtained by composing reflections about at most three mirrors.

Exercise 5.42. At the end of Section 5.1.3 we saw one case that the composition of two rotations is a translation and one where it is again a rotation. Use Exercise 5.40 to prove now that these are the only two possibilities in general for the composition of two rotations.

Moreover, show how to decide which case arises by proving:

1. The composition of two rotations, either both counter-clockwise or both clockwise, one of angle θ_1 and one of angle θ_2, about arbitrary centers, is a translation if and only if either $\theta_1 = \theta_2 = 0$ or $\theta_1 + \theta_2 = 2\pi$ (assume $0 \leq \theta_1, \theta_2 < 2\pi$).

2. The composition of two rotations, one counter-clockwise of angle θ_1 and the other clockwise of angle θ_2, about arbitrary centers, is a translation if and only if $\theta_1 = \theta_2$ (assume $0 \leq \theta_1, \theta_2 < 2\pi$).

Proposition 5.7. *Affine, Euclidean and rigid transformations of 2-space are related by the following inclusions, which are each proper:*

$$\textit{rigid transforms} \subset \textit{Euclidean transforms} \subset \textit{affine transforms}$$

Proof. The first inclusion follows from the definitions. It is proper because a reflection about any mirror is Euclidean but not rigid.

From Proposition 5.6 it follows that an Euclidean transformation is a composition of affine transformations (because translations, rotations and reflections are all affine) and, therefore, itself affine, proving the second inclusion. The inclusion is proper because a scaling by factors not all of unit magnitude is affine but not Euclidean. □

Remark 5.4. An interesting perspective on the proposition is to think of affine transformations as being made from translations, rotations, reflections and scalings; Euclidean transformations from translations, rotations and reflections; and rigid transformations from translations and rotations.

The worked example next says that Definition 5.5 about whether an Euclidean transformation reverses or preserves orientation is, in fact, independent of the choice of the three non-collinear points P, Q and R.

Example 5.10. Suppose that an affine transformation g of $\mathbb{R}^2$ maps *some* three non-collinear points P, Q and R in a manner that, looking at $\mathbb{R}^2$ from a fixed side, one of the sequences PQR and $g(P)g(Q)g(R)$ appears CW and the other CCW.

Show, then, that for *any* three non-collinear points X, Y and Z, one of the sequences XYZ and $g(X)g(Y)g(Z)$ appears CW and the other CCW, looking at $\mathbb{R}^2$ from the same side.

Answer: Use homogeneous coordinates to write $[g(W)\ 1]^T = M[W\ 1]^T$, where M is a fixed non-singular 3×3 matrix, and $W = [x\ y]^T$ is an arbitrary point of the plane.

Suppose that $P = [x_1\ y_1]^T$, $Q = [x_2\ y_2]^T$ and $R = [x_3\ y_3]^T$. Consider the equation

$$M \begin{bmatrix} x_1 & x_2 & x_3 \\ y_1 & y_2 & y_3 \\ 1 & 1 & 1 \end{bmatrix} = \begin{bmatrix} x_1' & x_2' & x_3' \\ y_1' & y_2' & y_3' \\ 1 & 1 & 1 \end{bmatrix}$$

in matrices, which gives the following

$$det(M) * \begin{vmatrix} x_1 & x_2 & x_3 \\ y_1 & y_2 & y_3 \\ 1 & 1 & 1 \end{vmatrix} = \begin{vmatrix} x_1' & x_2' & x_3' \\ y_1' & y_2' & y_3' \\ 1 & 1 & 1 \end{vmatrix}$$

relating determinants. Now, $[x_1'\ y_1'\ 1]^T = M[x_1\ y_1\ 1]^T = M[P\ 1]^T = [g(P)\ 1]^T$. Likewise, $[x_2'\ y_2'\ 1]^T = [g(Q)\ 1]^T$ and $[x_3'\ y_3'\ 1]^T = [g(R)\ 1]^T$. Therefore, the preceding equation can be written

$$det(M) * \begin{vmatrix} P & Q & R \\ 1 & 1 & 1 \end{vmatrix} = \begin{vmatrix} g(P) & g(Q) & g(R) \\ 1 & 1 & 1 \end{vmatrix}$$

Considering the signs of the three determinants above, and applying Lemma 5.1, one sees that PQR and $g(P)g(Q)g(R)$ appear differently oriented, from a fixed side of the plane, if and only if $det(M)$ is negative. But, then, by similar calculations, exactly the same would be true of XYZ and $g(X)g(Y)g(Z)$, for any points X, Y and Z.

5.3.3 Shear

With translations, rotations and scalings, and their compositions, we know that we "cover" all affine transformations. Shears, though, are a particularly distinctive kind of affine transformation that arise naturally from physical processes. For this reason they merit separate discussion. Roughly, a shear is the kind of distortion caused by placing a lump of putty between a pair of palms and then moving one palm parallel to the other.

A 2D shear s is uniquely determined by two parameters:

1. A directed line l called the *line of shear*.

2. An angle α called the *angle of shear*.

Here's how a point $P \in \mathbb{R}^2$ is mapped to the point P' by s (see Figure 5.16(a)):

(a) If P lies on l, then it is unchanged.

(b) If P lies a distance of h left of l, then it moves parallel to l in the positive direction of l a distance of $h \tan \alpha$.

(c) If P lies a distance of h right of l, then it moves parallel to l in the negative direction of l a distance of $h \tan \alpha$.

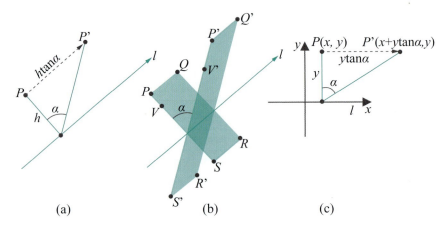

Figure 5.16: 2D shears: l is a directed line, α the angle of shear.

Note: Left or right is according to a viewer standing upright on the plane at a point of l, head pointing toward the positive z-axis (of a hypothetical right-handed coordinate system) and facing toward the direction of l.

Another way to think of the shear is as a force parallel to l which "bends" each perpendicular to it a fixed angle α. The rectangle $PQRS$ in Figure 5.16(b) is sheared into the parallelogram $P'Q'R'S'$. The farther points are from l, the proportionately more they travel under the shear; e.g., compare $V \mapsto V'$ and $P \mapsto P'$ in Figure 5.16(b). Figure 5.17 shows a sheared sheep (get it?).

If the directed line l of the shear is the x-axis, then it is particularly simple to determine its transformation equation. See Figure 5.16(c). The shear maps the point $P = [x \; y]^T$ to the point

Figure 5.17: Sheared
sheep.

$$P' = [x + y \tan \alpha \quad y]^T = \begin{bmatrix} 1 & \tan \alpha \\ 0 & 1 \end{bmatrix} P \qquad (5.22)$$

A shear along the x-axis, then, is a non-singular linear transformation given by Equation (5.22). Therefore, by Proposition 5.4, it is equivalent to a rotation about the origin, followed by a scaling, followed by another rotation about the origin. In fact, write

$$\begin{bmatrix} 1 & \tan \alpha \\ 0 & 1 \end{bmatrix} = \begin{bmatrix} \cos \phi & -\sin \phi \\ \sin \phi & \cos \phi \end{bmatrix} \begin{bmatrix} s_x & 0 \\ 0 & s_y \end{bmatrix} \begin{bmatrix} \cos \theta & -\sin \theta \\ \sin \theta & \cos \theta \end{bmatrix} \qquad (5.23)$$

It turns out that solving this equation is simpler than solving the more general (5.17). Indeed, it may be verified that the following four equations derive from (5.23):

$$\tan^2 \theta + \tan \alpha \tan \theta - 1 = 0 \qquad (5.24)$$
$$\phi = \theta - 90° \qquad (5.25)$$
$$s_x = \tan \theta \qquad (5.26)$$
$$s_y = \frac{1}{\tan \theta} \qquad (5.27)$$

The value of θ can then be calculated from (5.24) and those of ϕ, s_x and s_y subsequently from (5.25)-(5.27). In fact, it's interesting to visualize a shear along the x-axis as a rotation-scaling-rotation as in Figure 5.18.

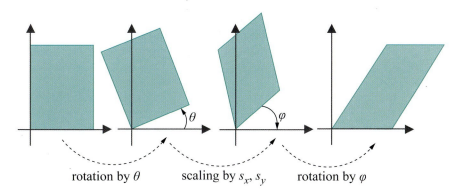

Figure 5.18: A shear as a rotation-scaling-rotation.

Exercise 5.43. Verify that the Equations (5.24)-(5.27) indeed follow from (5.23).

Example 5.11. Let s be the shear of angle $\alpha = 45°$ along the x-axis. Then $\tan \alpha = 1$ and Equation (5.24) in this case becomes

$$\tan^2 \theta + \tan \theta - 1 = 0$$

solving to (ignoring the negative root)

$$\tan \theta = \frac{-1 + \sqrt{5}}{2} \simeq 0.618034$$

so that

$$\theta \simeq 31.72°$$

Equations (5.25)-(5.27) give next:

$$\phi \simeq -58.28°$$
$$s_x \simeq 0.62$$
$$s_y \simeq 1.62$$

Therefore, s is equivalent to a rotation of $31.72°$ counter-clockwise about the origin, followed by scaling by factors 0.62, 1.62 along the x- and y-axes, respectively, followed by a rotation of $58.28°$ clockwise about the origin.

Exercise 5.44. Express the shear of angle $30°$ along the y-axis as a rotation-scaling-rotation.

Remark 5.5. We'll code shears in Section 5.4.8 following a discussion of their 3D version.

5.4 Geometric Transformations in 3-Space

Finally, the real world. Our discussions will mirror those of the previous section. In fact, extending translations, scalings and reflections from 2D to 3D is almost automatic. We'll pay our dues, though, for entering 3-space with a fair bit of work on rotations.

5.4.1 Translation

A translation is specified by a displacement vector $D = [d_x \; d_y \; d_z]^T$. The image of the point $P = [x \; y \; z]^T$ by this translation is $P' = [x + d_x \; y + d_y \; z + d_z]^T$ (see Figure 5.19).

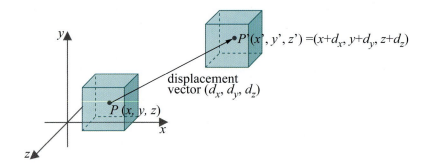

Figure 5.19: Translation.

Equivalently,

$$P' = P + D = \begin{bmatrix} 1 & 0 & 0 \\ 0 & 1 & 0 \\ 0 & 0 & 1 \end{bmatrix} P + D$$

which in homogeneous form, analogous to the 2D version (5.13), is

$$\begin{bmatrix} P' \\ 1 \end{bmatrix} = \begin{bmatrix} 1 & 0 & 0 & d_x \\ 0 & 1 & 0 & d_y \\ 0 & 0 & 1 & d_z \\ 0 & 0 & 0 & 1 \end{bmatrix} \begin{bmatrix} P \\ 1 \end{bmatrix} \tag{5.28}$$

For the record, the 4×4 matrix corresponding to translation by the displacement vector $[d_x \ d_y \ d_z]^T$ is denoted $T(d_x, d_y, d_z)$ and given by

$$T(d_x, d_y, d_z) = \begin{bmatrix} 1 & 0 & 0 & d_x \\ 0 & 1 & 0 & d_y \\ 0 & 0 & 1 & d_z \\ 0 & 0 & 0 & 1 \end{bmatrix} \tag{5.29}$$

Exercise 5.45. Write the 4×4 matrix corresponding to translation by the displacement vector $[3 \ 0 \ -1]^T$.

Exercise 5.46. Use Equation (5.28) to prove that the composition of translations is a translation and that the inverse of a translation is a translation as well.

Note: By default we're in 3-space from now on and all exercises and examples are in 3D.

5.4.2 Scaling

A scaling is specified by scaling factors s_x, s_y and s_z along the x-, y- and z-axis, respectively. The image of the point $P = [x \ y \ z]^T$ by this scaling is $P' = [s_x x \ s_y y \ s_z z]^T$ (see Figure 5.20).

Without further ado we write the 4×4 matrix corresponding to this scaling as (compare the 2D Equation (5.14))

$$S(s_x, s_y, s_z) = \begin{bmatrix} s_x & 0 & 0 & 0 \\ 0 & s_y & 0 & 0 \\ 0 & 0 & s_z & 0 \\ 0 & 0 & 0 & 1 \end{bmatrix} \tag{5.30}$$

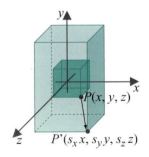

Figure 5.20: Scaling.

If any one or more of the scaling factors s_x, s_y and s_z is zero, the scaling is said to be *degenerate*; otherwise, it is *non-degenerate*. Clearly, a scaling is non-degenerate if and only if its matrix is non-singular. By a scaling we shall always mean a non-degenerate one, unless stated otherwise.

Exercise 5.47. Write the 4×4 matrix corresponding to scaling by the factors -1, 3 and 4 along the x-, y- and z-axis, respectively.

Exercise 5.48. Use (5.30) to prove that the composition of scalings is a scaling and that the inverse of a non-degenerate scaling is a non-degenerate scaling.

5.4.3 Rotation

Warning upfront: This section is much longer than the corresponding Section 5.1.3 on 2D rotations as there's much more magic to going around in 3D!

A rotation about a radial axis is specified by (a) a directed line l through the origin, which is the axis of rotation, and (b) the angle θ of the rotation.

We'll describe as a physical process how such a rotation maps a point P. First, if P lies on the axis l itself, then it does not move. Suppose, then, that P does not lie on l. Here's how it's mapped by the rotation (see Figure 5.21):

1. Drop the perpendicular from P to the point Q on l. Call the segment PQ as L. L lies on the plane h perpendicular to l through Q.

   Note that Figure 5.21 has Q and h on the positive side of l, but they could very well be on the other side, or even touching the origin, depending on where P is.

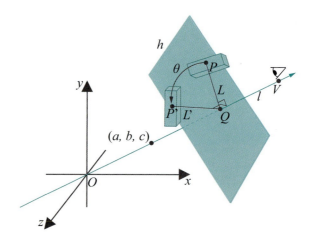

Figure 5.21: Rotation.

2. Locate a viewer at V far enough in the positive direction of l as to be able to see h when looking toward the origin.

3. Rotate the segment L about Q on h through an angle θ counterclockwise, as measured by the viewer.

4. If L' is the new position of L after rotation, then P is mapped to the corresponding endpoint P' of L'.

Remark 5.6. Giving a single point (a, b, c), not equal to the origin, is enough to specify the directed radial line l through it, as indicated in Figure 5.21. Therefore, all that remains to specify a rotation about l is the angle θ. This, of course, is exactly how the OpenGL command `glRotatef(`θ`, ` a`, ` b`, ` c`)` works, as described earlier in Section 4.1.3.

Rotation about the Coordinate Axes

The matrices corresponding to rotations in 3D about the coordinate axes are straightforwardly deduced from the 2D equation (5.3), reproduced below

$$\left[\begin{array}{c} x' \\ y' \end{array} \right] = \left[\begin{array}{cc} \cos\theta & -\sin\theta \\ \sin\theta & \cos\theta \end{array} \right] \left[\begin{array}{c} x \\ y \end{array} \right] \tag{5.31}$$

where $[x'\ y']^T$ is the image of $[x\ y]^T$ by a rotation on the xy-plane by an angle of θ about the origin, measured counter-clockwise by a viewer V on the positive side of the z-axis (Figure 5.22(a)).

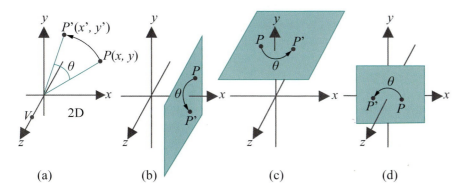

(a)   (b)   (c)   (d)

Figure 5.22: (a) 2D rotation on the xy-plane (b)-(d) 3D rotations about the coordinate axes.

In 3D, rotation about the x-axis by an angle θ (Figure 5.22(b)) maps a point $P = [x\ y\ z]^T$ to the point $P' = [x'\ y'\ z']^T$, where

(a) $x' = x$, because P travels parallel to the yz-plane, so its x value never changes.

(b) $[y\ z]^T \mapsto [y'\ z']^T$ is precisely as for a *2D rotation* by an angle θ CCW about the origin on the yz-plane, looking from the positive side of the x-axis.

Therefore, replacing x with y and y with z in (5.31), we have

$$\left[\begin{array}{c} y' \\ z' \end{array} \right] = \left[\begin{array}{cc} \cos\theta & -\sin\theta \\ \sin\theta & \cos\theta \end{array} \right] \left[\begin{array}{c} y \\ z \end{array} \right]$$

Therefore, the 4×4 matrix of 3D rotation about the x-axis is

$$R_x(\theta) = \left[\begin{array}{cccc} 1 & 0 & 0 & 0 \\ 0 & \cos\theta & -\sin\theta & 0 \\ 0 & \sin\theta & \cos\theta & 0 \\ 0 & 0 & 0 & 1 \end{array} \right] \tag{5.32}$$

(the first row serving to keep x unchanged).

Rotation about the y-axis by an angle θ (Figure 5.22(c)) maps a point $P = [x\ y\ z]^T$ to the point $P' = [x'\ y'\ z']^T$, where:

(a) $y' = y$

(b) $[x\ z]^T \mapsto [x'\ z']^T$ is as for a 2D rotation by an angle θ CCW about the origin on the xz-plane, looking from the positive side of the y-axis.

We have to be careful, though, in applying (5.31). For, compare Figure 5.22(a) with Figure 5.22(c) to observe that the role of x in the 2D figure is played by z in the 3D one, that of the 2D y by the 3D x and, of course, that of the 2D z (the viewer's axis) by the 3D y. (You can verify this by scratching out the current labels on the axes in Figure 5.22(a), relabeling them as just suggested, and then "mentally" turning the system to match Figure 5.22(c).) So, (5.31) gives

$$\begin{bmatrix} z' \\ x' \end{bmatrix} = \begin{bmatrix} \cos\theta & -\sin\theta \\ \sin\theta & \cos\theta \end{bmatrix} \begin{bmatrix} z \\ x \end{bmatrix}$$

or, equivalently,

$$\begin{bmatrix} x' \\ z' \end{bmatrix} = \begin{bmatrix} \cos\theta & \sin\theta \\ -\sin\theta & \cos\theta \end{bmatrix} \begin{bmatrix} x \\ z \end{bmatrix}$$

Finally, since y is fixed, we have

$$R_y(\theta) = \begin{bmatrix} \cos\theta & 0 & \sin\theta & 0 \\ 0 & 1 & 0 & 0 \\ -\sin\theta & 0 & \cos\theta & 0 \\ 0 & 0 & 0 & 1 \end{bmatrix} \tag{5.33}$$

We ask the reader to verify that the matrix of rotation about the z-axis by an angle θ (Figure 5.22(d)) is

$$R_z(\theta) = \begin{bmatrix} \cos\theta & -\sin\theta & 0 & 0 \\ \sin\theta & \cos\theta & 0 & 0 \\ 0 & 0 & 1 & 0 \\ 0 & 0 & 0 & 1 \end{bmatrix} \tag{5.34}$$

Exercise 5.49. Write the 4×4 matrix corresponding to rotation by an angle of $30°$ about the y-axis.

Rotation about an Arbitrary Radial Axis

It is a bit of work to find the matrix corresponding to rotation about an arbitrary axis through the origin. But it's important enough that we'll do it in two different ways. The first is mainly geometric and fairly intuitive. The second involves a bit of algebraic legerdemain, so it is a little less intuitive, but the final form it yields is more compact than that of the first.

Let the axis of rotation be specified as the directed line l through the origin O toward a point $P = (a, b, c)$ ($\neq O$), and the angle of rotation as θ. See Figure 5.23. To simplify computation we'll assume that P is a unit vector, i.e., $|P| = \sqrt{a^2 + b^2 + c^2} = 1$. There is no loss in generality because, if P is not of unit length, we can always divide it by $|P|$ to obtain a unit vector specifying the same rotation. Our goal is to compute the matrix, denote it $R_{a, b, c}(\theta)$, corresponding to this rotation.

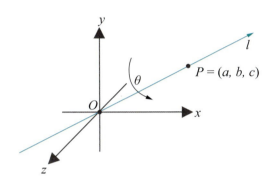

Figure 5.23: Rotating about an arbitrary radial axis.

Remark 5.7. To be honest, at this point we don't even know if a rotation about an arbitrary radial axis has a matrix representation at all, in other words, if it is a linear transformation!

Before we proceed, here's an exercise to nip in the bud a possible temptation.

Exercise 5.50. Can't we simply "add rotational axes" like vectors? For example, isn't it true, say, that glRotatef(90.0, 0.0, 1.0, 1.0) is the same as glRotatef(90.0, 0.0, 1.0, 0.0) followed by glRotatef(90.0, 0.0, 0.0, 1.0) or, maybe, the other way around?

If so, then writing the matrix corresponding to glRotatef(90.0, 0.0, 1.0, 1.0) would be simple: it is the product of the matrices corresponding to glRotatef(90.0, 0.0, 1.0, 0.0) and glRotatef(90.0, 0.0, 0.0, 1.0) in some order, both matrices easily written from what we know already about rotating about the coordinate axes themselves.

Prove that we *cannot*, in general, add rotational axes. In fact, show that glRotatef(90.0, 0.0, 1.0, 1.0) is neither glRotatef(90.0, 0.0, 1.0, 0.0) followed by glRotatef(90.0, 0.0, 0.0, 1.0), nor the other way around.

Hint: If, say, glRotatef(90.0, 0.0, 1.0, 1.0) were equal to the transformation glRotatef(90.0, 0.0, 1.0, 0.0) followed by glRotatef(90.0, 0.0, 0.0, 1.0), then the two would move all points identically. Consider the point $(0, 1, 1)$. How is it moved by glRotatef(90.0, 0.0, 1.0, 1.0)?

By glRotatef(90.0, 0.0, 1.0, 0.0) followed by glRotatef(90.0, 0.0, 0.0, 1.0)?

A Method to Compute the Rotation Matrix Which Is Mainly Geometric

Even though we can't quite add axes, the plan is still to express the rotation of θ about the radial axis l as a composition of rotations about the coordinate axes. We'll use the Trick. First we'll apply rotations to align l along one of the coordinate axes, then rotate by θ about that coordinate axis and, last, undo the initial rotations to bring l back where it was. For our plan to work, of course, the rotations to align l along a coordinate axis must themselves be about coordinate axes!

Here's a simple motivating experiment.

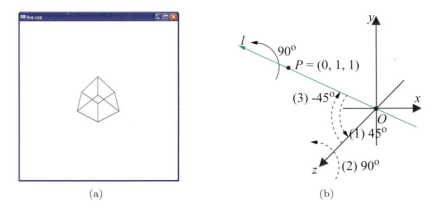

Figure 5.24: Experiment 5.1: (a) Output (b) Trick-based rotation scheme.

Experiment 5.1. Fire up box.cpp and insert a rotation command just before the box definition so that the transformation and object definition part of the drawing routine becomes:

```
// Modeling transformations.
glTranslatef(0.0, 0.0, -15.0);

glRotatef(90.0, 0.0, 1.0, 1.0);
glutWireCube(5.0); // Box.
```

The rotation command asks to rotate 90° about the line l from the origin through $(0, 1, 1)$. See Figure 5.24(a) for the displayed output.

Let's try now, instead, to use the strategy suggested above to express the given rotation in terms of rotations about the coordinate axes. Figure 5.24(b) illustrates the following simple scheme. Align l along the z-axis by rotating it 45° about the x-axis. Therefore, the given rotation should be equivalent to (1) a rotation of 45° about the x-axis, followed by (2) a rotation of 90° about the z-axis followed, finally, by a (3) rotation of −45° about the x-axis.

Give it a whirl. Replace the single rotation command `glRotatef(90.0, 0.0, 1.0, 1.0)` with a block of three as follows:

```
// Modeling transformations.
glTranslatef(0.0, 0.0, -15.0);

glRotatef(-45.0, 1.0, 0.0, 0.0);
glRotatef(90.0, 0.0, 0.0, 1.0);
glRotatef(45.0, 1.0, 0.0, 0.0);
glutWireCube(5.0); // Box.
```

Seeing is believing, is it not?!   **End**

Returning to the general problem, let's plan to rotate to align l along the z-axis in a manner that $P = (a, b, c)$ maps to the point $P'' = (0, 0, 1)$ on the positive side of the z-axis. We accomplish this by applying two successive rotations (see Figure 5.25):

(1) Rotate l an angle α about the x-axis onto a line l' on the xz-plane, taking P to P'.

(2) Rotate l' an angle $-\beta$ about the y-axis till it's aligned along the z-axis, taking P' to P'' (the minus sign in front of β is because the rotation is CW).

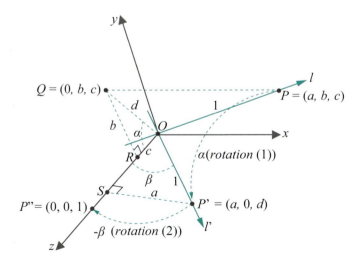

Figure 5.25: Aligning l along the z-axis.

Note: The choice of the z-axis to align l along was arbitrary – it could have been any of the three coordinate axes.

We must determine α and β. In fact, we'll simply determine the sine and cosine of both, which is sufficient to write the matrices $R_x(\alpha)$ and $R_y(-\beta)$ corresponding to the rotations (1) and (2), respectively.

Observe that the angle α that OP turns by rotation (1) about the x-axis is the same as the angle between its projection OQ on the yz-plane and the positive direction of the z-axis. In fact, imagine OQ as the "shadow" of OP cast on the yz-plane by a light shining down the x-axis – as OP turns so does its shadow, and by the same amount.

The coordinates of Q are $[0\ b\ c]^T$ as Q is the projection of $[a\ b\ c]^T$ on the yz-plane. Drop the perpendicular from Q to the point R on the z-axis. The angle QOR then is equal to α. We see from the coordinates of Q that $|OR| = c$ and $|RQ| = b$. Denoting $|OQ|$ by d, it follows from the right-angled triangle ORQ that $d = \sqrt{b^2 + c^2}$. Therefore, *assuming that $d \neq 0$,* we have $\sin\alpha = \frac{b}{d}$ and $\cos\alpha = \frac{c}{d}$.

Note: We don't lose any generality in assuming $d \neq 0$, because $d = 0$ means $Q = O$, which in turn means P is on the x-axis, implying that l lies along the x-axis as well, in which case we already know the matrix for rotation about l.

We can now use Equation (5.32) to write the matrix of rotation (1) as

$$R_x(\alpha) = \begin{bmatrix} 1 & 0 & 0 & 0 \\ 0 & \cos\alpha & -\sin\alpha & 0 \\ 0 & \sin\alpha & \cos\alpha & 0 \\ 0 & 0 & 0 & 1 \end{bmatrix} = \begin{bmatrix} 1 & 0 & 0 & 0 \\ 0 & c/d & -b/d & 0 \\ 0 & b/d & c/d & 0 \\ 0 & 0 & 0 & 1 \end{bmatrix}$$

After rotation (1), l coincides with l' on the xz-plane, and P with P'. The x coordinate of P' is a, the same as that of P, because rotation about the x-axis leaves this value unchanged; the z coordinate is d because rotation by α about the x-axis causes OQ, the shadow of OP whose length is d, to coincide with OS, the projection of OP' on the z-axis; the y coordinate, of course, is 0 as P' lies on the xz-plane.

Therefore, $P' = [a\ 0\ d]^T$, which means that in the right-angled triangle OSP', $|OS| = d$, $|SP'| = a$, and, therefore, $|OP'| = \sqrt{a^2 + d^2} = \sqrt{a^2 + b^2 + c^2} = 1$ (the latter evident as well from the fact that OP' is the unit vector OP rotated). Moreover, angle $P'OS = \beta$, where $-\beta$ is the angle turned by l' to align along the z-axis in rotation (2) above. Therefore, from the triangle OSP' we have $\sin\beta = a$ and $\cos\beta = d$.

We can now use Equation (5.33) to write the matrix of rotation (2) as

$$R_y(-\beta) = \begin{bmatrix} \cos(-\beta) & 0 & \sin(-\beta) & 0 \\ 0 & 1 & 0 & 0 \\ -\sin(-\beta) & 0 & \cos(-\beta) & 0 \\ 0 & 0 & 0 & 1 \end{bmatrix} = \begin{bmatrix} d & 0 & -a & 0 \\ 0 & 1 & 0 & 0 \\ a & 0 & d & 0 \\ 0 & 0 & 0 & 1 \end{bmatrix}$$

Returning to our original Trick-based plan, the first step of aligning l along the z-axis is accomplished, then, by the composition $R_y(-\beta)\, R_x(\alpha)$.

The next, of rotating by θ about the z-axis, is simply a matter of applying $R_z(\theta)$. Finally, the initial rotations aligning l along the z-axis are undone by the inverse transformation $(R_y(-\beta)\,R_x(\alpha))^{-1} = R_x(\alpha)^{-1}\,R_y(-\beta)^{-1} = R_x(-\alpha)\,R_y(\beta)$.

Putting everything together we have, finally,

$$
\begin{aligned}
R_{a,b,c}(\theta) \;=\;& R_x(-\alpha)\,R_y(\beta)\,R_z(\theta)\,R_y(-\beta)\,R_x(\alpha) \\[2mm]
=\;& \begin{bmatrix} 1 & 0 & 0 & 0 \\ 0 & c/d & b/d & 0 \\ 0 & -b/d & c/d & 0 \\ 0 & 0 & 0 & 1 \end{bmatrix}
\begin{bmatrix} d & 0 & a & 0 \\ 0 & 1 & 0 & 0 \\ -a & 0 & d & 0 \\ 0 & 0 & 0 & 1 \end{bmatrix} \\[2mm]
& \begin{bmatrix} \cos\theta & -\sin\theta & 0 & 0 \\ \sin\theta & \cos\theta & 0 & 0 \\ 0 & 0 & 1 & 0 \\ 0 & 0 & 0 & 1 \end{bmatrix}
\begin{bmatrix} d & 0 & -a & 0 \\ 0 & 1 & 0 & 0 \\ a & 0 & d & 0 \\ 0 & 0 & 0 & 1 \end{bmatrix} \\[2mm]
& \begin{bmatrix} 1 & 0 & 0 & 0 \\ 0 & c/d & -b/d & 0 \\ 0 & b/d & c/d & 0 \\ 0 & 0 & 0 & 1 \end{bmatrix}
\end{aligned}
\tag{5.35}
$$

The five matrices on the right side of the formula can be multiplied to give a single matrix, which would then be the value of $R_{a,b,c}(\theta)$. However, we'll not do so as the next method to calculate $R_{a,b,c}(\theta)$ gives a more concise form directly.

Exercise 5.51. Is rotation about an arbitrary radial axis a linear transformation? If so, is it always non-singular, or can it be singular?

Exercise 5.52. Use the Trick to write a rotation about an arbitrary axis l, not necessarily radial, as a seven-matrix product.

Example 5.12. Determine the 4×4 matrix corresponding to a $90°$ rotation about the radial axis l directed toward the point $[1\ 1\ 1]^T$ (which corresponds to the OpenGL command `glRotatef(90.0, 1.0, 1.0, 1.0)`).

Answer: The unit vector along l in the direction of $[1\ 1\ 1]^T$ is $P = [\frac{1}{\sqrt{3}}\ \frac{1}{\sqrt{3}}\ \frac{1}{\sqrt{3}}]^T$. Accordingly, keeping the notation used above,

$$
a = b = c = \frac{1}{\sqrt{3}}\,,\quad d = \sqrt{b^2 + c^2} = \frac{\sqrt{2}}{\sqrt{3}} \quad \text{and, of course,}\ \theta = \pi/2.
$$

Plugging these values into (5.35) we get the required matrix as

$$
R_{\overline{P}}(\theta) \;=\;
\begin{bmatrix}
1 & 0 & 0 & 0 \\
0 & \frac{1}{\sqrt{2}} & \frac{1}{\sqrt{2}} & 0 \\
0 & -\frac{1}{\sqrt{2}} & \frac{1}{\sqrt{2}} & 0 \\
0 & 0 & 0 & 1
\end{bmatrix}
\begin{bmatrix}
\frac{\sqrt{2}}{\sqrt{3}} & 0 & \frac{1}{\sqrt{3}} & 0 \\
0 & 1 & 0 & 0 \\
-\frac{1}{\sqrt{3}} & 0 & \frac{\sqrt{2}}{\sqrt{3}} & 0 \\
0 & 0 & 0 & 1
\end{bmatrix}
\begin{bmatrix}
0 & -1 & 0 & 0 \\
1 & 0 & 0 & 0 \\
0 & 0 & 1 & 0 \\
0 & 0 & 0 & 1
\end{bmatrix}
$$

$$
\begin{bmatrix}
\frac{\sqrt{2}}{\sqrt{3}} & 0 & -\frac{1}{\sqrt{3}} & 0 \\
0 & 1 & 0 & 0 \\
\frac{1}{\sqrt{3}} & 0 & \frac{\sqrt{2}}{\sqrt{3}} & 0 \\
0 & 0 & 0 & 1
\end{bmatrix}
\begin{bmatrix}
1 & 0 & 0 & 0 \\
0 & \frac{1}{\sqrt{2}} & -\frac{1}{\sqrt{2}} & 0 \\
0 & \frac{1}{\sqrt{2}} & \frac{1}{\sqrt{2}} & 0 \\
0 & 0 & 0 & 1
\end{bmatrix}
$$

$$
=\;
\begin{bmatrix}
\frac{1}{3} & \frac{1}{3}-\frac{1}{\sqrt{3}} & \frac{1}{3}+\frac{1}{\sqrt{3}} & 0 \\
\frac{1}{3}+\frac{1}{\sqrt{3}} & \frac{1}{3} & \frac{1}{3}-\frac{1}{\sqrt{3}} & 0 \\
\frac{1}{3}-\frac{1}{\sqrt{3}} & \frac{1}{3}+\frac{1}{\sqrt{3}} & \frac{1}{3} & 0 \\
0 & 0 & 0 & 1
\end{bmatrix}
$$

after some tedious computation.

Exercise 5.53. Determine the 4×4 matrix corresponding to a $90°$ rotation about the radial axis l directed toward the point $[0\ 1\ 1]^T$ (which corresponds to the OpenGL command `glRotatef(90.0, 0.0, 1.0, 1.0)`).

Exercise 5.54. Determine the 4×4 matrix corresponding to a $45°$ rotation about the radial axis l directed toward the point $[1\ -1\ 1]^T$ (which corresponds to the OpenGL command `glRotatef(45.0, 1.0, -1.0, 1.0)`).

Before discussing the second method to compute the matrix corresponding to rotation about an arbitrary axis, here are some facts about cross-products that we'll need. Skip this part if you are already familiar with cross-products of vectors.

Sidebar on Cross-Products

The *cross-product* (also called *vector product*) of two vectors u and v in $\mathbb{R}^3$ is another vector, denoted $u \times v$, defined as follows:

(a) If u and v are collinear, then $u \times v$ is the zero vector.

   Note: Two vectors are collinear if and only if any one is a scalar (positive, zero or negative) multiple of the other. Therefore, if at least one of the vectors is zero, the two are trivially collinear. (Figure 5.26(a) shows an example of three non-zero vectors, each pair being collinear.)

(b) If u and v are not collinear, then $u \times v$ is the vector whose (a) magnitude is $|u||v||\sin\theta|$, where θ is the angle between u and v, and (b) direction is perpendicular to the plane spanned by u and v, such that u, v and $u \times v$ form a right-handed system. See Figure 5.26(b).

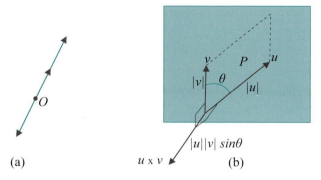

(a)

$u \times v$

(b)

Figure 5.26: (a) Non-zero collinear vectors drawn from the origin (b) Taking the cross-product.

Here's another way to think of the cross-product. The magnitude $|u||v|| \sin \theta|$ of the cross-product is nothing but the area of the parallelogram P with u and v as adjacent sides. The area of this parallelogram is, in fact, zero if and only if u and v are collinear. Consequently, the following is an alternate definition:

$u \times v$ is the vector whose magnitude is the area of the parallelogram with u and v as adjacent sides; if the magnitude is non-zero, then the direction of $u \times v$ is perpendicular to the plane spanned by u and v, such that u, v and $u \times v$ form a right-handed system.

If $u = [u_x \ u_y \ u_z]^T$ and $v = [v_x \ v_y \ v_z]^T$, a formula for the cross-product is the following:

$$u \times v = [u_y v_z - v_y u_z \quad v_x u_z - u_x v_z \quad u_x v_y - v_x u_y]^T \qquad (5.36)$$

A convenient way to remember this formula is with the help of a determinant, as you are asked to show in the following exercise.

Exercise 5.55. If

$$u = u_x \mathbf{i} + u_y \mathbf{j} + u_z \mathbf{k} \quad \text{and} \quad v = v_x \mathbf{i} + v_y \mathbf{j} + v_z \mathbf{k}$$

where $\mathbf{i}$, $\mathbf{j}$ and $\mathbf{k}$ are the unit vectors in the directions of the positive x-, y- and z-axes, show that

$$u \times v = \begin{vmatrix} \mathbf{i} & u_x & v_x \\ \mathbf{j} & u_y & v_y \\ \mathbf{k} & u_z & v_z \end{vmatrix} \qquad (5.37)$$

Example 5.13. Determine the cross-product $[2 \ 1 \ 0]^T \times [1 \ -2 \ 4]^T$.

Answer:

$$[2 \ 1 \ 0]^T \times [1 \ -2 \ 4]^T = \begin{vmatrix} \mathbf{i} & 2 & 1 \\ \mathbf{j} & 1 & 2 \\ \mathbf{k} & 0 & 4 \end{vmatrix} = 4\mathbf{i} - 8\mathbf{j} - 5\mathbf{k} = [4 \ -8 \ -5]^T$$

Exercise 5.56. Determine the cross-product $[3 \ -1 \ 2]^T \times [-1 \ 0 \ 3]^T$.

Exercise 5.57. Write the result of the cross-product of every ordered pair from the three vectors **i**, **j** and **k** (there are 9 such products if you include products of vectors with themselves).

Remark 5.8. An easy way to remember the answer to the preceding exercise is the following:

The cross-product of any of **i**, **j** and **k** with itself is the zero vector. For the product of two different ones from **i**, **j** and **k**, keep in mind the cyclic order **i** → **j** → **k** → **i**. Then, if two successive elements in this order are multiplied, the result is the next; if two successive elements are multiplied in reverse order, then the result is the negative of the next element. For example, $\mathbf{j} \times \mathbf{k} = \mathbf{i}$ and $\mathbf{k} \times \mathbf{j} = -\mathbf{i}$.

Exercise 5.58. Prove the following about cross-products, where u, v and w are any three vectors, and c an arbitrary scalar:

(a) u and v are collinear if and only if $u \times v = \mathbf{0}$ (*collinearity test*)

   Note: The "only if" direction follows from the definition of cross-product; "if" needs to be proved.

(b) $u \times u = \mathbf{0}$

(c) $u \times v = -(v \times u)$ (*cross-product is anti-commutative*)

(d) It may not be true that $(u \times v) \times w = u \times (v \times w)$ (*cross-product is not associative*). Give an example of u, v and w where it isn't true.

(e) $(cu) \times v = u \times (cv) = c(u \times v)$

(f) $u \times (v+w) = u \cdot v + u \cdot w$ and $(v+w) \times u = v \times u + w \times u$ (*cross-product distributes over a sum*)

Exercise 5.59. Prove that if u is a unit vector and v arbitrary, then the vector $(u \times v) \times u$ is the component of v perpendicular to u.

Interestingly, it turns out, as the next example shows, that a cross-product with one fixed vector is a linear transformation.

Example 5.14. Show for a fixed vector $u = u_x \mathbf{i} + u_y \mathbf{j} + u_z \mathbf{k}$ that the transformation of $\mathbb{R}^3$ defined by $v \mapsto u \times v$ is linear.

Answer: Check from the formula (5.36) for $u \times v$ that

$$u \times v = Mv$$

where

$$M = \begin{bmatrix} 0 & -u_z & u_y \\ u_z & 0 & -u_x \\ -u_y & u_x & 0 \end{bmatrix}$$

which proves that $v \mapsto u \times v$ is indeed linear, with defining matrix M.

A Method to Compute the Rotation Matrix Which Is Part Geometry and Part Algebra

The problem statement again:

The axis of rotation is the directed line l through the origin toward a point $P = [a \ b \ c]^T$, such that $|P| = 1$, and the angle of rotation is θ. The goal is to compute the matrix $R_{a,b,c}(\theta)$ corresponding to this rotation.

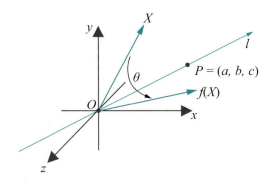

Figure 5.27: The vector $f(X)$ is obtained by rotating X an angle of θ about the radial line l.

Let the image of a vector X by the given rotation be $f(X)$. See Figure 5.27. First, split X as

$$X = X_1 + X_2$$

into components X_1 and X_2 parallel and perpendicular, respectively, to l. See Figure 5.28.

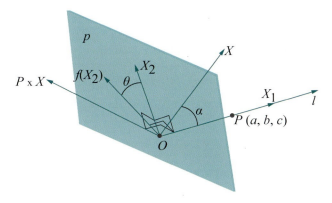

Figure 5.28: X_1 and X_2 are components of X parallel and perpendicular, respectively, to l; X_2, $f(X_2)$ and $P \times X$ all lie on the plane p through O perpendicular to l. X_2 and $P \times X$ are mutually perpendicular as well.

Since a rotation is a linear transformation, we have

$$f(X) = f(X_1) + f(X_2) \tag{5.38}$$

As X_1 lies on l, rotation about l leaves it unchanged. So

$$f(X_1) = X_1 \tag{5.39}$$

Now, X_2 lies on the plane p through O perpendicular to l and rotates by an angle θ about l, to $f(X_2)$. Therefore, $f(X_2)$ lies on p as well. We'll assume for now that X_2 is non-zero, meaning that X is not parallel to l, for, otherwise, $f(X) = f(X_1) = X_1$ and there's nothing more to do.

Observe that the vector $P \times X$, being perpendicular to l, lies on p; moreover, $P \times X$ is perpendicular to X_2 as it is perpendicular to the plane containing P and X, which contains X_2 as well. It follows, then, that the plane p is spanned by the two perpendicular vectors X_2 and $P \times X$. Consequently, these two specify coordinate axes on p. Let's determine the coordinates of $f(X_2)$ with respect to these axes, equivalently, the components of $f(X_2)$ parallel to X_2 and to $P \times X$.

The component of $f(X_2)$ parallel to X_2 has signed length $|f(X_2)| \cos \theta$. Moreover, the unit vector in the direction of X_2 is $X_2/|X_2|$. Therefore, the component of $f(X_2)$ parallel to X_2 is

$$|f(X_2)| \cos \theta \; X_2/|X_2| = X_2 \cos \theta \tag{5.40}$$

using the fact that $|f(X_2)| = |X_2|$, because $f(X_2)$ is X_2 rotated.

The component of $f(X_2)$ parallel to $P \times X$ has signed length

$$
\begin{aligned}
|f(X_2)| \sin \theta &= |X_2| \sin \theta \quad \text{(using again } |f(X_2)| = |X_2|) \\
&= |X \sin \alpha| \sin \theta \quad \text{(where } \alpha \text{ is the angle between } P \text{ and } X) \\
&= |P||X|| \sin \alpha| \sin \theta \quad \text{(as } |P| = 1) \\
&= |P \times X| \sin \theta \quad \text{(by definition of the cross-product)}
\end{aligned}
$$

The unit vector in the direction of $P \times X$ is $(P \times X)/|P \times X|$. It follows that the component of $f(X_2)$ parallel to $P \times X$ is

$$|P \times X| \sin \theta \; (P \times X)/|P \times X| = (P \times X) \sin \theta \tag{5.41}$$

Adding its components parallel to X_2 and $P \times X$ with help of (5.40) and (5.41), we conclude that

$$f(X_2) = X_2 \cos \theta + (P \times X) \sin \theta \tag{5.42}$$

Plugging the values from (5.39) and (5.42) into (5.38) we see that

$$
\begin{aligned}
f(X) &= X_1 + X_2 \cos \theta + (P \times X) \sin \theta \\
&= X_1 + (X - X_1) \cos \theta + (P \times X) \sin \theta \\
&= X \cos \theta + X_1(1 - \cos \theta) + (P \times X) \sin \theta \tag{5.43}
\end{aligned}
$$

Note: The preceding equation is valid even if X_2 is the zero vector, for, then, $X = X_1$ and $P \times X = 0$, so that the equation says $f(X) = X$, which is correct. So we're completely general from here on.

Use the results of Example 4.5 and Example 5.14 to replace X_1 and $P \times X$, respectively, with their equivalent matrix products:

$$
f(X) = \cos\theta \begin{bmatrix} 1 & 0 & 0 \\ 0 & 1 & 0 \\ 0 & 0 & 1 \end{bmatrix} X + (1 - \cos\theta) \begin{bmatrix} a^2 & ab & ac \\ ab & b^2 & bc \\ ac & bc & c^2 \end{bmatrix} X +
$$

$$
\sin\theta \begin{bmatrix} 0 & -c & b \\ c & 0 & -a \\ -b & a & 0 \end{bmatrix} X
$$

$$
= \left(\begin{bmatrix} a^2 & ab & ac \\ ab & b^2 & bc \\ ac & bc & c^2 \end{bmatrix} + \cos\theta \begin{bmatrix} 1 - a^2 & -ab & -ac \\ -ab & 1 - b^2 & -bc \\ -ac & -bc & 1 - c^2 \end{bmatrix} + \right.
$$

$$
\left. \sin\theta \begin{bmatrix} 0 & -c & b \\ c & 0 & -a \\ -b & a & 0 \end{bmatrix} \right) X \tag{5.44}
$$

Finally, adding the matrices in the parentheses and writing the result in homogeneous coordinates, we get a second form for the rotation matrix, different from the earlier geometrically-derived (5.35):

$R_{a,\,b,\,c}(\theta) =$

$$
\begin{bmatrix}
a^2(1 - \cos\theta) + \cos\theta & ab(1 - \cos\theta) - c\sin\theta & ac(1 - \cos\theta) + b\sin\theta & 0 \\
ab(1 - \cos\theta) + c\sin\theta & b^2(1 - \cos\theta) + \cos\theta & bc(1 - \cos\theta) - a\sin\theta & 0 \\
ac(1 - \cos\theta) - b\sin\theta & bc(1 - \cos\theta) + a\sin\theta & c^2(1 - \cos\theta) + \cos\theta & 0 \\
0 & 0 & 0 & 1
\end{bmatrix}
$$

$$
\tag{5.45}
$$

Remark 5.9. One can replace X_1 in (5.43) by $(X \cdot P)P$, as X_1 is the component of X parallel to the vector P of unit length (see Exercise 4.39(a)). The resulting equation

$$
f(X) = X\cos\theta + (X \cdot P)P(1 - \cos\theta) + (P \times X)\sin\theta \tag{5.46}
$$

is called *Rodrigues' rotation forumula.*

Exercise 5.60. Verify the result of Example 5.12 by computing the rotation matrix using Equation (5.45) instead of (5.35).

Exercise 5.61. Verify your answer to Exercise 5.53 by computing the rotation matrix using Equation (5.45) instead of (5.35).

$\mathcal{R}emark$ 5.10. An exercise which may have been conspicuous by its absence is to show that the composition of two rotations about radial axes is also a rotation about a radial axis. Unfortunately, though true, this is not easy to prove.

In fact, unlike its 2D counterpart, it's not even obvious that it is true. For example, is it evident that, say, a rotation of 45° about the x-axis followed by another, say, of 30° about the y-axis is a rotation about some axis in the first place? We'll prove that rotations do, in fact, compose to rotations using properties of rigid transformations in Section 5.4.5.

Whew, we told you this section was going to be long! It turned out fairly technical, too. We'll be coasting downhill the rest of the way and, believe it or not, get to see some code before long.

5.4.4 Reflection

The image of the point $P = [x \ y \ z]^T$ by reflection about a plane p, called the *mirror*, is $P' = [x' \ y' \ z']^T$ such that:

(a) if P lies on p, then $P' = P$;

(b) if P does not lie on p, then P' is the point on the other side of p such that PP' is perpendicular to p, and P' is the same distance from p as P. See Figure 5.29.

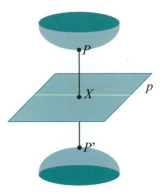

Figure 5.29: Reflection about plane p. $|XP| = |XP'|$.

Reflection about the xy-plane is simply scaling by the factors $s_x = 1$, $s_y = 1$ and $s_z = -1$. Its matrix, therefore, is

$$M = S(1, 1, -1) = \begin{bmatrix} 1 & 0 & 0 & 0 \\ 0 & 1 & 0 & 0 \\ 0 & 0 & -1 & 0 \\ 0 & 0 & 0 & 1 \end{bmatrix} \tag{5.47}$$

Exercise 5.62. Let p be an arbitrary plane mirror. Use the Trick to show that the matrix corresponding to reflection about p is of the form

$$M_1^{-1}M_2^{-1}S(1,1,-1)M_2M_1$$

where M_1 corresponds to a translation and M_2 to a rotation about a radial axis. You don't need to find exact values for M_1 and M_2.

Exercise 5.63. Write the 4×4 matrix corresponding to reflection about the plane $x - z = 0$.

Exercise 5.64. What is the result of composing reflections about the same mirror? What transformation is the result of composing reflections about two parallel mirrors? About two perpendicular mirrors? What is the inverse of a reflection?

Exercise 5.65. (Commutativity of transformations of 3-space)

(a) Do translations commute with each other?

(b) Do scalings commute with each other?

(c) Do rotations about the same radial axis commute with each other?

(d) Does a rotation about one radial axis commute with another about a different radial axis?

(e) Do translations and rotations commute?

(f) Do reflections about two different mirrors ever commute?

5.4.5 Affine Geometric Transformations

From Equations (5.29), (5.30), (5.45)and (5.47) and judicious applications of the Trick one sees that translations, rotations about arbitrary axes, scalings and reflections about arbitrary mirrors are all affine transformations of 3-space. Consequently, one has the following 3D analogue of Proposition 5.2 about their geometric niceness.

Proposition 5.8. *Let g be either a translation, a scaling, a rotation (about an arbitrary radial axis) or a reflection (about an arbitrary mirror). Then:*

(a) g maps straight lines to straight lines and planes to planes. Moreover, it maps parallel straight lines to parallel straight lines, intersecting straight lines to intersecting straight lines, parallel planes to parallel planes and intersecting planes to intersecting planes.

(b) g maps a convex set on one plane to a convex set on another plane. Moreover, it transforms the convex hull of a set of points $\{P_1, P_2, \ldots, P_k\}$ on one plane to the convex hull of $\{g(P_1), g(P_2), \ldots, g(P_k)\}$ lying on another plane.

Proof. Follows from Proposition 5.1 about affine transformations in general. ☐

Translations, rotations, scalings and reflections are all affine. In the opposite direction, the following analogues of the 2D Propositions 5.3 and 5.4 are true as well, though there seem to be no "low-level" proofs similar to the 2D ones. Fairly sophisticated linear algebra appears unavoidable. So at this time we'll only state Propositions 5.9 and 5.10, deferring the proof of the latter to later in this chapter as optional reading for the mathematically inclined. Mind that Proposition 5.9 follows straightforwardly from Proposition 5.10.

Proposition 5.9. *Any affine transformation of $\mathbb{R}^3$ is the composition in some order of translations, scalings and rotations about radial axes.*

In particular, an affine transformation $g : \mathbb{R}^3 \to \mathbb{R}^2$ can be factored into a composition $g = g_4 \circ g_3 \circ g_2 \circ g_1$, where g_1 is a rotation about a radial axis, g_2 a scaling, g_3 another rotation about a radial axis and g_4 a translation. ☐

Proposition 5.10. *Any non-singular linear transformation of $\mathbb{R}^3$ is the composition successively of a rotation about a radial axis, a scaling and another rotation about a radial axis.* ☐

OpenGL and Affine Transformations

The importance of Proposition 5.9, particularly to the design of an API like OpenGL, cannot be overestimated. The modeling transformations one creates using OpenGL are compositions of translations (`glTranslatef()`), scalings (`glScalef()`, excluding for the moment degenerate calls) and rotations about a radial axis (`glRotatef()`). The whole collection is affine as a composition of affine transformations is affine. Proposition 5.9 tells us that, conversely, any affine transformation of 3-space is a composition of translations, scalings and rotations about a radial axis and, therefore, may be implemented in OpenGL.

Conclusion: excluding degenerate scalings, *the modeling transformations one can create in OpenGL are precisely the affine transformations of 3-space and nothing else.* And, as we argued in Section 5.2.2, this is a welcome situation both from the API designer's point of view of being able to implement a simple rendering engine, and the application programmer's point of view of having a comprehensive set of transformations at her disposal.

Example 5.15. Express the affine transformation

$$f([x \ \ y \ \ z]^T) = [-y \ \ x \ \ z+2]^T$$

as a composition of OpenGL transformations.

Answer: The mapping by f is the composition

$$[x \ \ y \ \ z]^T \mapsto [-y \ \ x \ \ z]^T \mapsto [-y \ \ x \ \ z+2]^T$$

Now, the first map is easily seen to be a rotation of $\pi/2$ about the z-axis, while the second is a translation of 2 in the z-direction. We have, therefore, the required block of OpenGL transformations:

```
glTranslatef(0.0, 0.0, 2.0);
glRotatef(90.0, 0.0, 0.0, 1.0);
```

Exercise 5.66. Express each of the following affine transformations as a composition of OpenGL transformations:

(a) $f([x\ y\ z]^T) = [y\ x\ z]^T$

(b) $f([x\ y\ z]^T) = [y\ z\ x]^T$

(c) $f([x\ y\ z]^T) = [x - y\ \ x + y\ \ -z]^T$

Verifying the Matrices Generated by OpenGL

Appendix F of the red book lists the matrices which OpenGL generates for the modeling transformations.

The translation and scaling matrices are simple and seen to agree with Equations (5.29) and (5.30), respectively. We leave it to the reader to verify that the rotation matrix R which OpenGL generates for the rotation transformation `glRotate{fd}`(a, x, y, z) is equivalent to that of Equation (5.45); in fact, the red book expresses it in the form of the prior more expansive Equation (5.44).

Incidentally, it's clear now and worth emphasizing that the composition of modelview transformations is implemented in the OpenGL engine as 4×4 matrix multiplication. Almost all of the "action" in animation is, therefore, matrix multiplication. In fact, it's not an oversimplification to say that a graphics card animates as fast as it multiplies.

5.4.6 Accessing and Manipulating the Current Modelview Matrix

Finally, we surface from deep theory to see – *code*!

There are four commonly-used methods to access the current modelview matrix, i.e., the matrix at the top of the OpenGL modelview matrix stack, three to change its value and one to read it. After setting the matrix mode to GL_MODELVIEW with the command `glMatrixMode(GL_MODELVIEW)` if need be, the call

1. `glLoadIdentity()` sets the current modelview matrix to the identity matrix I_4.

2. `glLoadMatrix(`*matrixData*`)` sets the current modelview matrix to the 4×4 matrix whose elements are listed in the one-dimensional array pointed by *matrixData* in column-major order (which means elements of the first column are listed first in order of increasing row, then those of the second column and so on).

3. `glMultMatrix(`*matrixData*`)` multiplies the current modelview matrix on the right by the 4×4 matrix whose elements are listed in column-major order in the one-dimensional array pointed by *matrixData*.

4. `glGetFloatv(GL_MODELVIEW_MATRIX,` *modelviewMatrixData*`)` stores the 16 elements of the current modelview matrix in column-major order in the one-dimensional array pointed by *modelviewMatrixData*.

Figure 5.30: Screenshot from Experiment 5.2.

Experiment 5.2. Run `manipulateModelviewMatrix.cpp`. Figure 5.30 is a screenshot, although in this case we are really more interested in the transformations in the program rather than its visual output.

The `gluLookAt(0.0, 0.0, 10.0, 0.0, 0.0, 0.0, 0.0, 1.0, 0.0)` statement we understand to multiply the current modelview matrix on the right by the matrix of its equivalent modeling transformation. The current modelview matrix is changed again by the `glMultMatrixf(matrixData)` call, which multiplies it on the right by the matrix corresponding to a rotation of 45° about the z-axis, equivalent to a `glRotatef(45.0, 0.0, 0.0, 1.0)` call. It's changed one last time by `glScalef(1.0, 2.0, 1.0)`.

The current modelview matrix is output to the command window initially and then after each of the three modelview transformations. We'll discuss next if the four output values match our understanding of the theory. **End**

As expected, the first matrix output by the program is the identity

$$I_4 = \begin{bmatrix} 1 & 0 & 0 & 0 \\ 0 & 1 & 0 & 0 \\ 0 & 0 & 1 & 0 \\ 0 & 0 & 0 & 1 \end{bmatrix}$$

The `gluLookAt(0.0, 0.0, 10.0, 0.0, 0.0, 0.0, 0.0, 1.0, 0.0)` command is equivalent to `glTranslatef(0.0, 0.0, -10.0)`, whose matrix, from Equation (5.29), is

$$T(0, 0, -10) = \begin{bmatrix} 1 & 0 & 0 & 0 \\ 0 & 1 & 0 & 0 \\ 0 & 0 & 1 & -10 \\ 0 & 0 & 0 & 1 \end{bmatrix}$$

Therefore, after the `gluLookAt()` call the current modelview matrix should equal

$$\begin{bmatrix} 1 & 0 & 0 & 0 \\ 0 & 1 & 0 & 0 \\ 0 & 0 & 1 & 0 \\ 0 & 0 & 0 & 1 \end{bmatrix} \begin{bmatrix} 1 & 0 & 0 & 0 \\ 0 & 1 & 0 & 0 \\ 0 & 0 & 1 & -10 \\ 0 & 0 & 0 & 1 \end{bmatrix} = \begin{bmatrix} 1 & 0 & 0 & 0 \\ 0 & 1 & 0 & 0 \\ 0 & 0 & 1 & -10 \\ 0 & 0 & 0 & 1 \end{bmatrix}$$

which indeed is the second one output.

Next, the current modelview matrix is multiplied on the right by the matrix

$$M = \begin{bmatrix} X & -X & 0 & 0 \\ X & X & 0 & 0 \\ 0 & 0 & 1 & 0 \\ 0 & 0 & 0 & 1 \end{bmatrix}$$

where $X = 0.70710678 \simeq 1/\sqrt{2}$, so M corresponds to a rotation of $45°$ about the z-axis. The third matrix output then is, as expected,

$$\begin{bmatrix} 1 & 0 & 0 & 0 \\ 0 & 1 & 0 & 0 \\ 0 & 0 & 1 & -10 \\ 0 & 0 & 0 & 1 \end{bmatrix} \begin{bmatrix} X & -X & 0 & 0 \\ X & X & 0 & 0 \\ 0 & 0 & 1 & 0 \\ 0 & 0 & 0 & 1 \end{bmatrix} = \begin{bmatrix} X & -X & 0 & 0 \\ X & X & 0 & 0 \\ 0 & 0 & 1 & -10 \\ 0 & 0 & 0 & 1 \end{bmatrix}$$

The final matrix output is after the call `glScalef(1.0, 2.0, 1.0)`, a scaling by a factor of 2 along the y-axis. From Equation (5.30)

$$S(1, 2, 1) = \begin{bmatrix} 1 & 0 & 0 & 0 \\ 0 & 2 & 0 & 0 \\ 0 & 0 & 1 & 0 \\ 0 & 0 & 0 & 1 \end{bmatrix}$$

which multiplies the third matrix on the right to indeed give the final output matrix as

$$\begin{bmatrix} X & -X & 0 & 0 \\ X & X & 0 & 0 \\ 0 & 0 & 1 & -10 \\ 0 & 0 & 0 & 1 \end{bmatrix} \begin{bmatrix} 1 & 0 & 0 & 0 \\ 0 & 2 & 0 & 0 \\ 0 & 0 & 1 & 0 \\ 0 & 0 & 0 & 1 \end{bmatrix} = \begin{bmatrix} X & -2X & 0 & 0 \\ X & 2X & 0 & 0 \\ 0 & 0 & 1 & -10 \\ 0 & 0 & 0 & 1 \end{bmatrix}$$

Exercise 5.67. (Programming) Replace the `gluLookAt()` statement in `manipulateModelviewMatrix.cpp` with the following

```
gluLookAt(0.0, 10.0, 10.0, 0.0, 0.0, 0.0, 0.0, 1.0, 0.0);
```

Theoretically verify the correctness of the modelview matrix output by the program after the new `gluLookAt()` statement.
Hint: The new `gluLookAt()` statement is simulated as a translation by displacement vector $[0\ -10\ -10]^T$, followed by a rotation of $45°$ about the x-axis.

Exercise 5.68. (Programming) Verify your answer to Exercise 5.53 by comparing it with the output from an appropriately modified `manipulateModelviewMatrix.cpp`.

Exercise 5.69. (Programming) What is the current modelview matrix after the following piece of code in the drawing routine:

```
glMatrixMode(GL_MODELVIEW);
glLoadIdentity();
glScalef(1.0, 2.0, 2.0);
glTranslatef(2.0, 1.0, 0.0);
glRotatef(90.0, 1.0, 0.0, 0.0);
```

Find the answer theoretically by multiplying appropriate 4×4 matrices and then verify with the help of `manipulateModelviewMatrix.cpp`.

Exercise 5.70. (Programming) Verify your answers to Exercise 4.48(a)-(e) by using `manipulateModelviewMatrix.cpp` to find the matrix corresponding to the given `gluLookAt()` call, as well as that corresponding to the composed sequence of modeling transformations which you gave as being equivalent.

Exercise 5.71. (Programming) In Remark 4.12 about the viewing transformation being simulated by modeling transformations, we claimed that, following a `gluLookAt()` call, the current modelview matrix changes, but not the current projection matrix.

The part about the current modelview matrix is clearly true from what we have just seen in Experiment 5.2.

We ask the reader to verify the claim about the current projection matrix, particularly for the program `manipulateModelviewMatrix.cpp`, by reading the current projection matrix, both before and after the `gluLookAt()` statement, with the help of `glGetFloatv(GL_PROJECTION_MATRIX,` *projectionMatrixData*) calls.

5.4.7   Euclidean and Rigid Transformations

We have definitions analogous to the 2D case for Euclidean and rigid transformations of 3-space. Euclidean transformations are important because they preserve distance and, therefore, shape as well.

Definition 5.6. An *Euclidean transformation* (also called *isometry*) f of $\mathbb{R}^3$ is such that $|f(P)f(Q)| = |PQ|$ for any two points $P, Q \in \mathbb{R}^3$.

For the discussion of orientation coming up next, we need first to know when triples of vectors in 3-space are right-handed or left-handed.

Definition 5.7. An ordered triple of non-coplanar vectors $\{u, v, w\}$, each assumed originating from the same point O, is said to form a *right-handed* *system* (or, simply, be *right-handed*) if the rotation of u about O toward v, along the plane containing u and v, and along the smaller of the angles between the two, appears counter-clockwise to a viewer watching from the endpoint of w; otherwise, it is said to be *left-handed*. See Figures 5.31(a) and (b) for examples.

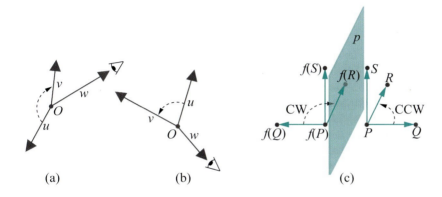

(a)   (b)   (c)

Figure 5.31: (a) $\{u, v, w\}$ is left-handed (b) $\{u, v, w\}$ is right-handed (c) The reflection f about the plane p is orientation-reversing, because the triple $\{PQ, PR, PS\}$ is right-handed, while the triple of images $\{f(P)f(Q), f(P)(R), f(P)(S)\}$ is left-handed.

Remark 5.11. We actually first discussed handedness in the context of coordinate systems way back in Section 2.2.

Remark 5.12. Another term for handedness – more scientific-sounding, but less used – is *chirality*.

Definition 5.8. An Euclidean transformation f of $\mathbb{R}^3$ is said to be *orientation-reversing* if there exist four non-coplanar points P, Q, R and S in $\mathbb{R}^3$ such that one of the two ordered triples of vectors $\{PQ, PR, PS\}$ and $\{f(P)f(Q), f(P)f(R), f(P)f(S)\}$ is right-handed, while the other is left-handed.

An Euclidean transformation that is not orientation-reversing is said to be *orientation-preserving*.

Orientation-preserving Euclidean transformations are called *rigid transformations*.

Remark 5.13. The property of the transformation f described in the first paragraph of the definition does not depend on the choice of the non-coplanar points: we'll see that if it is true for *some* four non-coplanar points P, Q, R and S in $\mathbb{R}^3$ that one of $\{PQ, PR, PS\}$ and $\{f(P)f(Q), f(P)f(R), f(P)f(S)\}$ is right-handed, while the other left-handed, then it is true for *any* four non-coplanar points.

Example 5.16. The reflection f of Figure 5.31(c) about the plane p is an orientation-reversing Euclidean transformation.

Exercise 5.72. Scalings in general are not Euclidean transformations, but with certain choices of scaling factors they are. List these choices and for each say if the scaling preserves or reverses orientation.

Exercise 5.73. Show that the composition of two Euclidean transformations is Euclidean and that of two rigid transformations is rigid.

Exercise 5.74. Show that the composition of two orientation-reversing Euclidean transformations is an orientation-preserving Euclidean transformation (in other words, rigid). Show that the composition of an orientation-preserving and an orientation-reversing Euclidean transformation is orientation-reversing.

The following result gives a way to decide if an ordered triple of vectors is right-handed or left-handed.

Lemma 5.2. *Assuming that the coordinate axes themselves form a right-handed system, then an ordered triple $\{u, v, w\}$ of non-coplanar vectors, where $u = [u_x\ u_y\ u_z]^T$, $v = [v_x\ v_y\ v_z]^T$ and $w = [w_x\ w_y\ w_z]^T$, is right-handed or left-handed according as the determinant*

$$\begin{vmatrix} u_x & v_x & w_x \\ u_y & v_y & w_y \\ u_z & v_z & w_z \end{vmatrix}$$

is greater or less than zero (it cannot be zero as $\{u, v, w\}$ is non-coplanar).

Proof. The proof is not difficult, but uses more linear algebra than we would like to assume at this time, so we ask the reader to refer to a text such as [7]. □

The next exercise, analogue of the 2D Example 5.10, says that Definition 5.8 about an Euclidean transformation reversing or preserving orientation is independent of the choice of the four non-coplanar points P, Q, R and S.

Exercise 5.75. Suppose that an affine transformation f of $\mathbb{R}^3$ maps some four non-coplanar points P, Q, R and S in $\mathbb{R}^3$ such that one of the two ordered triples of vectors $\{PQ, PR, PS\}$ and $\{f(P)f(Q), f(P)f(R), f(P)f(S)\}$ is right-handed, while the other is left-handed.

Show, then, that for any four non-coplanar points, P', Q', R' and S', one of the two ordered triples of vectors $\{P'Q', P'R', P'S'\}$ and $\{f(P')f(Q'), f(P')f(R'), f(P')f(S')\}$ is right-handed, while the other is left-handed.

Hint: Use the same approach as for Example 5.10. You will need as well to apply the preceding lemma.

Next is the 3D analogue of the 2D Proposition 5.5.

Proposition 5.11. *A translation or a rotation about an arbitrary axis is a rigid transformation of 3-space. A reflection about an arbitrary mirror is an orientation-reversing Euclidean transformation of 3-space.*

Proof. The proof that a 3D translation t is distance-preserving is exactly similar to the 2D version in the proof of Proposition 5.5. That t is orientation-preserving is even simpler, because the two ordered triples of vectors $\{PQ, PR, PS\}$ and $\{t(P)t(Q), t(P)t(R), t(P)t(S)\}$ are identical for any four points P, Q, R and S in $\mathbb{R}^3$.

We'll leave the reader to prove the claims for rotations and reflections. $\square$

The following is the 3D version of the 2D Proposition 5.6.

Proposition 5.12. *A rigid transformation of $\mathbb{R}^3$ keeping the origin fixed is a rotation about a radial axis, while an arbitrary rigid transformation of $\mathbb{R}^3$ is a composition of a rotation about a radial axis followed by a translation.*

An Euclidean transformation of $\mathbb{R}^3$ is a composition of a rotation about a radial axis followed by a translation, possibly followed again by a reflection.

Proof. The first statement of the proposition can be proved using linear algebra, but more interesting is to apply elementary arguments along the lines of Proposition 5.6. We ask the reader, who doesn't mind wallowing in a bit of geometry, to follow the approach suggested below. If you are not so inclined, it won't hurt to skip the proof altogether.

Suggested approach: Show first that a rigid transformation f of 3-space fixing the origin O is a rotation about a radial axis as follows:

If f is the identity, then it is trivially a rotation.

If f is not the identity, then suppose that P is a point such that $f(P) \neq P$. There are three possibilities:

(a) $f(f(P)) \neq P$.

Argue that the three points P, $f(P)$ and $f(f(P))$ cannot be collinear. Therefore, they belong to a unique plane p. Show that the line l through O perpendicular to p is the axis of f; further, the angle of rotation θ is the angle between the perpendiculars from P and $f(P)$ to l. See Figure 5.32(a).

(b) $f(f(P)) = P$ and the line l' joining P and $f(P)$ does not contain O.

Show in this case that the line l through O perpendicular to l' is the axis of f; furthermore, the angle of rotation is π. See Figure 5.32(b).

(c) $f(f(P)) = P$ and the line l' joining P and $f(P)$ does contain O.

In this case, let Q be a point not lying on l'.

If $f(Q) = Q$, then show that the line l through O and Q is the axis of f and the angle of rotation is π. See Figure 5.32(c).

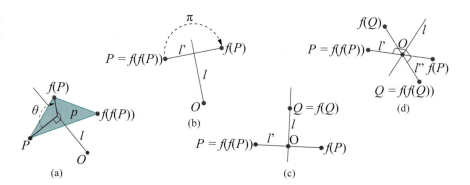

Figure 5.32: Finding the axis of a rigid transformation that fixes the origin.

If $f(Q) \neq Q$, then assume that $f(f(Q)) = Q$ and that the line l'' joining Q and $f(Q)$ contains O, for, if not, this case is equivalent to one of (a) or (b). Then show that the line l through O perpendicular to both l and l' is the axis and the angle of rotation is π. See Figure 5.32(d).

The rest of the proposition follows easily from the first part. □

Remark 5.14. The first part of Proposition 5.12, that a rigid transformation of $\mathbb{R}^3$ which keeps the origin fixed is a rotation about a radial axis, is often called Euler's Theorem. It is actually one of several theorems, proved by the great eighteenth century Swiss mathematician Leonhard Euler, bearing his name.

Finally, here's the 3D analogue of the 2D Proposition 5.7.

Proposition 5.13. *Affine, Euclidean and rigid transformations of 3-space are related by the following inclusions, which are each proper:*

$$\text{rigid transforms} \subset \text{Euclidean transforms} \subset \text{affine transforms}$$

Proof. We leave this to the reader. □

A proposition whose proof we kept putting off, because of its apparent difficulty, is now all of a sudden simple to prove:

Proposition 5.14. *The composition of two rotations about radial axes in 3-space is another such.*

Proof. Let f_1 and f_2 be rotations about radial axes in 3-space. By Proposition 5.11 they are rigid transformations and, moreover, both fix the origin. By Exercise 5.73 the composition $f_1 \circ f_2$ is rigid, and it obviously fixes the origin because both f_1 and f_2 do so. Proposition 5.12 then completes the proof. □

Exercise 5.76. Consider reflections through *points*. For example, the reflection through the origin corresponds to the transformation $[x \; y \; z]^T \mapsto$

$[-x \ -y \ -z]^T$. This transformation is clearly affine, in fact, linear. Is it Euclidean? Rigid? How about reflections through arbitrary points?

Sketch how the boy of Figure 4.7 of the last chapter would be transformed by reflection through the origin.

Proof of Proposition 5.10

This section is only for those with a good knowledge of linear algebra and may be safely skipped by others with no consequences to their learning of CG.

Lemma 5.3. *A special orthogonal transformation of $\mathbb{R}^3$, i.e., one whose matrix is orthogonal with determinant 1, is a rotation about a radial axis.*

Proof. It's easy to verify that a linear transformation f of $\mathbb{R}^3$ defined by an orthogonal matrix of determinant 1 preserves both distance and orientation. Therefore, it is rigid and the lemma follows from Proposition 5.12. □

The author learned the proof of the following lemma from T. K. Mukherjee [93].

Lemma 5.4. *For any non-singular real $n \times n$ matrix M, there exist special orthogonal matrices P and Q and a real diagonal matrix D with all non-zero entries such that*

$$M = PDQ$$

Proof. Consider the product MM^T. As it is symmetric, by a standard result of linear algebra there exists a real orthogonal matrix P such that

$$P^{-1}(MM^T)P = D'$$

where D' is a diagonal matrix.

Moreover, MM^T is positive definite, which implies that D' is as well. Therefore, each element of the diagonal of D' is positive. Let $D = \sqrt{D'}$. In particular, if $D' = [d_i']$, then $D = [d_i]$, where d_i is the positive square root of d_i'.

Accordingly,

$$P^{-1}MM^TP = D^2$$

It follows that

$$
\begin{aligned}
I & = D^{-1}\left(P^{-1}MM^TP\right)D^{-1} = \left(D^{-1}P^{-1}M\right)\left(M^TPD^{-1}\right) \\
& = \left(D^{-1}P^{-1}M\right)\left(D^{-1}P^{-1}M\right)^T
\end{aligned}
$$

as $(D^{-1})^T = D^{-1}$ and $(P^{-1})^T = P$.

Writing

$$Q = D^{-1}P^{-1}M$$

we have from the above that $QQ^T = I$, so Q is orthogonal

Now

$$M = PDQ$$

and we can assume that both P and Q are, in fact, special orthogonal, for, if either is not, then it can be multiplied by a diagonal matrix with determinant -1, viz.,

$$R = \begin{bmatrix} -1 & 0 & 0 & \ldots & 0 \\ 0 & 1 & 0 & \ldots & 0 \\ 0 & 0 & 1 & \ldots & 0 \\ & \ldots & \ldots & \ldots & \\ 0 & 0 & 0 & \ldots & 1 \end{bmatrix}$$

and, correspondingly, D multiplied by R^{-1} $(=R)$. □

The two lemmas combine to establish Proposition 5.10. □

5.4.8 Shear

Thinking back to the analogy made in Section 5.3.3, imagine again placing a lump of putty between your palms and then moving one palm parallel to the other. In proper 3D there are three choices to make: (1) how to initially align your palms in space with putty between them, (2) which direction to move, say, the upper palm, but keeping it parallel to the fixed lower one and, finally, (3) how far to move the upper palm. Accordingly, a 3D shear s is uniquely determined by three parameters:

1. A plane p called the *plane of shear*.

2. A directed line l called the *line of shear*, which is parallel to p.

3. An angle α called the *angle of shear*.

The action of s is equivalent to that of "infinitely many" 2D shears applied to parallel planes. Here's how (see Figure 5.33)(a)):

Given a point $P \in \mathbb{R}^3$, let q be the unique plane containing P which is both perpendicular to the plane p of shear s *and* parallel to the line l of shear s. Therefore, q intersects p in a directed line, denote it l', parallel to l. P, then, is transformed by s to the point P' *exactly* as it would by the particular 2D shear s', on the plane q, specified by the directed line l' and the angle α.

In other words, imagine 3D space "sliced" into infinitely many parallel planes, each perpendicular to p and parallel to l, and s as the "union of identical 2D shears" on each of these slices (Figure 5.33)(a) depicts two more slices parallel to q).

Example 5.17. Figure 5.33(b) depicts a 3D shear along the xz-plane whose line is the x-axis and angle α, shearing a cube into a parallelepiped. As the "slices" are all parallel to the xy-plane and the line of the shear in each

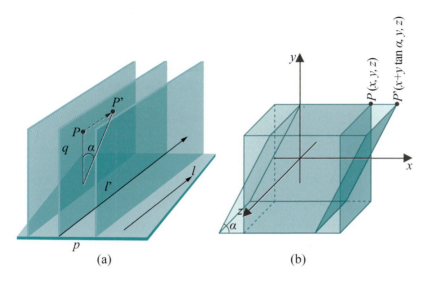

Figure 5.33: (a) A 2D slice of a 3D shear on the plane q and two more "copies" of q (b) A shear along the xz-plane whose line is the x-axis.

parallel to the x-axis, one can straightforwardly apply the 2D Equation (5.22) to write the equation of this shear as follows:

$$P' = \begin{bmatrix} 1 & \tan\alpha & 0 & 0 \\ 0 & 1 & 0 & 0 \\ 0 & 0 & 1 & 0 \\ 0 & 0 & 0 & 1 \end{bmatrix} P$$

Exercise 5.77. Verify that the 4×4 matrix of a 3D shear along the xz-plane, whose line is the z-axis and angle α, is:

$$\begin{bmatrix} 1 & 0 & 0 & 0 \\ 0 & 1 & 0 & 0 \\ 0 & \tan\alpha & 1 & 0 \\ 0 & 0 & 0 & 1 \end{bmatrix}$$

Exercise 5.78. What is the 4×4 transformation matrix of a 3D shear along the xy-plane, whose line is the x-axis and angle α?

Exercise 5.79. Prove that a shear along a radial plane is a non-singular linear transformation of 3-space, while an arbitrary 3D shear is an affine transformation. Conclude that any 3D shear is a composition of translations, scalings and rotations about radial axes.

When implementing a shear s in OpenGL, it's typically more efficient to compute the matrix M of s and then use a `glMultMatrix()` call to directly multiply the current modelview matrix by M, rather than expressing s as a composition of modeling transformations.

Exercise 5.80. (Programming) From Example 5.11 it follows that the 3D shear s along the xz-plane, whose line is the x-axis and angle $\alpha = 45°$, is the composition of a rotation of $31.72°$ about the z-axis, followed by scaling by factors 0.62, 1.62 and 1.0 along the x-, y- and z-axes, respectively, followed by a rotation of $-58.28°$ about the z-axis (up to floating point error).

Verify this by modifying `manipulateModelviewMatrix.cpp` to output the current modelview matrix after applying the preceding modeling transformations and separately computing the value of the shear matrix.

Experiment 5.3. Run `shear.cpp`. Press the left and right arrow keys to move the ball. The box begins to shear when struck by the ball. See Figure 5.34.

The shear matrix is explicitly computed in the code and multiplied from the right into the current modelview matrix via a `glMultMatrix()` call.

End

Figure 5.34: Screenshot of `shear.cpp`.

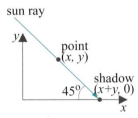

Figure 5.35: Shadow of a point cast by the sun at 45° in the sky.

Exercise 5.81. (Programming) Recall the program `ballAndTorus-Shadowed.cpp` from Experiment 4.34. Modify the shadow-drawing part of the program to cast shadows by the sun at $45°$ in the sky. Figure 5.35 indicates that the transformation to be implemented in the xy-plane is

$$(x, y) \mapsto (x + y, 0)$$

5.5 Summary, Notes and More Reading

In this chapter we opened up the graphics animation engine to find out what makes it tick. The short answer is 4×4 matrix multiplication and we learned why.

Each modeling transformation corresponds to an affine transformation of 3-space, which is represented using homogeneous coordinates by a 4×4 matrix. The composition of modeling transformations corresponds to multiplication of their matrices. Viewing transformations, being compositions of modeling transformations, each corresponds to a 4×4 matrix as well. We learned the particular matrix representations of basic geometric transformations such as translation, scaling, rotation and reflection.

We learned as well that the modeling transformations of OpenGL – translation, scaling and rotation – were chosen for good reasons by the designers of the API. Not only are these transformations affine, but they combine to generate all affine transformations.

We studied certain subclasses of affine transformations which are particularly useful. One was that of shape-preserving transformations, the Euclidean transformations, which in turn include rigid transformations that model the motion of rigid objects in space. Another special class of affine transformations which we studied was that of the shears.

We learned to access and manipulate OpenGL's modelview matrix stack as well, allowing us to program transformations directly into the stack if need be, rather than through calls to OpenGL's own modeling transformations.

Till Chapter 4 we were primarily interested in *using* OpenGL. Now we have an understanding of the *working* of this API as well. True, familiarity, say, with the functioning of an internal combustion engine does not necessarily make one a better driver; however, it certainly does help one better understand technical issues, more confidently deal with them and, generally, be a more informed consumer, which has its value.

One topic the knowledgeable reader might think missing from this chapter on transformation theory is a discussion of the so-called projection transformation in the graphics pipeline – which plays an important role in the shoot part of the shoot-and-print rendering paradigm from Chapter 2 – and how it is implemented by means of mathematical projective transformations. However, we thought it best to introduce the projection transformation as an application of projective spaces and their transformations later in the book in Chapter 18. The reason is that the choice of the particular projective transformations applied in the graphics pipeline is hard to motivate without some understanding of projective spaces, and we did not want pull it out of a hat at this stage.

For further reading about geometric transformations the reader is recommended to see Mortenson [91] and Yaglom's series [149, 150, 151]. Articles about transformations and their role in computer graphics, written in recreational style and yet very informative, can be found in the books by Blinn [16, 17] and Glassner [50, 51].

CHAPTER 6

Advanced Animation Techniques

The goal for this chapter is to learn techniques to cope with two issues that arise often in animation projects. The first is managing large worlds where the polygon count may painfully slow down the rendering pipeline. The programmer can help ease the logjam by pre-filtering polygons lying outside the camera's field of view. This process is called frustum culling and we describe how to do it by means of space partitioning in Section 6.1.

The next issue is that of animating the orientation of an object. In Chapter 4 we learned all about animating motion, coding balls and boxes that flew, fell, spun and revolved around one another. But how about animating orientation or pose? For example, an aircraft maneuvering in a dogfight or a camera tracking a scene. Changing orientation involves modeling transformations as well, particularly rotation, but first one must develop a method to *quantify* orientation, just as (x, y, z) quantify position. Only then comes the question of moving between two orientations.

In Section 6.2 we learn how to use Euler angles – which we first encountered in Section 4.6.3 – to quantify orientation in 3D. Animating between a pair of orientations given by their Euler angle tuples is possible but, as we shall see, potentially problematic. A more sophisticated approach is with the use of quaternions. This is the topic of Section 6.3, which begins with an introduction to the mathematics of quaternions, and then goes on to describe their applications to representing and animating orientation.

There is a fair amount of theory in this chapter but it has important practical applications and we back it all the way with code.

6.1 Frustum Culling by Space Partitioning

Frustum culling is *de rigueur* for game programmers or, for that matter, anyone creating scenes with large polygon counts. We'll motivate the proceedings by repeating Experiment 4.31.

Experiment 6.1. Run the program `spaceTravel.cpp` after increasing `ROWS` and `COLUMNS` both to 100 from 8 and 5, respectively. Figure 6.1 is a screenshot. The spacecraft now responds sluggishly to the arrow keys, at least on a typical desktop. You may have to increase even more the values of `ROWS` and `COLUMNS` if yours is exceptionally fast. **End**

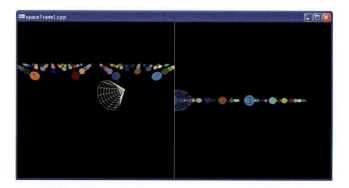

Figure 6.1: Screenshot of `spaceTravel.cpp` with a 100×100 array of asteroids.

Let's do a back-of-the-envelope calculation first. Assuming the viewable space of `spaceTravel.cpp` to be a box of sides 250, significantly larger, in fact, than the viewing frustum defined by the `glFrustum(-5.0, 5.0, -5.0, 5.0, 5.0, 250.0)` statement of the program, and noting that asteroids are 30 units apart in both the x and z directions, one deduces that at most 9*9=81 asteroids are viewable in either viewport at any given time. That's not a lot for OpenGL to draw. In fact, set ROWS to 9 and COLUMNS to 9 to find no perceptible slow-down! So what's going on? Why the slow-down in simply *creating* a larger asteroid field?

A moment's consideration of the rendering pipeline reveals the answer. At every redisplay, in other words, every arrow press, the `ROWS` × `COLUMNS` number of asteroids in `arrayAsteroids[ROWS][COLUMNS]` are *all* – in fact, their *collective polygons* are *all* – zapped first with the modelview matrix, then with the projection matrix, and *then* those that fall outside the viewing volume are clipped, and the rest projected to the viewing face and rendered. Specifically, if `ROWS` and `COLUMNS` are both 100, then the polygons of all 10,000 asteroids, several hundred thousand in total, enter the rendering pipeline before only those belonging to approximately 80 are drawn. That's more than 99% of the polygons, each incurring computational cost in the pipeline, ultimately not being drawn. Talk about waste!

It's not OpenGL's fault though. OpenGL finds out which polygons belong on-screen and off only *after* transforming and clipping, operations well into the pipeline. However, the programmer can help by pre-identifying as many objects as possible that do not intersect the viewing frustum, which means that they will end up being clipped, and not letting them into the pipeline in the first place. This is a process called *frustum culling*, which consists, then, of adding to the program routines to check if a polygon (or object) intersects the frustum and filtering through to the drawing routine only those which do or, equivalently, culling those which do not.

6.1.1 Space Partitioning

The most straightforward way to frustum cull is to test each object individually if it intersects the frustum. This may, in fact, give decent speed-up if the objects are simple enough that the combined cost of testing them all is still cheap.

However, frustum culling is typically more efficient if space is first partitioned in a hierarchical manner into cells which each contain only a few objects. Cost-effective partitioning must be driven by the distribution of the objects – subdividing into smaller ones only cells containing many objects. Once space is partitioned, frustum culling can be accomplished by hierarchically checking cells to determine if they intersect the frustum and passing to the drawing routine objects belonging only to those which do. This approach is based on a few premises:

(a) That partitioning space and determining the distribution of the objects in individual cells is primarily a one-time pre-processing cost, which is true if most objects in the scene are static. The few moving objects, in this case, can be passed mandatorily to the drawing routine.

(b) That the cells are of a shape easy to check for intersection with the frustum.

(c) That the hierarchical nature of the partition leads to efficiency because, if a cell is found to not intersect the frustum, then its sub-cells and the objects which they contain can all be eliminated from further consideration, a process called *pruning*.

(d) That the partitioning process is efficient in that, if a cell contains only a few objects, then it is not subdivided, so that the final partition reflects the distribution of the objects in space.

There's more than one way to hierarchically partition space, but intuitively simplest is the *octree* for 3-space and its analogue, the *quadtree*, for 2-space. We'll explain quadtree-based space partitioning using the scenario of `spaceTravel.cpp` as a running example because we have, in fact, an implementation in the program `spaceTravelFrustumCulled.cpp`.

6.1.2 Quadtrees

Note, first, that `spaceTravel.cpp` is essentially a 2D problem as the asteroids and frustum can all be projected onto the xz-plane – asteroids to discs and the frustum to a trapezoid – and intersection testing done on that plane. Therefore, a quadtree is appropriate for `spaceTravel.cpp` even though, nominally, the scene is 3D.

A quadtree partitions 2-space into axis-aligned squares, the collection having the hierarchy of a tree. In frustum culling applications the root node corresponds to a square large enough to contain all objects that might potentially be culled.

Terminology: We'll use the terms "node" and "square" interchangeably in the context of quadtrees.

Figure 6.2(a) illustrates a hypothetical projected scenario of `space-Travel.cpp` (an irregular distribution of asteroids is obtained by setting `FILL_PROBABILITY` to a value less than 100). The craft itself is ignored because it moves; therefore, it's always passed into the pipeline.

To begin with, the root square of the quadtree is chosen big enough to bound the entire asteroid field – see the setting of the `initialSize` variable of the `setup()` routine of `spaceTravelFrustumCulled.cpp`. (Note that we'll be running `spaceTravelFrustumCulled.cpp` soon enough, but, for now, let's see how the code is developed from incorporating frustum culling into `spaceTravel.cpp`.)

Subsequently, at each level, each square may be subdivided into four equal sub-squares (quadrants). The criterion to subdivide is generally determined by the programmer. In the particular case of `spaceTravelFrustumCulled-.cpp` we subdivide a square if it intersects more than one asteroid. If a square is subdivided, then its four quadrants become its children in the tree hierarchy and are denoted SW, NW, NE and SE according to their location in the parent square. Given our condition for when to subdivide, evidently leaf squares – which are not further subdivided – intersect either one or no asteroids.

The squares of the quadtree corresponding to the arrangement of asteroids in Figure 6.2(a) is shown in Figure 6.2(b) and the underlying tree structure in Figure 6.2(c).

Once the quadtree is built, culling is straightforward: check, starting at the root, for squares that intersect the frustum; if a non-leaf square intersects the frustum, then recursively process its children; if a leaf square intersects the frustum, then pass its asteroid (if any) to the drawing routine.

Let's verify the premises (a)-(d) for space partitioning, mentioned earlier, in the case of the `spaceTravelFrustumCulled.cpp` quadtree:

(a) The asteroids are all static while the spacecraft is the only object which is not, so a one-time quadtree is built for the asteroids, while the craft itself is always passed to the drawing routine.

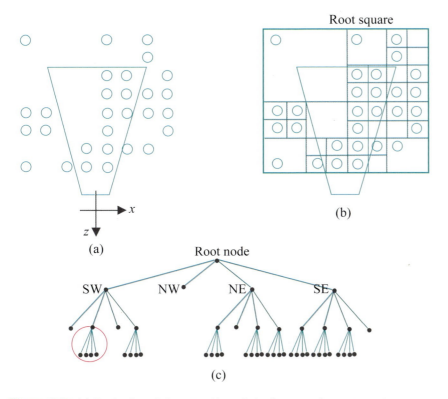

Figure 6.2: (a) Projection of the asteroids and the frustum of `spaceTravel.cpp` onto the xz-plane. (b) Corresponding quadtree squares (the root square is bold) (c) The tree structure with children at each node drawn SW, NW, NE, SE from left to right; the nodes in the red circle are *some* of those pruned.

In the second viewport the camera moves, which, as we know, is implemented by actually transforming the scene. However, in order to not have to update the quadtree structure, it's preferable to imagine the viewing frustum itself moving, attached to the front of the spacecraft, as in Figure 6.3.

(b) The cells are each a square, a shape easy to test for intersection with the trapezoidal projection of a frustum.

(c) Several of the nodes, even in the simple example of Figure 6.2, are pruned, e.g., the ones inside the red circle (and others).

(d) It can be seen from Figure 6.2 that the spatial distribution of the quadtree squares indeed tracks that of the asteroids.

Exercise 6.1. Indicate *all* the nodes of the tree of Figure 6.2 which are pruned by the quadtree-based frustum culling.

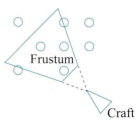

Figure 6.3: Spacecraft carrying a viewing frustum "attached" to its front.

6.1.3   Implementation

Experiment 6.2. Run `spaceTravelFrustumCulled.cpp`, which enhances `spaceTravel.cpp` with optional quadtree-based frustum culling. Pressing space toggles between frustum culling enabled and disabled. As before, the arrow keys maneuver the craft.

The current size of the asteroid field is 100×100. Dramatic isn't it, the speed-up from frustum culling?!

Important: Make sure to place the file `intersectionDetectionRoutines.cpp` in the same directory as `spaceTravelFrustumCulled.cpp`.

Note: When the number of asteroids is large, the display may take a while to come up because of pre-processing to build the quadtree structure.

End

We have already described the development of `spaceTravelFrustum-Culled.cpp`, which follows pretty much word for word the quadtree-based strategy described at the start of the section. Here are specifics.

The quadtree `asteroidsQuadtree` is an object of the `Quadtree` class containing nodes belonging to the `QuadtreeNode` class. The member function `numberAsteroidsIntersected()` of the class `QuadtreeNode` helps decide for each quadtree square if it is to be subdivided, while the member list `asteroidList` stores for each leaf square the asteroid (if any) intersecting it.

In addition to `checkSpheresIntersection()` from the original `space-Travel.cpp`, used in `asteroidCraftCollisiond()` to detect (approximately) intersection between the spacecraft and an asteroid, routines from the program `intersectionDetectionRoutines.cpp` are invoked for other intersection tests. In particular, `checkQuadrilateralsIntersection()` determines if the frustum in either viewport intersects a quadtree square, while `checkDiscRectangleIntersection()` if an asteroid intersects a quadtree square.

In Figure 6.2, asteroids lie either entirely inside or outside a square. This need not be the case in general, of course, and in our code an asteroid straddling the boundary of a quadtree square is associated with it.

With large numbers of asteroids, the speed-up gained to be had through frustum-culling is clearly enormous. Even so, our implementation `spaceTravelFrustumCulled.cpp` is minimal and there are further optimizations to be made. We ask the reader to explore a couple next.

Exercise 6.2. (Programming) A large quadtree costs both in RAM space and pre-processing time. Try the following two options in `space-TravelFrustumCulled.cpp` to control its size:

(a) The size of the quadtree tends to grow exponentially with its height. Accordingly, set a *cut-off depth* beyond which nodes cannot be partitioned.

(b) The criterion for subdividing a square, currently if it intersects more than one asteroid, can be made stricter by setting a larger threshold for the number to be intersected, again reducing the size of the quadtree.

6.1.4 More about Space Partitioning

Octrees are a straightforward generalization of quadtrees to 3-space – space is partitioned into a tree-like hierarchy of axis-aligned cubes. Each cube in an octree can be partitioned into 8 child octants (see Figure 6.4).

Quadtrees and octrees are not the only ways to partition space. There are more sophisticated data structures, such as kd-trees, range trees and BSP (Binary Space Partitioning) trees, which can be applied in two and higher dimensions. We'll be discussing BSP trees ourselves in Chapter 19 in the context of hidden surface removal.

Moreover, applications of space partitioning are not limited to frustum culling either. Another important one is collision detection. The principle is that two objects can intersect only if they belong to the same or adjacent cells; accordingly, one can pre-process and pass only "nearby" pairs to the, typically, costly intersection-checking routines.

Dynamic scenes with multiple mobile objects are a challenge for any space partitioning application. Options include predictively locating moving objects in cells if there is prior knowledge of their trajectories, followed, possibly, by a repartitioning of space or a redistribution of objects to cells.

Bottom line: There is overhead both in code and pre-processing in setting up a space-partitioning structure, but if the application is appropriate, e.g., frustum culling a fairly static scene, the bang for the buck is enormous.

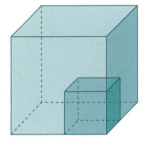

Figure 6.4: An octree cube and one of its 8 octants.

Exercise 6.3. (Programming) Currently, `spaceTravelFrustumCulled-.cpp` checks for intersection between the spacecraft and *every* asteroid. Use the quadtree to improve on this. In particular, check only for collision between the craft and each asteroid associated with a leaf square that the craft intersects.

Exercise 6.4. (Programming) Design a busy scene, maybe part of a game, and draw it using frustum culling.

6.2 Animating Orientation Using Euler Angles

The lead-in to this section is the discussion of viewing transformations in Section 4.6, particularly orientation and Euler angles in 4.6.3, so you might want to review this earlier material.

6.2.1 Euler Angles and the Orientation of a Rigid Body

Consider this for a second: it's no different to locate and orient a camera in 3-space than it is to locate and orient an aircraft, spacecraft or any other freely-movable rigid object. The `gluLookAt()` command happens to provide intuitive syntax to use, in particular, with a camera, namely, translate to (*eyex*, *eyey*, *eyez*), point at (*centerx*, *centery*, *centerz*) and turn about the line of sight according to the (*upx*, *upy*, *upz*) value. As you can see from Figure 6.5, the captain of a spacecraft could use similar syntax to steer her ship.

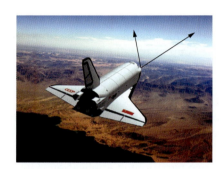

Figure 6.5: Spacecraft and video camera.

Accordingly, replace the camera with an arbitrary rigid object B. Assume that the location of B is fixed or, more precisely, that the location of a point P belonging B is fixed, say, at the origin (see Figure 6.6(a), where the point P at the end of the long leg of an L-shaped object B is fixed at the origin). Now we are exactly at the point in Section 4.6.3 that we considered a `gluLookAt()` command with zero translational component, i.e., (*eyex*, *eyey*, *eyez*) = (0, 0, 0). From discussions in that section, the orientation of B is specified by *Euler angles* α, β and γ such that it can be obtained from a fixed reference orientation – in the case of the OpenGL camera this is its default pose – by applying the following rotation sequence:

```
glRotatef(α, 1.0, 0.0, 0.0);
glRotatef(β, 0.0, 1.0, 0.0);
glRotatef(γ, 0.0, 0.0, 1.0);
```

Of the two orientations shown of B in Figure 6.6(a), the reference one is bold.

Remark 6.1. The yaw, pitch and roll of an aircraft to which pilots refer are nothing but Euler angles, the difference being that the axis system is carried by the aircraft itself (e.g, the roll axis is the line through the middle of the craft from tail to nose). See Figure 6.6(b).

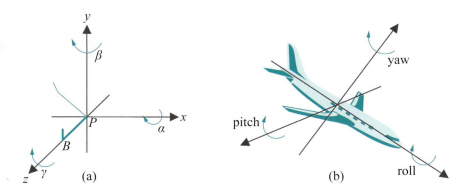

Figure 6.6: (a) Orienting an object in space with respect to fixed axes – the fixed reference orientation of the L is shown in bold (b) Orientation of an aircraft with respect to local axes "carried" by it.

6.2.2 Animating Orientation

It seems, then, that animating orientation is a matter simply of changing Euler angles. This is true.

E̶xperiment 6.3. Run `eulerAngles.cpp`, which shows an L, similar to the one in Figure 6.6(a), whose orientation can be interactively changed.

The original orientation of the L has its long leg lying along the z-axis and its short leg pointing up parallel to the y-axis. Pressing 'x/X', 'y/Y' and 'z/Z' changes the L's Euler angles and delete resets. The Euler angle values are displayed on-screen. Figure 6.7 is a screenshot of the initial configuration.

E̶nd

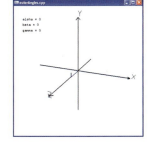

Figure 6.7: Screenshot of `eulerAngles.cpp`.

R̶emark 6.2. If the commands

```
glRotatef(Xangle, 1.0, 0.0, 0.0);
glRotatef(Yangle, 0.0, 1.0, 0.0);
glRotatef(Zangle, 0.0, 0.0, 1.0);
```

in `eulerAngles.cpp` to change the Euler angles look familiar, well, we've been using similar ones since the second chapter to rotate scenes.

It all seems easy enough so far. However, things can get rather strange with Euler angles. Run `eulerAngles.cpp`, or use paper and pencil, to determine the two orientations of the L specified by the following distinct tuples of Euler angles (all angles are in degrees in this section):

(a) $\alpha = 0$, $\beta = 90$, $\gamma = 0$

(b) $\alpha = -90$, $\beta = 90$, $\gamma = 90$

The two orientations are identical, both equal to the destination orientation of Figure 6.8, with the L's long leg along the x-axis and short leg pointing up parallel to the y-axis. Imagine now that the L is a spacecraft

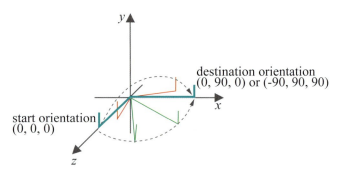

Figure 6.8: The bold blue start orientation is given by the Euler angle tuple $(0, 0, 0)$ and the bold blue destination one by either $(0, 90, 0)$ or $(-90, 90, 90)$. Intermediate orientations (green) in the linear interpolation between $(0, 0, 0)$ and $(0, 90, 0)$ all lie on the xz-plane, while those (red) between $(0, 0, 0)$ and $(-90, 90, 90)$ arc above it.

you want to take from its start orientation specified by the Euler angle tuple $(0, 0, 0)$ to the destination one specified by either of the tuples (a) or (b).

What comes to mind naturally is a linear interpolation between the start and destination orientations, exactly as if one had to translate the spacecraft from a start to a destination *location*, except that, instead of intermediate positions, one has now intermediate orientations. However, the ambiguity in representing the destination leads to a surprise we see next.

6.2.3 Problems with Euler Angles: Gimbal Lock and Ambiguity

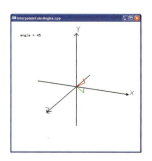

Figure 6.9: Screenshot of interpolateEuler-Angles.cpp.

Experiment 6.4. Run interpolateEulerAngles.cpp, which is based on eulerAngles.cpp. It simultaneously interpolates between the tuples $(0, 0, 0)$ and $(0, 90, 0)$ and between $(0, 0, 0)$ and $(-90, 90, 90)$. Press the left and right arrow keys to step through the interpolations (delete resets). For the first interpolation (green L) the successive tuples are $(0, angle, 0)$ and for the second (red L) they are $(-angle, angle, angle)$, *angle* changing by 5 at each step in both.

The paths are different! The green L seems to follow the intuitively straighter path by keeping its long leg always on the xz-plane as it rotates about the y-axis, while the red L arcs above the xz-plane, as diagrammed in Figure 6.8. Figure 6.9 is a screenshot of interpolateEulerAngles.cpp part way through the interpolation. **End**

Two different paths arise in the program, of course, because of the non-unique representation of the destination orientation by Euler angles. Are $(0, 90, 0)$ and $(-90, 90, 90)$ the only tuples of Euler angles that represent this particular destination? Emphatically no, as we see next!

Use eulerAngles.cpp to see that any tuple of the form $(-A, 90, A)$ does as well: the $-A$ rotation about the x-axis seems always to cancel the A rotation about the z-axis to make an equivalence to $(0, 90, 0)$! It's

the A rotation about the z-axis to make an equivalence to $(0, 90, 0)$! It's not hard to understand why. The first rotation that is applied, viz., `glRotatef(A, 0.0, 0.0, 1.0)`, twists the L about its long leg lying along the z-axis, the second `glRotatef(90, 0.0, 1.0, 0.0)` rotates the long leg to line it up with x-axis, so that the final `glRotatef(−A, 1.0, 0.0, 0.0)` again twists the L about its long leg, but equally and oppositely to the first rotation.

We've run into the problem of *gimbal lock* which afflicts the Euler angle representation of orientation. Let's get a better understanding of this phenomenon.

Experiment 6.5. Run `eulerAngles.cpp` again.

Press 'x' and 'X' a few times each – the L turns longitudinally. Reset by pressing delete. Press 'y' and 'Y' a few times each – the L turns latitudinally. Reset. Press 'z' and 'Z' a few times each – the L twists. There appear to be three physical *degrees of freedom* of the L derived from rotation about the three coordinate axes: longitudinal, latitudinal and "twisting".

Now, from the initial configuration of `eulerAngles.cpp` press 'y' till $\beta = 90$. Next, press 'z' or 'Z' – the L twists. Then press 'x' or 'X' – the L still twists! **End**

Even with β fixed in `eulerAngles.cpp` one would expect two degrees of freedom to remain, viz., twisting *and* longitudinal. However, because of the particular value of β, namely, 90, which takes the L from the z-axis to the x-axis, we seem to have lost the longitudinal one.

Well, physical space hasn't changed and all three degrees of freedom are obviously still there for the taking. It's simply because of the particular value of β that the other two Euler angles α and γ both "act on the same degree of freedom", disallowing access to the remaining one.

This *apparent* loss of a degree of freedom is precisely gimbal lock, and it's the reason as well for the multiple representations of the destination orientation of the L in `interpolateEulerAngles.cpp`: as both the rotations about the x- and z-axes act on the same degree of freedom when $\beta = 90$, one can be used to cancel the other.

Exercise 6.5. Show that gimbal lock also arises when $\beta = -90$.

Even though infinitely many different representations of an orientation occur only at the two gimbal lock values of $\beta = \pm 90$, there's actually instability near these values as well. In fact, two paths beginning and ending at nearby orientations which happen to be close to gimbal lock values can differ widely.

Exercise 6.6. (Programming) Verify the preceding remark by modifying `interpolateEulerAngles.cpp` to simultaneously interpolate between the Euler angle tuples $(0, 0, 0)$ and $(5, 95, 0)$ and the tuples $(0, 0, 0)$ and $(-95, 85, 95)$.

Moreover, there's another pesky problem with Euler angles: that of *non-unique*, in fact, *dual* representation of *every* orientation. To begin with, mentally visualize, or run `eulerAngles.cpp`, to see that the Euler angle tuple $(180, 180, 180)$ is equivalent to the Euler angle tuple $(0, 0, 0)$, both representing the initial orientation of the L.

This generalizes:

Proposition 6.1. *The Euler angle tuples* (α, β, γ) *and* $(\alpha + 180, -\beta + 180, \gamma + 180)$ *are equivalent in that they both represent the same orientation.* (Note the minus sign in front of the β in the second tuple.)

Proof. We'll not prove this formally, but ask the reader to "visually" verify it in the next exercise. $\square$

Exercise 6.7. (Programming) Verify the claim of the proposition by comparing the orientation of the L of `eulerAngles.cpp`, as specified by the Euler angle tuples (α, β, γ) and $(\alpha + 180, -\beta + 180, \gamma + 180)$, for a few different values of α, β and γ.

Note: If you happen to go outside the range of $-180°$ to $180°$ with α or β or γ, then you can get back in again by adding or subtracting $360°$.

Of course, an Euler angle tuple (α, β, γ) is always equivalent to the Euler angle tuple $(\alpha \pm 360, \beta \pm 360, \gamma \pm 360)$ for any choice of the pluses and minuses, simply because angular arithmetic is modulo 360. However, the equivalence of the Euler angle tuples (α, β, γ) and $(\alpha + 180, -\beta + 180, \gamma + 180)$ is *not* a consequence of angular arithmetic, but a true geometric *duality* intrinsic to Euler angles.

Exercise 6.8. (Programming) Modify `interpolateEulerAngles.cpp` to simultaneously interpolate between the initial orientation $(0, 90, 0)$ and the destination orientation represented dually by the distinct Euler angle tuples $(0, 0, 0)$ and $(180, 180, 180)$.

Again, the two paths are different.

All this is not the best news if one is in the business of animating the orientation of a camera or rigid body, as one then wants to be able to *unambiguously* choose representations of the start and destination orientations in order to interpolate between the two. It turns out that there is, in fact, a more efficient representation by means of mathematical thingies called quaternions, in which case there's never gimbal lock and, though, there's still an issue with ambiguous representation, it's one that can be elegantly resolved prior to interpolation. Quaternions and their application to orientation are the topic of the next section.

6.3 Quaternions

We are going to continue our discussion of animating orientation from the previous section. However, we'll now invoke quaternions for this particular application.

The air of mathematical mystique surrounding them has caused a fair bit of (mis)apprehension of quaternions among game programmers, which is particularly unfortunate as it is interactive applications like games that stand to benefit most from their use. Hopefully, this section will convince at least the reader that not only are quaternions fairly benign – about as hard as complex numbers as a matter of fact – they are not hard to apply either.

6.3.1 Quaternion Math 101

Quaternions were invented by the Irish mathematician William Hamilton in the mid-1900s as part of his investigation into 3D mechanics. Think of them as complex numbers on steroids. Whereas complex numbers extend the reals with one imaginary i, a square root of -1, quaternions add three such numbers i, j and k, all square roots of -1. Formally:

Definition 6.1. A *quaternion* q is a number of the form

$$q = w + xi + yj + zk$$

where w, x, y and z are real numbers

It's often written as $q = w + \mathbf{v}$, where w is the *real* or *scalar* part, while $\mathbf{v} = xi + yj + zk$ is the *vector* or *pure quaternion* part. A quaternion of the form $q = xi + yj + zk$, with a zero scalar part, is also called a pure quaternion.

Note: In this section vector parts will be denoted in bold to distinguish them from the scalar (except for i, j and k themselves, as there's little risk of ambiguity with the three).

The set of quaternions is commonly denoted $\mathbb{H}$ in honor of its inventor.

The real numbers are a subset of the quaternions, the real w being identified with the quaternion w $(= w + 0i + 0j + 0k)$ with a zero vector part.

Example 6.1. Some quaternions:

$$2 - 3.4i + 4.8j + 2k, \quad -i + \sqrt{2}k, \quad 10.9, \quad -6.3j, \quad 0$$

Quaternions are added component-wise, just as complex numbers:

Definition 6.2. The *sum* of two quaternions $q_1 = w_1 + x_1 i + y_1 j + z_1 k$ and $q_2 = w_2 + x_2 i + y_2 j + z_2 k$ is the quaternion

$$q_1 + q_2 = (w_1 + w_2) + (x_1 + x_2)i + (y_1 + y_2)j + (z_1 + z_2)k$$

Exercise 6.9. Add the quaternions

(a) $2 - 3.4i + 4.8j + 2k$ and $-6.3j$

(b) $-i + \frac{1}{\sqrt{2}}k$ and $\frac{1}{\sqrt{2}}k$

The "square root of -1" property kicks in when multiplying quaternions. Here are the rules to keep in mind:

$$i^2 = j^2 = k^2 = -1$$

$$ij = k \qquad ji = -k$$

$$jk = i \qquad kj = -i$$

$$ki = j \qquad ik = -j$$

They're not hard to remember. The square of i, j and k each is -1. For the rest, simply keep in mind the cyclic order $i \to j \to k \to i$. If two successive elements in this order are multiplied, the result is the next element; if two successive elements are multiplied in reverse order, the result is the negative of the next element. This second rule is a replica of that of taking the cross-product of two different ones from the three unit vectors **i**, **j** and **k** (see Remark 5.8).

When multiplying two quaternions, it's a matter of applying distributivity and the preceding rules. Here's an example:

Example 6.2.

$$
\begin{aligned}
(2 - 3i + 2j)(3 + i - k) &= 2(3 + i - k) - 3i(3 + i - k) + 2j(3 + i - k) \\
&= 2*3 + 2*i + 2*-k - 3i*3 - 3i*i \\
&\quad -3i*-k + 2j*3 + 2j*i + 2j*-k \\
&= 6 + 2i - 2k - 9i + 3 - 3j + 6j - 2k - 2i \\
&= 9 - 9i + 3j - 4k
\end{aligned}
$$

Note: We use the symbol $*$ only to make multiplication clear.

Exercise 6.10. Multiply

$$(4 + 2i + 2j - k)(1 - k)$$

Exercise 6.11. Prove the following:

(a) The addition of quaternions is commutative and associative. In particular,

$$q_1 + q_2 = q_2 + q_1 \quad \text{and} \quad (q_1 + q_2) + q_3 = q_1 + (q_2 + q_3)$$

for any three quaternions q_1, q_2 and q_3.

(b) The formula for the product of two quaternions $q_1 = w_1 + x_1 i + y_1 j + z_1 k$ and $q_2 = w_2 + x_2 i + y_2 j + z_2 k$ is

$$q_1 q_2 = (w_1 w_2 - x_1 x_2 - y_1 y_2 - z_1 z_2) + (w_1 x_2 + x_1 w_2 + y_1 z_2 - z_1 y_2) i$$
$$(w_1 y_2 + y_1 w_2 - x_1 z_2 + z_1 x_2) j + (w_1 z_2 + z_1 w_2 + x_1 y_2 - y_1 x_2) k$$
$$(6.1)$$

(c) The multiplication of quaternions is associative and, moreover, distributes both ways over addition. In particular,

$$(q_1 q_2) q_3 = q_1 (q_2 q_3), \quad q_1 (q_2 + q_3) = q_1 q_2 + q_1 q_3, \quad (q_2 + q_3) q_1 = q_2 q_1 + q_3 q_1$$

for any three quaternions q_1, q_2 and q_3. *Alert*: Not hard, but tedious!

(d) The multiplication of quaternions is *not* commutative. In particular, it need not be true that $q_1 q_2 = q_2 q_1$ for two quaternions q_1 and q_2. Give an example.

(e) The additive identity is 0 and the multiplicative identity 1. In particular,
$$q + 0 = 0 + q = q = q1 = 1q$$
for any quaternion q.

There is a useful shorter expression for the product of two quaternions in terms of vector operations:

Exercise 6.12. Prove that if $q_1 = w_1 + \mathbf{v_1}$ and $q_2 = w_2 + \mathbf{v_2}$, then

$$q_1 q_2 = w_1 w_2 - \mathbf{v_1} \cdot \mathbf{v_2} + w_1 \mathbf{v_2} + w_2 \mathbf{v_1} + \mathbf{v_1} \times \mathbf{v_2} \qquad (6.2)$$

where $\cdot$ and $\times$ represent the vector dot and cross-product, respectively.

Note: Here we treat the pure quaternion (vector) part $\mathbf{v} = xi + yj + zk$ of the quaternion $q + \mathbf{v}$ as the geometric vector $x\mathbf{i} + y\mathbf{j} + z\mathbf{k}$. It'll be clear from the context if we mean a pure quaternion or a geometric vector.

Similar to complex numbers, quaternions each have a magnitude and a conjugate.

Definition 6.3. The *magnitude* of the quaternion $q = w + xi + yj + zk$, denoted $|q|$, is the non-negative value of the square root

$$\sqrt{w^2 + x^2 + y^2 + z^2}$$

Therefore, if we write $q = w + \mathbf{v}$, then its magnitude $|q| = \sqrt{w^2 + |\mathbf{v}|^2}$, where $|\mathbf{v}|$ denotes as usual the magnitude of the vector $\mathbf{v}$. A *unit quaternion* is one with magnitude 1.

Definition 6.4. The *conjugate* of the quaternion $q = w + xi + yj + zk$ is the quaternion

$$\bar{q} = w - xi - yj - zk$$

In other words, if $q = w + \mathbf{v}$ then its conjugate $\bar{q} = w - \mathbf{v}$.

And, as with complex numbers, a quaternion and its conjugate and magnitude are related:

Exercise 6.13. Prove that $q\bar{q} = \bar{q}q = |q|^2$.

Definition 6.5. The *inverse* of a quaternion q, if it exists, is the quaternion q^{-1} such that

$$qq^{-1} = q^{-1}q = 1$$

Exercise 6.14. Use Exercise 6.13 to prove that a quaternion q has an inverse if and only if it is non-zero, in which case

$$q^{-1} = \frac{\bar{q}}{|q|^2}$$

Therefore, $q^{-1} = \bar{q}$ for a unit quaternion q.

Example 6.3. Determine the inverse of the quaternion $q = 1 + i + j + k$.

Answer:

$$q^{-1} = \frac{\bar{q}}{|q|^2} = \frac{1 - i - j - k}{1^2 + 1^2 + 1^2 + 1^2} = \frac{1}{4}(1 - i - j - k)$$

Exercise 6.15. Determine the inverses of the quaternions

(a) j

(b) $\sqrt{3}i + \sqrt{2}k$

(c) $-1 + i + 2j - k$

Exercise 6.16. Prove the following if q_1 and q_2 are quaternions and c a scalar:

(a) $\overline{(q_1 q_2)} = \bar{q_2}\, \bar{q_1}$

(b) $(cq_1)^{-1} = c^{-1}q_1^{-1}$, provided both c and q_1 are non-zero.

(c) $(q_1 q_2)^{-1} = q_2^{-1}q_1^{-1}$, provided both q_1 and q_2 are non-zero.

(d) $|q_1 q_2| = |q_1||q_2|$

6.3.2 Quaternions and Orientation

So what do quaternions have to do with orienting a rigid body? The answer comes by way of the rotation transformation. Recall from the last section that the orientation of a rigid object B is specified by Euler angles α, β and γ, so that the specified orientation can be obtained from a fixed reference orientation by applying the following sequence of rotations:

```
glRotatef(α, 1.0, 0.0, 0.0);
glRotatef(β, 0.0, 1.0, 0.0);
glRotatef(γ, 0.0, 0.0, 1.0);
```

Now, we know from Proposition 5.14 that the composition of rotations about radial axes is another such. Therefore, the three above can be combined into a *single* rotation

```
glRotatef(θ, x, y, z)
```

for some angle θ and some values of x, y and z.

It follows that, instead of Euler angles, one can represent an orientation in 3D by means of a single 3D rotation about a radial axis. Moreover, it turns out that quaternions, as we shall see, each represents a 3D rotation about a radial axis. The conclusion, then, is that quaternions each represents a 3D orientation. The next proposition says how a quaternion gives a 3D rotation.

Note: In what follows we'll often identify the quaternions i, j and k with their vector counterparts $\mathbf{i}$, $\mathbf{j}$ and $\mathbf{k}$, respectively. It should be clear from the context whether we mean a quaternion or vector. Accordingly, we'll not distinguish for the rest of this section $\mathbf{i}$, $\mathbf{j}$ and $\mathbf{k}$ with boldface as earlier.

Proposition 6.2. *Suppose the axis of a 3D rotation is specified by the directed line l through the origin O toward a point $P = [a\ b\ c]^T$. Assume that $|P| = 1$. Denote the unit vector $OP = a\mathbf{i} + b\mathbf{j} + c\mathbf{k}$ by $\mathbf{u}$. (See Figure 6.10.)*

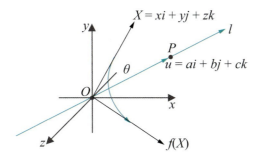

Figure 6.10: The vector $f(\mathbf{X})$ is obtained by rotating $\mathbf{X}$ about the line l.

If $\mathbf{X} = xi + yj + zk$ *is an arbitrary vector in* $\mathbb{R}^3$ *then the image, call it* $f(\mathbf{X})$, *of* $\mathbf{X}$ *by a rotation of angle* θ *about* l, *is given by*

$$f(\mathbf{X}) = q\,\mathbf{X}\,q^{-1} \qquad (6.3)$$

where q *is the unit quaternion*

$$q = \cos\frac{\theta}{2} + \mathbf{u}\sin\frac{\theta}{2} \qquad (6.4)$$

Let's put the proposition into context first. It gives one more way to determine the image $f(\mathbf{X})$ of a vector X by rotation of an angle θ about a radial axis l. The first, which we derived in Section 5.4.3, expressed $f(\mathbf{X})$ as the matrix product $R_{a,b,c}(\theta)\,\mathbf{X}$, where $R_{a,b,c}(\theta)$ was, in fact, given in a couple of different ways by the equations (5.35) and (5.45). Now, instead of a matrix, we manufacture a unit quaternion q such that the rotated vector $f(\mathbf{X})$ is $q\,\mathbf{X}\,q^{-1}$ (this operation of pre-multiplying by an element and then post-multiplying by its inverse is called an *inner automorphism* by that element).

The proof itself is a straight slog.

Proof. As q is a unit quaternion, its inverse is

$$q^{-1} = \bar{q} = \cos\frac{\theta}{2} - \mathbf{u}\sin\frac{\theta}{2}$$

Repeatedly applying the multiplication formula (6.2) we get the following equations:

$$
\begin{aligned}
q\,&\mathbf{X}\,q^{-1} \\
&= \left(\cos\frac{\theta}{2} + \mathbf{u}\sin\frac{\theta}{2}\right)\mathbf{X}\left(\cos\frac{\theta}{2} - \mathbf{u}\sin\frac{\theta}{2}\right) \\
&= \left(-(\mathbf{u}\cdot\mathbf{X})\sin\frac{\theta}{2} + \mathbf{X}\cos\frac{\theta}{2} + (\mathbf{u}\times\mathbf{X})\sin\frac{\theta}{2}\right)\left(\cos\frac{\theta}{2} - \mathbf{u}\sin\frac{\theta}{2}\right) \\
&= -(\mathbf{u}\cdot\mathbf{X})\sin\frac{\theta}{2}\cos\frac{\theta}{2} + (\mathbf{u}\cdot\mathbf{X})\sin\frac{\theta}{2}\cos\frac{\theta}{2} + (\mathbf{u}\times\mathbf{X})\cdot\mathbf{u}\sin^2\frac{\theta}{2} \\
&\quad + (\mathbf{u}\cdot\mathbf{X})\mathbf{u}\sin^2\frac{\theta}{2} + \mathbf{X}\cos^2\frac{\theta}{2} + (\mathbf{u}\times\mathbf{X})\sin\frac{\theta}{2}\cos\frac{\theta}{2} \\
&\quad + (\mathbf{u}\times\mathbf{X})\sin\frac{\theta}{2}\cos\frac{\theta}{2} - ((\mathbf{u}\times\mathbf{X})\times\mathbf{u})\sin^2\frac{\theta}{2}
\end{aligned}
$$

Of the eight terms in the final expression, the first two cancel, while the third is 0 because $\mathbf{u}\times\mathbf{X}$ is perpendicular to $\mathbf{u}$. Let $\mathbf{X} = \mathbf{X_1} + \mathbf{X_2}$, where $\mathbf{X_1}$ and $\mathbf{X_2}$ are the components of $\mathbf{X}$ parallel and perpendicular, respectively, to $\mathbf{u}$. Use the facts that $\mathbf{X_1} = (\mathbf{u}\cdot\mathbf{X})\mathbf{u}$ and $\mathbf{X_2} = \mathbf{X} - \mathbf{X_1} = (\mathbf{u}\times\mathbf{X})\times\mathbf{u}$, which we know from Exercises 4.39 and 5.59, respectively, to further simplify the final expression:

$$
\begin{aligned}
q\,\mathbf{X}\,q^{-1} &= \mathbf{X_1}\sin^2\frac{\theta}{2} + \mathbf{X}\cos^2\frac{\theta}{2} + (\mathbf{u}\times\mathbf{X})\sin\frac{\theta}{2}\cos\frac{\theta}{2} \\
&\quad + (\mathbf{u}\times\mathbf{X})\sin\frac{\theta}{2}\cos\frac{\theta}{2} - (\mathbf{X}-\mathbf{X_1})\sin^2\frac{\theta}{2} \\
&= \mathbf{X}(\cos^2\frac{\theta}{2} - \sin^2\frac{\theta}{2}) + 2\mathbf{X_1}\sin^2\frac{\theta}{2} + 2(\mathbf{u}\times\mathbf{X})\sin\frac{\theta}{2}\cos\frac{\theta}{2} \\
&= \mathbf{X}\cos\theta + \mathbf{X_1}(1-\cos\theta) + (\mathbf{u}\times\mathbf{X})\sin\theta \qquad (6.5)
\end{aligned}
$$

Comparing (6.5) with the formula (5.43) derived for $f(\mathbf{X})$ in Section 5.4.3 completes the proof. $\qquad\square$

So, indeed, we have the correspondence:

$$\text{orientation} \to \text{radial rotation} \to \text{quaternion}$$

Example 6.4. Let's verify the preceding proposition in the case of rotating the vector i by an angle of $\pi/2$ about the axis k.

It's easily checked by hand that this particular rotation takes i to j.

Next, to use the proposition, write first $\mathbf{u} = k$, $\mathbf{X} = i$ and $\theta = \pi/2$. Then the quaternion

$$q = \cos\frac{\theta}{2} + \mathbf{u}\sin\frac{\theta}{2} = \frac{1}{\sqrt{2}} + \frac{1}{\sqrt{2}}k$$

Therefore, by the proposition the image of i by the rotation of $\pi/2$ about axis k is

$$
\begin{aligned}
q\,\mathbf{X}\,q^{-1} &= (\frac{1}{\sqrt{2}} + \frac{1}{\sqrt{2}}k)\,i\,(\frac{1}{\sqrt{2}} - \frac{1}{\sqrt{2}}k) \\
&= (\frac{1}{\sqrt{2}}i + \frac{1}{\sqrt{2}}j)\,(\frac{1}{\sqrt{2}} - \frac{1}{\sqrt{2}}k) \\
&= \frac{1}{2}i + \frac{1}{2}j + \frac{1}{2}j - \frac{1}{2}i \\
&= j
\end{aligned}
$$

which, indeed, matches what we checked.

Exercise 6.17. Verify the proposition in the case of rotating the vector $i + k$ by an angle of π about j.

Proposition 6.2 says that every radial rotation corresponds to a unit quaternion. How about the other way around: does every unit quaternion correspond to a radial rotation in the sense that it gives that rotation by an inner automorphism? The answer is yes:

Proposition 6.3. *Let $q = w + xi + yj + zk$ be a unit quaternion. If $q = \pm 1$, then $\mathbf{X} \mapsto q\,\mathbf{X}\,q^{-1}$ is the identity transformation, i.e., a zero rotation about* *an arbitrary axis. If $q \neq \pm 1$, then there exists a unique pair $(\mathbf{u}, \theta)$, such that* $\mathbf{u}$ *is a unit vector and $\theta \in (0, 2\pi)$, and such that*

$$q = \cos \frac{\theta}{2} + \mathbf{u} \sin \frac{\theta}{2}$$

This implies that $\mathbf{X} \mapsto q \, \mathbf{X} \, q^{-1}$ is a rotation by angle θ about the radial axis directed along $\mathbf{u}$.

Proof. If $q = 1$ or $q = -1$ then it's obvious that $\mathbf{X} \mapsto q \, \mathbf{X} \, q^{-1}$ is the identity transformation.

Suppose, then, that $q \neq \pm 1$. As $w^2 + x^2 + y^2 + z^2 = |q|^2 = 1$, we must have $-1 \leq w \leq 1$. However, if $w = \pm 1$, then we would have $x = y = z = 0$, so that $q = \pm 1$, contradicting our assumption. One deduces, therefore, that $-1 < w < 1$, which implies that there is a unique $\theta/2 \in (0, \pi)$ – equivalently, a unique $\theta \in (0, 2\pi)$ – such that $\cos \frac{\theta}{2} = w$. Accordingly, $\sin \frac{\theta}{2} = \sqrt{1 - w^2}$, where the RHS is the positive square root. It follows that

$$\begin{aligned} q &= w + xi + yj + zk \\ &= w + \left(\frac{x}{\sqrt{1-w^2}} i + \frac{y}{\sqrt{1-w^2}} j + \frac{z}{\sqrt{1-w^2}} k \right) \sqrt{1-w^2} \\ &= \cos \frac{\theta}{2} + \mathbf{u} \sin \frac{\theta}{2} \end{aligned}$$

where

$$\mathbf{u} = \frac{x}{\sqrt{1-w^2}} i + \frac{y}{\sqrt{1-w^2}} j + \frac{z}{\sqrt{1-w^2}} k$$

is a unit vector because $x^2 + y^2 + z^2 = 1 - w^2$. The conclusion in the last line of the proposition now follows from an application of Proposition 6.2. □

Example 6.5. Determine the rotation corresponding to the unit quaternion

$$\frac{1}{\sqrt{3}} i + \frac{1}{\sqrt{3}} j + \frac{1}{\sqrt{3}} k$$

and write it in OpenGL form.

Answer: We want to express

$$q = \frac{1}{\sqrt{3}} i + \frac{1}{\sqrt{3}} j + \frac{1}{\sqrt{3}} k$$

in the form

$$\cos \frac{\theta}{2} + \mathbf{u} \sin \frac{\theta}{2}$$

Following the preceding proposition, write

$$q = w + xi + yj + zk \quad \text{where} \quad w = 0 \quad \text{and} \quad x = y = z = \frac{1}{\sqrt{3}}$$

Next, set

$$\cos \theta/2 = w = 0 \implies \theta/2 = \frac{\pi}{2} \implies \theta = \pi$$

and

$$\mathbf{u} = \frac{x}{\sqrt{1-w^2}}\,i + \frac{y}{\sqrt{1-w^2}}\,j + \frac{z}{\sqrt{1-w^2}}\,k = \frac{1}{\sqrt{3}}i + \frac{1}{\sqrt{3}}j + \frac{1}{\sqrt{3}}k$$

It follows that the given quaternion corresponds to the OpenGL rotation (up to round-off error)

```
glRotatef(180.0, 0.58, 0.58, 0.58)
```

Exercise 6.18. Determine the rotation corresponding to the unit quaternion

$$\frac{1}{\sqrt{2}} + \frac{1}{2}i + \frac{1}{2}k$$

and write it in OpenGL form.

We prove next a couple of useful facts related to Proposition 6.2:

Proposition 6.4.　*(a) If the rotation f_1 corresponds to the quaternion q_1 and the rotation f_2 to q_2, then the composed rotation $f_1 \circ f_2$ corresponds to the product $q_1 q_2$.*

In other words, rotations can be composed by multiplying their corresponding quaternions.

(b) If q is a unit quaternion and c an arbitrary non-zero scalar, then the transformation

$$\mathbf{X} \mapsto (cq)\,\mathbf{X}\,(cq)^{-1}$$

is equivalent to the rotation

$$\mathbf{X} \mapsto q\,\mathbf{X}\,q^{-1}$$

In other words, q and any non-zero scalar multiple cq give the same rotation by inner automorphism.

Proof. (a) For a vector $\mathbf{X}$,

$$
\begin{aligned}
(f_1 \circ f_2)(\mathbf{X}) &= f_1(f_2(\mathbf{X})) = q_1(q_2\,\mathbf{X}\,q_2^{-1})q_1^{-1} \\
&= (q_1 q_2)\,\mathbf{X}\,(q_2^{-1}q_1^{-1}) = (q_1 q_2)\,\mathbf{X}\,(q_1 q_2)^{-1}
\end{aligned}
$$

completing the proof.

(b) Since $(cq)^{-1} = c^{-1}q^{-1}$, the equalities

$$(cq)\,\mathbf{X}\,(cq)^{-1} = (cq)\,\mathbf{X}\,(c^{-1}q^{-1}) = (cc^{-1})q\,\mathbf{X}\,q^{-1} = q\,\mathbf{X}\,q^{-1}$$

complete the proof (note, for the second equality, that a scalar can be moved in and out of a product with impunity). □

Exercise 6.19. Let f_1 be a rotation of $60°$ about the x-axis and f_2 a rotation of $90°$ about the y-axis. Determine the composed rotations $f_1 \circ f_2$ and $f_2 \circ f_1$ (by giving their respective axis and amount of rotation).

Part answer: The plan is to go from rotation space to quaternion space, multiply and return to rotation space.

Proposition 6.2 tells us that rotation f_1 corresponds to the quaternion

$$q = \cos \frac{\theta}{2} + \mathbf{u} \sin \frac{\theta}{2}$$

where $u = i$ and $\theta = \pi/3$. In other words, f_1 corresponds to

$$\frac{\sqrt{3}}{2} + \frac{1}{2}i$$

Likewise, f_2 corresponds to

$$\frac{1}{\sqrt{2}} + \frac{1}{\sqrt{2}}j$$

By Proposition 6.4(a), $f_1 \circ f_2$ corresponds to

$$\left(\frac{\sqrt{3}}{2} + \frac{1}{2}i \right) \left(\frac{1}{\sqrt{2}} + \frac{1}{\sqrt{2}}j \right) = \frac{\sqrt{3}}{2\sqrt{2}} + \frac{1}{2\sqrt{2}}i + \frac{\sqrt{3}}{2\sqrt{2}}j + \frac{1}{2\sqrt{2}}k$$

Applying Proposition 6.3 to determine the rotation corresponding to the above quaternion, one finds after some calculation $f_1 \circ f_2$ to be `glRotatef(104.48, 0.45, 0.77, 0.45)` written as an OpenGL rotation up to round-off error.

The same method applies to $f_2 \circ f_1$.

Proposition 6.4 has a couple of interesting consequences.

To begin with, part (a) has an implication for computational efficiency. If a rotation is represented by a matrix, as in Equation (5.45), then the complexity of composing two rotations, equivalent to multiplying the corresponding matrices, is dominated by 27 scalar multiplications – treating the matrix of (5.45) as 3×3 because the fourth row and column don't add complexity and observing that each of the 9 entries of the product involves 3 multiplications.

On the other hand, if a rotation is represented by a quaternion, then composing two rotations, equivalent to multiplying the corresponding quaternions by Proposition 6.4(a), requires 16 scalar multiplications. The conclusion is that an efficient way to compose multiple rotations is via quaternion representation.

A consequence of part (b) of the proposition is a mathematically useful, though somewhat abstract, representation of 3D rotations. *You can safely skip this representation if you are not particularly inclined toward mathematical abstraction.*

Note, first, that the set $\mathbb{H}$ of quaternions is in one-to-one correspondence with 4D space $\mathbb{R}^4$ via the association $w + xi + yj + zk \leftrightarrow [w \ x \ y \ z]^T$. We can, therefore, identity $\mathbb{H}$ with $\mathbb{R}^4$ and refer to points of the latter as quaternions.

Now, Proposition 6.4(b) says that all non-zero quaternions on a given radial line in $\mathbb{R}^4$ correspond to the same rotation, as they differ one from another by a scalar multiple. It can also be verified that non-zero quaternions on distinct radial lines correspond to different rotations.

So here's a summary of the situation. Quaternions are in one-to-one correspondence with points of $\mathbb{R}^4$. Each non-zero quaternion also corresponds *uniquely* to the rotation of 3-space, which it gives by inner automorphism. Rotations, on the other hand, are not as "faithful" because each corresponds to infinitely many quaternions, in fact, a whole radial line's worth (except that the origin O does not correspond to a rotation). See Figure 6.11. Rotations of 3-space, therefore, are in one-to-one correspondence with the set of radial lines in $\mathbb{R}^4$.

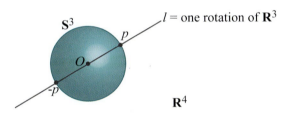

Figure 6.11: The unit sphere $\mathbb{S}^3$ in $\mathbb{R}^4$ with a radial line passing through a pair of antipodal points.

If one is uncomfortable with identifying rotations of $\mathbb{R}^3$ with lines in $\mathbb{R}^4$, then here's a way to identify rotations with points instead: since a radial line of $\mathbb{R}^4$ intersects $\mathbb{S}^3$, the unit sphere of $\mathbb{R}^4$, in two antipodal points, rotations are in 1-1 correspondence with the points of $\mathbb{S}^3$, *provided* one is willing to undertake the mental trick of identifying each antipodal pair as a single point. $\mathbb{S}^3$ with its antipodal points identified is actually the so-called projective 3-space $\mathbb{P}^3$, so one can identify the space of 3D rotations with $\mathbb{P}^3$.

Quaternion to Rotation Matrix

Proposition 6.3 enables us to find the unique rotation corresponding to a unit quaternion in terms of its axis and rotation angle. However, in various applications, e.g., when using OpenGL, the rotation matrix itself is more useful. We ask the reader to find the 4×4 rotation matrix corresponding to a given quaternion by completing the solution to the following exercise.

$\mathbf{Exercise}$ 6.20. Show that the 4×4 matrix representing the rotation corresponding to the unit quaternion

$$q = w + xi + yj + zk$$

is

$$\begin{bmatrix} w^2 + x^2 - y^2 - z^2 & 2xy - 2wz & 2xz + 2wy & 0 \\ 2xy + 2wz & w^2 - x^2 + y^2 - z^2 & 2yz - 2wx & 0 \\ 2xz - 2wy & 2yz + 2wx & w^2 - x^2 - y^2 + z^2 & 0 \\ 0 & 0 & 0 & 1 \end{bmatrix} \quad (6.6)$$

Part answer: Verify first the case when $q = \pm 1$ (keep in mind that this implies that $w = \pm 1$ and that $x = y = z = 0$). Suppose, next, that $q \neq \pm 1$. Proposition 6.3 says, then, that q gives, by inner automorphism, the rotation of angle θ about the unit vector **u** where

$$\cos \frac{\theta}{2} = w$$

and

$$\mathbf{u} = (x/\sqrt{1 - w^2})\,i + (y/\sqrt{1 - w^2})\,j + (y/\sqrt{1 - w^2})\,k$$

In Section 5.4.3 – see Equation (5.45) – we derived the following matrix corresponding to a rotation of angle θ about the radial axis l toward the point $P = [a\ b\ c]^T$, where $|P| = 1$.

$R_{a,b,c}(\theta) =$

$$\begin{bmatrix} a^2(1 - \cos\theta) + \cos\theta & ab(1 - \cos\theta) - c\sin\theta & ac(1 - \cos\theta) + b\sin\theta & 0 \\ ab(1 - \cos\theta) + c\sin\theta & b^2(1 - \cos\theta) + \cos\theta & bc(1 - \cos\theta) - a\sin\theta & 0 \\ ac(1 - \cos\theta) - b\sin\theta & bc(1 - \cos\theta) + a\sin\theta & c^2(1 - \cos\theta) + \cos\theta & 0 \\ 0 & 0 & 0 & 1 \end{bmatrix}$$

We ask the reader to finish the exercise by deriving the matrix (6.6), after plugging the following into the matrix expression above for $R_{a,b,c}(\theta)$:

$$
\begin{aligned}
a &= x/\sqrt{1 - w^2} \\
b &= y/\sqrt{1 - w^2} \\
c &= z/\sqrt{1 - w^2} \\
1 - \cos\theta &= 2\sin^2\frac{\theta}{2} = 2(1 - w^2) \\
\cos\theta &= 2\cos^2\frac{\theta}{2} - 1 = 2w^2 - 1 \\
\sin\theta &= 2\sin\frac{\theta}{2}\cos\frac{\theta}{2} = 2w\sqrt{1 - w^2}
\end{aligned}
$$

Spherical Linear Interpolation

Let's return to our original objective of applying quaternions to the animation of orientation. The strategy is to represent, initially, the start and end orientations corresponding, say, to the rotations f_1 and f_2 by unit quaternions, q_1 and q_2, respectively. Next is to interpolate between the two in quaternion space, traveling from q_1 to q_2 along a path of unit quaternions as well. The final step is to map this path back to a path in the space of rotations from f_1 to f_2, which, of course, is equivalent to a path in the space of orientations between the original start and end ones. Figure 6.12(a) is a conceptual diagram (numbers in parentheses indicate steps).

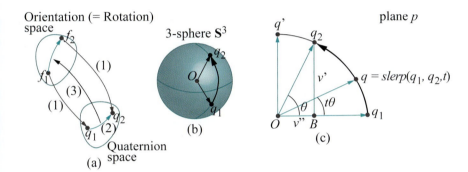

Figure 6.12: (a) Conceptual plan to use quaternion space to interpolate in orientation space (numbers in parentheses indicate steps) (b) The geodesic path from q_1 to q_2 on $\mathbb{S}^3$ (c) Slerping from q_1 to q_2.

Figure 6.12(b) shows the representing unit quaternions q_1 and q_2 as points on the unit sphere $\mathbb{S}^3$ of $\mathbb{R}^4$, the latter being identified with the set $\mathbb{H}$ of all quaternions. A shortest path on $\mathbb{S}^3$ between q_1 and q_2 is along a great circle – a *great circle* is the intersection of a radial plane with $\mathbb{S}^3$. A shortest path itself is called a *geodesic* path. If q_1 and q_2 are not antipodal, then there is a unique geodesic path joining them; otherwise, there are infinitely many (each half a great circle).

Remark 6.3. It helps the intuition to think of shortest paths on the surface of the Earth, where the situation is exactly the same, only one dimension lower.

Suppose, first, that q_1 and q_2 are not antipodal. We want to interpolate at a constant angular rate from q_1 to q_2 along the unique geodesic joining them. This is called *spherical linear interpolation*, or *slerp* for short. Even though q_1 and q_2 are points of $\mathbb{R}^4$, the slerp between them takes place entirely on the unique 2D plane p containing the two and the origin. See Figure 6.12(c).

Suppose that θ is the smaller of the angles between q_1 and q_2 on the great circle containing them. The point q, an angle of $t\theta$ from q_1 toward q_2 on that same circle, is denoted $slerp(q_1, q_2, t)$. As t varies from 0 to 1, $slerp(q_1, q_2, t)$ travels from q_1 to q_2. We seek, therefore, a formula for $slerp(q_1, q_2, t)$.

Let q' be the unit quaternion on the plane p on the same side of q_1 as q_2 and such that Oq' is perpendicular to Oq_1 (this assumes that $q_2 \neq q_1$, for slerping between them is trivial otherwise). Drop the perpendicular from q_2 to the point B on Oq_1 and denote the 4-vector Bq_2 by v' and OB by v''.

Observe, first, that

$$|v'| = |q_2| \sin \theta = \sin \theta \quad \text{and} \quad |v''| = |q_2| \cos \theta = \cos \theta$$

because $|q_2| = 1$. Moreover, $v'' = |v''|q_1$, because q_1 is the unit vector parallel to v'', which means $v'' = q_1 \cos \theta$. Therefore, we have

$$v' = q_2 - v'' = q_2 - q_1 \cos \theta$$

Since q' is of unit length and parallel to v', we have as well

$$q' = \frac{v'}{|v'|} = \frac{v'}{\sin \theta} = \frac{q_2 - q_1 \cos \theta}{\sin \theta}$$

(that q_1 and q_2 are neither equal nor antipodal ensures that θ and $\sin \theta$ are both non-zero). Therefore,

$$
\begin{aligned}
q = slerp(q_1, q_2, t) &= q_1 \cos(t\theta) + q' \sin(t\theta) \\
&\quad \text{(adding the components of } q \text{ along } q_1 \text{ and } q') \\
&= q_1 \cos(t\theta) + \frac{q_2 - q_1 \cos \theta}{\sin \theta} \sin(t\theta) \\
&= q_1 \left(\cos(t\theta) - \frac{\cos \theta}{\sin \theta} \sin(t\theta) \right) + q_2 \frac{\sin(t\theta)}{\sin \theta} \\
&= q_1 \frac{\sin((1-t)\theta)}{\sin \theta} + q_2 \frac{\sin(t\theta)}{\sin \theta} \qquad (6.7)
\end{aligned}
$$

which gives the formula sought for $slerp(q_1, q_2, t)$ (note that θ, the angle between them, is easily determined from q_1 and q_2).

Remark 6.4. Expectedly, linear interpolation has not been able to escape the ugly diminutive *lerp* in the CG literature, with the obvious formula

$$lerp(q_1, q_2, t) = (1 - t)q_1 + tq_2$$

Interpolating Orientations via Quaternions

We now have all the pieces in place to implement the following scheme to interpolate between two orientations corresponding to the rotations f_1 and f_2:

1. Go from rotation space to quaternion space by finding unit quaternions q_1 and q_2 corresponding to f_1 and f_2, respectively.

2. Observe that both q_2 and $-q_2$ represent the same rotation but one of the two makes an angle of at most $\pi/2$ with q_1 and the other an angle at least $\pi/2$. We want to interpolate to the one closer to q_1, in case the two are at different angular distances.

Accordingly, determine $q_1 \cdot q_2$. If its value is negative, then the angle between q_1 and q_2 is greater than $\pi/2$, in which case set $q_2 = -q_2$; otherwise, leave it as it is.

Note: It's in this last step that we resolve the problem from potentially ambiguous representation of a rotation. In fact, it is only when q_1 and q_2 are orthogonal that both q_2 and $-q_2$ are at the same angle of $\pi/2$ from q_1, and we *could* choose either. The last step above finesses this choice by leaving q_2 unchanged in this case. However, see Exercise 6.21 below.

Recall in this connection how much more troublesome were ambiguous Euler angle representations in the last section.

3. Compute $slerp(q_1, q_2, t)$, t varying from 0 to 1.

4. Return to rotation space by computing the rotation $f(t)$ corresponding to $slerp(q_1, q_2, t)$. Then $f(t)$ interpolates between the rotations f_1 and f_2 (equivalently, the corresponding orientations).

Exercise 6.21. Show that if q_1 and q_2 are orthogonal then ambiguity is *inherent*, in that the orientation corresponding to one is a rotation of 180° about some axis of the orientation corresponding to the other. In this case, rotating one way about that axis, to change one orientation to the other, is exactly symmetric to going the other way, and there is no procedure to prefer one over the other. For example, consider the two ways of going from the orientation AB to AB' of the solid straight arrow in Figure 6.13.

Figure 6.13: Changing the orientation from AB to AB' is inherently ambiguous.

Time for code.

Experiment 6.6. Run `quaternionAnimation.cpp`, which applies the preceding ideas to animate the orientation of our favorite rigid body, an L, with the help of quaternions. Press 'x/X', 'y/Y' and 'z/Z' to change the orientation of the blue L, whose current Euler angles are shown on the display. Its start orientation is the currently fixed red L. See Figure 6.14 for a screenshot.

Pressing enter at any time begins an animation of the red L from the start to the blue's current orientation. Press the up and down arrow keys to change the speed and delete to reset. **End**

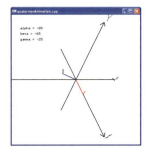

Figure 6.14: Screenshot of `quaternionAnimation.cpp`.

The routine `eulerAnglesToQuaternion()` of the program determines the unit quaternion corresponding to an orientation specified by three Euler angles, by computing first the unit quaternion corresponding to the rotation about a coordinate axis connoted by each Euler angle, and then multiplying the three.

The routine `slerp()` implements formula (6.7), except for the following "hack" to avoid problems with division by zero, or near-zero numbers, when θ is small. Applying the approximation $\sin\alpha \simeq \alpha$ for small α, formula (6.7) is modified as follows if $\theta \leq 0.000001$:

$$
\begin{aligned}
slerp(q_1, q_2, t) &= q_1 \frac{\sin((1-t)\theta)}{\sin\theta} + q_2 \frac{\sin(t\theta)}{\sin\theta} \\
&= q_2 \frac{(1-t)\theta}{\theta} + q_2 \frac{t\theta}{\theta} = (1-t)q_1 + tq_2
\end{aligned}
$$

(the value 0.000001 having been chosen arbitrarily).

The routine `quaternionToRotationMatrix()` gets us back to rotation space with help of (6.6) to find the 4×4 rotation matrix corresponding to a given quaternion.

Exercise 6.22. (**Programming**) Extend `quaternionAnimation.cpp` to animate the motion of a rigid body, say a spacecraft, from a start disposition to a user-specified target disposition which can differ *both* in location and orientation from the start. Transformation should be simultaneous in both position and orientation.

The utility of quaternions in interactive animation cannot be overemphasized and they should be in every game programmer's tool kit. For instance, try to do what `quaternionAnimation.cpp` does *without* using quaternions (good luck!).

6.4 Summary, Notes and More Reading

In this chapter we learned a few practically useful techniques. Particularly indispensable to programmers of heavily-populated environments is the method of frustum culling by space partitioning. For further reading, books on game programming, e.g., Lengyel [81] and Eberly [37], will, typically, contain descriptions of popular space partitioning methods such as quadtrees, octrees, kd-trees and BSP trees in the context of frustum culling and collision detection. An excellent computational geometry reference for space partitioning data structures, including range trees and kd-trees, is the one by de Berg et al. [10]. Samet [116] is a must for anyone seeking to learn about spatial data structures in depth. See Slater et al. [129] and the paper by Kumar et al. [79] for literature on the important problem of partitioning a space encompassing a dynamic scene.

We learned how to animate orientation with the use of both Euler angles and quaternions. This will come in handy in camera control and rigid body animation. The Euler angle representation, as we saw, suffers from certain problems, surmounted subsequently by the slick mathematics of quaternions. The books by Buss [21], Lengyel [81] and Watt [142], among others, contain discussions of Euler angles and quaternions and their relation to rigid-body kinematics. The ones by Hanson [59] and Kuipers [78] are all about quaternions and their applications.

Part IV

Geometry for the Home Office

CHAPTER 7

Convexity and Interpolation

I t's time now to get some of the geometric concepts underlying 3D modeling and lighting in place before we reach those particular chapters. We've seen programs where colors, defined at the vertices of a primitive, are mixed and spread throughout the primitive's interior. This is done by means of interpolation. In this chapter we'll study the exact mechanics of the interpolation process.

Section 7.1 motivates the process of interpolation with simple examples. Section 7.2 gets to the heart of the matter by showing first that line segment and triangle primitives are particularly suited for interpolation because of the property they share that any point of such a primitive can be uniquely represented as a so-called convex combination of its vertices. Section 7.3 shows precisely how this property is used by OpenGL to interpolate values such as color.

The geometric property which some objects have of being convex is closely related to interpolation. Section 7.4 defines convexity and the notion of the convex hull of a set of points and applies them to understanding if objects more complicated than line segments and triangles, e.g., polygons in general, can be equally easily interpolated. We see that the answer is no and that line segments and triangles are indeed special. We conclude in Section 7.5.

This chapter and the next two on triangulation and orientation, respectively, are intimately related and should be read one after another. The material is somewhat mathematical. However, the math involved is geometric, which means that it can be "seen to work", and not particularly abstract. The importance of these three chapters at the conceptual foundation of 3D computer graphics cannot be overemphasized. Having said this, it's true that this particular chapter will be fairly light reading for someone already familiar with linear interpolation and convexity, possibly from an earlier math class. If this is your case, then flip quickly through the pages – make sure the parts to do with OpenGL make sense – and move on.

7.1 Motivation

OpenGL has three favorites among its several drawing primitives: points, segments (by segment we'll always mean a straight line segment) and triangles. This is not owing to some idiosyncrasy of its specification as an API, but for a deeper reason. We've already seen that material values such as color, specified at a primitive's vertices, are apparently interpolated throughout its interior.

So here, briefly, is why points, segments and triangles are favored: they have the property that every point in each can be *unambiguously* (or, *uniquely*, same thing) represented in terms of its vertices. This makes it possible for values defined at the vertices to be *unambiguously* – therefore, automatically, by means of a program – interpolated throughout the primitive. We'll clarify all this soon, but to begin with here is a simple example.

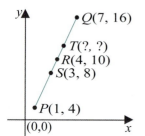

Figure 7.1: Points that split segment PQ.

Example 7.1. Using graph paper if you like, draw the segment PQ joining the point P, with coordinates $(1, 4)$, to the point Q, with coordinates $(7, 16)$. See Figure 7.1. Measure off the midpoint R of the segment. Verify that its coordinates are as indicated in the figure. Since the midpoint is halfway from either endpoint it does make sense that the coordinates of R are an exact average of those of P and Q, viz.,

$$\frac{1}{2} * (1,4) + \frac{1}{2} * (7,16) = \left(\frac{1}{2} * 1 + \frac{1}{2} * 7, \ \frac{1}{2} * 4 + \frac{1}{2} * 16\right) = (4, 10)$$

How about the point S a third of the way from P to Q? Again, measure it off and see if the coordinates shown in the figure are correct. S splits PQ in the ratio $\frac{2}{3} : \frac{1}{3}$, where it's $\frac{2}{3}$ toward P from Q and $\frac{1}{3}$ toward Q from P. Ergo, S's coordinates are

$$\frac{2}{3} * (1,4) + \frac{1}{3} * (7,16) = \left(\frac{2}{3} * 1 + \frac{1}{3} * 7, \ \frac{2}{3} * 4 + \frac{1}{3} * 16\right) = (3, 8)$$

Exercise 7.1. We ask you to calculate the coordinates of T, which is two-thirds of the way from P to Q, and verify by actual measurement.

Interestingly, P itself, which splits PQ in the ratio $1 : 0$, has coordinates

$$1 * (1,4) + 0 * (7,16) = (1,4)$$

while Q, which splits it in the ratio $0 : 1$, has coordinates

$$0 * (1,4) + 1 * (7,16) = (7,16)$$

It seems then that in the expression

$$X = c * (1, 4) + (1 - c) * (7, 16)$$

the variable c acts as a "dial" which can be turned from 1 to 0 to move the point X from P to Q.

Here's an exercise to get you thinking about using an expression like the last one to interpolate material properties.

Exercise 7.2. If the end vertex P of the segment in the preceding example is specified red (RGB $= (1, 0, 0)$) and Q green (RGB $= (0, 1, 0)$), then what should the colors be at the midpoint R? At the point S?

7.2 Convex Combinations

We said at the beginning of the last section that points, segments and triangles are favored by OpenGL. In fact, they are *the* building blocks of OpenGL. Even quadrilaterals and polygons in general, as we'll see, are first sub-divided into triangles before being processed. We informally described a property shared by the three primitives – that each point belonging to one has a unique expression in terms of its vertices – which makes unambiguous interpolation possible. Here is the formal statement in the case of a segment:

Proposition 7.1. *If P and Q are two points in $\mathbb{R}^3$, then a point V lies on the segment PQ if and only if it can be expressed as*

$$V = c_1 P + c_2 Q$$

where $0 \leq c_i \leq 1$, for both $i = 1$ and $i = 2$, and where $c_1 + c_2 = 1$.

Further, if P and Q are distinct – so that PQ does not degenerate to a point – then this expression for V is unique.

Before the proof, take a second to match the proposition with the example of the previous section: the points P, Q, R, S and T on the segment PQ could each indeed be expressed in the form $c_1 P + c_2 Q$, where c_1 and c_2 lie between 0 and 1, and add up to 1.

Proof. *Skip this proof if you start to get bogged down in the math. Just make sure to understand what the proposition says.*

For the first part of the proposition, suppose initially that $P \neq Q$ so that PQ is a non-degenerate segment. Consider a point V on this segment. The vector $V - P$ is parallel to the vector $Q - P$ (see Figure 7.2). Therefore, one is obtained from the other by multiplying by the ratio of their lengths:

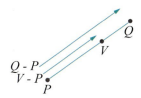

Figure 7.2: Illustration for the proof of Proposition 7.1.

$$V - P = \frac{|V - P|}{|Q - P|}(Q - P) \implies V - P = c(Q - P)$$

$$\implies V = (1 - c)P + cQ$$

where c denotes $\frac{|V-P|}{|Q-P|}$. Note then that $0 \le c \le 1$, as $|V-P| \le |Q-P|$. Writing $c_1 = 1 - c$ and $c_2 = c$ we have, indeed, that

$$V = c_1 P + c_2 Q$$

where $0 \le c_i \le 1$, for both $i = 1$ and $i = 2$, and, moreover, $c_1 + c_2 = 1$. This proves the "only if" direction of the first part, provided $P \ne Q$.

Conversely, for the "if" direction, assuming again $P \ne Q$, suppose that

$$V = c_1 P + c_2 Q$$

where $0 \le c_i \le 1$, for both $i = 1$ and $i = 2$, and $c_1 + c_2 = 1$. Then writing $c = c_2$ we have

$$V = (1-c)P + cQ = P + c(Q - P) \tag{7.1}$$

V is seen to be the point at a distance of $c|Q - P|$ from P in the direction of Q. As $0 \le c \le 1$, V indeed lies on the segment joining P and Q. This completes the proof of the first part of the proposition when $P \ne Q$. If $P = Q$, the first part is actually trivial to prove because the segment PQ degenerates to a point (so any point on it is P itself – we leave the rest to the reader).

For the second part regarding uniqueness, suppose that the point V on PQ can be expressed as both $V = c_1 P + c_2 Q$ and $V = d_1 P + d_2 Q$, where $c_1 + d_1 = d_1 + d_2 = 1$. Then

$$V = c_1 P + c_2 Q = d_1 P + d_2 Q \implies (c_1 - d_1)P = (d_2 - c_2)Q \tag{7.2}$$

From $c_1 + c_2 = d_1 + d_2 = 1$ we have that $c_1 - d_1 = d_2 - c_2$. Therefore, if these two equal quantities are not 0 we could multiply Equation (7.2) by $\frac{1}{c_1 - d_1} (= \frac{1}{d_2 - c_2})$ to deduce that $P = Q$, contradicting the hypothesis of the second part. We are led to conclude that $c_1 - d_1 = d_2 - c_2 = 0$, so that $c_1 = d_1$ and $c_2 = d_2$, proving that the expression for V in the proposition is indeed unique. □

$\mathcal{R}em\mathfrak{a}rk$ 7.1. To minimize notation we wrote the endpoints of the segment in the proposition as single variables P and Q. Of course, one can write out their coordinates as, say, $P = (p_x, p_y, p_z)$ and $Q = (q_x, q_y, q_z)$ and, correspondingly, the equation for V in the statement of the proposition as

$$(v_x, v_y, v_z) = c_1(p_x, p_y, p_z) + c_2(q_x, q_y, q_z) = (c_1 p_x + c_2 q_x, c_1 p_y + c_2 q_y, c_1 p_z + c_2 q_z)$$

For example, if $P = (1, 4, 3)$ and $Q = (2, 5, 2)$, then the proposition says that points on the segment PQ are of the form

$$(c_1 + 2c_2, 4c_1 + 5c_2, 3c_1 + 2c_2), \text{ where } 0 \le c_1, c_2 \le 1 \text{ and } c_1 + c_2 = 1.$$

$\mathcal{R}em\mathfrak{a}rk$ 7.2. We could have saved ourselves a variable and written $V = cP + (1-c)Q$, instead of $V = c_1 P + c_2 Q$, because $c_1 + c_2 = 1$, but chose not to in order to have separate variables for the coefficients of P and Q. This keeps our notation consistent with the proposition for triangles coming up.

Definition 7.1. A point V of the form

$$V = c_1 P + c_2 Q$$

where $0 \leq c_i \leq 1$, for both $i = 1$ and $i = 2$, and where $c_1 + c_2 = 1$, is said to be a *convex combination* – or *barycentric combination* – of P and Q. The scalars c_1 and c_2 are called the *barycentric coordinates* of V.

The following corollary then is just a rewrite of the first part of Proposition 7.1.

Corollary 7.1. *The segment joining two points P and Q consists of all their convex combinations.* □

A point $V = c_1 P + c_2 Q$ on the segment PQ can be usefully thought of as a *weighted sum* of P and Q, where the barycentric coordinates c_1 and c_2 are the weights – or influence, or ownership, if you will – of P and Q, respectively, on the location of V. The next exercise follows up on this idea.

Exercise 7.3. Suppose that $V = c_1 P + c_2 Q$ lies on the segment PQ.

(a) If $c_1 = c_2 = \frac{1}{2}$, *prove* that V is the midpoint of PQ; in other words, if the weights of P and Q on V are equal, then it's in the middle of the two. (We saw an illustration, though not proof, of this in Example 7.1.)

   Suggested approach:

   Without using vectors: Say $P = (p_x, p_y, p_z)$ and $Q = (q_x, q_y, q_z)$. Then

$$V = \frac{1}{2}(p_x, p_y, p_z) + \frac{1}{2}(q_x, q_y, q_z) = \left(\frac{p_x + q_x}{2}, \frac{p_x + q_x}{2}, \frac{p_x + q_x}{2} \right)$$

   Determine the distance between P and V and between Q and V using the formula for distance between a pair of points, and find that the two are equal.

   Note: The distance between the points (x, y, z) and (x', y', z') is $\sqrt{(x - x')^2 + (y - y')^2 + (z - z')^2}$.

   Using vectors: $V - P = (\frac{1}{2}P + \frac{1}{2}Q) - P = \frac{1}{2}Q - \frac{1}{2}P$. Similarly, determine $Q - V$ and find that it equals $V - P$.

(b) Prove generally that $PV : VQ$ equals $c_2 : c_1$.

(c) If $c_1 > c_2$, prove that V is closer to P than Q, and vice versa.

Exercise 7.4. Say $P \neq Q$ and that $V = c_1 P + c_2 Q$, where c_1 and c_2 are *any* real numbers such that $c_1 + c_2 = 1$ (the condition that the c_i's must lie between 0 and 1 is dropped). Show then that V may be any point on the (infinite) straight line through P and Q.
Hint: $V = (1 - c_2)P + c_2 Q = P + c_2(Q - P)$. How does this point change as c_2 varies?

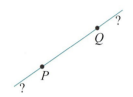

Figure 7.3: What are the conditions on c_1 and c_2 for $V = c_1P + c_2Q$ to lie on either side of PQ?

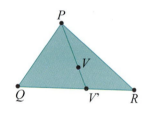

Figure 7.4: Illustration for the proof of Proposition 7.2.

Exercise 7.5. The previous exercise says that points on the whole straight line through P and Q (given $P \neq Q$) are of the form $c_1P + c_2Q$, where $c_1 + c_2 = 1$. We already know that points on this line *between* P and Q additionally satisfy $0 \leq c_1, c_2 \leq 1$. How about those on either side of PQ – what are the conditions on c_i? See Figure 7.3.

Statements analogous to Proposition 7.1 can be proved for triangles as well:

Proposition 7.2. *If P, Q and R are three points in $\mathbb{R}^3$, then a point V lies on the triangle PQR if and only if it can be expressed as*

$$V = c_1P + c_2Q + c_3R$$

where $0 \leq c_i \leq 1$, for $1 \leq i \leq 3$, and where $c_1 + c_2 + c_3 = 1$.

Further, if P, Q and R are not collinear – so that PQR does not degenerate to a segment or a point – then this expression for V is unique.

Proof. *Skip this proof if you start to get bogged down in the math. Just make sure to understand what the proposition says.*

For the first part of the proposition, suppose initially that the triangle PQR is non-degenerate, in other words, that P, Q and R are not collinear. Consider a point V on this triangle.

If $V = P$, then, of course, $V = 1P + 0Q + 0R$, which is an expression of the form required.

If $V \neq P$, suppose that the straight line from P through V intersects the edge QR at V' (see Figure 7.4). As V lies on the segment joining P and V' the previous proposition gives an expression

$$V = c_1'P + c_2'V' \tag{7.3}$$

where $0 \leq c_i' \leq 1$, for $i = 1$ and $i = 2$, and $c_1' + c_2' = 1$.

Again, by the previous proposition, since V' lies on the segment joining Q and R,

$$V' = c_1''Q + c_2''R \tag{7.4}$$

where $0 \leq c_i'' \leq 1$, for $i = 1$ and $i = 2$, and $c_1'' + c_2'' = 1$.

We have by using both (7.3) and (7.4) that

$$\begin{aligned} V &= c_1'P + c_2'(c_1''Q + c_2''R) \\ &= c_1'P + c_2'c_1''Q + c_2'c_2''R \end{aligned}$$

Writing $c_1 = c_1'$, $c_2 = c_2'c_1''$ and $c_3 = c_2'c_2''$, we see that

$$V = c_1P + c_2Q + c_3R \tag{7.5}$$

where it may be verified that $0 \leq c_i \leq 1$, for $1 \leq i \leq 3$, and, moreover, $c_1 + c_2 + c_3 = c_1' + c_2'c_1'' + c_2'c_2'' = c_1' + c_2'(c_1'' + c_2'') = c_1' + c_2' * 1 = 1$. This proves the "only if" direction of the first part, provided PQR is not degenerate; we leave the proof in case PQR is degenerate (so, either a point or segment) to the reader.

We leave the proof of the "if" direction and of the uniqueness of the expression in the case that PQR is non-degenerate, to the reader as well. □

Definition 7.2. A point V of the form

$$V = c_1 P + c_2 Q + c_3 R$$

where $0 \leq c_i \leq 1$, for $1 \leq i \leq 3$, and where $c_1 + c_2 + c_3 = 1$, is said to be a *convex combination* – or *barycentric combination* – of P, Q and R. The scalars c_1, c_2 and c_3 are called the *barycentric coordinates* of V.

The following corollary is a rewrite of the first part of Proposition 7.2.

Corollary 7.2. *The triangle with vertices at P, Q and R consists of all their convex combinations.* □

Similarly as for a segment, a point $V = c_1 P + c_2 Q + c_3 R$ on the triangle PQR can be thought of as a weighted sum of P, Q and R, where the barycentric coordinates c_1, c_2 and c_3 are the respective weights.

Exercise 7.6. Suppose that $V = c_1 P + c_2 Q + c_3 R$ lies on the triangle PQR. Where is V when

(a) one of the c_i's is 1 and the other two 0?

(b) one of the c_i's is 0 and the other two equal to $\frac{1}{2}$?

(c) all the c_i's are equal to $\frac{1}{3}$?

Example 7.2. If $P = (0,0,0)$, $Q = (20,0,0)$ and $R = (20,30,0)$, does the point $V = (10,20,0)$ lie on the triangle PQR? If so, express V as a convex combination of the three vertices.

Answer: Let's try to solve the equations

$$\begin{aligned} V &= c_1 P + c_2 Q + c_3 R \\ c_1 + c_2 + c_3 &= 1 \end{aligned}$$

The first gives

$$(10,20,0) = c_1(0,0,0) + c_2(20,0,0) + c_3(20,30,0)$$

Equating the values in each coordinate on either side of the above equation we get the following (the equation in the z coordinate is trivial and not written):

$$\begin{aligned} 20c_2 + 20c_3 &= 10 \\ 30c_3 &= 20 \end{aligned}$$

With the additional

$$c_1 + c_2 + c_3 = 1$$

one solves to find that

$$c_1 = \frac{1}{2}, \qquad c_2 = -\frac{1}{6}, \qquad c_3 = \frac{2}{3}$$

As the c_i's do not all lie between 0 and 1 we conclude that V is not a convex combination of P, Q and R and, therefore, does not lie on the triangle PQR.

Exercise 7.7. If $P = (0,0,0)$, $Q = (20,0,0)$ and $R = (20,30,0)$, does the point $V = (15,15,0)$ lie on the triangle PQR? If so, express V as a convex combination of the three vertices.

Exercise 7.8. What if $V = c_1P + c_2Q + c_3R$, where c_i, $1 \le i \le 3$, are *any* real numbers such that $c_1 + c_2 + c_3 = 1$ (the condition that the c_i's must lie between 0 and 1 is dropped)? Where does V lie?

Exercise 7.9. If $P = (30,50,45)$, $Q = (40,20,5)$ and $R = (30,20,0)$, which of the points $V = (35,25,20)$, $V' = (35,25,15)$ and $V'' = (28,80,89)$ lie on the triangle PQR? Which of them lie on the plane containing P, Q and R, but not on the triangle PQR?

Exercise 7.10. In Figure 7.5, points D, E and F are midpoints of the edges PQ, QR and RP, respectively. Are points in the triangle DEF a special kind of convex combination of P, Q and R? Precisely, if $V \in DEF$ is expressed as $V = c_1P + c_2Q + c_3R$, what restrictions, if any, are there on the values that the c_i can have?

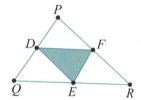

Figure 7.5: A triangle DEF with vertices at the midpoints of the edges of a larger triangle.

7.3 Interpolation

It is straightforward now to explain how OpenGL interpolates property values over its favorite primitives. Let's begin with a non-degenerate triangle $P_1P_2P_3$.

Suppose that the RGB color tuples at the vertices P_1, P_2 and P_3 are specified by the programmer to be (R_1, G_1, B_1), (R_2, G_2, B_2) and (R_3, G_3, B_3), respectively. Let V be any point of $P_1P_2P_3$. See Figure 7.6. By Proposition 7.2, there is a unique expression

$$V = c_1P_1 + c_2P_2 + c_3P_3 \tag{7.6}$$

of V as a convex combination of the vertices P_1, P_2 and P_3. OpenGL, in fact, determines this expression, particularly, the values of c_1, c_2 and c_3. The color at V is then set to

$$c_1(R_1, G_1, B_1) + c_2(R_2, G_2, B_2) + c_3(R_3, G_3, B_3) \tag{7.7}$$
$$= (c_1R_1 + c_2R_2 + c_3R_3, \; c_1G_1 + c_2G_2 + c_3G_3, \; c_1B_1 + c_2B_2 + c_3B_3)$$

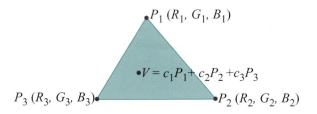

$P_1 (R_1, G_1, B_1)$

$V = c_1P_1 + c_2P_2 + c_3P_3$

$P_3 (R_3, G_3, B_3)$ $P_2 (R_2, G_2, B_2)$

Figure 7.6: Color values specified at the vertices V_1, V_2 and V_3 of a triangle are interpolated at V.

Simply put, OpenGL uses the weight of each vertex on the location of V as its weight on the color of V as well. As the weights c_1, c_2 and c_3 are unique, the interpolation process is *unambiguous* (and, therefore, *programmable*).

OpenGL performs an exactly similar computation to interpolate color values specified at the end vertices of a non-degenerate segment. In the case of the third of OpenGL's favorite primitives, the point, there is obviously nothing to interpolate.

Remark 7.3. What we call interpolation is often (more accurately) referred to as *linear interpolation* because the interpolation parameters enter as linear variables – i.e., with power one – in the expression for the interpolated value; e.g., the c_i's in the expression for V in Equation (7.6), viz., $V = c_1P_1 + c_2P_2 + c_3P_3$.

Remark 7.4. Color values aren't the only ones to be interpolated. In fact, any attribute specified *numerically* at a triangle's vertices can be interpolated through its interior. For example, in Phong's shading model, normal vector values defined at a triangle's vertices are interpolated.

Experiment 7.1. Replace the polygon declaration part of our old favorite square.cpp with the following (Block 1*):

```
glBegin(GL_TRIANGLES);
    glColor3f(1.0, 0.0, 0.0);
    glVertex3f(20.0, 20.0, 0.0);
    glColor3f(0.0, 1.0, 0.0);
    glVertex3f(80.0, 20.0, 0.0);
    glColor3f(0.0, 0.0, 1.0);
    glVertex3f(80.0, 80.0, 0.0);
glEnd();
```

Observe how OpenGL interpolates vertex color values throughout the triangle. Figure 7.7 is a screenshot.   **End**

Figure 7.7: Screenshot of Experiment 7.1.

Exercise 7.11. For the triangle of Experiment 7.1 calculate the RGB colors at the point $(70.0, 50.0, 0.0)$. Verify your answer by drawing a point with those colors at $(70.0, 50.0, 0.0)$ – the point should be invisible!

---

*To cut-and-paste you can find the block in text format in the file chap7codeModifications.txt in the directory Code/CodeModifications.

Exercise 7.12. Show by computation that the interpolated color value at the *centroid* (whose barycentric coordinates are all equal) of the triangle of Experiment 7.1 is a darkish gray. Again, verify your answer by drawing a point of that color at the centroid.

Exercise 7.13. If the vertices of a triangle at $P = (0, 0, 0)$, $Q = (20, 0, 0)$ and $R = (20, 30, 0)$ are colored red, cyan and magenta, respectively, what is the color at the point $(15, 15, 0)$?

Exercise 7.14. If the vertices of a triangle at $P = (10, 10, 0)$, $Q = (40, 10, 0)$ and $R = (30, 30, 0)$ are colored white, black and white, respectively, what is the color at the point $(25, 20, 0)$?

Remark 7.5. It's clear now from the per-triangle interpolation process that the computation involved in rendering a scene is proportional to its triangle count (or polygon count). In animated applications, in particular, where the scene is repeatedly re-rendered, an important objective then is to minimize this count without compromising visual quality.

Experiment 7.2. Run `interpolation.cpp`, which shows the interpolated colors of a movable point inside a triangle with red, green and blue vertices. The triangle itself is drawn white. See Figure 7.8 for a screenshot.

As the arrow keys are used to move the large point, the height of each of the three vertical bars on the left indicates the weight of the respective triangle vertex on the point's location. The color of the large point itself is interpolated (by the program) from those of the vertices. **End**

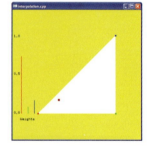

Figure 7.8: Screenshot of `interpolation.cpp`.

Exercise 7.15. (Programming) Replace the triangle declaration of `interpolation.cpp` with (Block 2):

```
glBegin(GL_TRIANGLES);
   glColor3f(1.0, 0.0, 0.0);
   glVertex3f(20.0, 20.0, 0.0);
   glColor3f(0.0, 1.0, 0.0);
   glVertex3f(80.0, 20.0, 0.0);
   glColor3f(0.0, 0.0, 1.0);
   glVertex3f(80.0, 80.0, 0.0);
glEnd();
```

The movable large point is no longer visible except when it pokes out of the triangle. Why?

Exercise 7.16. (Programming) The interpolation procedure described above requires the triangle or segment to be non-degenerate. Find out by writing code how OpenGL draws a degenerate segment or triangle. *Hint:* It doesn't!

Exercise 7.17. It is easy to test a segment with vertices at $P = (x_1, y_1, z_1)$ and $Q = (x_2, y_2, z_2)$ for degeneracy: simply check if the end vertices are identical, i.e., if $x_1 = x_2$ and $y_1 = y_2$ and $z_1 = z_2$.

How about a triangle? How does one test if the triangle with vertices at $P = (x_1, y_1, z_1)$, $Q = (x_2, y_2, z_2)$ and $R = (x_3, y_3, z_3)$ is degenerate? *Hint*: PQR is degenerate if and only if at least one of the two vectors $Q - P$ and $R - P$ is zero, *or*, if they are parallel. It's easy to test if one of them is zero; if not, their being parallel implies that each is a multiple of the other.

Exercise 7.18. If you know about determinants, then write a succinct condition for the degeneracy of a triangle lying on the xy-plane with vertices at $P = (x_1, y_1)$, $Q = (x_2, y_2)$ and $R = (x_3, y_3)$, using a single determinant.

Remark 7.6. A practical application of interpolation to rendering must take into account the fact that screen space is not actually a 2D continuum but, in practice, a rectangular array, called a *raster*, of finitely many pixels. Each pixel is not a point either but a square of non-zero size.

We know – see the discussion of shoot-and-print in Section 2.2 – that a primitive object, such as a triangle t, drawn in the viewing volume is projected to the volume's front and, then, scaled to its image t' on the OpenGL window. See the left and middle diagrams of Figure 7.9. The scaled t' is actually *rendered* by a set of pixels in the OpenGL window – the shaded ones in the raster on the right of Figure 7.9 (admittedly at a rather lousy resolution). A part of the print process, called *rasterization* or *scan conversion*, in fact, consists of choosing and coloring the pixels to render t'.

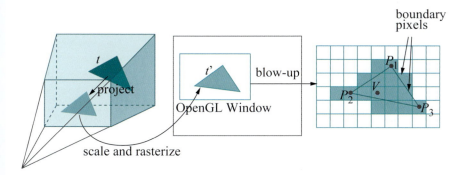

Figure 7.9: Project, scale and rasterize.

We'll be studying rasterization algorithms in depth later on, but it's worth noting a couple of issues at this time in relation to interpolation. Since a pixel is a square that contains not one point, but infinitely many, OpenGL must pick a representative one at which to interpolate the color values from the vertices and then set the entire pixel RGB to those particular values.

For a pixel in the interior of the primitive, a valid choice is its center point, e.g., V on the right of Figure 7.9 for the pixel to which it belongs.

The color of that pixel is then determined by formula (7.7), viz.,

$$c_1(R_1, G_1, B_1) + c_2(R_2, G_2, B_2) + c_3(R_3, G_3, B_3)$$

where $V = c_1 P_1 + c_2 P_2 + c_3 P_3$, and the programmer-specified color at P_i is (R_i, G_i, B_i), for $1 \leq i \leq 3$.

Coloring boundary pixels is more complicated, as both foreground and background colors need to be taken into account. In practice, depending on if effects such as antialiasing are in force, weights may be decided by the area of the pixel occupied by the primitive and the background, respectively. For example, compare the two boundary pixels pointed at in the right of Figure 7.9: the lower one should give greater weight to the foreground color (that of the triangle) than the background, while the opposite is the case for the upper pixel.

The reader may have been wondering the following for a while now. Points, segments and triangles seem to be nicely interpolatable, but how about quadrilaterals and, in general, polygons with four or more vertices? Isn't it true that points belonging to such figures have unique expressions in terms of their vertices as well? The answer is no, as we'll see. First, though, let's learn about convexity and the convex hull, which will lead to the answer and more.

7.4 Convexity and the Convex Hull

There's nothing about convex combinations in Definitions 7.1 and 7.2 that's specific to two or three points. They can be defined for arbitrary numbers of points:

Definition 7.3. If $F = \{P_1, P_2, \ldots, P_k\}$ is a set of k points in $\mathbb{R}^2$, then a point V of the form

$$V = c_1 P_1 + c_2 P_2 + \ldots + c_k P_k$$

where $0 \leq c_i \leq 1$, for $1 \leq i \leq k$, and where $c_1 + c_2 + \ldots + c_k = 1$, is said to be a *convex combination* – or *barycentric combination* – of F. The scalars c_i, $1 \leq i \leq k$, are the *barycentric coordinates* of V.

$Remark$ 7.7. We restrict to $\mathbb{R}^2$ as we'll be computing convex combinations only on a plane. Definitions and results can all be generalized to $\mathbb{R}^3$ and higher.

Corollaries 7.1 and 7.2 tell us that the convex combinations of two points form the segment joining them and those of three points the triangle with corners at these points. How about an arbitrary set $F = \{P_1, P_2, \ldots, P_k\}$ of points on the plane? What object is formed by its convex combinations? To answer the question we need first to define convexity.

Definition 7.4. A non-empty set S of points in $\mathbb{R}^2$ is said to be *convex* if, for any two points $P, Q \in S$, it is true that the segment $PQ \subset S$; in other words, if it is true that, if the endpoints of a segment are in S, then the segment itself is contained in S.

See Figure 7.10 for examples of both convex and non-convex subsets of $\mathbb{R}^2$. Intuitively, convexity ensures that the object has neither holes nor depressions.

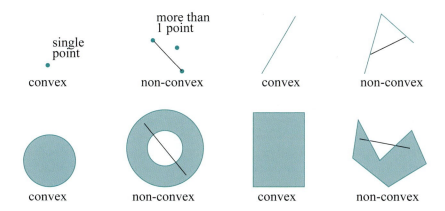

Figure 7.10: Convex and non-convex subsets of $\mathbb{R}^2$ – non-convexity is indicated by an example of a black line with endpoints in the object but that itself is not contained in it.

Exercise 7.19. Prove that points, segments and triangles are always convex.

Exercise 7.20. Prove that the entire plane $\mathbb{R}^2$ is itself convex. What about a *half-plane*, i.e., the part of the plane to only one side of a straight line? Consider both a closed half-plane (i.e., one which includes its bounding straight line) and an open one (which does not).

Exercise 7.21. Prove that the intersection of (any number of) convex sets is again convex. Is the union of two convex sets necessarily convex?

A polygon which happens to be a convex set is, of course, a *convex polygon*. Convex polygons are particularly important in OpenGL. As we noted even in the second chapter, a programmer should ensure that polygons he draws with GL_QUADS, GL_QUAD_STRIP or GL_POLYGON calls are each convex; otherwise, rendering is unpredictable (we'll see why in the next chapter).

Exercise 7.22. Show that it's possible to tell if a (plane) polygon is convex by measuring the internal angle at each vertex.
Hint: Compare the two polygons at the right of Figure 7.10, one of which is convex and the other not.

Exercise 7.23. For the four non-convex figures shown in Figure 7.10, fill them out *minimally* to make them convex; in particular, for each, shade in on the page itself an additional area as small as possible which, together with the original, forms a convex figure.

Part answer: See Figure 7.11.

Figure 7.11: Part answer to Exercise 7.23.

The previous exercise leads to the consideration of the *smallest* possible convex set which contains a given planar set F – obtained by "filling out the holes and depressions" of F. If F is convex already then there's nothing to do as, obviously, F is the smallest convex set containing itself. But does there always exist a smallest convex set containing an arbitrary F?

Consider the collection of *all* convex sets containing a given set F. This collection includes certainly the whole plane itself and, possibly, infinitely many other sets. Now, the intersection X of this collection surely contains F as each member does. And Exercise 7.21 tells us that X is convex as well. Moreover, X is no bigger than any convex set C containing F, because C was one of the collection that was intersected to derive X in the first place. So the answer is yes, there always exists a smallest convex set containing a given planar set F: it is simply the intersection of all convex sets containing F.

Definition 7.5. The smallest convex set containing a set F of points on the plane is called its *convex hull*, denoted $ch(F)$.

Remark 7.8. Again, the restriction to the plane is for our purposes only. It can all be made to work in higher dimensions as well.

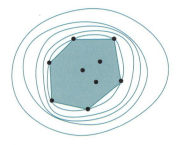

Figure 7.12: A rubber band snapping to bound the convex hull of the nails.

The intersection of infinitely many convex sets is admittedly a rather abstract notion. It's more intuitive to think of $ch(F)$ as the "limit" of a shrinking sequence of convex sets containing F. The process is equivalent to sticking a nail at each point of F, stretching a rubber band around all the nails, then releasing it. See Figure 7.12. When the rubber band becomes taut it bounds the convex hull of F. Figure 7.13 shows the convex hulls of a few small sets of points.

Figure 7.13: Convex hulls.

Experiment 7.3. Run `convexHull.cpp`, which shows the convex hull of 8 points on a plane. Use the space bar to select a point and the arrow keys to move it. Figure 7.14 is a screenshot.

Note: The program implements a very inefficient (but easily coded) algorithm to compute the convex hull of a set F as the union of all triangles with vertices in F. **End**

Example 7.3. What is the convex hull of the set consisting of two opposite edges of a parallelogram? How about that consisting of two adjacent edges?

Answer: The whole (filled) parallelogram. The triangle on the two edges.

Figure 7.14: Screenshot of `convexHull.cpp`.

Exercise 7.24. What are the convex hulls of the figures $+$ and $\times$?

Exercise 7.25. Earlier, in Exercise 7.20, you may have shown that a closed half-plane H is convex. Can you find two straight lines S and T, not necessarily finite, such that $H = ch(S \cup T)$?

We relate next convex hulls to convex combinations.

Proposition 7.3. *Given a set* $F = \{P_1, P_2, \ldots, P_k\}$ *of* k *points in* $\mathbb{R}^2$, $ch(F)$ *is exactly the set of convex combinations of* F.

Proof. *Skip this proof if you start to get bogged down in the math. Just make sure to understand what the proposition says.*

We'll prove first that the set X of convex combinations of F is a convex set. As X obviously contains F, a consequence will be that $ch(F) \subset X$, because $ch(F)$ is the smallest convex set containing F.

Accordingly, given $V, V' \in X$, we have to show that the segment VV' lies in X. As V and V' are convex combinations of F, they can be written as

$$V = \sum_{i=1}^{k} c_i P_i \quad \text{and} \quad V' = \sum_{i=1}^{k} c_i' P_i$$

where $0 \leq c_i, c_i' \leq 1$, for each i, and $\sum_{i=1}^{k} c_i = \sum_{i=1}^{k} c_i' = 1$. A point W on the segment VV' is of the form

$$W = dV + d'V'$$

where $0 \leq d, d' \leq 1$ and $d + d' = 1$. Therefore,

$$W = d \sum_{i=1}^{k} c_i P_i + d' \sum_{i=1}^{k} c_i' P_i = \sum_{i=1}^{k} (dc_i + d'c_i') P_i$$

It's easily verified that $0 \leq dc_i + d'c_i' \leq 1$, for each i. Moreover, $\sum_{i=1}^{k}(dc_i + d'c_i') = d\sum_{i=1}^{k} c_i + d'\sum_{i=1}^{k} c_i' = d.1 + d'.1 = 1$. One concludes that W is a convex combination of F as well and, therefore, belongs to X. As W was an arbitrary point of VV', we have proved that VV' indeed lies in X and, therefore, X is convex.

Next, we'll prove that any convex set Y containing F contains X as well. This will imply that $ch(F) \supset X$, which, together with the proof above that $ch(F) \subset X$, will complete the proof of the proposition.

In fact, we'll prove by induction that Y contains all convex combinations of the sets $\{P_1, P_2, \ldots, P_r\}$, for $r = 1, 2, \ldots, k$ (the case $r = k$ will, of course, prove that Y contains X). Starting the induction at $r = 1$ is trivial as the only convex combination of $\{P_1\}$ is P_1, which belongs to F and, therefore, to Y.

Suppose, inductively, that Y contains all convex combinations of the set $\{P_1, P_2, \ldots, P_r\}$, for some r, $1 \leq r \leq k - 1$. Let $V = \sum_{i=1}^{r+1} c_i P_i$ be a convex combination of $\{P_1, P_2, \ldots, P_r, P_{r+1}\}$. We'll prove that $V \in Y$, completing the induction. We can assume that $c_{r+1} < 1$, because, if $c_{r+1} = 1$, then $V = P_{r+1}$ which trivially belongs to Y. Now,

$$V = \sum_{i=1}^{r} c_i P_i + c_{r+1} P_{r+1}$$

$$= c \left(\sum_{i=1}^{r} \frac{c_i}{c} P_i \right) + c_{r+1} P_{r+1}$$

writing $c = \sum_{i=1}^{r} c_i$, which is not 0 as $c_{r+1} < 1$.

Since $\sum_{i=1}^{r} \frac{c_i}{c} = \frac{\sum_{i=1}^{r} c_i}{c} = 1$, apply the inductive hypothesis to conclude that the point

$$V' = \sum_{i=1}^{r} \frac{c_i}{c} P_i$$

is in Y. Next, use the preceding expression for V' to rewrite the earlier equation for V as

$$V = cV' + c_{r+1}P_{r+1}$$

which is a convex combination of V' and P_{r+1}, because $c + c_{r+1} = 1$, and so lies on the segment $V'P_{r+1}$. Since Y is a convex set containing both V' and P_{r+1}, it contains the segment $V'P_{r+1}$ and, therefore, V as well.

The proof of the proposition is complete. □

Extreme Points

The reader, contemplating again Figure 7.13, will note that some of the points of the two sets on the right lie at the corners of their respective hulls, while others are inside or elsewhere on the boundary. Here's a definition that classifies points accordingly.

Definition 7.6. If a point P of a set F of points on the plane is such that the convex hulls of F and $F - \{P\}$ (i.e., F with P deleted) are the same, then P is said to be a *non-extreme* point of F; otherwise, it is an *extreme point* of F.

Remark 7.9. Colloquially, non-extreme points are "expendable" in that the elimination of any one doesn't affect the convex hull. However, removing an extreme point will change the hull; it will, in fact, become smaller.

Exercise 7.26. What are the extreme points of a set of 10 points chosen arbitrarily from a given circle.

Exercise 7.27. What are the extreme points of a closed disc (mind that a closed disc contains its interior and bounding circle)?

Interpolation and Convexity

We've covered a fair amount of the theory of convexity so far. Let's pause to take stock of how it impacts our understanding of the interpolation process.

We asked at the end of the last section if it's true that points in a polygon with four or more vertices have unique expressions in terms of these vertices. Since the expressions that we seek are convex combinations and since the convex hull of the vertex set consists precisely of its convex combinations, a fair question to ask first is if each point in the convex hull of a set with at least four points has a unique expression as a convex combination of these points.

The answer to this question is *without an exception* no. Figure 7.15 indicates a counter-example in the case of a four-point set $\{P_1, P_2, P_3, P_4\}$ whose members are at the corners of a square. The convex hull of the set, of course, is the square itself. Observe now that the point V has at least two distinct expressions as a convex combination of P_1, P_2, P_3 and P_4:

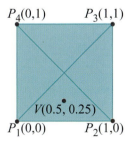

Figure 7.15: V has at least two expressions as a convex combination of P_1, P_2, P_3 and P_4.

(i) $V = 0.25P_1 + 0.5P_2 + 0.0P_3 + 0.25P_4$, obtained by computing for V as belonging to the triangle $P_1P_2P_4$ (so the coefficient of P_3 is 0).

(ii) $V = 0.5P_1 + 0.25P_2 + 0.25P_3 + 0.0P_4$, obtained by computing for V as belonging to the triangle $P_1P_2P_3$ (so the coefficient of P_4 is 0).

In such a situation, there is no unambiguous way in which to interpolate values, such as color, which are defined at the vertices. For example, which set of weights would one use to interpolate at V: $(0.25, 0.5, 0.0, 0.25)$ or $(0.5, 0.25, 0.25, 0.0)$? This problem is actually general and arises in the case of any convex polygon with at least four vertices, as we ask the reader to show.

Exercise **7.28.** Show that in any convex polygon with four or more vertices a point can be found which has more than one expression as a convex combination of its vertices.
Hint: If the polygon is $P_1P_2 \ldots P_n$, a point V in the intersection of the triangles $P_1P_2P_3$ and $P_1P_2P_4$ will do.

Exercise **7.29.** Could it be that some strange plane *non-convex* polygon has the property that all its points are uniquely expressible as convex combinations of its vertices?

Suggested approach: If you did Exercise 7.22 you know that a plane non-convex polygon $P_1P_2 \ldots P_n$ has at least one vertex, say P_r, where the internal angle is greater than $180°$. Show, in fact, that P_r is a non-extreme member of the set of vertices $\{P_1, P_2, \ldots, P_n\}$ and, moreover, contained in a triangle with corners at some three other vertices, say P_{i_1}, P_{i_2} and P_{i_3} (can you find such a triangle for the lone non-extreme vertex of the non-convex polygon on the right of Figure 7.10?). Then P_r itself has more than one expression as a convex combination of the vertices $P_1, P_2, \ldots, P_n$.

We conclude that any polygon with four or more vertices, be it convex or not, contains points with ambiguous representation as a convex combination of its vertices. It is with good reason, therefore, that OpenGL has chosen points, segments and triangles, and *none other*, as its fundamental primitives.

Planarity

One final note is that we have been making the tacit assumption that polygons under consideration are all planar, as evident from the fact that our definitions of convex combination and convexity have been for a plane. This assumption is valid because a polygon is *not even properly defined* if its vertices do not lie on one plane! The following example illustrates the problem.

Example **7.4.** A quadrilateral $q = P_1P_2P_3P_4$ is made starting from a square of sides 2 units on the xy-plane and then lifting one vertex a unit in the z-direction. Specifically, the vertices of q are $P_1 = (1, 1, 0)$, $P_2 = (3, 1, 0)$, $P_3 = (3, 3, 1)$ and $P_4 = (1, 3, 0)$. Now, which is q: is it the union of the two triangles $P_1 P_2 P_3$ and $P_1 P_4 P_3$ (Figure 7.16(a)) or is it the union of the two triangles $P_2 P_1 P_4$ and $P_2 P_3 P_4$ (Figure 7.16(b))? These two shapes are completely disjoint except for the shared boundary!

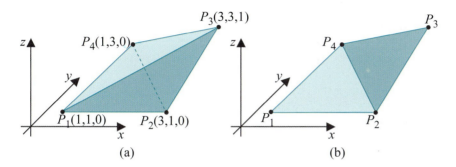

(a)   (b)

Figure 7.16: A non-planar quad q drawn in two different ways.

As a set of at most three vertices is always planar, no such problem as in the preceding example can arise for a point, a segment or a triangle – yet another reason for OpenGL to favor these three.

7.5 Summary, Notes and More Reading

In this chapter we learned that points, segments and triangles are preferred by a drawing API such as OpenGL for the reason that each of their points can be unambiguously interpolated from their vertices. This led to exploring the definition of a convex combination and the notions of convexity and a convex hull.

To learn more about convexity and related algorithms the place to look is the computational geometry literature. The introductory computational geometry texts by de Berg et al. [10] and O'Rourke [103] are good starting points. The mathematical introduction to computer graphics by Buss [21] has a chapter on averaging and interpolation and an interesting discussion of convexity as well.

CHAPTER 8

Triangulation

To create a 3D object in CG one assembles its surface, or a good likeness, with help of 2D primitives. For example, the (seemingly 3D) solid ball and torus in Figure 8.1 are depicted by the surface sphere and surface torus which bound them, respectively (note that the calls glut*Solid*Sphere() and glut*Solid*Torus() in ballAndTorusShadowed.cpp refer to *filled* triangles, not solid 3D objects). Boris's head in Figure 8.2 is a mesh of quads.

The triangle, as we learned in the previous chapter, is the preferred of the 2D primitives. It turns out, though, that simply cobbling together a collection of triangles that resembles the desired object may not be good enough. In order to avoid problems at the time of rendering the collection must follow certain rules. The goal of this chapter is to formulate these rules and understand their importance, particularly in the context of OpenGL.

Section 8.1 begins by listing the rules that make a collection of triangles a so-called triangulation. After several examples of triangulations and non-triangulations it then explains the logic behind these rules, in particular why they make a collection behave predictably at the time of rendering. The next section is a brief discussion of so-called Steiner vertices and how they can be included to improve the quality of a triangulation. Section 8.3 explains OpenGL's own somewhat simple-minded triangulation mechanism and the consequent importance of making sure that all polygons specified in a program are convex. Section 8.4 discusses the automatic triangulation routines, tesselators as they are called, available from the GLU (OpenGL Utility Library). Section 8.5 concludes.

This short chapter, together with its sisters Chapter 7 on convexity and interpolation and Chapter 9 on orientation, goes to the geometric core of CG.

Figure 8.1: Screenshot of `ballAndTorus-Shadowed.cpp`.

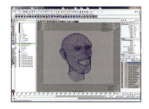

Figure 8.2: Mesh of Boris's head (courtesy of Sateesh Malla at `www.sateeshmalla.com`).

293

8.1 Definition and Motivation

Definition 8.1. Suppose $\mathcal{T}$ is a collection of triangles whose union is an object X. Then $\mathcal{T}$ is said to be a *triangulation* of X if, given any two triangles t_1 and t_2 from $\mathcal{T}$, exactly one of the following three is true:

1. t_1 and t_2 are disjoint, i.e., do not intersect at all.

2. t_1 and t_2 intersect in a vertex of both.

3. t_1 and t_2 intersect in an edge of both.

Informally, triangles in a triangulation are asked to intersect "nicely" or not at all. If $\mathcal{T}$ is a triangulation of X then X is said to be *triangulated* by $\mathcal{T}$.

A collection of triangles may be such that its union is X, but without being a triangulation of X according to the preceding rules – such a collection is called an *invalid triangulation* of X.

Remark 8.1. We should probably call triangulations "valid triangulations" to make the contrast with invalid ones clear, but that would be cumbersome.

Figure 8.3(a) shows the triangulation of a simple non-convex polygon. In Figure 8.3(b), $\{ABC, ADC\}$ is a triangulation of the rectangle $ABCD$.

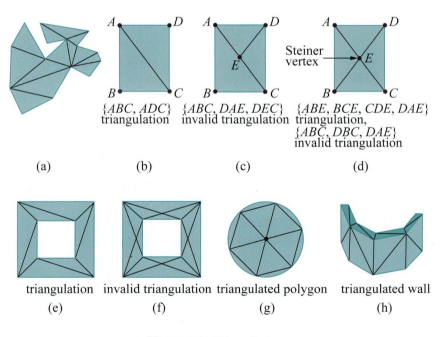

Figure 8.3: Triangulations.

In Figure 8.3(c), $\{ABC, DAE, DEC\}$ is an invalid triangulation of $ABCD$ as ABC intersects DAE in AE, which is an edge of DAE, but *not* of *ABC* (it is only *part* of the edge *AC*). Generally, if the vertex of one triangle lies on another, then it should be a vertex of the second triangle as well, or there is a violation of the triangulation rules.

In Figure 8.3(d), $\{ABE, BCE, CDE, DAE\}$ is a triangulation of *ABCD*, while $\{ABC, DBC, DAE\}$ is an invalid triangulation because *ABD* and *DBC* intersect in a triangle *EBC*, which, of course, is neither an edge nor vertex of either.

Figure 8.3(e) is a familiar triangulation of a square annulus, while Figure 8.3(f) is an often drawn invalid triangulation. Figure 8.3(g) is a triangulated polygon approximating a disc. The approximation may be made closer by increasing the number and decreasing the size of the sides of the polygon. Likewise, the wall of Figure 8.3(h) may be made to appear smoothly curved by increasing the number of flat panels and making them narrower.

Triangulation is not a unique process and the same object may have multiple triangulations.

Exercise 8.1. Draw a different triangulation of the polygon of Figure 8.3(a). Does the number of triangles change?

Exercise 8.2. Can you give a formula for the number of triangles that any triangulation of a simple polygon with n vertices must have? Assume that there is no triangle vertex other than those from the original polygon (e.g., unlike Figure 8.3(c) where vertex *E* is doesn't belong to the input rectangle).

Why the rules for triangulation? As long as a collection of triangles looks like the desired object why should one care if it happens to be a triangulation according to Definition 8.1 or not?

To answer this question consider the invalid triangulation $\{ABC, DBC, DAE\}$ of the rectangle in Figure 8.3(d). Say the programmer has specified color values at the vertices *A-E*. These are interpolated separately through the three triangles *ABC*, *DBC* and *DAE*. What happens though in the region of overlap *EBC* of the two triangles *ABC* and *DBC*? The color of *EBC* is determined by the one of triangles *ABC* and *DBC* which appears *later* in the code, as its color values overwrite those of the earlier one. This exact situation is implemented in the following experiment.

Experiment 8.1. Run `invalidTriangulation.cpp`, which implements exactly the invalid triangulation $\{ABC, DBC, DAE\}$ of the rectangle in Figure 8.3(d). Colors have been arbitrarily fixed for the five vertices *A-E*. Press space to interchange the order that *ABC* and *DBC* appear in the code. Figure 8.4 shows the difference. **End**

Exercise 8.3. (Programming) Theoretically, in Figure 8.3(c) a similar ambiguity arises in the coloring of the segment *AC*, depending on the order of the triangles *ABC*, *DAE* and *DEC* in the code. Try to write code like `invalidTriangulation.cpp` to demonstrate this.

(a)   (b)

Figure 8.4: Screenshots of `invalidTriangulation.cpp` with two different drawing orders: (a) *ABC* drawn first in code, *DBC* next (b) *DBC* drawn first, *ABC* next.

Interestingly, you will find that the expected ambiguity does not arise in practice. The reason is that, at the time of rasterization, pixels on the edge shared between two abutting polygons are assigned uniquely to one or the other, independent of code order, by the polygon rasterizing algorithm (we'll see how in Section 14.4). Accordingly, color values of these border pixels are obtained each from its "owning" polygon and there is no ambiguity.

Generally, it is not desirable that the image of an object be sensitive to the order in which the collection of triangles composing it *happens* to appear in the code. From a designer's perspective it should be enough to simply specify (a) a set of vertices, possibly, with color and other data at each, and (b) a collection of triangles with corners among these vertices. It should *not* be necessary to additionally specify a particular order on the collection. We have now the following proposition which the reader is asked next to prove.

Proposition 8.1. *If the collection of triangles composing an object satisfies the properties of a triangulation, then its image is independent of the order in which the triangles are rendered, i.e., independent of their code order.*

Exercise 8.4. Prove the proposition just stated.
Suggested approach: Let t_1 and t_2 be two triangles of a triangulation. If they don't intersect then, of course, they cannot "conflict" in coloring a region and code order between them does not matter. If they intersect in a vertex v of both then there cannot be a conflict to be resolved by code order either, because colors are specified per-vertex. If they intersect in an edge...

Remark 8.2. Even if a 2D collection is of quads, or a mix of quads and triangles, rules exactly as in Definition 8.1 can be applied and with exactly the same benefit as stated in the preceding proposition.

8.2 Steiner Vertices and the Quality of a Triangulation

A vertex used in the triangulation of an input object which is not a vertex of the object itself is called a *Steiner vertex*. For example, E is a Steiner vertex of the triangulation $\{ABE, BCE, CDE, DAE\}$ of the rectangle $ABCD$ of Figure 8.3(d). Even though they may not be necessary per se in order to triangulate, Steiner vertices are often inserted to improve the "quality" of a triangulation.

Roughly, a good triangulation is one where there are few long and thin triangles, called *slivers*, and where most triangles are of nearly equal size and relatively small with respect to the entire object. We'll not try to formalize any further the notion of the quality of a triangulation, but here's an example to explain how it can impact rendering.

Consider the long rectangular floor $ABCD$, triangulated as in the top of Figure 8.5. Located at either end are two lamps emitting white light – one can create light sources in OpenGL, as we'll see. The vertices A, B, C and D are clearly well-lit; say the color tuple computed at each is $(0.9, 0.9, 0.9)$. Recalling that OpenGL interpolates over a triangle the values evaluated at its vertices, the color at *every* point of $ABCD$ turns out to be $(0.9, 0.9, 0.9)$, even though, realistically, a point in its interior, such as P, should appear darker.

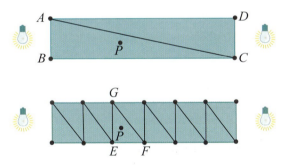

Figure 8.5: Triangulations of different quality.

Unfortunately, the sliver ABC causes what should be a local brightness to propagate globally. Such a problem can often be alleviated by improving the triangulation – e.g., the lower one of Figure 8.5 which uses Steiner vertices. There, the color intensities computed at E, F and G will be less owing to the increased distance from the light sources, so the interpolated values at P less too.

Exercise 8.5. Typical triangulations of a disc that come to mind are shown in Figures 8.6(a) and (b), but in either the problem with slivers becomes severe with an increasing number of edges. Can you suggest a better quality of triangulation? Maybe so that most of the disc is covered with "good" triangles and only a small area with slivers.

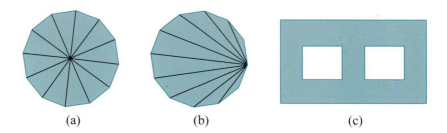

(a)   (b)   (c)

Figure 8.6: (a) and (b) Triangulations of a disc (c) Double annulus.

Exercise 8.6. Triangulate the double annulus depicted in Figure 8.6(c) using exactly one triangle strip, with the help of Steiner vertices.

8.3 Triangulation in OpenGL and the Trouble with Non-Convexity

We'll now resolve the mystery that arose in Experiment 2.19 of Chapter 2. Here's the experiment again.

Experiment 8.2. Replace the polygon declaration of `square.cpp` with (Block 1*):

```
glBegin(GL_POLYGON);
   glVertex3f(20.0, 20.0, 0.0);
   glVertex3f(80.0, 20.0, 0.0);
   glVertex3f(40.0, 40.0, 0.0);
   glVertex3f(20.0, 80.0, 0.0);
glEnd();
```

Display it *both* filled and outlined using appropriate `glPolygonMode` calls – you see a non-convex quadrilateral in either case (see Figure 8.7(a)).

Next, keeping the *same* cycle of vertices as above, list them starting with `glVertex3f(80.0, 20.0, 0.0)` instead (Block 2):

```
glBegin(GL_POLYGON);
   glVertex3f(80.0, 20.0, 0.0);
   glVertex3f(40.0, 40.0, 0.0);
   glVertex3f(20.0, 80.0, 0.0);
   glVertex3f(20.0, 20.0, 0.0);
glEnd();
```

*To cut-and-paste you can find the block in text format in the file `chap8codeModifications.txt` in the directory `Code/CodeModifications`.

Make sure to display it both filled and outlined. When filled it's a triangle, while outlined it's a non-convex quadrilateral identical to the one output earlier (see Figure 8.7(b))! Because the cyclic order of the vertices is unchanged, shouldn't it be as in Figure 8.7(a) both filled and outlined? **End**

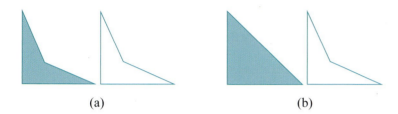

(a)   (b)

Figure 8.7: Experimental outputs.

Here is what's happening. When OpenGL is asked to draw a *filled* polygon P with n vertices $v_0, v_1, \ldots, v_{n-1}$, it renders a fan of $n-2$ triangles around the first vertex, exactly as though the call was made using the primitive `GL_TRIANGLE_FAN` instead of `GL_POLYGON`; in particular, the triangles of the fan are $v_0 v_1 v_2$, $v_0 v_2 v_3$, $\ldots$, $v_0 v_{n-2} v_{n-1}$.[*]

Now, if the polygon $P = v_0 v_1 \ldots v_{n-1}$ is convex then the fan around v_0 is a triangulation of P. In fact, if P is convex then it does not matter how the cycle of vertices is listed, i.e., at which vertex it starts: the fan around the first vertex, or any vertex for that matter, is *always* a triangulation of P. For example, Figures 8.8(a) and (b) show the triangulation fans corresponding to cyclically rotated listings of the vertices of a convex pentagon.

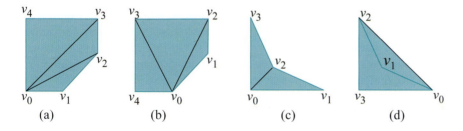

(a)   (b)   (c)   (d)

Figure 8.8: Triangle fans.

Remark 8.3. That OpenGL seeks to triangulate the polygon before rendering is understandable given, from what we learned in Chapter 7, that it can then unambiguously interpolate property values from the vertices through individual triangles.

---

[*]This triangulation is implementation-dependent but all implementations that we are aware of behave as described.

However, if the polygon $P = v_0 v_1 \ldots v_{n-1}$ is not convex then there is no guarantee that the fan $v_0 v_1 v_2$, $v_0 v_2 v_3$, ..., $v_0 v_{n-2} v_{n-1}$ is a triangulation of P. For example, the listing $v_0 v_1 v_2 v_3$ as in Figure 8.8(c) of the vertices of the non-convex quadrilateral of the preceding experiment does, in fact, give the fan triangulation $\{v_0 v_1 v_2, v_0 v_2 v_3\}$. However, the fan from the listing as in Figure 8.8(d) not only does not give a triangulation, it does not even give an invalid triangulation, as the union of the triangles $v_0 v_1 v_2$ and $v_0 v_2 v_3$ is $v_0 v_2 v_3$ itself, which is larger than the input quadrilateral $v_0 v_1 v_2 v_3$! This explains the differing filled outputs of Experiment 8.2.

It's accordingly vital to ensure that each filled polygon specified by GL_QUADS, GL_QUAD_STRIP and GL_POLYGON is convex; otherwise, rendering is unpredictable and may even be incorrect depending on the vertex listing.

When asked to draw a polygon in outline, OpenGL draws a line loop using the given vertex sequence, which is always valid however the vertices are listed. This explains the correctness of the outlined outputs in both cases in Experiment 8.2.

Why doesn't OpenGL automatically pre-triangulate a non-convex polygon? Then the problem of Experiment 8.2 would not arise. In fact, OpenGL does have triangulation routines (*tesselators*, as they are called) in its Utility Library GLU. However, it's left to the user to call them if she so desires – we'll discuss how in the next section – rather than OpenGL automatically imposing such a call on its own. There are two reasons for this:

(a) OpenGL's mission is to afford an efficient and portable 3D API, for which reason it's restricted to specifying low-level operations close to the hardware. Complex operations built on top of these are not mandated, but left to the GLU and other extensions.

(b) Recall that triangulations are not unique, may differ in quality, and that one can choose to add Steiner vertices. Only the programmer can know her particular requirements. On its own, therefore, OpenGL attempts no more than the least possible, leaving the programmer to implement more sophisticated routines if need be.

Exercise 8.7. (Programming) Replace the polygon declaration of square.cpp with (Block 3):

```
glBegin(GL_POLYGON);
   glColor3f(0.0, 0.0, 0.0);
   glVertex3f(80.0, 20.0, 0.0);
   glColor3f(1.0, 0.0, 0.0);
   glVertex3f(40.0, 40.0, 0.0);
   glColor3f(0.0, 0.0, 0.0);
   glVertex3f(20.0, 80.0, 0.0);
   glVertex3f(20.0, 20.0, 0.0);
glEnd();
```

It's actually the second listing of the polygon of Experiment 8.2 with all vertices, except the second, colored black, while the second vertex itself is colored red. The rendered figure is all black with no sign of red at all. Why?

Exercise 8.8. (Programming) Replace the polygon declaration of square.cpp with the following piece of code specifying a non-convex pentagon (Block 4):

```
glBegin(GL_POLYGON);
   glVertex3f(50, 10, 0);
   glVertex3f(40, 50, 0);
   glVertex3f(10, 60, 0);
   glVertex3f(90, 60, 0);
   glVertex3f(60, 50, 0);
glEnd();
```

Sketch the pentagon on graph paper first and then predict the filled output each time as you rotate the vertices cyclically.

Even in the case of a convex polygon, different triangulations may lead to different renderings as the following exercise illustrates.

Exercise 8.9. (Programming) Replace the polygon declaration of square.cpp with the following to make a colored pentagon (Block 5):

```
glBegin(GL_POLYGON);
   glColor3f(0.0, 0.0, 0.0);
   glVertex3f(20.0, 20.0, 0.0);
   glVertex3f(50.0, 20.0, 0.0);
   glVertex3f(80.0, 50.0, 0.0);
   glVertex3f(80.0, 80.0, 0.0);
   glColor3f(1.0, 0.0, 0.0);
   glVertex3f(20.0, 80.0, 0.0);
glEnd();
```

All the vertices are black except the last one listed, which is red. Next, cyclically rotate the vertices, *preserving* their colors (Block 6):

```
glBegin(GL_POLYGON);
   glColor3f(0.0, 0.0, 0.0);
   glVertex3f(50.0, 20.0, 0.0);
   glVertex3f(80.0, 50.0, 0.0);
   glVertex3f(80.0, 80.0, 0.0);
   glColor3f(1.0, 0.0, 0.0);
   glVertex3f(20.0, 80.0, 0.0);
   glColor3f(0.0, 0.0, 0.0);
   glVertex3f(20.0, 20.0, 0.0);
glEnd();
```

Explain the difference in rendering. Verify your understanding by calculating the color of the point $(50.0, 70.0, 0.0)$ in either pentagon and actually drawing a point of that color at $(50.0, 70.0, 0.0)$ (which should then be invisible).

Hint: See Figures 8.8(a) and (b).

Exercise 8.10. Raising the first vertex of `square.cpp` from `glVertex3f-(20.0, 20.0, 0.0)` to `glVertex3f(20.0, 20.0, 1.5)` causes the square – actually, the new figure which is not a square any more – to be clipped. If, instead, the second vertex is raised from `glVertex3f(80.0, 20.0, 0.0)` to `glVertex3f(80.0, 20.0, 1.5)`, then the figure is clipped too, *but* very differently from when the first vertex is raised. Why? Should not the results be similar by symmetry?

Remark 8.4. Worse than non-convexity can happen if a polygon has more than three vertices: the specified vertices may not be coplanar, in which case the polygon is not even properly defined, as Example 7.4 of the last chapter makes clear.

Points, lines and triangles are always plane and always convex. *Try to use only these!* In fact, our recommendation is to *avoid* GL_QUADS, GL_QUAD_STRIP and GL_POLYGON *altogether* (if a quadrilateral or polygon with more than three vertices does, in fact, arise in your design, then simply triangulate it).

Remark 8.5. OpenGL ES, the lean and mean version of OpenGL for small devices, has done away with quadrilateral and polygon primitives altogether.

Exercise 8.11. (Programming) Draw the objects in Figure 8.9 after first triangulating them. Allow rendering both filled and wireframe.

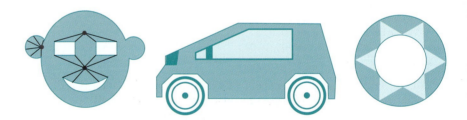

Figure 8.9: Objects to draw.

Make true holes for the eyes and mouth of the mask, which is a flat object. In addition to the vertices you'll need on the circular arcs in order to approximate them, it may be useful to situate Steiner vertices in the interior of the mask as well (possible strategic locations are indicated in the figure).

Make the car simple and boxy. Keep in mind that it is 3D and depicted in Figure 8.9 is just a side view. In fact, copy code from `hemisphere.cpp` to rotate an object so that the car may be viewed from all angles.

For the flat decorated annulus make sure that the triangles in the darkly shaded part are separate from those in the lightly shaded part.

8.4 OpenGL Tesselation

We have already learned that OpenGL polygons and quadrilaterals that one draws with GL_POLYGON and GL_QUAD* primitives should be convex; otherwise, rendering may not be correct. Consequently, 2D figures which happen not to be convex polygons, such as the two in Figure 8.10, must be decomposed into convex pieces – preferably triangles – prior to drawing. Till now we had always been triangulating "by hand" (i.e., by programming in our own triangulation of the object being made), which is not a bad thing as we retain control over the quality and, moreover, can choose to include Steiner vertices. However, the OpenGL Utility Library (GLU) does provide routines for automatic triangulation or, as they call it, *tesselation*, if one wishes to avoid the chore.

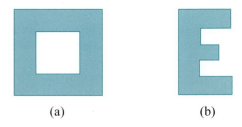

(a)   (b)

Figure 8.10: (a) The square annulus is a non-simple non-convex polygon (non-simple because it has two boundary components) (b) The 'E' is a simple non-convex polygon.

Experiment 8.3. Run tesselateAnnulus.cpp, which uses GLU tesselation routines to triangulate a square annulus, a non-simple non-convex polygon. Press the space bar to see the triangulation (Figure 8.11(a)). Compare with the triangulation we did ourselves of a square annulus in squareAnnulus1.cpp. *End*

There is an amount of managerial overhead, particularly in creating a new tesselation object of type GLUtesselator with a gluNewTess() call and specifying callback routines with gluTessCallback() commands. The description of the square annulus itself of tesselateAnnulus.cpp is contained within a gluTessBeginPolygon() and gluTessEndPolygon() pair. Moreover, each of its two boundary components is described by a sequence of gluTessVertex() statements within a gluTessBeginContour() and gluTessEndContour() pair.

Remark 8.6. GLU tesselation routines never generate Steiner vertices.

The following is an example of the tesselation of a simple non-convex polygon.

Experiment 8.4. Run tesselateE.cpp. Press space ((Figure 8.11(b)). *End*

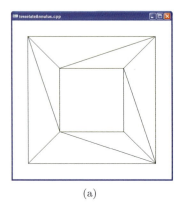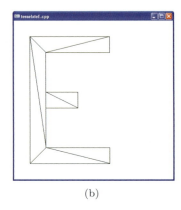

(a)　　　　　　　　　　　　　　　　　(b)

Figure 8.11: Screenshots of (a) `tesselateAnnulus.cpp` and (b) `tesselateE.cpp`.

We'll not go further into the GLU tesselation routines and refer the interested reader to the red book. Our experience though is that there is rarely occasion to invoke them as, mostly, polygons arising in practice either have few edges, or have numerous symmetries, programmer-coded triangulation in either case not being difficult.

8.5　Summary, Notes and More Reading

In this chapter we learned the importance of triangulation and the potential problems arising from an invalid one. We learned as well of OpenGL's somewhat cavalier default attitude toward triangulation and the actually good reasons for it.

For more on the topic, computational geometry literature, in particular, deals extensively with theory of triangulation – and tetrahedralization, its 3D equivalent – as well as practical algorithms for both. There are practical algorithms to triangulate a simple plane polygon with n vertices in $O(n \log n)$ time. You will find them described in de Berg et al. [10] and O'Rourke [103], both good introductions to computational geometry in general. The CGAL [27] library is a marvelous source of ready-to-use algorithms for various geometric applications, including triangulation.

Mesh generation, as triangulation is often called, is obviously key to object creation in computer graphics. We'll be seeing much more of this process as we go along and Chapter 10 on design is mostly devoted to the topic.

An advanced text on mesh generation is by Edelsbrunner [39]. The proceedings of the annual Meshing Roundtable Conference organized by Sandia National Laboratories [70] is a source for the latest developments in the field.

CHAPTER 9

Orientation

T he notion of orientation is vitally important in CG when drawing 3D scenes but, unfortunately, often confusing for the beginner. OpenGL itself makes critical use of orientation to determine the visible side of a surface. Note that the word "orientation" in the current context relates to handedness, e.g., clockwise or counter-clockwise, as we shall see, and has nothing to do with the orientation of a camera as discussed in Section 4.6.3, where the word meant pose or arrangement. The goal for this chapter is an understanding of orientation and its utility in CG.

The first section motivates the concept of orientation with a benign thought experiment. Section 9.2 describes how OpenGL applies orientation to determine the particular side of a 2D primitive which the viewer sees and then renders it with that side's specified material properties. If an object is specified as a collection of triangles, as in a triangulation, the question then arises of consistently orienting the collection. This is the topic of Section 9.3. Section 9.4 describes how OpenGL can make use of orientation to improve the efficiency of its rendering pipeline by culling certain triangles belonging to a closed surface, a procedure called back-face culling. In Section 9.5 we see how geometric transformations affect the perceived orientation of a primitive. We conclude in Section 9.6.

Although the three are conceptual in nature without a lot of excitement by way of programming, this chapter and the two preceding ones form a good part of the geometric core of CG.

9.1 Motivation

A thought experiment:

You and your friend, environmentally-conscious types both, are headed separately toward a meeting of the Tree Huggers' Union. The meeting is out in the open in a field with, well, lots of trees and no other landmarks. There is, though, a triangle of long helium-filled balloons with the letters T, H and U at the corners floating high above the meeting site. See Figure 9.1 (ignore superman with a spray can and the sheet in the middle for now).

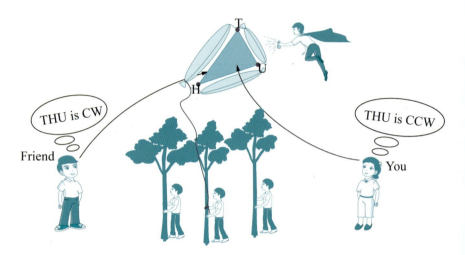

Figure 9.1: Meeting of the Tree Huggers' Union.

Now, you want to meet up with your friend before running into the crowd. So while walking you call him on his cell phone to try to figure out how he is currently situated with respect to you. How do you do this?

As both can see the balloons, a start is to determine if you are on the same side or not. Unfortunately, the letters at the corners (carefully chosen, of course!) are of no help as they each look the same from either side.

What you can do, though, is ask your friend, "Does the vertex sequence THU – that's T→H→U – appear CW (clockwise) or CCW (counter-clockwise) from where you are?" If the orientation appears the same for both, then you are on the same side of the balloons; if not, you are on opposite sides.

OpenGL, as well, must determine for each triangle if the viewer currently sees one side or the other. And, as we'll see, it does so in an exactly similar manner. We'll understand as well the reason for this (seemingly) roundabout method. Why does OpenGL need to distinguish sides in the first place? Because they may have properties (e.g., outlined/filled, color, etc.) specified differently by the programmer and OpenGL is obliged to display accordingly.

Figure 9.2: A bowl of two colors.

For example, if the inside of a triangulated bowl is green and the outside red, then the two sides of every triangle composing it are colored differently as well. Given a viewpoint, OpenGL must determine the visible side of each triangle and render it with the appropriate color. See Figure 9.2. From the current (reader's) viewpoint the red side of triangle t_1 and the green side of triangle t_2 are visible. If the viewpoint travels $180°$ around the bowl, then the visible side is reversed for both.

9.2 OpenGL Procedure to Determine Front and Back Faces

Here then is the procedure that OpenGL follows.

(1) First, it obtains the vertex orders of each 2D primitive from the code. For example, the declaration

```
glBegin(GL_TRIANGLES);
    v0; v1; v2; v3; v4; v5;
glEnd();
```

specifies the order of the vertices of the first triangle as v0, v1, v2 and that of the second as v3, v4, v5 (these orders are part of the GL_TRIANGLES definition; see Section 2.6). The declaration

```
glBegin(GL_TRIANGLE_STRIP);
    v0; v1; v2; v3; v4; v5;
glEnd();
```

specifies the vertex orders of the four successive triangles in the strip as v0, v1, v2 and v1, v3, v2 and v2, v3, v4 and v3, v5, v4. And, similarly, for the other 2D primitives.

(2) Second, OpenGL determines for each component primitive if the order of its vertices as determined in Step (1) is *perceived* as CW or CCW by the viewer. This is said to be the *orientation* of the primitive with respect to the viewer (keep in mind that orientation as just defined has nothing to do with the identical word used to describe the pose of a camera in Section 4.6.3).

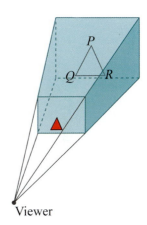

OpenGL can make this determination because it knows both the location of the viewer – at the origin in case of perspective projection and at some point on the viewing face (it doesn't matter which) in case of orthographic projection – and those of the primitive's vertices. For example, in Figure 9.3, if the vertex order of the triangle is P, Q, R, then it is perceived as CCW by the viewer. We'll see later in this section a specific algorithm to output the orientation given these respective inputs.

Figure 9.3: PQR is oriented CCW to the viewer, so it's rendered according to properties for the front face.

(3) Finally, those component primitives whose orientation the viewer perceives as CCW are presumed to be *front-facing*, i.e., the viewer is presumed by OpenGL to see their front faces, while those whose

orientation is perceived as CW *back-facing*. This is actually the default, which can be flipped with a `glFrontFace(GL_CW)` call. Front-facing components are rendered with properties specified for their front faces, while back-facing ones with those for their back faces.

For example, if the vertex order of triangle t_1 in Figure 9.2 happens to be P, Q, R and the viewer is the reader, then OpenGL determines that this triangle is oriented CCW with respect to the viewer, who sees, therefore, the front face. Accordingly, t_1 is rendered red, assuming that the code indeed specifies that front-facing triangles are red. In Figure 9.3 we show the red rendering on the viewing face itself, pretending that it is the OpenGL window.

The reader may wonder at this point why one needs to invoke a *particular* viewpoint to distinguish sides. In real life the inside of the bowl (which is *absolute* and does not depend on the location of any viewer) is painted green and the outside (absolute as well) red. Subsequently, a viewer's perception is determined simply by the laws of nature, in particular, how light from the bowl travels to her eyes.

Why doesn't OpenGL try and simulate this phenomenon? The answer is that, yes, it is true that the inside and outside of the bowl are absolute irrespective of the viewer, but *only after the entire bowl has been created*! If one breaks off a tiny piece of the bowl – a tiny flat triangle, if you will – and shows it to someone who has never seen the whole, then it is not possible for that person to decide which side of the piece originally lay on the bowl's inside and which the outside (Figure 9.4). OpenGL has no global notion of objects either as it simply draws them triangle by triangle, and, therefore, requires direction from the programmer as to which side of each triangle is which.

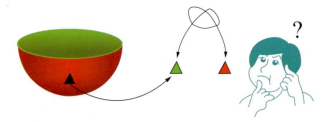

Figure 9.4: Was the inside of the bowl red or green?

The three-step procedure described above provides exactly a mechanism for such direction. Let's return to the thought experiment at the start of the section and assume that there is a giant triangular sheet of paper attached to the balloons (as in Figure 9.1) which you know is colored differently on either side. Then, of course, you could ask your friend, "What color do you see up there?" instead of "Does the vertex sequence THU – that's T→H→U – appear CW or CCW from where you are?" The point is that the two questions are exactly equivalent in that those who perceive a particular orientation see a particular side and vice versa.

Continuing with this line of thought, suppose you are walking that you notice a man high up about to spray-paint the sides of the triangular paper and he has a cell phone which you can call. You could then either ask him to arbitrarily paint one side green and the other red, which would at least serve the purpose of locating your friend, or you could ask him to paint your side green and the other red, which allows you (the programmer) to dictate that "CCW-seers" see red and "CW-seers" green.

There are three points worth emphasizing:

(a) First, "front-facing" and "back-facing" are merely terms to call one side and the other. There is no *intrinsic* front or back of an OpenGL triangle or other 2D primitive. If we didn't use these terms, we would have to say things like "the side which the viewer sees when the order v0v1v2 appears clockwise from the origin".

(b) A real-life 2D object (like a piece of paper) actually has two physical sides regardless of which an observer sees. This is not true of OpenGL, whose objects are all, of course, virtual. An OpenGL 2D primitive such as a triangle consists simply of data, e.g., vertex coordinates, color values, etc., residing inside the computer.

When asked to draw, OpenGL determines *if* the viewer is *supposed* to see the front or the back face according to the procedure described earlier and then *displays* the primitive with properties specified for that face. And, what it displays, of course, is simply a set of colored pixels in the OpenGL window (which has only one side!).

(c) OpenGL draws primitives *one by one* as they occur in the code. It has no global understanding of the objects formed by these primitives *together*.

Exercise 9.1. If a triangle t is specified by

```
glBegin(GL_TRIANGLES); v0; v1; v2; glEnd();
```

where the vertices are as below, in each case determine which side of t, front or back, a viewer at the origin sees, assuming the default of glFrontFace(GL_CCW):

(a) $v_0 = (1, 0, 0)$, $v_1 = (0, 1, 0)$, $v_2 = (0, 0, 1)$

Answer:

The back face because $v_0 v_1 v_2$ appears CW from O. See Figure 9.5.

(b) $v_0 = (0, 1, 0)$, $v_1 = (1, 0, 0)$, $v_2 = (0, 0, 1)$

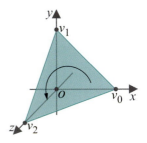

Figure 9.5: $v_0 v_1 v_2$ appears CW from O.

(c) $v_0 = (-1, 0, 0)$, $v_1 = (0, -1, 0)$, $v_2 = (0, 0, -1)$

(d) $v_0 = (1, 1, 1)$, $v_1 = (1, 1, -2)$, $v_2 = (-1, 1, -2)$

Exercise 9.2. A tacit assumption in all of the preceding discussion is that a viewer at a particular location sees, in fact, *only one* side – front or back – of a 2D primitive. For example, if a viewer could see both sides of a triangle, then is it front or back facing (or both)? Moreover, how then would one reconcile the situation with the three-step procedure at the start of the section, which purports to determine a unique orientation for the primitive?

So, is the assumption that only one side is visible a valid one?

Hint: A triangle is always *flat* (planar), while a quad or polygon should be specified to be so.

Definition 9.1. Two orders of the vertices of a polygon are said to be *equivalent* if one can be *cyclically rotated* into the other.

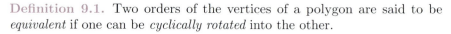

It follows that the sequence of vertices around any given polygon can be written in exactly *two* inequivalent orders. For example, the sequence of vertices around the quadrilateral q of Figure 9.6 can be written in eight different ways:

Figure 9.6: A quadrilateral.

$$v_0 v_1 v_2 v_3 \qquad v_1 v_2 v_3 v_0 \qquad v_2 v_3 v_0 v_1 \qquad v_3 v_0 v_1 v_2$$

$$v_0 v_3 v_2 v_1 \qquad v_3 v_2 v_1 v_0 \qquad v_2 v_1 v_0 v_3 \qquad v_1 v_0 v_3 v_2$$

The orders on the top line are all equivalent to each other, while those on the second all to each other as well, and none on the first equivalent to any on the second. The notion of equivalence is important precisely because of the fact that a viewer on one side of a polygon perceives equivalent orders of vertices as either all CW or all CCW.

Experiment 9.1. Replace the polygon declaration part of `square.cpp` with (Block 1*):

```
glPolygonMode(GL_FRONT, GL_LINE);
glPolygonMode(GL_BACK, GL_FILL);
glBegin(GL_POLYGON);
    glVertex3f(20.0, 20.0, 0.0);
    glVertex3f(80.0, 20.0, 0.0);
    glVertex3f(80.0, 80.0, 0.0);
    glVertex3f(20.0, 80.0, 0.0);
glEnd();
```

This simply adds the two `glPolygonMode()` statements to the original `square.cpp`. In particular, they specify that front-facing polygons are to be drawn in outline and back-facing ones filled. Now, the order of the vertices

---

*To cut-and-paste you can find the block in text format in the file `chap9codeModifications.txt` in the directory `Code/CodeModifications`.

is (20.0, 20.0, 0.0), (80.0, 20.0, 0.0), (80.0, 80.0, 0.0), (20.0, 80.0, 0.0), which appears CCW from the viewing face. Therefore, the square is drawn in outline.

Next, rotate the vertices cyclically so that the declaration becomes (Block 2):

```
glPolygonMode(GL_FRONT, GL_LINE);
glPolygonMode(GL_BACK, GL_FILL);
glBegin(GL_POLYGON);
   glVertex3f(20.0, 80.0, 0.0);
   glVertex3f(20.0, 20.0, 0.0);
   glVertex3f(80.0, 20.0, 0.0);
   glVertex3f(80.0, 80.0, 0.0);
glEnd();
```

As the vertex order remains equivalent to the previous one, the square is still outlined.

Reverse the listing next (Block 3):

```
glPolygonMode(GL_FRONT, GL_LINE);
glPolygonMode(GL_BACK, GL_FILL);
glBegin(GL_POLYGON);
   glVertex3f(80.0, 80.0, 0.0);
   glVertex3f(80.0, 20.0, 0.0);
   glVertex3f(20.0, 20.0, 0.0);
   glVertex3f(20.0, 80.0, 0.0);
glEnd();
```

The square is drawn filled as the vertex order now appears CW from the front of the viewing box. **End**

Exercise 9.3. (Programming) If the polygon declaration part of `square.cpp` is replaced with the following piece of code (Block 4), then is an outlined or filled triangle seen? Try to answer first without running the program.

```
glFrontFace(GL_CW);
glPolygonMode(GL_FRONT, GL_LINE);
glPolygonMode(GL_BACK, GL_FILL);
glBegin(GL_TRIANGLES);
   glVertex3f(80.0, 10.0, -1.0);
   glVertex3f(90.0, 75.0, 1.0);
   glVertex3f(15.0, 10.0, 0.5);
glEnd();
```

Remark 9.1. Before we get to Chapter 11 and learn about material properties and how to color the sides of an object differently, we'll have to do with distinguishing them by the unglamorous means of drawing one in outline and the other filled.

Algorithm to Decide the Orientation Perceived by a Viewer

An algorithmic question, that we did not address then, arose earlier in this section in Step (2) of OpenGL's procedure to determine the side of a primitive a viewer sees: given a viewpoint and a primitive with its vertices ordered, how to decide if the given order appears CW or CCW? We invite the reader to answer this for herself in the following exercise, with a fair amount of input from our end.

Exercise 9.4. Assume that the viewpoint is at the origin O and that the vertices of a triangle are $P = (x_1, y_1, z_1)$, $Q = (x_2, y_2 z_2)$ and $R = (x_3, y_3, z_3)$. See Figure 9.7. Determine if the viewer at O perceives the order PQR as CCW or CW.

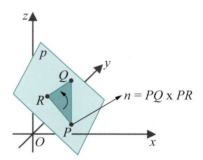

Figure 9.7: The plane p contains the triangle PQR: the orientation of PQR depends on which side of p the viewer is located.

If you don't do the exercise do at least read the conclusion below in terms of the determinant D.

Suggested approach: Supposing, first, that P, Q and R are not collinear, i.e., PQR is a non-degenerate triangle, determine the equation $ax + by + cz + d = 0$ of the unique plane p containing P, Q and R.

A point (x, y, z) lies on p if $ax + by + cz + d = 0$. A point lies in one *half-space* of p, i.e., on one side of p or the other, depending on whether $ax + by + cz + d < 0$ or $ax + by + cz + d > 0$. In particular, one half-space consists of all points (x, y, z) such that $ax + by + cz + d < 0$, and the other such that $ax + by + cz + d > 0$. (To be particular, we are talking of open half-spaces excluding the plane itself.)

Observe that a viewer located on p sees triangle PQR "edge-on", in other words, as a line and not a triangle, so the question of orientation does not arise. Moreover, the orientation of PQR perceived by a viewer not on p depends on the half-space he is in: all viewers in one half-space perceive CCW, while those in the other CW.

Therefore, the perception at viewpoint O, in particular, depends on whether O lies on p or, if not, on which side.

Finally, conclude the following:

Let D be the determinant

$$\begin{vmatrix} x_1 & x_2 & x_3 \\ y_1 & y_2 & y_3 \\ z_1 & z_2 & z_3 \end{vmatrix}$$

1. If $D = 0$, then either (a) P, Q and R are collinear, in which case PQR is a degenerate triangle and the question of an orientation of PQR does not arise, or (b) O lies on the plane p containing P, Q and R, so that the viewer at O sees triangle PQR edge-on and, again, the question of orientation does not arise.

2. If $D > 0$, then the viewer at O perceives the order PQR as CW.

3. If $D < 0$, then the viewer at O perceives the order PQR as CCW.

Another approach is with the use of cross-products, by observing that $n = PQ \times PR$ is normal to the plane p and, in fact, points to the half-space where observers perceive PQR as CCW. Therefore, if the eye direction vector PO makes an angle of less than $90°$ with n – placing it in the same half-space as n – then the viewer at O perceives the order PQR as CCW; if greater, then as CW (in the configuration depicted in the figure the angle is, in fact, greater than $90°$). Whether the angle between the two vectors n and PO is greater or less than $90°$ can be decided from the sign of the dot product $n \cdot PO$.

Note: If you're not familiar with the dot or cross-product of vectors, we have short sidebars Sections 4.6.1 and 5.4.3, respectively.

$\mathbf{E}$xercise 9.5. Does a viewer at the origin perceive the order PQR of the points $P = (-1, 2, 0)$, $Q = (3, 2, 2)$ and $R = (-3, -8, 6)$ as CW or CCW?

$\mathbf{E}$xercise 9.6. Does a viewer at the point $O' = (1, 3, 2)$ perceive the order PQR of the points $P = (3, 7, 5)$, $Q = (4, 1, 2)$ and $R = (0, 1, 2)$ as CW or CCW?
Hint: Translate all points by $(-1, -3, -2)$ to bring O' to the origin and then apply the result of Exercise 9.4.

$\mathbf{E}$xercise 9.7. Relate Lemma 5.1 to the answer to Exercise 9.4.

9.3 Consistently Oriented Triangulation

The notion of orientation gets more interesting when one considers a collection of triangles, as in a triangulation. The issue arises then of *consistency*. We have the following definition:

Definition 9.2. Suppose an order is given of the vertices of each triangle belonging to some triangulation $\mathcal{T}$ of an object X. $\mathcal{T}$ is said to be *consistently oriented* if any two triangles of $\mathcal{T}$ which share an edge order the shared edge oppositely; otherwise, $\mathcal{T}$ is *inconsistently oriented*.

Figure 9.8(a) shows a consistently oriented triangulation. For example, the edge shared by the two triangles $v_0 v_1 v_2$ and $v_1 v_3 v_2$ is ordered $v_1 v_2$ by the first and $v_2 v_1$ by the second. The triangulation of Figure 9.8(b) is not consistently oriented as the edge shared by the two leftmost triangles is ordered $v_1 v_2$ by both.

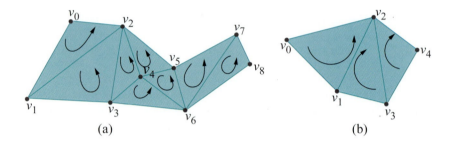

Figure 9.8: (a) Consistently oriented triangulation (b) Inconsistently oriented triangulation.

Intuitively, triangles in a consistently oriented triangulation of X appear oriented either all CW or all CCW "looking at one side of X". What exactly does this mean?

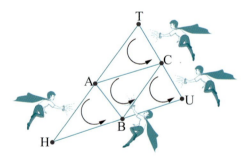

Figure 9.9: OpenGL spray-painting bots.

Let's return again to the earlier thought experiment at the point when you were about to call the painter. Looking up again you make out that the large triangular sheet is actually composed of four smaller ones and that there's a painter for each, so you'll have to call them separately (Figure 9.9). Moreover, all that you are allowed to specify to each is the order of his triangle's vertices – e.g., you can specify to the painter at the top his vertex order as either C→A→T or T→A→C – for these painters are nothing but OpenGL bots that have been programmed to do the following:

Determine if you perceive the order that you just called in as CCW or CW; if CCW then paint your side red, if not green.

Clearly, the onus then is on you to call in the four orders so that the small triangles are consistently oriented or else your side of the large triangle will be colored disparately.

Are we saying that an observer at a given position can see only one side of a consistently oriented surface? Not at all. For example, the man in Figure 9.10 can see parts of both sides of the consistently oriented triangulated wall. However, he sees a change in side, according to the CW/CCW rule, only across boundary edges, never across an internal edge – which is physically authentic. If the wall were not consistently oriented, though, then this would not be the case. For example, the reader using the CW/CCW rule would believe herself to be seeing two different sides of the polygon of Figure 9.8(b) along the edge $v_1 v_2$.

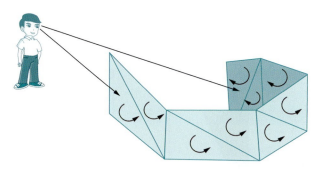

Figure 9.10: Man looking at both sides of a consistently oriented wall.

Recall again the bowl of Figure 9.2 with its inside green and outside red. If it's created in OpenGL as a triangulation, the programmer should then (a) specify that all front faces are of one color and back faces of the other, *and* (b) ensure consistent orientation of the triangulation so that the entire inside and entire outside appear of the desired colors, respectively.

In fact, the preceding rule should apply to all surfaces that we create. Here's what can happen if it doesn't.

Experiment 9.2. Replace the polygon declaration part of `square.cpp` with (Block 5)

```
glPolygonMode(GL_FRONT, GL_LINE);
glPolygonMode(GL_BACK, GL_FILL);
glBegin(GL_TRIANGLES);
    // CCW
    glVertex3f(20.0, 80.0, 0.0);
    glVertex3f(20.0, 20.0, 0.0);
```

```
glVertex3f(50.0, 80.0, 0.0);

//CCW
glVertex3f(50.0, 80.0, 0.0);
glVertex3f(20.0, 20.0, 0.0);
glVertex3f(50.0, 20.0, 0.0);

// CW
glVertex3f(50.0, 20.0, 0.0);
glVertex3f(50.0, 80.0, 0.0);
glVertex3f(80.0, 80.0, 0.0);

// CCW
glVertex3f(80.0, 80.0, 0.0);
glVertex3f(50.0, 20.0, 0.0);
glVertex3f(80.0, 20.0, 0.0);
glEnd();
```

The specification is for front faces to be outlined and back faces filled, but, as the four triangles are not consistently oriented, we see both outlined and filled triangles (Figure 9.11(a)). **End**

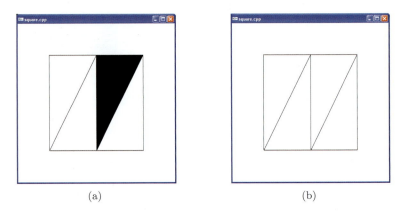

(a)   (b)

Figure 9.11: Screenshots for (a) Experiment 9.2 and (b) Experiment 9.3.

Experiment 9.3. Continuing the previous experiment, next replace the polygon declaration part of square.cpp with (Block 6):

```
glPolygonMode(GL_FRONT, GL_LINE);
glPolygonMode(GL_BACK, GL_FILL);
glBegin(GL_TRIANGLE_STRIP);
    glVertex3f(20.0, 80.0, 0.0);
    glVertex3f(20.0, 20.0, 0.0);
    glVertex3f(50.0, 80.0, 0.0);
    glVertex3f(50.0, 20.0, 0.0);
```

```
        glVertex3f(80.0, 80.0, 0.0);
        glVertex3f(80.0, 20.0, 0.0);
    glEnd();
```

The resulting triangulation is the same as before, but, as it's consistently oriented, we see only outlined front faces. (Figure 9.11(b)). **End**

In the next experiment we see an example of a consistently oriented object, both sides of which are visible.

Experiment 9.4. Run `squareOfWalls.cpp`, which shows four rectangular walls enclosing a square space. The front faces (the outside of the walls) are filled, while the back faces (the inside) are outlined. Figure 9.12(a) is a screenshot.

The triangle strip of `squareOfWalls.cpp` consists of eight triangles which are consistently oriented, because triangles in a strip are *always* consistently oriented. **End**

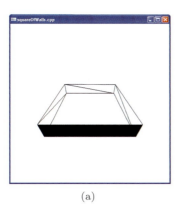
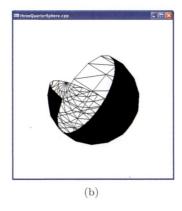

(a) (b)

Figure 9.12: Screenshots of (a) `squareOfWalls.cpp` and (b) `threeQuarterSphere.cpp`.

Experiment 9.5. Run `threeQuarterSphere.cpp`, which adds one half of a hemisphere to the bottom of the hemisphere of `hemisphere.cpp`. The two polygon mode calls ask the front faces to be drawn filled and back ones outlined. Turn the object about the axes by pressing 'x', 'X', 'y', 'Y', 'z' and 'Z'.

Unfortunately, the ordering of the vertices is such that the outside of the hemisphere appears filled, while that of the half-hemisphere outlined. Figure 9.12(b) is a screenshot. Likely, this would not be intended in a real design application where one would, typically, expect a consistent look throughout one side.

Such mixing up of orientation is not an uncommon error when assembling an object out of multiple pieces. Fix the problem in the case of `threeQuarterSphere.cpp` in four different ways:

(a) Replace the loop statement

```
for(i = 0; i <= p/2; i++)
```

of the half-hemisphere with

```
for(i = p/2; i >= 0; i--)
```

to reverse its orientation.

(b) Interchange the two **glVertex3f()** statements of the half-hemisphere, again reversing its orientation.

(c) Place the additional polygon mode calls

```
glPolygonMode(GL_FRONT, GL_LINE);
glPolygonMode(GL_BACK, GL_FILL);
```

before the half-hemisphere so that its back faces are drawn filled.

(d) Call

```
glFrontFace(GL_CCW)
```

before the hemisphere definition and

```
glFrontFace(GL_CW)
```

before the half-hemisphere to change the front-face default to be CW-facing for the latter.

Of the four, either (a) or (b) is to be preferred because they go to the source of the problem and repair the object, rather than hide it with the help of state variables, as do (c) and (d). **End**

It is not hard to orient consistently when creating objects in OpenGL because the primitives themselves tend to help. Verify from the definition of the drawing primitives in Section 2.6 that the set of triangles created by a call to GL_TRIANGLE_STRIP or GL_TRIANGLE_FAN is, in fact, consistently oriented. Therefore, a GL_TRIANGLE_STRIP or a GL_TRIANGLE_FAN call guarantees consistent orientation, at least for that particular set of triangles, so it's a good idea to use as many such as possible.

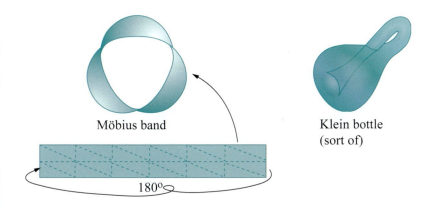

Möbius band

Klein bottle
(sort of)

180°

Figure 9.13: Non-orientable surfaces.

Non-Orientable Surfaces

Before concluding this section, mention must be made of non-orientability. There do exist surfaces which can be triangulated but *never* consistently oriented. The most famous two, the Möbius band and Klein bottle, are depicted in Figure 9.13. Such surfaces are said to be *non-orientable*. Surfaces for which consistently oriented triangulations do exist are *orientable*.

Experiment 9.6. Make a Möbius band as follows.

Take a long and thin strip of paper and draw two equal rows of triangles on one side to make a triangulation of the strip as in the bottom of Figure 9.13. Turn the strip into a Möbius band by pasting the two end edges together after twisting one 180°. The triangles you drew on the strip now make a triangulation of the Möbius band.

Try next to orient the triangles by simply drawing a curved arrow in each, in a manner such that the entire triangulation is consistently oriented. Were you able to?! **End**

We have less to worry about with the Klein bottle, at least as far as real-world applications are concerned, because it cannot be created in 3-space. It needs at least 4D space to hold it properly.

Further formalization of the notion of orientability requires knowledge of topology, but what we have discussed so far is ample from the point of view of first-level computer graphics. By the way, in case non-orientability looks like a potential can of worms, rest assured you will almost never encounter a non-orientable surface in practical applications.

9.4 Culling Obscured Faces

Consider a *closed* surface such as a sphere, cube or torus, i.e., a surface that bounds a solid. See Figure 9.14. If the surface is opaque, then a viewer outside of it sees only one side, no matter where she is located, while a viewer inside sees the other.

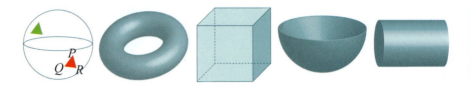

Figure 9.14: First three closed surfaces, next two non-closed (the sphere is not shaded to reveal the inside). The green back face of a triangle is not visible from outside the sphere.

Such a situation is replicated in OpenGL by a consistently oriented triangulation of the given closed surface. For example, suppose the outside of the sphere of Figure 9.14 is painted red, while the inside green. Suppose, too, it's consistently triangulated so that the orientation of the triangle PQR appears CCW, as shown in the figure, to a viewer outside the sphere (e.g., the reader). Then *any* viewer outside the sphere sees only front-facing (assuming the default of CCW = front-facing) triangles and never any back-facing ones (e.g., the green back face of the other triangle in the figure) because, for such a viewer, all back-facing triangles are hidden behind front-facing ones. The precise opposite is true for viewers inside the sphere who see only back-facing triangles.

Now, OpenGL cannot know if a surface is closed or not because this is a global decision to be made *after* the *entire* surface has been drawn (e.g., if even one triangle were missing from the sphere then it would no longer be closed). Closedness cannot, therefore, be determined by an API which simply draws one triangle after another. As a result, what happens, for example, in the case of the sphere above with the viewer outside, is that OpenGL processes *every* triangle and then ends up discarding back-facing ones at the time of hidden surface removal, because it's only then that OpenGL discovers back-facing triangles to be obscured by front-facing ones.

Therefore, knowing that a viewer located outside the closed sphere can see only front-facing triangles, the programmer can help OpenGL be more efficient by directing it to not process any further a triangle once it's been determined to be back-facing. This is called *back-face culling* or *polygon culling*.

Experiment 9.7. Run `sphereInBox1.cpp`, which draws a green ball inside a red box. Press up or down arrow keys to open or close the box. Figure 9.15(a) is a screenshot of the box partly open.

Ignore the statements to do with lighting and material properties for now. The command `glCullFace(`*face*`)` where *face* can be `GL_FRONT`, `GL_BACK` or `GL_FRONT_AND_BACK`, is used to specify if front-facing or back-facing or all polygons are to be culled. Culling is enabled with a call to `glEnable(GL_CULL_FACE)` and disabled with `glDisable(GL_CULL_FACE)`.

You can see at the bottom of the drawing routine that back-facing triangles of the sphere are indeed culled, which makes the program more efficient because these triangle are hidden in any case behind the front-facing ones.

Comment out the `glDisable(GL_CULL_FACE)` call and open the box. Oops, some sides of the box have disappeared, as you can see in Figure 9.15(b). The reason, of course, is that the state variable `GL_CULL_FACE` is set when the drawing routine is called the first time so that all back-facing triangles, including those belonging to the box, are eliminated on subsequent calls.

$\textbf{End}$

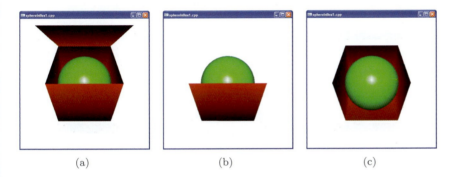

| (a) | (b) | (c) |

Figure 9.15: Screenshots for (a) Experiment 9.7 (b) Experiment 9.7 (disable culling commented out) (c) Experiment 9.8.

$\textbf{Experiment}$ **9.8.** Here's a trick often used in 3D design environments like Maya and Studio Max to open up a closed space. Suppose you've finished designing a box-like room and now want to work on objects inside it. A good way to do this is to remove only the walls that obscure your view of the inside and leave the rest, but the obscuring walls are either *all* front-facing or *all* back-facing, so a cull will do the trick.

Insert the pair of statements

```
glEnable(GL_CULL_FACE);
glCullFace(GL_FRONT);
```

in the drawing routine of `sphereInBox1.cpp` just before `glDrawElements()`. The top and front sides of the box are not drawn, leaving its interior visible. Figure 9.15(c) is a screenshot.

$\textbf{End}$

9.5 Transformations and the Orientation of Geometric Primitives

We know now how OpenGL uses the vertex order to determine the orientation of a primitive perceived by a viewer and, accordingly, the face seen, front or back. A reader, recollecting the theory of transformations, particularly Section 5.4.7 about orientation-preserving Euclidean transformations (i.e., rigid transformations) and orientation-reversing ones, may have already thought about and guessed the answer to the following question: how do these transformations affect the perceived orientation of a geometric primitive?

Answer: An orientation-preserving Euclidean transformation preserves the viewer's perceived orientation of the primitive, while an orientation-reversing one reverses it. An experiment will help make this clear.

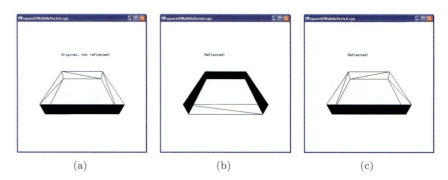

(a)   (b)   (c)

Figure 9.16: Screenshots from Experiment 9.9: (a) Original (b) Wrongly reflected (c) Correctly reflected.

Experiment 9.9. Run `squareOfWallsReflected.cpp`, which is `square-OfWalls.cpp` with the following additional block of code, including a `glScalef(-1.0, 1.0, 1.0)` call, to reflect the scene about the yz-plane.

```
// Block to reflect the scene about the yz-plane.
if (isReflected)
{
    ...
    glScalef(-1.0, 1.0, 1.0);
    // glFrontFace(GL_CW);
}
else
{
    ...
    // glFrontFace(GL_CCW);
}
```

The original walls are as in Figure 9.16(a). Press space to reflect. Keeping in mind that front faces are filled and back faces outlined, it seems that `glScalef(-1.0, 1.0, 1.0)` not only reflects, but turns the square of walls inside out as well, as you can see in Figure 9.16(b)

Well, of course! The viewer's (default) agreement with OpenGL is that if she perceives a primitive's vertex order as CCW, then she is shown the front, if not the back. Reflection about the yz-plane, an orientation-reversing Euclidean transformation, flips all perceived orientations, so those primitives whose front the viewer used to see now have their back to her, and vice versa.

We likely want the reflection to transform the primitives but not simultaneously change their orientation. This is easily done by revising the viewer's agreement with OpenGL with a call to `glFrontFace(GL_CW)`. Accordingly, uncomment the two `glFrontFace()` statements in the reflection block. Now the reflection looks right, as shown in Figure 9.16(c). The primitives are clearly still being reflected about the yz-plane, but front and back stay same. **End**

9.6 Summary, Notes and More Reading

In this chapter we learned how OpenGL uses orientation to determine which side of a 2D primitive is visible and to render it accordingly. We saw as well the importance of consistently orienting a triangulation in order to avoid disparate rendering. The technique of back-face culling to improve efficiency in rendering a closed surface was a useful addition to our repertoire. We learned as well how orientation-preserving and orientation-reversing transformations impact the orientation of a primitive.

Although our discussion of orientation at the elementary level is ample for the practical programmer, a fairly sophisticated mathematical setting is required to formalize the concept of the orientability of a surface. The interested reader is urged to look up an introductory topology text. The two by Munkres [94, 95], as well as the one by Singer & Thorpe [128], are classics. Incidentally, the mathematically-inclined student of CG will find many things of use in topology. One has only to scan the latest ACM SIGGRAPH papers [125] to see the heavy application of topological ideas in cutting-edge CG.

Part V

Making Things Up

CHAPTER 10

Modeling in 3D Space

T he goal for this chapter is to systematically study the modeling
of objects in 3D space in order to be able to populate the movies,
games and other scenes that we create.

As OpenGL has only straight and flat drawing primitives, curved objects
must necessarily be approximated. We'll develop general strategies to
manufacture approximations of both curves and surfaces. We'll examine in
depth certain special classes of curves and surfaces especially important in
applications. Particular attention will be paid to Bézier primitives because
of their utility, as well as the easy-to-use OpenGL syntax available to code
them. Another popular class we'll study is that of fractals. Although we'll
delve into some of the mathematics underlying curves and surfaces, we'll
never be far from practical code: throughout this chapter are numerous
illustrative programming examples and exercises.

We begin in Section 10.1 with the modeling of curves. The first two
subsections, 10.1.1 and 10.1.2, describe how a curve is specified by equations,
either implicitly or parametrically. A strategy to draw a curve as a polyline
approximation is the topic of 10.1.3. We discuss polynomial and rational
parametrizations of curves in 10.1.4, as they are computationally more
efficient than other kinds. The conic sections, including parabolas, ellipses
and hyperbolas, comprise a very important and commonly-occurring class
of curves that we investigate briefly in 10.1.5. Section 10.1.6 is a short
introduction to the mathematics of curves, particularly giving a rigorous
definition of what it means to be a curve, and discussing continuity and
regularity. This section can and probably should be skipped on a first
reading.

We move on to surfaces in Section 10.2. We present the following 2D
primitives in an informal order of increasing drawing complexity: polygons,
meshes, planar surfaces and general surfaces. Subsections 10.2.1-10.2.3

describe the first three, which are straightforward to draw. The next two subsections, 10.2.4 and 10.2.5, discuss the specification of a general surface and how to model one as a mesh approximation.

The powerful technique of making a surface by sweeping a curve is the topic of 10.2.6. In 10.2.7 we pause to apply our newly-acquired skills in a bunch of modeling projects. We continue our study of surfaces in 10.2.8, discussing a special class of swept surfaces, called ruled surfaces. This class includes bilinear patches and generalized cones and cylinders. The generalization of conic sections to 3D, the quadric surfaces, is described in 10.2.9. Objects of the GLU library which are somewhat inappropriately called the GLU quadrics are introduced in 10.2.10. The beautifully symmetric regular polyhedra, or Platonic solids as they are often called, are presented in 10.2.11. Section 10.2.12 parallels 10.1.6 in a formal discussion of surfaces and the properties of continuity and regularity – and the same recommendation applies that it be skipped on a first reading.

Although we'll be discussing Bézier theory in depth in a later chapter, it turns out that a fair amount of design with Bézier curves and surfaces can be accomplished even with limited theoretical understanding. Therefore, in keeping with our aim in this chapter of equipping the reader with as many practical modeling techniques as possible, Section 10.3 introduces Bézier design – curves in 10.3.1 and surfaces in 10.3.2.

Fractal curves, ubiquitous in nature, and so often used to create surreal shapes by designers, are the topic of Section 10.4.

10.1 Curves

One-dimensional objects are unions of straight and curved segments. See Figure 10.1. Parts composed of straight segments can be drawn exactly – in an OpenGL environment one would invoke the GL_LINES, GL_LINE_STRIP and GL_LINE_LOOP primitives. Curved segments, on the other hand, have to be approximated.

Terminology: The term "curve" can mean any segment, curved or straight.

We'll formalize the process of approximating a curve with a polygonal line. However, let's first see how to mathematically specify a curve.

10.1.1 Specifying Plane Curves

We begin with *plane curves*, which are those that lie on a plane (or 2-space, mathematically). There are two ways to specify such a curve, implicit and parametric.

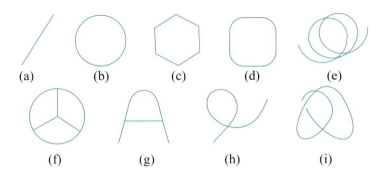

(a) (b) (c) (d) (e)

(f) (g) (h) (i)

Figure 10.1: One-dimensional objects.

Implicit

A plane curve c is specified *implicitly* by the equation

$$F(x, y) = 0 \qquad (10.1)$$

if the points of c are those whose coordinates (x, y) satisfy this equation. $F(x, y) = 0$ is said to be the *implicit equation* of c and the curve c the *graph* of this equation. An implicit equation, therefore, gives a *Boolean condition* for points on the curve to satisfy: a point (a, b) lies on the curve $F(x, y) = 0$ if $F(a, b) = 0$; it doesn't if $F(a, b) \neq 0$.

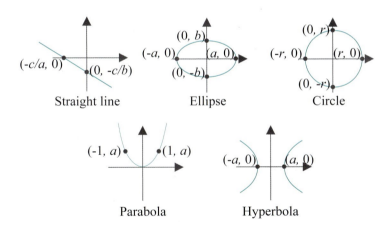

Figure 10.2: Graphs of familiar plane curves (curves are fairly accurate sketches but not exact plots).

Example 10.1. Here are examples of implicit equations of curves. Five familiar ones first (see Figure 10.2, which shows a few points on the graph of each as well):

(a) Straight line: $ax + by + c = 0$

(b) Ellipse: $\frac{x^2}{a^2} + \frac{y^2}{b^2} = 1$

(c) Circle (special case of ellipse): $x^2 + y^2 = r^2$

(d) Parabola: $y = ax^2$

(e) Hyperbola: $\frac{x^2}{a^2} - \frac{y^2}{b^2} = 1$

Remark 10.1. An implicit equation is often written in the form $F(x, y) = G(x, y)$ with the RHS not necessarily equal to 0, but, of course, it can be rearranged as $F(x, y) - G(x, y) = 0$.

The following two exotic curves (Figure 10.3) may not be as familiar:

(f) Witch of Agnesi: $y(x^2 + 4) = 8$

(g) Lemniscate of Bernoulli: $(x^2 + y^2)^2 = x^2 - y^2$

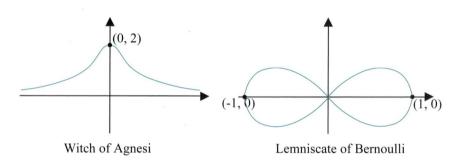

Witch of Agnesi   Lemniscate of Bernoulli

Figure 10.3: A couple of exotic curves.

Parametric

A plane curve c is specified *parametrically*, or *explicitly*, by the two equations

$$x = f(t), \ y = g(t), \ \text{where } t \in T \qquad (10.2)$$

if the points of c are those whose coordinates (x, y) satisfy $x = f(t)$ and $y = g(t)$, for some value of $t \in T$. Mathematically, $c = \{(f(t), g(t)) : t \in T\}$.

The functions f and g are called *parameter functions*, t the *parameter variable* and T the *parameter space* (also *parameter domain*). Typically, T is an interval of the real line $\mathbb{R}$, bounded or unbounded.

Example 10.2. Here are parametrizations of the curves earlier given implicitly in Example 10.1:

(a) Straight line: $x = t, \ y = -\frac{a}{b}t - \frac{c}{b}, \ t \in (-\infty, \infty)$, assuming $b \neq 0$.

(b) Ellipse: $x = a \cos t$, $y = b \sin t$, $t \in [-\pi, \pi]$.

Observe that the two different parameter values $t = -\pi$ and $t = \pi$ map to the same point $(-a, 0)$ on the ellipse, which is by no means illegal. A larger parameter space, e.g., $(-\infty, \infty)$, would cause even more overlap of parameter images. The half-open parameter interval $[-\pi, \pi)$ causes no overlap, but is a bit ungainly.

There's nothing special about $[-\pi, \pi]$ other than that it's symmetric about the origin. Any other closed interval of size 2π would do as well, e.g., $[0, 2\pi]$.

(c) Circle: $x = r \cos t$, $y = r \sin t$, $t \in [-\pi, \pi]$.

(d) Parabola: $x = t$, $y = at^2$, $t \in (-\infty, \infty)$.

(e) Hyperbola: $x = a \sec t$, $y = b \tan t$, $t \in [-\pi, -\pi/2) \cup (-\pi/2, \pi/2) \cup (\pi/2, \pi]$.

The parameter space is $[-\pi, \pi]$ *minus* the two values $\pm \pi/2$, where sec and tan become "infinite".

(f) Witch of Agnesi: $x = 2t$, $y = \frac{2}{1+t^2}$, $t \in (-\infty, \infty)$. Alternately, $x = 2 \tan t$, $y = 2 \cos^2 t$, $t \in (-\frac{\pi}{2}, \frac{\pi}{2})$.

(g) Lemniscate of Bernoulli: $x = \frac{\cos t}{1+\sin^2 t}$, $y = \frac{\cos t \sin t}{1+\sin^2 t}$, $t \in [-\pi, \pi]$.

Exercise 10.1. Prove that parametrizations of Example 10.2 are indeed those of the curves given implicitly in Example 10.1.
Hint: Plug the parametric forms for x and y into the implicit equation and verify the equality, e.g., for the circle

$$x^2 + y^2 = (r \cos t)^2 + (r \sin t)^2 = r^2(\cos^2 t + \sin^2 t) = r^2$$

Whereas an implicit equation $F(x, y) = 0$ for a curve gives a Boolean check for points "aspiring" to be on the curve, a parametric representation is more functional. If a curve c is given parametrically by the equations $x = f(t)$ and $y = g(t)$, where $t \in T$, one can think of the image $(f(t), g(t))$ of the parameter value t as traveling on c, as t travels in the parameter space. For example, as t traverses the real line, the point $(t, -\frac{a}{b}t - \frac{c}{b})$ traverses the straight line $ax + by + c = 0$, and as t goes from $-\pi$ to π, the point $(a \cos t, b \sin t)$ sweeps around the ellipse $\frac{x^2}{a^2} + \frac{y^2}{b^2} = 1$ from $(-a, 0)$ and back again.

In summary, while an implicit specification $F(x, y) = 0$ is ideal for *verifying* if a given point lies on a curve c, it's not as useful for the purpose of *generating* points on c; it is exactly the opposite in the case of a parametric specification such as $x = f(t)$, $y = g(t)$, $t \in T$.

Example 10.3.

(a) Verify if the points $(1,0)$ and $(1,-1)$ lie on the Lemniscate of Bernoulli.

(b) Generate three distinct points on the Lemniscate of Bernoulli.

Answer: (a) Plugging $(1,0)$ and $(1,-1)$ successively into the implicit equation $(x^2 + y^2)^2 = x^2 - y^2$, one sees that the first point lies on the curve, while the second doesn't.

(b) Plugging $t = 0$, $\pi/4$ and $\pi/2$ successively into the parametric equations

$$x = \frac{\cos t}{1 + \sin^2 t} \quad y = \frac{\cos t \sin t}{1 + \sin^2 t}$$

one gets the points $(1,0)$, $(\frac{\sqrt{2}}{3}, \frac{1}{3})$ and $(0,0)$ on the curve.

A parametric representation is to be preferred to an implicit for drawing as it enables the programmer to efficiently generate sample points on the curve. Going from one kind of representation to another often requires a bit of mathematical dexterity. The following two exercises are not particularly difficult though.

Exercise 10.2. An astroid, a curve traced by a fixed point on a circle rolling inside another circle of four times the diameter (see Figure 10.4), is given by the implicit equation

$$x^{\frac{2}{3}} + y^{\frac{2}{3}} - 1 = 0$$

Find parametric equations.

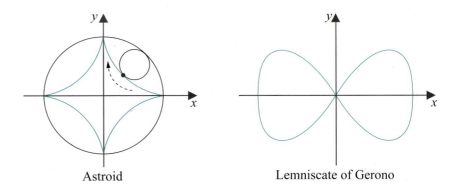

Astroid   Lemniscate of Gerono

Figure 10.4: More exotic curves.

Exercise 10.3. Parametric equations for another exotic curve, the Lemniscate of Gerono (see Figure 10.4), are

$$x = \cos t, \ y = \cos t \sin t, \ t \in [-\pi, \pi]$$

Find an implicit equation.

Remark 10.2. Neither the implicit nor parametric specification of a curve is ever unique. For example, the unit circle can be written implicitly as both $x^2 + y^2 = 1$ and $2x^2 + 2y^2 = 2$, or parametrically as $x = \cos t$, $y = \sin t$, $t \in [-\pi, \pi]$ and $x = \cos(t/2)$, $y = \sin(t/2)$, $t \in [-2\pi, 2\pi]$.

10.1.2 Specifying Space Curves

The extra dimension they have in which to move makes curves in 3-space – the real world – more interesting than their plane counterparts. Such curves are called *space curves*.

Implicit

The implicit specification of a space curve requires two equations:

$$\begin{aligned} F(x, y, z) &= 0, \\ G(x, y, z) &= 0 \end{aligned} \qquad (10.3)$$

The reason for two equations rather than the one as in the case of a plane curve is as follows. $\mathbb{R}^3$ itself – unconstrained by any equations – is of dimension 3. However, each additional equation imposed reduces the resulting object's dimension by one. For example, points of $\mathbb{R}^3$ satisfying the one equation

$$x^2 + y^2 + z^2 - 1 = 0$$

make a sphere, a surface of dimension 2. Adding the equation of, say, the plane $x + y + z - 1 = 0$, one obtains the circle

$$\begin{aligned} x^2 + y^2 + z^2 - 1 &= 0, \\ x + y + z - 1 &= 0 \end{aligned}$$

which is a curve of dimension 1 at the intersection of the two (Figure 10.5). Generally, two equations imposed on $\mathbb{R}^3$ give an object of dimension $3 - 2 = 1$, a curve.

If the implicit equation of a plane curve is already known to be $F(x, y) = 0$, then it can be written as a space curve by means of the two equations

$$\begin{aligned} F(x, y) &= 0, \\ z &= 0 \end{aligned}$$

Figure 10.5: A sphere and a plane intersect in a circle.

Remark 10.3. We have used the term "dimension" without defining it formally. We'll not do so at this time as it would take us too far afield, but think intuitively of an object's dimension as the number of "independent directions of movement" – "degrees of freedom" would be apt as well – on it.

For example, there is only one independent direction of movement on a curve (mind that forward and backward are not independent, but merely the negative of one another). A surface allows two independent directions of movement. True, there are infinitely many directions of movement from any given point on a surface, but at most any two are independent. For example, take a point on a sphere – using only latitude and longitude one can represent any direction starting from it.

Exercise 10.4. What space curve is

$$\begin{aligned} x^2 + y^2 &= 1, \\ x + z &= 1 \end{aligned}$$

Describe or sketch the curve.

Exercise 10.5. What space curve is

$$\begin{aligned} x^2 + y^2 + z^2 &= 1, \\ x^2 + y^2 + z^2 - 2x &= 0 \end{aligned}$$

Describe or sketch the curve.

Parametric

The parametric, or explicit, specification of a space curve is similar to that of a plane one except, as one would expect, another parameter function is required to determine the z coordinate value:

$$x = f(t), \ y = g(t), \ z = h(t), \ t \in T \tag{10.4}$$

Any plane curve is a space curve as well, of course, and its parametric equations in 2-space are extended to 3-space by adding $z = 0$.

Example 10.4. Parametric equations for a helix whose axis is along the z-axis (Figure 10.6) are

$$x = r \cos t, \ y = r \sin t, \ z = t, \ t \in \mathbb{R}$$

which, in fact, we used (slightly modified) to draw one in Section 2.8.2.

Example 10.5. Parametric equations for a general straight line in space are

$$x = a_1 t + b_1, \ y = a_2 t + b_2, \ z = a_3 t + b_3, \ t \in \mathbb{R}$$

where the $a_i, b_i, 1 \leq i \leq 3$, are constants.

Example 10.6. Give implicit equations for the helix of Example 10.4.

Answer:

$$\begin{aligned} x - r \cos z &= 0 \\ y - r \sin z &= 0 \end{aligned}$$

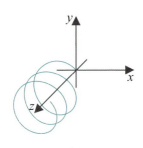

Figure 10.6: Helix.

Exercise 10.6. Give parametric equations for the infinite straight line through (p_x, p_y, p_z) and (q_x, q_y, q_z). How is the parameter space restricted if we are interested only in the finite segment *between* the two points?

Exercise 10.7. Sketch using pencil and paper the curve, called a *conical helix*, specified parametrically by

$$x = t \cos t, \ y = t \sin t, \ z = t, \ t \in (-\infty, \infty)$$

10.1.3 Drawing Curves

Drawing a curve from its parametric equations is straightforward. We did this in Chapter 2 for a few particular curves, in particular, the circle, parabola and helix. We'll describe next the procedure for a general space curve c given parametrically by

$$x = f(t), \ y = g(t), \ z = h(t), \ t \in T \tag{10.5}$$

Assume that T is $[a, b]$, a closed interval. The point $(f(t), g(t), h(t))$ on c is denoted $c(t)$.

It's useful to imagine T as being part of the real line and to imagine c as lying in a "separate" 3-space. The parameter equations together $c(t) = (f(t), g(t), h(t))$ can then be thought to map the former to the latter or, more vividly, to lift and shape T into c. See Figure 10.7.

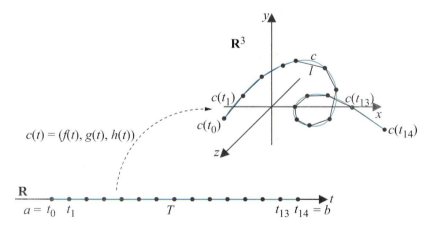

Figure 10.7: Parameter space $T = [a, b]$ mapped to a curve c. The sample grid on T, as well as its corresponding mapped sample on c, has 15 points (not all labeled). The polyline l connecting the mapped sample approximates c.

A sample

$$a = t_0 < t_1 < \ldots < t_n = b$$

of $n+1$ points from $[a, b]$ is called a *sample grid* on $[a, b]$ (note that the end points a and b are always included in the sample). It maps to a sample

$$c(t_0), c(t_1), \ldots, c(t_n)$$

of $n+1$ points of c, called the *mapped sample*.

The polyline l joining, successively, $c(t_0), c(t_1), \ldots, c(t_n)$ is an approximation of c. Each individual segment $c(t_{i-1})c(t_i)$, $1 \le i \le n$, of l approximates the arc of c between $c(t_{i-1})$ and $c(t_i)$. The sample grid and its corresponding mapped sample in Figure 10.7 contain 15 points each.

Keep in mind that a sample which is uniformly spaced along $[a, b]$ may *not* map to one which is uniformly spaced along c. For, the length of the arc of c between $c(t_{i-1})$ and $c(t_i)$ depends not only on the length of the interval $[t_{i-1}, t_i]$, but also on the "speed" of $c(t)$ with respect to t. We see next an example of this.

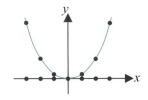

Example 10.7. Figure 10.8 shows the parabola $y = x^2$. The seven sample points on the real line are uniformly spaced, but their images on the curve are not because the rate of change of y with respect to x – equaling $\frac{dy}{dx} = 2x$ for the parabola – increases away from the origin.

Figure 10.8: A uniformly sampled parabola $y = x^2$.

Experiment 10.1. Compare the outputs of `circle.cpp`, `helix.cpp` and `parabola.cpp`, all drawn in Chapter 2.

The sample is chosen uniformly from the parameter space in all three programs. The output quality is good for both the circle – after pressing '+' a sufficient number of times for a dense enough sample – and the helix. The parabola, however, shows a difference in quality between its curved bottom and straighter sides, the sides becoming smoother more quickly than the bottom. In curves such as this, one may want to sample non-uniformly, in particular, more densely from parts of greater curvature. **End**

Here's another simple curve-drawing program.

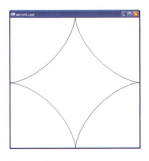

Experiment 10.2. Run `astroid.cpp`, which was written by modifying `circle.cpp` to implement the parametric equations

$$x = \cos^3 t, \ y = \sin^3 t, \ z = 0, \ 0 \le t \le 2\pi$$

for the astroid of Exercise 10.2. Figure 10.9 is a screenshot. **End**

Figure 10.9: Screenshot of `astroid.cpp`.

Exercise 10.8. (Programming) Draw the Lemniscate of Bernoulli with the help of the parametric equations given in Example 10.2(g).

Exercise 10.9. (Programming) Draw the conical helix of Exercise 10.7.

Exercise 10.10. (Programming) Draw a curve with a repeating pattern as close as possible to that of Figure 10.10(a). The two arcs of the "shark's fin" could be parts of circles. Your program should allow the user to specify the number of repetitions.

(a)   (b)

Figure 10.10: (a) Curve with a repeating pattern (b) Axe head.

Exercise 10.11. (Programming) Draw an axe head as in Figure 10.10(b).

Exercise 10.12. (Programming) The *twisted cubic* is a space curve given parametrically by the equations

$$x = t, \; y = t^2, \; z = t^3$$

Draw a part of it near the origin.

Exercise 10.13. (Programming) Animate the drawing of an astroid, as described in Exercise 10.2, as the curve traced by a point of a circle rolling inside another four times as large. The popular children's drawing toy called Spirograph can draw an astroid, among other curves.

Superellipses

A class of plane curves that generalizes both the ellipse and the astroid is that of the *superellipses*, invented by Lamé in 1818, given by the implicit equation:

$$\left| \frac{x}{a} \right|^n + \left| \frac{y}{b} \right|^n = 1$$

where a, b and n each is a positive constant.

Note: Because the exponent n can be fractional, the modulus signs are to avoid imaginaries if x or y is negative.

Figure 10.11 shows a few of the curves for $a = b = 1$ and different values of n. Generally, if $a = b$, the superellipse is called a *supercircle*. When $n = 1$ the supercircle is a square, when $n > 1$ it's convex outwards, and when $n < 1$ concave outwards.

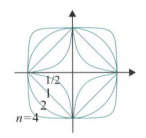

Figure 10.11:
Supercircles
$|x|^n + |y|^n = 1$, for
$n = 1/2, 1, 2, 4$.

Exercise 10.14. Justify the claim that superellipses generalize both ellipses and astroids.

Exercise 10.15. Deduce the parametric equations of a superellipse.

Exercise 10.16. (Programming) Write a program to draw a supercircle $|x|^n + |y|^n = 1$, allowing the user to choose n.

10.1.4 Polynomial and Rational Parametrizations

A *rational function* is the ratio of two polynomials. For example,

$$\frac{1 + 2t}{1 - t + t^2 - t^3} \quad \text{and} \quad \frac{x}{1 + x} \tag{10.6}$$

are rational functions of t and x, respectively. A *polynomial function* is, of course, simply a special case of a rational function where the denominator is 1, e.g.,

$$1 + 2t - 3t^2 + 4t^3 \quad \text{and} \quad x^3 \tag{10.7}$$

Unlike polynomial functions, rational functions may become undefined, which happens when their denominator vanishes. The first rational function of (10.6) is undefined at $t = 1$ and the second at $x = -1$.

If the parameter functions of a curve c are all rational, then it is said to be a *rational curve* and to have a *rational parametrization*; if they are, in fact, all polynomial, then c is a *polynomial curve* with a *polynomial parametrization*. For example, Examples 10.2 (a) and (d) give polynomial parametrizations, while the first part of (f) a rational one.

The parametrization

$$x = r \cos t, \ y = r \sin t, \ t \in [-\pi, \pi]$$

of Example 10.2(c) of the circle $x^2 + y^2 = r^2$ uses trigonometric functions and is called, of course, a *trigonometric parametrization*. It turns out that there's an alternate rational parametrization of the circle, as the reader is asked to show in the following exercise.

Exercise 10.17. Show, first, that

$$x = r\frac{1 - t^2}{1 + t^2} \quad \text{and} \quad y = r\frac{2t}{1 + t^2}$$

satisfy $x^2 + y^2 = r^2$ for all values of t.

By plotting a few values of (x, y) for values of $t = 0, \pm 1, \pm 2, \ldots$ (see Figure 10.12) convince yourself that

$$x = r\frac{1 - t^2}{1 + t^2}, \ y = r\frac{2t}{1 + t^2}, \ t \in (-\infty, \infty)$$

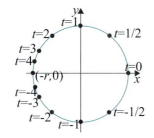

Figure 10.12: Points on a circle from a rational parametrization.

is a parametrization of the entire circle of radius r centered at the origin *minus* the point $(-r, 0)$. The point $(-r, 0)$, which the parametrization "cannot reach," is called a *singularity* of the curve (with respect to this particular parametrization).

Polynomial and rational parametrizations are often preferred to trigonometric ones in applications because the former can be computed exactly (up to round-off error) while a trigonometric function can at best be approximated by a series. For example, the power series $1 - x^2/2! + x^4/4! - x^6/6! + \ldots$ must be summed to some desired degree of accuracy to determine $\cos x$.

Exercise 10.18. (Programming) Draw a unit circle centered at the origin with the help of a rational parametrization. Avoid the problem of the singularity at $(-r, 0)$ in the previous exercise, as well as the infinite parameter domain there, by using the equations

$$x = \frac{1 - t^2}{1 + t^2}, \quad y = \frac{2t}{1 + t^2}, \quad t \in [-1, 1]$$

to draw the right half semi-circle and

$$x = -\frac{1 - t^2}{1 + t^2}, \quad y = \frac{2t}{1 + t^2}, \quad t \in [-1, 1]$$

to draw the left half.

Exercise 10.17 gave a rational parametrization of a circle. Is there a polynomial parametrization which might, in fact, be computationally better because it does not require the expensive division operation? The answer is no, as is shown in the following.

Example 10.8. Prove that the circle $x^2 + y^2 = r^2$ has no polynomial parametrization.

Answer: Suppose, if possible, that $x = f(t)$ and $y = g(t)$ is a polynomial parametrization of the circle. Then $f(t)$ and $g(t)$ are polynomial functions such that

$$f(t)^2 + g(t)^2 = r^2 \tag{10.8}$$

If $f(t)$ and $g(t)$ are both constants, i.e., neither contains a power of t, then $(f(t), g(t))$ is just one point and certainly cannot represent a circle. Therefore, either one or both of the two functions must contain a power of t. Let t^m be the highest power of t in $f(t)$ or $g(t)$. Write

$$f(t) = a_m t^m + a_{m-1} t^{m-1} + \ldots a_0 \quad \text{and} \quad g(t) = b_m t^m + b_{m-1} t^{m-1} + \ldots b_0$$

where at least one of a_m and b_m is non-zero. Then

$$f(t)^2 + g(t)^2 = (a_m^2 + b_m^2) t^{2m} + \quad \text{lower powers of } t$$

where the coefficient of t^{2m} is non-zero, in fact, positive, because at least one of a_m and b_m is non-zero.

We see, then, that the LHS of (10.8) contains a non-zero power of t, but, in this case, it cannot equal an RHS which is only scalar, proving that circle $x^2 + y^2 = r^2$ indeed has no polynomial parametrization.

10.1.5 Conic Sections

The ellipse, parabola and hyperbola are well-known members of a special class of plane curves called *conic sections* or, simply, *conics*. A conic is nothing but the graph on the plane of a quadratic equation in two variables, typically written:

$$Ax^2 + Bxy + Cy^2 + Dx + Ey + F = 0 \qquad (10.9)$$

(At least one of A, B and C should not be zero, or the equation is no longer that of a quadratic.)

Conditions on the coefficients determine the type of conic. If the quantity $B^2 - 4AC$, called the *discriminant* of the conic, is less than zero, then the conic is an ellipse, if it is zero then a parabola, and if it is greater than zero then a hyperbola. If $A = C$ are non-zero and $B = 0$ we get a circle, which is a special case of the ellipse. However, in all cases, there are *degenerate* instances when the equation is that of a point, straight line(s), or nothing at all, as the reader is asked to find for herself next.

Exercise 10.19. Show that the following are equations of degenerate conics by determining their graphs:

$$x^2 + 2y^2 + 1 = 0, \qquad x^2 + y^2 = 0, \qquad x^2 + 2xy + y^2 = 0, \qquad x^2 - y^2 = 0$$

In each case say as well if it is a degenerate ellipse, circle, parabola or hyperbola.

Conics arise frequently in design applications. Consequently, it's useful that they all have polynomial or rational parametrizations. In fact, any non-degenerate conic can be transformed by translation and rotation to one of the following four particular normalized forms (pictured in Figure 10.13):

Conic	Implicit	Polynomial or Rational Parametrization	Singularity
Ellipse	$\frac{x^2}{a^2} + \frac{y^2}{b^2} = 1$	$x = a\frac{1-t^2}{1+t^2}, y = b\frac{2t}{1+t^2}, t \in (-\infty, \infty)$	$(-a, 0)$
Circle	$x^2 + y^2 = r^2$	$x = r\frac{1-t^2}{1+t^2}, y = r\frac{2t}{1+t^2}, t \in (-\infty, \infty)$	$(-r, 0)$
Parabola	$y = ax^2$	$x = t, y = at^2, t \in (-\infty, \infty)$	None
Hyperbola	$\frac{x^2}{a^2} - \frac{y^2}{b^2} = 1$	$x = a\frac{1+t^2}{1-t^2}, y = b\frac{2t}{1-t^2}, t \in (-\infty, \infty) - \{-1, 1\}$	$(-a, 0)$

Geometric Construction

There is a rather neat geometric construction of the conics which, in fact, explains why they are called conic sections. Consider the double cone C formed from all the lines through the origin intersecting a a circle c centered some distance vertically above the origin. See Figure 10.14(a).

Now, the section of C by a non-radial plane p aligned as in Figure 10.14(b) is a hyperbola (a *non-radial* line or plane is one that does not pass through the origin). A hyperbola is not the only curve that can be sectioned off a double cone, as the reader is asked to show next.

Exercise 10.20. Using paper and pencil draw three non-radial planes so that their intersections with a double cone are a circle, ellipse and parabola, respectively.

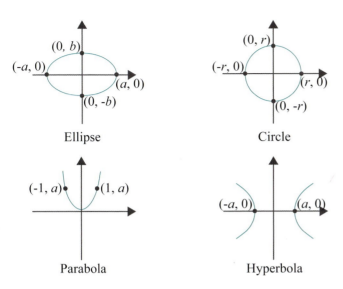

Ellipse

Circle

Parabola

Hyperbola

Figure 10.13: Conic sections.

Let's try to determine precisely the section of the double cone C by some given plane p. Assume that the half-angle at the vertex of C is θ and that the angle between p and the axis of C is ϕ. A typical cross-sectional view is drawn in Figure 10.14(c).

First, suppose that p is non-radial, as in Figure 10.14(c). We have then the following: *the section of C by p is an ellipse, parabola or hyperbola according as $\theta < \phi$, $\theta = \phi$ or $\theta > \phi$.* We'll leave the reader to convince herself of this fact by mentally rotating the plane p of Figure 10.14(c), where, in fact, currently $\theta < \phi$.

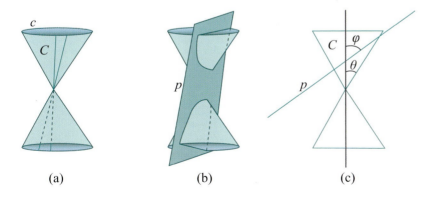

(a) (b) (c)

Figure 10.14: (a) A double cone C showing two of the lines through the origin lying on it (b) A hyperbolic section of C by a non-radial plane p (c) Cross-sectional view of a non-radial plane p intersecting C.

Exercise 10.21. Suppose now that p is *radial*. Determine the three different *degenerate* conic sections that arise, again according as $\theta <=> \phi$.

10.1.6 Curves More Formally

This section may be safely skipped on a first reading.

As we make our living in computer graphics drawing curves and surfaces, it's reasonable to try and understand some of their underlying mathematical formalism. We'll make a start with curves in this section. The theory of curves is a vast area within mathematics. Our objective in contrast is modest: to bring across a few definitions and results we believe most relevant to graphics applications, and in as intuitive a manner as possible.

We assume that you have some basic calculus. In other words, statements such as the function $f(x) = x^2$ is continuous and derivable (derivable, differentiable, same thing), that its derivative is the function $f'(x) = 2x$ and that its tangent at the point $(1, 1)$ has gradient 2 all make sense to you. Good!

Moving on, we'll first examine some "holes" in the rules given earlier to specify a curve. We'll then try to fix these and motivate in the process a more rigorous definition.

An implicit equation of the form $F(x, y) = 0$ on the plane may well specify an object that does not agree with our notion of what a curve should be. For example, $x^2 - y^2 = 0$ specifies two intersecting straight lines. See Figure 10.15(a). Writing $x^2 - y^2 = 0$ as $(x - y)(x + y) = 0$ explains the graph. And $x^2 + y^2 = 0$ defines just the single point $(0, 0)$, as in Figure 10.15(b)!

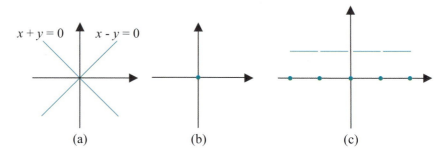

Figure 10.15: Non-curves: (a) $x^2 - y^2 = 0$ (b) $x^2 + y^2 = 0$ (c) $y = 0$, if x is an integer, 1 otherwise (gaps in the blue line indicate missing points).

Parametric equations may not fare better. The following is contrived certainly but makes the point:

$$x = t, \quad y = \begin{cases} 0, & t \text{ is an integer} \\ 1, & t \text{ is not an integer} \end{cases}, \quad t \in \mathbb{R}$$

define a disconnected union of points and straight segments (Figure 10.15(c)).

It seems, therefore, that the definition earlier of implicit and parametric curves simply by sets of equations might not be enough to agree consistently with our intuition at least of what a curve should be. To motivate a better definition, contemplate again the one-dimensional objects in Figure 10.1. All, except (f) and (g), seem to match the notion of a curve as being the *trajectory of a continuously-moving point*. We'll build on this simple observation.

In fact, we begin with the following definition of so-called C^0-continuity:

Definition 10.1. A real-valued function f defined on a closed interval T is said to be C^0-*continuous* or, simply, C^0, if it is continuous on T.

Remark 10.4. Yes, the definition simply re-christens what we know already as continuous. The reason to do this is that we'll soon be encountering so-called higher orders of continuity, to be called C^1, C^2, etc.

Remark 10.5. As the only functions that we consider are real-valued we won't explicitly say this any more.

Remark 10.6. A closed interval is either bounded of the form $[a, b]$ or unbounded in one direction of the form $(-\infty, b]$ or $[a, \infty)$ or unbounded in both when it can only be $(-\infty, \infty)$.

The C^0-continuity of functions leads to the definition of C^0 curves:

Definition 10.2. Three C^0 functions f, g and h defined on a closed interval T give the following C^0-*parametrization* of a space curve c:

$$x = f(t), \ y = g(t), \ z = h(t), \ t \in T$$

The curve c itself is the set of all image points $\{(f(t), g(t), h(t)) : t \in T\}$. If a curve c has a C^0-parametrization, then it is said to be C^0-*continuous* or, simply, C^0.

Remark 10.7. An equivalent definition of a C^0 plane curve is obtained by dropping the "$z = h(t)$" term. Henceforth, we'll stick to 3D and give definitions only for space curves.

Example 10.9. The parametrization

$$x = t, \quad y = \begin{cases} 0, & t \text{ is an integer} \\ 1, & t \text{ is not an integer} \end{cases}, \quad t \in \mathbb{R}$$

given earlier is not C^0 as y is not a continuous function of t.

Example 10.10. The single point $(0, 0)$ defined implicitly by $x^2 + y^2 = 0$ is, strangely enough, a C^0 curve. For, it has the C^0-parametrization

$$x = 0, \ y = 0, \ t \in (-\infty, \infty)$$

on the plane, for the constant functions $x = 0$ and $y = 0$ are, in fact, continuous.

Example 10.11. All the parametrizations given in Example 10.2, except for the hyperbola, are C^0. The problem with the hyperbola is that its parameter space is not a (single) closed interval as we require. See the next exercise.

Exercise 10.22. Parametrize either of the two wings of the hyperbola of Example 10.2(e) so that each is a C^0 curve defined on a closed interval.

Example 10.12. Even though there are no actual parametrizations given there to go by, it's believable that, except for (f) and (g), the one-dimensional objects of Figure 10.1 are each a C^0 curve. The problem with (f) and (g) is that both seem composed of more than one trajectory.

Exercise 10.23. How about the astroid of Exercise 10.2 and the Lemniscate of Gerono of Exercise 10.3? Are they C^0?

Exercise 10.24. Is the graph of the following function C^0?

$$y = \begin{cases} 0, & x \leq 0 \\ 1, & x > 0 \end{cases}$$

Exercise 10.25. Is the graph of the function $y = |x|$ C^0?

Because of the continuity conditions, a C^0 curve is at least minimally well-behaved. However, it is not even guaranteed to possess a tangent at any given point (e.g., the one-point curve $x^2 + y^2 = 0$, or the hexagon of Figure 10.1(c) at its corners, do not seem to have meaningful tangents). In fact, it's often desirable that a curve be "smooth" in that, not only does it possess a tangent at every point, but the tangent turns continuously along the curve as well.

Note: We use the term "smooth" now as an informal descriptor. There are technical definitions of a "smooth function" and a "smooth curve" which will come up shortly.

The following two definitions are formulated to impose smoothness:

Definition 10.3. A function f defined on a closed interval T is said to be C^1-*continuous* or, simply, C^1 if its derivative f' exists and is continuous on T; equivalently, if f' exists and is C^0 on T.

Definition 10.4. Three C^1 functions f, g and h defined on a closed interval T give the following C^1-*parametrization* of a curve c:

$$x = f(t), \ y = g(t), \ z = h(t), \ t \in T$$

The curve c itself is the set of all image points $\{(f(t), g(t), h(t)) : t \in T\}$. If a curve c has a C^1-parametrization, then it is said to be C^1-*continuous* or, simply, C^1.

If, additionally, the three derivatives f', g' and h' never vanish together at any point of $[a, b]$, then the parametrization is said to be *regular*, and c is said to be a *regular curve*.

Because derivability implies continuity, C^1 curves and regular curves are C^0 as well. Regularity is "nice enough" for most CG applications. Why? Because regularity assures the smoothness of a curve c in the following sense:

(a) The tangent line to c at any point $c(t) = (f(t), g(t), h(t))$ exists. In fact, it is parallel to the vector $c'(t) = (f'(t), g'(t), h'(t))$, which is non-zero because the derivatives of the parameter functions do not vanish simultaneously (see Figure 10.16(a)). The existence of a tangent line everywhere on c means that it has a well-defined *direction* at every point.

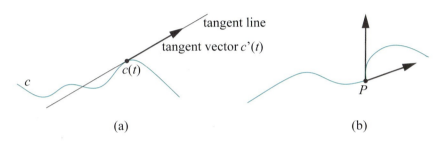

(a)   (b)

Figure 10.16: (a) A smooth curve (b) A non-smooth curve with a corner at P where the tangent changes direction abruptly.

(b) Since f', g' and h' are continuous, $c'(t) = (f'(t), g'(t), h'(t))$ is continuous along c, which means that, not only does the tangent line exist at all points of c, it turns continuously along c as well (intuitively, this means that c cannot have a corner where its direction changes abruptly as in Figure 10.16(b)).

The upshot is that a regular curve appears smooth when drawn.

It's precisely the non-vanishing property of $c'(t)$ that is not guaranteed by mere C^1-continuity, as opposed to regularity, as we see in the following example.

Example 10.13. The astroid of Exercise 10.2 has parametric equations

$$x = \cos^3 t, \ y = \sin^3 t, \ t \in [0, 2\pi]$$

As $x = \cos^3 t$ and $y = \sin^3 t$ are continuous functions of t, the astroid is C^0. Now

$$\frac{dx}{dt} = -3\cos^2 t \sin t, \quad \frac{dy}{dt} = 3\sin^2 t \cos t$$

which are continuous functions of t as well, proving that the astroid is C^1 as well. However, as both $\frac{dx}{dt}$ and $\frac{dy}{dt}$ vanish at $t = 0$, $\frac{\pi}{2}$, π and $\frac{3\pi}{2}$, the astroid is not regular. In fact, it has cusps at precisely these four parameter values. Intuitively, as well, one sees that a point traveling along the astroid has to abruptly reverse direction on reaching a cusp.

Figure 10.17: Tangent
vectors to a circle.

Exercise 10.26. How about the Lemniscate of Gerono of Exercise 10.3? Is it C^0, C^1, regular?

The (non-zero) vector $c'(t) = (f'(t), g'(t), h'(t))$ is called a *tangent vector* to a regular curve $c(t) = (f(t), g(t), h(t))$. We say *a* tangent vector because any non-zero multiple of $c'(t)$, i.e., a vector collinear with it, is tangent at $c(t)$ as well. For example, Figure 10.17 shows tangent vectors, intentionally drawn of varying lengths and inconsistently oriented, at three points of a circle.

Example 10.14. Consider the helix given by the parametrization

$$x = \cos t, \ y = \sin t, \ z = t, t \in \mathbb{R}$$

A tangent vector at the point $(\cos t, \sin t, t)$ of the helix is by differentiation $(-\sin t, \cos t, 1)$, which never vanishes, so the helix is regular.

Exercise 10.27. The stationary curve

$$x = 0, \ y = 0, \ t \in (-\infty, \infty)$$

which is just the point $(0, 0)$, we saw earlier to be C^0. Is it C^1? Regular?

Exercise 10.28. What about the hexagon of Figure 10.1(c)? Is it C^0? C^1? Imagine a parametrization for it by "stringing" together parametric equations for its straight sides.

Exercise 10.29. What is a tangent vector to the graph of the function $y = f(x)$ on the xy-plane at the point $(x, f(x))$? Assume f to be differentiable.

Exercise 10.30. Is the graph of the function

$$y = \left\{ \begin{array}{ll} x^3, & x < 0 \\ x^2, & x \geq 0 \end{array} \right.$$

C^0? C^1? Regular? The point of interest is obviously the origin.

Exercise 10.31. (Programming) Animate the non-regularity of the astroid. In particular, draw the asteroid and animate its tangent vector, moving along the curve with changing t, as an arrow from $(\cos^3 t, \sin^3 t)$ to $(\cos^3 t - 3 \cos^2 t \sin t, \ \sin^3 t + 3 \sin^2 t \cos t)$.

The tangent vector shrinks to zero at each cusp and grows again as it leaves the cusp.

We can loosen up the definition of regularity to be a bit more inclusive.

Definition 10.5. A curve c is said to be *piecewise regular* if it can be made by sequentially joining a finite number of regular curves *end to end*.

In other words, a piecewise regular curve is regular except, possibly, for a finite number of corners inside it.

Example 10.15. Of those 1D objects depicted in Figures 10.1, the hexagon (c) is piecewise regular but not regular, while (f) and (g) are not even piecewise regular. The others are all regular curves.

Exercise 10.32. What about the curves of Figures 10.3 and 10.4? Identify those that are piecewise regular but not regular.

We define next a regular one-dimensional object as composed of pieces that are each a regular curve, except that they are not required to be joined end to end as for a piecewise regular curve.

Definition 10.6. A *regular one-dimensional object* is a finite union of regular curves.

In other words, even though composed of pieces that are regular, a regular one-dimensional object may not have the property of a curve that it can be continuously traversed end to end.

Example 10.16. Figures 10.1(f) and (g) are regular one-dimensional objects, but not piecewise regular curves. So are the letters 'K', 'Q' and 'X'.

We have the obvious proper inclusions

regular curves $\subset$ piecewise regular curves $\subset$ regular 1D objects

Remark 10.8. Curves arising in real life – strings, wires, rubber bands, edges of a car or aircraft – are almost invariably piecewise regular, if not regular. Likewise, one-dimensional objects we see around around us are almost all regular one-dimensional.

It is important to keep in mind that a curve is regular (or C^0, or C^1) *if* there is *some* parametrization according to the respective definition. For example, the curve c on the plane given by the parametric equations

$$x = y = t, \quad t \in (-\infty, \infty) \qquad (10.10)$$

is regular as $c'(t) = (1, 1)$ for all t. (Yes, it's the straight line $y = x$ drawn in Figure 10.18.)

However, the same straight line is defined by the cubic equations

$$x = y = t^3, \quad t \in (-\infty, \infty) \qquad (10.11)$$

but now regularity is "lost" at the origin (verify)!

In fact, to preempt the issue of finding the "best" parametrization, mathematical texts often define a curve to be the set of parameter functions itself, rather than the image, so, e.g., (10.10) and (10.11) would actually represent different curves, thus avoiding ambiguity over continuity.

If the reader is now wondering, no, there is no parametrization which will make either the hexagon or the astroid regular, though we'll not try to prove these facts.

The following intuitive proposition suggests how to join two regular curves end to end so that their union is regular.

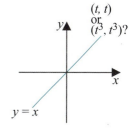

Figure 10.18: Good and bad parametrizations.

Proposition 10.1. *If two regular curves c_1 and c_2 meet at the point P, which is an endpoint of both, and if their tangent vectors at P are collinear, then the curve $c_1 \cup c_2$ is regular as well (Figure 10.19).*

Proof. We leave it to the reader. □

Hint: One must find a regular parametrization for the union $c_1 \cup c_2$. Change the parametrization of one of the curves, say c_1, from $t \mapsto c_1(t)$ to $t \mapsto \alpha t \mapsto c_1(\alpha t)$, choosing the constant α so that c_1's tangent vector at P becomes identical to that of c_2 (such rescaling of the parameter necessitates that the parameter interval be resized as well). Finally, "move" one parameter interval to abut the other.

Exercise 10.33. Prove that the two helixes

$$x = \cos t, \ y = \sin t, \ z = t, \ t \in [0, 10\pi]$$

and

$$x = 2 + \cos t, \ y = -\sin t, \ z = t - \pi, \ t \in [\pi, 10\pi]$$

meet at their common endpoint $(1, 0, 0)$, corresponding to $t = 0$ of the first curve and $t = \pi$ of the second, and share there a tangent vector. Can you come up with a single parametrization for the union of the two?

As expected, there are higher orders of continuity (C^0 is said to be zero-order, C^1 first-order) one can define as well, pretty much in the obvious manner:

Definition 10.7. A function f defined on a closed interval T is said to be C^m-*continuous* or, simply, C^m, where $m \geq 1$, if all of its derivatives of order m and less exist and are continuous on T.

A function f that is C^m-continuous for all m is said to be C^∞-*continuous* or, simply, C^∞. C^∞ functions are also called *smooth*.

Definition 10.8. Three C^m functions (m can be ∞ as well) f, g and h defined on a closed interval T give the following C^m-*parametrization* of a curve c:

$$x = f(t), \ y = g(t), \ z = h(t), \ t \in T$$

The curve c itself is the set of all image points $\{(f(t), g(t), h(t)) : t \in T\}$. If a curve c has a C^m-parametrization, then c is said to be C^m-*continuous* or, simply, C^m. If it has a C^∞-parametrization, then c is said to be *smooth*.

Remark 10.9. C^m-continuity implies C^n continuity for any $n < m$.

Remark 10.10. It is usual to assume regularity in addition to C^m-continuity, if $m \geq 1$.

Exercise 10.34. How continuous is a polynomial curve, in other words, what is the maximum order of continuity it possesses?

C^2 is about the highest order of continuity which can be distinguished visually. Even so, the lack of C^2-continuity is not as easy for the eye to catch as the lack of C^0 or C^1-continuity, which is, typically, obvious. The labeled continuity of all the curves of Figure 10.20, except the third, is probably easy to understand.

not C^0 C^0, not C^1 C^1, not C^2 C^∞

Figure 10.20: Various orders of continuity.

The third one is C^1-continuous but loses C^2-continuity at the two points P and Q where a half-circle meets straight segments, because the tangent stops turning. More explicitly, the tangent vector which is rotating at a uniform speed along the half-circle "suddenly" stops rotating altogether when it crosses into one of the straight segments, so that its rate of change drops from uniform to zero, giving rise to C^2-discontinuity.

An interesting application of C^2-continuity arises in planning the motion of a camera. We ask the reader to see this for herself in the following two exercises.

Exercise 10.35. Verify that the graph of the function (encountered earlier in Exercise 10.30)

$$y = \begin{cases} x^3, & x < 0 \\ x^2, & x \geq 0 \end{cases}$$

is not C^2 because of a second-order discontinuity at the origin.

Exercise 10.36. (Programming) Use `gluLookAt()` to simulate the view of a simple scene (populated, say, by spheres) from a camera moving along the graph of the preceding exercise and pointing always along its tangent. See Figure 10.21. Move the camera by uniformly incrementing its x-value at each time step.

The viewer perceives a jolt as the camera passes the origin. The reason is as follows. Even though the *path* of the camera is smooth, in that it is C^1, the *direction* of the camera, which is along the *tangent* to this path, does not turn smoothly past the origin because of the C^2-discontinuity there. Moreover, the location of the camera and its direction together determine the scene, so a discontinuity in either is echoed in the animation.

Bottom line: (Regular) C^1 is good enough for a curve to appear smooth, while smooth camera movement should be (regular) C^2.

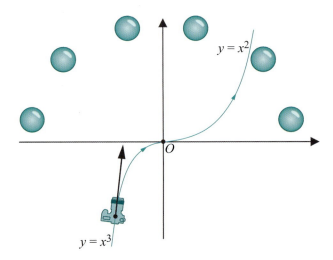

Figure 10.21: Camera moving along a path with a C^2-discontinuity at the origin O.

10.2 Surfaces

Two-dimensional objects are composed of surfaces. Analogously to the situation for one-dimensional objects, pieces of a two-dimensional object that cannot be drawn exactly using OpenGL's two-dimensional drawing primitives – triangles, quadrilaterals and polygons – must be approximated.

We'll discuss two-dimensional objects in an informal taxonomy ordered by increasing complexity from the point of view of drawing.

10.2.1 Polygons

The simplest two-dimensional object is the familiar polygon. By a polygon we shall always mean, unless stated otherwise, a *simple planar polygon*, i.e., one whose boundary lies on a plane and consists of a single component which is a non-self-intersecting line loop. All the polygons in Figure 10.22 are planar. However, (a) and (b) are non-simple polygons, while (c) and (d) are simple. A simple planar polygon can be equivalently described as a connected planar surface bounded by straight edges, and without any holes.

A convex polygon, e.g., Figure 10.22(c), can be drawn as a single GL_POLYGON primitive. A non-convex polygon should be drawn after decomposing it into convex pieces, otherwise (recall from Section 8.3) it may not render correctly. Figure 10.22(d) is an example of a non-convex polygon decomposed into triangles, in other words, *triangulated*. It's recommended, in fact, that *all* polygons, convex or otherwise, be first triangulated and then drawn using GL_TRIANGLE* primitives.

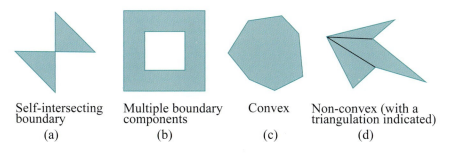

Self-intersecting boundary
(a)

Multiple boundary components
(b)

Convex
(c)

Non-convex (with a triangulation indicated)
(d)

Figure 10.22: (a) and (b) Non-simple planar polygons (c) and (d) Simple planar polygons (what we call polygons).

10.2.2 Meshes

The next simplest kind of two-dimensional object is a *polygonal mesh* or, simply, *mesh*, also called a *polyhedral surface*. A mesh is a union of polygons satisfying the following two conditions:

1. Any two polygons in the union are either disjoint or intersect in a vertex of both or intersect in an edge of both.

    Note: This is a repetition of the condition for a collection of triangles to be a triangulation (see Definition 8.1) and motivated likewise to ensure deterministic rendering.

2. The "neighborhood around each vertex is sheet-like".

    We said the taxonomy would be informal. We'll not try to define what it means exactly for a neighborhood around a vertex to be sheet-like, leaving it instead to the reader's intuition with a few suggestive examples coming next.

Figure 10.23(a) is part of a hexagonal tiling of the plane with the shaded piece in the middle missing. Figure 10.23(b) is the surface of a glass with a hexagonal base and six rectangular walls. Figure 10.23(c) is the surface of an octahedron. Evident from the drawings themselves, all three objects are unions of polygons satisfying the first condition above. Moreover, for any vertex V belonging to any one of them, one can imagine fitting a small rubber sheet exactly onto the surface in an area around V, for an informal verification of the second condition above. Therefore, Figures 10.23(a)-(c) are meshes.

Both Figures 10.23(d) and (e) clearly satisfy the first condition to be a mesh as well. However, Figure 10.23(d) fails the second condition because it's crimped at U – no rubber sheet, no matter how pliable, can be squeezed to a point. Figure 10.23(e), which consists of three rectangles sharing an edge, is not a mesh either because it has multiple panels around W. One would have to tear a sheet into pieces to cover a neighborhood of W.

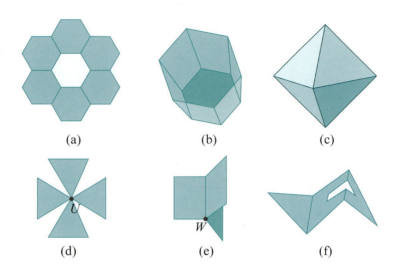

Figure 10.23: (a), (b) and (c) Meshes (d) and (e) Not meshes: parts around vertices V and W are not sheet-like (f) Your call.

Exercise 10.37. Is Figure 10.23(f) a mesh?

Remark 10.11. If you must know, in mathematical terms condition 2 above is for a mesh to be a so-called topological manifold, which guarantees a certain respectability to shapes that a mesh can take.

The polygons comprising a mesh are its *faces*. The *boundary* of a mesh consists of those edges with a polygon on only one side. The mesh of Figure 10.23(a) has two boundary components (one inside bordering the shaded missing piece and the other on the outside); that of Figure 10.23(b) has one boundary component (the rim of the glass); and that of Figure 10.23(c) has no boundary. A mesh with no boundary, a *closed mesh* as such is called, bounds a solid figure. For example, the closed mesh of Figure 10.23(c) bounds an octahedron.

It's usual to require that the faces of a mesh be convex polygons. In fact, from an OpenGL point of view it's best they all be triangles. A mesh, all of whose faces are triangles, is a *triangular mesh*. Of course, any mesh can be made triangular by triangulating every face.

Drawing a mesh is, of course, simply a matter of drawing each of its polygonal faces. Again, the recommendation is that each face be triangulated first, if it's not already a triangle, then drawn using GL_TRIANGLE* primitives. Because of the first condition for a mesh, the rendering is consistent whatever order the faces be drawn (for why, see the discussion in Section 8.1 of the reasons for the rules for a triangulation).

In fact, a moment's consideration of its primitives indicates that the *only* 2D objects OpenGL can draw *exactly* are meshes and unions of meshes. Others have to be approximated. So for the remaining two classes of 2D objects we will be trying to find good mesh approximations.

Example 10.17. Figures 10.23(d) and (e) are not meshes as we have seen. But are they a union of meshes?

Answer: Easily yes. For example, Figure 10.23(d) is the union of four meshes, each consisting of a single triangle.

10.2.3 Planar Surfaces

A planar surface is a generalization of a polygon. A polygon is a connected planar surface bounded by straight edges, and with no holes, while a general planar surface has no such restrictions. It may comprise multiple components and contain holes, while its boundary may be composed of both straight and curved parts. See Figure 10.24.

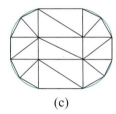

(a)　　　　　　(b)　　　　　　(c)　　　　　　(d)

Figure 10.24: Planar surfaces (the last one has two components). The black edges belong to approximating meshes.

The drawing of a general planar surface can be reduced to that of drawing an approximating mesh by the following approach:

1. Apply the technique of Section 10.1.3 to make polyline approximations of the curved edges on the surface's boundary. Together with the existing straight edges, these polylines then bound a possibly non-simple planar polygon (or a union of such if there are multiple components) approximating the surface.

2. Triangulate the approximating polygon(s). The result is a triangular mesh approximation of the surface.

The dashed edges in Figures 10.24(a)-(d) clarify the approach.

Exercise 10.38. (Programming) Draw a rounded rectangle as in Figure 10.24(c).

Exercise 10.39. (Programming) Draw a likeness in wireframe of the chair in Figure 10.25.

First of all, assume all the panels to be of zero thickness. So, for example, the seat is a flat rounded quad, while all four legs are bow-shaped planar

Figure 10.25: Wooden chair.

surfaces. The only part that's evidently not planar, even at zero thickness, is the back rest, so, in fact, replace it with a rounded rectangle. Be sure to use the symmetries: the two front legs are identical, as are the two back legs, so make only one of each.

10.2.4 General Surfaces

More general surfaces, which may be neither planar nor a union of polygons, are drawn by approximation by triangular meshes. But, let's see how a general surface s is specified in the first place. An *implicit specification* consists of an equation

$$F(x, y, z) = 0 \qquad (10.12)$$

such that points of s are those whose coordinates (x, y, z) satisfy this equation. A *parametric*, or *explicit*, *specification* consists of three equations

$$x = f(u, v), \ y = g(u, v), \ z = h(u, v), \ \text{where } (u, v) \in W \qquad (10.13)$$

In this case, the points of s are those whose coordinates (x, y, z) satisfy $x = f(u, v)$, $y = g(u, v)$ and $z = h(u, v)$, for some value of $(u, v) \in W$. The parameter space W itself is a subset of the plane $\mathbb{R}^2$. There are two parameter variables for a surface, versus one for a curve, because it is of dimension two.

The point $(f(u, v), g(u, v), h(u, v))$ on the surface s is often denoted $s(u, v)$.

Example 10.18. The (infinitely long circular) cylinder with its axis along the z-axis, and a circular cross-section of radius 1, is given by the implicit equation

$$x^2 + y^2 = 1$$

It's also given by the parametric equations

$$x = \cos u, \ y = \sin u, \ z = v, \quad (u, v) \in [-\pi, \pi] \times (-\infty, \infty)$$

where the parameter space is an infinitely long rectangular subset of the plane.

Figure 10.26 shows a finite part – the rectangle bounded by the lines $u = \pm\pi$ and $v = \pm 1$ – of the parameter space, as well as the corresponding piece of the cylinder.

If the parameter variables are u and v, then the image on the surface s, of a straight line $v = \beta$ in the parameter space, is a curve, denoted $s(v = \beta)$, and called a *u-parameter curve* of s; in other words, a u-parameter curve is traced on s by fixing the parameter v and varying u. Similarly, the image $s(u = \alpha)$ on s, of the line $u = \alpha$, is called a *v-parameter curve*.

The u-parameter curves of the cylinder of the preceding example are circles, while the v-parameter curves are vertical straight lines. One curve of either class is shown on the right of Figure 10.26.

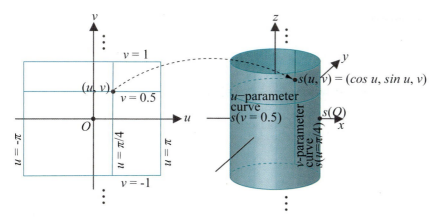

Figure 10.26: Parametric mapping of a circular cylinder s.

One can imagine the surface of s either as the union of its u-parameter curves $s(v = \beta)$ as β varies, or the union of its v-parameter curves $s(u = \alpha)$ as α varies, each over its respective range. For example, the infinite cylinder of the preceding example is the union of its u-parameter circles as the corresponding parameter β varies in $(-\infty, \infty)$, while the finite part depicted in Figure 10.26 as β varies in $[-1, 1]$. More graphically, the infinite cylinder is *swept* by its u-parameter circle as β changes from $-\infty$ to ∞. Complementarily, it's swept by its v-parameter straight lines as α varies from $-\pi$ and π.

Exercise 10.40. The implicit equation $x^2 + y^2 = 1$ of the cylinder of the preceding example did not involve z at all. In fact, it's the equation of a circle on the xy-plane simply applied to 3-space. Generally, if the implicit equation of a plane curve is $F(x, y) = 0$, then what surface is represented by the same equation in 3-space?

10.2.5 Drawing General Surfaces

The strategy to approximate a surface is similar to that for a curve. However, instead of straight line segments to approximate sub-arcs of a curve, triangles are used to approximate small patches of the surface and a triangular mesh the entire surface. We make the assumption that the parametric specification of a surface s is given as

$$x = f(u, v), \ y = g(u, v), \ z = h(u, v), \ \ (u, v) \in W \qquad (10.14)$$

where the parameter space W is the plane rectangle $[a, b] \times [c, d]$. Think then of the parametric equations as lifting and shaping the rectangle W from uv-space into s in xyz-space. As an example, Figure 10.26 illustrates the mapping of the rectangle $[-\pi, \pi] \times [-1, 1]$ to a circular cylinder by the parametric equations

$$x = \cos u, \ y = \sin u, \ z = v, \ \ (u, v) \in [-\pi, \pi] \times [-1, 1]$$

Sample, next, W at the $(p + 1)(q + 1)$ points

$$(u_i, v_j), \ 0 \le i \le p, \ 0 \le j \le q$$

of a *rectangular sample grid* where

$$a = u_0 < u_1 < \ldots < u_p = b$$
$$c = v_0 < v_1 < \ldots < v_q = d$$

The *mapped sample*

$$s(u_i, v_j), \ 0 \le i \le p, \ 0 \le j \le q$$

of $(p + 1)(q + 1)$ points of s are used as vertices of a triangular mesh approximation of s. This mesh consists of the following $2pq$ triangular faces:

$$\triangle \, s(u_i, v_j) \, s(u_{i+1}, v_j) \, s(u_i, v_{j+1}) \ \text{ and } \ \triangle \, s(u_i, v_{j+1}) \, s(u_{i+1}, v_j) \, s(u_{i+1}, v_{j+1}),$$

for $0 \le i \le p - 1$, $0 \le j \le q - 1$. (Note that $\triangle ABC$ denotes the triangle with vertices at A, B and C.)

Each face approximates a patch of the surface. In particular, the face with corners at the images of the vertices of a grid triangle approximates the image of that triangle on the surface. Spelling this out for the two triangles listed above: the face $\triangle \, s(u_i, v_j) \, s(u_{i+1}, v_j) \, s(u_i, v_{j+1})$ approximates the patch $s(\triangle \, (u_i, v_j) \, (u_{i+1}, v_j) \, (u_i, v_{j+1}))$ which is the image on s of the grid triangle $\triangle \, (u_i, v_j) \, (u_{i+1}, v_j) \, (u_i, v_{j+1})$; likewise, the face $\triangle \, s(u_i, v_{j+1}) \, s(u_{i+1}, v_j) \, s(u_{i+1}, v_{j+1})$ approximates the patch $s(\triangle \, (u_i, v_{j+1}) \, (u_{i+1}, v_j) \, (u_{i+1}, v_{j+1}))$.

It's easiest to understand this visually. Let's use again the circular cylinder

$$x = \cos u, \ y = \sin u, \ z = v, \ \ (u, v) \in [-\pi, \pi] \times [-1, 1]$$

from the previous example. Refer to Figure 10.27. Sample points in the parameter space are seen at the upper left and the corresponding mapped sample points on the cylinder at the upper right (here, $p = 6$ and $q = 4$). The triangles of the mesh along a band of the cylinder are shown at the upper right as well.

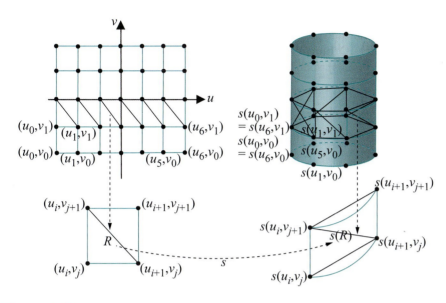

Figure 10.27: Triangular mesh approximation of a circular cylinder. *Upper*: A uniform sample grid on the parameter rectangle and its corresponding mapped sample on the cylinder. Only a few points are labeled. Vertices of a triangle strip on the rectangle maps to those of a strip approximating a band of the cylinder. *Lower*: A map from a grid rectangle to a patch of the cylinder.

The lower left and right diagrams are blow-ups, respectively, of a grid rectangle R, with corners at (u_i, v_j), (u_{i+1}, v_j), (u_{i+1}, v_{j+1}) and (u_i, v_{j+1}), and the patch $s(R)$ of the cylinder which is its image. The two triangles $s(u_i, v_j)s(u_{i+1}, v_j)s(u_i, v_{j+1})$ and $s(u_i, v_{j+1})s(u_{i+1}, v_j)s(u_{i+1}, v_{j+1})$ of the approximating mesh together approximate $s(R)$.

Cylinder

Experiment 10.3. Run `cylinder.cpp`, which shows a triangular mesh approximation of a circular cylinder, given by the parametric equations

$$x = f(u, v) = \cos u, \qquad y = g(u, v) = \sin u, \qquad z = h(u, v) = v,$$

for $(u, v) \in [-\pi, \pi] \times [-1, 1]$. Pressing arrow keys changes the fineness of the mesh. Press 'x/X', 'y/Y' or 'z/Z' to turn the cylinder itself. Figure 10.28 is a screenshot. **End**

The approximating mesh of `cylinder.cpp` is constructed according to the method above. However, a minor technicality is that the parameter space of the program is taken to be the square $[0, 1] \times [0, 1]$, rather than the parameter rectangle $[-\pi, \pi] \times [-1, 1]$ of the definition, so the former has first to be scaled to the latter. In fact, see the definitions of the functions `f`, `g` and `h` in the program:

Figure 10.28: Screenshot of `cylinder.cpp`.

```
float f(int i, int j)
{
    return ( cos( (-1 + 2*(float)i/p) * PI ) );
}

float g(int i, int j)
{
    return ( sin( (-1 + 2*(float)i/p) * PI ) );
}

float h(int i, int j)
{
    return ( -1 + 2*(float)j/q );
}
```

The expression returned by f first applies the mapping $u \mapsto (-1 + 2u)\pi$ to scale $[0, 1]$ to $[-\pi, \pi]$, then applies cos; likewise, the expression returned by g applies $u \mapsto (-1 + 2u)\pi$, then sin; the expression returned by h applies $v \mapsto -1 + 2v$ to scale $[0, 1]$ to $[-1, 1]$. Figure 10.29 indicates the scheme.

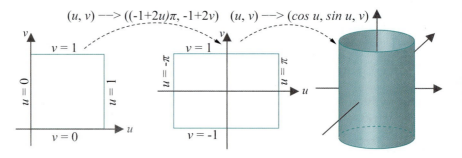

Figure 10.29: The composed mapping implemented in cylinder.cpp: first the parameter space is scaled, then mapped to the cylinder.

The most important part of cylinder.cpp's implementing the drawing strategy of Section 10.2.5 is that the coordinate values $(i/p, j/q)$ run over a uniformly-spaced $(p + 1) \times (q + 1)$ grid of sample points in $[0, 1] \times [0, 1]$, as the integer argument i runs from 0 to p, and j from 0 to q. Correspondingly, $(f(i, j), g(i, j), h(i, j))$ run over the mapped sample points on the cylinder itself. These mapped sample coordinate values are written into a vertex array by the fillVertexArray() routine. Triangles corresponding to each row of the grid in parameter space are drawn as a single triangle strip in the drawing routine, so that there are q triangle strips consisting of $2p$ triangles each.

The right and left arrow keys are programmed to increase and decrease p, respectively; the up and down arrow keys increase and decrease q, respectively.

Let's revisit next a surface we had first drawn in Experiment 2.26 of Chapter 2 to put it into the perspective of our general drawing strategy.

Exercise 10.41. (Programming) Run `hemisphere.cpp`, which draws a triangulated hemisphere. Figure 10.30 is a screenshot. Press 'p/P' and 'q/Q' to coarsen or refine the triangulation. The parametric equations of the hemisphere implemented in the program are

$$x = R\cos\phi\cos\theta, \quad y = R\sin\phi, \quad z = R\cos\phi\sin\theta,$$

for $0 \le \theta \le 2\pi$ and $0 \le \phi \le \pi/2$.

Does the program use the strategy above? *Yes, it does.* Accordingly, alter only the `f`, `g` and `h` function definitions of `cylinder.cpp` to obtain a program equivalent to `hemisphere.cpp`.

Figure 10.30: Screenshot of `hemisphere.cpp`.

You can write down *any* set of parametric equations you like and implement the corresponding surface with the help of the template of `cylinder.cpp`. All that changes in the program are the function definitions `f`, `g` and `h`. If you have a shape in mind then, of course, first deduce appropriate functions.

Helical Pipe

Experiment 10.4. Without really knowing what to expect (honestly!) we tweaked the parametric equations of the cylinder to the following:

$$x = \cos u + \sin v, \ y = \sin u + \cos v, \ z = u, \quad (u, v) \in [-\pi, \pi] \times [-\pi, \pi]$$

It turns out the resulting shape looks like a helical pipe – run `helical-Pipe.cpp`. Figure 10.31 is a screenshot.

Functionality is the same as for `cylinder.cpp`: press the arrow keys to coarsen or refine the triangulation and 'x/X', 'y/Y' or 'z/Z' to turn the pipe.

Looking at the equations again, it wasn't too hard to figure out how this particular surface came into being. See the next exercise. **End**

Figure 10.31: Screenshot of `helicalPipe.cpp`.

Exercise 10.42. Why do the parametric equations of the preceding experiment create a helical pipe?
Hint: The equation of the surface is

$$s(u, v) = (\cos u + \sin v, \ \sin u + \cos v, \ u) = (\cos u, \ \sin u, \ u) + (\sin v, \ \cos v, \ 0)$$

Note now that $u \mapsto (\cos u, \ \sin u, \ u)$ gives a helix, while $v \mapsto (\sin v, \ \cos v, \ 0)$ a circle.

Exercise 10.43. (Programming) Changing only the functions `f`, `g` and `h` of `cylinder.cpp`, draw wireframe surfaces resembling those in Figure 10.32.

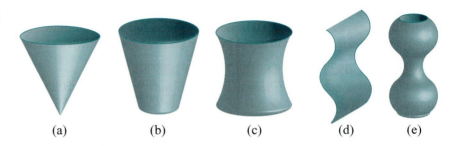

Figure 10.32: Draw these by modifying `cylinder.cpp`.

10.2.6 Swept Surfaces

A powerful design method to create surfaces is *sweeping* a curve. For example, consider the circular cylinder of Figure 10.33(a). One can think of it as the surface swept by a circle moving up, its center traveling along a vertical line. The curve that sweeps the surface is called its *profile curve* or, simply, *profile*. The path followed by the profile is the *trajectory*. The trajectory is actually the path of a point on the profile, or some point fixed with respect to it, such as the center of the circle sweeping the cylinder. The surface itself is the *swept surface*.

A torus (Figure 10.33(b)) is swept by a circular profile itself traveling along a circular trajectory. When the trajectory is a circle, the swept surface is called a *surface of revolution*. A cone (Figure 10.33(c)) is a surface of revolution swept by a straight segment profile in a circular trajectory about the cone's axis.

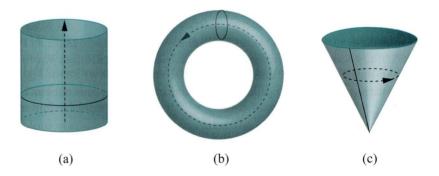

Figure 10.33: Swept surfaces: trajectories dashed arrows, profiles solid black.

When the trajectory is a straight segment, as in the case of the cylinder swept by a circle, the resulting surface is said to be an *extrusion*, or *extruded surface*, obtained by *extruding* the profile curve. In this case the profile is often called the *base curve*.

Exercise 10.44. Our description of a cylinder was as an extrusion of a circle. Can it be conceived of as a surface of revolution as well? If so, what are the profile and trajectory curves?

Exercise 10.45. How about a sphere, a hemisphere, an ellipsoid (egg shape) and (the surface of) a cube? Are they surfaces of revolution, extrusions, ...?

The advantage of being able to describe a surface as a swept surface is that its parametric equations are often, then, easy to deduce from those of its profile and trajectory curves. We see a few examples of this next.

Torus

Example 10.19. Let's compute the parametric equations of a torus. The profile is a circle c of radius r, whose center revolves along a circular trajectory C. C itself is of radius R, centered at the origin O and lying on the xy-plane. Each configuration of c, as it revolves, lies on a plane containing the z-axis and a radius of C. See Figure 10.34, where both the torus and a section through it are drawn.

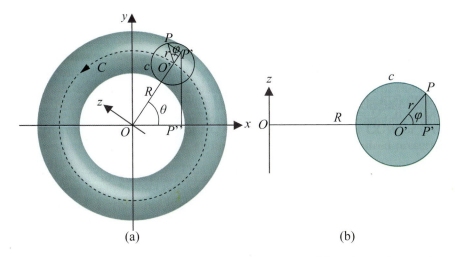

Figure 10.34: Computing parametric equations of a torus: (a) Profile circle c revolves along trajectory circle C (b) Sectional view of the left diagram along the plane containing the z-axis and OO'.

A point P on the torus is specified by two angles θ and ϕ as follows:

(a) θ is the angle made with the x-axis by the radius OO' of C from its center (the origin) to the center of the configuration of c containing P.

(b) ϕ is the angle made by $O'P$, a radius of the configuration of c containing P, with the extension of OO'.

Let P' be the projection of P on the xy-plane and P'' be the projection of P' on the x-axis.

The x coordinate of P is

$$
\begin{aligned}
OP'' &= OP' \cos\theta = (OO' + O'P')\cos\theta = (OO' + O'P\cos\phi)\cos\theta \\
&= (R + r\cos\phi)\cos\theta
\end{aligned}
$$

The y coordinate of P is

$$
\begin{aligned}
P'P'' &= OP' \sin\theta = (OO' + O'P')\sin\theta = (OO' + O'P\cos\phi)\sin\theta \\
&= (R + r\cos\phi)\sin\theta
\end{aligned}
$$

The z coordinate of P is

$$
P'P = O'P\sin\phi = r\sin\phi
$$

These give the parametric equations of the torus as

$$
x = (R + r\cos\phi)\cos\theta, \quad y = (R + r\cos\phi)\sin\theta, \quad z = r\sin\phi, \quad -\pi \le \theta, \phi \le \pi \tag{10.15}
$$

Figure 10.35: Screenshot of `torus.cpp`.

$\mathbf{E}$xperiment 10.5. Run `torus.cpp`, which applies the parametric equations deduced above in the template of `cylinder.cpp` (simply swapping new `f`, `g` and `h` function definitions into the latter program). The radii of the circular trajectory and the profile circle are set to 2.0 and 0.5, respectively. Figure 10.35 is a screenshot.

Functionality is the same as for `cylinder.cpp`: press the arrow keys to coarsen or refine the triangulation and 'x/X', 'y/Y' or 'z/Z' to turn the torus. $\mathbf{End}$

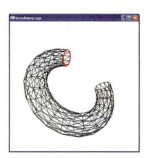

Figure 10.36: Screenshot of `torusSweep.cpp`.

$\mathbf{E}$xperiment 10.6. Run `torusSweep.cpp`, modified from `torus.cpp` to show the animation of a circle sweeping out a torus. Press space to toggle between animation on and off. Figure 10.36 is a screenshot part way through the animation. $\mathbf{End}$

$\mathbf{E}$xercise 10.46. ($\mathbf{P}$rogramming) Plump a *toroidal helix* – which is a helix coiling around a torus (Figure 10.37(a)) – into a pipe (Figure 10.37(b)). Allow the user to choose the number of times the pipe coils before closing. No need to draw the torus itself.

Suggested approach: Begin by using the parametric equations of the torus itself as determined in Example 10.19, namely,

$$
x = (R + r\cos\phi)\cos\theta, \quad y = (R + r\cos\phi)\sin\theta, \quad z = r\sin\phi, \quad -\pi \le \theta, \phi \le \pi
$$

to find those for the toroidal helix.

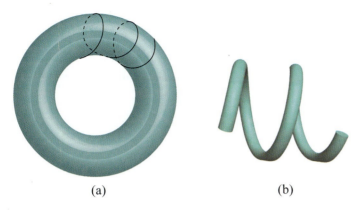

Figure 10.37: (a) Part of a toroidal helix (b) Part of a toroidal helix pipe.

The torus, being a surface, has the two degrees of freedom represented by θ and ϕ. For a curve lying on it, ϕ should depend on θ, leaving only a single degree of freedom. The function $\phi = n\theta$ yields a helix coiling n times around the torus before closing. Substituting, then, for ϕ in the equation above, one gets the parametric equations of a toroidal helix:

$$x = (R + r\cos(n\theta))\cos\theta,\ y = (R + r\cos(n\theta))\sin\theta,\ z = r\sin(n\theta),\ -\pi \le \theta \le \pi$$

To plump the toroidal helix into a pipe, observe that the pipe is swept by a circle c traveling along the toroidal helix.

Table

We'll draw next a table as the surface of revolution swept by revolving a profile curve c about the y-axis. The profile c is a polygonal line composed of seven segments lying on the xy-plane, starting at the point A and ending at B, as shown in Figure 10.38(a) (where, the 0 z-coordinates are not written).

We'll parametrize c first by using the length t along c measured from A to P as the parameter value for a point P on c. Then the x coordinate $x_c(t)$ of a point with parameter value t is given below, as can be verified from a straightforward reading of Figure 10.38(a).

$$x_c(t) = \begin{cases} t, & 0 \le t \le 4 \\ 4, & 4 \le t \le 5 \\ 9 - t, & 5 \le t \le 8 \\ 1, & 8 \le t \le 22 \\ t - 21, & 22 \le t \le 31 \\ 10, & 31 \le t \le 32 \\ 42 - t, & 32 \le t \le 42 \end{cases}$$

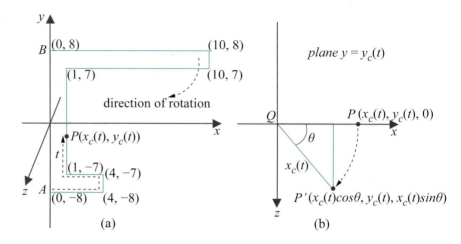

(a) (b)

Figure 10.38: (a) The profile curve for a table on the xy-plane, with the z-coordinates, all 0, not written (b) A point on the profile curve after a rotation θ CW about the y-axis.

Likewise, the y coordinate $y_c(t)$ is given by:

$$
y_c(t) = \begin{cases}
-8, & 0 \le t \le 4 \\
t - 12, & 4 \le t \le 5 \\
-7, & 5 \le t \le 8 \\
t - 15, & 8 \le t \le 22 \\
7, & 22 \le t \le 31 \\
t - 24, & 31 \le t \le 32 \\
8, & 32 \le t \le 42
\end{cases}
$$

When c revolves an angle of θ clockwise about the y-axis from its start configuration on the xy-plane, then a point P on c with coordinates $(x_c(t),\ y_c(t),\ 0)$ rotates an angle of θ on the plane $y = y_c(t)$ about the point $Q = (0, y_c(t), 0)$ to a new point P'. See Figure 10.38(b). Since $|QP'| = |QP| = x_c(t)$, the coordinates of P' are $(x_c(t) \cos\theta,\ y_c(t),\ x_c(t) \sin\theta)$.

Therefore, parametric equations for the table are

$$
x = x_c(t) \cos\theta, \quad y = y_c(t), \quad z = x_c(t) \sin\theta, \quad 0 \le t \le 42 \text{ and } -\pi \le \theta \le \pi
\tag{10.16}
$$

$\mathbf{E}$xperiment 10.7. These equations are implemented in `table.cpp`, again using the template of `cylinder.cpp`. Press the arrow keys to coarsen or refine the triangulation and 'x/X', 'y/Y' or 'z/Z' to turn the table. See Figure 10.39 for a screenshot of the table.

Figure 10.39: Screenshot of `table.cpp`.

Note that the artifacts at the edges of the table arise because sample points may not map exactly to corners $(0, -8)$, $(4, -8)$, ..., $(0, 8)$ of the profile drawn in Figure 10.38(a) – which can be avoided by including always t values 0, 4, 5, 8, 22, 31, 32 and 42 in the sample grid.

$\mathbf{End}$

Exercise 10.47. (Programming) Modify `table.cpp` to eliminate the artifacts at the edges in the manner suggested above.

Doubly-Curled Cone

Next is an experiment where the alignment of the profile curve varies as it travels along its trajectory. We want to make a "doubly-curled" cone, much like a cone made by curling a sheet of paper so that the edges don't meet, but that one wraps inside the other.

Let's write first the parametric equations for a plain-vanilla cone obtained by revolving a straight segment profile c of length 1 about the z-axis, with one end of c fixed at the origin, and with c making an angle of A with the xy-plane. We'll leave the reader to use Figure 10.40(a) to verify that the coordinates of the point P at a distance t from the origin along c, after the latter has revolved an angle of θ CCW from an original configuration on the xz-plane, are given by:

$$x = t \cos A \cos \theta, \ y = t \cos A \sin \theta, \ z = t \sin A, \quad 0 \leq t \leq 1 \text{ and } 0 \leq \theta \leq 2\pi$$

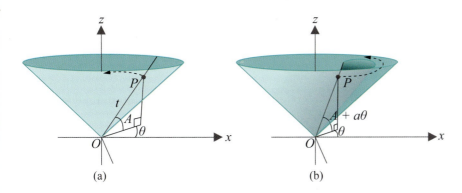

Figure 10.40: (a) A cone and (b) a doubly-curled cone as swept surfaces.

To make the cone doubly-curled we'll bring the profile c in toward the z-axis as it rotates, by increasing its angle with the xy-plane. Moreover, we'll rotate c twice about the z-axis to make a double curl.

A simple way to bring c in uniformly is to increment A, the angle that it makes with the xy-plane, by a multiple $a\theta$ of the amount of c's rotation. Figure 10.40(b) indicates the plan and it's straightforward to modify the parametric equations of the plain cone to write those of the doubly-curled:

$$x = t \cos(A + a\theta) \cos \theta, \ y = t \cos(A + a\theta) \sin \theta, \ z = t \sin(A + a\theta),$$

for $0 \leq t \leq 1$ and $0 \leq \theta \leq 4\pi$.

Experiment 10.8. The plan above is implemented in `doublyCurled-Cone.cpp`, again using the template of `cylinder.cpp`, with the value of A set to $\pi/4$ and a to 0.05. Press the arrow keys to coarsen or refine the triangulation and 'x/X', 'y/Y' or 'z/Z' to turn the cone. Figure 10.41 is a screenshot. **End**

Figure 10.41: Screenshot of `doublyCurledCone.cpp`.

Exercise 10.48. (Programming) Modify `doublyCurledCone.cpp` to change as well the length of the revolving segment c as it sweeps the cone.

Exercise 10.49. A *superellipsoid* is given generally by the implicit equation

$$\left|\frac{x}{a}\right|^n + \left|\frac{y}{b}\right|^n + \left|\frac{z}{c}\right|^n = 1$$

where a, b, c and n are each a positive constant. It's an extension to 3D of the superellipse (see Section 10.1.3).

In fact, a special type of superellipsoid is obtained as a surface of revolution by simply revolving a superellipse about either the x- or y-axis. Deduce the parametric equations of the particular superellipsoid obtained by revolving the superellipse

$$\left|\frac{x}{a}\right|^n + \left|\frac{y}{b}\right|^n = 1$$

about the y-axis.

Exercise 10.50. (Programming) Draw a superellipsoid, allowing the user to choose parameters.

Extruded Helix

For the record, here's a simple example of extrusion.

Experiment 10.9. Run `extrudedHelix.cpp`, which extrudes a helix, using yet again the template of `cylinder.cpp`. The parametric equations of the extrusion are

$$x = 4cos(10\pi u), \ y = 4sin(10\pi u), \ z = 10\pi u + 4v, \ \ 0 \le u, v \le 1$$

the constants being chosen to size the object suitably. As the equation for z indicates, the base helix is extruded parallel to the z-axis. Figure 10.42 is a screenshot. **End**

Figure 10.42: Screenshot of `extrudedHelix.cpp`.

Exercise 10.51. (Programming) Can you extrude the panels of the chair of Exercise 10.39, all of which you were then asked to draw flat, to make them now truly solid? Make the back rest curved too.

10.2.7 Drawing Projects

Here are real-life 3D projects for your drawing pleasure.

Exercise 10.52. (Programming) Draw the objects depicted in Figure 10.43: wine glass, vase, helmet with visor, extruded 'A', arch. Draw in wireframe.

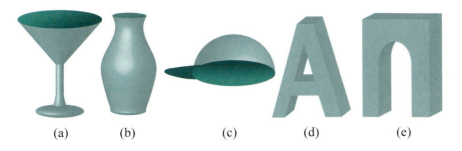

(a)   (b)   (c)   (d)   (e)

Figure 10.43: Stuff to draw.

Exercise 10.53. (Programming) Draw your name in 3D text.

Exercise 10.54. (Programming) Draw the six different chess pieces.

Exercise 10.55. (Programming) You have a chair from Exercise 10.39 (or Exercise 10.51) and a table from Experiment 10.7. Can you place these into `animateMan1.cpp` of Experiment 4.32, and get the man to walk up to the chair, sit and lean forward with his elbows on the table?

Figure 10.44: Model these? You gotta be kidding!

Exercise 10.56. (Programming) Make a likeness of an interesting structure at the place where you live, e.g., building, bridge, multi-level highway crossing, train station, or a famous one, e.g., the Eiffel Tower, Taj Mahal, of which images are available (Figure 10.44). Ignore details. Try to be as faithful as possible to the large-scale geometry.

Hint: The Eiffel Tower and Taj Mahal may seem daunting at first, but there are multiple symmetries in each which can be exploited to simplify the design process.

Consider the Eiffel Tower. The base has four identical panels reminiscent of the arch of Exercise 10.52 a little earlier. As for the tower above the base, it has fourfold symmetry. Once a suitable profile curve has been chosen for one of the four identical edges of the tower – its designer Gustave Eiffel used an exponential equation in order for the structure to be able to withstand severe wind forces – it's a matter of placing this curve four times, rotated 90° each time, and filling identical wireframes between successive pairs. Don't forget display lists from Section 3.2 in your design process.

The Taj Mahal has arches as well and numerous surfaces of revolution.

You can skip the rest of Section 10.2 on a first reading and go directly now to Section 10.3 on Bézier curves and surfaces. The reason is that Sections 10.2.8-10.2.11 deal with special classes of surfaces which you can explore later at leisure, while 10.2.12 is about the theory of surfaces which can be deferred as well.

10.2.8 Ruled Surfaces

A *ruled surface* is a swept surface whose profile curve is a straight line. In other words, a ruled surface is traced by a straight line traveling through space. Each instance of the sweeping line is called a *ruling*. See Figure 10.45.

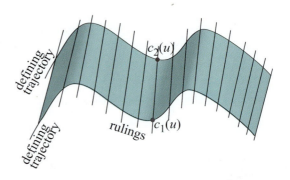

Figure 10.45: A ruled surface showing several rulings and two defining trajectories.

The parametrization of a ruled surface is particularly simple. Say that the paths of two distinct points on the profile line are $c_1(u)$ and $c_2(u)$,

$u \in [a, b]$, respectively, each called a *defining trajectory* of the surface. A parametrization then is

$$s(u, v) = (1 - v)c_1(u) + vc_2(u), \quad u \in [a, b], \ v \in (-\infty, \infty) \quad (10.17)$$

where u varies over the defining trajectories, and v over the (infinite) straight line through a pair of corresponding points on the two. If we want only the part of the surface *between* the defining trajectories, then we have to restrict the parameter space as follows:

$$s(u, v) = (1 - v)c_1(u) + vc_2(u), \quad u \in [a, b], \ v \in [0, 1] \quad (10.18)$$

Various surfaces arise as ruled surfaces. Here are three interesting ones.

Bilinear Patches

A *bilinear patch* is a ruled surface whose defining trajectories c_1 and c_2 are straight line segments both and which lies between the two trajectories. See Figure 10.46. Suppose the endpoints of c_1 are p_1 and q_1, so that it can be parametrized $c_1(u) = (1 - u)p_1 + uq_1$, $0 \le u \le 1$, while the endpoints of c_2 are p_2 and q_2, and it is parametrized $c_2(u) = (1 - u)p_2 + uq_2$, $0 \le u \le 1$. Plugging these equations into (10.18) we get the equation of the bilinear patch:

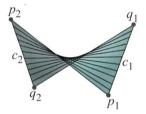

$$s(u, v) = (1 - u)(1 - v)\, p_1 + u(1 - v)\, q_1 + (1 - u)v\, p_2 + uv\, q_2, \quad u, v \in [0, 1] \quad (10.19)$$

Figure 10.46: Bilinear patch.

Counter-intuitively, even though a bilinear patch is made of a family of straight segments joining points again on two straight segments c_1 and c_2, it need not be flat! In fact, it's flat only when c_1 and c_2 are coplanar; otherwise, it is a curved surface. Interestingly, it turns out, as we'll see in upcoming Section 10.2.9, that, generally, a bilinear patch is nothing but the familiar saddle surface.

Experiment 10.10. Run `bilinearPatch.cpp`, which implements precisely Equation (10.19). Press the arrow keys to refine or coarsen the wireframe and 'x/X', 'y/Y' or 'z/Z' to turn the patch. Figure 10.47 is a screenshot. $\qquad$ End

Figure 10.47: Screenshot of `bilinearPatch.cpp`.

Generalized Cones

A *generalized cone* is a ruled surface one of whose defining trajectories is an arbitrary curve c, while the other is stationary, in other words, a single point p (not belonging to c). The cone is said to be *over* c with *apex* at p. See Figure 10.48 for three examples. Unless the qualifier "generalized" is used, the curve c is typically presumed closed. Often, informally meant by the term cone is a *right circular cone*, which is a cone over a circle c whose apex

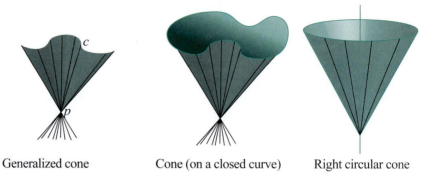

Generalized cone   Cone (on a closed curve)   Right circular cone

Figure 10.48: Generalized cones: (a) over a non-closed curve (b) over a closed curve (c) Right circular cone. Only the part between the two trajectories is drawn.

is located on the line which is through the center of c and perpendicular to its plane.

The equation of the part of the generalized cone between its trajectories – a part like those depicted in Figure 10.48 – is obtained by plugging $c_1(u) = p$ for the trajectory stationary at the apex p, and $c_2(u) = c(u)$ for the other, into (10.18):

$$s(u,v) = (1-v)p + vc(u), \quad u \in [a,b], \ v \in [0,1] \tag{10.20}$$

Evidently, all rulings pass through the apex p, at $v = 0$.

Exercise 10.57. (Programming) Draw a cone over the astroid of `astroid.cpp`.

Generalized Cylinders

A *generalized cylinder* is a ruled surface whose defining trajectories c_1 and c_2 are translates of one another. See Figure 10.49. Colloquially, cylinder typically means the familiar *right circular cylinder*, where c_1 and c_2 are circles whose centers are joined by a line perpendicular to the plane of both.

The equation for the generalized cylinder is obtained by writing the equation of one trajectory as $c(u)$ and the other as $c(u) + d$, where $u \in [a,b]$, and d is the vector translating the first trajectory to the second. Plugging these equations into (10.18) we get the generalized cylinder as

$$s(u,v) = (1-v)c(u)+v(c(u)+d) = c(u)+vd, \quad u \in [a,b], \ v \in [0,1] \tag{10.21}$$

The rulings are evidently all parallel to the vector d.

Exercise 10.58. Are generalized cylinders the same as extrusions?

Exercise 10.59. (Programming) Draw a generalized cylinder using an astroid as a trajectory.

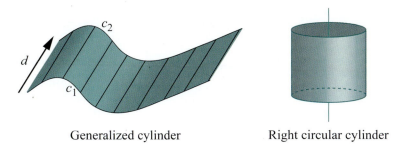

Generalized cylinder   Right circular cylinder

Figure 10.49: A generalized cylinder and a special case.

10.2.9   Quadric Surfaces

As we saw in 10.1.5, conics are curves on a plane given by a quadratic equation in two variables. *Quadric surfaces* or, simply, *quadrics*, are their generalization to one dimension higher. They are surfaces in 3D space given by a quadratic equation in *three* variables:

$$Ax^2 + By^2 + Cz^2 + Dyz + Ezx + Fxy + Px + Qy + Rz + H = 0$$

Excluding degenerate instances (e.g., $x^2 + y^2 + z^2 = 0$, which gives a single point) a quadric is of one of the nine kinds shown in Figure 10.50. In fact, any non-degenerate quadric can be transformed by translation and rotation to one of the normalized forms in the following table, corresponding each to one of those pictured in Figure 10.50.

Quadric	Implicit Equation
Ellipsoid	$\frac{x^2}{a^2} + \frac{y^2}{b^2} + \frac{z^2}{c^2} = 1$
Elliptic Paraboloid	$\frac{x^2}{a^2} + \frac{y^2}{b^2} - z = 0$
Hyperbolic Paraboloid	$\frac{x^2}{a^2} - \frac{y^2}{b^2} - z = 0$
Hyperboloid (1 sheet)	$\frac{x^2}{a^2} + \frac{y^2}{b^2} - \frac{z^2}{c^2} = 1$
Hyperboloid (2 sheets)	$\frac{x^2}{a^2} - \frac{y^2}{b^2} - \frac{z^2}{c^2} = 1$
Elliptic Cone	$\frac{x^2}{a^2} + \frac{y^2}{b^2} - \frac{z^2}{c^2} = 0$
Elliptic Cylinder	$\frac{x^2}{a^2} + \frac{y^2}{b^2} = 1$
Parabolic Cylinder	$y = ax^2$
Hyperbolic Cylinder	$\frac{x^2}{a^2} - \frac{y^2}{b^2} = 1$

A sphere is, of course, a special case of an ellipsoid. The hyperbolic paraboloid, for an obvious reason, is often called a *saddle surface*. The three cylindrical quadrics along the bottom row are probably the least interesting, as they are merely extrusions of plane conics. Parametrization, both trigonometric and rational, of the quadrics are not hard to derive.

$\mathbf{Example}$ **10.20.** Find both trigonometric and rational parametrizations of the ellipsoid.

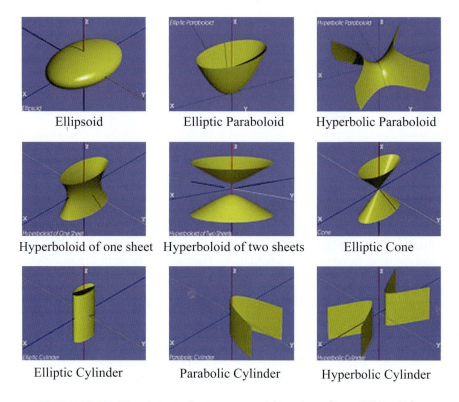

Ellipsoid   Elliptic Paraboloid   Hyperbolic Paraboloid

Hyperboloid of one sheet   Hyperboloid of two sheets   Elliptic Cone

Elliptic Cylinder   Parabolic Cylinder   Hyperbolic Cylinder

Figure 10.50: The nine non-degenerate quadric surfaces (from Wikimedia).

Answer: We begin with the ellipsoid's implicit equation

$$\frac{x^2}{a^2} + \frac{y^2}{b^2} + \frac{z^2}{c^2} = 1$$

A trigonometric parametrization is

$$x = a\cos\theta\cos\phi, \ y = b\sin\theta\cos\phi, \ z = c\sin\phi, \quad \theta \in [-\pi, \pi], \ \phi \in [-\pi/2, \pi/2]$$

while a rational one is

$$x = a\frac{1 - u^2 + v^2}{1 + u^2 + v^2}, \quad y = b\frac{2uv}{1 + u^2 + v^2}, \quad z = c\frac{2u}{1 + u^2 + v^2}, \quad u, v \in (-\infty, \infty)$$

with a singularity at $(-a, 0, 0)$.

Exercise **10.60.** Find trigonometric and rational parametrizations of the elliptic paraboloid.

Remark 10.12. If $a = b$ in the equation of the elliptic paraboloid – the equation becomes $\frac{x^2}{a^2} + \frac{y^2}{a^2} - z = 0$ in this case – then it's actually a surface of revolution obtained from revolving a parabola around its axis. This special case of an elliptic paraboloid is called a *circular paraboloid*. It is the shape used in mirrors behind headlamps because light from a bulb placed at its focal point reflects in a parallel beam.

Drawing the quadrics is simple. We'll use a trigonometric parametrization to draw next the hyperboloid of one sheet.

Experiment 10.11. Run `hyperboloid1sheet.cpp`, which draws a triangular mesh approximation of a single-sheeted hyperboloid with the help of the parametrization

$$x = \cos u \sec v, \ y = \sin u \sec v, \ z = \tan v, \quad u \in [-\pi, \pi], \ v \in (-\pi/2, \pi/2)$$

Figure 10.51(a) is a screenshot. In the implementation we restrict v to $[-0.4\pi, 0.4\pi]$ to avoid $\pm\pi/2$ where sec is undefined.   **End**

It's interesting that a few of the non-degenerate quadrics are, in fact, ruled surfaces and, therefore, traced by a straight line traveling through space. The ones on the bottom row of Figure 10.50 are evidently so. We'll prove a less obvious case.

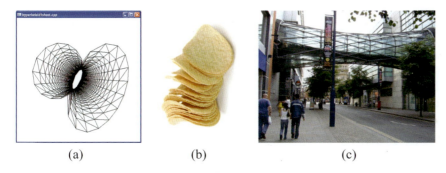

(a)   (b)   (c)

Figure 10.51: (a) Screenshot of `hyperboloid1sheet.cpp` (b) Edible hyperbolic paraboloids (c) Hyperboloid footbridge over Corporation Street in Manchester in England supported by its rulings (courtesy of Richard Litherland).

Example 10.21. Show that the hyperbolic paraboloid is a ruled surface.

Answer: We'll work with the instance s given by the implicit equation

$$x^2 - y^2 = z$$

as setting the coefficients all equal to 1 simplifies calculations (without costing in generality). Write the equation as

$$(x + y)(x - y) = z$$

Setting $u = x + y$ and $v = x - y$ then leads to the following parametrization of s:

$$x = \frac{u+v}{2}, \quad y = \frac{u-v}{2}, \quad z = uv, \quad u, v \in (-\infty, \infty) \qquad (10.22)$$

Now, a u-parameter curve of s is obtained by fixing $v = \beta$:

$$x = \frac{u+\beta}{2}, \quad y = \frac{u-\beta}{2}, \quad z = \beta u, \quad u \in (-\infty, \infty)$$

which is a straight line. Therefore, s is swept by a straight line profile, particularly the u-parameter curve for $v = \beta$, as β varies, proving it is indeed ruled. Evidently, it's *doubly-ruled*, the u-parameter curves and v-parameter curves defining distinct symmetric families of rulings.

In fact, it gets even more interesting! Two defining trajectories for s can be obtained as the v-parameter curves corresponding to a couple of distinct values of u, because they intersect each u-parameter curve in a distinct pair of points. Accordingly, set u equal to 0 and 1 in Equation (10.22) to get, respectively, the equations

$$x = \frac{v}{2}, \quad y = -\frac{v}{2}, \quad z = 0, \quad v \in (-\infty, \infty)$$

and

$$x = \frac{1+v}{2}, \quad y = \frac{1-v}{2}, \quad z = v, \quad v \in (-\infty, \infty)$$

which are both straight lines. So a hyperbolic paraboloid is a ruled surface with straight-line defining trajectories, which means it's a bilinear patch. This justifies an earlier remark that bilinear patches are saddle surfaces in general.

Remark 10.13. People often snack on hyperbolic paraboloids! See Figure 10.51(b).

Exercise 10.61. Prove that the single-sheeted hyperboloid is doubly-ruled. Figure 10.51(c) illustrates how this fact is applied to build a bridge – note the two sets of steel rulings and how they intersect in a grid. You likely have seen baskets woven in the shape of single-sheeted hyperboloids as well.

Exercise 10.62. (**Programming**) Animate a straight line segment sweeping out a single-sheeted hyperboloid.

10.2.10 GLU Quadric Objects

We are already familiar with several GLUT library objects such as spheres, cubes and cones which are ready-to-use for 3D drawing. The OpenGL Utility Library GLU provides additional routines to create four kinds of so-called quadric objects: sphere, tapered cylinder, annular disc and partial annular disc. See Figure 10.52.

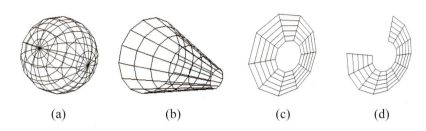

Figure 10.52: GLU quadrics: (a) Sphere (b) Tapered cylinder (c) Annular disc (d) Partial annular disc.

Experiment 10.12. Run `gluQuadrics.cpp` to see all four GLU quadrics. Press the left and right arrow keys to cycle through the quadrics and 'x/X', 'y,Y' and 'z/Z' to turn them. The images in Figure 10.52 were, in fact, generated by this program. **End**

Remark 10.14. It's a bit unfortunate that OpenGL chooses to render the quadrics quadrilateralized, rather than triangulated.

Here's how the syntax works (refer to Figure 10.53):

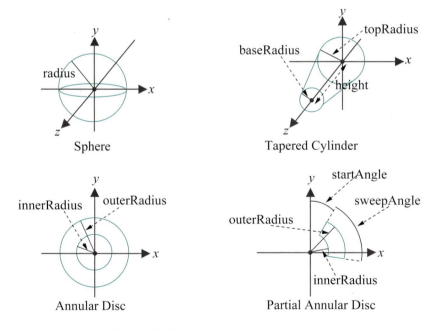

Figure 10.53: Defining the GLU quadrics.

1. gluSphere(*qobj*, *radius*, *slices*, *stacks*)

   Draws a *sphere* of radius *radius* centered at the origin. The parameters *slices* and *stacks* determine the fineness of the quadrilateralization.

Note: The parameter *qobj* in this case, and in the following, points to a quadric object.

2. **gluCylinder(***qobj, *baseRadius*, *topRadius*, *height*, *slices*, *stacks***)**

   Draws a *tapered cylinder* with its axis along the z-axis, whose base is a circle of radius *baseRadius* lying on the $z = 0$ plane and whose top a circle of radius *topRadius* lying on the $z = height$ plane. If either *baseRadius* or *topRadius* is zero then the object is a cone. The parameters *slices* and *stacks* determine the fineness of the quadrilateralization.

3. **gluDisk(***qobj, *innerRadius*, *outerRadius*, *slices*, *rings***)**

   Draws an *annular disc* centered at the origin and lying on the $z = 0$ plane, whose inner boundary is of radius *innerRadius* and outer boundary of radius *outerRadius*. The parameters *slices* and *rings* determine the fineness of the quadrilateralization.

4. **gluPartialDisk(***qobj, *innerRadius*, *outerRadius*, *slices*, *rings*, *startAngle*, *sweepAngle***)**

   Draws a *partial annular disc*: precisely, the sector of the annular disc defined by **gluDisk(***qobj, *innerRadius*, *outerRadius*, *slices*, *rings***)**, starting from angle *startAngle* and ending at *startAngle* + *sweepAngle*, where either angle is measured clockwise (looking from the $+z$-direction) along the xy-plane starting from the y-axis.

GLU calls of the form **gluQuadric*(***qobj, ***)** determine various properties of the quadric. For example, the call **gluQuadricDrawStyle(qobj, GLU_LINE)** in **gluQuadrics.cpp** causes the quadric to be rendered in wireframe.

Remark 10.15. The GLU quadrics are somewhat ambitiously named. Although they are each a part of one of the mathematical quadric surfaces described in the preceding section, they will hardly help in drawing the more complex ones.

10.2.11 Regular Polyhedra

To begin with here's a definition of a particular kind of polygon which should be familiar:

Definition 10.9. A *regular polygon* is a simple planar polygon whose sides are of equal length and which has equal interior angles at its vertices.

A regular polygon with n sides is convex and its vertices are spaced equally along a circle, called its *circumscribed circle*, at an angle of $2\pi/n$ apart (Figure 10.54). The larger n the more closely the polygon approximates its circumscribed circle.

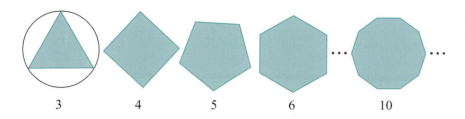

3   4   5   6   10

Figure 10.54: Regular polygons with number of sides indicated. The triangle shows its circumscribed circle.

Exercise **10.63.** What is the interior angle at a vertex of an n-sided regular polygon?

Exercise **10.64.** Show that if the condition "and which has equal interior angles at its vertices" is dropped from the definition of a regular polygon, then we could make one not belonging to the family of Figure 10.54.

Regular polyhedra are a generalization of regular polygons to three dimensions. Here first is the definition of a polyhedron based on that of a polygonal mesh (described in Section 10.2.2).

Definition 10.10. A *polyhedron* is a solid object whose boundary is a polygonal mesh (in other words, a solid whose faces are all polygons).

Definition 10.11. A *regular polyhedron* is a polyhedron all of whose faces are identical regular polygons.

Because of the symmetry constraints on their faces, there exist only five different regular polyhedra, ignoring difference in size. They are the *tetrahedron*, *hexahedron* (familiarly, *cube*), *octahedron*, *dodecahedron* and *icosahedron* in order of increasing number of faces. See Figure 10.55.

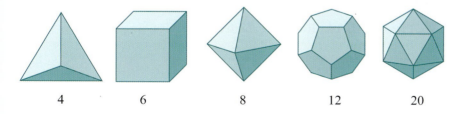

4   6   8   12   20

Figure 10.55: The five regular polyhedra with the number of faces indicated.

Geometric data for the five in numerical form is collected in the following table.

	Faces	Edges	Vertices	Edges of a face	Faces at a vertex
Tetrahedron	4	6	4	3	3
Hexahedron (cube)	6	12	8	4	3
Octahedron	8	12	6	3	4
Dodecahedron	12	30	20	5	3
Icosahedron	20	30	12	3	5

The hexahedron is bounded by squares, the dodecahedron by regular pentagons and the remaining three by equilateral triangles. The value (m, n) for a regular polyhedron, where m is the number of edges of a face and n the number of faces meeting at a vertex – the items in the last two columns of the table – is called its *Schläfli symbol*. The five different Schläfli symbols are $(3, 3), (4, 3), (3, 4), (5, 3)$ and $(3, 5)$, and each uniquely identifies a regular polyhedron. It's not an accident, as we'll see, that the reverse of each Schläfli symbol is another.

Remark 10.16. Regular polyhedra are also called *Platonic solids* because they were known to Plato, as recorded in his Timaeus dialogues. In fact, archeological finds suggest that these beautiful shapes were familiar to even earlier people.

Modeling Regular Polyhedra

There's an easy way to draw regular polyhedra using OpenGL: call them from the GLUT library! All five are available as GLUT objects.

Experiment 10.13. Run `glutObjects.cpp`, a program we originally saw in Chapter 3. Press the left and right arrow keys to cycle through the various GLUT objects and 'x/X', 'y/Y' and 'z/Z' to turn them. Among other objects you see all five regular polyhedra, both in solid and wireframe. **End**

However, in case you are the hardy do-it-yourself type, here's what you need to know.

Figure 10.56: Screenshot of `tetrahedron.cpp`.

Experiment 10.14. Run `tetrahedron.cpp`. The program draws a wireframe tetrahedron of edge length $2\sqrt{2}$ which can be turned using the 'x/X', 'y/Y' and 'z/Z' keys. Figure 10.56 is a screenshot. **End**

The coordinates of the vertices of the tetrahedron of `tetrahedron.cpp`, as well as the indices of the vertices comprising each of its triangular faces, are listed in the following two global arrays:

```
// Vertex coordinate vectors for the tetrahedron.
  static float vertices[] =
  {
      1.0,  1.0,  1.0, // V0
```

```
     -1.0,  1.0, -1.0, // V1
      1.0, -1.0, -1.0, // V2
     -1.0, -1.0,  1.0  // V3
};

// Vertex indices for the four triangular faces.
static int triangleIndices[4][3] =
{
    {1, 2, 3}, // F0
    {0, 3, 2}, // F1
    {0, 1, 3}, // F2
    {0, 2, 1}  // F3
};
```

For example, the face $F0$ is a triangle with corners at the vertices $V1$, $V2$ and $V3$.

Here's similar data for a cube of edge length 2:

Cube

Vertex	coordinates	Face	Vertices
$V0$	$(1,1,1)$	$F0$	$(V3,\ V0,\ V1,\ V2)$
$V1$	$(1,1,-1)$	$F1$	$(V2,\ V1,\ V5,\ V6)$
$V2$	$(1,-1,-1)$	$F2$	$(V6,\ V5,\ V4,\ V7)$
$V3$	$(1,-1,1)$	$F3$	$(V7,\ V4,\ V0,\ V3)$
$V4$	$(-1,1,1)$	$F4$	$(V1,\ V0,\ V5,\ V4)$
$V5$	$(-1,1,-1)$	$F5$	$(V3,\ V2,\ V6,\ V7)$
$V6$	$(-1,-1,-1)$		
$V7$	$(-1,-1,1)$		

You may be wondering why we bothered at all with the totally trivial cube. The reason is that a cube sets up modeling an octahedron by way of the beautiful relationship of *duality* between regular polyhedra.

Duality

The *dual* of a regular polyhedron P is the polyhedron P' *inscribed* in P as follows:

(a) For each face f of P there is a vertex of P', called f's dual, located at the center of f.

(b) For each edge e of P there is an edge of P', called e's dual, joining the dual of the two faces of P adjacent to e.

(c) For each vertex v of P there is a face of P', called v's dual, with vertices at the duals of the faces of P that meet at v.

See Figure 10.57. Fascinatingly enough, it turns out that the dual of a regular polyhedron is another regular polyhedron. Cubes and octahedrons are duals of one another, as are dodecahedrons and icosahedrons, while tetrahedrons are self-dual.

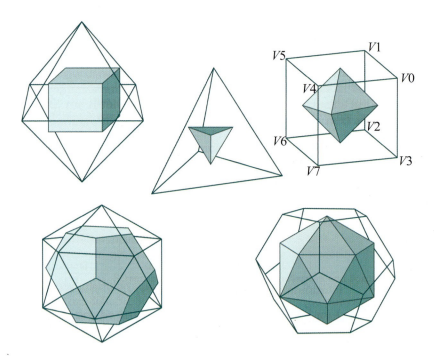

Figure 10.57: The five regular polyhedra each containing its inscribed dual (the cube is labeled to help with Exercise 10.65).

It's clear from the construction that a regular polyhedron and its dual have the same number of edges, while the number of vertices of one equals the number of faces of the other. Moreover, their Schläfli symbols are flips one of the other.

Returning to the drawing of regular polyhedra, it's easy now to compute the data for the octahedron dual to the cube whose data we listed earlier. In fact, we leave the reader to verify data for the dual octahedron in the next exercise.

Exercise 10.65. Verify the data for the octahedron, dual to the cube whose data was listed earlier, as given in the two tables below. Note that the vertex $V'i$ of the octahedron is the dual of the face Fi of the cube, while the face $F'j$ of the octahedron dual of the face Vj of the cube. Moreover, the edge length of this particular octahedron is $\sqrt{2}$.

Octahedron

Vertex	coordinates
$V'0$	$(1, 0, 0)$
$V'1$	$(0, 0, -1)$
$V'2$	$(-1, 0, 0)$
$V'3$	$(0, 0, 1)$
$V'4$	$(0, 1, 0)$
$V'5$	$(0, -1, 0)$

Face	Vertices
$F'0$	$(V'0, V'4, V'3)$
$F'1$	$(V'4, V'0, V'1)$
$F'2$	$(V'5, V'1, V'0)$
$F'3$	$(V'5, V'0, V'3)$
$F'4$	$(V'2, V'3, V'4)$
$F'5$	$(V'1, V'2, V'4)$
$F'6$	$(V'2, V'1, V'5)$
$F'7$	$(V'5, V'3, V'2)$

Part answer: In addition to the data for the cube, it's helpful as well to refer to the diagram of the octahedron inscribed in the labeled cube in Figure 10.57.

$V'0$, dual of the face $F0$, is located at the center of $F0$. It's coordinates, therefore, are

$$\frac{1}{4}(V3 + V0 + V1 + V2) = \frac{1}{4}((1, -1, 1) + (1, 1, 1) + (1, 1, -1) + (1, -1, -1))$$
$$= (1, 0, 0)$$

$F'0$, dual of the vertex $V'0$, has vertices that are the duals of the faces of the cube that contain $V0$. From the cube's table the faces containing $V0$ are $F0$, $F3$ and $F4$. Therefore, $F'0$ has vertices $V'0$, $V'3$ and $V'4$.

Use the data for an icosahedron in the two tables below to solve the following two problems.

Exercise 10.66. (Programming) Draw an icosahedron.

Exercise 10.67. (Programming) Use duality to compute the data for a dodecahedron from that of an icosahedron. In fact, write a short program for this purpose which takes as input the icosahedron data.

10.2.12 Surfaces More Formally

The material in this section is fairly theoretical though we do our best to motivate it practically. We suggest skipping it on a first reading of the book and returning later. If it proves not to your taste at all then you can skip it altogether without affecting your CG skills.

In addition to calculus 101, a basic understanding of partial derivatives is required in order to formalize the notion of surfaces, particularly that of the regular surfaces. If you're not familiar with partial derivatives then a math class or calculus book, e.g., Stewart [131] or Schaum's Outlines [5, 145], is the place to pick the stuff up. We have a handy primer ourselves in Section 11.10 (independent of the rest of that chapter).

Icosahedron

Vertex	coordinates
$V0$	$(0, 1, X)$
$V1$	$(0, 1, -X)$
$V2$	$(1, X, 0)$
$V3$	$(1, -X, 0)$
$V4$	$(0, -1, -X)$
$V5$	$(0, -1, X)$
$V6$	$(X, 0, 1)$
$V7$	$(-X, 0, 1)$
$V8$	$(X, 0, -1)$
$V9$	$(-X, 0, -1)$
$V10$	$(-1, X, 0)$
$V11$	$(-1, -X, 0)$

Face	Vertices
$F0$	$(V6,\ V2,\ V0)$
$F1$	$(V2,\ V6,\ V3)$
$F2$	$(V3,\ V6,\ V5)$
$F3$	$(V6,\ V7,\ V5)$
$F4$	$(V0,\ V7,\ V6)$
$F5$	$(V8,\ V2,\ V3)$
$F6$	$(V1,\ V2,\ V8)$
$F7$	$(V2,\ V1,\ V0)$
$F8$	$(V10,\ V0,\ V1)$
$F9$	$(V9,\ V10,\ V1)$
$F10$	$(V8,\ V9,\ V1)$
$F11$	$(V4,\ V8,\ V3)$
$F12$	$(V3,\ V5,\ V4)$
$F13$	$(V11,\ V4,\ V5)$
$F14$	$(V10,\ V11,\ V7)$
$F15$	$(V0,\ V10,\ V7)$
$F16$	$(V4,\ V11,\ V9)$
$F17$	$(V8,\ V4,\ V9)$
$F18$	$(V11,\ V5,\ V7)$
$F19$	$(V10,\ V9,\ V11)$

Note: The constant $X = (\sqrt{5} - 1)/2$ is the reciprocal of the golden ratio. It's value is approximately 0.618.

Recall that a C^0 curve was defined as the continuous image of a closed interval. Defining a surface as the continuous image of, say, a rectangle seems then a reasonable thing to do.

This does indeed pass muster for simple surfaces. See the objects of Figure 10.58. It's straightforward to map the rectangle W continuously onto the disc. The cylinder requires the parametric functions to roll W – mapping an edge onto the opposite one. The torus can be made in an additional step from the cylinder, by mapping the cylinder's opposite ends onto each other. How about the double torus though, which is certainly a surface? Is it apparent how to map W onto a double torus? Or, consider something as simple as the punctured rectangle, also a surface. How can one map the (unpunctured) rectangle W onto a punctured one in a continuous manner?

It's not quite clear, then, if the view of a surface as simply the image of a rectangle can be successful. Well, even if it might not succeed *globally*, it does *locally*. Huh?

Here's a thought experiment to explain what we mean. Straighten and bring the fingers of your right hand together so that it looks like an ellipse.

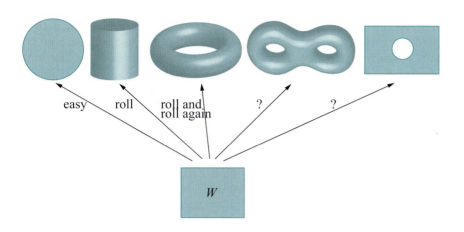

Figure 10.58: Mapping a rectangle onto surfaces.

Actually, let's pretend that it's the rectangle W of Figure 10.58. A question now: if you coated your palm and fingers with gray ink could you color each of the surfaces of Figure 10.58 gray by patting repeatedly? Pats are allowed to overlap. Ignore size constraints as well – think of your palm or any of the surfaces to be as small or large as you like. After a couple of minutes, then, the double torus may look like Figure 10.59.

Well ...? We're hoping your answer is yes, that you could pat each of the surfaces fully gray. What would that mean then? Exactly that the surfaces can each be covered by patches, each of which is a continuous image of a rectangle – continuous in the sense that you'll probably have to bend and squeeze your palm a lot, but not do anything drastic like poke a hole through it! In other words, per patch (locally!) the surface is indeed the continuous image of a rectangle.

Figure 10.59: Patting gray a double torus.

We're close to a definition of a surface. First, though, we have to formalize the notion of a so-called C^0 coordinate patch.

Definition 10.12. A C^0 *coordinate patch* in $\mathbb{R}^3$ is specified by three real-valued C^0 functions f, g and h, all defined on a closed rectangle $W = [a, b] \times [c, d]$ on the plane, such that the function

$$(u, v) \mapsto (f(u, v),\ g(u, v),\ h(u, v))$$

from W to its image B is one-to-one. The image set

$$B = \{(f(u, v), g(u, v), h(u, v)) : (u, v) \in W\}$$

itself is called a C^0 coordinate patch in $\mathbb{R}^3$.

Remark 10.17. The one-to-one condition ensures that B is topologically equivalent to W, which is stronger than if B were merely a continuous image of W. The examples next clarify this.

Example 10.22. Assume W to be the rectangle $[-1, 1] \times [-1, 1]$ on the plane. Refer to Figure 10.60 for diagrams of the following functions from W.

(a)
$$(u, v) \mapsto (u, v, u^2 + v^2)$$

specifies a coordinate patch that covers the bottom part of a paraboloid.

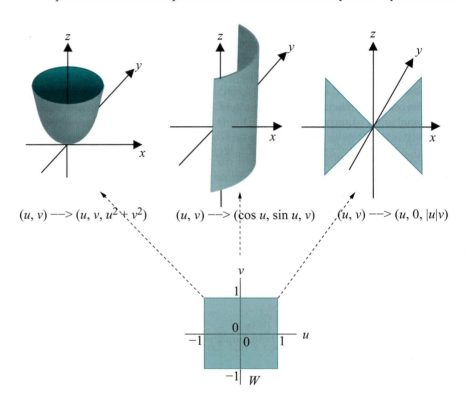

$(u, v) \dashrightarrow (u, v, u^2 + v^2)$ $\qquad$ $(u, v) \dashrightarrow (\cos u, \sin u, v)$ $\qquad$ $(u, v) \dashrightarrow (u, 0, |u|v)$

Figure 10.60: Functions $(u, v) \mapsto (f(u, v),\ g(u, v),\ h(u, v))$ and their images.

(b)
$$(u, v) \mapsto (\cos u, \sin u, v)$$

specifies a coordinate patch that covers part of a cylinder.

(c)
$$(u, v) \mapsto (u, 0, |u|v)$$

continuously maps the rectangle onto the union of two triangles *but* is not one-to-one – the entire segment $\{0\} \times [-1, 1]$ of W is mapped to the single point $(0, 0, 0)$ – so does not specify a coordinate patch.

The union cannot be patted gray either, as you can squeeze your palm as thin as you like, but never to a point.

We employ coordinate patches next to define surfaces.

Definition 10.13. A subset s of $\mathbb{R}^3$ is a C^0 *surface* if there is a collection $\mathcal{B}$ of C^0 coordinate patches in $\mathbb{R}^3$ such that

(a) the union of the coordinates patches in $\mathcal{B}$ equals s, and

(b) for each point $P \in s$, a sufficiently small neighborhood of P – i.e., consisting of points within a distance δ of P for some small positive δ – lies inside a single coordinate patch belonging to $\mathcal{B}$.

Remark 10.18. Patches in $\mathcal{B}$ can overlap.

Remark 10.19. $\mathcal{B}$ can be an infinite collection.

Remark 10.20. A C^0 surface is said to be a *two-dimensional topological manifold* or *topological surface*.

The first condition of the definition formalizes the intuition of a surface being covered by "rectangle-like" patches. The second one is to eliminate the sort of pathology shown in Figure 10.61. The object s consisting of two intersecting planes should not qualify as a surface – even though it can evidently be covered by patches – as it's not of a single sheet. In fact, a neighborhood of P, no matter how small, consists of two intersecting fragments, which can never lie in a single coordinate patch.

Example 10.23. What is the minimum number of coordinate patches required to cover the cylinder

$$x = \cos u, \; y = \sin u, \; z = v, \; (u,v) \in [-\pi, \pi] \times [-1, 1]$$

to prove that it's a C^0 surface according to Definition 10.13? You don't have to write equations for the coordinate patches. Just sketch them on the cylinder.

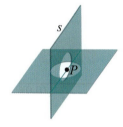

Figure 10.61: Any neighborhood of P will consist of two intersecting fragments, which cannot lie in one coordinate patch.

Answer: Two coordinate patches are sufficient. Figure 10.62(a) indicates one patch covering a sector of more than 180°. Another patch that's a mirror image of this one would cover the rest of the cylinder. The two would overlap, of course.

One coordinate patch by itself will never do. Although this is fairly evident given the shape of the cylinder, a proof requires topology (in particular, the fact that the cylinder is not "homeomorphic" to a rectangle). Thus, two patches is the minimum.

Exercise 10.68. The two coordinate patches of the preceding example overlapped significantly. To avoid "waste" suppose we had two patches that each spanned 180°, covering exactly one half of the cylinder, one a mirror image of the other. Their intersection would then consist of two line segments.

Would these two patches do to prove a cylinder to be a C^0 surface?

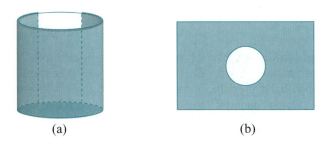

(a)   (b)

Figure 10.62: (a) One coordinate patch wrapping almost all the way around a cylinder (b) A punctured square.

Exercise 10.69. How about the following punctured square lying on the xy-plane?

$$[-1, 1] \times [-1, 1] \times \{0\} - \{(x, y, 0) : x^2 + y^2 < 0.5\}$$

See Figure 10.62(b). Note that it's closed with two boundary components – an outside one bounding the square and an inside one bounding the missing disc. How many coordinate patches does one need to prove that it's a C^0 surface? Answer with a sketch.

Exercise 10.70. Consider the following *open* disc (i.e., missing its boundary) lying on the xy-plane:

$$\{(x, y, 0) : x^2 + y^2 < 1\}$$

How does one cover it with coordinate patches to prove that it's a C^0 surface?

Suggested approach: A finite number of coordinate patches will not do. Find, first, a continuous mapping of a closed rectangle W to the closed disc

$$D_r = \{(x, y, 0) : x^2 + y^2 \leq 1 - 1/r\}$$

for any $r \geq 2$, to make D_r a coordinate patch. Consider, then, the union $\cup_{r=2}^{\infty} D_r$.

Defining C^m surfaces for values of m greater than zero is the next natural step and not hard. One must first define the C^m continuity of a function of more than one variable. Not surprisingly, this involves partial derivatives. Compare the following with Definition 10.7 of the C^m-continuity of a function of a single variable.

Definition 10.14. A function f defined on a closed rectangle $W = [a, b] \times [c, d]$ on the plane is said to be C^m-*continuous* or, simply, C^m, where $m \geq 1$, if all of its partial derivatives of order m and less exist and are continuous on W.

C^m coordinate patches will obviously invoke C^m functions of two variables:

Definition 10.15. A C^m *coordinate patch* in $\mathbb{R}^3$, where $m \geq 1$, is specified by three real-valued C^m functions f, g and h, all defined on a closed rectangle $W = [a, b] \times [c, d]$ on the plane, such that the function

$$(u, v) \mapsto (f(u, v),\ g(u, v),\ h(u, v))$$

from W to its image B is one-to-one. The image set

$$B = \{(f(u, v), g(u, v), h(u, v)) : (u, v) \in W\}$$

itself is called a C^m *coordinate patch*.

If, additionally, the two vectors

$$\begin{bmatrix} \dfrac{\partial f}{\partial u} & \dfrac{\partial g}{\partial u} & \dfrac{\partial h}{\partial u} \end{bmatrix}^T \quad \text{and} \quad \begin{bmatrix} \dfrac{\partial f}{\partial v} & \dfrac{\partial g}{\partial v} & \dfrac{\partial h}{\partial v} \end{bmatrix}^T$$

are linearly independent – i.e., if they are both non-zero and are not collinear – for every $(u, v) \in W$, then the coordinate patch is said to be *regular*.

It's usual to consider regular C^m coordinate patches, when $m \geq 1$, rather than just C^m. Accordingly, here's the definition of a surface covered by such patches:

Definition 10.16. A subset s of $\mathbb{R}^3$ is a *regular C^m surface*, where $m \geq 1$, if there is a collection $\mathcal{B}$ of regular C^m coordinate patches in $\mathbb{R}^3$ such that

(a) the union of the coordinate patches in $\mathcal{B}$ equals s, and

(b) for each point $P \in s$, all points of s sufficiently close to P lie in a single coordinate patch belonging to $\mathcal{B}$.

A regular C^m surface, $m \geq 1$, is often simply called a *regular surface*. A regular surface that is C^m for any m is regular C^∞, also called *smooth*.

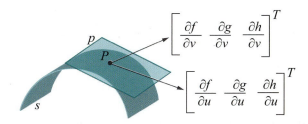

Figure 10.63: The non-zero linearly independent tangent vectors $\begin{bmatrix} \frac{\partial f}{\partial u} & \frac{\partial g}{\partial u} & \frac{\partial h}{\partial u} \end{bmatrix}^T$ and $\begin{bmatrix} \frac{\partial f}{\partial v} & \frac{\partial g}{\partial v} & \frac{\partial h}{\partial v} \end{bmatrix}^T$ span the tangent plane p at P.

Remark 10.21. Recall that the regularity condition in the case of a curve – that the tangent vector never vanishes – ensures a meaningful tangent direction at each of its points. The regularity condition for a surface ensures likewise that a meaningful *tangent plane* exists at each point. See Figure 10.63.

Figure 10.64 shows four surfaces of various orders of continuity. All labels should be clear except maybe for the second one, which is a cylinder capped by a hemisphere. This surface is regular C^1 but not regular C^2, for precisely the same reason that the third curve of Figure 10.20 is C^1 and not C^2 (we'll leave the reader to revisit the explanation there if need be).

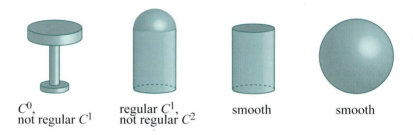

C^0,
not regular C^1 regular C^1,
not regular C^2 smooth   smooth

Figure 10.64: Various orders of surface continuity.

The table in Figure 10.64 is not regular, though it is composed of regular pieces. It would be nice to have a definition of piecewise regularity to apply to such surfaces. However, formulating an analogue of Definition 10.5 of piecewise regular curves is not straightforward, as it isn't clear what it means to join a number of surfaces end to end. The following definition finesses the problem.

Definition 10.17. A *piecewise regular surface* in $\mathbb{R}^3$ is a C^0 surface that is the union of finitely many regular surfaces.

The requirement of C^0-continuity in the definition assures the sheet-like nature of the union. The table is then piecewise regular, but the intersecting planes of Figure 10.61 are not as they don't form a C^0 surface.

Exercise 10.71. Is the surface of a regular polyhedron C^0, C^1, piecewise regular, ...?

We finish up with a definition of regular two-dimensional objects analogous to Definition 10.6 of regular one-dimensional objects, which does apply to the intersecting planes of Figure 10.61, and pretty much everything else one is likely to run into in 3D graphics.

Definition 10.18. A *regular two-dimensional object* is a finite union of regular surfaces.

And for the record we have the proper inclusions

regular surfaces $\subset$ piecewise regular surfaces $\subset$ regular 2D objects

10.3 Bézier Phrase Book

Bézier and NURBS (Non-Uniform Rational B-Spline) curves and surfaces are two special classes of curves and surfaces widely used in 3D design. Their utility to the applications programmer lies in the ability to sculpt a primitive in an intuitive manner by manipulating so-called control points, rather than by devising equations (the equations do exist but are created and managed transparently by the API). Several 3D modeling systems, in fact, allow the user to *interactively* design Bézier and NURBS primitives in a WYSIWYG environment. OpenGL, however, offers both in a sparser code-it-yourself manner.

There is a fair amount of theory underlying both, which is the reason for their effectiveness in the first place, and it's important that designers have a reasonable understanding in order to use them effectively. We'll study polynomial Bézier and NURBS theory in depth in Chapters 15-16. The rational version of both classes of primitives is developed in Chapter 18.

Although it is most convenient to design Bézier and NURBS primitives in a WYSIWYG environment, nevertheless, with a little effort fairly complex designs can be coded in straight OpenGL. Polynomial Bézier curves and surfaces, in particular, are quite intuitive and it's perfectly possible to learn their OpenGL syntax and begin design even before fully grasping the theory. Unfortunately, such is not the case with the B-splines, polynomial or rational or, for that matter, rational Bézier primitives, as it's difficult to make sense of their OpenGL syntax without some theoretical understanding.

In keeping, therefore, with the goal of this chapter to acquaint the reader with as many 3D design techniques as possible, we'll discuss polynomial OpenGL Bézier primitives – both curves and surfaces – without, for now, much of the theory.

Remark 10.22. There are plug-ins and third party software available for most popular 3D modeling systems to export their designs to OpenGL, so if you are willing to invest the time – and money, as commercial systems can be expensive – to learn to use one of them, then you can enjoy the convenience of WYSIWYG interactivity even for your OpenGL productions. However, it's our opinion that the new programmer gains most by cutting her teeth in the minimalist OpenGL environment.

10.3.1 Curves

A *Bézier curve c* is specified by a sequence $P_0, P_1, \ldots, P_n$ of *control points* in 3-space, whose number $n + 1$ is called the *order* of c. The curve starts at

Figure 10.65: Screenshot of bezierCurves.cpp with six control points, showing both the Bézier curve and its control polygon.

Figure 10.66: Screenshot of bezierCurveWithEvalCoord.cpp.

the first control point P_0, ends at the last P_n and approaches, but does not necessarily pass through, the intermediate ones. Think of the intermediate control points as "attractors" which mold the shape of c. See Figure 10.65, a screenshot of the program bezierCurves.cpp, which we'll be discussing momentarily, showing six control points and their Bézier curve (the curvy line).

Specifying the control points defines c in a parametric form

$$x = f(t), \; y = g(t), \; z = h(t)$$

where t belongs to a user-specified interval $[t1, t2]$. We'll leave discussing how the functions f, g and h are obtained to Chapter 15.

Experiment 10.15. Run bezierCurves.cpp. Press the up and down arrow keys to select an order between 2 and 6 on the first screen. Press enter to proceed to the next screen where the control points initially lie on a straight line. Press space to select a control point and then the arrow keys to move it. Press delete to start over. Figure 10.65 is a screenshot for order 6.

In addition to the black Bézier curve, drawn in light gray is its *control polygon*, the polyline through successive control points. Note how the Bézier curve tries to mimic the shape of its control polygon. **End**

We'll explain the syntax of the OpenGL Bézier curve with the help of the following two simpler programs.

Experiment 10.16. Run bezierCurveWithEvalCoord.cpp, which draws a fixed Bézier curve of order 6. See Figure 10.66 for a screenshot. **End**

The pair of statements

```
glMap1f(GL_MAP1_VERTEX_3, 0.0, 1.0, 3, 6, controlPoints[0]);
glEnable(GL_MAP1_VERTEX_3);
```

in the initialization routine of bezierCurveWithEvalCoord.cpp specify and enable the Bézier curve. The command

$$\text{glMap1f}(target, \; t1, \; t2, \; stride, \; order, \; *controlPoints)$$

defines what OpenGL calls a *one-dimensional Bézier evaluator*. Depending on how the parameter *target* is specified, the evaluator can be used to generate data for position, color, texture or normal direction. For now, we'll use it only to generate positional data and, accordingly, set *target* to GL_MAP1_VERTEX_3: the '1' is the dimension of the evaluator, while '3' calls for x, y and z coordinate values.

The parameters $t1$ and $t2$ specify the endpoints of the parameter interval of the curve. The parameter *order* specifies the number of control points, the coordinate values of which are to be found in the array pointed by *controlPoints*. The parameter *stride* is the number of floating point values between the start of the data set for one control point and that of the next in the array.

An evaluator of the form `glMap1f`(*target*, ...) must be enabled with a corresponding `glEnable`(*target*) command.

The Bézier curve itself of `bezierCurveWithEvalCoord.cpp` is drawn as a line strip joining vertices returned by calls to `glEvalCoord1f-((GLfloat)i/50.0)`. Generally, `glEvalCoord1f`(*t*) evaluates the coordinates of the point on the Bézier curve corresponding to the value *t* in the parameter interval.

Exercise 10.72. (Programming) Guess what will be displayed if the line strip definition in the drawing routine of `bezierCurveWithEval-Coord.cpp` is changed to either of the two below:

(a)
```
glBegin(GL_LINE_STRIP);
    for (i = 0; i <= 25; i++) glEvalCoord1f( (float)i/50.0 );
glEnd();
```

(b)
```
glBegin(GL_LINE_STRIP);
    for (i = 0; i <= 4; i++) glEvalCoord1f( (float)i/4.0 );
glEnd();
```

As Bézier curves are most often sampled evenly through the parameter interval, OpenGL provides a convenient way to do so, as we see in the next experiment.

Experiment 10.17. Run `bezierCurveWithEvalMesh.cpp`. This program is the same as `bezierCurveWithEval.cpp` except that, instead of calls to `glEvalCoord1f()`, the pair of statements

```
glMapGrid1f(50, 0.0, 1.0);
glEvalMesh1(GL_LINE, 0, 50);
```

are used to draw the approximating polyline.

The call `glMapGrid1f`(*n*, *t1*, *t2*) specifies an *evenly-spaced* grid of $n + 1$ sample points in the parameter interval, starting at *t1* and ending at *t2*. The call `glEvalMesh1`(*mode*, *p1*, *p2*) works in tandem with the `glMapGrid1f`(*n*, *t1*, *t2*) call. For example, if *mode* is `GL_LINE`, then it draws a line strip through the mapped sample points, starting with the image of the *p1*th sample point and ending at the image of the *p2*th one, which is a polyline approximation of part of the Bézier curve. **End**

End Tangents

Not only does a Bézier curve pass through its first and last control points, the tangent at the first control point is along the straight line joining the first two control points. In other words, it lies along the first segment of the control polygon. Likewise, the tangent at the other end lies along the last control polygon segment. This makes it possible to smoothly join two Bézier

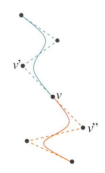

Figure 10.67: Two Bézier curves, one blue and one red, meet smoothly at an endpoint, as their control polygons meet smoothly (because v', v and v'' are collinear).

curves which meet at a common end control point v, by arranging v and its adjacent control points in either curve so that all three are on one straight line. See Figure 10.67.

Experiment **10.18.** Run `bezierCurveTangent.cpp`. The blue curve may be shaped by selecting a control point with the space key and moving it with the arrow keys. Visually verify that the two curves meet smoothly when their control polygons meet smoothly. Figure 10.68 is a screenshot of such a configuration. **End**

10.3.2 Surfaces

From Bézier curves to Bézier surfaces is straightforward. A *Bézier surface* (also called *Bézier patch*) s is specified by an $(n + 1) \times (m + 1)$ array of control points P_{ij}, $0 \le i \le n$, $0 \le j \le m$. The surface passes through the four "corner" control points $P_{00}, P_{n0}, P_{0m}, P_{nm}$, but not necessarily the others which, nevertheless, act as attractors. Let's continue the discussion with live code in front.

Figure 10.68: Screenshot of `bezier-CurveTangent.cpp`.

Experiment **10.19.** Run `bezierSurface.cpp`, which allows the user herself to shape a Bézier surface by selecting and moving control points originally in a 6×4 grid. Drawn in black actually is a 20×20 quad mesh approximation of the Bézier surface. Also drawn in light gray is the *control polyhedron*, which is the polyhedral surface with vertices at control points.

Press the space and tab keys to select a control point. Use the left/right arrow keys to move the selected control point parallel to the x-axis, the up/down arrow keys to move it parallel to the y-axis, and the page up/down keys to move it parallel to the z-axis. Press 'x/X', 'y/Y' and 'z/Z' to turn the surface. Figure 10.69 is a screenshot. **End**

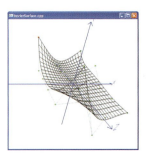

Specifying the control points array causes a Bézier surface s to be defined in a parametric form

$$x = f(u, v), \ y = g(u, v), \ z = h(u, v)$$

where u belongs to the interval $[u1, u2]$ and v to the interval $[v1, v2]$. We'll see in Chapter 15 how the functions f, g and h are obtained.

The statement

```
glMap2f(GL_MAP2_VERTEX_3, 0, 1, 3, 4, 0, 1, 12, 6,
        controlPoints[0][0]);
```

Figure 10.69: Screenshot of `bezierSurface.cpp`, showing both the surface mesh and its control polyhedron.

in the drawing routine of `bezierSurface.cpp` specifies the Bézier surface, while

```
glEnable(GL_MAP2_VERTEX_3);
```

enables it.

The syntax of the command

```
glMap2f(target, u1, u2, ustride, uorder, v1, v2, vstride,
        vorder, *controlPoints)
```

defining a *two-dimensional Bézier evaluator*, or Bézier surface, is an extension of that for a one-dimensional evaluator, taking into account the extra dimension. Like its one-dimensional counterpart, a two-dimensional evaluator can be used to generate data for position, color, texture or normal direction. We'll restrict ourselves to positional data for the present, setting *target* to GL_MAP2_VERTEX_3 (indicating a 2D surface in 3D space).

The values $u1$ and $u2$ specify the endpoints of the u-parameter interval and $v1$ and $v2$ those of the v-parameter interval. The parameter *uorder* is $m+1$, the number of columns of the control points array P_{ij}; *vorder* is $n+1$, the number of rows.

The coordinate values of the control points are located in the array pointed by *controlPoints*. The parameter *ustride* is the number of floating point values between the starts of the data sets for control points P_{ij} and $P_{i,j+1}$; *vstride* is the number of floating point values between the starts of the data sets for control points P_{ij} and $P_{i+1,j}$.

The pair of statements

```
glMapGrid2f(20, 0.0, 1.0, 20, 0.0, 1.0);
glEvalMesh2(GL_LINE, 0, 20, 0, 20);
```

are analogous in functionality to `glMapGrid1f()` and `glEvalMesh1()` discussed in the context of drawing Bézier curves. In particular,

```
glMapGrid2f(numberU, u1, u2, numberV, v1, v2)
```

specifies a $(numberU + 1) \times (numberV + 1)$ grid of sample points in the parameter rectangle, evenly spaced along both rows and columns, each row starting with u-value $u1$ and ending with u-value $u2$, and each column starting with v-value $v1$ and ending with v-value $v2$. The call

```
glEvalMesh2(mode, i1, i2, j1, j2)
```

works in tandem with the `glMapGrid1f(nu, u1, u2, nv, v1, v2)` call. For example, if *mode* is GL_LINE, then a stack of outlined quad strips is drawn with vertices at the mapped sample points, making a quadrilateral mesh approximation of the Bézier surface; if *mode* is GL_FILL then the strips are drawn filled.

Remark 10.23. It's a minor design flaw of OpenGL that a Bézier surface is approximated with a stack of quad strips, rather than triangle strips.

Remark 10.24. The u-parameter curves and v-parameter curves of a Bézier surface are (no surprise) Bézier curves. Think of the parameter u as being associated with (i.e., varying along) the columns of the control points array, and v with the rows. Accordingly, for a fixed i, the points $P_{i0}, P_{i1}, \ldots, P_{im}$ on the ith row are the control points of a u-parameter curve. The order of a u-parameter curve, therefore, is $m + 1$, which is the parameter *uorder* of `glMap2f()`.

Likewise, for a fixed j, $P_{0j}, P_{1j}, \ldots, P_{nj}$ are the control points of a v-parameter curve, whose order is $n + 1$, the parameter *vorder* of `glMap2f()`.

$\mathcal{R}em\mathit{ark}$ 10.25. There is a `glEvalCoord2f()` call available as well, analogous to `glEvalCoord1f()`, to evaluate the coordinates of a point on the surface corresponding to parameter point (u, v).

Next is an example of how to make a target shape by manipulating the control points of a Bézier surface.

$\mathcal{E}$xperiment 10.20. Run `bezierCanoe.cpp`. Repeatedly press the right arrow key for a design process that starts with a rectangular Bézier patch, and then edits the control points in each of three successive steps until a canoe is formed. The left arrow reverses the process. Press 'x/X', 'y/Y' and 'z/Z' to turn the surface.

The initial configuration is a 6×4 array of control points placed in a rectangular grid on the xz-plane, making a rectangular Bézier patch.

The successive steps are:

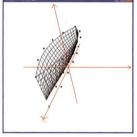

(1) Lift the two end columns of control points up in the y-direction and bring them in along the x-direction to fold the rectangle into a deep pocket.

(2) Push the middle control points of the end columns outwards along the x-direction to plump the pocket into a "canoe" with its front and back open.

(3) Bring together the two halves of each of the two end rows of control points to stitch closed the erstwhile open front and back. Figure 10.70 is a screenshot after this step.

End

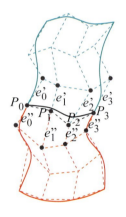

Figure 10.71: Two bicubic Bézier patches and their control polyhedrons, one pair blue and one red. The patches meet smoothly along a shared boundary curve which, together with its control polygon, is black.

Bicubic Bézier Patches and How to Join Them

Most often invoked in design are Bézier surfaces specified by a 4×4 array of control points, called *bicubic Bézier patches*. Complex shapes can be made by connecting multiple such patches. A similar principle applies to joining two bicubic patches – in fact, arbitrary Bézier patches – smoothly, as applies to joining two Bézier curves smoothly. First, two patches are contiguous if they share a common end row or end column of control points, in which case their control polyhedrons abut along that row or column. For the two patches to join smoothly, one further requires every pair of edges, one from either control polyhedron, meeting at a vertex of the shared border but not lying on the border itself, to be collinear.

For example, in Figure 10.71 two bicubic patches meet along a common boundary curve specified by their shared control points P_0, P_1, P_2 and P_3, which also specify the shared border of their control polyhedrons. The edges of the control polyhedrons on either side that meet at the border, viz., the pairs e_i' and e_i'', for $0 \leq i \leq 3$, are collinear, so the patches join smoothly.

Utah Teapot

Probably the most famous object ever made from bicubic Bézier patches is the *Utah Teapot*, created originally by Martin Newell, then at the University of Utah, in 1975. It rapidly became an iconic benchmark model within the CG community for the testing of rendering algorithms. The GLUT library's wireframe version is shown on the left of Figure 10.72.

Figure 10.72: GLUT library's version of the Utah teapot and Martin Newell's original porcelain Melitta model (from Wikimedia).

Newell's original design consisted of 28 patches and had neither a bottom nor a rim for the lid to rest on. The current incarnation, available from the GLUT library, has both and consists of 32 patches and a total of 306 different control points – 12 patches specify the body of the teapot, 4 the handle, 4 the spout, 8 the lid and 4 the bottom. Patches that meet obviously share control points and those composing the same part of the teapot join smoothly based on the principle described above.

Newell modeled an actual porcelain teapot that he owned, manufactured by the Melitta company. It's now an exhibit at the Computer History Museum in California. A picture is on the right of Figure 10.72. The rendered version, in fact, is squatter than the original. For the reason why and an entertaining account of the teapot's evolution read Crow's article [31]. Crow gives the patch control points data as well.

Torpedo

Experiment 10.21. Run `torpedo.cpp`, which shows a torpedo composed of a few different pieces, including bicubic Bézier patch propeller blades:

(i) Body: GLU cylinder.

(ii) Nose: hemisphere.

(iii) Three fins: identical GLU partial discs.

(iv) Backside: GLU disc.

(v) Propeller stem: GLU cylinder.

(vi) Three propeller blades: identical bicubic Bézier patches (control points arranged by trial-and-error).

Press space to start the propellers turning. Press 'x/X', 'y/Y' and 'z/Z' to turn the torpedo. Figure 10.73 is a screenshot.   End

Figure 10.73: Screenshot of `torpedo.cpp`.

Exercise 10.73. (Programming) Emulate Newell by modeling an everyday object using multiple bicubic Bézier patches. First, modify `bezierSurface.cpp` for an editable bicubic patch over a 4 × 4 array of control points.

Exercise 10.74. (Programming) Animate a river scene with many boats. Modify the canoe from `bezierCanoe.cpp` for a couple of different kinds of boats and put them in display lists and place (scaled and colored) copies on the river. Give a split screen view, one global from the bank and one from a particular boat.

Exercise 10.75. (Programming) Model the aircraft (Figure 10.74) and make it fly. Use bicubic patches. Parts like the wings, tail and fins can be panels of zero thickness, as can be the jet engine cases. Ignore logos and details. These can be textured in later. Focus on large-scale geometry.

Figure 10.74: Aircraft and express train.

Exercise 10.76. (Programming) Model a running version of the express train (Figure 10.74) using bicubic patches.

Exercise 10.77. (Programming) Make a recognizable replica of some familiar automobile using bicubic patches.

10.4 Fractals

Fractals are fun, but not essential to design. You can safely skip this section if you are in a hurry to progress through the book.

A *fractal* shape, or just fractal for short, is one that possesses the characteristic of *self-similarity*. A canonical example of a fractal in nature is a coastline. The sketch on the left of Figure 10.75 is that of a hypothetical stretch of a hundred miles of coastline, as it may appear from an aircraft, outlining the shape of a country's border with the sea. To its right is a view zoomed in on a part of maybe about ten miles, seen from a low-flying aircraft, showing bays and inlets. Rightmost is an even closer zoom-in on a stretch of one mile of beach showing its own features. One notices the similarity across different levels of resolution – self-similarity – in the undulations of the coastline. Clouds, trees and neural systems of animals are among a multitude of other naturally occurring fractals.

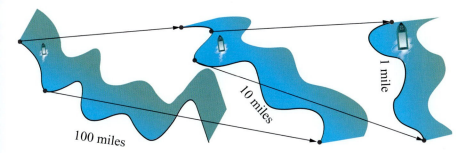

Figure 10.75: A coastline at increasing degrees of resolution: pairs of arrows indicate a blow-up.

Self-similarity lends itself immediately to programming by recursion. We'll give a semi-formal definition of fractal curves which makes self-similarity explicit. Our goal is not mathematical rigor, but a reasonable framework within which to write recursive code. The running example we'll use is a classic fractal curve – the *Koch curve* – invented by the Swedish mathematician Helge von Koch.

The first step in defining a fractal curve is to specify a *source* curve s. The location and orientation of s are not specified: it can be placed freely anywhere on the plane if we are drawing in 2D or space if in 3D. In the case of the Koch curve, s is a straight line segment on a fixed plane. See the top diagram in Figure 10.76.

The next step is to specify a rule to generate a *sequel* curve s' from the source curve. The sequel is rigidly associated with the source in that, if the location and orientation of s are fixed, then so are those of s'. The sequel for a Koch curve, in particular, is obtained by deleting the middle one-third segment of the source s and replacing it with the two opposite edges of an equilateral triangle, one of whose edges is the now-deleted middle third. See

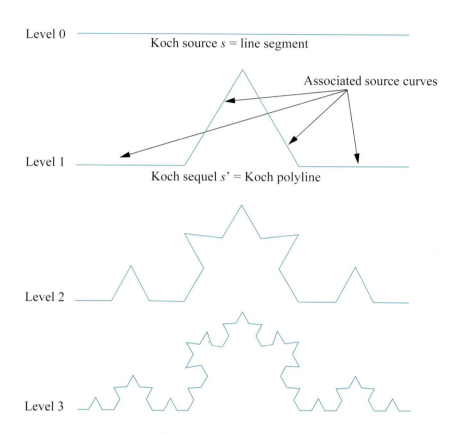

Figure 10.76: Koch curves.

the second diagram from the top in Figure 10.76. This particular 4-segment polyline sequel is called the *Koch polyline*.

The third and final step is to specify a rule to allow iterative reproduction of the sequel curves. In particular, for a sequel s' this rule specifies a finite set $\{s_1, s_2, \ldots, s_n\}$ of curves that are each a "transformed" copy of the source s. We'll not try to be precise as to how the transformations must be specified. It's best to imagine them being such that the s_i each are similar in shape to the source s, differing only in scale and location, which is the most common situation. Moreover, each s_i, $0 \leq i \leq n$, is rigidly associated with s' in that, if the location and orientation of s' are fixed, then so are those of s_i. The s_i, $0 \leq i \leq n$, are said to be the source curves *associated* with the sequel. In the case of the Koch curve, the associated source curves are simply the four segments of the Koch polyline, as indicated in Figure 10.76, each obviously a scaled version of the Koch source.

We are now ready to recursively produce the fractal curve to any desired level. Level 0 is a fixed copy s of the source. Level 1 is s replaced with its sequel s'. Level 2 is s' replaced with the union of the sequels of its associated sources $s_1, s_2, \ldots, s_n$, with the proviso that transformed copies of the source each generate equally transformed copies of the sequel. Level 3 and higher are obtained by repeating the procedure. Figure 10.76 shows the Koch curves till level 3.

The following program demonstrates the flexibility afforded by the framework just described.

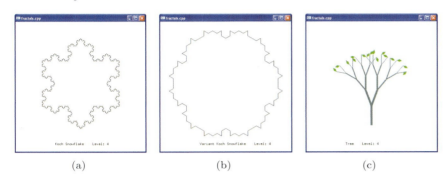

(a) (b) (c)

Figure 10.77: Screenshots from `fractals.cpp`: (a) Koch snowflake (b) Variant Koch snowflake (c) Tree.

$\mathbf{Experiment}$ **10.22.** Run `fractals.cpp`, which draws three different fractal curves – a Koch snowflake, a variant of the Koch snowflake and a tree – all within the framework above, by simply switching source-sequel specs! Press the left/right arrow keys to cycle through the fractals and the up/down arrow keys to change the level of recursion. Figure 10.77 shows all three at level 4. $\mathbf{End}$

The first curve of `fractals.cpp` drawn is the so-called *Koch snowflake* which consists of three Koch curves, as described earlier, each starting with one edge of an equilateral triangle as its level 0 source.

The variant Koch snowflake is produced with a single change from the definition of the Koch snowflake: instead of specifying the four segments of a sequel Koch polyline as its associated sources, the two segments joining the end vertices of the polyline to its middle are specified. See the upper two diagrams in Figure 10.78. Observe that it's not necessary that the associated sources be parts of the sequel.

The source for the fractal tree is a straight line segment as well, the initial copy being vertical. The sequel is a V-shaped two-segment polyline located atop the source, the length of each segment being a specified fraction (the constant `RATIO` in the program) of the length of the corresponding source, with a specified angle (the constant `ANGLE`) between them. The sources associated with the sequel are its two segments. See the bottom diagram of Figure 10.78.

The tree is produced by drawing the original vertical source line segment, as also the succeeding sequels at *all* levels, till the highest level of recursion.

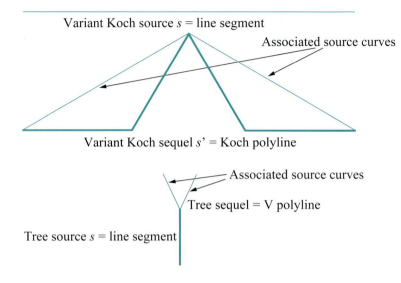

Figure 10.78: The variant Koch curve and fractal tree.

Note the difference here with the Koch curve and its variant, *only* the highest-level sequels being drawn in the case of the latter two. Moreover, sequels at successive levels of the fractal tree are drawn thinner – not part of the fractal definition. The final drawing is also embellished with leaves – not part of the fractal structure either – which are quadrilaterals at random angles at the ends of the top-level V's.

The program combines all three fractals by providing different member functions for each in the classes `Source` and `Sequel`.

Exercise 10.78. (Programming) Create a non-uniform tree by adding randomization so that not all sequels are drawn.

Exercise 10.79. (Programming) The variant of the Koch snowflake drawn in `fractals.cpp` self-intersects at high enough levels. Draw an "interesting" variation of the snowflake which doesn't self-intersect.

Exercise 10.80. (Programming) Draw a fractal cloud.

Exercise 10.81. (Programming) Draw a fractal flower.

10.5 Summary, Notes and More Reading

In this chapter we learned how to create a range of objects in 3D space. Of course, we had already been creating objects earlier, but in this chapter we studied 3D drawing techniques systematically. These included polygonal line approximation of curves and mesh approximation of surfaces. Special classes of curves and surfaces particularly useful in drawing, including conics and quadrics, swept and ruled surfaces, regular polyhedra, Bézier curves and surfaces, and fractals were discussed. We were introduced at an elementary level to the mathematical theory underlying curves and surfaces.

The dictum that practice makes perfect applies particularly to drawing. The more drawing projects one completes the better one will know how to proceed on the next one, not to mention the reusable object parts and code one will begin to accumulate.

A great thing about drawing scenes for movies and games is that if it walks like a duck and talks like a duck, then it *is* a duck. We have already taken advantage of this notion in faking curved objects with the help of straight and flat primitives. There's more to it though as the following example shows.

(a)

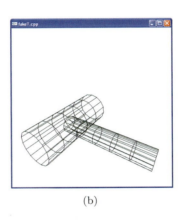

(b)

Figure 10.79: A T-pipe is simulated by sticking one GLU cylinder into another.

On the left of Figure 10.79 is a picture of what seems a perfectly respectable T-pipe of a sort one might find at a plumbing goods store. The picture on the right though reveals how it's drawn – by pushing one GLU cylinder into another so that an end of it protrudes inside (run `fakeT.cpp`). This is unlikely to be an acceptable industrial design of a T-pipe, but for the purposes of 3D graphics it is. Consider the saving in complexity. Authentic industrial design would require a hole whose boundary is a non-planar loop to be excised in one cylinder and an end of the other pared to match – neither trivial operations at all!

Figure 10.80: Sheared? But, what about the other side?

One is reminded of the amusing story of a farmer and a mathematics professor traveling together on a train through the countryside. The farmer looking out of the window remarks to the professor, "The sheep look like they've just been sheared." The ever-precise professor replies, "Well, we can say only that the side that we see has been sheared. Nothing can be concluded about the other side!" In CG one can indeed get away with shearing only the sides that the viewer sees. This liberation from real-world rectitude (or even rationality!) throws the doors wide open to creativity. If you are working on a game or a movie, now is the time to enhance it with various objects. And, if you are getting a bit tired now of wireframe, it won't be long – less than a page as a matter of fact – before we begin to paint and illuminate!

All introductory texts on 3D graphics have parts devoted to drawing curves and surfaces in 3-space. Books containing a more specialized treatment of modeling include Farin [42], the two by Mortenson [90, 92] and Rogers & Adams [113]. They all include discussions of the Bézier primitives as well. In Simmons [127]) the reader will find more about conics and quadrics. The author's own paper [58] is a mathematical investigation of a rather curious irregularity of regular polyhedra.

Keep in mind that we are not at all done with modeling ourselves. This chapter laid the groundwork. More of the theory of Bézier curves and surfaces comes in Chapter 15. Chapter 16 is about B-spline curves and surfaces, which are staple in modern design. That chapter discusses the polynomial version of B-splines, while NURBS – non-uniform rational B-splines, the most general version – are a topic of the later Chapter 18. Chapter 17 is about Hermite curves and patches, which interpolate – i.e., actually pass through – their control points, rather than merely approximate.

The reader interested in the mathematical theory of curves and surfaces, especially those wishing at some point to get into the research end of 3D graphics, should refer to math books such as Lipschutz [84] and Pressley [110] for a fairly soft introduction to differential geometry, while the books by Do Carmo [35], Kreyszig [77], O'Neill [102] and Singer & Thorpe [128] are written at a higher level.

Our account of fractals, though basic, has probably enough for the person who primarily wants to draw them. There are several excellent books to learn more about this popular topic. In addition to Mandelbrot's classic [87], which had seminal influence in formalizing fractals and attracting popular interest, a couple of more recent ones are by Barnsley [8] and Falconer [41].

Part VI

Lights, Camera, Equation

CHAPTER 11

Color and Light

O ur objects so far have mostly looked as if they plan to spend the afternoon home watching the game. It's time now to dress up and go party. The goal for this chapter is to learn how to use light sources to illuminate a scene and complementarily define material properties of objects to determine how they appear when lit.

We begin with a brief discussion of the theory of vision and color models in Section 11.1, learning particularly about the RGB color model so important in CG, as well as a few other models which pop up occasionally, such as CMY, CMYK and HSV. In Section 11.2 we study Phong's lighting model and how it conceives of light coming off an object as comprised of three components – ambient, diffuse and specular – based on the nature of their reflectance. This section concludes with a formula to derive the RGB intensities of the light reflected at a vertex based on Phong's model.

We move on to OpenGL in Section 11.3 and see how faithfully it implements Phong's model. And, we begin extensively to experiment and code. In Section 11.4 we describe OpenGL's so-called lighting model – not to be confused with Phong's lighting model – which sets certain environmental parameters. Directional light sources, located far from the scene, and positional lights, located in or near it, are discussed in Section 11.5, as is the related notion of attenuation of light over distance. Spotlights are the topic of Section 11.6. At this point we have all the parts needed to formulate in Section 11.7 the famous lighting equation that OpenGL actually implements to calculate color intensities at a vertex.

We discuss the two so-called shading models OpenGL offers, smooth and flat, in Section 11.8. The former familiarly interpolates the vertex colors through a primitive, while the latter is a somewhat idiosyncratic discrete coloring scheme. Animation of light sources is the topic of Section 11.9.

Specifying appropriate surface normals is critical to good lighting.

OpenGL can sometimes help with automatic normals, but often the user is on her own and the task can require a fair amount of calculation. Before we begin with normal computation proper, we have an optional introduction in Section 11.10 to the calculus of partial derivatives, to the extent required to calculate tangent planes and normals to the kinds of surfaces typical in CG. We recommend it be skipped at first and consulted subsequently if need be.

The long Section 11.11 is devoted to computing and applying surface normals to lighting. It begins by following the informal taxonomy of 2D objects introduced in Section 10.2, and moves on to Bézier and quadric surfaces for which automatic normals are available.

Section 11.12 contains a discussion of an alternate shading model proposed by Phong, which is more sophisticated (and more computation-intensive) than OpenGL's smooth shading. We conclude in Section 11.13.

11.1 Vision and Color Models

We begin with a little of the physics and biology underlying color and its perception.

Electromagnetic (EM) radiation consists of oscillating electric and magnetic fields moving through space. It is produced by the motion of electrically charged particles. From a physics point of view, EM radiation can be treated dually as waves or a stream of massless particles called photons traveling through a vacuum at the speed of light. EM radiation is characterized by its frequency or, equivalently, wavelength, the inverse of frequency. The EM spectrum consists of EM radiation of all possible frequencies, of which visible light is a very small part. See Figure 11.1.

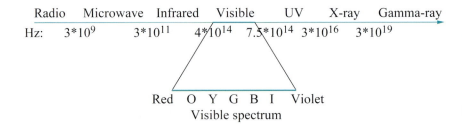

Figure 11.1: EM spectrum indicating approximate frequency ranges in Hz.

Visible light emitted from a source is rarely pure, i.e., of one particular frequency. Rather, there is an intensity distribution across the entire visible spectrum, and the perceived color depends on the particular distribution. Light from a source with an intensity distribution, for example, as in Figure 11.2(a), would be perceived as blue, as this color's intensity dominates, while one with the distribution of Figure 11.2(b) would appear white, because white is a mix of all colors in the visible spectrum.

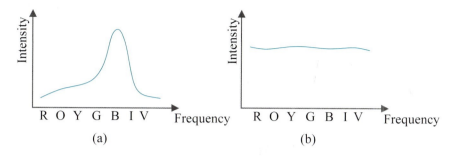

Figure 11.2: Intensity distributions across the visible spectrum: (a) appears blue (b) appears white.

We humans can see because of millions of light-sensitive cells embedded in the retinas of our eyes (see Figure 11.3 for a simplified anatomy). These cells are of two kinds, rod and cone. Rod cells are sensitive to low-intensity light, but not its frequency, which accounts for our night vision, as well as the fact that we have particular difficulty distinguishing colors in the dark. Cone cells, on the other hand, are stimulated only by fairly bright light, but can efficiently distinguish frequencies in the visible light spectrum, enabling us to perceive color. In fact, there are three kinds of cones – red, green and blue – according to the color of the light that most stimulates them. This is the basis of the *tristimulus* theory of human vision that perceived color is the net effect of the stimulation of these three kinds of cells.

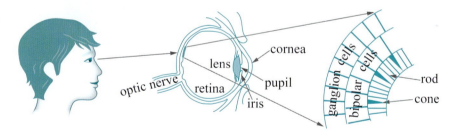

Figure 11.3: The eye.

11.1.1 RGB Color Model

A consequence of the tristimulus theory is the ubiquitous *RGB color model*: each color is represented as a sum of the three *primary colors*, red, green and blue, and each with a certain intensity, typically a value between 0 and 1 (for this reason RGB is called an *additive color model*). A color is denoted by a *color tuple* (r, g, b), where each component is the respective primary color's intensity.

Note: An intensity distribution curve, as those in Figure 11.2, one corresponding to each primary color, has been standardized by the International Commission on Illumination (CIE, from its French name Commission Internationale de L'Éclairage), as also a standard to convert intensity distributions across the visible spectrum to RGB triplets.

The RGB color space can be depicted as a cube, called a *color cube*, with axes corresponding to R, G and B values (see Figure 11.4(a)). The origin $(0, 0, 0)$ of the cube corresponds to black, while its diagonally opposite vertex $(1, 1, 1)$ to white, which, of course, is the maximal equal mix of red, green and blue. The other three diagonally opposite pairs each corresponds to a primary color and its complement (the complement of a color being that which with it combines to produce white). The straight line segment from black to white, each point (x, x, x) of which has equal parts of the primary colors, represents the gray scale. Figure 11.4(b) is a popularly drawn Venn diagram, where discs correspond to primary colors and their intersections are colored according to the mixing of the primaries.

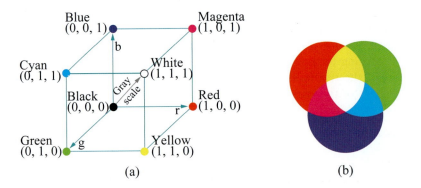

(a)   (b)

Figure 11.4: (a) The RGB color cube (b) Venn diagram combining colors.

The mechanics of the "addition" of colors in the RGB model is interesting. The color cube, for instance, indicates that an equal mix of red (which is $(1, 0, 0)$) and green $((0, 1, 0))$ produces yellow $((1, 1, 0))$. The reason for this is that the sensation produced in the human eye by a mix of two lights, one whose red frequency dominates and another whose green dominates, is similar to that produced by a single light dominant in the yellow frequency. This is a consequence of how our optic nerves react to the stimulation of particular combinations of cone cells, and *not* because the frequencies of red and green combine according to some law of physics to produce that of yellow! The RGB model, therefore, rests more on the biology of human vision than the physics of light.

RGB Color Model and Computer Graphics

The RGB model is implemented in millions of color display units around us, including computer monitors, both CRT and LCD. A CRT (cathode-ray tube) monitor has phosphors of the three primary colors located at each one of a rectangular array of pixels, and three electron guns that each fires a beam at phosphors of one color. A mechanism to aim and control their intensities causes the beams to travel together row by row, striking successive pixels in order to excite the RGB phosphors at each to values specified for it in the color buffer. See Figure 11.5(a).

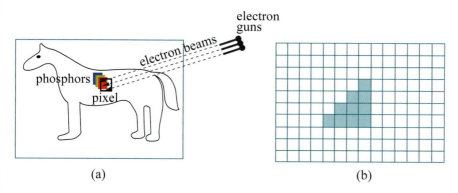

(a)   (b)

Figure 11.5: (a) Color CRT monitor with electron beams aimed at a pixel with phosphors of the 3 primaries (b) A raster of pixels showing a rasterized triangle.

Pixels in an LCD (liquid crystal display) monitor, on the other hand, each consist of three subpixels made of liquid crystal molecules, which separately filter through light of only one primary color. The amount of primary color emerging through each subpixel is controlled by an electric charge, whose intensity in turn is determined by the corresponding value in the color buffer.

From the point of view of OpenGL and, indeed, most CG theory, what matters most is that the pixels in a monitor, CRT or LCD, are, in fact, arranged in a rectangular array, called a *raster* (as depicted in Figure 11.5(b)). The number of rows and columns in the raster determines the monitor's resolution. For, this layout is the basis of the lowest-level CG algorithms, the so-called raster algorithms, which actually select and color the pixels to represent user-specified shapes such as lines and triangles on the monitor. Figure 11.5(b), for example, shows the rasterization of a right-angled triangle (with terrible jaggies because of the low resolution). We'll be studying raster algorithms in fair depth ourselves in Chapter 14.

Furthermore, a memory location called the *color buffer*, either in the CPU or graphics card, contains, typically, 32 bits of data per raster pixel – 8 bits for each of RGB and 8 for the alpha value (used for blending). It is the RGB values in the color buffer which determine the corresponding raster pixel's color intensities. The values in the color buffer are read by the raster – at which time the raster is said to be refreshed – at a rate called the monitor's refresh rate. Beyond this, the technology underlying a particular display device matters little practically in computer graphics.

11.1.2 CMY and CMYK Color Models

The *CMY color model*, whose augmentation CMYK is typically used in color printing, is a *subtractive color model*. CMY stands for cyan, magenta and yellow, and they are, respectively, the complements of red, green and blue. For example, cyan reflects blue and green but absorbs (or subtracts) red. Likewise, magenta and yellow subtract green and blue, respectively. Accordingly, cyan, magenta and yellow are referred to as the *subtractive primaries*. The color cube and Venn diagram for the CMY color model are depicted in Figure 11.6.

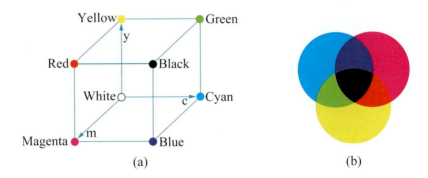

Figure 11.6: (a) The CMY color cube (b) Venn diagram of the CMY model.

Going between the RGB and CMY color spaces is simple:

$$\begin{bmatrix} c \\ m \\ y \end{bmatrix} = \begin{bmatrix} 1 \\ 1 \\ 1 \end{bmatrix} - \begin{bmatrix} r \\ g \\ b \end{bmatrix} \quad \text{and} \quad \begin{bmatrix} r \\ g \\ b \end{bmatrix} = \begin{bmatrix} 1 \\ 1 \\ 1 \end{bmatrix} - \begin{bmatrix} c \\ m \\ y \end{bmatrix} \quad (11.1)$$

Ink of the color of each of the three subtractive primaries is coated as a grid of dots (called a *screen*) on a sheet of paper during printing. The relative proportions of CMY ink at each dot determines the amounts of light of various frequencies subtracted there; the remaining light emerges through the ink layers and imparts to the dot its perceived color.

However, in practice, the combination of CMY ink to produce RGB color does not work as well as the equations (11.1) might suggest. The CMY pigments in printer toner cartridges are never pure enough that an equal mix produces 100% black or even proper shades of gray. In addition to this technical problem there is an economic one too: making blacks and grays, by far the most common colors in printing, by mixing colored inks is expensive.

Modern color printers, accordingly, supplement their CMY inks with a black ink to directly produce both blacks and the gray scale, the resulting process called *four-color printing*.

The CMY model augmented with black is called the *CMYK color model* (for reasons to do with printing terminology black is denoted by K rather than B). Conversion formulae between the CMYK color space and the RGB and CMY color spaces are more complicated than those between the latter two and we'll not discuss them here. However, drawing and image editing programs, such GNU's GIMP (freeware [49]), which offer both RGB and CMYK models will automatically convert between the two.

A practical point to keep in mind is that mapping from RGB to CMYK is often device-dependent and rarely 100% accurate, which is why CMYK print-outs are frequently significantly different from the original RGB display. Moreover, the space of colors representable in the RGB and CMYK color models – their *gamuts* – are not identical either, so some colors simply cannot be transferred exactly from monitor to paper (or vice versa).

11.1.3 HSV (or HSB) Color Model

The RGB color model, though pretty much ingrained into applications around us, is not particularly intuitive for the mixing of colors. For example, what RGB values should an artist combine for a jungle green, sunset orange, ocean blue, ...? The *HSV color model* was created by Alvy Ray Smith (one of the co-founders of Pixar Corporation) in 1978 as a more user-friendly alternative for designers. HSV is the abbreviation for hue, saturation and value. The model is also called *HSB*, where B stands for brightness.

The HSV model gets past the problem of having to numerically mix primaries by allowing the designer to choose a color's "coloredness" (that which we perceive as jungle green, sunset orange, ocean blue, etc.) directly with a *single* parameter, the *hue*. The hue parameter space is circular and often called the *color wheel*.

Here's how the color wheel is derived. See Figure 11.7. Begin with a triangle with corners representing the red, green and blue hues. Double the number of vertices to make a hexagon and fill in the middle hues, yellow, cyan and magenta (e.g., yellow is an equal mix of red and green, so midway between them). Again double to a dodecagon and add new hues by interpolating between previous ones. Continuing the process leads to a continuum of hues in a circle. A position on this circle – or, the color wheel as it's called – thus represents a particular hue. Typically, red, green and blue are located at $0°$, $120°$ and $240°$, respectively.

The hue parameter by itself is insufficient to specify a color. Two other parameters are required as well. The *saturation* of a color, typically given as a percentage, represents its purity. The higher the saturation the richer and more vibrant the color appears; conversely, less saturated colors (called *desaturated*) appear faded and grayish. The final parameter is *value*, given

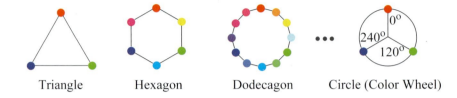

Triangle   Hexagon   Dodecagon   Circle (Color Wheel)

Figure 11.7: Hues on a triangle, hexagon, dodecagon and circle (color wheel).

as a percentage as well, representing a color's intensity or brightness.

The saturation and value amounts of a color are often specified by positioning a point on an equilateral triangle inscribed in the color wheel, with a vertex of that triangle located at and turning with the color's hue position on the wheel. As indicated in Figure 11.8(a), value changes parallel to the edge opposite the hue vertex, while saturation increases with distance from the opposite edge. Figure 11.8(b) shows GIMP's color dialog box setting a 100% blue in HSV mode.

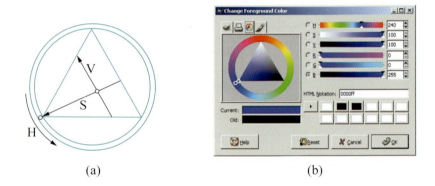

(a)   (b)

Figure 11.8: (a) The outer small hollow circle is positioned on the colored wheel to set the H value and the inner one inside the triangle to set S and V values. (b) The Gimp color dialog box making a 100% blue using the HSV color wheel option.

11.1.4   Summary of the Models

In drawing applications the predominant model is RGB and we'll really not have use for any other through the rest of this book. It's useful, though, to have a nodding acquaintance with models that may occasionally pop up elsewhere. With CMY, CMYK and HSV we have covered the ones the user is most likely to encounter. CMYK does, in fact, become particularly important when one goes from drawing to printing. There are other color models not used as much, such as Lab (an option in Adobe's Photoshop) and HLS (for hue, lightness, saturation).

The gold standard among color models was established by the CIE in 1931. It's called the *CIE XYZ model* (also the *CIE 1931 model*) where the X, Y and Z parameters represent, respectively, three theoretical primaries, each corresponding to a particular intensity distribution standardized by the CIE. Although not seen in practical interfaces, the CIE XYZ color model is used to calibrate implementations of the other ones.

11.2 Phong's Lighting Model

A model of interaction between light sources and objects is called a *lighting model* (or *reflection model*, or *illumination model*). In 1975 Vietnamese computer scientist Phong Bui Tuong [105] invented a particular lighting model, now known by his name, which is currently the one most widely used in practice. Despite the subsequent development of more authentic lighting models, e.g., Cook-Torrance [28], ray tracing, etc., Phong's has endured in popularity, especially because it delivers realistic lighting at moderate computational cost. OpenGL implements Phong's model. But, before we start coding up light let's first get an understanding of the model.

11.2.1 Phong Basics

In Phong's model the light reflected off an object O is the sum of three components – *ambient*, *diffuse* and *specular* – based on the *reflectance* properties of its surface. We'll describe each component next before explaining how to specify and calculate them.

Ambient: Ambient reflectance models O's reflection of background light that strikes it from multiple directions. Ambient light is scattered equally in all directions from the surface of O as well because of fine-scale graininess. See Figure 11.9.

Of the light sources in the environment – e.g., lamps and the sun – a part of the light from each is presumed ambient in that it's scattered by minute particles such as dust in the environment, effectively becoming part of background light before striking O. The direction of the light's source, therefore, is lost in that part of it which is ambient. Neither does it matter where the viewer is located because of the scattered reflection from the surface of O, presumed equal in all directions. In practical terms, the ambient component models that part of light which supplies constant illumination throughout a scene. An example of a familiar light source which is mostly ambient is a tube lamp recessed behind a frosted panel.

In addition to the ambient parts of each light source, there is presumed to be a *global ambient light* as well, from no identifiable source (i.e., "true" background light). For example, when modeling a scene inside a building, we can adjust the global ambient to account for light coming in from outside through doors and windows, without trying to model every possible light source such as the sun and lights on the street, which would be very complex indeed.

The total ambient component of the light reflected from an object O is the sum of what it reflects of the ambient parts from each source, plus what it reflects of the global ambient. Informally:

$$\text{ambient reflectance from } O = \sum (\text{reflectance of ambient part from each light source}) + \text{reflectance of global ambient}$$

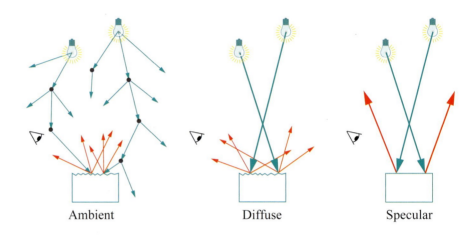

| Ambient | Diffuse | Specular |

Figure 11.9: Ambient, diffuse and specular reflectance: incident light drawn blue, reflected red.

Diffuse: Diffuse reflectance specifically models the fine-scale graininess of the surface: the diffuse part of light from a particular source travels in a coherent beam toward O and then is scattered equally in all directions by diffuse reflectance from the surface of O. Therefore, the direction of the light source does matter in case of diffuse reflectance, but not that of the viewer. Practically then, the diffuse component models the "soft" part of the light with little focus, e.g., that reflected off polished wood or silky fabric.

The total diffuse component of light reflected from O is the sum of the reflectances of the diffuse parts from each source:

$$\text{diffuse reflectance from } O = \sum (\text{reflectance of diffuse part from each light source})$$

Specular: Specular reflectance models the shininess of the surface: the specular part of light from a particular source travels in a coherent beam toward O and then is reflected in mirror-like manner, again in a coherent beam, by specular reflectance from the surface of O. Both the direction of the light source and the viewer matter in the case of specular reflectance. Specular light is "hard" light with a focus, e.g., that from a beam bouncing off a polished metal surface.

The total specular component of the light reflected from O is the sum of the reflectances of the specular parts from each source:

$$\text{specular reflectance from } O = \sum (\text{reflectance of specular part from each light source})$$

Remark 11.1. Because specular reflection is mirror-like, while the ambient and diffuse reflections are due to scattering from the surface of the object, the color of specularly reflected light depends primarily on that of the source itself, while those of the ambient and diffusely reflected on the native color of the object, as well as the light source.

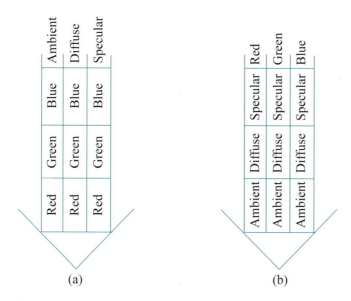

Figure 11.10: Orthogonal splitting of light: (a) Reflectance followed by color (b) Color followed by reflectance.

Important: The split of light into the three components of ambient, diffuse and specular according to reflectance is *independent* of the split into the primary color components of red, green and blue, in that each of the ambient, diffuse and specular components has RGB subcomponents and all nine subcomponents can be independently set. Or, one can equivalently say that each of RGB has ambient, diffuse and specular subcomponents which can all be independently set. In other words, you can think of light as being split as in Figure 11.10(a) or Figure 11.10(b) – it does not matter. Yet another way this is often phrased is by saying that color and reflectance splits are *orthogonal*.

A final component of light emerging from O is not reflected.

Emissive: The emissive component of light from an object O is that which is "manufactured" at O and unrelated to external light sources or the global ambient light. An example of an emissive object would be a lamp or the headlight of an automobile.

It is extremely important to keep in mind that, in OpenGL implementations, emissive light is perceived *only* by the viewer and does *not* illuminate other objects – it does *not* make O a light source for the rest of the environment.

11.2.2 Specifying Light and Material Values

OpenGL allows several light sources to be specified – the exact number depending on the implementation. For each of the N light sources L^i, $0 \leq i \leq N - 1$, the RGB intensities of each of its ambient, diffuse and specular components can be set to between 0 and 1, for nine values altogether per light source. These are written, typically, in a 3×3 *light properties matrix*:

$$\begin{bmatrix} L^i_{amb, R} & L^i_{amb, G} & L^i_{amb, B} \\ L^i_{dif, R} & L^i_{dif, G} & L^i_{dif, B} \\ L^i_{spec, R} & L^i_{spec, G} & L^i_{spec, B} \end{bmatrix} \tag{11.2}$$

Similarly, for each object O or, more precisely, each vertex V of O, one can set *scaling factors* between 0 and 1 to determine how much of each component of the incident light is reflected, for again nine values, contained in a 3×3 *material properties matrix*:

$$\begin{bmatrix} V_{amb, R} & V_{amb, G} & V_{amb, B} \\ V_{dif, R} & V_{dif, G} & V_{dif, B} \\ V_{spec, R} & V_{spec, G} & V_{spec, B} \end{bmatrix} \tag{11.3}$$

These so-called reflectance values represent the object's color: the higher one is, the more of the corresponding incoming light is reflected, and the more of that color the object appears to be.

The RGB values of the global ambient light are contained in a 3-vector called the *global ambient light vector*:

$$[globAmb_R \quad globAmb_G \quad globAmb_B] \tag{11.4}$$

The RGB values of the emissive light from a vertex V is a 3-vector called the *emissive light vector*:

$$[V_{emit, R} \quad V_{emit, G} \quad V_{emit, B}] \tag{11.5}$$

11.2.3 Calculating the Reflected Light

We come now to calculating each component of the reflected light.

Ambient

Calculating the ambient component emerging from a vertex V owing to a particular light source consists simply of scaling the light's ambient intensity by V's ambient reflectance. If the original intensity of the ambient light of some primary color from a source L (or the global ambient) is I, then that of its reflection from the surface at V is

$$I * material\ ambient\ scaling\ factor \qquad (11.6)$$

The *material ambient scaling factor* is the fraction of the incident ambient light that the material reflects. It is nothing but the ambient reflectance value $V_{amb,\ X}$, where X may be R, G or B, in the first row of the material properties matrix. An example will clarify use of the equation.

Example 11.1. Say the intensities of the ambient light from source L are given by

$$L_{amb,\ R} = 0.4, \quad L_{amb,\ G} = 0.9, \quad L_{amb,\ B} = 0.2$$

and the ambient reflectances of V by

$$V_{amb,\ R} = 0.9, \quad V_{amb,\ G} = 0.9, \quad V_{amb,\ B} = 0.1$$

Then the part of the red light emanating from V owing to the L ambient is

$$L_{amb,\ R} * V_{amb,\ R} = 0.36$$

and the part of the green light emanating from V owing to the L ambient is

$$L_{amb,\ G} * V_{amb,\ G} = 0.81$$

and the part of the blue light emanating from V owing to the L ambient is

$$L_{amb,\ B} * V_{amb,\ B} = 0.02$$

Exercise 11.1. If the global ambient light vector is

$$[0.2\ 0.2\ 0.2]$$

and all the ambient reflectances of a vertex V are as in the preceding example, calculate the parts of the RGB light emanating from V owing to the global ambient.

Diffuse

Calculation of the diffuse component of the light reflected from V is more complex than that of the ambient as, not only must the incident light be scaled by the reflectance at V, but its direction taken into account as well. The latter is done by measuring the angle between the direction of the light source and the normal to the surface at V.

Remark 11.2. A line l is *normal* to a surface s at the point P if it is perpendicular to the tangent plane p of s at P. Any non-zero vector n parallel to l is a *normal vector* to s at P. See Figure 11.11. (Think intuitively of the tangent plane as a hard board pressed to touch s at P.)

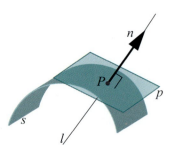

Figure 11.11: A normal vector n to a surface s at P lies along the normal line l there and is perpendicular to the tangent plane p at P.

The light source L is modeled as a point. Further, the surface of the object O around the illuminated vertex V is assumed flat; in fact, it's taken to coincide with its own tangent plane at V. See Figure 11.12(a). Diffuse light is reflected in all directions from V.

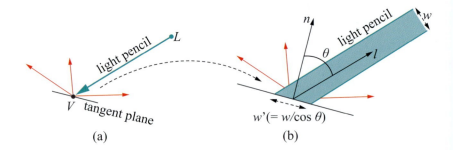

Figure 11.12: Calculating the diffuse component: (a) A light pencil from a point source L hits the surface, represented by it tangent plane at V (b) A blow-up of the pencil showing the normal vector n and the light direction vector l.

One observes from the blow-up in Figure 11.12(b) that a pencil of light of cross-sectional width w from L illuminates an area of width w', which is, typically, greater than w.

In this figure, θ is the angle between a direction vector l from V to the light source L, called the *light direction vector* and an outward normal vector n at V. The angle θ is called the *angle of incidence* of the light. We ask the reader to show next, by elementary trigonometry in Figure 11.12(b), that the width $w' = w/\cos\theta$.

Exercise 11.2. Verify the claim just made about the width of the area illuminated by a light of width w being $w' = w/\cos\theta$.

Since the area illuminated is greater by a factor of $1/\cos\theta$ than the cross-sectional area of the light pencil, the intensity of the light is diminished by a factor of $1/\cos\theta$ from I to $1/(1/\cos\theta) * I = \cos\theta * I$. Accordingly, if the original intensity of the diffuse light of some primary color emanating from the light source L is I, then that of its reflection from the surface of O at V is

$$\cos\theta * I * material\ diffuse\ scaling\ factor \qquad (11.7)$$

The *material diffuse scaling factor*, given by the values $V_{dif,\,X}$, where X is R, G or B, in the second row of material properties matrix, determines the fraction of the incident diffuse light the material reflects.

Example 11.2. Say the intensities of the diffuse light from source L are given by

$$L_{dif,\,R} = 0.3, \quad L_{dif,\,G} = 1.0, \quad L_{dif,\,B} = 1.0$$

and the diffuse reflectances of a vertex V by

$$V_{dif,\,R} = 0.8, \quad V_{dif,\,G} = 1.0, \quad V_{dif,\,B} = 0.8$$

and that the angle θ of incidence at V is $60°$.

Then the part of the red light emanating from V owing to the L diffuse is

$$\cos\theta * L_{dif,\,R} * V_{dif,\,R} = 0.5 * 0.3 * 0.8 = 0.12$$

Likewise, the part of the green light emanating from V owing to the L diffuse is

$$\cos\theta * L_{dif,\,G} * V_{dif,\,G} = 0.5$$

and the part of the blue light emanating from V owing to the L diffuse is

$$\cos\theta * L_{dif,\,B} * V_{dif,\,B} = 0.4$$

$Remark$ 11.3. The relationship that the intensity of the reflected lighted varies as the cosine of the angle of incidence is known as *Lambert's law*. It is Lambert's law which explains why early mornings and late evenings, when the sun is lower in the sky, are cooler and darker than mid-day.

Specular

For specular light, as in the case of diffuse light, the light source L is modeled as a point, and so too the eye E. And, again, the surface of O is identified with its tangent plane at the illuminated vertex V. An outward normal vector to the surface of O at V is n. Let s be a vector, call it a *halfway vector*, which bisects the angle between a light direction vector l from V toward the light source L and an eye direction vector e from V toward the eye E. See Figure 11.13(a) (ignore r for now).

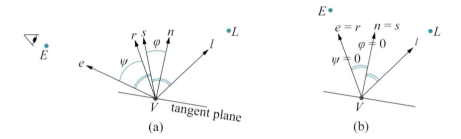

Figure 11.13: Calculating the specular component: (a) The light direction vector l, eye direction vector e, halfway vector s, normal vector n and reflection vector r (b) The special case when reflection is in the direction of the eye. (Double arcs indicate equal angles.)

We'll next state a relationship between the intensity of the reflected specular light and that of the incident which may seem unintuitive at first, but motivation will soon be apparent:

If the intensity of the specular light of some primary color emanating from the light source L is I, then that of its reflection from the surface of O at V is

$$\cos^f \phi * I * material\ specular\ scaling\ factor \tag{11.8}$$

where ϕ is the angle between halfway vector s and the normal vector n, $f \geq 0$ is a scalar, called the *shininess exponent* and the *material specular scaling factor*, a value read from the material properties matrix, determines the fraction of the incident specular light the material reflects.

Here's what's happening. If the surface of O is *perfectly* mirror-like, then a ray of light from L to V reflects according to the laws of reflection, which say that the normal to O at V, the incident ray and the reflected ray all lie on the same plane and, moreover, that the incident ray and the reflected ray make the same angle with the normal. In this particular case, if the eye E is located in the direction of reflection, given by the *reflection vector* r, then it perceives all the incident light, if not no light at all.

Say ψ is the angle the reflection vector makes with the eye direction vector e, as in Figure 11.13(a). Figure 11.13(b) shows a particular case of the general Figure 11.13(a), where the eye is actually situated in the direction of reflection, so that $\psi = 0$. Observe, in this case, that the halfway vector is aligned with the normal, in other words, $\phi = 0$ as well.

Most real surfaces, however, are not perfectly mirror-like and do not reflect light only along the direction of reflection, but, rather, spread it *about* that direction with an intensity which diminishes with increasing angle. In other word, maximum intensity is perceived by the viewer in a configuration as in Figure 11.13(b); nevertheless, even in a general configuration as in Figure 11.13(a), the eye receives light, though, with intensity inversely related to ψ.

This suggests that the intensity of specular reflection be modeled by the formula

$$\text{angular attenuation factor} * I * \text{material specular scaling factor} \qquad (11.9)$$

where the *angular attenuation factor* is, in fact, the factor in inverse relationship with ψ.

Phong suggested the angular attenuation factor $\cos^f \psi$, where the exponent f is larger the more mirror-like (i.e., shiny) the surface is. His considerations were empirical rather than based on actual physics. In particular, $\cos \psi$ is a function of ψ which is at its maximum of 1 when $\psi = 0$ and drops off as ψ increases, behavior expected of the angular attenuation factor. The function $\cos^f \psi$ also shows the same behavior, but more markedly, as f increases. In particular, the larger the value of f the more rapid the drop from the value of 1 as ψ increases from 0. See Figure 11.14. Intuitively, the shinier the surface the more rapidly the light diminishes away from the direction of reflection.

The angle ψ is often replaced by ϕ, the angle between the halfway vector s and the normal vector n, because ϕ is easier to compute, and because it is legitimate to do so given the following linear relation between the two.

Example 11.3. Show that $\psi = 2\phi$.

Answer: Label the angles from the tangent plane to the light direction, the reflection and eye direction vectors θ_1, θ_2 and θ_3, respectively, as in Figure 11.15.

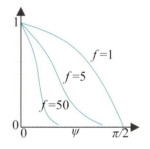

Figure 11.14: Graphs of $\cos^f \psi$ for different values of f (not exact plots).

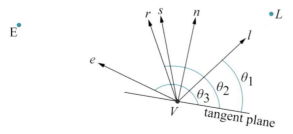

Figure 11.15: Proving that $\psi = 2\phi$.

The angle to the halfway vector, then, is $(\theta_1 + \theta_3)/2$, implying that the angle between the halfway vector and the normal is $\phi = (\theta_1 + \theta_3)/2 - \pi/2$.

The angle between the light vector and the normal, which is $\pi/2 - \theta_1$, is the same as the angle between the normal and the reflection vector, by laws of reflection. Therefore, $\theta_2 = \pi/2 + (\pi/2 - \theta_1) = \pi - \theta_1$. This, then, implies that the angle between the eye direction vector and the reflection vector is $\psi = \theta_3 - \theta_2 = \theta_3 - (\pi - \theta_1) = \theta_1 + \theta_3 - \pi$. That $\psi = 2\phi$ now follows.

Given the relationship between ϕ and ψ contained in the preceding example, substituting $\cos^f \phi$ for $\cos^f \psi$ as the angular attenuation factor in Equation (11.9) makes no qualitative difference. The result of the substitution, in fact, is Equation (11.8), which is now fully justified.

Example 11.4. Give a formula for the halfway vector s in terms of the light direction vector l and the eye direction vector e from V, which are, of course, the two vectors that s bisects. Assume that both l and e are of unit length. Give s as a unit vector as well.

Answer: See Figure 11.16, where $l = \overrightarrow{OA}$, $e = \overrightarrow{OB}$, and where $l + e$ is drawn with the help of the parallelogram law of addition of vectors. Since $|l| = |e| = 1$, all four sides of the parallelogram $OACB$ are of unit length as well. A consequence is that corresponding sides of the triangles OAC and OBC are of equal lengths. The two triangles are, therefore, congruent, so $\angle AOC = \angle BOC$. One concludes that the vector $l + e$ bisects l and e.

Accordingly, the unit halfway vector

$$s = \frac{l + e}{|l + e|} \quad \text{(provided that } l + e \text{ is not the zero vector)}$$

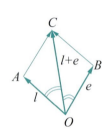

Figure 11.16: The vector $l = e$ bisects l and e.

Remark 11.4. When a vector u is used to represent a direction, so that its magnitude is not of importance, it is often convenient to scale it to unit length, a step called *normalizing u*. Normalization of a non-zero vector u (note that a vector representing a direction cannot be zero) consists simply of dividing it by its length, in other words, replacing u by $u/|u|$.

Exercise 11.3. Give a simple formula, in an OpenGL setting, for the eye direction vector from a vertex V whose position vector is v. Accordingly, rewrite the formula for the halfway vector of the preceding example in terms of l and v.

Example 11.5. Say the intensities of the specular light from source L are given by

$$L_{spec,\,R} = 1.0, \quad L_{spec,\,G} = 1.0, \quad L_{spec,\,B} = 1.0$$

and the specular reflectances of a vertex V by

$$V_{spec,\,R} = 0.0, \quad V_{spec,\,G} = 1.0, \quad V_{spec,\,B} = 0.6$$

and that the angle ϕ between the halfway vector and the outward normal vector at V is $60°$ and that the shininess exponent is 2.0.

Then the part of the red light emanating from V owing to the L specular is

$$\cos^f \phi * L_{spec,\, R} * V_{spec,\, R} = 0.25 * 1.0 * 0.0 = 0.0$$

Likewise, the part of the green light emanating from V owing to the L specular is

$$\cos^f \phi * L_{spec,\, G} * V_{spec,\, G} = 0.25$$

and the part of the blue light emanating from V owing to the L specular is

$$\cos^f \phi * L_{spec,\, B} * V_{spec,\, B} = 0.15$$

It's interesting that calculation of the reflected light never actually required determination of the reflection vector r itself. It's not hard though to find r, as we see next.

Exercise 11.4. Suppose that n is the unit (outward) normal vector and l the unit light direction vector at a vertex V. Prove that the unit vector r in the direction of reflection is given by the equation

$$r = 2(n \cdot l)n - l$$

Part answer: According to the laws of reflection we have to verify that r lies on the plane of l and n and makes the same angle with n as l. We must also prove that r is of unit length.

That r lies on the plane of l and n follows from its formula above, because of the linear dependence of r on l and n. Now

$$|r|^2 = r \cdot r = (2(n \cdot l)n - l) \cdot (2(n \cdot l)n - l) = 4(n \cdot l)^2 - 4(n \cdot l)^2 + l \cdot l = |l|^2 = 1$$

proving that r indeed is a unit vector.

We'll leave the reader to prove that r makes the same angle with n as l by computing its dot product with n.

11.2.4   First Lighting Equation

Our formulae from the last section straightforwardly combine into a single so-called *lighting equation*. Assume that we are given the values of the lighting properties matrix (11.2) for each light source L^i, $0 \le i \le N - 1$, the material properties matrix (11.3) for the vertex V, the global ambient light vector (11.4), as well as the emissive light vector (11.5) at V. Further, denote the normalized light direction and halfway vectors corresponding to light source L^i at vertex V by l^i and s^i, respectively. Denote the normalized outward surface normal vector at V by n and its shininess exponent by f.

Here then is the lighting equation giving the color intensity V_X at V, where X may be any of RGB:

$$
\begin{aligned}
V_X \;=\; & V_{emit,\,X} \;+ \\
& globAmb_X * V_{amb,\,X} \;+ \\
& \sum_{i=0}^{N-1} \Big(L^i_{amb,\,X} * V_{amb,\,X} \;+ \\
& \qquad \max\{l^i \cdot n,\, 0\} * L^i_{dif,\,X} * V_{dif,\,X} \;+ \\
& \qquad (\max\{s^i \cdot n,\, 0\})^f * L^i_{spec,\,X} * V_{spec,\,X} \Big)
\end{aligned}
\qquad (11.10)
$$

Note: The dot product of two unit vectors gives the cosine of the angle between them.

Note: If the RHS sums to more than 1, for any of X equal to R, G or B, then it is clamped to 1.

The lighting equation simply collects the components we have already discussed separately. The first summand on the RHS is the emissive component, while the second the global ambient scaled by the ambient reflectance. The third summand is a summation over the n light sources of

(a) The incident ambient component scaled by the ambient reflectance (Equation (11.6)).

(b) The incident diffuse component scaled by the diffuse reflectance and the cosine of the incident angle (Equation (11.7)).

(c) The incident specular component scaled by the specular reflectance and the angular attenuation factor (Equation (11.8)).

The reason for the $\max\{*, 0\}$ terms is to not allow a negative multiplier, which would imply the physically impossible phenomenon of light being subtracted. For example, $l^i \cdot n$ is negative when the angle between l^i and n is greater than $\pi/2$, which means that the light source L^i is behind the surface on which V is located, contributing zero light, rather than negative light.

Equation (11.10) is actually a first draft. The final lighting equation of OpenGL, which we'll see soon, enhances it by taking into account the attenuation of light over distance, as well as the spotlight effect, where light from a source emerges as a cone, rather than in all directions.

Exercise 11.5. There are two light sources L^0 and L^1, the respective values of whose lighting properties matrices are

$$
\begin{bmatrix}
0.0 & 0.0 & 0.0 \\
0.7 & 0.1 & 0.1 \\
0.7 & 0.1 & 0.1
\end{bmatrix}
\quad \text{and} \quad
\begin{bmatrix}
0.0 & 0.0 & 0.0 \\
0.1 & 0.7 & 0.1 \\
0.1 & 0.7 & 0.1
\end{bmatrix}
$$

The material properties matrix at a vertex V is

$$\begin{bmatrix} 0.1 & 0.8 & 0.9 \\ 0.1 & 0.8 & 0.9 \\ 1.0 & 1.0 & 1.0 \end{bmatrix}$$

Furthermore, the shininess exponent of the surface at V is 2.0, there is no emission from V, and the unit outward normal vector at V is

$$[0.0\ 1.0\ 0.0]^T$$

The position vectors of L^0, L^1, V and the eye are, respectively,

$$[0.0\ 5.0\ 5.0]^T, \quad [5.0\ 5.0\ 5.0]^T, \quad [0.0\ 0.0\ 5.0]^T \text{ and } [0.0\ 0.0\ 0.0]^T$$

The global ambient light vector is

$$[0.1\ 0.1\ 0.1]^T$$

Compute the color vector at V using the lighting equation (11.10).

Remark 11.5. It is important to realize that Phong's lighting model is *local*: the color at each vertex V depends only on the interaction between the external light properties and the material properties at V *itself*. No account is taken of whether V is obscured from a light source by another object (shadows), or of light that strikes V not directly from a light source but having bounced off other objects (reflection and secondary lighting). Colloquially, object-object light interaction is not considered, only light-object. We will discuss two global lighting models, ray tracing and radiosity, where shadows, reflections and other secondary effects are captured, in a later chapter.

11.3 OpenGL Light and Material Properties

The mapping from Phong's lighting model to OpenGL syntax is pretty much one-to-one. For each light source the user defines the values in the lighting properties matrix (11.2), as also the values in the material properties matrix (11.3) for each vertex. The global ambient vector (11.4) is user-defined as well. The user, too, defines the shininess exponent f, the emission color vector (11.5) and, very importantly, the normal vector at each vertex.

If you are beginning to worry that that's a lot of values to specify to light a scene, don't! Remember that OpenGL is a state machine, so material properties – which are state variables – persist in their current setting until explicitly changed, making it convenient for the programmer to apply the same properties to all vertices of a single object. Moreover, OpenGL has sensible defaults for values the programmer doesn't care to define.

Time to look at code.

Experiment 11.1. Run again `sphereInBox1.cpp`, which we ran the first time in Section 9.4. Press the up-down arrow keys to open or close the box. Figure 11.17 is a screenshot of the box partly open. We'll use this program as a running example to explain much of the OpenGL lighting and material color syntax. **End**

Figure 11.17: Screenshot of `sphereInBox1.cpp`.

11.3.1 Light Properties

Properties of light sources are set by statements of the form:

`glLight*(`*light, parameter, value*`)`

where *light* is the label of the light (viz., `GL_LIGHT0`, `GL_LIGHT1`, ...) and its particular *parameter* set to *value*.

The properties of the single light source of `sphereInBox1.cpp` are specified by the following statements in the `setup()` routine:

```
glLightfv(GL_LIGHT0, GL_AMBIENT, lightAmb);
glLightfv(GL_LIGHT0, GL_DIFFUSE, lightDifAndSpec);
glLightfv(GL_LIGHT0, GL_SPECULAR, lightDifAndSpec);
glLightfv(GL_LIGHT0, GL_POSITION, lightPos);
```

Typically, the diffuse and specular color vectors, i.e., the values of the `GL_DIFFUSE` and `GL_SPECULAR` parameters, respectively, are set identically to values perceived as the actual color of the light source. So, that of `sphereInBox1.cpp` is a bright white.

It's simplifying, as well, to consolidate all light source ambients – their `GL_AMBIENT` values – into the global ambient; in other words, set light source ambient colors all to 0.0 and adjust the one global ambient light vector. We follow this approach in `sphereInBox1.cpp`, as in all our lit programs. The fourth component, the alpha value, of each of the three color vectors – ambient, diffuse and specular – should always be 1.0 for a light source.

The value $\{x, y, z, w\}$ of `GL_POSITION` specifies the location $[x\ y\ z\ w]^T$ of the light source in homogeneous coordinates. If $w \neq 0$ then the light source is said to be positional and is located at world coordinates

$$[x/w\ \ y/w\ \ z/w]^T$$

The single positional light source of `sphereInBox1.cpp` is at $[0.0\ 1.5\ 3.0]^T$, which is just above and some ways in front of the box. We'll discuss what happens if $w = 0$ in Section 11.5 when we discuss directional light sources.

Note that *no visible object* is created by OpenGL at the location of a light source! This location is simply a point used for the purpose of lighting calculation. If you want the light to appear to be from a lamp or car headlight or such object you'll have to model the object and position it yourself.

Global ambient light in `sphereInBox1.cpp` is set with the statement

`glLightModelfv(GL_LIGHT_MODEL_AMBIENT, globAmb);`

in the **setup()** routine, where the second parameter points to the global ambient vector. Don't forget that lighting calculation is enabled with the call **glEnable(GL_LIGHTING)** and individual lights with calls to **glEnable(GL_LIGHTi)**.

Exercise 11.6. Show that nothing, in fact, is lost according to the first lighting equation (11.10) by setting all light source ambient colors to 0.0. In particular, prove that, however the light source ambients are initially set, they can all be reset to 0.0 and the global ambient adjusted accordingly so that the color computed at each vertex by (11.10) remains unchanged.

11.3.2   Material Properties

Material properties at a vertex are set by statements of the form:

```
glMaterial*(face, parameter, value)
```

where the *parameter* of *face* is set to *value*. The value of face can be **GL_FRONT**, **GL_BACK** or **GL_FRONT_AND_BACK** for both faces.

Material properties of (each vertex of) the box of **sphereInBox1.cpp** are specified by the following statements in the **drawScene()** routine:

```
glMaterialfv(GL_FRONT_AND_BACK, GL_AMBIENT_AND_DIFFUSE, matAmbAndDif1);
glMaterialfv(GL_FRONT_AND_BACK, GL_SPECULAR, matSpec);
glMaterialfv(GL_FRONT_AND_BACK, GL_SHININESS, matShine);
```

Typically, the ambient and diffuse color vectors are set identically to values perceived as an object's native color. OpenGL makes it convenient to do so via the **GL_AMBIENT_AND_DIFFUSE** parameter; however, they can be set separately as well using **GL_AMBIENT** and **GL_DIFFUSE**. As specular light is obtained from reflection from the light source, it's reasonable to set an object's **GL_SPECULAR** value either to white $\{1.0, 1.0, 1.0, 1.0\}$, fully reflecting the incident specular light, or a shade of gray $\{\gamma, \gamma, \gamma, 1.0\}$, *equally* scaling each color component. Ignore, for now, the fourth, or alpha, component of the material color vectors, all currently set to 1 – the alpha value pertains to blending, which is discussed in a later chapter.

The value of **GL_SHININESS** is, of course, the shininess exponent f of the first lighting equation (11.10). Its value must be in the range $[0.0, 128.0]$. The default is 0.0, which causes no angular attenuation of specular reflectance.

The emissive color at a vertex can be set using the **GL_EMISSION** parameter, but we choose to go with the default of $\{0.0, 0.0, 0.0, 1.0\}$, in other words, no emission, for each vertex in **sphereInBox1.cpp**.

Exercise 11.7. (Programming) What are the material specular values and the shininess exponent of the sphere of **sphereInBox1.cpp**?

11.3.3 Experimenting with Properties

The two programs `lightAndMaterial1.cpp` and `lightAndMaterial2.cpp` allow the user to experiment with various material and light properties. Both show a blue ball lit by two lights, one white and one green, whose positions are indicated by small wire spheres. Figure 11.18 shows screenshots of both the programs.

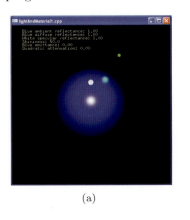
(a)

(b)

Figure 11.18: Screenshots of (a) `lightAndMaterial1.cpp` (b) `lightAndMaterial2.cpp`.

Using the first program one can change material properties of the blue ball, as well as move it. The second program, on the other hand, allows properties of the white light to be controlled, as also of the global ambient, and enables the user to rotate the white light. Text messages show property values. Let's take a quick tour of the two before experimenting with properties.

Experiment 11.2. Run `lightAndMaterial1.cpp`.

The ball's current ambient and diffuse reflectances are identically set to a maximum blue of $\{0.0, 0.0, 1.0, 1.0\}$, its specular reflectance to the highest gray level $\{1.0, 1.0, 1.0, 1.0\}$ (i.e., white), shininess to 50.0 and emission to zero $\{0.0, 0.0, 0.0, 1.0\}$.

Press 'a/A' to decrease/increase the ball's blue **A**mbient and diffuse reflectance. Pressing 's/S' decreases/increases the gray level of its **S**pecular reflectance. Pressing 'h/H' decreases/increases its s**H**ininess, while pressing 'e/E' decreases/increases the blue component of the ball's **E**mission.

The program has further functionalities which we'll explain as they become relevant. **E**nd

Experiment 11.3. Run `lightAndMaterial2.cpp`.

The white light's current diffuse and specular are identically set to a maximum of $\{1.0, 1.0, 1.0, 1.0\}$ and it gives off zero ambient light. The green light's attributes are fixed at a maximum diffuse and specular of $\{0.0, 1.0, 0.0, 1.0\}$, again with zero ambient. The global ambient is a low intensity gray at $\{0.2, 0.2, 0.2, 1.0\}$.

Press 'w' or 'W' to toggle the **W**hite light off and on. Pressing 'g' or 'G' toggles the **G**reen light off and on. Press 'd/D' to decrease/increase the gray level of the white light's **D**iffuse and specular intensity (the ambient intensity never changes from zero). Pressing 'm/M' decreases/increases the gray intensity of the global a**M**bient. Rotate the white light about the ball by pressing the arrow keys.

The program has more functionality too which we'll need later. **End**

Experiment 11.4. Run `lightAndMaterial1.cpp`.

Reduce the specular reflectance of the ball. Both the white and green highlights begin to disappear, as it's the specular components of the reflected lights which appear as specular highlights. **End**

Exercise 11.8. (**P**rogramming) The specular highlight is sharpened or blunted, respectively, by increasing or decreasing the shininess exponent. Why?

Hint: The higher the shininess exponent the more rapidly the specular light diminishes as the vertex normals turn away from the eye direction (recall the definition of the angular attenuation factor in Section 11.2.3).

Experiment 11.5. Restore the original values of `lightAndMaterial1.cpp`.

Reduce the diffuse reflectance gradually to zero. The ball starts to lose its roundness until it looks flat as a disc. The reason for this is that the ambient intensity, which does not depend on eye or light direction, is uniform across vertices of the ball and cannot, therefore, provide the sense of depth that obtains from a contrast in color values across the surface. Diffuse light, on the other hand, which varies across the surface depending on light direction, can provide an illusion of depth.

Even though there is a specular highlight, sensitive to both eye and light direction, it's too localized to provide much contrast. Reducing the shininess spreads the highlight but the effect is not a realistic perception of depth.

Moral: Diffusive reflectance lends three-dimensionality. **End**

Experiment 11.6. Restore the original values of `lightAndMaterial1.cpp`.

Now reduce the ambient reflectance gradually to zero. The ball seems to shrink! This is because the vertex normals turn away from the viewer at the now hidden ends of the ball, scaling down the diffuse reflectance there (recall the $\cos\theta$ term in the diffusive reflectance equation (11.7)). The result is that, with no ambient reflectance to offset the reduction in diffuse, the ends of the ball are dark.

Moral: Ambient reflectance provides a level of uniform lighting over a surface. **End**

Experiment 11.7. Restore the original values of `lightAndMaterial1.cpp`.

Reduce both the ambient and diffuse reflectances to nearly zero. It's like the cat disappearing, leaving only its grin! Specular light is clearly for highlights and not much else. **End**

Exercise 11.9. (Programming) Restore the original values of light-AndMaterial1.cpp.

Reduce all three of the ball's diffuse, ambient and specular reflectances and raise its emissive light intensity. It does appear to glow but also appears flat. Why?

Experiment 11.8. Run lightAndMaterial1.cpp with its original values.

With it's current high ambient, diffuse and specular reflectances the ball looks a shiny plastic. Reducing the ambient and diffuse reflectances makes for a heavier and less plastic appearance. Restoring the ambient and diffuse to higher values, but reducing the specular reflectance makes it a less shiny plastic. Low values for all three of ambient, diffuse and specular reflectances give the ball a somewhat wooden appearance.   **End**

Experiment 11.9. Run lightAndMaterial2.cpp.

Reduce the white light's diffuse and specular intensity to 0. The ball becomes a flat dull blue disc with a green highlight. This is because the ball's ambient (and diffuse) is blue and cannot reflect the green light's diffuse component, losing thereby three-dimensionality.

Raising the white global ambient brightens the ball, but it still looks flat in the absence of diffusive light.   **End**

Exercise 11.10. (Programming) When the white light is switched off in lightAndMaterial2.cpp, the only evidence of green on the ball is the specular highlight; moreover, if the ambient is tamped down as well then the ball begins to disappear altogether.

However, this is not so in the opposite situation, when the white light is switched on and the green off – a sector of the ball is clearly visible no matter how low the ambient. Why?

Experiment 11.10. Nate Robins has a bunch of great tutorial programs at the site [96]. This is a good time to run his lightmaterial tutorial, which allows the user to control a set of parameters as well.   **End**

11.3.4   Color Material Mode

Remember glColor*() which we used to set color in the dark days before there were light sources? Now that we have light and glMaterial*() allows us to set all sorts of material properties, it seems there's no use any more for glColor*(). Well, it turns out that the good folk who designed OpenGL found a way to keep it on the payroll.

Here's how. Suppose you're in the not uncommon situation coloring a scene where only a particular color attribute, say the ambient and diffuse reflectances of the front faces, changes from one object to the next, other attributes remaining constant. What you can do in this case, instead of repeatedly calling glMaterialfv(GL_FRONT, GL_AMBIENT_AND_DIFFUSE, *value*), is to:

1. Enable the so-called *color material mode* with a call to `glEnable(GL_COLOR_MATERIAL)`.

2. Call `glColorMaterial(GL_FRONT, GL_AMBIENT_AND_DIFFUSE)`, which tells OpenGL to use the current color, set by `glColor*()`, to determine the front-face ambient and diffuse color values.

   Generally, the `glColorMaterial()` call can be of the form `glColorMaterial(face, parameter)` where *face* can be `GL_FRONT`, `GL_BACK` or `GL_FRONT_AND_BACK`, and *parameter* one of `GL_AMBIENT`, `GL_DIFFUSE`, `GL_AMBIENT_AND_DIFFUSE`, `GL_SPECULAR` or `GL_EMISSION`.

3. Make a call to `glColor*()` to set the front-face ambient and diffuse color from one object to the next.

This method may, in fact, be more efficient with certain implementations of OpenGL, not to mention the convenience of not having to change a programming habit if one is used to coloring with `glColor*()`.

Experiment 11.11. Run `spotlight.cpp`. The program is primarily to demonstrate spotlighting, the topic of a forthcoming section. Nevertheless, press the page-up key to see a multi-colored array of spheres. Figure 11.19 is a screenshot.

Currently, the point of interest in the program is the invocation of the color material mode for the front-face ambient and diffuse reflectances by means of the last two statements in the initialization routine, viz.,

```
glEnable(GL_COLOR_MATERIAL);
glColorMaterial(GL_FRONT, GL_AMBIENT_AND_DIFFUSE);
```

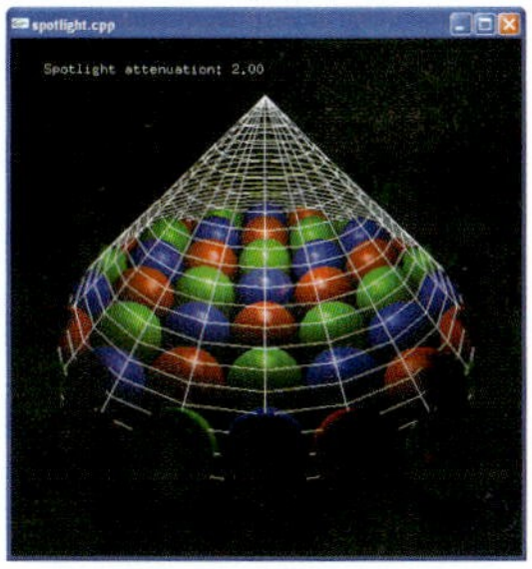

Figure 11.19: Screenshot of `spotLight.cpp`.

and subsequent coloring of the spheres in the drawing routine by `glColor4f()` statements. **End**

11.4 OpenGL Lighting Model

The so-called OpenGL *lighting model* sets certain environmental parameters. The terminology, even though used in the red book, is somewhat unfortunate as it may suggest laws of interaction between light and objects, or a relation with Phong's model – neither of which is true. The four parameters the OpenGL lighting model sets are the following:

1. The global ambient light with the statement

   ```
   glLightModel*(GL_LIGHT_MODEL_AMBIENT, globAmb)
   ```

   where *globAmb* is the global ambient light vector. This we've seen already.

431

2. Whether to use a local or infinite viewpoint for lighting calculation.

See again the lighting equation (11.10). The halfway vector s^i at a vertex, one for each light source, is the unit vector bisecting the angle between the direction vector l^i to the light source L^i and the direction vector e to the eye.

The OpenGL eye being fixed at the origin $[0 \; 0 \; 0]^T$, evidently $e = -V$, where V is the vertex's position vector, which changes from one to another. However, it simplifies lighting computation to keep e constant, particularly $e = [0 \; 0 \; 1]^T$, equivalent to *assuming* an eye that is infinitely far up the z-axis and so, effectively, in the same direction from every vertex. See Figure 11.20. This simplification, often, still gives adequately authentic lighting.

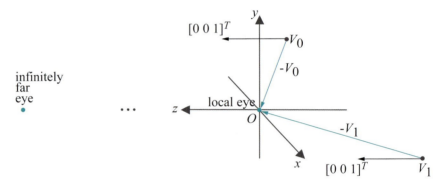

Figure 11.20: Local versus infinite viewpoint: the direction vector from each vertex toward the infinite viewpoint is black, while that toward the local viewpoint – i.e., the eye vector – is blue.

Remark 11.6. The direction vector l^i to the light source, too, changes from vertex to vertex if the source is a positional one, i.e., if $w \neq 0$ in the value $[x \; y \; z \; w]^T$ of the source's GL_POSITION parameter. Moreover, a simplification exactly similar to that of assuming an infinite viewpoint can be achieved, not by tweaking the OpenGL lighting model, but by making the light directional by setting $w = 0$. We'll discuss this in the next section.

The OpenGL default viewpoint, in fact, is infinite. For lighting calculation to be done using a local viewpoint instead – with the eye at the origin, that is – call

　　glLightModel*(GL_LIGHT_MODEL_LOCAL_VIEWER, GL_TRUE)

which is what we do in the setup() routines of both sphereIn-Box1.cpp and lightAndMaterial1.cpp, while lightAndMaterial2.cpp provides an option. The local viewpoint is more realistic at the expense of greater computation.

Remark 11.7. The chosen light model viewpoint is used *only for lighting calculations*. The viewing frustum or box stays unchanged – therefore, in the case of a frustum, for example, we still *see the scene* from the eye at the origin.

Exercise 11.11. (Programming) Press 'l' or 'L' to toggle between the **L**ocal and the infinite viewpoint in `lightAndMaterial2.cpp`. The change seems to be only in the highlights, in other words, only the specular reflectances. Why?

3. Whether to enable two-sided lighting.

   The OpenGL default is to perform lighting calculations for each polygon based on its specified `GL_FRONT` face parameter values and its specified vertex normals, regardless of if it is front or back facing. As the user likely sets material properties and normal values with the front faces of polygons in mind, results tend to be unrealistic for those whose back faces happen to be visible. So, when back faces might be visible, the command to use is

   ```
   glLightModel*(GL_LIGHT_MODEL_TWO_SIDE, GL_TRUE)
   ```

   which causes OpenGL to

   (a) use the `GL_BACK` (or `GL_FRONT_AND_BACK`) parameter values to color back-facing polygons, and

   (b) *reverse* the specified vertex normal for back-facing polygons.

Experiment 11.12. Run `litTriangle.cpp`, which draws a single triangle, whose front is coded red and back blue, initially front-facing and lit two-sided. Press the left and right arrow keys to turn the triangle and space to toggle two-sided lighting on and off. See Figure 11.21 for screenshots.

Notice how the back face is dark when two-sided lighting is disabled – this is because the normals are pointing oppositely of the way they should be. End

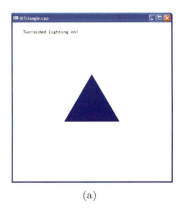

(a)   (b)

Figure 11.21: Screenshots of `litTriangle.cpp` showing the back face with (a) two-sided lighting on (b) two-sided lighting off.

4. Whether to apply specular light before or after texturing.

   The following remarks will be more meaningful after the discussion of textures in the next chapter.

   The OpenGL default is to apply textures after all lighting calculations, which can cause specular highlights to be smothered. However, the command

   ```
   glLightModel*(GL_LIGHT_MODEL_COLOR_CONTROL,
                 GL_SEPARATE_SPECULAR_COLOR)
   ```

   makes OpenGL

   (a) separately produce two colors at each vertex: a primary color calculated from all incoming non-specular components and a secondary color from all incoming specular components,

   (b) combine only the primary color with texture color at the time of texture mapping and, finally,

   (c) add in the secondary color to the result of the previous step, which assures the specular highlights.

11.5   Directional Lights, Positional Lights and Attenuation of Intensity

Directional and Positional Light Sources

We know that the value of the `GL_POSITION` parameter of a light source L specifies its location $[x\ y\ z\ w]^T$ in homogeneous coordinates.

If $w \neq 0$, then the light source is called *positional*, or *local*, and located at world coordinates $[x/w \ y/w \ z/w]^T$. This is the kind of source we have used so far. If $w = 0$, then the light source is *directional* and assumed located at an infinite distance in the direction of $[x \ y \ z]^T$ from the origin. See Figure 11.22.

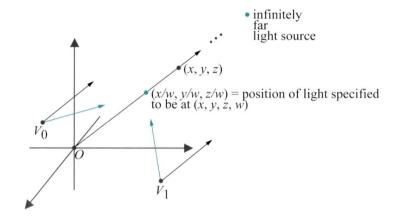

Figure 11.22: Directional versus positional light: the direction vector from each vertex toward the directional light is black and parallel to the direction of the directional light, while that toward the positional light is blue.

A positional light is located within the environment, e.g., a car headlight, while a directional light is far removed, e.g., the sun. From the point of view of lighting calculation, the difference is that the light direction vector l from a vertex V to the light source L depends on the coordinates of V if L is positional, while it is constant for all vertices if L is directional. Evidently, lighting calculation is cheaper for directional sources.

The default value for GL_POSITION is $[0 \ 0 \ 1 \ 0]^T$, which defines a directional light shining down from high up the z-axis.

$\mathbf{Experiment}$ 11.13. Press 'p' or 'P' to toggle between **P**ositional and directional light in lightAndMaterial2.cpp.

The white wire sphere indicates the positional light, while the white arrow the incoming directional light. $\mathbf{End}$

Attenuation of Light

In the real world, the intensity of light from a source diminishes with distance from the source following an inverse square law. This phenomenon, called *distance attenuation*, can be modeled in OpenGL as well by a multiplicative *distance attenuation factor*

$$\frac{1}{k_c + k_l d + k_q d^2}$$

where d is the distance from the light source and k_c, k_l and k_d are the values of the light parameters GL_CONSTANT_ATTENUATION, GL_LINEAR_ATTENUATION and GL_QUADRATIC_ATTENUATION, respectively. These values are set by statements of the form

```
glLightf(GL_LIGHTi, GL_CONSTANT_ATTENUATION, kc);
glLightf(GL_LIGHTi, GL_LINEAR_ATTENUATION, kl);
glLightf(GL_LIGHTi, GL_QUADRATIC_ATTENUATION, kq);
```

The default values are $k_c = 1$ and $k_l = k_q = 0$, which imply no attenuation over distance at all. Attenuating the intensity of a directional light over distance is not meaningful as it's already infinitely far from every vertex; therefore, default values for the attenuation parameters cannot be changed for such a source.

Experiment 11.14. Run lightAndMaterial1.cpp. The current values of the constant, linear and quadratic attenuation parameters are 1, 0 and 0, respectively, so there's no attenuation. Press 't/T' to decrease/increase the quadratic aTtenuation parameter. Move the ball by pressing the up/down arrow keys to observe the effect of attenuation.   End

11.6   Spotlights

The default for a light source is that it's *regular*, emitting light in all directions. This can be altered by turning it into a *spotlight*, in which case the emitted light is in the shape of a cone. Figures 11.23(a) and (b) show, respectively, plane sections of the light from both a regular and a spotlight.

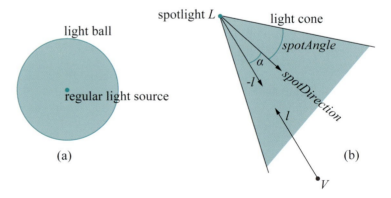

(a)   (b)

Figure 11.23: Sections of (a) Regular light (b) Spotlight. Shown for the spotlight are the light direction vector l from vertex V toward light source L, the inverse vector $-l$ from the light source toward the vertex, the direction vector *spotDirection* of the cone's axis and the half-angle *spotAngle* at its apex.

Experiment 11.15. Run `spotlight.cpp`, which shows a bright white spotlight illuminating a multi-colored array of spheres. A screenshot was shown earlier in Figure 11.19.

Press the page up/down arrows to increase/decrease the angle of the light cone. Press the arrow keys to move the spotlight. A white wire mesh is drawn along the light cone boundary. End

All statements pertaining to the spotlight properties of the single light source of `spotlight.cpp` are located in the `drawScene()` routine. The first step to turning a light source L into a spotlight is to specify the half-angle at the apex of the light cone, called the *cone angle*, with the command

  `glLightf(GL_LIGHT0, GL_SPOT_CUTOFF,` *spotAngle*`)`

which, in fact, sets the cone angle to the value *spotAngle*. This should be between 0.0 and 90.0. The default is the special value of 180.0, meaning that L is not a spotlight, but a regular source emitting in all directions.

The next step is to specify the direction of the spotlight or, more specifically, that of the axis of its cone with a command

  `glLightfv(GL_LIGHT0, GL_SPOT_DIRECTION,` *spotDirection*`)`

which sets the axis in a direction parallel to the vector

$$spotDirection = [x \ y \ z]^T$$

The default value of `GL_SPOT_DIRECTION` is $[0 \ 0 \ -1]^T$, aiming the spotlight down the negative z-axis.

A final spotlight parameter is `GL_SPOT_EXPONENT`, whose value is called the *spotlight attenuation factor* and which controls the distribution of intensity through the light cone. If the value of `GL_SPOT_EXPONENT` is h and the angle between the axis of the light cone and the direction from source L toward vertex V is α, then the intensity of light at V is attenuated by the multiplicative factor $\cos^h \alpha$. This, of course, presumes that V lies within the light cone in the first place; if not, no light reaches V from L at all. Note that, as depicted in Figure 11.23, a vector from L toward V is, simply, $-l$, the negative of a light direction vector.

The motivation behind the spotlight attenuation factor is similar to that for the angular attenuation factor in the calculation of specular reflection in Equation (11.8) – so that the greater the value of h, the more rapidly the intensity of the spotlight attenuates away from the cone's axis. Equivalently, the greater h the more "concentrated" the spotlight. The default value of `GL_SPOT_EXPONENT` is 0, implying no attenuation at all.

Experiment 11.16. Run again `spotlight.cpp`. The current value of the spotlight's a**T**tenuation is 2.0, which can be decreased/increased by pressing 't/T'. Note the change in visibility of the balls near the cone boundary as the attenuation changes. End

Exercise 11.12. A spotlight should always be positional. Why?

For use in the upcoming final OpenGL light equation, let's write a single complete formula for a *spotlight attenuation factor* or, briefly, *saf*, at a vertex V, for a given light source L. Denote the unit vector along the spotlight axis – the normalized value of `GL_SPOT_DIRECTION` – by d and assume that l, the light direction vector from V, is normalized as well (see Figure 11.23). Then:

$$saf = \begin{cases} 1 & , & \text{if } spotAngle = 180° \\ 0 & , & \text{if } -l \cdot d < \cos(spotAngle) \\ (-l \cdot d)^h & , & \text{otherwise} \end{cases} \qquad (11.11)$$

Here's how to parse the formula.

The first line is the case when L is not a spotlight, so there's no attenuation.

For the second line, recall that $-l$ is the unit vector from L toward V. Therefore, $-l \cdot d = \cos\alpha$, where α is the angle between the axis of the light cone and the direction of V from L. Now, if $\cos\alpha < \cos(spotAngle)$, then $\alpha > spotAngle$, which means that V lies outside the light cone and gets zero light. This explains the second line.

The third line, of course, gives the angular attenuation factor.

Exercise 11.13. Why isn't it necessary to write $(\max\{-l \cdot d, 0\})^h$, instead of $(-l \cdot d)^h$, in Equation (11.11) in a manner similar to the first lighting equation (11.10)?

Exercise 11.14. (Programming) In addition to the spotlight attenuation, light from a spotlight source can be distance attenuated as well. Additionally, allow distance attenuation to be controlled in `spotlight.cpp`. Add vertical motion capability to the light source to in order to accentuate the effect of distance attenuation. And while you're at it, why not make the light emerge from a well at the bottom of a flying saucer?!

11.7 OpenGL Lighting Equation

We now have the two additional pieces needed to enhance the first lighting equation (11.10) to the form that is, in fact, used by OpenGL to calculate RGB color intensities at a vertex V, namely, distance attenuation and spotlight attenuation. The enhancement is straightforward.

All symbols from the first lighting equation retain the same meaning. Additionally, d^i denotes the distance of V from the ith light source; k_c^i, k_l^i and k_q^i denote, respectively, the constant, linear and quadratic attenuation parameters for the ith light source; and saf^i is the spotlight attenuation factor for the ith light source at the vertex V, as given by Equation (11.11).

So finally, here it is, the grand ole lighting equation of OpenGL:

$$V_X = V_{emit, X} +$$
$$globAmb_X * V_{amb, X} +$$
$$\sum_{i=0}^{n-1} \frac{1}{k_c^i + k_l^i d^i + k_q^i (d^i)^2} * sa f^i *$$
$$\left(L_{amb, X}^i * V_{amb, X} + \right.$$
$$\max\{l^i \cdot n, 0\} * L_{dif, X}^i * V_{dif, X} +$$
$$\left. (\max\{s^i \cdot n, 0\})^f * L_{spec, X}^i * V_{spec, X} \right) \quad (11.12)$$

where V_X is the color intensity at V, X being any of RGB.

The additions to the first lighting equation (11.10) are exactly the two multiplicative terms on the third line of the current equation, representing distance attenuation and spotlight attenuation, respectively.

Remark 11.8. It's really Phong's lighting equation, but, given the context, we'll more often than not call it the OpenGL lighting equation.

Remark 11.9. We must revisit Exercise 11.6 at this time. Its implication that all individual light source ambients can be consolidated into the global ambient is not true any more if one uses Equation (11.12) instead of Equation (11.10), because the same light source can contribute different amounts of ambient light to different vertices owing to distance and spotlight attenuation.

Nevertheless, the simplification of setting all individual light source ambients to zero, and adjusting only the global, is probably still authentic enough for most applications.

Exercise 11.15. If there is a single directional light source in an OpenGL program, which is not distance attenuated, which of the three – ambient, diffuse and specular – reflectance components at its vertices is changed by *translating* an object?

Exercise 11.16. If there is a single positional light source in an OpenGL program, which is not a spotlight and not distance attenuated, which of the three – ambient, diffuse and specular – reflectance components at an object's vertex can change by moving the light source? By translating the object?

11.8 OpenGL Shading Models

A *shading model* is a method to shade, or color, the interiors of primitives. Keep in mind that Phong's lighting model, as implemented through the OpenGL lighting equation, determines colors *only* at the vertices of primitives, but says nothing about how to spread them inside. OpenGL's default shading model, called *smooth shading* or *Gouraud* shading, is to interpolate color values computed at its vertices through a primitive's interior. We discussed in Section 7.2 the mechanics of interpolation by computing barycentric coordinates of interior points.

An alternate shading model, called *flat shading*, is available, as well, in OpenGL. It is specified by a call to

```
glShadeModel(GL_FLAT)
```

The default of smooth shading is restored by calling

```
glShadeModel(GL_SMOOTH)
```

When flat shading, even if the color values differ across the vertices of a primitive, OpenGL chooses *one* of them and applies its color to the entire primitive. For example, the first vertex (according to the order in the code) of a polygon is used. In a triangle strip, the ith triangle is painted with the color of the $i+2$th vertex. The reader is referred to the red book for a full listing of which vertex it is whose color is used for a given primitive.

Flat shading can be a reasonable alternative in the absence of lighting. Computationally it's, of course, far less expensive than smooth shading. One interesting application of flat shading is in applying "discrete" color schemes, which, often, is difficult with smooth shading. The following experiment is an illustration.

Experiment 11.17. Run `checkeredFloor.cpp`, which creates a checkered floor drawn as an array of flat shaded triangle strips. See Figure 11.24. Flat shading causes each triangle in the strip to be painted with the color of the last of its three vertices, according to the order of the strip's vertex list.

End

Exercise 11.17. (Programming) Try and replicate the checkered floor of the preceding experiment using smooth shading instead of flat.

Figure 11.24: Screenshot of `checkeredFloor.cpp`.

In Section 11.12 we'll see yet another shading model – Phong's shading model – which is more sophisticated than smooth shading and computationally more expensive as well. Phong's shading model is not available in first-generation OpenGL, i.e., OpenGL 1.x, but can be user-coded in the second generation of the API.

11.9 Animating Light

There are three ways that a light source can be animated by changing spatial properties:

1. By moving its position.

2. By changing its direction if it's a spotlight.

3. By changing the light cone angle if it's a spotlight.

We've already seen light animation in two programs in this chapter: lightAndMaterial2.cpp and spotlight.cpp.

The things to keep in mind are:

(a) A light source's position vector, specified by a glLightfv(*light*, GL_POSITION, *lightPos*) statement, is transformed by the value of the current modelview matrix by multiplication from the left. (See Section 4.2 if you need to review modelview matrices.)

Effectively, modelview transformations in the code prior to the glLightfv(*light*, GL_POSITION, *lightPos*) statement apply to a light's position, exactly as those prior to a glVertex3f() statement, defining a vertex, apply to that vertex.

(b) Likewise, a spotlight source's direction vector, specified by the glLightfv(*light*, GL_SPOT_DIRECTION, *spotDirection*) statement is transformed by the value of the current modelview matrix by multiplication from the left.

For example, as the light source of sphereInBox1.cpp is positioned by the glLightfv(GL_LIGHT0, GL_POSITION, lightPos) statement in the initialization routine setup(), it is unaffected by any modelview transformations in drawScene().

However, both lights of lightAndMaterial1.cpp are positioned in the display routine following the viewing command gluLookAt(), so their positions are, in fact, transformed by gluLookAt(), which effectively means that the lights stay static relative to the scene, no matter if the viewpoint is changed. The position of the lights in lightAndMaterial2.cpp is similarly transformed by its own gluLookAt().

Note: The push-pop pairs surrounding the code to position the lights in both programs are to isolate the transformations applied to the spheres that depict the light sources.

The spotlight of spotlight.cpp is positioned in the display routine after the viewing transformation *and* a user-specified translation; moreover, its cone angle can be changed by the user too. We ask you next to look into changing its direction.

Exercise 11.18. (Programming) Consider this an extension of Exercise 11.14 – add capability to aim the spotlight of spotlight.cpp.

Remark 11.10. We've discussed only animating the spatial attributes of a light source. Color values can easily be animated as well.

Exercise 11.19. (Programming) Cause the color of the balls of spotlight.cpp to brighten, fade and change.

Exercise 11.20. (Programming) Make the ball of `ballAndTorus.cpp` carry a spotlight, which is aimed always at the torus, and whose cone angle and color change as the ball travels. You may want to copy some light and material properties from `ballAndTorusShadowed.cpp`, but ignore the shadows.

11.10 Partial Derivatives, Tangent Planes and Normal Vectors 101

This section is an introduction to the calculus sometimes required to calculate normals to surfaces. It is not mandatory reading. We suggest you skip this section initially and consult it later if need be.

Actually, if you know how to compute derivatives of a function of a single variable, e.g., $f(x) = x^2$ or $f(x) = \sin x$, as we'll assume you do, you already know how to compute partial derivatives. Because. . .

Definition 11.1. Suppose that f is a function of more than one variable $x, y, \ldots$ The *partial derivative* of f with respect to one of these variables, say x, is the derivative of f as a function *only* of x, assuming the other variables all fixed. The partial derivative of f with respect to x is denoted $\frac{\partial f}{\partial x}$.

Example 11.6. Evaluate the partial derivatives of

$$f(x, y) = x^2 + y^2$$

at the point $(1, 2)$.

Answer: We have

$$\frac{\partial f}{\partial x}(x, y) = 2x, \qquad \frac{\partial f}{\partial y}(x, y) = 2y$$

Therefore,

$$\frac{\partial f}{\partial x}(1, 2) = 2, \qquad \frac{\partial f}{\partial y}(1, 2) = 4$$

Remark 11.11. Often $\frac{\partial f}{\partial x}(x, y)$ is simply written $\frac{\partial f}{\partial x}$, e.g., the first two equations of the preceding answer could be written

$$\frac{\partial f}{\partial x} = 2x, \qquad \frac{\partial f}{\partial y} = 2y$$

Example 11.7. Evaluate the partial derivatives of

$$f(x, y) = x^2 \sin y$$

at the point $(1, \pi/2)$.

Answer: We have

$$\frac{\partial f}{\partial x} = 2x\sin y, \qquad \frac{\partial f}{\partial y} = x^2\cos y$$

Therefore,

$$\frac{\partial f}{\partial x}(1,\,\pi/2) = 2, \qquad \frac{\partial f}{\partial y}(1,\,\pi/2) = 0$$

Example 11.8. Evaluate the partial derivatives of

$$f(x,y,z) = xz + \sin x \cos y \cos z + y$$

at the point $(\pi/2, \pi, 0)$.

Answer: We have

$$\frac{\partial f}{\partial x} = z + \cos x \cos y \cos z, \quad \frac{\partial f}{\partial y} = 1 - \sin x \sin y \cos z, \quad \frac{\partial f}{\partial z} = x - \sin x \cos y \sin z$$

Therefore,

$$\frac{\partial f}{\partial x}(\pi/2,\,\pi,\,0) = 0, \quad \frac{\partial f}{\partial y}(\pi/2,\,\pi,\,0) = 1, \quad \frac{\partial f}{\partial y}(\pi/2,\,\pi,\,0) = \pi/2$$

Exercise 11.21. Evaluate the partial derivatives of

$$f(x,y) = xy$$

at the point $(2,3)$.

Exercise 11.22. Evaluate the partial derivatives of

$$f(x,y,z) = x\cos y + y\cos z + z\cos x$$

at the point $(\pi/2, 0, \pi/2)$.

The reader may wonder that if the partial derivative $\frac{\partial f}{\partial x}$, for example, is obtained by differentiating f with respect to the single variable x, assuming the others fixed, then why those other variables occasionally pop up again in the expression for $\frac{\partial f}{\partial x}$? Here's the reason.

Consider the function $f(x,y) = x^2\sin y$ of Example 11.7 above. Fixing y at, say, the value $\pi/6$ gives the function $f(x,\pi/6) = x^2/2$, while fixing y at $\pi/2$ gives the function $f(x,\pi/2) = x^2$. Both $f(x,\pi/6)$ and $f(x,\pi/2)$ are functions of the one variable x, but they are *different* functions because y's been fixed at two *different* values.

Moreover,

$$\frac{\partial f}{\partial x}(x,\pi/6) = \frac{d}{dx}(x^2/2) = x \quad \text{and}$$

$$\frac{\partial f}{\partial x}(x, \pi/2) = \frac{\mathrm{d}}{\mathrm{d}x}(x^2) = 2x$$

are different as well, as they are derivatives of different functions. This is why $\frac{\partial f}{\partial x}$ depends on y, as well as on x.

So far so good. At least calculating partial derivatives is no different from calculating ordinary derivatives. But what do partial derivatives mean? For example, we understand the *geometric meaning* of ordinary derivatives along curves specified both implicitly and parametrically:

(a) *Implicit*: Suppose a curve is given by the equation

$$y = f(x)$$

Then the value of

$$\frac{\mathrm{d}f}{\mathrm{d}x}$$

at $x = a$ is the gradient of the tangent line to the curve at the point $(a, f(a))$.

For example, the gradient of the tangent line l at the point $(1, 1)$ of the parabola

$$y = x^2$$

is 2 as

$$\frac{\mathrm{d}}{\mathrm{d}x}(x^2) = 2x$$

which equals 2 when x is 1. See Figure 11.25(a).

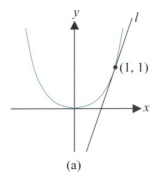
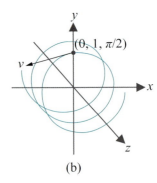

Figure 11.25: Tangents: (a) Tangent line l to the parabola $y = x^2$ at $(1, 1)$ (b) Tangent vector v to the helix $c(t) = (\cos t, \sin t, t)$ at $(0, 1, \pi/2)$.

(b) *Parametric*: Suppose a curve is given by

$$c(t) = (f(t),\ g(t),\ h(t))$$

Then the value of the vector

$$c'(t) = \begin{bmatrix} \dfrac{df}{dt} & \dfrac{dg}{dt} & \dfrac{dh}{dt} \end{bmatrix}^T$$

at $t = a$ is a tangent vector (provided it's non-zero) to the curve at the point $(f(a),\ g(a),\ h(a))$.

For example, a tangent vector v to the helix

$$c(t) = (\cos t,\ \sin t,\ t)$$

at the point $(0,\ 1,\ \pi/2)$, corresponding to $t = \pi/2$, is $[-1\ 0\ 1]^T$, as

$$\begin{bmatrix} \dfrac{d}{dt}(\cos t) & \dfrac{d}{dt}(\sin t) & \dfrac{d}{dt}(t) \end{bmatrix}^T = [-\sin t\ \cos t\ 1]^T$$

which equals $[-1\ 0\ 1]^T$ when $t = \pi/2$. See Figure 11.25(b).

It turns out that, just as the computation of partial derivatives is based on computing ordinary derivatives, their geometric significance obtains from that of ordinary derivatives too. Here's how:

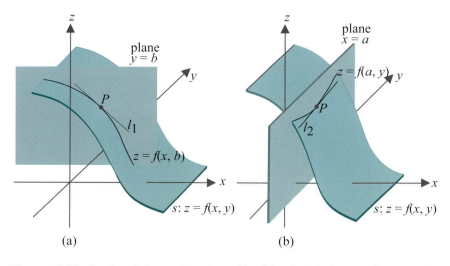

(a)   (b)

Figure 11.26: Section of the graph s of $z = f(x, y)$ by the (a) plane $y = b$, giving the tangent line l_1 at $P = (a,\ b,\ f(a, b))$, (b) plane $x = a$, giving the tangent line l_2 at P.

(a) *Implicit*: Consider $z = f(x, y)$, a function of two variables. It defines a surface s, called the graph of f. See Figure 11.26(a).

Now, if we fix y at, say, the value b, then $z = f(x, b)$ gives a curve s. In fact, this curve is the section of s by the plane $y = b$.

We know that the value of $\frac{\partial f}{\partial x}$ at (a, b) is the value at a of the ordinary derivative $\frac{\mathrm{d}}{\mathrm{d}x} f(x, b)$. This helps find geometric meaning for the partial derivative as follows.

The value of

$$\frac{\partial f}{\partial x}$$

at (a, b) is the gradient of the tangent line l_1 to the sectional curve $z = f(x, b)$ at the point $P = (a,\ b,\ f(a, b))$.

Likewise, the value of

$$\frac{\partial f}{\partial y}$$

at (a, b) is the gradient of the tangent line l_2, at the point $P = (a,\ b,\ f(a, b))$, to the curve $z = f(a, y)$, which is the section of s by the plane $x = a$ (Figure 11.26(b)).

(b) *Parametric*:

Consider next the surface s specified by the parametric equations

$$x = f(u, v),\ y = g(u, v),\ z = h(u, v),\ (u, v) \in W$$

where $W = [u_1, u_2] \times [v_1, v_2]$ is a rectangle in uv parameter space. The function $(u, v) \mapsto s(u, v) = (f(u, v),\ g(u, v),\ h(u, v))$ maps W to the surface s in 3-space. See Figure 11.27.

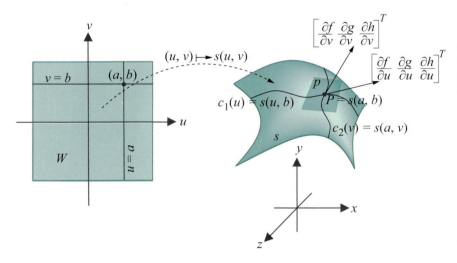

Figure 11.27: The surface s is the image of a parameter rectangle W by the map $(u, v) \mapsto s(u, v) = (f(u, v),\ g(u, v),\ h(u, v))$. Tangents to the parameter curves on s at the point $P = s(a, b)$ span the tangent plane p at P.

Fix a point $(a, b) \in W$. The image of the line $v = b$ by s is the u-parameter curve c_1 with equation

$$c_1(u) = (f(u, b), \ g(u, b), \ h(u, b)), \quad u \in [u_1, u_2]$$

The tangent vector to this curve is

$$
\begin{aligned}
c_1'(u) &= \left[\frac{d}{du} f(u, b) \quad \frac{d}{du} g(u, b) \quad \frac{d}{du} h(u, b) \right]^T \\
&= \left[\frac{\partial f}{\partial u}(u, b) \quad \frac{\partial g}{\partial u}(u, b) \quad \frac{\partial h}{\partial u}(u, b) \right]^T
\end{aligned}
$$

at $u \in [u_1, u_2]$. Therefore, the value of the vector

$$\left[\frac{\partial f}{\partial u} \quad \frac{\partial g}{\partial u} \quad \frac{\partial h}{\partial u} \right]^T$$

at the point (a, b) is a tangent vector (provided it's non-zero) to the u-parameter curve

$$c_1(u) = s(u, b)$$

at the point $s(a, b)$.

Likewise, the value of the vector

$$\left[\frac{\partial f}{\partial v} \quad \frac{\partial g}{\partial v} \quad \frac{\partial h}{\partial v} \right]^T$$

at the point (a, b) is a tangent vector (provided it's non-zero) to the v-parameter curve

$$c_2(v) = s(a, v)$$

at the point $s(a, b)$.

Definition 11.2. If the tangent vectors

$$\left[\frac{\partial f}{\partial u} \quad \frac{\partial g}{\partial u} \quad \frac{\partial h}{\partial u} \right]^T$$

and

$$\left[\frac{\partial f}{\partial v} \quad \frac{\partial g}{\partial v} \quad \frac{\partial h}{\partial v} \right]^T$$

to the two parameter curves through the point $P = s(a, b)$ are linearly independent – in other words, they are not collinear – then they span a plane p, called the *tangent plane* to the surface s at P. This is the case in Figure 11.27.

Any line l on p through P is said to be a *tangent line* to s at P and any non-zero vector v lying on p is said to be a *tangent vector* to s at P (v is usually drawn emanating from P). See Figure 11.28. The line perpendicular to p through P is said to be the *normal line* to s at P and any non-zero vector lying on this line a *normal vector* to s at P.

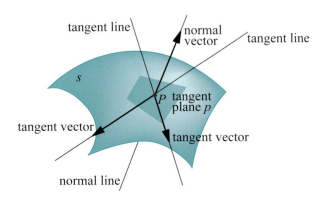

Figure 11.28: Two tangent lines and vectors on them, normal line and a normal vector to the surface s at P.

A tangent plane to a surface is precisely the geometric analogue of a tangent line to a curve. A thin straight stick pressed to a plane wire curve aligns itself along the tangent line at the point of contact; similarly, a thin flat board pressed to a surface in 3-space aligns itself along the tangent plane at the point of contact.

Example 11.9. Determine the tangent plane and a normal vector to the paraboloid

$$z = x^2 + y^2$$

at the point $(1, 2, 5)$.

Answer: It's easy first to write the given implicit equation in the parametric form

$$x = u, \quad y = v, \quad z = u^2 + v^2 \tag{11.13}$$

Differentiating,

$$\left[\frac{\partial x}{\partial u} \ \frac{\partial y}{\partial u} \ \frac{\partial z}{\partial u} \right]^T = [1 \ 0 \ 2u]^T$$

$$\left[\frac{\partial x}{\partial v} \ \frac{\partial y}{\partial v} \ \frac{\partial z}{\partial v} \right]^T = [0 \ 1 \ 2v]^T \tag{11.14}$$

The point $(1, 2, 5)$ corresponds to the parameter values $u = 1$ and $v = 2$ in (11.13). Therefore, two tangent vectors to the paraboloid at $(1, 2, 5)$ are obtained by substituting these particular parameter values into the general expressions (11.14) above for tangent vectors at arbitrary points. Particularly, these two vectors are $[1 \ 0 \ 2]^T$ and $[0 \ 1 \ 4]^T$, which are evidently linearly independent. Therefore, the tangent plane to the paraboloid at $(1, 2, 5)$ is spanned by $[1 \ 0 \ 2]^T$ and $[0 \ 1 \ 4]^T$. See Figure 11.29.

A normal vector to the paraboloid at the point $(1, 2, 5)$ is perpendicular to its tangent plane there and, therefore, to both spanning vectors $[1 \ 0 \ 2]^T$

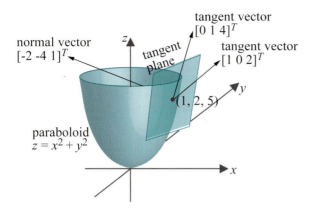

Figure 11.29: Tangent vectors, tangent plane and normal vector at the point $(1, 2, 5)$ to the paraboloid $z = x^2 + y^2$.

and $[0\ 1\ 4]^T$. It is obtained, then, as the cross-product of the latter (cross-products of vectors were reviewed in Section 5.4.3), viz.,

$$[1\ 0\ 2]^T \times [0\ 1\ 4]^T = [-2\ -4\ 1]^T$$

Remark 11.12. Computing the tangent plane at a point of a surface and computing a normal vector there are equivalent.

Exercise 11.23. Determine the tangent plane and a normal vector to the circular cylinder

$$x = \cos u, \ y = \sin u, \ z = v$$

at the point corresponding to the parameter values $(u, v) = (\pi/4, 3)$.

Exercise 11.24. Determine the tangent plane and a normal vector to the saddle-shaped surface (hyperbolic paraboloid is the mathematical name)

$$z = xy$$

at the point $(2, 3, 6)$.

Exercise 11.25. (Programming) Draw the paraboloid of Example 11.9 and its tangent plane at some point. The paraboloid should be wireframe and the tangent plane a finely meshed rectangle. Allow the user to press the arrow keys to slide the tangent plane over the paraboloid.

Normals from Function Gradients

Definition 11.2 of a tangent plane assumes a parametric representation of the surface. There's, however, a neat way to compute directly a normal vector at a point of a surface given implicitly.

If a surface s is specified implicitly by an equation of the form

$$F(x, y, z) = 0$$

then a normal vector to the surface at the point (a, b, c) is given by the value of the so-called *gradient* of F, denoted $grad(F)$, at that point, provided this value is not the zero vector. The gradient is defined by

$$grad(F) = \begin{bmatrix} \dfrac{\partial F}{\partial x} & \dfrac{\partial F}{\partial y} & \dfrac{\partial F}{\partial z} \end{bmatrix}^T$$

We'll not try to prove that $grad(F)$ is indeed normal to the surface $F(x, y, z) = 0$, but simply assume so for the purpose of computation. For the actual proof and more about the gradient, as well as its related functions *divergence* and *curl*, the reader is referred to books on vector calculus, e.g., Schey [118] and Spiegel [130].

Example 11.10. Determine a normal vector to the paraboloid

$$z = x^2 + y^2$$

at the point $(1, 2, 5)$.

Answer: Write the implicit equation in the form

$$F(x, y, z) = z - x^2 + y^2 = 0$$

Then

$$grad(F) = \begin{bmatrix} \dfrac{\partial F}{\partial x} & \dfrac{\partial F}{\partial y} & \dfrac{\partial F}{\partial z} \end{bmatrix}^T = \begin{bmatrix} -2x & -2y & 1 \end{bmatrix}^T$$

Therefore, a normal vector at the point $(1, 2, 5)$ is $\begin{bmatrix} -2 & -4 & 1 \end{bmatrix}^T$, which is obtained from putting $x = 1$ and $y = 2$ in the preceding equation. This result checks with Example 11.9.

Exercise 11.26. Verify your answer to Exercise 11.23 by finding a normal vector to the cylinder using the *grad* function. You must write an implicit equation for the cylinder first.

11.11   Computing Normals and Lighting Surfaces

Look carefully at the OpenGL lighting equation (11.12) once more. Outside of a bunch of user-specified color properties, the only data needed to compute the color intensities at a vertex V of an object O consists of the position of V, the positions of the light sources *and* the normal vector n at V.

The position of V is, of course, part of O's design. As for the light sources, they are usually few, and the user is free to locate them as he pleases. Remaining is the normal vector n, which the user is free to set as well. However, for authentic lighting it should actually be perpendicular to the surface of O at V or at least nearly so. For example, the choice of the normal vector n at the vertex V of the sphere in Figure 11.30 seems good, though either of the other two vectors drawn there could conceivably have been picked as well.

We'll discuss computing surface normals following the informal taxonomy of 2D objects in Section 10.2 before moving on to Bézier and quadric surfaces for which OpenGL provides automatic normals.

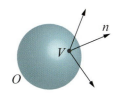

Figure 11.30: Three vectors at a vertex on a sphere, one of which has been chosen as the normal.

11.11.1 Polygons and Planar Surfaces

Polygons in particular, and planar surfaces in general, are the simplest. The normal at each vertex is simply normal to the plane itself containing the surface. In particular, unit vertex normals are all identical across a given side of the surface.

So how does one determine the normal direction to a plane p? If two non-collinear vectors u and v are known to lie on p, then the cross-product $u \times v$ is normal to p (cross-products were reviewed in Section 5.4.3). For example, any two adjacent edges of a polygon determine non-collinear vectors u and v spanning the plane p containing the polygon; therefore, $u \times v$ is normal to p. In Figure 11.31, $n = (P_1 - P_0) \times (P_4 - P_0)$ is normal to p.

Exercise 11.27. Determine a normal to the plane p of the triangle with vertices at

$$P_0 = [0\ 3\ 5]^T, \qquad P_1 = [1\ -2\ 0]^T, \qquad P_2 = [3\ 3\ 3]^T$$

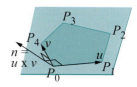

Figure 11.31: Vector n is normal to the plane p.

11.11.2 Meshes

Polygonal meshes are of interest next. Let's work with real examples.

Experiment 11.18. Run again sphereInBox1.cpp. The normal vector values at the eight box vertices of sphereInBox1.cpp, placed in the array normals[], are

$$[\pm 1/\sqrt{3}\ \ \pm 1/\sqrt{3}\ \ \pm 1/\sqrt{3}]^T$$

each corresponding to one of the eight possible combinations of signs. End

The choice of the normals in sphereInBox1.cpp is easily motivated. The box being situated symmetrically about the origin, the normal values are chosen as unit vectors along the lines from the origin to each of the eight vertices, which indeed give the values above. The box is depicted in Figure 11.32(a), where only the normal vector at the lower-right vertex V

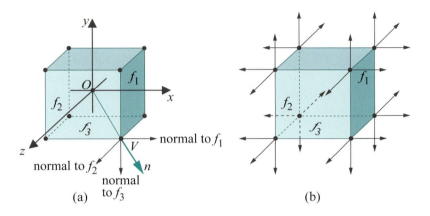

Figure 11.32: (a) The box of `sphereInBox1.cpp` with the averaged normal vector n at vertex V, together with the normals to the three faces that meet at V (f_1 right face, f_2 front face, f_3 bottom face) (b) The unaveraged normals of `sphereInBox2.cpp`.

of the front face is shown: it is the arrow n drawn by extending OV a unit distance from V.

In fact, probably a better rationale for this particular choice of normals – which would still hold if the same box happened to be drawn not centered at the origin, but elsewhere – is that the one at each vertex is the normalized *average* of the unit outward normals to the three faces meeting at that vertex. For example, in Figure 11.32(a) the unit outward normals to f_1, f_2 and f_3 are $[1\ 0\ 0]^T$, $[0\ 0\ 1]^T$ and $[0\ -1\ 0]^T$, respectively, whose average is $[1/3\ \ -1/3\ \ 1/3]^T$, which normalizes to $[1/\sqrt{3}\ \ -1/\sqrt{3}\ \ 1/\sqrt{3}]^T$, which one can verify from the code is indeed the value of the normal at V in `sphereInBox1.cpp`.

Although they possess the virtue of symmetry, it's clear, nevertheless, the box normals of `sphereInBox1.cpp` are not nearly actually perpendicular to the surface of the box, in particular, not to any of its faces. This consideration leads to another approach – to set the normal at each vertex of a face as a normal to that face itself. This is implemented as an option in `sphereInBox2.cpp`.

$\mathrm{Experiment}$ 11.19. Run `sphereInBox2.cpp`, which modifies `sphereIn-Box1.cpp`. Press the arrow keys to open or close the box and space to toggle between two methods of drawing normals.

The first method is that of `sphereInBox1.cpp`, specifying the normal at each vertex as an average of incident face normals. The second creates the box by first drawing one side as a square with the normal at each of its four vertices specified to be the unit vector perpendicular to the square, then placing that square in a display list and, finally, drawing it six times appropriately rotated. Figure 11.32(b) shows the vertex normals to three faces. Figure 11.33 shows screenshots of the box created with and without averaged normals.

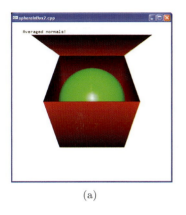

(a)

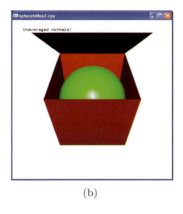

(b)

Figure 11.33: Screenshot of `sphereInBox2.cpp`: (a) Averaged box normals (b) Unaveraged box normals.

The contrast in output between the two ways of defining box normals in `sphereInBox2.cpp` is clear and the reason not hard to understand. The first method softens the edges because the averaged normal at each vertex is shared by all its three adjacent faces. Consequently, the interpolation of color values in each face's interior continues smoothly across its boundary.

The second method is significantly different. As each face is drawn separately with the normals at all its four vertices equal and perpendicular to the face itself, interpolation in the interior results in the entire face being colorized as if with that one normal value throughout. Moreover, this normal value turns abruptly by 90° from one face to the next. The upshot is that there is a significant difference in color intensities, as well, from one face to the next, throwing the edges between them into sharp relief. Which approach to choose depends on the effect desired.

Remark 11.13. Using the second method, colors at pixels along an edge are defined differently by its two adjacent faces, while pixel colors at a vertex are defined, in fact, by its three adjacent faces. At these pixels, therefore, code order determines which color prevails. This is not desirable, but it is not a serious issue because such "ambiguous" pixels lie only along edges and not in the interior of faces which constitute the bulk of the figure.

Versions of the averaging approach implemented sometimes to achieve greater realism use a *weighted* average rather than a straight one. Two possibilities are:

(a) Weight each adjacent face normal with the angle of that face at the vertex. In Figure 11.34, five faces meet at the vertex V subtending angles $\theta_1, \theta_2, \ldots, \theta_5$, respectively. The angle-weighted average value of the normal at V is:

$$n = \frac{\theta_1 n_1 + \theta_2 n_2 + \theta_3 n_3 + \theta_4 n_4 + \theta_5 n_5}{\theta_1 + \theta_2 + \theta_3 + \theta_4 + \theta_5}$$

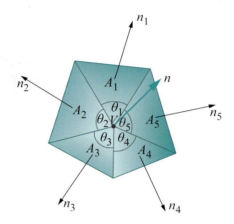

Figure 11.34: Weighted average of normals: θ_i are angles, A_i area, n_i face normals and n a weighted average normal at V.

(b) Weight each adjacent face normal with the area of that face. The areas of the five faces in Figure 11.34 meeting at V are $A_1, A_2, \ldots, A_5$, respectively. The area-weighted average value of the normal at V is then:

$$n = \frac{A_1 n_1 + A_2 n_2 + A_3 n_3 + A_4 n_4 + A_5 n_5}{A_1 + A_2 + A_3 + A_4 + A_5}$$

Important: Whatever approach you adopt to compute normals, make sure, as a last step, to normalize each to unit length (easy enough – just divide each by its length). The reason is that OpenGL uses the dot product to compute the cosine of the angle between two vectors (see Equation (11.12)), which is correct *only if* they are of unit length.

Example 11.11. For the trash can mesh whose vertices are given in Figure 11.35, compute the unit normals to the three faces adjacent to the vertex V. Then compute the (unweighted) average of these three normals and normalize to unit length.

Answer: The three edge vectors emanating from V are:

$$\begin{aligned}
u_1 &= [1 \ -1 \ -1]^T - [1 \ -1 \ 1]^T = -2\mathbf{k} \\
u_2 &= [1.2 \ 1 \ 1.2]^T - [1 \ -1 \ 1]^T = 0.2\mathbf{i} + 2\mathbf{j} + 0.2\mathbf{k} \\
u_3 &= [-1 \ -1 \ 1]^T - [1 \ -1 \ 1]^T = -2\mathbf{i}
\end{aligned}$$

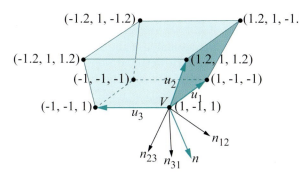

Figure 11.35: Trash can of five quadrilateral sides. The vectors n_{12}, n_{23} and n_{31} from V are normals to V's adjacent faces, while n is the averaged normal.

Therefore, the outward unit normal to the face with edges u_1 and u_2 is

$$n_{12} = (u_1 \times u_2) \, / \, |u_1 \times u_2| = (4\mathbf{i} - 0.4\mathbf{j}) \, / \, \sqrt{4^2 + 0.4^2} \simeq 0.995\mathbf{i} - 0.0995\mathbf{j}$$

and that to the face with edges u_2 and u_3 is

$$n_{23} = (u_2 \times u_3) \, / \, |u_2 \times u_2| = (-0.4\mathbf{j} + 4\mathbf{k}) \, / \, \sqrt{4^2 + 0.4^2} \simeq -0.0995\mathbf{j} + 0.995\mathbf{k}$$

while the outward unit normal to the face with edges u_3 and u_1, the bottom face, is easily seen to be

$$n_{31} = -\mathbf{j}$$

The normalize average of these normals is

$$
\begin{aligned}
n &= (n_{12} + n_{23} + n_{31}) \, / \, |n_{12} + n_{23} + n_{31}| \\
&\simeq (0.995\mathbf{i} - 1.199\mathbf{j} + 0.995\mathbf{k}) \, / \, \sqrt{0.995^2 + 1.199^2 + 0.995^2} \\
&\simeq 0.538\mathbf{i} - 0.649\mathbf{j} + 0.538\mathbf{k}
\end{aligned}
$$

Exercise 11.28. (Programming) Use data from the preceding example to replace the box of `sphereInBox2.cpp` with a trash can. Omit the sphere. Let the user choose between averaged and unaveraged normals. Allow the can to be rotated keeping the light source fixed.

11.11.3   General Surfaces

As a general surface is drawn by approximating it with a polygonal mesh, the thought comes to mind to simply use the methods of the preceding section to find normals. Precisely, (a) formulate a mesh approximation of the surface and (b) specify the normal at each vertex as an average of those of its adjacent faces (we really want to use an average here, especially if the original surface is smooth, to avoid color discontinuities between adjacent mesh faces).

This approach is perfectly reasonable if the surface is known to the user only by its mesh approximation. However, if one knows, say, a parametric representation of the surface, why not get the normals from the "horse's mouth" – that being the parametrization itself? In other words, use the parametrization to *analytically* compute the normals at the mesh vertices. This makes for stable normals independent of the vagaries of the particular mesh approximation, not to mention those of the averaging process (possibly, angle-weighted or area-weighted). For example, working from the mesh approximation of the surface s in Figure 11.36, normals to the six faces adjacent to vertex V must be averaged to determine the normal n at V. However, knowledge of s itself could enable a direct computation.

So let's see how to compute normals analytically. We're going to assume in the following that you know that a tangent plane at the point $s(u, v)$ to a surface s given parametrically by the equations

$$x = f(u, v), \ y = g(u, v), \ z = h(u, v)$$

is spanned by the two vectors

$$\left[\frac{\partial f}{\partial u} \ \frac{\partial g}{\partial u} \ \frac{\partial h}{\partial u}\right]^T \quad \text{and} \quad \left[\frac{\partial f}{\partial v} \ \frac{\partial g}{\partial v} \ \frac{\partial h}{\partial v}\right]^T$$

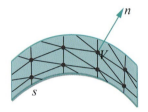

Figure 11.36: Normal vector n to the surface s at a vertex V of its mesh approximation.

evaluated at (u, v) (provided they are not collinear). Moreover, a normal vector to s at $s(u, v)$ is the cross-product

$$\left[\frac{\partial f}{\partial u} \ \frac{\partial g}{\partial u} \ \frac{\partial h}{\partial u}\right]^T \times \left[\frac{\partial f}{\partial v} \ \frac{\partial g}{\partial v} \ \frac{\partial h}{\partial v}\right]^T \tag{11.15}$$

evaluated at (u, v). If you need to brush up, Section 11.10 has a review of the needed calculus.

Denote the normalized value of the vector (11.15) – obtained by dividing it by its magnitude – by

$$[f_n(u, v) \quad g_n(u, v) \quad h_n(u, v)]^T \tag{11.16}$$

which, therefore, is a unit normal to s at $s(u, v)$.

Finally, we'll specify either $[f_n(u, v) \quad g_n(u, v) \quad h_n(u, v)]^T$ or its reverse, $[-f_n(u, v) \quad -g_n(u, v) \quad -h_n(u, v)]^T$, as the unit normal at $s(u, v)$ depending on which direction is appropriate for front-facing triangles. There's not much to worry about making a wrong choice, as it'll be plenty clear from the viewable output! Let's get to work on a benign surface first.

Cylinder

Example 11.12. Consider the circular cylinder $s(u, v)$ with parametric equations

$$x = \cos u, \ y = \sin u, \ z = v, \ \text{where} \ (u, v) \in [-\pi, \pi] \times [-1, 1]$$

We drew it using these equations in `cylinder.cpp` of Experiment 10.3. To color and light, let's do normal calculations. The vectors spanning the tangent plane at $s(u, v)$ are

$$\left[\frac{\partial(\cos u)}{\partial u} \quad \frac{\partial(\sin u)}{\partial u} \quad \frac{\partial v}{\partial u} \right]^T = [-\sin u \quad \cos u \quad 0]^T$$

and

$$\left[\frac{\partial(\cos u)}{\partial v} \quad \frac{\partial(\sin u)}{\partial v} \quad \frac{\partial v}{\partial v} \right]^T = [0 \ 0 \ 1]^T$$

so a normal vector is

$$[-\sin u \quad \cos u \quad 0]^T \times [0 \ 0 \ 1]^T = [\cos u \quad \sin u \quad 0]^T$$

which happens to be normalized already. So, in the terminology of (11.16), for the cylinder,

$$f_n(u, v) = \cos u, \quad g_n(u, v) = \sin u, \quad h_n(u, v) = 0$$

We'll add this normal data to `cylinder.cpp` next.

Experiment 11.20. Run `litCylinder.cpp`, which builds upon `cylinder.cpp` using the normal data calculated above, together with color and a single directional light source. Press 'x/X', 'y/Y' and 'z/Z' to turn the cylinder. The functionality of being able to change the fineness of the mesh approximation has been dropped. Figure 11.37 is a screenshot. **End**

Compare the two programs `cylinder.cpp` and `litCylinder.cpp` – it's not really a lot of code from the first to the second. Essentially, the additions are (a) the `fn()`, `gn()` and `hn()` normal component functions as calculated above, (b) the `fillNormalArray()` function to fill the array `normals[]`, and (c) a bunch of routine code specifying light and material properties, which can be kept similar across most programs with lighting.

So the extra code arising from analytic normal computation is really in (a) and (b), about 20 lines all told. Not too bad, huh? And it gets better. As we used the template of `cylinder.cpp` to draw various surfaces, simply swapping in new `f()`, `g()` and `h()` functions according to the given parametrization, so we can use `litCylinder.cpp` for lit applications, additionally swapping in new `fn()`, `gn()` and `hn()` functions.

Figure 11.37: Screenshot of `litCylinder.cpp`.

Exercise 11.29. (Programming) Reverse the normals of `litCylinder.cpp` by changing their specification in the `fillNormalArray()` routine as follows:

```
normals[k++] = -fn(i,j);
normals[k++] = -gn(i,j);
normals[k++] = -hn(i,j);
```

Not good! As we remarked earlier, wrongly-oriented normals are easy to spot. Can you fix the problem caused by the normal values above by a minimal amount of code change *only* in the drawing routine?

Hint: Think orientation, in particular, reversing the orientation of the strip triangles.

Exercise 11.30. It's a bit late now, but do we really need partial derivatives, as in Example 11.12, to determine the normal to the cylinder at the point $V = (\cos u, \sin u, v)$?

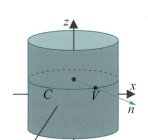

The outward normal to the cylinder at V evidently lies along a radius of the circle C which is the section of the cylinder through V by a plane perpendicular to its axis. See Figure 11.38. Use this to compute the parametric equation for a unit normal vector to the cylinder without any calculus.

Figure 11.38: Normal n to a cylinder.

Often, as in the preceding exercise, normals to a surface can be determined by elementary geometric considerations. Unfortunately, this does not seem to be the case with the doubly-curled cone of Experiment 10.8.

Doubly-curled Cone

Next, we light the doubly-curled cone of `doublyCurledCone.cpp`. Its parametric equations are

$$x = t \cos(A + a\theta) \cos \theta, \ y = t \cos(A + a\theta) \sin \theta, \ z = t \sin(A + a\theta),$$

where $0 \leq t \leq 1$ and $0 \leq \theta \leq 4\pi$. A somewhat tedious calculation gives a normal to the cone as

$$\left[\frac{\partial x}{\partial \theta} \ \frac{\partial y}{\partial \theta} \ \frac{\partial z}{\partial \theta} \right]^T \times \left[\frac{\partial x}{\partial t} \ \frac{\partial y}{\partial t} \ \frac{\partial z}{\partial t} \right]^T$$

$$= [-at \sin \theta + t \sin(A + a\theta) \cos(A + a\theta) \cos \theta,$$
$$at \cos \theta + t \sin(A + a\theta) \cos(A + a\theta) \sin \theta,$$
$$- t \cos^2(A + a\theta)]^T \qquad (11.17)$$

Moreover, the length of this normal is

$$t\sqrt{a^2 + \cos^2(A + a\theta)} \qquad (11.18)$$

Dividing the normal (11.17) by its length (11.18) gives a unit normal to the cone.

Experiment 11.21. The program `litDoublyCurledCone.cpp`, in fact, applies the preceding equations for the normal and its length. Press 'x/X', 'y/Y', 'z/Z' to turn the cone. See Figure 11.39 for a screenshot.

Figure 11.39: Screenshot of `litDoublyCurled-Cone.cpp`.

As promised, `litDoublyCurledCone.cpp` is pretty much a copy of `litCylinder.cpp`, except for the different `f()`, `g()`, `h()`, `fn()`, `gn()` and `hn()` functions, as also the new `normn()` to compute the normal's length.

End

Exercise 11.31. Verify Equations (11.17) and (11.18) for the normal and its magnitude of the doubly-curled cone.

Exercise 11.32. (Programming) The doubly-curled cone would probably benefit from at least one more light source, particularly to brighten the inside. Code this in.

Exercise 11.33. (Programming) Color and light the table of Experiment 10.7. You don't need any calculus in order to compute the normals to the various component surfaces – which happen each to be either cylindrical or flat. Make sure to choose normals so that edges appear sharp.

Exercise 11.34. (Programming) Color and light the helical pipe of Experiment 10.4.

Exercise 11.35. (Programming) Color and light the pipe of Exercise 10.46, which coils around a torus.

Exercise 11.36. (Programming) Color and light the single-sheeted hyperboloid of Experiment 10.11.

Exercise 11.37. Which of the three components – ambient, diffuse and specular – of light reflected from a vertex V are affected if the normal at V is altered?

11.11.4 Bézier and Quadric Surfaces

Good news! All one has to do is type in the command `glEnable(GL_AUTO_-NORMAL)` for OpenGL to automatically calculate unit normals at the vertices of a Bézier surface which has been created using `glMap2f(GL_MAP2_VERTEX_3, ... )` and `glEnable(GL_MAP2_VERTEX_3)`.

Canoe

Experiment 11.22. Run `litBezierCanoe.cpp`. Press 'x/X', 'y/Y', 'z/Z' to turn the canoe. You can see a screenshot in Figure 11.40.

This program illuminates the final shape of `bezierCanoe.cpp` of Experiment 10.20 with a single directional light source. Other than the expected command `glEnable(GL_AUTO_NORMAL)` in the initialization routine, an important point to notice about `litBezierCanoe.cpp` is the reversal of the sample grid along the u-direction. In particular, compare the statement

```
glMapGrid2f(20, 1.0, 0.0, 20, 0.0, 1.0)
```

of `litBezierCanoe.cpp` with

```
glMapGrid2f(20, 0.0, 1.0, 20, 0.0, 1.0)
```

Figure 11.40: Screenshot of `litBezierCanoe.cpp`.

of `bezierCanoe.cpp`. This change reverses the directions of one of the tangent vectors evaluated at each vertex by OpenGL and, correspondingly, that of the normal (which is the cross-product of the two tangent vectors).

Modify `litBezierCanoe.cpp` by changing

```
glMapGrid2f(20, 1.0, 0.0, 20, 0.0, 1.0);
```

back to `bezierCanoe.cpp`'s

```
glMapGrid2f(20, 0.0, 1.0, 20, 0.0, 1.0);
```

Wrong normal directions! The change from `bezierCanoe.cpp` is necessary. Another solution is to leave `glMapGrid2f()` as it is in `bezierCanoe.cpp`, instead making a call to `glFrontFace(GL_CW)`. **End**

The lesson to take from this is that if you obtain normals automatically from OpenGL, then you might have to subsequently alter their orientation for authenticity, which is not unreasonable because OpenGL cannot know which you intend to be the front face of a primitive.

Remark 11.14. If the user wishes to define her own normals for a Bézier surface, she can do so with a `glMap2f(GL_MAP2_NORMAL, ...)` call. We'll not have occasion to use this call ourselves.

Quadrics are even simpler. The call

```
gluQuadricNormals(qobj, GLU_SMOOTH)
```

automatically generates a normal at each vertex of the quadric pointed by `qobj`.

The next program we'll look at is a fairly substantial animation which invokes both `glEnable(GL_AUTO_NORMAL)` for Bézier surface normals and `gluQuadricNormals(qobj, GLU_SMOOTH)` for quadric surfaces.

Movie with a Ship and Torpedo

Experiment 11.23. Run `shipMovie.cpp`. Pressing space start an animation sequence which begins with a torpedo traveling toward a moving ship and which ends on its own after a few seconds. Figure 11.41 is a screenshot as the torpedo nears the ship.

There are a few different objects. The hull of the ship is obviously inspired by the Bézier canoe of the previous experiment. The deck is a flat Bézier surface – all its control point y-values are identical – which is designed to fit the hull. Each of the ship's three storeys is a cylindrical quadric, as is its chimney.

The torpedo should be familiar from the program `torpedo.cpp` of Experiment 10.21. Each of the four grayish boats in the background is a couple of quads, while the sea itself is a solid blue cube.

The smoke from the chimney is a simple-minded *particle system*. In particular, we render a sequence of quadric discs in point mode and hack for it a coloring and animation scheme. **End**

Exercise 11.38. (Programming) The program `shipMovie.cpp` bears a lot of improvement. Try at least the following:

(a) Add detail to the ship.

(b) Make the water more realistic, possibly by adding movement, variation in color, etc.

(c) Put stars and a moon in the sky.

(d) Improve the smoke particle system.

(e) Make a particle system to simulate water spray from the torpedo's propeller.

Exercise 11.39. (Programming) Fill, paint and light the character of `animateMan*.cpp` in surroundings less bland than a plane with a ball. Make an animation sequence.

11.11.5 Transforming Normals

Normals are transformed by modelview transformations, but not as straightforwardly as vertices are by multiplication from the left by the transformation matrix. Let's see how they are transformed by each of the fundamental transformations – translation, rotation and scaling.

1. Translation:

   A translation leaves a normal vector at a vertex unchanged because the normal simply translates parallely (see Figure 11.42(a)).

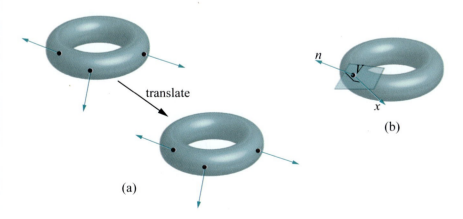

Figure 11.42: (a) Vertex normals translate parallely as the torus is translated (b) The normal n at V is perpendicular to any vector x which lies on the tangent plane at V.

2. Rotation and Scaling:

These cases are not as simple and require a bit of calculation.

A given rotation or non-degenerate scaling, say t, is a non-singular linear transformation, with, therefore, a non-singular defining matrix, say N. Suppose, as well, that n is a normal vector at a vertex V of an object O. Therefore, n is perpendicular to an arbitrary vector, say x, tangent to the surface of O at V (see Figure 11.42(b)).

Now, if we apply t, it will transform all the vertices of O, as well as vectors tangent to O's surface, by multiplication on the left by N.

Note: To convince yourself that tangent vectors are transformed identically with vertices, think of a tangent vector as connecting two vertices infinitesimally close together on the surface of O. Therefore, these two vertices "carry" the tangent vector with them.

So x is transformed to Nx. We would, therefore, like to transform n to a vector perpendicular to Nx. Since n is perpendicular to x, we already have $n \cdot x = 0$, which is equivalent to $n^T x = 0$, the latter being a matrix equation. It follows that

$$n^T (N^{-1}N)x = n^T x = 0$$

Therefore,

$$0 = n^T (N^{-1}N)x = (n^T N^{-1})(Nx) = ((N^{-1})^T n)^T (Nx)$$

(invoking rules of matrix algebra).

One sees that $((N^{-1})^T n) \cdot Nx = 0$, so $(N^{-1})^T n$ is indeed perpendicular to Nx. The conclusion, then, is that the appropriate transformation to apply to the normal vector n, under a rotation or non-degenerate scaling corresponding to the matrix N, is left multiplication by $(N^{-1})^T$, i.e., $n \mapsto (N^{-1})^T n$.

OpenGL actually transforms normals as just described. If the current modelview matrix is

$$M = \begin{bmatrix} a_{11} & a_{12} & a_{13} & a_{14} \\ a_{21} & a_{22} & a_{23} & a_{24} \\ a_{31} & a_{32} & a_{33} & a_{24} \\ a_{41} & a_{42} & a_{43} & a_{44} \end{bmatrix}$$

then "erasing" the translational part, which, as we know, has no impact on the normal, leaves the 3×3 matrix

$$N = \begin{bmatrix} a_{11} & a_{12} & a_{13} \\ a_{21} & a_{22} & a_{23} \\ a_{31} & a_{32} & a_{33} \end{bmatrix}$$

and, in fact, the matrix $(N^{-1})^T$ to use to transform normals, called the *normal matrix*, is stored in a state variable (which can be accessed by the user via the Shading Language in OpenGL 2.0 and higher). It should be noted that, correspondingly, the OpenGL normal is a 3-vector (recall that it was only to accommodate translations into the matrix multiplication scheme that real world 3-vectors were homogenized to length 4).

Exercise 11.40. We gave above a general formula for how a normal vector is transformed by a rotation or non-degenerate scaling in terms of its defining matrix. Ignoring the formula for a moment, can you deduce from elementary considerations what should happen in the particular case of a rotation? Then relate your answer to the formula.

11.11.6 Normalizing Normals

Normalizing a (non-zero) vector means dividing it by its magnitude to obtain a vector with the same direction, but of unit length. We've already seen that it's important to specify normalized normals because OpenGL uses the dot product to compute the cosine of the angle between two vectors, which is correct only if they are both of unit length.

Here's a simple modification of `litTriangle.cpp` to show what can happen if one is careless.

Experiment 11.24. Run `sizeNormal.cpp` based on `litTriangle.cpp`.

The ambient and diffuse colors of the three triangle vertices are set to red, green and blue, respectively. The normals are specified separately as well, initially each of unit length perpendicular to the plane of the triangle.

However, pressing the up/down arrow keys changes (as you can see) the size, but not the direction, of the normal at the red vertex. Observe the corresponding change in color of the triangle. Figure 11.43 is a screenshot.

End

Figure 11.43: Screenshot of `sizeNormal.cpp`.

There are, typically, two reasons why normals turn out not normalized:

(a) The user does not specify them of unit length in the first place.

(b) Even if they are specified of unit length, a subsequent application of a scaling transformation changes the length.

If the user is not inclined to write code to ensure normals of unit length, there's a way to ask OpenGL's help. Calling `glEnable(GL_NORMALIZE)` causes OpenGL to normalize all normal vectors before lighting calculation. Beware, though, it's not a particularly efficient call and should be avoided if possible.

Experiment 11.25. Run `sizeNormal.cpp` after placing the statement `glEnable(GL_NORMALIZE)` at the end of the initialization routine. Press the up/down arrow keys. The triangle no longer changes color (though the white arrow still changes in length, of course, because its size is that of the program-specified normal).   **End**

There's a cheaper renormalization call, `glEnable(GL_RESCALE_NORMAL)`, which can be used if you originally did provide unit normals that were subsequently all changed by the *same* scaling transformation.

11.12   Phong's Shading Model

An alternate shading model first proposed by Phong, though computationally far more intensive, significantly improves the realism of a rendered image.

Note: Phong's shading model should not be confused with his lighting model, which we know already that OpenGL implements.

Instead of computing light values only at primitives' vertices and then interpolating through its interior as in Gouraud shading – or smooth shading as it's also called, the OpenGL default – Phong suggested to (a) interpolate the vertex normal values through the primitive, and then (b) compute light values at each pixel using the interpolated normals.

Figure 11.44 illustrates the idea. Unit normals n_0, n_1 and n_2 are specified by the programmer at the vertices V_0, V_1 and V_2, respectively, of triangle t. These normals are then interpolated, and normalized, throughout t. For example, if the barycentric coordinates of the point V are given by

$$V = c_0 V_0 + c_1 V_1 + c_2 V_2$$

then the normal value n at V is computed to be

$$n = (c_0 n_0 + c_1 n_1 + c_2 n_2) / |c_0 n_0 + c_1 n_1 + c_2 n_2| \qquad (11.19)$$

(provided the denominator is not zero).

The color values of a pixel which happens to be centered at V are then computed in Phong's model using the lighting equation (11.12), where, now, the normal value n applied is from (11.19) above, the color values $V_{*,X}$ are interpolated from the vertices as well, while the light direction and halfway vectors l^i and s^i are determined from the coordinates of V itself.

OpenGL, as we know, offers only flat and Gouraud shading as options. However, the OpenGL Shading Language, part of the version 2.0 specification, allows individual pixels to be programmed, which means the programmer herself can code in Phong shading. We'll be doing precisely this as an application, after learning the Shading Language ourselves in Chapter 20.

Remark 11.15. Phong lighting calculation at each vertex followed by Gouraud shading, OpenGL's default process, is often called *per-vertex* lighting to contrast it with the *per-pixel* lighting of Phong's shading model. (More appropriate might have been per-vertex and per-pixel *shading*, but the given usage is common.)

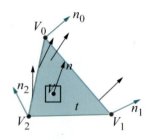

Figure 11.44: Normals n_0, n_1 and n_2 at the vertices of the triangle are programmer-specified. Shown also are (black) normalized interpolated normals at a few points and a pixel centered at V.

11.13 Summary, Notes and More Reading

In Chapters 4, 5 and 6 we learned to animate objects, in Chapter 10 to draw them, and now we have begun to "dress them up" with color and light. In this chapter we learned the underlying color and lighting models which OpenGL implements, the related syntax, and how to use them to specify light sources and material properties, as well as related environmental parameters. The technical issue of normal computation was an important part of our program too. We'll continue this theme in the next chapter when we learn of yet another technique to decorate an object, texturing.

For a further reference on coloring models, the somewhat encyclopedic Wyszecki and Stiles [147] is frequently called the bible of color science. The books by Berns [11] and Jackson et al. [71] are probably easier to read though.

Since the publication of Phong's model in 1975 [105] several other lighting models, both local and global, have been proposed. Local models like Phong's do not consider object-object light interaction, while global ones do, thereby displaying secondary effects such as shadows and reflections. Lighting models are often used in an application-specific manner, certain models being more realistic in rendering particular material properties and finishes.

A few of the local models which appeared after Phong's are Blinn [14], Cook-Torrance [28], He et al. [63, 64], Nayar-Oren [97], Poulin-Fournier [108] and Schlick [120]. However, the only local model that we discuss or use in this book is Phong's.

The two most commonly implemented global models are ray tracing [4, 143] and radiosity [54], which as a matter of fact complement each other. Global models, though much more realistic than local ones, are notoriously computation-intensive, so rarely apt for interactive applications. However, they are often used when frames can be created off-line, as in movies. We discuss both ray tracing and radiosity in Chapter 19.

The theory of lighting models necessarily involves a fair amount of physics and mathematics. The reader interested in learning more is best advised to start with advanced books such as those by Akenine-Möller, Haines & Hoffman [1], Buss [21] and Watt [142] and then proceed to original research papers, as the area is particularly active. The canonical source for the latest in CG research in general is the annual ACM SIGGRAPH conference [125].

CHAPTER 12

Texture

W e continue to explore methods to attire objects and enhance realism, which we began in the last chapter with color and light. The topic of this chapter is texturing. Textures are a vital part of the wardrobe available to designers. Texturing makes it possible to create lifelike scenes at acceptable costs. It's an enormously important technique in modern-day CG. We'll examine how texturing is implemented in OpenGL and various aspects of texturing in practice. Textures can be combined, as well, with color and light to good effect, as we'll see.

We cover the basics of loading and applying textures in Section 12.1, as well as the so-called texture map that specifies how a texture is painted onto an object. Sections 12.2-12.3 discuss setting various texturing parameters including the very important ones to control filtering. Specifying texture coordinates to determine the texture map is the topic of Section 12.4. We learn how to combine texture with light and color in Section 12.5 and, finally, conclude in Section 12.6.

12.1 Texture Basics and the Texture Map

Texturing a surface consists essentially of painting a picture onto it, a process which can be used to two great advantages when programming graphics:

(a) *Authenticity*: Realistically depicting objects, e.g., a soda can or a box of cereal, which happen to be painted in real life, requires painting the surface that models the object as well.

(b) *Illusion of geometric detail*: Instead of reproducing the geometry of an object, painting a picture of it in a scene can achieve a realistic result at a fraction of the cost in the number of polygons. For example, instead of modeling the individual bricks in a wall, paint a picture of a brick wall onto a single rectangle; instead of modeling individual trees in the backdrop of a scene, paint pictures of trees onto appropriately located polygons.

There are various ways to go about applying a texture in OpenGL. We'll discuss the one which is most common: where the texture is a rectangular array of pixels and applied to a polygonal surface. The pixels in a texture are called *texels*, each texel storing color values, such as 24-bit RGB or 32-bit RGBA, just as their counterpart pixels in the frame buffer.

The texture itself can be an external image which is imported into an OpenGL program or one created in the program itself. The former is called an *external* texture while the latter a *procedural*, or *synthetic*, texture. Once loaded though there is no difference between the two. Let's run a program using both kinds.

Experiment 12.1. Run `loadTextures.cpp`, which loads an external image of a shuttle launch as a texture and generates internally a chessboard image as another.

Notes:

1. The `Textures` folder must be placed in the same one as the program is in.

2. Because our programs all use the particular routine `getBMPData()` to read image files, textures applied by them have all to be in the uncompressed 24-bit `bmp` format for which this routine is written. Files in other formats have first to be converted. You can use image-editing software like Windows Paint, GIMP and Adobe Photoshop for this purpose.

3. OpenGL requires that the width and height of a texture be powers of two (in particular, for *unbordered* textures, which is the only kind we'll use). A further requirement is that both dimensions be at least 64; the maximum possible value depends on the implementation. Image files of dimensions not satisfying these conditions have to be resized accordingly. Again, most image-editing software have the capability to do this.

The program paints both the external and the procedural texture onto a square. Figure 12.1 shows the two. Press space to toggle between them, the left and right arrow keys to turn the square and delete to reset it.   **End**

Let's start to understand the texture-related OpenGL commands in `loadTextures.cpp`. A texture index array `texture` of size two is created first, in readiness to hold two textures, by the call `glGenTextures(2, texture)` in the initialization routine. The `loadExternalTextures()` routine then loads the image `launch.bmp` with the statements

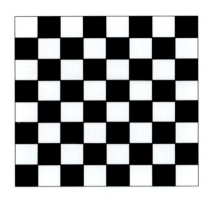

Figure 12.1: The two textures of `loadTextures.cpp`: shuttle launch (external, from NASA) and chessboard (synthetic).

```
BitMapFile *image[1];
image[0] = getBMPData("Images/launch.bmp");
```

Next, ignoring, for now, the four parameter-setting commands of the form `glTexParameteri()` in the `loadExternalTextures()` routine, the statements

```
glBindTexture(GL_TEXTURE_2D, texture[0]);
...
glTexImage2D(GL_TEXTURE_2D, 0, GL_RGB, image[0]->sizeX, image[0]->sizeY,
             0, GL_RGB, GL_UNSIGNED_BYTE, image[0]->data);
```

bind the image of the launch to the texture index `texture[0]`.

Likewise, the two statements

```
glBindTexture(GL_TEXTURE_2D, texture[1]);
...
glTexImage2D(GL_TEXTURE_2D, 0, GL_RGB, 64, 64, 0, GL_RGB,
             GL_UNSIGNED_BYTE, chessboard);
```

in the routine `loadProceduralTextures()` bind the image of a chessboard to `texture[1]`. The chessboard image itself is generated by the `createChessboard()` routine and stored in a 64×64 array `chessboard[64][64][3]` of RGB values. Each square of the board is represented by an $8 \times 8 \times 3$ subarray of `chessboard`, consisting either of all black or all white color values.

Figure 12.2: A striped board.

Exercise 12.1. (Programming) Replace the image of the launch with others downloaded from the web. Remember to first resize appropriately.

Exercise 12.2. (Programming) Write a routine `createStripedBoard()` that generates the image of a striped board, depicted in Figure 12.2, in a 64×64 RGB array.

A texture, once loaded, occupies the unit square with corners at (0,0), (1,0), (1,1) and (0,1) of an imaginary plane called *texture space*. This is regardless of whether the original rectangle of texels itself is equal-sided or not. If it is not, then it is scaled to fit the square, as illustrated in Figure 12.3. The axes of texture space are usually denoted s and t.

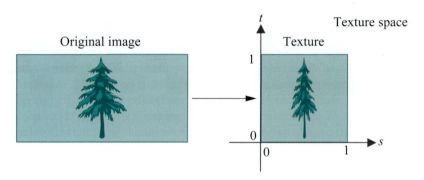

Figure 12.3: An image stored as a texture in a unit square of texture space.

The four statements within the `glBegin(GL_POLYGON)`–`glEnd()` pair of the following piece of code, from the drawing routine of `loadTextures.cpp`, each maps the vertex of a (square) polygon to a point in texture space.

```
glBegin(GL_POLYGON);
    glTexCoord2f(0.0, 0.0); glVertex3f(-10.0, -10.0, 0.0);
    glTexCoord2f(1.0, 0.0); glVertex3f(10.0, -10.0, 0.0);
    glTexCoord2f(1.0, 1.0); glVertex3f(10.0, 10.0, 0.0);
    glTexCoord2f(0.0, 1.0); glVertex3f(-10.0, 10.0, 0.0);
glEnd();
```

The first statement, for example, maps the vertex at $(-10.0, -10.0, 0.0)$ of world space to the point $(0.0, 0.0)$ of texture space. The coordinates of the mapped point in texture space are called the *texture coordinates* of the vertex. The mapping of the polygon vertices to texture space is interpolated throughout the polygon to obtain the so-called *texture map*, which, therefore, is a map from a part of world space (that occupied by the polygon) to texture space. The texture, finally, is painted onto the polygon by applying to each point of it the RGB color values of its image by the texture map.

Exercise 12.3. In `loadTextures.cpp`, what are the texture coordinate of the following points of the world-space square?

(a) $(0.0, 0.0, 0.0)$

(b) $(5.0, 5.0, 0.0)$

(c) $(10.0, 0.0, 0.0)$

Part answer: (a) $(0.5, 0.5)$, as the midpoint of the world-space square maps to the midpoint of the texture square by linearity.

Experiment 12.2. Replace every 1.0 in each `glTexCoord2f()` command of `loadTextures.cpp` with 0.5 so that the polygon specification is (Block 1*):

```
glBegin(GL_POLYGON);
   glTexCoord2f(0.0, 0.0); glVertex3f(-10.0, -10.0, 0.0);
   glTexCoord2f(0.5, 0.0); glVertex3f(10.0, -10.0, 0.0);
   glTexCoord2f(0.5, 0.5); glVertex3f(10.0, 10.0, 0.0);
   glTexCoord2f(0.0, 0.5); glVertex3f(-10.0, 10.0, 0.0);
glEnd();
```

The lower left quarter of the texture is interpolated over the square (Figure 12.4(a)). Make sure to see both the launch and chessboard textures! **End**

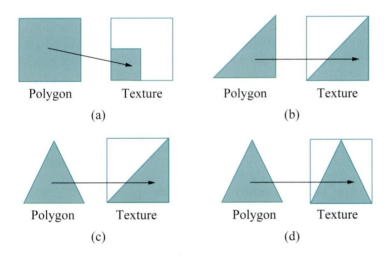

Figure 12.4: Texture maps.

Experiment 12.3. Restore the original `loadTextures.cpp` and delete the last vertex from the polygon so that the specification is that of a triangle (Block 2):

```
glBegin(GL_POLYGON);
   glTexCoord2f(0.0, 0.0); glVertex3f(-10.0, -10.0, 0.0);
   glTexCoord2f(1.0, 0.0); glVertex3f(10.0, -10.0, 0.0);
   glTexCoord2f(1.0, 1.0); glVertex3f(10.0, 10.0, 0.0);
glEnd();
```

*To cut-and-paste you can find the block in text format in the file `chap12codeModifications.txt` in the directory `Code/CodeModifications`.

Exactly as expected, the lower-right triangular half of the texture is interpolated over the world-space triangle (Figure 12.4(b)).

Change the coordinates of the last vertex of the world-space triangle (Block 3):

```
glBegin(GL_POLYGON);
    glTexCoord2f(0.0, 0.0); glVertex3f(-10.0, -10.0, 0.0);
    glTexCoord2f(1.0, 0.0); glVertex3f(10.0, -10.0, 0.0);
    glTexCoord2f(1.0, 1.0); glVertex3f(0.0, 10.0, 0.0);
glEnd();
```

Interpolation is clearly evident now. Parts of both launch and chessboard are skewed by texturing, as the triangle specified by texture coordinates is not similar to its world-space counterpart (Figure 12.4(c)).

Continuing, change the texture coordinates of the last vertex (Block 4):

```
glBegin(GL_POLYGON);
    glTexCoord2f(0.0, 0.0); glVertex3f(-10.0, -10.0, 0.0);
    glTexCoord2f(1.0, 0.0); glVertex3f(10.0, -10.0, 0.0);
    glTexCoord2f(0.5, 1.0); glVertex3f(0.0, 10.0, 0.0);
glEnd();
```

The textures are no longer skewed as the triangle in texture space is similar to the one being textured (Figure 12.4(d)). **End**

Expe**r**imen**t** 12.4. Restore the original `loadTextures.cpp` and replace `launch.bmp` with `cray2.bmp`, an image of a Cray 2 supercomputer. View the original images in the `Textures` folder and note their sizes: the launch is 512×512 pixels while the Cray 2 is 512×256. As you can see, the Cray 2 is scaled by half width-wise to fit the square polygon. **End**

Exerci**se** 12.4. (**P**rogrammin**g**) Change the polygon specs so that the Cray 2 is not distorted.

Expe**r**imen**t** 12.5. Restore the original `loadTextures.cpp` and then change the coordinates of only the third world-space vertex of the textured polygon (Block 5):

```
glBegin(GL_POLYGON);
    glTexCoord2f(0.0, 0.0); glVertex3f(-10.0, -10.0, 0.0);
    glTexCoord2f(1.0, 0.0); glVertex3f(10.0, -10.0, 0.0);
    glTexCoord2f(1.0, 1.0); glVertex3f(20.0, 0.0, 0.0);
    glTexCoord2f(0.0, 1.0); glVertex3f(-10.0, 10.0, 0.0);
glEnd();
```

The launch looks odd. The rocket rises vertically, but the flames underneath are shooting sideways! Toggle to the chessboard and it's instantly clear what's going on. Figure 12.5 shows both textures.

(a)

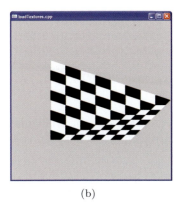
(b)

Figure 12.5: Screenshots from Experiment 12.5.

The polygon and the texture have been triangulated *equivalently* – in particular, triangles in the triangulation of one correspond to those in the other via the texture map. Corresponding triangle in this case, though, evidently differ in shape. Subsequently, either triangle of the texture has been *separately* interpolated over the corresponding triangle of the polygon, causing the perceived distortion. See Figure 12.6. **E**nd

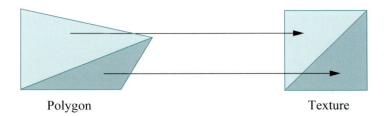

Polygon   Texture

Figure 12.6: Each of the two texture triangles is interpolated over the corresponding polygon triangle.

When we had said a little earlier that the texture map is obtained by interpolating over the entire polygon the mapping from its vertices to points in texture space, we had not taken into account the fact that there is no unambiguous way to do this if the polygon has more than three sides (recall discussions to this effect in Section 7.4). We know now how OpenGL gets past the problem. It interpolates not over the polygon but, after triangulation, over each triangle separately.

Exercise **12.5. (P**rogramming**)** Change the polygon specification in `loadTextures.cpp` to map a five-sided polygon in world space to a five-sided polygon in texture space (Block 6):

```
glBegin(GL_POLYGON);
```

```
    glTexCoord2f(0.0, 0.0); glVertex3f(-10.0, -10.0, 0.0);
    glTexCoord2f(1.0, 0.0); glVertex3f(10.0, -10.0, 0.0);
    glTexCoord2f(1.0, 0.5); glVertex3f(20.0, 0.0, 0.0);
    glTexCoord2f(0.5, 1.0); glVertex3f(0.0, 10.0, 0.0);
    glTexCoord2f(0.0, 1.0); glVertex3f(-10.0, 0.0, 0.0);
glEnd();
```

Can you make out the triangulations in world and texture space, as well as the correspondence between triangles?

12.2 Repeating and Clamping Textures

So far we've been careful to keep texture coordinates in the range $[0, 1]$, along both the s and t axes. What happens if we slip outside? Let's find out.

Figure 12.7: Screenshot of Experiment 12.6.

Experiment 12.6. Restore the original `loadTextures.cpp` and change the texture coordinates of the polygon as follows (Block 7):

```
glBegin(GL_POLYGON);
    glTexCoord2f(-1.0, 0.0); glVertex3f(-10.0, -10.0, 0.0);
    glTexCoord2f(2.0, 0.0); glVertex3f(10.0, -10.0, 0.0);
    glTexCoord2f(2.0, 2.0); glVertex3f(10.0, 10.0, 0.0);
    glTexCoord2f(-1.0, 2.0); glVertex3f(-10.0, 10.0, 0.0);
glEnd();
```

It seems that the texture space is *tiled* using the texture. See Figure 12.7.

In particular, the texture seems repeated in every unit square of texture space with integer vertex coordinates. As the world-space polygon is mapped to a 3×2 rectangle in texture space, it is painted with six copies of the texture, each scaled to an aspect ratio of 2:3. The scheme itself is indicated Figure 12.8. **End**

Experiment 12.7. Change the texture coordinates again by replacing each -1.0 with -0.5 (Block 8):

```
glBegin(GL_POLYGON);
    glTexCoord2f(-0.5, 0.0); glVertex3f(-10.0, -10.0, 0.0);
    glTexCoord2f(2.0, 0.0); glVertex3f(10.0, -10.0, 0.0);
    glTexCoord2f(2.0, 2.0); glVertex3f(10.0, 10.0, 0.0);
    glTexCoord2f(-0.5, 2.0); glVertex3f(-10.0, 10.0, 0.0);
glEnd();
```

Again it's apparent that the texture space is tiled with the specified texture and that the world-space polygon is painted over with its rectangular image in texture space. **End**

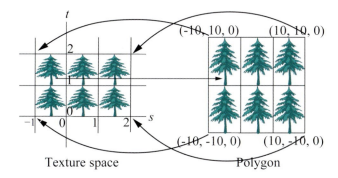

Texture space   Polygon

Figure 12.8: Tiling of texture space. The curved bold black arrows indicate the texture map. The straight arrow indicates the painting of one tile onto a sub-rectangle of the polygon; other tiles similarly paint corresponding sub-rectangles.

That the texture space is tiled with the texture is because of the following two statements in both the `loadExternalTextures()` and `loadProcedural-Textures()` routines of `loadTextures.cpp`, specifying the *wrapping mode* to be *repeated* along both s and t directions:

```
glTexParameteri(GL_TEXTURE_2D, GL_TEXTURE_WRAP_S, GL_REPEAT);
glTexParameteri(GL_TEXTURE_2D, GL_TEXTURE_WRAP_T, GL_REPEAT);
```

There is another option for the wrapping mode, instead of repeated, namely, *clamped*.

Experiment 12.8. Restore the original `loadTextures.cpp` and then change the texture coordinates as below, which is the same as in Experiment 12.6 (Block 7):

```
glBegin(GL_POLYGON);
    glTexCoord2f(-1.0, 0.0); glVertex3f(-10.0, -10.0, 0.0);
    glTexCoord2f(2.0, 0.0); glVertex3f(10.0, -10.0, 0.0);
    glTexCoord2f(2.0, 2.0); glVertex3f(10.0, 10.0, 0.0);
    glTexCoord2f(-1.0, 2.0); glVertex3f(-10.0, 10.0, 0.0);
glEnd();
```

Next, replace the **GL_REPEAT** parameter in the

```
glTexParameteri(GL_TEXTURE_2D, GL_TEXTURE_WRAP_S, GL_REPEAT);
```

statement of both the `loadExternalTextures()` and `loadProcedural-Textures()` routines with **GL_CLAMP** so that it becomes

```
glTexParameteri(GL_TEXTURE_2D, GL_TEXTURE_WRAP_S, GL_CLAMP);
```

This causes the wrapping mode to be clamped in the s-direction. It's probably easiest to understand what happens in this mode by observing in particular the chessboard texture: see Figure 12.9. Texture s coordinates

Figure 12.9: Screenshot from Experiment 12.8.

475

greater than 1 are clamped to 1, those less than 0 to 0. Precisely, instead of the texture space being tiled with the texture, points with coordinates (s, t), where $s > 1$, obtain their color values from the point $(1, t)$, while those with coordinates (s, t), where $s < 0$, obtain them from $(0, t)$. **End**

Experiment 12.9. Continue the previous experiment by clamping the texture along the t-direction as well. In particular, replace the `GL_REPEAT` parameter in the

```
glTexParameteri(GL_TEXTURE_2D, GL_TEXTURE_WRAP_T, GL_REPEAT);
```

statement with `GL_CLAMP`. We leave the reader to parse the output. **End**

The repeating option is appropriate to tile the surface of an object with a particular color pattern, e.g., a wall with a brick pattern, a table with a wood grain pattern, the ground with a grass pattern and so on, while the clamping option is appropriate to paint on a single copy of the texture, e.g., the facade of a building onto a rectangle which is situated next to a street scene.

12.3 Filtering

Experiment 12.10. Run `fieldAndSky.cpp`, where a grass texture is tiled over a horizontal rectangle and a sky texture clamped to a vertical rectangle. There is the added functionality of being able to transport the camera over the field by pressing the up and down arrow keys. Figure 12.10 shows a screenshot.

Figure 12.10: Screenshot of `fieldAndSky.cpp`.

As the camera travels, the grass seems to *shimmer* – *flash* and *scintillate* are terms also used to describe this phenomenon. This is our first encounter with the *aliasing* problem in texturing. Any visual artifact that arises owing to the finite resolution of the display device and the correspondingly "large" size of the individual pixels – at least to the extent that individual ones are discernible by the human eye – is said to be caused by aliasing. **End**

Let's try and understand why shimmer is caused by aliasing. Recall that the texture map is obtained by interpolating through each triangle the map from its vertices to points in texture space specified by texture coordinates. Subsequently, each point of the triangle is colored with the values of its mapped texture point. However, a technicality arises at this stage we did not consider earlier. Color values in the computer are *not* associated per *point*, either in texture space or the polygon. In reality, they are associated one set (RGB or RGBA) per *pixel* in the display, as also one set per *texel* in the texture.

Now, once the polygon has been rasterized – i.e., its set of corresponding pixels determined – the texture map is unlikely to map pixels to texels in a one-to-one manner. The situation, more typically, is as depicted in

Figure 12.11, where the triangle $v_1v_2v_3$ in world space is mapped to the raster triangle $v_1'v_2'v_3'$ – or screen space triangle, if you like, as it is the one to be displayed – via the rasterization process and to the texture space triangle $v_1''v_2''v_3''$ via the texture map. These two maps (downwards in the figure) induce the "real" texture map from raster to texture space (left to right) which takes pixels to texels.

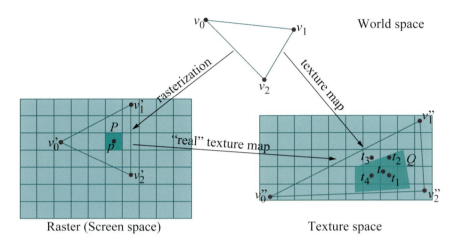

Figure 12.11: The aliasing problem in texture mapping. A single pixel P is mapped to a quadrilateral Q covering many texels (minification).

The bold pixel P in the raster maps to the bold quadrilateral Q in texture space (mind that the texture map need not preserve shape). As Q intersects multiple texels, how should OpenGL choose color values for P? Particularly, which texel should OpenGL pick to apply its particular color values to P?

Here's a reasonable solution: if the texture map takes the center p of P to the point t in texture space, then choose the texel whose center is nearest to t. In Figure 12.11, then, the chosen texel is centered at t_1, so this texel's colors would be applied P. In fact, this is precisely the so-called *filtering* option specified by the GL_NEAREST value in the following two parameter-setting commands in the statement blocks that bind the grass and sky textures of the loadExternalTextures() routine of fieldAndSky.cpp:

```
glTexParameteri(GL_TEXTURE_2D, GL_TEXTURE_MIN_FILTER, GL_NEAREST);
glTexParameteri(GL_TEXTURE_2D, GL_TEXTURE_MAG_FILTER, GL_NEAREST);
```

(We'll discuss the difference between MIN and MAG filters momentarily.)

The reason for the shimmer observed in fieldAndSky.cpp is now not hard to grasp. See again Figure 12.11. Suppose that the object in world space moves a small distance so that its rasterization changes a small amount as well, causing the map from raster to texture space to map p to a new point just to the left of t, closer to the pixel center t_4 than t_1. Correspondingly, according to the GL_NEAREST filtering principle, there is a switch in the color values at pixel P, obtained now from the texel centered at t_4, rather than the one at t_1. It's exactly these relatively large discrete changes in pixel colors arising from minute movements of the object which cause shimmer.

Figure 12.11 itself suggests a way to ameliorate the problem: instead of obtaining color values from just the one texel centered at t_1, take an average of the values at the four texels whose centers (t_1, t_2, t_3 and t_4) surround t. This smooths the color transitions and, in fact, is offered by OpenGL as the GL_LINEAR filtering option.

Remark 12.1. The basis for linear filtering is exactly the same as for moving averages in statistics. For example, as part of analyzing the stock market, one may chart average values over a sliding window of size one week or month, instead of daily values, in order to smooth out near-term fluctuations.

Experiment 12.11. Change to linear filtering in fieldAndSky.cpp by replacing every GL_NEAREST with GL_LINEAR. The grass still shimmers though less severely. The sky seems okay with either GL_NEAREST or GL_LINEAR. We'll see next even more powerful filtering options that almost eliminate the problem. **End**

The process of selecting color values for pixels based on the texture map is called filtering. OpenGL offers a few different filtering options, in addition to GL_NEAREST and GL_LINEAR, allowing the user to trade between speed and output quality. OpenGL makes a distinction, as well, between *minification* and *magnification* when filtering. Minification occurs when a pixel is mapped onto multiple texels as in Figure 12.11, while magnification is when many pixels map onto a single texel as in Figure 12.12.

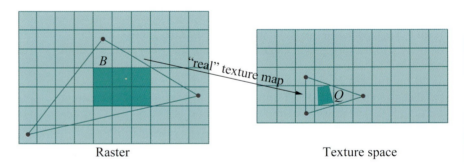

Raster   Texture space

Figure 12.12: A block B of many pixels is mapped to a quadrilateral Q inside a single texel (magnification).

Remark 12.2. The term minification arises because the phenomenon of a pixel being mapped to many texels occurs when a painted surface moves into the distance to occupy a smaller part of the screen or, equivalently, when the viewer zooms out. For example, as an aircraft flies away from the camera, the texels comprising its logo occupy an increasingly smaller region of the screen, until, when the craft is far enough, multiple texels occupy single pixels (which is equivalent to texture map taking a single pixel to multiple texels). Magnification, related to zooming in, is, of course, the inverse phenomenon.

The statement

```
glTexParameteri(GL_TEXTURE_2D, case, filter)
```

causes the filtering option *filter* to be applied in the case of minification or magnification, according as the value of *case* is GL_TEXTURE_MIN_FILTER or GL_TEXTURE_MAG_FILTER. With the commands

```
glTexParameteri(GL_TEXTURE_2D, GL_TEXTURE_MIN_FILTER, GL_NEAREST);
glTexParameteri(GL_TEXTURE_2D, GL_TEXTURE_MAG_FILTER, GL_NEAREST);
```

fieldAndSky.cpp asks for the GL_NEAREST filter in the case of either minification or magnification.

In the case of minification, particularly, OpenGL offers an assortment of efficient filtering options based on *pre-assigning* a set of textures to be used at different levels of minification. The idea, conceived by Lance Williams [144], is clever yet simple. Starting with the original texture, the *base texture* as it is called, a set of textures of progressively lower resolution, called *mipmaps*, is prepared. These mipmaps are either computed by OpenGL by an averaging process – we'll see momentarily how – or supplied by the programmer. Subsequently, during run-time, OpenGL maps a geometric primitive, based upon the size it occupies in the raster, to that particular mipmap which affords a nearly one-to-one correspondence between pixels and texels, rather than the one-to-many which would occur if the base texture were used. This (a) saves on run-time filtering computation, and (b) assures quality (provided mipmaps are initially well-chosen).

Figure 12.13 illustrates the idea with an idealized example. The base texture is of resolution 8×4 with a single scalar color value at each texel. Mipmaps of successively lower resolution till 2×1 are computed by averaging the color values in 2×2 squares of texels; finally, the 1×1 mipmap is computed by averaging the two color values in the 2×1 mipmap. For example, the value 3 in the bold texel in the 4×2 mipmap is the average of the four values in the bold square of texels in the 8×4 mipmap.

Now, when the triangle primitive shown is mapped to the base texture, minification causes the shaded pixel to map to the shaded square of texels, which requires run-time linear filtering to return the color value 3. On the other hand, if it is mapped to the 4×2 mipmap, then there is no minification and the color value of 3 is returned *without* run-time filtering. For this reason mipmaps are often called *pre-filtered* textures.

The idealized situation described above is not far removed from what actually happens in OpenGL. If a base texture of resolution $2^m \times 2^n$ is

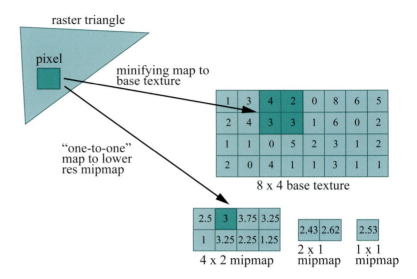

Figure 12.13: Mipmapping.

to be mipmapped, then OpenGL requires mipmaps of resolution $2^{m-1} \times 2^{n-1}$, $2^{m-2} \times 2^{n-2}, \ldots$, obtained by halving both width and height, until one of the dimensions becomes 1; subsequently, if the other dimension is still greater than 1, then it must be repeatedly halved and mipmaps provided for each resolution down to 1×1.

Exercise 12.6. If the base texture is 4×8 with color values at the texels as follows

1	0	4	2
3	2	1	5
0	1	2	6
8	2	7	7
2	3	1	2
6	4	3	8
7	3	6	1
3	5	0	2

then find all the mipmaps down to the one of lowest resolution.

Example 12.1. What is the total space required to store all the mipmaps for a base texture of resolution $2^m \times 2^n$? What is the ratio of this space to that required for only the base texture?

Answer: Suppose without loss of generality that $m \geq n$. The total number of texels in all the mipmaps is

$$2^m \times 2^n + 2^{m-1} \times 2^{n-1} + \ldots + 2^{m-n+1} \times 2 + 2^{m-n} \times 1$$
$$+ 2^{m-n-1} + 2^{m-n-2} + \ldots + 1$$

(the first line above bringing one dimension down to 1, the second the next)

$$= 2^{m-n}(1 + 2^2 + \ldots + 2^{2n}) + (1 + 2 + \ldots + 2^{m-n-1})$$
$$= 2^{m-n}(2^{2n+2} - 1)/3 + 2^{m-n} - 1$$
$$= \frac{4}{3} 2^{m+n} + \frac{2}{3} 2^{m-n} - 1$$

The space required is, therefore, the above quantity multiplied by the number of bits per texel.

Now, the number of texels in the base texture is $2^m \times 2^n = 2^{m+n}$ and

$$\frac{4}{3} 2^{m+n} + \frac{2}{3} 2^{m-n} - 1 < \frac{4}{3} 2^{m+n} + \frac{2}{3} 2^{m+n} = 2 \times 2^{m+n}$$

so the ratio of the space required for all the mipmaps to that for the base texture is less than 2.

The preceding example says that mipmapping offers efficiency and quality at a cost of only twice the amount of space. Once mipmaps have been set, OpenGL has four filtering options, in addition to GL_NEAREST and GL_LINEAR, available for use in case of minification. In order of increasing quality *and* computational cost they are:

(1) GL_NEAREST_MIPMAP_NEAREST: Applies the mipmap that's a closest fit resolution-wise to the rasterized primitive and then uses the GL_NEAREST filtering option within that mipmap.

(2) GL_LINEAR_MIPMAP_NEAREST: Applies the mipmap that's a closest fit resolution-wise to the rasterized primitive and then the GL_LINEAR filtering option within that mipmap.

(3) GL_NEAREST_MIPMAP_LINEAR: Finds the two mipmaps that are closest resolution-wise to the rasterized primitive, then uses the GL_NEAREST filtering option within either mipmap to produce two sets of color values and, finally, takes a weighted average of the two sets.

(4) GL_LINEAR_MIPMAP_LINEAR: Finds the two mipmaps that are closest resolution-wise to the rasterized primitive, then uses the GL_LINEAR filtering option within either mipmap to produce two sets of color values and, finally, takes a weighted average of the two sets.

Mipmaps are *not* used in the case of magnification because, with the viewer zooming in, one wants only the highest resolution, namely, the base texture. Accordingly, the only two filters available in the case of magnification are GL_NEAREST and GL_LINEAR.

Remark 12.3. If there is ambiguity as to which filter to apply – mappings can arise with aspects of both minification and magnification – OpenGL applies an internal algorithm to make a choice between the two.

It's time to see mipmapping in action. It's most efficient to use the GLU library routine `gluBuild2DMipmaps()` to automatically generate mipmaps, as in the following program.

Experiment 12.12. Run `fieldAndSkyFiltered.cpp`, identical to `field-AndSky.cpp` except for additional filtering options. Press the up/down arrow keys to move the camera and the left/right ones to cycle through filters for the grass texture. A message at the top left tells the current filter.

Using `gluBuild2DMipmaps()` is straightforward. Simply replace the `glTexImage2D()` command with

```
gluBuild2DMipmaps(GL_TEXTURE_2D, GL_RGB, image[0]->sizeX,
                  image[0]->sizeY, GL_RGB, GL_UNSIGNED_BYTE,
                  image[0]->data);
```

to generate all the mipmaps for the base texture in `image[0]`.

The `loadExternalTextures()` routine loads the same grass image as six different textures with the min filter ranging from `GL_NEAREST` to `GL_LINEAR_MIPMAP_LINEAR`. The mag filter used is `GL_NEAREST` when the min filter is `GL_NEAREST` as well; otherwise, it's `GL_LINEAR`. The sky texture is not mipmapped.

As one sees, the more expensive filters do nearly eliminate shimmering, but at the same time tamp down possibly desirable sharpness. For example, blades of grass can be distinguished in Figure 12.14(a), where the weakest filter is applied, but not in Figure 12.14(b), which applies the strongest.

End

(a)

(b)

Figure 12.14: Screenshots of `fieldAndSkyFiltered.cpp`: (a) Weakest filter (b) Strongest filter.

$\mathcal{R}emark$ 12.4. The `gluBuild2DMipmaps()` routine does not require the width and height of its input image to be powers of two. However, it's still recommended to scale texture images to dimensions that are.

We have a couple more programs for you to experiment with mipmaps and filters.

Figure 12.15: Screenshot of `compareFilters.cpp` initially.

$\mathcal{E}xperiment$ 12.13. Run `compareFilters.cpp`, where one sees side-by-side identical images of a shuttle launch bound to a square. Press the up and down arrow keys to move the squares. Press the left arrow key to cycle through filters for the image on the left and the right arrow key to do likewise for the one on the right. Messages at the top say which filters are currently applied. Figure 12.15 is a screenshot of the initial configuration.

Compare, as the squares move, the quality of the textures delivered by the various min filters. If one of the four mipmap-based min filters – `GL_NEAREST_MIPMAP_NEAREST` through `GL_LINEAR_MIPMAP_LINEAR` – is applied, then the particular mipmaps actually used depend on the screen space occupied by the square. *End*

$\mathcal{E}xercise$ 12.7. ($\mathbf{P}$rogramming) Replace the launch image of `compare-Filters.cpp` with other ones.

$\mathcal{E}xperiment$ 12.14. Run `mipmapLevels.cpp`, where mipmaps are supplied by the program, rather than computed automatically with use of `gluBuild2DMipmaps()`. The mipmaps are very simple: just differently colored square images, starting with blue, from size 64×64 down to 1×1, created by the routine `createMipmaps()`. Commands of the form

```
glTexImage2D(GL_TEXTURE_2D, level, GL_RGB, width, height,
             0, GL_RGB, GL_UNSIGNED_BYTE, image);
```

each binds a *width* $\times$ *height* mipmap image to the current texture index, starting with the highest resolution image with *level* parameter 0, and with each successive image of lower resolution having one higher *level* all the way up to 6.

Move the square using the up and down arrow keys. As it grows smaller a change in color indicates a change in the currently applied mipmap. Figure 12.16 is screenshot after the first change. As the min filter setting is GL_NEAREST_MIPMAP_NEAREST, a unique color, that of the closest mipmap, is applied to the square at any given time. **End**

Remark 12.5. OpenGL offers options in addition to those that we have discussed to fine-tune mipmapping. The interested reader is referred to the red book.

Remark 12.6. Mipmapping is one of a class of LOD (*level-of-detail*) methods, or *multiresolution* methods as they are also called, which are important in graphics from the point of view of run-time efficiency.

Figure 12.16: Screenshot of `mipmapLevels.cpp`.

Representing objects by polygonal meshes of varying levels of refinement is another practically important LOD application and one related to the drawing methods that we studied in Chapter 10. For example, if the camera is close to a spacecraft, then one may want a "base mesh" of thousands, or even millions, of triangles to be rendered. However, after the ship has flown some distance off to occupy a smaller portion of the screen, a mesh with fewer triangles may not only be visually adequate, but, in fact, desirable both for quicker rendering and to avoid aliasing artifacts. Accordingly, it's often advantageous to pre-compute a set of meshes for a moving object, exactly as mipmaps for a texture, with varying numbers of triangles.

Instead of a spacecraft, we have a cow at three different levels of resolution in Figure 12.17, starting from the highest at left, and then simplified twice with the help of mesh simplification software [23] co-developed by the author.

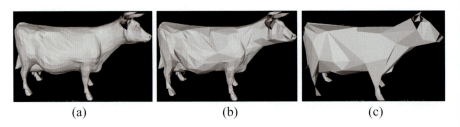

(a)   (b)   (c)

Figure 12.17: Cow at 3 different resolutions: (a) 5804 (b) 1772 (c) 328 triangles.

There's an extensive literature on LOD methods. A recent book is by Luebke et al. [85].

12.4 Specifying Texture Coordinates

Our programs so far have been simple from the point of view of specifying the texture map. The surfaces textured were all polygons, so that we simply had to specify texture coordinates at the corners. How about more general surfaces? It's actually surprisingly straightforward if the surface is parametrized – we can leverage the parametrization to derive texture coordinates.

12.4.1 Parametrized Surfaces

Experiment 12.15. Run `texturedTorus.cpp`, which shows a synthetic (red-black) chessboard texture mapped onto a torus. Figure 12.18 is a screenshot. Press space to see animation: the texture scrolls around the torus. The pace of the animation can be changed by pressing the up and down arrow keys. End

Figure 12.18: Screenshot of `texturedTorus.cpp`.

The program `texturedTorus.cpp` is based on `torus.cpp` of Experiment 10.5. As i runs from 0 to p and j from 0 to q, $(i/p,\ j/q)$ runs over sample points in $[0, 1] \times [0, 1]$, and $(\ f(i, j),\ g(i, j),\ h(i, j)\ )$ – see the corresponding function definitions `f()`, `g()` and `h()` in the program – over mapped sample points on the torus. Since the (i, j)th entry in the vertex array is the image of the point $(i/p, j/q)$ of the parameter rectangle $[0, 1] \times [0, 1]$, an obvious texture map is to associate this very same image with the point $(i/p, j/q)$ of texture space, effectively identifying the parameter rectangle with the texture! See Figure 12.19. This is exactly what's done in `texturedTorus.cpp` by the routine `fillTextureCoordArray()`, which fills values into a texture coordinates array.

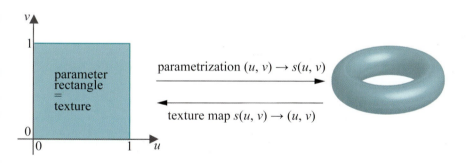

Figure 12.19: Texturing a torus by identifying the parameter rectangle with the texture.

The animation is effected by cycling the variable `shift` of `fillTexture-CoordArray()` through the values $0, 0.01, 0.02, \ldots, 1$, thereby cycling the t-values of the texture map as well within the interval $[0, 1]$.

Exercise 12.8. (Programming) The length of each red and black sector on the surface of the torus of `texturedTorus.cpp` is much more than its width. Make the aspect ratio more even.

Exercise 12.9. (Programming) If you did Exercise 12.2 to create the synthetic striped board texture of Figure 12.2, apply it now to the torus of `texturedTorus.cpp`.

Exercise 12.10. (Programming) Texture the helical pipe of Experiment 10.4 to give it an appearance of rusted metal.

Exercise 12.11. (Programming) Texture the table of Experiment 10.7 with a wood grain texture.

Exercise 12.12. (Programming) Animate the sky of `fieldAndSky.cpp` to make the clouds move. A possibility is to map the vertical world-space rectangle to only a part of a unit square in the sky's texture space and then move it slowly around in that square.

12.4.2 Bézier and Quadric Surfaces

It's fairly simple to use OpenGL to automatically generate texture coordinates for Bézier and quadric surfaces, as is demonstrated in the next experiment.

Experiment 12.16. Run `texturedTorpedo.cpp`, which textures parts of the torpedo of `torpedo.cpp` – from Experiment 10.21 – as you can see in the screenshot in Figure 12.20.   **End**

Figure 12.20: Screenshot of `texturedTorpedo.cpp`: propeller blades textured with the chessboard, body with stripes.

The texturing of the propeller blade Bézier surface of `textured-Torpedo.cpp` is most important. We want to do this in the same manner as the torus of Experiment 12.15 – by identifying its parameter rectangle in particular with the synthetic chessboard texture. However, because of the way OpenGL is set up for texture coordinate generation we are forced to a slightly roundabout approach – first, we have to create a Bézier surface s' in *texture space*. The statement

```
glMap2f(GL_MAP2_TEXTURE_COORD_2, 0, 1, 2, 2, 0, 1, 4, 2,
        texturePoints[0][0])
```

in the display list for the propeller blade does just this. The syntax of this statement is similar to that of the `glMap2f(GL_MAP2_VERTEX_3, ...)` we are already familiar with to create a world-space Bézier surface. Control points of s' stored in the `texturePoints` array are $(0,0)$, $(0,1)$, $(1,0)$ and $(1,1)$, making s' the rectangle $[0,1] \times [0,1]$. See Figure 12.21. Moreover, the bilinearity of the parametric mapping of s' (it's of order 2, or degree 1, along both u and v) implies that it is simply the identity map from the rectangle $[0,1] \times [0,1]$ in parameter space to s' in texture space.

Now, OpenGL takes the coordinates of the image on the texture-space Bézier surface of each parameter point (u, v) to be the texture coordinates of (u, v)'s image on the world-space surface, as depicted in Figure 12.21. Given our particular texture Bézier surface s', texture coordinates (u, v) are

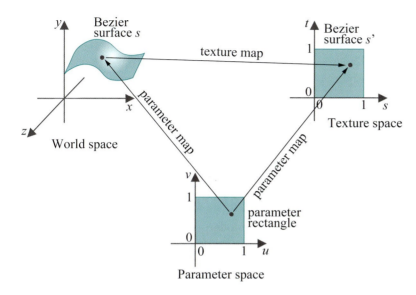

Figure 12.21: Texture mapping a Bézier surface via a Bézier surface in texture space. The parameter map on the right from parameter to texture space is the identity in case of `texturedTorpedo.cpp`.

assigned to the image on the world Bézier surface of the parameter point (u, v), effectively identifying the texture with the parameter rectangle as we set out to do.

The two statements

```
glEnable(GL_MAP2_TEXTURE_COORD_2);
glMapGrid2f(5, 0.0, 1.0, 5, 0.0, 1.0);
```

in the propeller blade's display list actually instigate texture coordinate generation at the image on the world-space Bézier surface – the propeller blade in this case – of an evenly-spaced 5×5 grid of sample points in the parameter domain.

OpenGL's particular mechanism to generate texture coordinates for a Bézier surface offers flexibility. For example, by changing the control points of the texture-space surface and so its extent, one can paint the real-world surface with a different region of texture space. The next exercise asks you to try this.

Exercise 12.13. (Programming) Make the red and blue squares of the chessboard pattern on the propeller blades of `texturedTorpedo.cpp` twice as big are they are currently, without changing the original texture as generated by the program.

Getting OpenGL to generate texture coordinates for a quadric is even simpler. The statement gluQuadricTexture(qobj, GL_TRUE)

in the `drawScene()` routine of `texturedTorpedo.cpp` is all that it takes to automatically generate texture coordinates for the quadric torpedo body.

Exercise 12.14. (Programming) Texture the parts of the torpedo of `texturedTorpedo.cpp` which are still wire mesh.

12.5 Lighting Textures

By now the reader may be wondering if color and light can coexist with texture or if the two are mutually exclusive ways of adorning an object. The answer is that OpenGL, in fact, offers a few different options to combine them. An option is selected using the following texture function statement, most often located in the initialization routine:

glTexEnvf(GL_TEXTURE_ENV, GL_TEXTURE_ENV_MODE, *parameter*)

If *parameter* is **GL_REPLACE**, as it has been in our programs thus far, then the colors used to render each primitive are derived *solely* from its texture. The primitive's own material color as well as light sources in the environment are ignored.

The most common way to combine color and light with texture is by setting *parameter* to **GL_MODULATE**, in which case OpenGL does the following:

(a) Computes RGB values at a primitive's vertices using OpenGL's lighting equation (11.12) and interpolates these through its interior – assuming that smooth shading is on – to determine the RGB values at each of its pixels.

(b) Uses the texture map to obtain RGB values from the texture at each of the primitive's pixels.

(c) Determines the final RGB values at each pixel as the *product* of the corresponding values from the preceding two steps.

In short, OpenGL *separately* computes RGB values for color and light as if there were no texture and RGB values for texture as if there were no color and light and, finally, scales one with the other.

Example 12.2. If the RGB tuple at a pixel P is $(0.5, 0.75, 0.1)$ as obtained by interpolation from vertex RGB values computed after lighting, while that determined at P from the texture via the texture map is $(0.4, 0.5, 1.0)$, then the final color applied to P using the **GL_MODULATE** option is $(0.5 \times 0.4, 0.75 \times 0.5, 0.1 \times 1.0) = (0.2, 0.375, 0.1)$.

Experiment 12.17. Run `fieldAndSkyLit.cpp`, which applies lighting to the scene of `fieldAndSky.cpp` with help of the `GL_MODULATE` option. The light source is directional – imagine the sun – and its direction controlled using the left and right arrow keys, while its intensity can be changed using the up and down arrow keys. A white line indicates the direction and intensity of the sun. Figure 12.22(a) is a screenshot.

The material colors are all white, as is the light. The normal to the horizontal grassy plane is vertically upwards. Strangely, we use the same normal for the sky's vertical plane, because using its "true" value toward the positive z-direction has the unpleasant, but expected, consequence of a sky that doesn't darken together with land. **End**

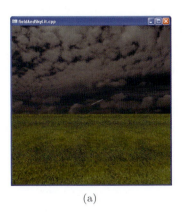
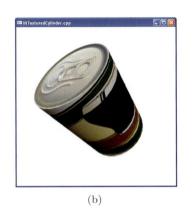

(a)   (b)

Figure 12.22: Screenshots of (a) `fieldAndSkyLit.cpp` and (b) `litTextured-Cylinder.cpp`.

Experiment 12.18. Run `litTexturedCylinder.cpp`, which adds a label texture and a can top texture to `litCylinder.cpp`. Figure 12.22(b) is a screenshot.

Most of the program is routine – the texture coordinate generation is, in fact, a near copy of that in `texturedTorus.cpp` – except for the following lighting model statement which we're using for the first time:

```
glLightModeli(GL_LIGHT_MODEL_COLOR_CONTROL, GL_SEPARATE_SPECULAR_COLOR)
```

We had briefly encountered this statement as an OpenGL lighting model option in Section 11.4. It causes a modification of OpenGL's `GL_MODULATE` procedure: the specular color components are separated and not multiplied with the corresponding texture color components, as are the ambient and diffuse, but added in after. The result is that specular highlights are preserved rather than blended with the texture. **End**

Exercise 12.15. (Programming) Close off the bottom of the cylinder of `litTexturedCylinder.cpp` with a metal-textured disc.

Exercise 12.16. (Programming) Animate a lit textured flag fluttering in the wind.

Hint: A surface whose section is of the form $y = \sin(x + t)$, where t depends on time, in other words a "moving" sine curve, simulates fluttering.

Exercise 12.17. (Programming) Continue improvement of `ship-Movie.cpp`, which you began in Exercise 11.38, now with the help of textures. There are numerous possibilities of which a few are:

(a) Texture the black back plane with the image of a night-time city skyline.

(b) Paint the surface of the sea with a water texture, possibly animated.

(c) Add detail to the ship by texturing it with images of parts of real ships.

(d) Texture the background boats. You may want to strategically place additional light sources.

Exercise 12.18. (Programming) Continue with Exercise 11.39 where you enhanced `animateMan*.cpp` with color and light, now using texture.

12.6 Summary, Notes and More Reading

Texturing, the process of painting an image onto the surface of an object, is of great practical importance in computer graphics. In this chapter we learned the basics of how to apply a texture, underlying principles of the texturing process and a fair number of techniques to effectively manage textures, one being how to combine them with lighting.

The seminal reference for textures in CG is Heckbert [66]. The article by Haeberli and Segal [61] is easily-readable and informative as well. More advanced CG books, e.g., the ones by Akenine-Möller, Haines & Hoffman [1] and Watt [142], all have sections on texturing that will take the reader beyond what she has learned in this chapter. The book by Reynolds and Blythe [88] is a good reference for texturing in OpenGL in particular. Texturing techniques are a field of active research and, as for CG research in general, the place to visit for the latest developments is the annual ACM SIGGRAPH conference [125].

CHAPTER 13

Special Visual Techniques

nd now for special effects! Special visual techniques are the topic of the seven sections of this chapter. These include blending, fogging, billboarding, antialiasing, environment mapping, stencil buffer techniques and bump mapping.

The goal of blending objects by combining their color values, the topic of Section 13.1, is primarily to engender translucency; however, blending can also be used in effects such as reflection and morphing. Section 13.2 shows how to use fog to cue the viewer to the distance of an object. This imparts realism to a scene, as does the technique of billboarding introduced in Section 13.3, a cost-effective way to create the illusion of a 3D object by means of its 2D image. We'll learn in Section 13.4 how to antialias straight lines in order to remove jaggedness in their rendering. Environment mapping, discussed in Section 13.5, enables a shiny object to reflect its surroundings, again making more realistic the rendering of things such as gleaming teapots and rocket nose cones. The stencil buffer, described in Section 13.6, which allows drawing only to selected regions of the display, is useful in creating certain interesting effects. Bump mapping, which we'll study in Section 13.7, is a technique to add the illusion of detail to an object by altering its normals, but without changing its geometry. Section 13.8 concludes the chapter.

13.1 Blending

We spend a little time first assimilating the theory of blending before putting it into practice.

13.1.1 Theory

Our plan is to understand first how OpenGL operates without blending, and then with.

No Blending

Consider the following piece of pseudo-code and how OpenGL would process it without blending:

```
glClear(GL_COLOR_BUFFER_BIT | GL_DEPTH_BUFFER_BIT);
draw triangle T;
draw quad Q;
```

See Figure 13.1. The z-values in the depth buffer are each initialized to a very large value by the `glClear()` command, as are values in the color buffer all to the clearing color white. This initial configuration is depicted in the upper left grid, where pixels only in a region of interest are labeled with their z-values, unlabeled pixels all having ∞ z-value as well.

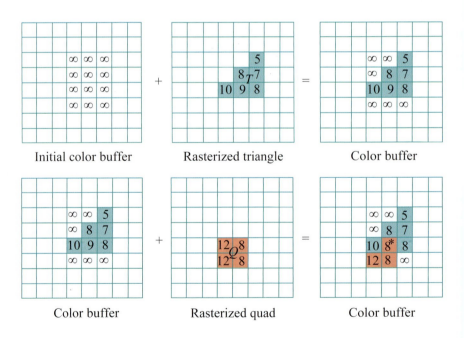

Initial color buffer Rasterized triangle Color buffer

Color buffer Rasterized quad Color buffer

Figure 13.1: Assuming depth testing on: T rasterized and rendered (upper row), followed by Q rasterized and rendered (lower row). The starred pixel is considered in an example below.

The triangle T is then *rasterized*, i.e., the set of pixels corresponding to T determined, and color and z-values computed for each – the upper middle grid using hypothetical z-values. A raster pixel, together with its color values and z-value, is called a *fragment*.

OpenGL next *renders* T to the color buffer, which will be viewed when flushed to the screen, according to sets of rules depending on whether depth testing is on or not.

If *depth testing is enabled*, then the process is two-step:

1. The z-value of each of T's fragments, called a *source fragment*, is compared with that in the corresponding pixel, called the *destination pixel*, in the color buffer.

2. If the source fragment's z-value is less than that of the destination pixel, then its color values *overwrite* the current color values of the destination pixel and its z-value overwrites the current z-value, as well; if the source fragment's z-value is not less than that of the destination pixel, then its color values and z-value are discarded.

    Note: That its z-value be less than that in the destination pixel is the default test that a source fragment has to pass in order to overwrite the destination. Other tests can be invoked with a call to `glDepthFunc()` (see the red book for a listing of possible tests).

If *depth testing is not enabled*, then the process is a simple single step:

1. Each one of T's fragments unconditionally overwrites the color values of its destination pixel. The z-values are not relevant and never change.

Given depth testing as enabled, the result of rendering T is depicted in the upper right grid in Figure 13.1, as also in the lower left. Following T, Q is rasterized (lower middle) and rendered (lower right) in an identical manner, Q replacing T in the two-step procedure above. One sees that three of Q's fragments pass the depth test and overwrite their destination pixels.

The Difference with Blending

Next, let's understand what happens *with* blending. The differences are precisely the following:

1. In all the cases above – without blending, that is – where a source fragment's color values are supposed to overwrite those of its destination pixel, OpenGL instead *combines* the two sets and applies the result to the destination pixel. We'll discuss momentarily how color values are, in fact, combined.

2. In all the cases above, where the source fragment does not overwrite its destination pixel – e.g., if depth testing is on and its z-value is greater – then it's discarded as before.

3. As for the z-values, only if depth testing is on does a lower source z-value replace that of the destination, *regardless* of blending.

Here's how OpenGL combines color values in order to blend. Say the source fragment's color values are $(source_R, source_G, source_B)$, while those of the destination $(dest_R, dest_G, dest_B)$, a total of six scalar values. Based upon the programmer's specifications, OpenGL assigns a *blending factor* to each of these six scalars, which we'll denote with an additional superscript b (e.g., the blending factor of $source_R$ is denoted $source_R^b$). OpenGL then determines the final color values of the destination pixel using the following *blending equation*:

$$dest_X = source_X^b * source_X + dest_X^b * dest_X \qquad (13.1)$$

where X may be any of RGB. In other words, for each primary color a weighted sum of the source and destination values is used, weights being the respective blending factors. If a color value computed by the blending equation exceeds 1, then it is clamped to 1.

Example 13.1. Consider the starred pixel in the lower right grid of Figure 13.1. Say its RGB color values prior to the rendering of Q, in particular, those of the corresponding pixel in the triangle T in the lower left grid, are $(0.6, 0.4, 0.2)$, while those of the corresponding pixel of the quad in the lower middle are $(0.5, 0.5, 0.5)$. Suppose as well that

$$source_R^b = source_G^b = source_B^b = 0.3 \quad \text{and} \quad dest_R^b = dest_G^b = dest_B^b = 0.7$$

and that depth testing is on. What are the final color values of the starred pixel if blending is not enabled? If it is?

Answer: Suppose, first, that blending is off. The source fragment's z of 8 being less than the destination's 9, it overwrites the destination's colors. Final color values of the starred pixel are, therefore, $(0.5, 0.5, 0.5)$.

Next, suppose blending is enabled. Since the source would overwrite the destination colors if blending were off, given that blending is, in fact, on, their color values are combined instead. The blending equation gives the resulting values as

$$(0.3*0.5+0.7*0.6, \; 0.3*0.5+0.7*0.4, \; 0.3*0.5+0.7*0.2) = (0.57, 0.43, 0.29)$$

The resulting z of 8 for the starred pixel is the same, though, both in blended and unblended applications, as it is determined only by competition in the depth buffer.

Exercise 13.1. How about the pixel just to the left of the starred one? What are its color values with blending enabled, assuming that the color values in the corresponding pixels in the left and middle grids are the same as for the starred one in the preceding example and that blending factors are identical as well? Consider when depth testing is both on and off.

Alpha

It's in assigning the blending factors that the programmer can make use of the *alpha* values, the A in RGBA. For example, the *blend function* command

```
glBlendFunc(GL_SRC_ALPHA, GL_ONE_MINUS_SRC_ALPHA)
```

causes OpenGL to set

$$source_X^b = source_A \quad \text{and} \quad dest_X^b = 1 - source_A$$

for X equal to each of RGB, where $source_A$ denotes the source fragment's alpha value. The consequence of this particular blend function is that the greater the source's alpha the more its contribution to the final color or, intuitively, the more *opaque* it is.

In general, blending factors are set by calling

```
glBlendFunc(srcFactor, destFactor)
```

where the values of the parameters *srcFactor* and *destFactor* tell OpenGL how to determine the source and destination blending factors, respectively. We've already seen that the value GL_SRC_ALPHA chooses the source alpha value $source_A$ for each of RGBA, while GL_ONE_MINUS_SRC_ALPHA chooses $1 - source_A$. GL_DST_ALPHA and GL_ONE_MINUS_DST_ALPHA, likewise, choose $dest_A$ and $1 - dest_A$, respectively. The reader is referred to the red book for a full list of possible values for *srcFactor* and *destFactor*. Any of these values may be used for either parameter, e.g., even GL_SRC_ALPHA for *destFactor*, which would be an odd choice indeed.

It is interesting that the alpha value $dest_A$ of the destination pixel is computed, too, from those of the source and destination using an equation similar to (13.1):

$$dest_A = source_A^b * source_A + dest_A^b * dest_A \qquad (13.2)$$

where $source_A^b$ and $dest_A^b$ are the source and destination alpha blending factors, respectively. Moreover, the blend function command

```
glBlendFunc(GL_SRC_ALPHA, GL_ONE_MINUS_SRC_ALPHA)
```

causes OpenGL to set

$$source_A^b = source_A \quad \text{and} \quad dest_A^b = 1 - source_A$$

as well.

Remark 13.1. The blending equation (13.1), though the default and most commonly used, isn't the only option OpenGL provides to combine source and destination color values. Other formulae are available as well, which can be chosen by a call to glBlendEquation(). Refer to the red book for a list.

13.1.2   Experiments

Experiment 13.1. Run `blendRectangles1.cpp`, which draws two translucent rectangles with their alpha values equal to 0.5, the red one being closer to the viewer than the blue one. The *code* order in which the rectangles are drawn can be toggled by pressing space. Figure 13.2 shows screenshots of either order.   **End**

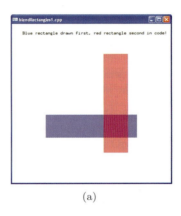

(a)   (b)

Figure 13.2: Screenshot of `blendRectangles1.cpp` with (a) the blue rectangle first in code (b) the red rectangle first in code.

Blending is enabled in `blendRectangles1.cpp` with the call

    glEnable(GL_BLEND)

and the blend function used is

    glBlendFunc(GL_SRC_ALPHA, GL_ONE_MINUS_SRC_ALPHA)

As depth testing is currently disabled, if there were no blending, then the rectangle second in the code would overwrite the one drawn first. Therefore, from our understanding of theory now, with blending on, their colors are combined.

As $source_A = 1 - source_A = 0.5$, no matter which rectangle is drawn first, the blending equation is

$$dest_X = 0.5 * source_X + 0.5 * dest_X$$

which is symmetric in source and destination. So one may wonder why the one drawn second dominates. The reason is that the rectangle drawn first is blended with the background black, diluting its positive color value (blue for the blue rectangle, red for the red) to 0.5, while the second-drawn rectangle comes in at "full strength".

Exercise 13.2. (Programming) Change the alpha values of both rectangles in `blendRectangles1.cpp` successively to 0.0, 0.25, 0.75 and 1.0. Explain what you see.

Exercise 13.3. (Programming) Although there is a depth buffer in `blendRectangles1.cpp`, depth testing has been disabled. Enable it by replacing the call `glDisable(GL_DEPTH_TEST)` with `glEnable(GL_DEPTH_TEST)`. Explain what you observe.

Keeping depth testing enabled, disable blending by replacing the call `glEnable(GL_BLEND)` with `glDisable(GL_BLEND)`. Again, explain what you observe.

13.1.3 Opaque and Translucent Objects Together

Experiment 13.2. Run `blendRectangles2.cpp`, which draws three rectangles at different distances from the eye. The closest one is vertical and a translucent red ($\alpha = 0.5$), the next one is angled and opaque green ($\alpha = 1$), while the farthest is horizontal and a translucent blue ($\alpha = 0.5$). Figure 13.3(a) is a screenshot of the output.

The scene is clearly not authentic as no translucency is evident in either of the two areas where the green and blue rectangles intersect the red. The fault is not OpenGL's as it is rendering as it's supposed to with depth testing.

End

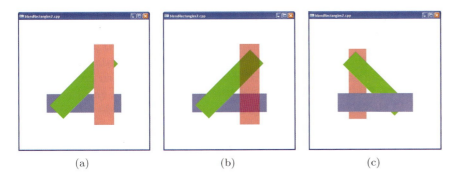

      (a)   (b)   (c)

Figure 13.3: Screenshots of `blendRectangles1.cpp`: (a) Original (b) With rectangles re-ordered to blue, green, red in the code (c) New ordering seen from the $-z$-direction.

Example 13.2. Verify the claim just made that the program is doing as it's supposed to, using our understanding of the rendering process.

Answer: The program's drawing order is:

```
drawRedRectangle(); // Red rectangle closest to viewer, translucent.
drawGreenRectangle(); // Green rectangle second closest, opaque.
drawBlueRectangle(); // Blue rectangle farthest, translucent.
```

Once the red rectangle, the closest, is drawn, the green and blue both fail the depth test where they intersect the red, so there's no translucency apparent in those two regions (keep in mind that only if the source fragment passes the depth test does blending kick in).

Returning to the experiment, disabling depth testing doesn't help either, as the green blocks out the red (which it shouldn't as it's farther away), while it doesn't block out the blue (which it should as it's closer). See Figure 13.4.

In fact, trying all 6 (=3!) possible orders to draw the rectangles, it's seen that the only one producing an authentic rendering is:

Figure 13.4: Screenshot of `blendRectangles1.cpp` with depth testing disabled.

```
drawBlueRectangle(); // Blue rectangle farthest, translucent.
drawGreenRectangle(); // Green rectangle second closest, opaque.
drawRedRectangle(); // Red rectangle closest to viewer, translucent.
```

See Figure 13.3(b). As one would expect, both green and blue rectangles can be seen through the red, while the green blocks out the blue. The order that the primitives are drawn in code happens to be according to their distance from the viewer, starting with the farthest; moreover, with this order it doesn't matter if depth testing is on or off.

Exercise 13.4. Explain why the (farthest-to-nearest) blue-green-red drawing order is successful, regardless whether depth testing is enabled or not.

Unfortunately, this method is not a particularly robust way to produce an authentic scene, as the farthest-to-nearest order depends on the viewpoint. For example, keeping the same drawing order, replace the viewing transformation

```
gluLookAt(0.0, 0.0, 3.0, 0.0, 0.0, 0.0, 0.0, 1.0, 0.0)
```

with

```
gluLookAt(0.0, 0.0, -3.0, 0.0, 0.0, 0.0, 0.0, 1.0, 0.0)
```

to view the rectangles from the $-z$-direction, which means now the blue is closest, the green next and the red farthest away. Oops! See Figure 13.3(c). The result is no longer authentic as the closest translucent blue rectangle now blocks out both red and green, which, of course, it shouldn't. The reason for the breakdown is clearly that what used to be farthest-to-nearest rendering is now no longer.

So, what is one to do? Precisely, the following:

1. Enable depth testing.

2. Draw first the opaque objects. Because of depth testing they block one another out according to distance from viewpoint, as one would want.

3. Make the depth buffer read-only with a call to `glDepthMask(GL_FALSE)`.

4. Draw next the translucent objects. As depth testing is still on, translucent objects are blocked out by nearer opaque ones, again as one would want. *However*, as they can no longer update the z-buffer to their own z-values, closer translucent objects don't block out farther ones – which is what would happen if there z-values were recorded – but blend instead. In fact, all translucent objects, which are closer than the closest opaque one, blend into one another. Exactly what the doctor ordered.

5. Restore the writability of the depth buffer with a call to `glDepthMask-(GL_TRUE)`.

Note: The depth buffer is writable by default.

Try it.

$E_{xperiment}$ **13.3.** Rearrange the rectangles and insert two `glDepth-Mask()` calls in the drawing routine of `blendRectangles2.cpp` as follows:

```
// Draw opaque objects.
drawGreenRectangle(); // Green rectangle second closest, opaque.

glDepthMask(GL_FALSE); // Make depth buffer read-only.

// Draw translucent objects.
drawBlueRectangle(); // Blue rectangle farthest, translucent.
drawRedRectangle(); // Red rectangle closest to viewer, translucent.

glDepthMask(GL_TRUE); // Make depth buffer writable.
```

Try both `gluLookAt(0.0, 0.0, 3.0, ...)` and `gluLookAt(0.0, 0.0, -3.0, ...)`. Interchange the drawing order of the two translucent rectangles as well. The scene is authentic in every instance. E_{nd}

$E_{xperiment}$ **13.4.** Run `sphereInGlassBox.cpp`, which makes the sides of the box of `sphereInBox2.cpp` glass-like by rendering them translucently. Only the unaveraged normals option of `sphereInBox2.cpp` is implemented. Press the up and down arrow keys to open or close the box and 'x/X', 'y/Y' and 'z/Z' to turn it.

The opaque sphere is drawn first and then the translucent box sides, after making the depth buffer read-only. A screenshot is Figure 13.5(a). E_{nd}

$E_{xercise}$ **13.5. ($P_{rogramming}$)** Inscribe a glass dodecahedron inside a glass icosahedron.

$E_{xercise}$ **13.6. ($P_{rogramming}$)** Make a solid ball travel through a glass helical pipe (see Experiment 10.4 for the helical pipe).

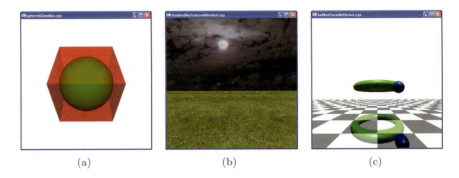

Figure 13.5: Screenshots of blended effects: (a) `sphereInGlassBox.cpp` (b) `fieldAnd-SkyTexturesBlended.cpp` (c) `ballAndTorusReflected.cpp`.

13.1.4   Blending Textures

There's no reason why textures cannot be blended – it's simply a matter of one or both of the source fragment and destination pixel obtaining color values from a texture map. Here's a fun program doing just that.

Experiment 13.5. Run `fieldAndSkyTexturesBlended.cpp`, which is based on `fieldAndSkyLit.cpp`. Press the arrow keys to move the sun. As the sun rises the night sky morphs into a day sky. Figure 13.5(b) shows late evening. The program's a fairly straightforward application of alpha blending. We point out a few interesting features:

(a) The sky rectangle is no longer lit as in `fieldAndSkyLit.cpp` because the night texture itself causes the sky to darken.

(b) Source blending factors all 1 (`GL_ONE`) and destination blending factors all 0 (`GL_ZERO`) enable the grass and night sky textures to initially paint their respective rectangles without dilution.

(c) The statements

```
if (theta <= 90.0) alpha = theta/90.0;
else alpha = (180.0 - theta)/90.0;
glColor4f(1.0, 1.0, 1.0, alpha);
```

in the drawing routine link the **alpha** value to the angle **theta** of the sun in the sky, so that the former increases from 0 to 1 as the sun rises from the horizon to vertically above.

(d) The day sky is blended into the night sky because both textures paint the same rectangle and because the prior disabling of depth testing allows an incoming fragment to write to a destination pixel, even if its z-value is equal to the current one (with depth testing on it has to be less in order to do so). The call $\texttt{glBlendFunc(GL\_SRC\_ALPHA,}$ $\texttt{GL\_ONE\_MINUS\_SRC\_ALPHA)}$ in the drawing routine sets the source blending factor equal to $\texttt{alpha}$ and the destination blending factor to $1 - \texttt{alpha}$. **End**

Remark 13.2. The simple-minded alpha-morph just described should work fairly well if the transition required is mainly in the colors of a scene and not the geometry, because it is a straight linear interpolation between corresponding source and destination color values. For the same reason, one should not expect much from it by way of morphing *shapes*, as in the Terminator movies.

Exercise 13.7. (Programming) Morph a day image of a static scene (e.g., city skyline, mountainous landscape, etc.) to a night image.

13.1.5 Creating Reflections

Another neat application of blending is to simulate reflection.

Experiment 13.6. Run $\texttt{ballAndTorusReflected.cpp}$, which builds on $\texttt{ballAndTorusShadowed.cpp}$. Press space to start the ball traveling around the torus and the up and down arrow keys to change its speed.

The reflected ball and torus are obtained by drawing them scaled by a factor of -1 in the y-direction, which creates their reflections in the xz-plane, and then blending the floor into the reflection. Figure 13.5(c) shows a screenshot. **End**

Exercise 13.8. (Programming) Draw the reflection of the ship of $\texttt{shipMovie.cpp}$ in the sea.

Exercise 13.9. (Programming) Make the character of $\texttt{animate-Man1.cpp}$ walk along a shiny reflective floor to a window. The camera should then move to the character's point of view as he looks down at a city scene which is really a single textured image.

13.2 Fog

Fog is an *atmospheric effect* that OpenGL offers ready to use. What fog does is blend objects with a programmer-specified fog color so that the farther away an object is from the viewer the more the fog color dominates, the effect being of objects fading into the distance.

We'll explain how fog is implemented in OpenGL using the program $\texttt{fieldAndSkyFogged.cpp}$ as a running example.

Experiment 13.7. Run `fieldAndSkyFogged.cpp`, which is based on our favorite workhorse program `fieldAndSky.cpp`, adding to it a movable black ball and controllable fog. Figure 13.6 is a screenshot. There's interaction as well, which we'll describe after discussing next the code and how fog is implemented. **End**

All fog-related calls in the program are near the top of the drawing routine. Fog is enabled or disabled by a call to `glEnable(GL_FOG)` or `glDisable(GL_FOG)`, respectively. The *fog color* is specified by the statement

    glFogfv(GL_FOG_COLOR, *fogColor*)

Figure 13.6: Screenshot of `fieldAndSkyFogged.cpp` with exponential fogging.

where *fogColor* points to the fog color values (a medium gray in the program).

The user can set the *fog mode* to be one of three – GL_LINEAR, GL_EXP and GL_EXP2 – by assigning that value to the parameter *fogMode* in the statement

    glFogi(GL_FOG_MODE, *fogMode*)

It is the fog mode which, together with a few associated parameters, determines the "thickness" of the fog. Here's how.

OpenGL invokes the fog mode and the z-distance of an incoming fragment from the eye to compute a number f, called the *fog factor*, which is used to blend the fragment with the fog color. The equation which determines the fog factor f depends on the fog mode as follows:

$$\begin{aligned}
\text{GL\_LINEAR:} \qquad f &= \frac{fogEnd - z}{fogEnd - fogStart} \\
\text{GL\_EXP:} \qquad f &= e^{-(fogDensity * z)} \\
\text{GL\_EXP2:} \qquad f &= e^{-(fogDensity * z)^2}
\end{aligned}$$
(13.3)

The values of the parameters *fogStart*, *fogEnd* and *fogDensity* are user-specified as well in the following statements:

    glFogf(GL_FOG_START, *fogStart*);
    glFogf(GL_FOG_END, *fogEnd*);
    glFogf(GL_FOG_DENSITY, *fogDensity*);

Their default values are 0, 1 and 1, respectively. One sees from Equations (13.3) that if fog mode is GL_LINEAR, then *fogStart* and *fogEnd* give the two endpoints of a linear ramp along the z-axis along which f decreases from 1 to 0; moreover, if fog mode is GL_EXP or GL_EXP2 then *fogDensity* controls the (exponential or doubly exponential) rate of diminishment of f with increasing z – the greater *fogDensity* the more rapidly f diminishes. See Figure 13.7 for sketches of how f changes with z.

After it's computed from (13.3), the fog factor f is clamped in the range $[0, 1]$. OpenGL then blends the fog color tuple *fogColor* with the incoming fragment's color tuple C_{in}, using the equation

$$C_{dest} = (1 - f)\, fogColor + f\, C_{in}$$

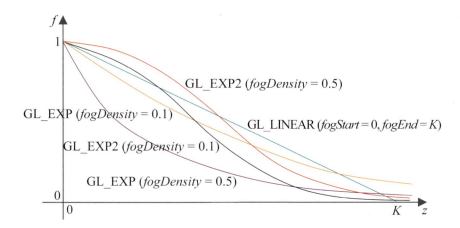

Figure 13.7: f versus z for various parameter values – the graphs are *not* mathematically exact.

to determine the color tuple C_{dest} of the destination pixel. The smaller the fog factor, therefore, the more the fog dominates and the fragment fades. One sees, as well, from the equations in (13.3) that fog modes GL_LINEAR, GL_EXP and GL_EXP2 in that order create increasingly thicker fog in general, though the constants in the equations have to be taken into account as well.

Interaction: Let's return now to `fieldAndSkyFogged.cpp` to describe the interaction. Press the up and down arrow keys to move the ball. Press the space bar to cycle through the different fog modes. If the fog mode is linear, then the left and right arrow keys change the *fogEnd* parameter, while if it's exponential or doubly exponential, they change the *fogDensity* parameter. The *fogStart* parameter is fixed at 0. Messages on the display indicate the fog mode and parameter values. Press delete at any time to reset the ball. Observe how even a mild fog *depth cues* the ball, adding realism as it travels away.

Note: The rendition of the ball suffers from the lack of lighting – loss of three-dimensionality in particular – but we wanted to keep the program simple.

Note: If you move the ball far enough it suddenly disappears altogether. That's not because of the fog, but because it's gone behind the sky rectangle!

The statement

```
glHint(GL_FOG_HINT, GL_NICEST)
```

in the drawing routine is a run-time advisory to OpenGL to use the highest-quality (and computationally most expensive) option available. Instead of GL_NICEST one could also pass as parameter values GL_FASTEST (computationally least expensive) or GL_DONT_CARE (no particular preference).

Exercise 13.10. (Programming) Exercise 12.17 was about enhancing shipMovie.cpp. Fog can be used to good effect in the scene as well to depth cue the traveling torpedo and ship.

13.3 Billboarding

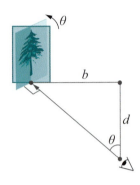

Figure 13.8: Billboarding: the original placement of the billboard (bold border) is rotated so its plane is normal to the direction of the viewer.

The technique of *billboarding* is to simulate a 3D object in a scene by placing an image of it as a texture on a rectangle, the *billboard*. The latter is then continuously rotated to keep it normal always to the direction of the viewer, giving the latter an illusion of the real object. See Figure 13.8.

As long as the object is a peripheral one, e.g., a background tree, a piece of furniture, a road sign, the device of holding up a 2D image is often authentic enough, thereby saving on the geometry required to make a "real" 3D version.

Experiment 13.8. Run billboard.cpp, where an image of two trees is textured onto a rectangle. Press the up and down arrow keys to move the viewpoint and the space bar to turn billboarding on and off. See Figure 13.9 for screenshots.   End

(a)

(b)

Figure 13.9: Screenshots of billboard.cpp: (a) Billboarding off (b) Billboarding on.

The billboard rectangle of billboard.cpp is located in the scene by drawing it first on the xy-plane centered about the z-axis and then translating it d units down the z-direction and b units left (Figure 13.8). Therefore, the angle θ that the billboard must be rotated about the vertical line through its center to keep it normal to the viewer's direction is given by

$$\theta = \tan^{-1}(b/d)$$

This particular rotation is implemented if billboarding is on; otherwise, the billboard remains parallel to the xy-plane (the position indicated by the dashed rectangle in Figure 13.8). The effect of billboarding is marked as the viewer travels "into" the scene by pressing the up arrow key.

Remark 13.3. One way to seamlessly fit a billboard into a scene is to paint its background texels a color matching the scene's backdrop (e.g., both white in the case of `billboard.cpp`). Another is to make the billboard's background texels transparent by setting their alpha values to 0 and then blend the billboard onto the backdrop.

13.4 Antialiasing

A straight line segment s specified, say, to be one pixel wide is rasterized by selecting a set of fragments that best approximates it and setting each to the color specified for s, while unselected fragments remain of the background color. See Figure 13.10(a) for a particularly low res example. Such discrete on/off rasterization protocols are computationally inexpensive, but tend to give poor visual quality at certain alignments of the segment owing to the jaggedness of the rasterization, the so-called *jaggies*.

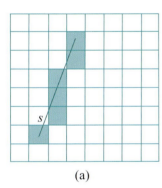
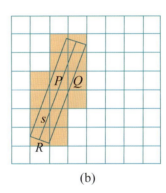

(a)   (b)

Figure 13.10: (a) Dark fragments represent a rasterization of a line segment s, specified to be one pixel wide (b) Shaded fragments are those that are intersected by the one-pixel wide rectangle R centered on s: the area that R covers of individual fragments, e.g., P and Q, varies.

Jaggies are another example of *aliasing*, a visual artifact that arises because of the limited resolution of the display device. Since a line segment not parallel to one of the axes can at best be approximated in a raster, one cannot hope to eliminate jaggies altogether. However, there are techniques to attenuate their visual impact, not surprisingly at the cost of extra computation.

OpenGL, in particular, offers an *antialiasing* method for line segments based on so-called *coverage* values. Consider again the line segment s. A one-pixel wide rectangle R centered on s intersects the 14 fragments shaded in

Figure 13.10(b). However, the area that R covers of each, called its coverage of that fragment, varies. For example, the coverage value of fragment P is more than twice that of Q. When antialiasing is enabled, OpenGL draws the line by setting each of the 14 intersected fragment's color to that of s, then multiplying the alpha value of each with its coverage and, finally, using the resulting weighted alpha to blend the fragment with the corresponding pixel already in the color buffer. The amount of the segment's color, therefore, blended into a destination pixel is proportional to the area of its source fragment covered by R. This has the effect of smoothing out the jaggies.

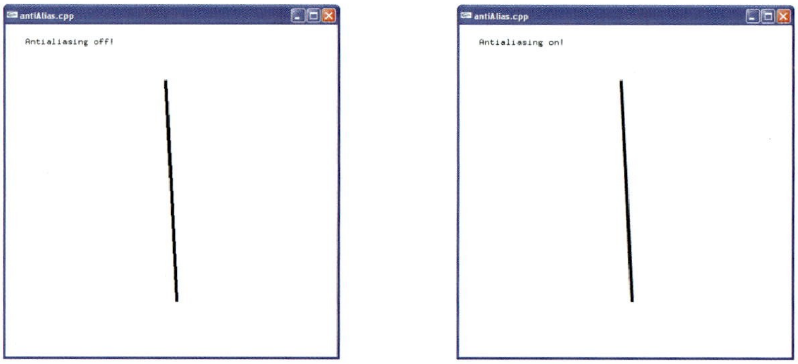

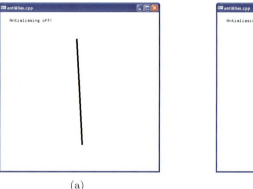

(a)   (b)

Figure 13.11: Screenshots of `antiAlias.cpp`: (a) Antialiasing off (b) Antialiasing on.

Experiment 13.9. Run `antiAlias.cpp`, which draws a straight line segment which can be rotated with the arrow keys and whose width changed with the page up/down keys. Press space to toggle between antialiasing off and on. Figure 13.11 shows screenshots of antialiasing both off and on.

Antialiasing is simple to implement in OpenGL. One has to first enable blending. The blending factors `GL_SRC_ALPHA` and `GL_ONE_MINUS_SRC_ALPHA`, used in `antiAlias.cpp`, are the best choice. Antialiasing itself is enabled with a call to

    glEnable(GL_LINE_SMOOTH)

A final point to note in the program is that we ask for the best possible antialiasing with the call

    glHint(GL_LINE_SMOOTH_HINT, GL_NICEST)

End

$Remark$ 13.4. OpenGL offers another antialiasing method, based on so-called *multisampling*, where color values computed at a sample of points in a pixel area are combined to apply to that pixel. Multisampling is particularly effective in antialiasing the vicinity of a polygon boundary. The reader is referred to the red book for details of its implementation and syntax.

Exercise 13.11. The unaided human eye can resolve to a minimum size of approximately 0.1 mm. (about 0.004 inch). So, what resolution levels must a 22 inch desktop monitor reach in terms of number of pixels by number of pixels for aliasing problems (and antialiasing algorithms) to go the way of the floppy disc? What's your best guess as to how long it will take for technology to get there?

13.5 Environment Mapping

The goal of *environment mapping* is to simulate an object reflecting its surrounding, e.g., a shiny kettle reflecting the kitchen or a well-polished car reflecting the street. As the environment can be seen by reflection off the object, it is said to be mapped onto the object. An approach to environment mapping originally invented by Blinn and Newell [18] in the seventies is still popular today because of its ease of implementation. The Blinn-Newell method makes clever use of textures, but the basic idea is not hard. An image of the environment (presumed static) is captured in a texture or multiple textures. Subsequently, the particular texture and texture coordinates used to paint a point on the environment-mapped object are determined from the position of the viewer relative to the object.

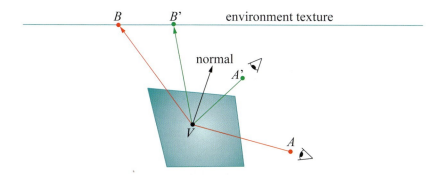

Figure 13.12: Blinn-Newell environment mapping principle: texture coordinates for a vertex V on an environment-mapped surface are obtained from the point on the texture image struck by the reflected ray originating from the eye.

Figure 13.12 illustrates the principle. The texture coordinates at the vertex V of an environment-mapped quad are determined by the point of the environment – more precisely, the corresponding point of the environment texture – seen by the viewer by reflection off the object. For example, when the viewer is at A, V is painted with the color values at B (red in the figure); when she moves to A', those of B' are used (green). The crux of the Blinn-Newell approach then is to *dynamically* compute texture coordinates, based on the laws of reflection, as the viewpoint changes.

OpenGL provides support for two methods of environment mapping: *sphere mapping* and *cube mapping*. Both are based on the Blinn-Newell approach, the difference being in the way that the environment is captured on texture and that texture coordinates are computed. OpenGL provides *automatic* texture coordinate computation for either method. We'll discuss sphere mapping in fair detail.

13.5.1 Sphere Mapping

Getting It to Work

Implementing sphere mapping using OpenGL is simple, as the following program shows.

Experiment 13.10. Run `sphereMapping.cpp`, which shows the scene of a shuttle launch with a reflective rocket cone initially stationary in the sky in front of the rocket. Press the up and down arrow keys to move the cone. As the cone flies down, the reflection on its surface of the launch image changes. Figure 13.13 is a screenshot as it's about to crash to the ground. **End**

The two commands

Figure 13.13: Screenshot of `sphereMapping.cpp`.

```
glTexGeni(GL_S, GL_TEXTURE_GEN_MODE, GL_SPHERE_MAP);
glTexGeni(GL_T, GL_TEXTURE_GEN_MODE, GL_SPHERE_MAP);
```

in the initialization routine of `sphereMapping.cpp` ask OpenGL to use functions from its library to generate the s and t texture coordinates for sphere mapping.

The pair of commands

```
glEnable(GL_TEXTURE_GEN_S);
glEnable(GL_TEXTURE_GEN_T);
```

and its inverse

```
glDisable(GL_TEXTURE_GEN_S);
glDisable(GL_TEXTURE_GEN_T);
```

in the drawing routine, bracketing the drawing of the cone, enable and disable the use of these functions. That's pretty much all there is to implementing a sphere map using OpenGL! Note that at the time sphere mapping is activated the currently bound texture is the launch image, which, of course, is why it is reflected in the cone.

Now, a reader watching the cone as it zooms down may be wondering how authentic actually is the reflection. Good question, and it leads us to investigate how OpenGL computes sphere-mapped texture coordinates.

How It Works

This part is fairly mathematical. If your interest is practical and limited to using the technique, you can safely skip it and jump to the part on preparing the environment texture.

Here's how sphere-mapped texture coordinates are generated at a vertex V. See Figure 13.14. The unit vector u from the eye (the origin O in OpenGL) toward V is $v/|v|$, where v is the position vector of V, assuming, of course, that $v \neq 0$. The unit eye direction vector from V then is $-u$. The unit normal n at V is user-provided.

OpenGL computes the reflection vector r, the unit vector in the direction that a hypothetical ray from the eye is reflected at V, with the help of the following equation (obtained by replacing light direction vector l with eye direction vector $-u$ in the formula of Exercise 11.4):

$$r = u - 2(n \cdot u)n$$

Suppose, then, OpenGL finds that $r = (r_x, r_y, r_z)$. Computed next is the quantity

$$m = 2\sqrt{r_x^2 + r_y^2 + (r_z + 1)^2}$$

Finally, the texture coordinates at V are calculated as

$$s = \frac{r_x}{m} + \frac{1}{2} \quad \text{and} \quad t = \frac{r_y}{m} + \frac{1}{2}$$

Whew!

If we parse the expressions for s and t carefully, though, it'll not be hard to understand the game plan. Using the expression for m above, write

$$s = \frac{1}{2}\frac{r_x}{\sqrt{r_x^2 + r_y^2 + (r_z + 1)^2}} + \frac{1}{2} \quad \text{and} \quad t = \frac{1}{2}\frac{r_y}{\sqrt{r_x^2 + r_y^2 + (r_z + 1)^2}} + \frac{1}{2}$$

or

$$s = \frac{1}{2}R_x + \frac{1}{2} \quad \text{and} \quad t = \frac{1}{2}R_y + \frac{1}{2} \tag{13.4}$$

where the variables

$$R_x = \frac{r_x}{\sqrt{r_x^2 + r_y^2 + (r_z + 1)^2}} \quad \text{and} \quad R_y = \frac{r_y}{\sqrt{r_x^2 + r_y^2 + (r_z + 1)^2}} \tag{13.5}$$

Once we understand what the mapping

$$(r_x, r_y, r_z) \mapsto (R_x, R_y)$$

does geometrically the rest will be straightforward.

The reflection vector $r = (r_x, r_y, r_z)$ is the position vector of some point, say P, on the unit sphere S centered at the origin. See Figure 13.15(a).

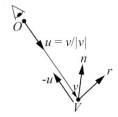

Figure 13.14: The vectors involved in generating texture coordinates.

Now, the position vector of P with respect to the *south pole* $(0, 0, -1)$ of S is $r' = (r_x, r_y, r_z + 1)$. And r' normalized is the vector

$$r'' = \frac{1}{\sqrt{r_x^2 + r_y^2 + (r_z + 1)^2}} (r_x, r_y, r_z + 1) = \left(R_x, R_y, \frac{r_z + 1}{\sqrt{r_x^2 + r_y^2 + (r_z + 1)^2}} \right)$$

(13.6)

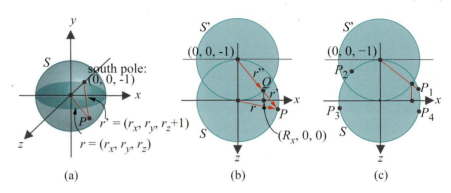

Figure 13.15: Determining R_x from r.

In fact, r'' itself is the position vector, with respect to the south pole, of the point Q of intersection of the line from the south pole to P with the unit sphere S' centered at the pole. S' is not drawn in Figure 13.15(a), but Figure 13.15(b) shows both S and S' in section along the xz-plane (for this particular drawing we assume that P lies on this section). R_x being the x-value of Q by (13.6), the projection of Q on the x-axis in Figure 13.15(b) is $(R_x, 0, 0)$ (Q's y-value R_y is 0, of course, as it's on the xz-plane). Here's an exercise to reinforce your understanding of the preceding construction to find R_x from r.

Exercise 13.12. For each point P_i, $1 \le i \le 4$, in Figure 13.15(c), use a ruler and pencil to draw the corresponding point $(R_x, 0, 0)$ on the x-axis.

Part answer: Red lines indicate the construction for P_1.

The reader may now agree that, at least as P varies over the xz-section of S, $(R_x, 0, 0)$ varies between $(-1, 0, 0)$ and $(1, 0, 0)$ and, correspondingly, R_x between -1 and 1. Moreover, the closer P gets to the south pole the closer is R_x to -1 or 1, depending on which side of the pole P is. However, P should never be *at* the south pole, for, otherwise, the construction to determine R_x breaks down. It follows that R_x itself reaches neither value -1 nor 1. In fact, considering now all of the sphere S, not just its xz-section, it's not hard to see that R_x varies over the open interval $(-1, 1)$ as P varies over S minus its south pole.

The mapping from P to R_y is similar. Therefore, as P moves over S minus its south pole, (R_x, R_y) moves within the interior of the square $[-1, 1] \times [-1, 1]$.

For an even better understanding, let's determine analytically the dependence of (R_x, R_y) on P.

Choose a Z in $-1 < Z \leq 1$. The plane $z = Z$ intersects S in a latitudinal circle

$$x^2 + y^2 + Z^2 = 1 \quad \text{or} \quad x^2 + y^2 = 1 - Z^2$$

Now, from (13.5) we have that

$$R_x^2 + R_y^2 = \frac{r_x^2 + r_y^2}{r_x^2 + r_y^2 + (r_z + 1)^2}$$

Therefore, if P lies on the latitudinal circle $x^2 + y^2 = 1 - Z^2$, so that $r_x^2 + r_y^2 = 1 - Z^2$ and $r_z = Z$, then the preceding equation says that (R_x, R_y) lies on the circle

$$R_x^2 + R_y^2 = \frac{1 - Z^2}{1 - Z^2 + (1 + Z)^2} = \frac{1 - Z^2}{2 + 2Z} \tag{13.7}$$

Now, we can see how (R_x, R_y) varies with $r = (r_x, r_y, r_z)$ as we had set out to. In fact, we'll draw a picture. See the two diagrams on the left of Figure 13.16.

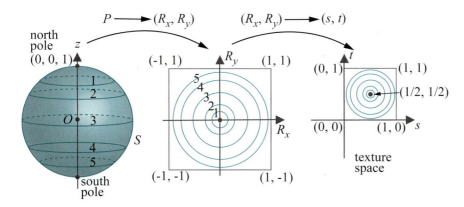

Figure 13.16: The maps $P \mapsto (R_x, R_y)$ and $(R_x, R_y) \mapsto (s, t)$.

Keep in mind that r is P's position vector, the latter varying over S. Latitudinal circles on S (now drawn upright at left with the north pole at the top to better see these circles) each maps to a circle centered at the origin and in the square $[-1, 1] \times [-1, 1]$ in $R_x R_y$-space (drawn in the middle). In particular, the north pole maps to the origin, and latitudinal circles from the north pole downward map to increasingly larger circles inside $[-1, 1] \times [-1, 1]$. Five pairs of corresponding circles have been drawn and labeled similarly in the two diagrams. As the latitudinal circles approach the south pole, the mapped circles draw nearer and nearer to the containing square.

Exercise 13.13. What is the radius of the circle to which the equator maps? What are the radii of the images of the latitudinal circles 45°N and 60°S?

Hint: Equation (13.7) gives the radius of the circle in $R_x R_y$-space, which is the image of the latitudinal circle at $z = Z$. For example, the latitudinal circle 45°N has z-value $\sin 45° = 1/\sqrt{2}$, so plug $Z = 1/\sqrt{2}$ into (13.7) to find the radius of its mapped circle in $R_x R_y$-space.

Exercise 13.14. How do longitudinal great circles on S map?

Hint: Straight lines through the origin in $R_x R_y$-space...

The final transformation from $R_x R_y$-space to st-space (texture space) is simple. See the righthand two diagrams of Figure 13.16. (R_x, R_y) is mapped to $(\frac{1}{2}R_x + \frac{1}{2}, \frac{1}{2}R_y + \frac{1}{2})$ via Equations (13.4), which linearly transform the square $[-1, 1] \times [-1, 1]$ in $R_x R_y$-space to the unit square $[0, 1] \times [0, 1]$ in texture space. The images in texture space of the five circles in $R_x R_y$-space are shown as well in the rightmost diagram.

Bottom Line

Time for a wrap-up in plain and simple English. If vertex V were on a perfect mirror and the environment around it arranged along the unit sphere S centered at V, then the eye would see the point P where S is intersected by the reflection of the line of sight at V (see Figure 13.17). However, OpenGL's only knowledge of the environment is from a user-provided texture occupying a unit square in texture space. So what it does is this: if the eye wants to see the point P in the spherical environment, OpenGL shows it instead the point (s, t) in texture space to which P is mapped as described above by $P \mapsto (R_x, R_y) \mapsto (s, t)$.

The calculations above tell exactly what happens in physical terms. If the eye asks to see the north pole of the environment, then it's shown instead the center of the texture. As the eye travels to see points farther and farther from the north pole, it's shown points farther and farther from the center of the texture. Precisely, latitudinal circles in the environment are replaced for viewing by circles in the texture centered at its middle.

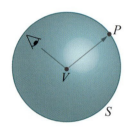

Figure 13.17: The environment S around a perfect mirror vertex V.

Preparing the Environment Texture

Given this sphere-mapped scheme to present the environment to the viewer via a texture, what is the right way to prepare the texture? Practically speaking, how then should one photograph the environment in order to create the texture image? Comparing the left and right diagrams of Figure 13.16 suggests an answer. The camera should be located at the origin O pointing up the z-axis toward the north pole and have a *very* wide-angle lens; in fact, it would be helpful if the field of view were nearly 360°! Of course, this is impossible, but a fairly wide-angle picture taken with a camera located in the vicinity of the object to be environment mapped, focused up the z-axis of world space, should be good.

Remark 13.5. Since the texel used depends only on the value of the reflection vector at a vertex, and not the vertex's location, reflections in parallel directions appear the same at all vertices. Practically, this means that the environment-mapped object should be small compared to its surroundings for authenticity.

Remark 13.6. Some practitioners advocate the application of filters to the texture prior to sphere mapping. For example, NeHe [98] suggests using the *spherizing* filter (available, e.g., in Adobe's Photoshop software).

Exercise 13.15. (Programming) Sphere map the torpedo in `ship-Movie.cpp`.

It's fun experimenting with sphere mapping – as with any special effect – as long as you remember that in CG authenticity lies in the eyes of the beholder. If it looks real then it *is* real. Which still leaves unanswered the question that started our mathematical investigation in the first place: how authentic is the reflection in the cone of `sphereMapping.cpp`? But perhaps the reader at this point has an opinion already!

13.5.2 Cube Mapping

Cube mapping is even simpler than sphere mapping and often more authentic. The only difficulty with implementing it is that the environment has to be captured on multiple textures. In fact, the environment is imagined to be a cube and the user asked to provide an image of each of the six faces. Figure 13.18 gives the idea.

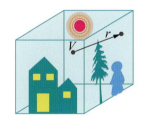

Figure 13.18: Cube mapping.

In the case of cube mapping, given the reflection vector r, first must be computed which of the six faces is struck by the reflected ray, and then the actual point, to extract color values from the texel at that location. As for sphere mapping, there is support in OpenGL for cube mapping. OpenGL automatically generates cube-mapped texture coordinates once the user loads the six environment textures. We'll leave it at this and refer the reader to the red book for more.

13.6 Stencil Buffer Techniques

A space in memory reserved for pixel-related data, equal space being assigned to each pixel, is called a *buffer*. A computer system can have numerous buffers for different purposes, and different buffers may have differing amounts of space reserved per pixel. We'll briefly introduce all the system buffers OpenGL supports and their uses before focusing in particular on the stencil buffer.

13.6.1 OpenGL Buffers

A particular OpenGL system can support the following buffers:

- Color buffers

- Depth buffer

- Stencil buffer

- Accumulation buffer

We are already familiar with the first two kinds and they are, in fact, the most commonly used of the buffers.

A *color buffer* stores primary RGB color values and the alpha A value, typically to a precision of 8 bits each, for a total of 32 bits per pixel. Color buffers are the only ones of the buffers the user can directly draw into, and it is the final RGB values in some particular color buffer which are *flushed* to the screen for viewing.

In fact, there may be more than one color buffer. A double-buffered system has at least two – front (viewable) and back (drawable) – critical to smooth animation, as we learned in Section 4.5.1. If stereoscopic viewing is supported, there will be left and right color buffers, possibly even front-left, back-left, front-right and back-right, if combined with double buffering. The left buffers are shown to the left eye and the right ones to the right eye, typically with the user wearing special glasses. If the images in the left and right buffers are of the same scene projected toward two points slightly offset one from the other, as are human eyes, a perception of 3D is created.

Additionally, a system may have so-called auxiliary color buffers, usually in hardware, to store intermediate steps of a complicated rendering process.

We are also familiar with the *depth buffer* (or *z-buffer* as it's popularly called) which stores depth information, usually a 24-bit integer, one for each pixel in the color buffer. When depth testing is enabled, the depth buffer helps sort out objects according to their depth from the viewer along lines of sight, permitting nearer objects to obscure further ones in the rendering phase.

Less familiar to the reader might be the *stencil buffer*. This is a buffer used to tag pixels in the color buffer. The stencil buffer most often contains 8 bits, called tags, for each color buffer pixel. The typical way to use the stencil buffer is to set the tags in a first phase when nothing is drawn to the screen and then employ these tags as controls in a second phase when actual drawing takes place. The tag values allow the user to constrain drawing to limited portions of the screen, making possible various creative applications. We'll be studying the stencil buffer in fair detail shortly.

The *accumulation buffer* is yet another buffer used for special effects. Think of the accumulation buffer as a giant color buffer. It contains RGBA values per pixel as well, but often to a much higher precision. Typical accumulation buffers dedicate 16 bits to each of RGBA, for a total of 64 bits per pixel.

The primary use of the accumulation buffer is to composite several drawings into one. The accumulation buffer cannot be directly drawn into, nor can it be directly displayed. Rather, drawings are made to color buffers and combined one by one into the accumulation buffer. Successive incoming drawings can be combined with the one currently resident in the accumulation buffer in various ways, e.g., added or multiplied, the latter explaining the need for higher precision. The final composited drawing is returned from the accumulation buffer to a color buffer for display.

The collection of all buffers in a system is called its *frame buffer*. Figure 13.19(a) shows a frame buffer containing all four kinds of buffers supported by OpenGL. The size of each constituent buffer – which is its *width × height* as an array of bit strings – matches that of the display device. This size is also called the *resolution* of the frame buffer.

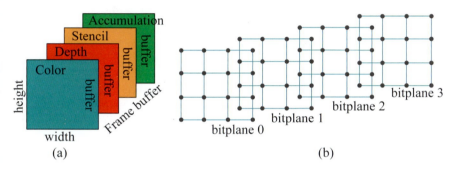

Figure 13.19: (a) A complete frame buffer (b) A 4 × 4 buffer with 4-bit precision as a stack of four bitplanes: points represent bits.

The number of bits per pixel in a buffer determines the buffer's *precision*. Bits in the buffer in some particular bit position, between 0 to *precision*−1, form an array called a *bitplane*. So, of course, a buffer's precision is identical to its number of bitplanes. Figure 13.19(b) depicts a 4 × 4 buffer of 4-bit precision as a stack of bitplanes.

$Exercise$ **13.16.** Suppose we want a graphics card which has four 32-bit precision color buffers for stereoscopic viewing with double-buffering, an auxiliary color buffer of similar precision, a 24-bit depth buffer, an 8-bit stencil buffer and a 64-bit accumulation buffer, all to support a 1024 × 768 resolution display. How much on-card memory are we asking for?

$Remark$ 13.7. A given OpenGL implementation may not support all the possible buffers. One can determine which are supported, as well as the number of bitplanes in each, with the help of `glGet*v()` calls, e.g., `glGetIntegerv(GL_DEPTH_BITS, *pointer)` returns the number of bitplanes in the depth buffer. Check the blue book for the specs for such calls.

13.6.2   Using the Stencil Buffer

Applications using the stencil buffer usually set the *stencil bits*, or *tags* as they are called – typically there being 8 for each pixel – in a first phase, by means of a *stencil test* applied to each incoming fragment. Stencil testing is enabled by calling glEnable(GL_STENCIL_TEST). The particular test applied depends on a glStencilFunc() call. The stencil test is applied in the graphics pipeline just before the depth test and only if a fragment passes the stencil test does it proceed to the depth test. If a fragment fails either test then it is discarded from the drawing pipeline.

How the incoming fragments set tags depends on a glStencilOp() call, paired with the glStencilFunc() call. Incoming fragments that fail the stencil test, those that pass the stencil test but fail the depth test and those that pass both tests can set tags differently. The color buffer is often disabled during the tag-setting first phase so that nothing is actually rendered to the display.

Once the tags have been set, actual drawing to the display occurs in the second phase, when tags are read per incoming fragment and a stencil test applied to determine if it is to continue on down the pipeline. We'll describe next the mechanics of the twin commands glStencilFunc() and glStencilOp() before putting everything together in a program.

The glStencilFunc(*func*, *ref*, *mask*) command sets a comparison function *func* to use in the stencil test, as well as a reference value *ref* to compare the stencil tag with. For example, if *func* is GL_LESS, then the test passes if *ref* is less than the value of the stencil tag. The reference value is clamped to the range $[0, 2^k - 1]$ if the stencil buffer contains k bitplanes and the stencil tag, too, interpreted as an integer in the same range. However, prior to comparison, the *mask* is bitwise ANDed with both the reference value and the stencil tag. Effectively, therefore, the two integers made from bits in the reference value and stencil tag, respectively, at positions corresponding to the 1-bits in the mask are compared.

E_xample 13.3. If the call is glStencilFunc(GL_EQUAL, 0xFF, 0x3F) and the stencil tag corresponding to a fragment is 0xBF, then the fragment passes the stencil test because only the lower six bits of the mask are 1, and the reference value and the given stencil tag, in fact, agree in each of these positions.

We will henceforth use only a mask value of 1 (=00000001, assuming an 8-bit stencil buffer), which means that the lowest bit of the reference value is compared with the lowest bit of the stencil tag.

E_xample 13.4. The call glStencilFunc(GL_EQUAL, 1, 1) allows a fragment to pass the stencil test only if the lowest bit in its corresponding stencil tag equals 1. Suppose this, in fact, is the call prior to drawing the square R consisting of four fragments, as shown in Figure 13.20(a), and that the contents of the stencil buffer are as in Figure 13.20(b) (only the lowest bit of each tag is drawn). Then the left two fragments of R pass the stencil test and proceed on to the depth test, while the right two fail and are ejected from the pipeline. Ignore the right grid for now.

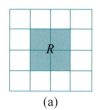

1	1	0	0
1	1	0	0
1	1	0	0
1	1	0	0

1	1	0	0
1	1	1	0
1	1	1	0
1	1	0	0

(a)   (b)   (c)

Figure 13.20: (a) Square R, which is drawn after calls to `glStencilFunc(GL_EQUAL, 1, 1)` and `glStencilOp(GL_REPLACE, GL_REPLACE, GL_REPLACE)` (b) Stencil buffer configuration before R is drawn (c) Stencil buffer configuration before R is drawn. Only each lowest bit in the stencil buffer is shown.

The call `glStencilOp(`*fail, zfail, zpass*`)`, paired with `glStencil-Func()`, specifies how a stencil tag is updated following a stencil test. The values of *fail, zfail* and *zpass* determine the update action in case the fragment fails the stencil test, passes the stencil test but fails the ensuing depth test, and passes both tests, respectively. Figure 13.21 indicates the scheme. Values for each of the three parameters that we'll use are `GL_KEEP`, `GL_REPLACE` and `GL_INVERT` (for a full list of possible values see the red book).

Example 13.5. The call `glStencilOp(GL_REPLACE, GL_REPLACE, GL_-REPLACE)` causes the stencil tag to be replaced with the reference value in all cases. If this call were indeed paired with `glStencilFunc(GL_EQUAL, 1, 1)` prior to drawing the square R of Figure 13.20(a), then the stencil buffer would be updated as in Figure 13.20(c).

Exercise 13.17. Determine how the stencil buffer would be updated in the situation of Figure 13.20 if the call were `glStencilOp(GL_INVERT, GL_REPLACE, GL_KEEP)` instead of `glStencilOp(GL_REPLACE, GL_REPLACE, GL_REPLACE)`, and it was known that rectangle fragments all fail the depth test if they come to it. Assume that the call `glStencilFunc(GL_EQUAL, 1, 1)` remains.

Note: `GL_KEEP` keeps the stencil tag bits unchanged, while `GL_INVERT` inverts them.

Drawing reflections in a constrained area is a canonical application of the stencil buffer which we illustrate next.

Experiment 13.11. Run `ballAndTorusStenciled.cpp`, based on `ball-AndTorusReflected.cpp`. The difference is that in the earlier program the entire checkered floor was reflective, while in the current one the red floor is

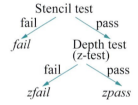

Figure 13.21: Potential outcomes for a fragment through the stencil and depth tests.

non-reflective except for a mirror-like disc lying on it. Pressing the arrow keys moves the disc and pressing the space key starts and stops the ball moving. As you can see in the screenshot Figure 13.22, the ball and torus are reflected only in the disc and nowhere else. **End**

Figure 13.22: Screenshot of ballAndTorus-Stenciled.cpp.

The effect of restricted reflection in `ballAndTorusStenciled.cpp` is obtained using four successive pairs of `glStencilFunc()` and `glStencilOp()` calls. The first pair

```
glStencilFunc(GL_ALWAYS, 1, 1);
glStencilOp(GL_REPLACE, GL_REPLACE, GL_REPLACE);
```

causes the next drawing statement `drawReflectiveDisc(xVal, zVal)` to create a "mask" of the disc (nothing to do with the mask parameter of `glStencilFunc()`!) in the stencil buffer with 1's at positions corresponding to a disc fragment and 0's elsewhere. The reason for this is that the GL_ALWAYS parameter value of `glStencilFunc()` ensures that every fragment of the disc passes the stencil test, while the second two GL_REPLACE values of `glStencilFunc()`, for *zfail* and *zpass*, respectively, ensure that the stencil tag corresponding to a disc fragment is always replaced with the reference value 1, regardless the result of the depth test. The remaining stencil tags remain at their clearing value 0 set by the call `glClearStencil(0)` in the initialization routine. Both the color and depth buffers are disabled prior to the drawing statement, so that only the stencil buffer is updated and nothing is drawn to the screen by the first `drawReflectiveDisc(xVal, zVal)`.

Remark 13.8. It's because the stencil buffer itself cannot be directly drawn into, that the mask of the disc has to be created indirectly with the help of the buffer-manipulating command `glStencilOp()`.

Once the mask of the disc in the stencil buffer has been created, actual drawing to the window commences. The color and depth buffers are accordingly enabled next. The second pair of stencil-buffer manipulating calls

```
glStencilFunc(GL_EQUAL, 1, 1);
glStencilOp(GL_KEEP, GL_KEEP, GL_KEEP);
```

causes the subsequent drawing statement `drawFlyingBallAndTorus()` to draw the reflected ball and torus in the mask area of the disc, because only fragments corresponding to this area pass the stencil test. The contents of the stencil buffer are kept unchanged.

The next drawing statement, `drawReflectiveDisc(xVal, zVal)`, actually draws (or, rather, blends) the disc onto the reflected ball and torus.

The third pair of stencil-buffer manipulating calls

```
glStencilFunc(GL_NOTEQUAL, 1, 1);
glStencilOp(GL_KEEP, GL_KEEP, GL_KEEP);
```

prevents the red quad drawn next from erasing the disc by allowing drawing only outside the area corresponding to the disc.

The final pair of stencil-buffer manipulating calls

```
glStencilFunc(GL_ALWAYS, 1, 1);
glStencilOp(GL_KEEP, GL_KEEP, GL_KEEP);
```

allows the real ball and torus to be drawn by a `drawFlyingBallAndTorus()` statement regardless of the stencil tags.

Exercise 13.18. (Programming) Reverse the roles of the floor and disc in `ballAndTorusStenciled.cpp`. In particular, make the floor reflective, while a movable red disc blocks reflection.

Exercise 13.19. (Programming) Draw the glass door of a roadside building reflecting a passing vehicle – all shapes in your scene being simple and boxy.

For more about drawing shadows and reflections with the help of the stencil buffer read Mark Kilgard's tutorial [74].

13.7 Bump Mapping

Blinn [15] developed an ingenious method, called *bump mapping*, to give the illusion of geometric detail on a surface, e.g., making it appear ridged or dimpled, by means of perturbing the surface normals, but *without* actually changing any geometry. The idea is to re-align the normals to the original surface so that light reflects from it *as if* it were detailed. A one-dimensional example will make matters clear.

Consider the straight line c of Figure 13.23(a). The unit normals $n(u)$ at points $c(u)$ of c are identical vectors perpendicular to c (drawn in the figure as a discrete sequence of points).

Next, suppose that one wants to wrinkle c to make it look like the blue curve c' of Figure 13.23(b). Actually wrinkling c entails replacing it with a multi-segment polyline approximation of c', an object substantially more complex than c. The bump mapping approach is to leave c as it is, but, instead, to redefine the normal at each point $c(u)$ so that it equals $n'(u)$, the normal at the corresponding point $c'(u)$ of c'. Figure 13.23(b) shows $n'(u)$ at one point $c'(u)$ of c', while Figure 13.23(c) shows the so-called *bump-mapped* c with its perturbed normals.

The premise of bump mapping is that c with normals redefined to match those of c' will resemble c' when lit, because the reflection of light from a surface depends on the normals there. We describe next Blinn's method to compute the perturbed normals.

Suppose that s is a surface in 3-space defined parametrically on some parameter domain W by

$$s(u, v) = (f(u, v), g(u, v), h(u, v))$$

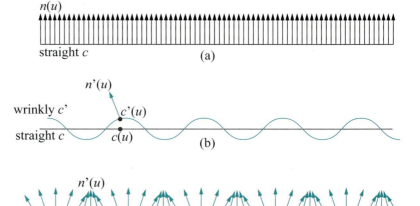

$n(u)$

straight c (a)

$n'(u)$

wrinkly c' $c'(u)$

straight c $c(u)$ (b)

$n'(u)$

bump-mapped c (c)

Figure 13.23: Bump mapping: (a) The original curve c and its true unit normals $n(u)$ (b) The wrinkled curve c' and its unit normal $n'(u)$ at a single point $c'(u)$ (c) Bump-mapped c with redefined normals $n'(u)$.

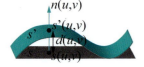

$n(u,v)$

$s'(u,v)$

$d(u,v)$

$s(u,v)$

s'

s

Figure 13.24: The bumped surface s' is obtained from s by displacing each point $s(u.v)$ a distance $d(u, v)$ along the normal $n(u, v)$ at $s(u, v)$.

and that $n(u, v)$ is a unit normal vector to s at $s(u, v)$.

Suppose, as well, that the desired (hypothetical) detailed surface s' is obtained from s by displacing each point $s(u, v)$ a distance $d(u, v)$ along $n(u, v)$. See Figure 13.24. The scalar-valued function $d(u, v)$ giving this displacement is called the *bump map*.

Accordingly,

$$s'(u, v) = s(u, v) + d(u, v)n(u, v)$$

which we write more simply by dropping the arguments as

$$s' = s + dn \qquad (13.8)$$

A normal n' to s' is given by

$$n' = \frac{\partial s'}{\partial u} \times \frac{\partial s'}{\partial v} \qquad (13.9)$$

because the partial derivatives of s' with respect to u and v span the tangent plane to that surface at $s'(u, v)$, so that their cross-product is normal (see Section 11.10 for more about partial derivatives and their application to finding tangents and normals to a surface). We evaluate n' next.

First, from (13.8),

$$\begin{aligned}
\frac{\partial s'}{\partial u} &= \frac{\partial s}{\partial u} + \frac{\partial d}{\partial u}n + d\frac{\partial n}{\partial u} \\
&\simeq \frac{\partial s}{\partial u} + \frac{\partial d}{\partial u}n
\end{aligned}$$

where the approximation in the second line is made by dropping the term $d\frac{\partial n}{\partial u}$, which is negligibly small on the assumptions that

(a) the displacement d is small, and

(b) $\frac{\partial n}{\partial u}$ is small as well, from the reasonable premise that the original surface s lacked detail and was fairly smooth, meaning that its normal direction changed only slowly with u.

Likewise, one writes

$$\frac{\partial s'}{\partial v} \simeq \frac{\partial s}{\partial v} + \frac{\partial d}{\partial v}n$$

Plugging the preceding two approximations into (13.9) one gets

$$\begin{aligned}
n' &\simeq (\frac{\partial s}{\partial u} + \frac{\partial d}{\partial u}n) \times (\frac{\partial s}{\partial v} + \frac{\partial d}{\partial v}n) \\
&= (\frac{\partial s}{\partial u} \times \frac{\partial s}{\partial v}) + \frac{\partial d}{\partial u}(n \times \frac{\partial s}{\partial v}) - \frac{\partial d}{\partial v}(n \times \frac{\partial s}{\partial u}) \\
&= n + \frac{\partial d}{\partial u}(n \times \frac{\partial s}{\partial v}) - \frac{\partial d}{\partial v}(n \times \frac{\partial s}{\partial u}) \quad\quad\quad (13.10)
\end{aligned}$$

which expresses the perturbed normal n' in terms of the original surface s, the original normal n and the bump map d. Finally, the new normal function for the bump mapped s is obtained by normalizing n' to unit length (provided that it's not zero).

Bump mapping comes into its own in the per-pixel lighting of Phong's shading model, as the programmer can then apply Equation (13.10) to compute normal values at *each pixel* for subsequent use in the lighting equation. This cannot be achieved with first-generation OpenGL, where the user is restricted to specifying normals at vertexes only. It is possible, though, using the Shading Language, which is part of the second-generation of the language. Without per-pixel programming bump, mapping is at best awkward, but we do have a simple proof-of-concept program.

Experiment 13.12. Run bumpMapping.cpp, where a plane is bump mapped to make it appear corrugated. Press space to toggle between bump mapping turned on and off. Figure 13.25 shows screenshots. End

The equation of the plane in bumpMapping.cpp is

$$s(u, v) = (u, 0, -v)$$

(the minus sign in front of v is so that the normal $\frac{\partial s}{\partial u} \times \frac{\partial s}{\partial v}$ to the plane points in the upward y-direction) and that of the bump map

$$d(u, v) = \sin(2u)$$

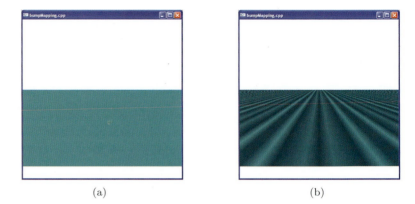

Figure 13.25: Screenshots of `bumpMapping.cpp`: (a) Bump mapping off (b) Bump mapping on.

We leave it to the reader to verify that with these equations for s and d, (13.10) gives

$$n'(u, v) = (2\cos(2u),\ 1,\ 0)$$

This formula for the perturbed normals to the plane is, in fact, implemented in `bumpMapping.cpp` when bump mapping is turned on (together with a call to `glEnable(GL_NORMALIZE)` to normalize the normals).

In Section 20.3 of the last chapter we'll see this same plane bump mapped with the use of per-pixel lighting courtesy the Shading Language. The difference in quality from what we have now will be significant.

13.8   Summary, Notes and More Reading

In this chapter we learned a few different visual techniques to help embellish our scenes, games and movies. It's worth emphasizing that visual techniques are as much an art as a science. Experience counts a lot in knowing how to get the "right effect", a subjective notion in the first place. Collect code to save re-inventing the wheel and for inspiration. A lot of people out there are doing amazingly creative stuff. Although much of it is commercial, still there's plenty of free stuff to be found on the net. The OpenGL site [99] has numerous pointers. Check out Nehe [98] as well.

It's worth noting that we've just crossed a milestone in the progression of this book. With the only two significant exceptions of NURBS and programmable shaders, both to come in later chapters, *we've mostly covered all in this book to do with coding OpenGL*. There's *much more* to CG than OpenGL, of course, and topics we've yet to see such as rasterization, Bézier, B-spline and NURBS theory, Hermite interpolation, projective spaces, ray tracing and radiosity, among others, are extremely important for a solid understanding of the field. Nevertheless, it's heartening to realize how far we have come since the first chapter, particularly, from the point of view of practical programming.

Part VII

Pixels, Pixels, Everywhere

CHAPTER 14

Raster Algorithms

I n this chapter we are going to be traveling almost all the way from one end of the graphics pipeline to the other – from world space to just behind screen space – to understand some of the low-level processes which take place at the time primitives are transformed to pixels in the raster. In particular, the goal for this chapter is to learn algorithms to clip and rasterize lines and polygons. These are operations a user cannot herself call directly or interact with via a high-level API such as OpenGL, which is not a bad thing as she can devote herself then to modeling and animation. Nevertheless, it's useful to have a grasp overall of the functioning of the pipeline. We'll not attempt here a comprehensive coverage of raster algorithms, but focus on four which are commonly implemented and fairly representative.

First, comes clipping, which is the process of determining the part of a primitive within some restricted area. We're already familiar with the functionality of OpenGL's clipping to a viewing volume and in this chapter we'll learn how the operation is implemented in two 2D cases, namely, those of a straight line segment and a convex polygon, both clipped to a rectangle. In particular, the Cohen-Sutherland line clipper is the topic of Section 14.1 and the Sutherland-Hodgeman polygon clipper of Section 14.2. Both clippers can be straightforwardly extended to 3D to clip a line segment or convex polygon against a box, the version actually implemented in the rendering pipeline.

Next, we'll investigate rasterization, the process of selecting and coloring pixels from the raster to represent a given primitive. We'll again limit ourselves to the two cases of straight segments and polygons. Moreover, we shall only be choosing pixels to comprise a primitive, leaving the problem of coloring them to a later chapter. Section 14.3 presents Bresenham's line rasterizer, actually as an improvement over the DDA (Digital Differential

Analyzer) line rasterizing algorithm. Section 14.4 next discusses scan-based polygon rasterization. Section 14.5 concludes.

14.1 Cohen-Sutherland Line Clipper

Straight line segments traveling down the OpenGL graphics pipeline are clipped to within an axis-aligned 3D viewing box prior to rasterization. Cohen-Sutherland is the classic algorithm for this purpose. Our exposition of Cohen-Sutherland, though, will be in 2D – clipping a segment to a rectangle – for the sake of simplicity. However, extension of the 2D version to three dimensions, where a segment is clipped to a box, is fairly straightforward.

Remark 14.1. How about clipping to a viewing volume which is not a box, but a frustum, as when the projection statement is `glFrustum()` or `gluPerspective`? It turns out that in a stage in the pipeline, prior to clipping, the frustum is "straightened" into a box by a so-called projective transformation, so one need clip only to a box.

The inputs to (2D) Cohen-Sutherland are, then, the endpoints p_i and p_j of a straight line segment S on a plane, and an axis-aligned rectangle R on the same plane bounded by the lines $x = a$, $x = b$, $y = c$ and $y = d$. See Figure 14.1(a) for a diagram with multiple input segments. R is called the *clipping rectangle*.

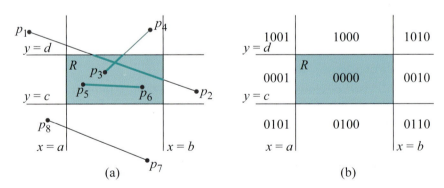

(a)   (b)

Figure 14.1: (a) A clipping rectangle R and four straight line segments with their parts clipped to R colored (b) Nine regions of the plane by outcode.

The output consists of the endpoints p'_i and p'_j of the intersection $S \cap R$ of S with R if it is non-empty, and *empty* otherwise. The output is said to be the segment S clipped to R. The parts of the segments $p_1 p_2$, $p_3 p_4$ and $p_5 p_6$ clipped to R are indicated with color in Figure 14.1(a), while the output for $p_7 p_8$ is *empty* (the end points of the colored segments are not all labeled to avoid clutter).

Outcodes and Trivial Termination

Critical to Cohen-Sutherland is a classification of points of the plane, according to their disposition with respect to the input rectangle R, by means of so-called outcodes. A point $p = (x, y)$ is said to have *outcode* the 4-bit string $k_3 k_2 k_1 k_0$, whose value is determined by comparing the x- and y-values of p with those of the edges of R as follows:

- $k_0 = 0$, if $x \geq a$; $k_0 = 1$, if $x < a$

- $k_1 = 0$, if $x \leq b$; $k_1 = 1$, if $x > b$

- $k_2 = 0$, if $y \geq c$; $k_2 = 1$, if $y < c$

- $k_3 = 0$, if $y \leq d$; $k_3 = 1$, if $y > d$

It's easily seen that the four infinite straight lines, viz., $x = a$, $x = b$, $y = c$ and $y = d$, bounding R divide the plane into nine regions by outcode, as indicated in Figure 14.1(b). All points in each region have the outcode with which the region is labeled. For each of the lines $x = a$, $x = b$, $y = c$ and $y = d$, call the side of it containing R the *inside*, the other the *outside*. The line itself is included on its inside (so, e.g., the inside of $x = a$ is the closed half-plane $x \geq a$, while its outside is the open half-plane $x < a$). The rules above say, then, that each of the four lines determines an outcode bit, which is 0 if the point is on its inside, 1 if it's outside.

Cohen-Sutherland begins with the first step of determining the outcodes o_i and o_j of the endpoints p_i and p_j, respectively, of the input segment S, using the rules above. The next step is to determine if one of the following two cases applies, when the algorithm can immediately return the answer and terminate:

(a) S lies entirely inside R. In this case, both outcodes o_i and o_j are 0000, which can be verified by performing the logical bitwise operation $o_i \vee o_j$ and checking that the result is false, i.e., 0000.

(b) S lies entirely outside one of the four straight lines bounding R. In this case, both outcodes must have a 1-bit in the same position, which can be verified by performing the operation $o_i \wedge o_j$ and checking that the result is true, i.e., *not* 0000.

If (a) holds, the algorithm *trivially accepts*, returning the endpoints of S itself; if (b) holds, it *trivially rejects*, returning *empty*.

Recursion

If the outcome of the second step is neither a trivial accept nor a trivial reject, then Cohen-Sutherland proceeds recursively as follows.

There exists a bit position where o_i and o_j differ for, otherwise, one of the cases (a) and (b) above would have occurred. Scan the bits of o_i and

o_j from right to left to find the first (i.e., lowest) bit position, say the rth, where they differ. It follows that p_i and p_j lie on opposite sides of the infinite straight line bounding R, call it L, corresponding to the rth bit. Therefore, S intersects L and the algorithm calculates the point of intersection q. As they lie on opposite sides of L, exactly one of p_i and p_j lies inside L. The algorithm, then, calls itself recursively on the segment $p_i q$ or $p_j q$, according as p_i or p_j is inside L.

The next example illustrates Cohen-Sutherland.

Example 14.1. Apply the Cohen-Sutherland algorithm to the segment $p_1 p_2$ in Figure 14.1(a).

Answer: The rectangle R and segment $p_1 p_2$ of Figure 14.1(a) are drawn separately in Figure 14.2. The outcodes of p_1 and p_2 are 1001 and 0010, respectively, and neither of the two conditions that terminate Cohen-Sutherland trivially holds.

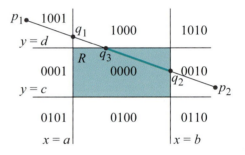

Figure 14.2: Cohen-Sutherland, called on segment $p_1 p_2$, recursively calls itself on $p_2 q_1$, $q_1 q_2$ and $q_2 q_3$, successively.

Scanning the bits of the two outcodes from right to left, they are seen to differ in the rightmost bit k_0, corresponding to the line $x = a$. Accordingly, the intersection q_1 of $p_1 p_2$ with $x = a$ is determined. As p_2 lies inside $x = a$, a recursive call is made on the segment $p_2 q_1$.

The outcodes of p_2 and q_1 are, respectively, 0010 and 1000. Again, there is no trivial termination. The rightmost bit at which the outcodes differ is k_1, so the intersection q_2 of $p_2 q_1$ with $x = b$ is computed. As q_1 lies inside $x = b$, a recursive call is made next on the segment $q_1 q_2$.

The outcodes of q_1 and q_2 are, respectively, 1000 and 0000. There is no trivial termination. The rightmost bit the outcodes differ at is k_3, so the intersection q_3 of $q_1 q_2$ with $y = d$ is computed. As q_2 lies inside $y = d$, a recursive call is made on the segment $q_2 q_3$.

The outcodes of q_2 and q_3 are both 0000 and the call terminates trivially, returning q_2 and q_3.

Exercise 14.1. Apply Cohen-Sutherland to the other three segments in Figure 14.1(a).

Exercise 14.2. Show that the maximum number of times Cohen-Sutherland can recursively call itself on a single input segment is four. Give an example of a segment where, in fact, four calls are made.

Exercise 14.3. (Programming) Animate Cohen-Sutherland using OpenGL. Draw a fixed clipping rectangle R, but allow the user to specify the endpoints of an arbitrary segment. Subsequently, highlight its subsegments as they are recursively processed.

The following two exercises extend Cohen-Sutherland.

Exercise 14.4. Extend Cohen-Sutherland to handle *semi-infinite* segments. A semi-infinite segment S is specified by one finite endpoint p_1 and another point p_2 in the *direction* of which it is infinite. See Figure 14.3.

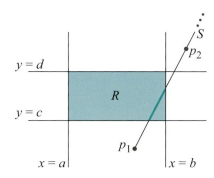

Figure 14.3: Clipping a semi-infinite segment to a rectangle.

Exercise 14.5. Extend Cohen-Sutherland to 3D: pseudo-code a 3D version to clip a straight-line segment S in 3-space against an axis-aligned box. See Figure 14.4. This is the actual clipper invoked in a 3D synthetic-camera rendering pipeline, like OpengGL's.

Questions to ponder: How does one define outcodes in 3D? How many bits are there in an outcode? Into how many regions is 3-space divided by outcode? Beyond these, need there be a significant difference between the 2D and 3D algorithms?

Extend your 3D clipper to handle semi-infinite segments as well.

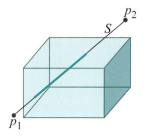

Figure 14.4: Clipping a line segment to an axis-aligned box.

Complexity

The complexity of Cohen-Sutherland lies mainly in the intersection computation resulting if the logical operation on the outcodes in the second step fails to terminate the algorithm trivially. Suppose the endpoints of the input segment S are $p_1 = (x_1, y_1)$ and $p_2 = (x_2, y_2)$ as in Example 14.1,

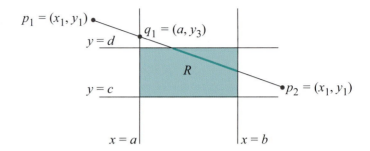

Figure 14.5: Cohen-Sutherland intersection computation.

drawn again in Figure 14.5. The first intersection to be calculated is $q_1 = (x_3, y_3)$, where S and $x = a$ meet.

The slope-intercept form of the equation of the straight line on which S lies is

$$y = \frac{y_2 - y_1}{x_2 - x_1} x + x_1$$

Obviously, $x_3 = a$, so y_3 can be found by plugging $x = a$ into this equation to get

$$y_3 = \frac{y_2 - y_1}{x_2 - x_1} a + x_1$$

Intersections with $x = b$, $y = c$ and $y = d$ can be found similarly as required.

If the input segments are pre-processed to determine their slope-intercept form, e.g., in the case above this entails pre-computing $\frac{y_2 - y_1}{x_2 - x_1}$, then, evidently, one floating point multiplication and one addition is performed per intersection finding.

Note: When pre-processing the slope-intercept form, one has to be careful to check for vertical line segments for which this form is not defined. However, once detected, vertical segments are obviously easy to clip to any rectangle.

We know from Exercise 14.2 that a call to Cohen-Sutherland to clip an input segment spawns at most four more recursive calls. The conclusion then is that a call to Cohen-Sutherland may require four floating point multiplications and four additions per input segment beyond pre-processing, in the worst case, which is fairly expensive. Refinements of Cohen-Sutherland, e.g., Liang-Barsky [82], invest in greater pre-processing in order to reduce subsequent intersection computation.

14.2 Sutherland-Hodgeman Polygon Clipper

The Sutherland-Hodgeman strategy for clipping a convex polygon P to a rectangle R is to successively clip off parts of P lying outside the four straight lines bounding R – the outside being the side of the straight line not containing R. The process can be conceived of as a pipeline of four clippers, as in Figure 14.6. The clipping rectangle is drawn only in outline in this section.

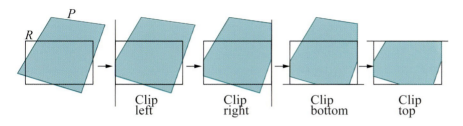

Clip left   Clip right   Clip bottom   Clip top

Figure 14.6: A pipeline of clippers.

Remark 14.2. There is no particular merit to the left-right-bottom-top ordering in Figure 14.6 – any other ordering could have been chosen.

The implementation of the four clippers is similar and we explain in detail only the leftmost one, which clips off outside the line along the left of R. In general, the input to this clipper is an ordered list $\{v_0, v_1, \ldots, v_n\}$ of the vertices of a convex polygon P and a vertical straight line L. See Figure 14.7. The output is an ordered list of vertices of the polygon P' resulting from clipping P off to the left of L.

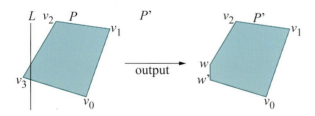

Figure 14.7: The left clipper in action: input $= \{v_0, v_1, v_2, v_3\}$, output $= \{v_0, v_1, v_2, w, w'\}$.

Output Rules for Left Clipper

The input list $\{v_0, v_1, \ldots, v_n\}$ of vertices is processed, in fact, in successive pairs, plus a final pair containing the first and last vertices, in particular, $v_0 v_1, v_1 v_2, \ldots, v_n v_0$. Equivalently, processing is edge by edge around the polygon P. Each edge outputs zero, one or two vertices to the output list. The output of an edge depends on the respective disposition of its end vertices with respect to L, in particular, if either is inside or outside L.

There is, therefore, a total of four possible dispositions. The four output rules, one corresponding to each disposition, are listed below. Refer to Figure 14.8 as you read (take v_{i+1} to be v_0, if v_i is v_n).

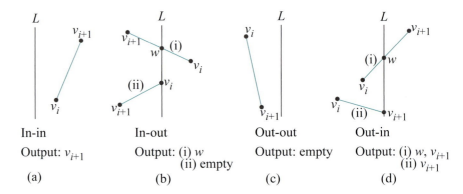

Figure 14.8: Output of a pair of successive vertices (equivalently, edge) entering the left clipper. Note both in-out and out-in dispositions have a special case, labeled (ii), where the vertex on the inside actually lies *on* L.

(a) In-in: Both v_i and v_{i+1} are inside L. Output is v_{i+1}.

(b) In-out: v_i is inside, while v_{i+1} outside L. There are two subcases, according as v_i lies strictly right of L or on it:

   (i) v_i is strictly right of L. Output is w, the point where the segment $v_i v_{i+1}$ intersects L.

   (ii) v_i lies on L. Output is *empty*.

(c) Out-out: Both v_i and v_{i+1} are outside L. Output is *empty*.

(d) Out-in: v_i is outside, while v_{i+1} inside L. There are two subcases, according as v_{i+1} lies strictly right of L or on it:

   (i) v_{i+1} is strictly right of L. Output is w, v_{i+1}, where w is the point where the segment $v_i v_{i+1}$ intersects L (this is the only case when two vertices are output).

   (ii) v_{i+1} lies on L. Output is v_{i+1}.

Observe that, if L is $x = a$, it is easy to determine if a point is inside or outside L by simply comparing its x-value with a. Therefore, it's easy as well to decide which of the four rules above to apply to each successive pair of vertices. We ask the reader next to determine the new vertex w, in case it arises, in the following exercise.

Exercise 14.6. Suppose $v_i = (x_i, y_i)$, $0 \le i \le n$, and L is the line $x = a$. Observe, from the rules above, that if a new vertex – one, that is, not belonging to the original input sequence – is at all output, then there is only one such. Give a formula to determine the new vertex.

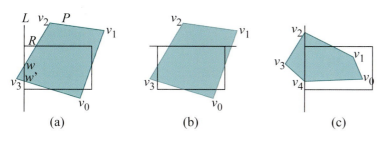

Figure 14.9: Applying the clipping rules.

Example 14.2. Let's apply the rules above to the initial polygon P, at the leftmost of Figure 14.6, and the vertical straight line L along the left edge of the rectangle R. See Figure 14.9(a). The output of successive vertex pairs is as follows:

$$
\begin{aligned}
v_0 v_1 &\Rightarrow v_1 &\text{(in-in)} \\
v_1 v_2 &\Rightarrow v_2 &\text{(in-in)} \\
v_2 v_3 &\Rightarrow w &\text{(in-out (i))} \\
v_3 v_0 &\Rightarrow w', v_0 &\text{(out-in (i))}
\end{aligned}
$$

Accordingly, the vertex list returned by the left clipper is $\{v_1, v_2, w, w', v_0\}$, which indeed is the sequence of vertices around the polygon resulting from clipping off the part of P outside L.

Exercise 14.7. Clip the initial polygon P of Figure 14.6 to the top, in other words, apply the top clipper to it first. See Figure 14.9(b). Rules, exactly as for those given already for the left clipper, apply to the top.

Exercise 14.8. Clip the polygon P of Figure 14.9(c) to the left.

Pipelining

Now that we understand the implementation of its individual clippers, we come to the beauty of the Sutherland-Hodgeman algorithm: that it can be *pipelined* with all four clippers running in *parallel*, each one after the first using as input the output of its predecessor. This follows from observing that each clipper *incrementally* produces its output list as it *incrementally* consumes its input vertices. Therefore, the next clipper in the sequence does not have to wait till its predecessor completes processing – it can begin to operate *as soon* as it receives the first two vertices output by its predecessor.

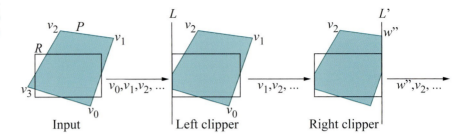

Figure 14.10: The right clipper consumes the first two vertices v_1 and v_2 entering into it from the left clipper to output w'' and v_2.

For example, we saw in Example 14.2 that the successive vertices output by the left clipper operating on P are $\{v_1, v_2, w, w', v_0\}$. Observe, now, that as soon as v_1 and v_2 enter the next clipper in the sequence – the right one according to the scheme in Figure 14.6 – the latter can process them to output w'', v_2. See Figure 14.10, where the disposition of v_1 and v_2 with respect to L' is case (i) of out-in.

The Sutherland-Hodgeman pipeline of four clippers is often implemented in hardware, the clippers being identical but separate modules.

Exercise 14.9. Assuming that each clipper takes unit time to perform the operation of applying one of the rules of Figure 14.8 to a pair of successive vertices and that vertices move from one clipper to the next in zero time, how long does it take for the clipping pipeline of Figure 14.6 to process the particular polygon P in that figure?

Exercise 14.10. (Programming) Code and creatively animate the Sutherland-Hodgeman clipping pipeline.

Exercise 14.11. Extend Sutherland-Hodgeman to 3D to clip a convex polygon against an axis-aligned box. This, in fact, is simpler than generalizing the Cohen-Sutherland line clipper to three dimensions, which we considered in Exercise 14.5, because the Sutherland-Hodgeman output rules go through pretty much verbatim in one higher dimension, except that the part of the polygon to one side of a plane, rather than a line, is clipped off.

14.3 DDA and Bresenham's Line Rasterizers

Rasterization of a straight line segment consists of picking and coloring the pixels to comprise that segment, given its start and end vertices and their color attributes. Both the DDA and Bresenham's algorithm actually only pick pixels. Coloring them is a straightforward application of linear interpolation that we learned in Chapter 7, except for the added twist of "perspective correction" needed in case of perspective projection. We'll fill out details of the coloring process when we examine the synthetic-camera pipeline in Chapter 19.

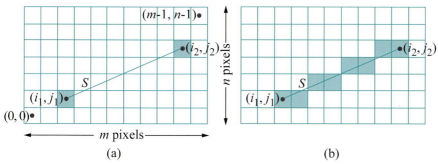

Figure 14.11: An $m \times n$ raster of pixels and a rasterized straight line segment.

Let's take the raster to be a rectangular $m \times n$ grid of square pixels, each of side length one, located axis-aligned on a plane, so that the pixel centers have integer coordinates (i, j), $0 \le i \le m - 1$ and $0 \le j \le n - 1$. See Figure 14.11(a).

We'll assume that both end vertices of an input segment S have already been projected and scaled onto the raster – shot on film and printed according to the analogy in the second chapter – so that they lie at the pixels centered at (i_1, j_1) and (i_2, j_2), respectively, as shown in Figure 14.11(a).

Note: Even though S is actually a segment in world space, we'll call its image on the raster S as well – which of the two we mean should be clear from the context.

The rasterization task then is to choose the pixels between its two ends to represent S, as, say, in Figure 14.11(b). We can assume that one end of S lies strictly to the right of the other because, otherwise, either both ends are at the same pixel or one is vertically above the other, both cases being trivial to rasterize. Suppose, without loss of generality then, that (i_2, j_2) is to the right of (i_1, j_1), meaning $i_2 > i_1$. The line equation of S is:

$$\frac{y - j_1}{x - i_1} = \frac{j_2 - j_1}{i_2 - i_1} \qquad \text{or} \qquad y = m(x - i_1) + j_1 \qquad (14.1)$$

where $m = \frac{j_2 - j_1}{i_2 - i_1}$ is the slope of S.

DDA Algorithm – Floating Point Heavy

We'll warm up with the very simple DDA (Digital Differential Analyzer) rasterization algorithm which Bresenham subsequently improves.

Remark 14.3. The rather fancy name Digital Differential Analyzer comes first from Differential Analyzer, a class of mechanical machines invented in the 1800s to solve differential equations. A Digital Differential Analyzer, or DDA, is a digital version of the Differential Analyzer. The DDA line rasterizer implements an incrementing loop borrowed from the DDA, hence the name.

Suppose, first, that its slope m lies between -1 and 1, so that the input segment S makes an angle of at most $45°$ with the x-axis (as in Figure 14.11(a)). It's clear in this case that a rasterization of S should contain exactly one pixel per x-value within the x-span of S. Moreover, the equation on the right of (14.1) implies that the y-value along S increases by m as the x-value increases by 1. This motivates the following algorithm:

```
// DDA Line Rasterizer
// Assume i2 > i1 and -1 <= m <= 1
float y = j1;
float m = (j2 - j1)/(i2 - i1);
for (int x = i1; x <= i2; x++)
{
    pickPixel(x, round(y));
    y += m;
}
```

Note: Due to obvious typesetting constraints we write variables such as i_1 and j_1 as i1 and j1 in the code snippet.

The algorithm starts with the pixel centered at (i_1, j_1), then increases the x-value by 1 and the y-value by m at each step, until x equals i_2. The pixel chosen at each step by the pickPixel() function follows rounding of the y-value to an integer.

Experiment 14.1. Run DDA.cpp, which is pretty much a word for word implementation of the DDA algorithm above. A point of note is the *simulation* of the raster by the OpenGL window: the statement gluOrtho2D(0.0, 500.0, 0.0, 500.0) identifies pixel-to-pixel the viewing face with the 500×500 OpenGL window.

There's no interaction and the endpoints of the line are fixed in the code at $(100, 100)$ and $(300, 200)$. Figure 14.12 is a screenshot. **End**

Figure 14.12: Screenshot of DDA.cpp.

Exercise 14.12. (Programming) The restriction on the slope m in the DDA rasterizer can be removed by noting that a rasterization of S should contain exactly one pixel per y-value in its y-span, if it makes an angle of more than $45°$ with the x-axis.

Accordingly, rewrite DDA.cpp so that there are no restrictions at all and that it interactively accepts arbitrary input end vertices (i_1, j_1) and (i_2, j_2).

The problem with the DDA algorithm is that it invokes two floating point operations per loop iteration – an addition and a rounding – floating point operations always being computationally expensive. We'll see next how Bresenham manages to avoid floating point.

Eliminating Floating Points – Bresenham's Algorithm

Bresenham's algorithm avoids floating point operations altogether. Here's the idea in a nutshell. The algorithm incrementally picks one pixel after another starting with the left end vertex of S, the smarts being in using the location of the current pixel to cost-effectively deduce that of the next.

Assume first that the slope of S satisfies $0 < m < 1$; we'll handle other cases later. Given this constraint, there is exactly one pixel of S for each x-value within its x-span.

Suppose that pixel (i, j) has just been picked (pixel (i, j) means, of course, the pixel centered at (i, j)). See Figure 14.13(a), where only pixel centers are shown in a grid. One can assume that the point $p = (p_x, p_y)$ of the intersection of S with the line $x = i$ is at least as close to (i, j) as it is to the centers of either pixel above or below it, for, otherwise, pixel (i, j) would not have been chosen. In other words, $j - \frac{1}{2} \leq p_y \leq j + \frac{1}{2}$. These inequalities, together with the condition $0 < m < 1$ on the gradient of S, imply that S intersects $x = i + 1$ at the point $q = (q_x, q_y)$, where $j - \frac{1}{2} < q_y < j + \frac{3}{2}$, as we'll show next.

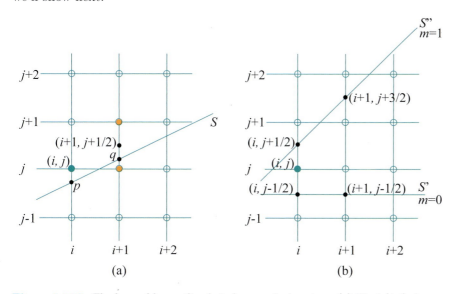

(a)   (b)

Figure 14.13: The larger blue-outlined circles are pixel centers. (a) Pixel (i, j) shown filled blue has just been chosen, the two candidate pixels for the next step are filled orange (b) Diagram for Example 14.3.

Example 14.3. Prove the inequalities claimed on q_y in the preceding statement.

Answer: Let the line S' go through $(i, j - \frac{1}{2})$ with gradient 0 and the line S'' go through $(i, j + \frac{1}{2})$ with gradient 1. See Figure 14.13(b), where S' and S'' are shown but not S. Since $j - \frac{1}{2} \leq p_y \leq j + \frac{1}{2}$ and $0 < m < 1$, S intersects

$x = i + 1$ at a point $q = (q_x, q_y)$ strictly between the points where S' and S'' intersect $x = i + 1$. It's straightforward geometry to see that S' intersects $x = i + 1$ at $(i + 1, j - \frac{1}{2})$ and S'' intersects $x = i + 1$ at $(i + 1, j + \frac{3}{2})$. It follows that, indeed, $j - \frac{1}{2} < q_y < j + \frac{3}{2}$.

As $j - \frac{1}{2} < q_y < j + \frac{3}{2}$, the pixel chosen when x is incremented to $i + 1$ is either $(i + 1, j)$ or $(i + 1, j + 1)$, depending on which of the two is closer to q. Now, q is closer to $(i + 1, j)$ if the midpoint $(i + 1, j + \frac{1}{2})$ between the two candidate pixels is above S (the situation depicted in Figure 14.13(a)); it's closer to $(i + 1, j + 1)$ if the midpoint is below S.

Evidently, we need to determine the disposition of q with respect to S. So, how does one generally determine the disposition of some given point (x, y) with respect to S? Rewrite the line equation of S, on the left of Equation (14.1), as

$$(i_2 - i_1)y - (j_2 - j_1)x + i_1 j_2 - i_2 j_1 = 0$$

Denote the LHS of the preceding equation by $D'(x, y)$, i.e.,

$$D'(x, y) = (i_2 - i_1)y - (j_2 - j_1)x + i_1 j_2 - i_2 j_1 \qquad (14.2)$$

Therefore, if (x, y) satisfies $D'(x, y) = 0$, then the point lies on the straight line containing S. In fact, it lies on S itself if, additionally, its x-value lies within the x-span of S. Furthermore, (x, y) lies above that straight line if $D'(x, y) > 0$ and below if $D'(x, y) < 0$. See Figure 14.14.

Exercise 14.13. Verify the claim made in the last statement.

Hint: Consider a point (x, y) lying on the straight line containing S, so that $D'(x, y) = 0$. If only its y-value is increased to raise it above the line, as indicated in Figure 14.14, then the value of $D'(x, y)$ increases because the coefficient of y in the formula (14.2) for $D'(x, y)$ is positive.

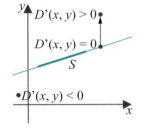

Figure 14.14: Discriminating the position of a point with respect to a segment: S is bold, while the straight line through it is thin.

For a reason which will soon be apparent, we'll use

$$D(x, y) = 2D'(x, y) = 2(i_2 - i_1)y - 2(j_2 - j_1)x + 2(i_1 j_2 - i_2 j_1) \qquad (14.3)$$

whose sign is always the same as that of $D'(x, y)$, as the *discriminant* to determine the disposition of (x, y) with respect to the straight line on S.

Returning to the question of which of the candidate pixels $(i + 1, j)$ or $(i + 1, j + 1)$ to choose when $x = i + 1$, we see the answer now as:

- Choose $(i + 1, j)$ if $D(i + 1, j + \frac{1}{2}) > 0$ (when the midpoint $(i + 1, j + \frac{1}{2})$ between the candidate pixels lies above S, implying that S is closer to the center of the lower one).

- Choose $(i + 1, j + 1)$ if $D(i + 1, j + \frac{1}{2}) < 0$ (complementary to the first case).

- Choose $(i+1, j+1)$ if $D(i+1, j+\frac{1}{2}) = 0$ (this choice being made arbitrarily as S is equidistant from the centers of both candidates).

We near the crux of Bresenham's algorithm, which is to use the old value of D to compute its new value after x is incremented once again to $i+2$. There are two cases:

(a) $D(i+1, j+\frac{1}{2}) > 0$, which implies, according to the pixel-choosing process just described, that $(i+1, j)$ was chosen when $x = i+1$.

Repeating the reasoning that took us from $x = i$ to $x = i+1$, we see that the two candidate pixels when $x = i+2$ are $(i+2, j)$ and $(i+2, j+1)$; moreover, $(i+2, j)$ is chosen if $D(i+2, j+\frac{1}{2}) > 0$ and $(i+2, j+1)$ if $D(i+2, j+\frac{1}{2}) \leq 0$. As for the new value of the discriminant, we ask the reader to verify from (14.3) that it can be calculated from its previous value by:

$$D(i+2, j+\frac{1}{2}) = D(i+1, j+\frac{1}{2}) - 2(j_2 - j_1) \qquad (14.4)$$

(b) $D(i+1, j+\frac{1}{2}) \leq 0$, which implies that $(i+1, j+1)$ was chosen when $x = i+1$.

The two candidate pixels when $x = i+2$ are now $(i+2, j+1)$ and $(i+2, j+2)$; moreover, $(i+2, j+1)$ is chosen if $D(i+2, j+\frac{3}{2}) > 0$ and $(i+2, j+2)$ if $D(i+2, j+\frac{3}{2}) \leq 0$. Again, we ask the reader to verify that:

$$D(i+2, j+\frac{3}{2}) = D(i+1, j+\frac{1}{2}) + 2(i_2 - i_1 - j_2 + j_1) \qquad (14.5)$$

We have in hand now the crux of Bresenham's algorithm: Equations (14.4) and (14.5) together say how D changes as x is incremented by 1. It only remains to get the algorithm started by initializing D. The first pixel picked is at the left endpoint (i_1, j_1) of S. Accordingly, the candidate pixels for the next value of x, namely, $x = i_1 + 1$, are $(i_1 + 1, j_1)$ and $(i_1 + 1, j_1 + 1)$, whose midpoint is $(i_1, j_1 + \frac{1}{2})$. Therefore, the first value of D is

$$D(i_1 + 1, j_1 + \frac{1}{2}) = i_2 - i_1 - 2(j_2 - j_1) \qquad (14.6)$$

It's here that taking $D(x, y) = 2D'(x, y)$, rather than $D'(x, y)$ itself, pays off. For, $D'(i_1 + 1, j_1 + \frac{1}{2}) = \frac{1}{2}(i_2 - i_1) - (j_2 - j_1)$, which may be fractional (remember, we want only integers to stay away from floating point calculation).

With (14.4)-(14.6) all pieces now are in place to implement Bresenham's strategy: initialize D using (14.6) and then successively increment it with the help of (14.4)-(14.5), picking at each step the next pixel based on the sign of D. Here's code:

```
// Bresenham's Line Rasterizer
// Assume i2 > i1 and 0 < m < 1
int y = j1;
int diff1 = -2*(j2 - j1);
int diff2 =  2*(i2 - i1 - j2 + j1);
int D = i2 - i1 - 2*(j2 - j1);
for (int x = i1; x <= i2; x++)
{
    pickPixel(x, y);
    if (D > 0) D += diff1;
    else y++; D += diff2;
}
```

Compare this with the DDA algorithm earlier which required floating point rounding and addition – there's nary a floating point value in sight in Bresenham's procedure! Bresenham is extremely efficient and, often, implemented in the graphics card itself for even higher rendering speeds.

Exercise 14.14. Use Bresenham's Line Rasterizer as coded above to pick the pixels on the straight line segment joining pixel centers $(9, 17)$ and $(25, 25)$.

Part answer: We'll get the reader started. Set $(i1, j1) = (9, 17)$ and $(i2, j2) = (25, 25)$ in the code above. Accordingly,

$$\begin{aligned} diff1 &= -2 * (j2 - j1) = -16 \\ diff2 &= 2 * (i2 - i1 - j2 + j1) = 16 \\ D &= i2 - i1 - 2 * (j2 - j1) = 0 \quad \text{(initially)} \end{aligned}$$

The first pixel picked is the left endpoint $(9, 17)$. As $D = 0$, the next pixel picked is $(10, 18)$ and D changes to $D + diff2 = 16 > 0$. Therefore, the next pixel picked is $(11, 18)$ and D changes to $D + diff1 = 0, \ldots$

The reader should sketch the rasterization on graph paper and verify that it indeed starts and ends at the given endpoints.

Exercise 14.15. (Programming) We ask the reader now to implement Bresenham's algorithm, simulating the raster using the OpenGL window, as in DDA.cpp. Remove, as well, the assumptions in our formulation of Bresenham's algorithm. In particular, allow the two input end vertices (i_1, j_1) and (i_2, j_2) to be arbitrary.

Moreover, ensure that your program is not sensitive to the order that the end vertices are input. In other words, the rasterization should be identical if (i_1, j_1) and (i_2, j_2) are swapped in the routine. This is as a user would expect: it should not matter in what order the end vertices happen to appear in the segment definition in code.

14.4 Scan-Based Polygon Rasterizer

We'll describe a commonly-implemented scan-based rasterization algorithm to draw a polygon, or *fill* it, as is commonly said, because pixels comprising the interior of the polygon are selected. The notion of scanning arises from CRT technology where an electron gun repeatedly scans the screen, pixel row by pixel row. One row of pixels is called a *scan line*. Although rasterization algorithms are implemented when filling pixel values in the frame buffer and have really nothing to do with the front-end display technology, this particular algorithm for polygon rasterization happens to mimic the scan process, hence the CRT reference.

A central part of the scan-based algorithm is to use the so-called *parity test*, also called the *inside-outside test*, to determine if a point lies inside or outside a polygon P.

Parity Test

Suppose P is a simple planar polygon, i.e., one which lies on a plane and whose boundary is a single line loop which doesn't self-intersect. Let q be a point known to lie outside P (possibly, by choosing q's x or y value to be either very large or very small). Now, suppose we wish to determine if another given point p lies inside or outside P. Consider the ray R from q to p. The ray, obviously, starts from outside P. Upon its first intersection (if any) with the boundary of P, R enters P; upon its second intersection with the boundary of P, it exits P; upon its third intersection, it re-enters P; and so on. See Figure 14.15.

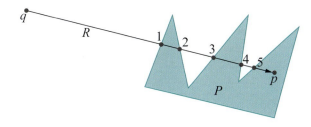

Figure 14.15: Intersections of R with the boundary of P are labeled with their respective ranks.

This leads to the following test to decide if p lies inside P or outside.

Parity Test: If the ray R from a point q, known to lie outside P, to some given point p, intersects the boundary of P an odd number of times, then p lies inside P, while if it intersects the boundary of P an even number of times, then P lies outside P. We assume, when applying the parity test, that p itself does not lie *on* the boundary of P, i.e., it is either inside or outside.

The parity of the number of intersections of R with the boundary of P, odd or even, is often called the *parity* of p. So the parity test may be rephrased to say that points of odd parity lie inside P, while those of even parity lie outside.

Note: If R does not intersect the boundary of P at all, then P's parity is that of zero, which is even, and, of course, p lies outside P.

Exercise 14.16. For the five points p_1, p_2, p_3, p_4, p_5 in Figure 14.16, use the parity test to verify where they lie with respect to P. The rays from an outside point q are already drawn. Consider how you would handle the particular singularities seen in case of p_3 and p_4 without revising the statement of the parity test itself (we'll be discussing this matter in some detail momentarily).

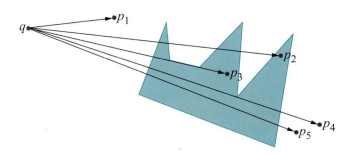

Figure 14.16: Testing parity: the points p_3 and p_4 require handling a singularity.

See Figure 14.17. The integer label beneath a segment of the semi-infinite ray R from q indicates the number of intersections of R with the boundary of P before it reaches that segment. Therefore, all points in the *interior* of a segment share the same parity (the restriction to the interior is to avoid endpoints lying on the boundary of P, whose parity is not defined). For example, points in the interior of qp_1 have even parity, those in the interior of p_1p_2 odd parity and so on. The segments w and w' are not labeled because they each lie entirely on the boundary.

One has to be careful, evidently, when counting intersections, to take into account edges such as w and w', as well as points such as p_6, where the ray touches the polygon without properly intersecting it. Here are the conventions to follow, and which have been followed in Figure 14.17, to ensure that these singularities are handled correctly.

Conventions for Singularities:

(a) When the ray R passes through a vertex v, but neither of the edges adjacent to v lie along R, there are two cases:

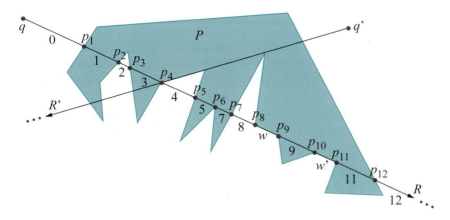

Figure 14.17: Applying the parity test: the integer label beneath a segment of R indicates the number of intersections of R with the boundary of P prior to reaching that segment. The reason for the change from 5 to 7 after p_6 and from 9 to 11 after edge w' is explained in the text. The edges w and w' are not labeled because they lie on the boundary.

  (i) if these two edges lie on opposite sides of R, then v is a proper intersection point and counted once (p_4 in Figure 14.17);

  (ii) otherwise, R *touches* P at v, and v is counted as two intersections (p_6, explaining the difference in count on either side of it).

(b) When the ray R passes through a vertex v and one of the edges, say e, adjacent to v lies along R (we'll assume that successive edges of P are never collinear, so that both edges adjacent to v cannot lie on R):

  (i) if the two edges adjacent to e lie on opposite sides of R, then count the entire edge e as one intersection (w in Figure 14.17);

  (ii) otherwise, count the entire edge e as two intersections (edge w').

In other words, case (b) is the same as (a) if one imagines e as one giant vertex.

Exercise 14.17. Label the segments of the ray R' emanating from the point q' in Figure 14.17 in a manner similar to that of segments along R.

Exercise 14.18. (Programming) Implement the parity test. Allow the user to select a polygon P, as well as a test point p, by clicking on the OpenGL window. Either ask the user to, or automatically, choose a point q outside P. Then display the working of the test with some creative animation. For example, you might show the ray from q approaching p, highlighting and labeling intersections and segments on the way.

It's easy now to explain the strategy of the scan-based polygon-filling algorithm. Treating each scan line as a horizontal ray starting from some point far to the left of the screen, pixels along it are selected to fill a polygon P according to the parity test. Figure 14.18 shows one such scan line. In particular, each scan line is split into segments, each with endpoints at intersections with the boundary of P. Moreover, pixels in one segment all have the same parity. The result is alternate runs of pixels filling and not filling P.

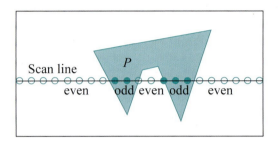

Figure 14.18: The scan line as a ray with parities along segments indicated. Pixels fill the polygon according to the parity test: pixels in the polygon are drawn solid, others hollow.

Before we proceed to describe an algorithm for scan-based polygon filling, we need first to resolve how to deal with pixels which happen to lie on the boundary of a polygon, whose parity, therefore, is indeterminate. We do this next.

Ownership of Boundary Pixels

One must be cautious in scan-based filling about pixels lying *on* the boundary of a polygon. For example, given a couple of *abutting* polygons – ones whose edges overlap, e.g., P and Q in Figure 14.19 – one has to decide ownership of the pixels along the overlapping part, which, in turn, decides the coloring of these pixels. Clearly, it is desirable to do this in a consistent manner, independent of the order in which the polygons happen to appear in the code.

For example, if each polygon is awarded ownership of all its boundary pixels, then the color of shared boundary pixels depends, in fact, on the order in which the abutting polygons appear in the code, the color of the one appearing last prevailing. At the other extreme, if boundary pixels are excluded altogether from a polygon, then there will arise gaps in the rendering between abutting polygons.

Remark 14.4. If one follows the rules of triangulation as described in Chapter 8, then there will be no anomalies, as we saw then, if every triangle simply owns its boundary. However, the intention here is to set up rules robust enough to hold *even* given an invalid triangulation, as in Figure 14.19.

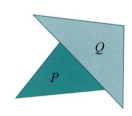

Figure 14.19: Abutting polygons.

A simple and oft-implemented method to resolve the ownership of a polygon's boundary and, accordingly, boundary pixels is by means of the following:

Edge Ownership Rule: Among its non-horizontal edges, a polygon owns only the *left* edges, while, among its horizontal edges, it owns only the *bottom* ones. A vertex shared by a left and right edge is owned only if both incident edges extend above the vertex; it is not owned otherwise.

Note: A left edge of a polygon is one such that its interior lies to the right of the edge; the edge does not have to be physically located at a left extreme of the polygon. Similar remarks apply to right, bottom and top edges. Figure 14.20 labels the edges of the polygon P accordingly.

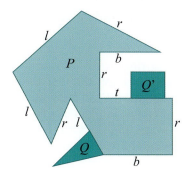

Figure 14.20: The non-horizontal edges of P are labeled left or right, while the horizontal ones top or bottom. P owns the part of the boundary it shares with Q; P does not own the part of the boundary it shares with Q'.

Remark 14.5. The rule above does not resolve the ownership question in every case of a vertex being incident to one edge which is owned by the polygon, and and another which is not. Generally, this depends on the particular filling algorithm implemented.

In Figure 14.20, from the edge ownership rule, P owns the pixels on the part of the edge it shares with Q, while those on the part of the edge it shares with Q' are owned by the latter. The two pixels in Figure 14.18 on the boundary of the polygon P have been processed according to this rule as well.

Remark 14.6. The edge ownership rule explains why the expected ambiguity in edge color did not arise in Exercise 8.3.

14.4.1 Algorithms

Assume that P is a simple polygon input as the list of its edges, each edge specified by the coordinates of its end vertices. Here's a first cut at an algorithm to rasterize P:

<u>Scan-based Polygon-filling Algorithm (Version 1)</u>
for each scan line s
{

    1. for each non-horizontal edge e of P intersecting s
           determine the intersection point between s and e;

    2. sort the points from the preceding step from left to right along s
       in a list $p_1, \ldots, p_k$;

    3. fill, as belonging to P, pixels *strictly* between each of the pairs
       p_1 and p_2, p_3 and p_4, $\ldots$;

    4. if a pixel coincides with a p_i, which means it's a boundary pixel,
       fill it or not according as i is odd or even;

    5. for each horizontal edge e intersecting s
           include/exclude e according as it is a bottom/top edge;

}

The reader will agree that this algorithm is at least a forthright attempt to implement the scan-based strategy: the first three statements inside the bracketed **for** loop apply the parity test to select alternate runs of pixels to fill, while the fourth statement implements the disambiguation rule for ownership of left and right edges and the fifth that for bottom and top edges.

However, upon a more careful examination, Version 1 is seen to have three significant flaws:

(1) When a scan line passes through a vertex whose adjacent edges are non-horizontal and lie on either side of the scan line, that vertex is listed twice as an intersection point, in violation of clause (i) of convention (a) for singularities listed earlier.

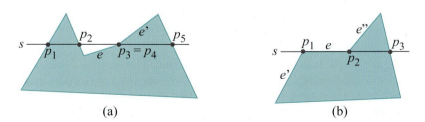

(a)   (b)

Figure 14.21: Problems with Version 1.

For example, in Figure 14.21(a), the sorted list of intersection points for scan line s is $\{p_1, p_2, p_3, p_4, p_5\}$, where p_3 ($= p_4$) appears twice on the list, once as an intersection with the edge e and once as an intersection with edge e', leading to pixels between p_4 and p_5 being wrongfully left unfilled by statement 3 of Version 1.

(2) If a pixel coincides with more than one p_i – in particular, if it's at a vertex adjacent to two non-horizontal edges – how do we decide to include/exclude it when executing statement 4?

For example, in Figure 14.21(a), do we treat the pixel at the vertex shared by e and e' as a p_i with an odd subscript (i.e., p_3), or even subscript (p_4), at the time of processing statement 4?

(3) There may be ambiguity in rendering pixels along a horizontal edge because its vertices are those as well of its non-horizontal neighbors and will be processed as such for the scan line upon which it lies.

For example, in Figure 14.21(b), pixels in the interior of the horizontal edge e (strictly between p_1 and p_2) are filled when processing scan line s at statement 3, because s's sorted list of intersections is $\{p_1, p_2, p_3\}$. However, when processing e itself at statement 5, we find that these pixels should be excluded, as e is a top edge.

It looks like rescuing Version 1 will be tricky, but it turns out, in fact, that a fix is not hard at all. Just add in the following two rules:

(i) For a non-horizontal edge, don't list the intersection of a scan line with the upper endpoint of that edge.

(ii) Moreover, don't process horizontal edges at all.

That's it! Here's the amended algorithm with the few changes from Version 1 highlighted with comments:

Scan-based Polygon-filling Algorithm (Version 2)
for each scan line s
{

    1. for each non-horizontal edge e of P intersecting s
    {

        determine the intersection point between s and e;
        ignore this intersection point if it is the upper endpoint
        of e; // New line.

    }
    2. sort the points from the preceding step from left to right along s in a list $p_1, \ldots, p_k$;
    3. fill, as belonging to P, pixels *strictly* between each of the pairs p_1 and p_2, p_3 and p_4, $\ldots$;
    4. if a pixel coincides with a *single* p_i, which means it's a boundary pixel, fill it or not according as i is odd or even; // Modified.
    5. if a pixel coincides with *more than one* p_i, which also means it's a boundary pixel, fill it; // New line. Processing of horizontal edges
                    // is now omitted.

}

We'll soon see why this version is correct, but let's take it for a spin first. Figure 14.22 shows how many times each intersection point along the boundary of a polygon P appears in the sorted list for the scan line containing it according to statements 1 and 2 of Version 2. Based on these labels is an exercise next to run the new version (with a part answer).

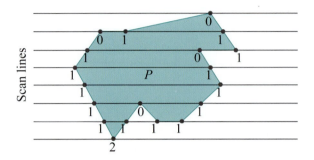

Figure 14.22: Counting intersections according to Version 2: shown are eight scan lines, their intersections with the polygon boundary and the number of times each intersection point appears in the sorted list for that scan line, according to statements 1 and 2 of algorithm Version 2. Polygon vertices and intersection points are not named to avoid clutter.

Exercise 14.19. Describe how pixels are filled along each of the eight scan lines drawn in Figure 14.22 by Version 2.

Part answer: We'll assume first that all the vertices of P are located exactly at pixels.

Bottom scan line: Denote the single vertex on this scan line by p_1 (note that the labels p_i are not drawn in the figure to avoid clutter). Then the sorted list on this scan line, after statements 1 and 2, is $\{p_1, p_1\}$. Only the pixel at p_1 is filled (by statement 5); all others on the scan line are not.

Second scan line: Denote the four intersection points along this scan line from left to right by p_1, p_2, p_3, p_4. Then the sorted list (statements 1 and 2) is $\{p_1, p_2, p_3, p_4\}$. All pixels strictly between p_1 and p_2 and strictly between p_3 and p_4 are filled (statement 3). If there is a pixel at p_1, it is filled (statement 4). The pixel at p_3 is filled (statement 4 again). All other pixels on the scan line are not filled.

Third scan line: Denote the three intersection points along this scan line from left to right by p_1, p_2, p_3. Then the sorted list (statements 1 and 2) is $\{p_1, p_3\}$. All pixels strictly between p_1 and p_3 are filled (statement 3). Note that the pixel at p_2 is filled. If there is a pixel at p_1, it is filled (statement 4). All other pixels on the scan line are not filled.

As the reader may have surmised from completing the exercise, the new version gets past the three problems of the earlier one as follows:

(1) Ignoring upper endpoints eliminates the first flaw.

(2) Noting that a pixel coincides with more than one p_i only when it is the bottom endpoint of two adjacent edges – upper endpoints are ignored remember – and, therefore, must be included removes the second flaw.

(3) Noting that the pixels on a horizontal edge lie as well between the endpoints of the two non-horizontal edges adjacent to it suggests that we can leave off separately processing horizontal edges altogether. Version 2 does so, resolving the last flaw.

Exercise 14.20. How are the problems with Version 1, as indicated particularly in Figures 14.21(a) and (b), resolved in Version 2?

14.4.2 Optimizing Using Edge Coherence – Active Edge List

Version 2 is pretty much ready to go as long as a couple of statements that require efficient implementation in code are kept in mind: in particular, statements 1 and 2 to determine and sort the intersection points of a scan line with polygon edges. These two tasks need not be done from scratch per scan line – which would be not be efficient at all – if one exploits so-called *edge coherence*.

This simple but very useful concept is illustrated in Figure 14.23, which shows an edge e straddling a run of scan lines. Now, if scan lines are processed in order, say from bottom to top, then e first appears in the intersection list for scan line s, remains in the intersection list for each scan line until s' and then disappears forever. Moreover, the intersection of e with successive scan lines clearly travels uniformly along the x-direction.

Consideration of edge coherence leads to the creation and maintenance of a particular dynamic data structure, the *active edge list* (AEL). The AEL is a linked list of records, one for each non-horizontal edge e intersecting the *current* scan line s (hence the qualifier "active"). The record for an edge e contains three data items:

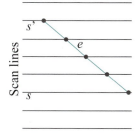

Figure 14.23: Edge coherence.

1. The y-value of the upper endpoint of e.

2. The reciprocal $1/m$ of e's gradient (as e can't be horizontal, $m \neq 0$).

3. The x-value of the intersection of e with s.

The reason for choosing these particular items will become apparent as we learn to process the AEL. The AEL, additionally, is sorted according to the left to right order of the intersections of its member edges with s, a left edge (of the polygon) preceding a right one if their intersections with s coincide. Observe that the first two items of each record are static, depending only on the particular edge e, while the value of the third item varies as s sequences through successive scan lines.

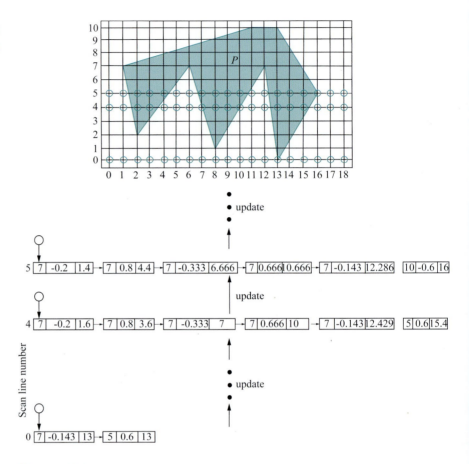

Figure 14.24: The AEL values for scan lines 0, 4 and 5 for a polygon P in a 19×11 raster. Pixels are drawn as hollow circles only for these three scan lines.

See Figure 14.24, which shows a polygon P in a 19×11 raster, as well as the AEL values for scan lines 0, 4 and 5. For example, the data values in the first record of the bottom AEL arise because it corresponds to the edge joining vertices at $(12, 7)$ and $(13, 0)$, the reciprocal of whose gradient, therefore, is $-1/7 \simeq -0.143$, and which intersects scan line 0 at x-value 13.

Exercise 14.21. Verify the data values in each record of each of the three AEL values shown in Figure 14.24.

Given the value of the AEL for a scan line s, it's straightforward to apply Version 2 to compute the runs of filled pixels along s by traversing the AEL from left to right, reading off the third data value from successive records.

Initialization – Edge Table

All that's left now to do is initialize the AEL data structure and say how to
update it from scan line to scan line. To this end, a bit of pre-processing
first is needed to put the non-horizontal edges of P into a data structure
called the *edge table* (ET).

The ET is an array of linked lists, one for each scan line. The linked
list for a scan line consists of one record for each non-horizontal edge that
has a lower endpoint on that scan line. Each record is similar in structure
to an AEL record and, in fact, the first two data items are identical, while
the third contains the x-value of the lower endpoint of the edge. Moreover,
each list is sorted according to the left to right order of its lower endpoints,
with a left polygon edge again preceding a right one if their lower endpoints
coincide.

Figure 14.25 shows the ET for the polygon P of Figure 14.24. The ET is
evidently a static data structure, consisting, essentially, of a bucket sort of
the non-horizontal edges keyed on their lower y-value.

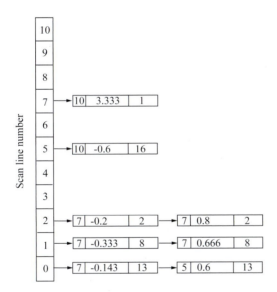

Figure 14.25: Edge table for the polygon P of Figure 14.24.

Processing the AEL

With the ET in hand, processing the AEL is straightforward. Set $j = -1$, the
number of an imaginary scan line just below the drawing area and initialize
the AEL to empty. To update the AEL from scan line j to scan line $j + 1$,
do the following:

Updating the AEL

for each record in the AEL

{

1. if the first data value (i.e., the y-value of the upper endpoint of the corresponding edge) equals $j + 1$, delete the record;

2. else, update the third data value (i.e., the x-value of the intersection of the edge with the scan line) by adding to it the second data value (i.e., the reciprocal of the edge's gradient);

}

if the ET list for scan line $j + 1$ is not empty, merge that list with the AEL using the third data value as the key;

The reader is asked next to prove the one fact required to assure the validity of the AEL update procedure.

Exercise 14.22. Show that the third data value is correctly computed for each record in the new AEL by the update procedure.

Exercise 14.23. Refer to Figure 14.24. Apply the update procedure to compute the AEL values for scan lines 1, 2, 3 and 6.

Exercise 14.24. Incorporate into the Scan-based Polygon-filling Algorithm (Version 2) both initialization of the edge table and subsequent AEL-driven processing.

If you're coming to this polygon-filling algorithm fresh from Bresenham's line rasterizer, then alarm bells are probably going off in your head right now about floating point computations. And rightly so. The update procedure as just described does require a floating point addition to update a record – which can be avoided by optimizing the implementation. We'll let you figure this out in the next exercise.

Exercise 14.25. To cut floating points from Version 2, observe that, generally, a run of filled pixels along a given scan line s will (1) start at at the intersection of a left edge e with s if there is a pixel at the intersection or, if not, at the pixel just after the intersection, *and* (2) the run will end at the pixel before the intersection of a right edge e with s, even if there is a pixel at the intersection. See Figure 14.26(a). The one exceptional case – easily detected – is shown in Figure 14.26(b).

Therefore, instead of the floating point x-value, call it X, of the intersection of a segment e with a scan line s, one can store in AEL records the smallest integer greater than or equal to X if e is a left edge, or the largest integer strictly less than X if it is a right edge.

Accordingly, suggest a new form of the AEL with no floating point data items and a method to update it.

Remark 14.7. Polygon rasterization in OpenGL is particularly simple, as it consists only of rasterizing triangles in case of polygon drawing calls –

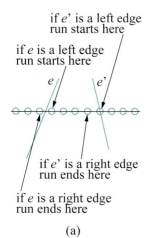

if e' is a left edge run starts here

if e is a left edge run starts here

e e'

if e' is a right edge run ends here

if e is a right edge run ends here

(a)

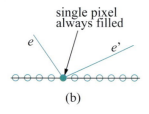

single pixel always filled

e e'

(b)

Figure 14.26: (a) Typical dispositions of a left and right edge with respect to a run of pixels (b) The one exceptional case where a left and a right edge meet at a common lower endpoint.

recall that OpenGL polygons are always pre-triangulated – or axis-aligned rectangles in case of glRectf() calls.

Section 14.4
SCAN-BASED POLYGON
RASTERIZER

Flood-fill

The scan-based polygon rasterization algorithm does not explicitly draw a polygon P's boundary given its vertices. Rather, its output is directly a set of pixels filling an area corresponding to P. Another approach to rasterizing P is, in fact, to draw first its boundary by, say, repeatedly applying Bresenham's line rasterizer for each edge, and then fill its interior.

Once the boundary of P has been drawn, a particularly intuitive algorithm to fill its interior is the so-called *flood-fill* algorithm. Here's how flood-fill works:

Start with a pixel p known to be in P's interior, found, possibly, by following a ray through the raster and applying the parity test. Fill p. Then examine four of p's neighboring pixels, in particular, the ones to its north, south, east and west. Of these pixels, which are said to be *4-adjacent* to p, fill those that don't belong to the boundary and have not yet been filled. Next, examine the pixels 4-adjacent to the ones just filled and, again, fill those that don't belong to the boundary and have not yet been filled. Continue in this manner until no more pixels can be filled. See Figure 14.27 for the initial configuration and the next two steps of a flood-fill.

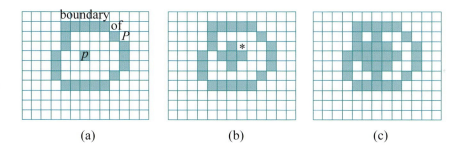

(a)   (b)   (c)

Figure 14.27: Flood-fill: (a) Initially (b) Fill pixels 4-adjacent to p (c) Fill pixels 4-adjacent to the ones filled in the previous step. The starred pixel of (b) is examined by both its south and west neighbors at this step.

Exercise 14.26. In how many more steps after Figure 14.27(c) does the flood-fill algorithm terminate?

Coding flood-fill is simple and we ask the reader to do this next.

Exercise 14.27. Assume that initially all pixels in the raster are of background_color, that a line rasterizer has subsequently been invoked on the edges of an input polygon P to set its boundary pixels to foreground_color and that the pixel at location (x, y) is known to be in the interior of P.

Pseudo-code a recursive flood-fill procedure to set pixels in the interior of P to foreground_color as well.

Now, the flood-fill algorithm will fill the interior of a polygon P provided that it is *4-connected*, in that any pair of pixels in the interior can be joined by a path of pixels, each consecutive pair of which is 4-adjacent. The interior of the polygon P of Figure 14.27 is 4-connected and flood-fill indeed terminates successfully after filling the whole interior. On the other hand, the boundary of a polygon Q where flood-fill will not succeed is shown in Figure 14.28. Evidently, Q's interior is not 4-connected and, whichever of its two pixels is chosen to start with, flood-fill will terminate after filling only that one.

It might seem, then, that one need only enhance flood-fill to examine all eight neighbors of the current pixel, which are said to be 8-adjacent to it, instead of only the 4-adjacent ones. However, care is needed. For, consider the two interior pixels of Q in Figure 14.28. Both are 8-adjacent to pixels belonging neither to Q's boundary nor its interior. Therefore, a simple-minded enhancement of flood-fill could "leak" and fill the whole raster. We'll leave the reader to ponder an appropriate enhancement.

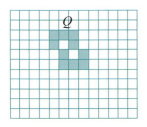

Figure 14.28: Flood-fill fails on the polygon Q whose boundary is drawn.

Exercise 14.28. Enhance flood-fill to examine all 8-adjacent neighbors of the current pixel, instead of only the 4-adjacent ones. Make sure your procedure does not leak.

Flood-fill is not efficient, particularly compared to scan-based filling. The primary reason is that flood-fill examines the same pixel repeatedly as a neighbor of different ones. For example, the starred pixel in Figure 14.27(b) is examined twice in the second step, once by its south neighbor and once by its west neighbor. Nevertheless, flood-fill is simple to implement and, moreover, can be applied even to curved non-polygonal shapes, once the shape's boundary has been identified. For this reason many paint programs allow the user to flood-fill the interior of a "blob".

14.5 Summary, Notes and More Reading

In this chapter we learned a few important algorithms that are applied deep in the graphics pipeline. These included line and polygon clipping, as well as the raster-conversion of these primitives. This knowledge will hardly impact our programming of graphics, particularly because high-level API's like OpenGL allow no access to these algorithms. Still, it's good to have some understanding of what goes on at the far end of the pipeline. And, in Section 19.1, when we assemble the entire synthetic-camera rendering pipeline, clipping and rasterization algorithms will be incorporated into its final stage.

The algorithms in this chapter are all from the foundational period of modern CG, particularly the sixties and early seventies. The Cohen-Sutherland line clipper is from Sketchpad [133], the pioneering interactive CG system Sutherland developed in 1963. The Sutherland-Hodgeman polygon clipper [134] is from that period as well, as is Bresenham's line rasterizer [20], the latter originally being proposed as a method to plot lines on real paper using a computer-controlled pen. The scan-based filling techniques migrated from pen-plotters to raster displays as well.

The classic books by Foley et al. [44] and Rogers [112] contain extensive discussions of various raster algorithms, while the two by Akenine-Möller, Haines & Hoffman [1] and Watt [142] describe modern-day implementations.

Part VIII

Anatomy of Curves and Surfaces

CHAPTER 15

Bézier

T he goal for this chapter is an understanding of the theory underlying Bézier primitives. We are already familiar with much of their practical aspects. In Section 10.3 of the chapter on drawing we saw how to specify Bézier curves and surfaces and incorporate them into our designs. This was possible then – even before theory – as an intuitive understanding of control points and their role as attractors in shaping Bézier primitives is sufficient to grasp the OpenGL syntax. We went even further in Sections 11.11.4 and 12.4.2, learning how to light and texture Bézier surfaces.

We'll restrict ourselves in this chapter to *polynomial* Bézier primitives. The more general form is *rational* – a rational function being the ratio of two polynomials. We'll postpone the discussion of the rational primitives to Chapter 18, as an application of projective spaces, which are the natural setting for these primitives.

Several 3D modeling systems support rational Bézier primitives – in fact, often, the even more general class of NURBS primitives – in a WYSIWYG environment where users create primitives interactively by manipulating control points. OpenGL is often the front-end of such modelers, itself offering both rational Bézier and NURBS primitives in a low-level "code-it-yourself" manner. Of course, a system supporting rational primitives supports as well its polynomial subclass.

There is a bit of math in the development of Bézier theory, but behind it always is the fairly intuitive "mechanics" of Bézier primitives, which we'll try to make as apparent as possible. Illustrative code is interspersed throughout this chapter as well.

We begin with Bézier curves in Section 15.1. First, de Casteljau's procedural approach to defining linear and quadratic Bézier curves is explained in Sections 15.1.1-15.1.2. The reader is asked to do most of the work for cubic Bézier curves in 15.1.3. Section 15.1.4 generalizes the development to Bézier curves of arbitrary order and we see as well a host of properties that make Bézier curves so useful in design. From Bézier curves to Bézier surfaces in Section 15.2 is a fairly intuitive progression. Section 15.3 concludes the chapter.

15.1 Bézier Curves

Suppose a programmer specifies a sequence $P_0, P_1, \ldots, P_n$ of $n+1$ *control points*, asking for a curve not necessarily passing through them, but, rather, whose shape is molded by the control points. In other words, the control points are expected to act as "attractors", each exerting a pull on the curve. The generated curve is said to *approximate* the control points. Figure 15.1 is illustrative of the situation.

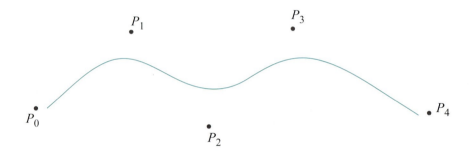

Figure 15.1: A curve approximating five control points.

Circa 1960, two French automotive designers, Bézier [12, 13] and de Casteljau [33, 34], independently invented a particular method to approximate a sequence of control points. It bears the name of Bézier because his publications had earlier circulation in the design community. However, de Casteljau's approach to Bézier curves is actually the more intuitive and it is what we'll first describe.

15.1.1 Linear Bézier Curves

Let's start with the simplest case, where there are only two control points P_0 and P_1. The Bézier curve c approximating P_0 and P_1 is, simply, the straight line segment joining the two. We write the parametric equation of c as follows:

$$c(u) = (1 - u)P_0 + uP_1 \qquad (0 \le u \le 1) \qquad (15.1)$$

See Figure 15.2(a). This Bézier curve is said to be *linear*, or of *degree one*, or of *second order*, order being the number of control points.

If the ambient space is $\mathbb{R}^2$, and $P_0 = [x_0 \ y_0]^T$ and $P_1 = [x_1 \ y_1]^T$, we can write (15.1) as

$$c(u) = [(1-u)\,x_0 + u\,x_1 \quad (1-u)\,y_0 + u\,y_1]^T \quad (0 \le u \le 1) \quad (15.2)$$

$\mathbf{Exercise}$ 15.1. Write an equation analogous to (15.2) if the ambient space is $\mathbb{R}^3$.

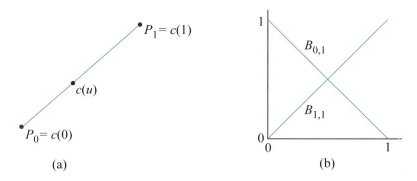

Figure 15.2: (a) Bézier curve of degree 1 (b) Bernstein polynomials of degree 1: $B_{0,1}(u) = 1 - u, \quad B_{1,1}(u) = u.$

$\mathcal{R}em\mathbf{ark}$ 15.1. As far as the theory goes, the control points can belong to a real space of arbitrary dimension, but practical applications are in $\mathbb{R}^2$ or $\mathbb{R}^3$.

$\mathcal{R}em\mathbf{ark}$ 15.2. It is evident from either (15.1) or (15.2) that a linear Bézier curve linearly interpolates between its two control points (recall linear interpolation from Section 7.2).

$\mathbf{Example}$ 15.1. What is the equation of the linear Bézier curve c with control points $[5 \ 1]^T$ and $[-1 \ 0]^T$? What are the points on c corresponding to the values 0, 0.3 and 1 of the parameter u?

Answer: The equation of c is

$$c(u) = (1-u)[5 \ 1]^T + u[-1 \ 0]^T = [5 - 6u \quad 1 - u]^T \quad (0 \le u \le 1)$$

The point corresponding to $u = 0$ is $[5 \ 1]^T$, the first control point.
The point corresponding to $u = 0.3$ is $[3.2 \ 0.7]^T$.
The point corresponding to $u = 1$ is $[-1 \ 0]^T$, the second control point.

$\mathbf{Exercise}$ 15.2. What is the equation of the linear Bézier curve c with control points $[0 \ -2 \ -4]^T$ and $[3 \ 8 \ 0]^T$? What are the points on c corresponding to the values 0, 0.3 and 1 of the parameter u?

We make the following observations, all straightforward, for a linear Bézier curve c:

1. The parametric equation (15.1) for c is linear in u, which, of course, is why it's called a linear Bézier curve.

2. The *point* $c(u)$ of the curve c is a weighted sum of the control points P_0 and P_1, the weights of P_0 and P_1 being the *values* of $1 - u$ and u, respectively. Accordingly, the *curve* c can be thought of as a weighted sum of P_0 and P_1, where the weights of P_0 and P_1 are the *functions* $1 - u$ and u, respectively.

   These functions are called the *blending functions* of the respective control points (the term *basis function* is used as well).

   The blending functions $1 - u$ (of the first control point) and u (of the second control point) are known as the *Bernstein polynomials* of degree 1. They are denoted $B_{0,1}(u)$ and $B_{1,1}(u)$, respectively. Figure 15.2(b) shows their graphs. Accordingly, Equation (15.1) can be written as

   $$c(u) = B_{0,1}(u)P_0 + B_{1,1}(u)P_1 \qquad (0 \le u \le 1) \qquad (15.3)$$

   The blending function $B_{0,1}(u)$ of the first control point decreases from 1 to 0 as u goes from 0 to 1, while exactly the opposite is true of that of the second control point.

3. Because

   (a) $B_{0,1}(u)$ and $B_{1,1}(u)$ both lie between 0 and 1, and

   (b) $B_{0,1}(u) + B_{1,1}(u) = 1$,

   for each u in $0 \le u \le 1$, every point of c is a convex combination (recall Definition 7.3) of the control points P_0 and P_1 and, therefore, lies in their convex hull, which is actually pretty obvious in this simple case.

4. c starts at the first control point P_0, when $u = 0$, and ends at the second one P_1, when $u = 1$.

   If an approximating curve passes through a control point, then it is said to *interpolate* it. So a linear Bézier curve interpolates both its control points.

15.1.2   Quadratic Bézier Curves

Consider next three control points P_0, P_1 and P_2. We want to construct the Bézier curve approximating these control points by means of a process of linear interpolation. Is there, though, an evident way to linearly interpolate a curve between three control points "simultaneously"? What does this even mean?

A possibility, of course, is to linearly interpolate between P_0 and P_1 and then between P_1 and P_2, to get the two-segment polyline $P_0P_1P_2$, which is not particularly attractive because of the corner at P_1 (see Figure 15.3(a)). De Casteljau's method, however, succeeds by adding a third interpolation step to "amalgamate" the two segments P_0P_1 and P_1P_2, smoothening thereby the corner. Here's how it works.

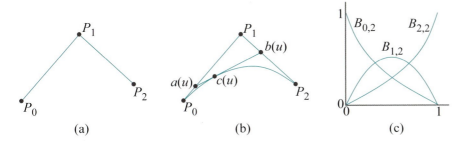

Figure 15.3: (a) An "unhappy" way of approximating three control points (b) $c(u)$ describes a Bézier curve of degree 2 interpolating P_0, P_1 and P_2 after a "triple" interpolation (c) Bernstein polynomials of degree 2: $B_{0,2}(u) = (1-u)^2$, $B_{1,2}(u) = 2(1-u)u$, $B_{2,2}(u) = u^2$.

Given a u, $0 \leq u \leq 1$ (see Figure 15.3(b)):

1. First interpolate between P_0 and P_1 to find the point
$$a(u) = (1-u)P_0 + uP_1$$

2. Next interpolate between P_1 and P_2 to find the point
$$b(u) = (1-u)P_1 + uP_2$$

3. Finally, interpolate between $a(u)$ and $b(u)$ to determine the point
$$c(u) = (1-u)a(u) + ub(u)$$

Substituting the expressions for $a(u)$ and $b(u)$ into that for $c(u)$, one obtains the parametric equation for a curve c:

$$c(u) = (1-u)^2 P_0 + 2(1-u)uP_1 + u^2 P_2 \qquad (0 \leq u \leq 1) \qquad (15.4)$$

As u varies from 0 to 1, $c(u)$ describes the *quadratic*, or *degree two*, or *third-order*, Bézier curve approximating three control points P_0, P_1 and P_2, which is indeed smooth.

Note: Curves drawn in this chapter are fairly accurate sketches, but not necessarily exact plots of their equations.

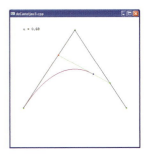

Figure 15.4: Screenshot of deCasteljau3.cpp.

Experiment 15.1. Run `deCasteljau3.cpp`, which shows an animation of de Casteljau's method for three control points. Press the left or right arrow keys to decrease or increase the curve parameter u. The interpolating points $a(u)$, $b(u)$ and $c(u)$ are colored red, green and blue, respectively. Figure 15.4 is a screenshot. End

If the ambient space is $\mathbb{R}^2$, and $P_0 = [x_0 \; y_0]^T$, $P_1 = [x_1 \; y_1]^T$ and $P_2 = [x_2 \; y_2]^T$, one can write (15.4) as

$$c(u) = [(1-u)^2 x_0 + 2(1-u)u \, x_1 + u^2 x_2$$
$$(1-u)^2 y_0 + 2(1-u)u \, y_1 + u^2 y_2]^T \quad (0 \leq u \leq 1) \; (15.5)$$

Example 15.2. What is the equation of the third-order Bézier curve c with control points $[0 \; -1]^T$, $[1 \; 2]^T$ and $[5 \; -1]^T$?

Answer: The equation of c is

$$c(u) = (1-u)^2[0 \; -1]^T + 2(1-u)u[1 \; 2]^T + u^2[5 \; -1]^T$$
$$= [2u + 3u^2 \quad -1 + 6u - 6u^2]^T \quad (0 \leq u \leq 1)$$

Exercise 15.3. What is the equation of the third-order Bézier curve c with control points $[-2 \; 2 \; 2]^T$, $[0 \; 3 \; 5]^T$ and $[-6 \; 0 \; 2]^T$? What are the points on c corresponding to the values 0, 0.5 and 1 of the parameter u?

Experiment 15.2. Run `bezierCurves.cpp`, which allows the user to choose a Bézier curve of order 2-6 and move each control point.

You can choose an order in the first screen by pressing the up and down arrow keys. Select 3. Press enter to go to the next screen to find the control points initially on a straight line. Press space to select a control point – the selected one is red – and then arrow keys to move it. Delete resets to the first screen. Figure 15.5 is a screenshot.

The polygonal line joining the control points, called the *control polygon* of the curve, is drawn in light gray. Evidently, the Bézier curve "mimics" its control polygon, but smoothly, avoiding a corner. **End**

Figure 15.5: Screenshot of `bezierCurves.cpp` with three control points, showing both the Bézier curve and its control polygon.

Compare the following observations for a third-order Bézier curve c with the corresponding ones, made earlier, for a second-order curve:

1. c is quadratic in u.

2. c is a weighted sum of the control points P_0, P_1 and P_2, where the weights of P_0, P_1 and P_2 are the blending functions $(1-u)^2$, $2(1-u)u$ and u^2, respectively. These blending functions are called the Bernstein polynomials of degree 2, and denoted $B_{0,2}(u)$, $B_{1,2}(u)$ and $B_{2,2}(u)$, respectively. Figure 15.3(c) shows their graphs. Accordingly, Equation (15.4) can be written as

$$c(u) = B_{0,2}(u)P_0 + B_{1,2}(u)P_1 + B_{2,2}(u)P_2 \quad (0 \leq u \leq 1) \quad (15.6)$$

It's useful to think of the value of a control point's blending function at $c(u)$ as the amount of its "attraction" (or "pull", or "weight") on that point of the Bézier curve and of u as a dial propelling $c(u)$ along the curve by altering these attractions.

The blending function $B_{0,2}(u)$ of the first control point P_0 decreases from 1 to 0 as u goes from 0 to 1; the blending function $B_{1,2}(u)$ of the middle control point P_0 starts and ends at 0, reaching a maximum value of $\frac{1}{2}$ at $u = \frac{1}{2}$; finally, the blending function $B_{2,2}(u)$ of the last control point P_0 increases from 0 to 1. So, e.g., the attraction of the middle control point is greatest on the point of the curve corresponding to $u = \frac{1}{2}$.

3. Every point of c is a convex combination of the control points P_0, P_1 and P_2, because

   (a) $B_{0,2}(u)$, $B_{1,2}(u)$ and $B_{2,2}(u)$ all lie between 0 and 1, and

   (b) $B_{0,2}(u) + B_{1,2}(u) + B_{2,2}(u) = (1-u)^2 + 2(1-u)u + u^2 = 1$,

   for each u in $0 \leq u \leq 1$. It follows that the entire curve c is contained in the convex hull of P_0, P_1 and P_2.

   Colloquially, (a) and (b) say that the attraction of each control point is between 0 and 1 and that the total attraction of all three is always 1.

4. c interpolates the first and last control points, but not necessarily the middle one.

Exercise 15.4. What is the attraction of each of the control points P_0, P_1 and P_2 on $c(0.2)$?

Exercise 15.5. Verify that Equation (15.4) can be written in the matrix form:

$$c(u) = [P_0 \ P_1 \ P_2] \begin{bmatrix} 1 & -2 & 1 \\ -2 & 2 & 0 \\ 1 & 0 & 0 \end{bmatrix} [u^2 \ u \ 1]^T \qquad (15.7)$$

15.1.3 Cubic Bézier Curves

Consider, now, four control points P_0, P_1, P_2 and P_3. We're going to ask you, dear reader, to do most of the work. Perform step-by-step interpolation, similarly to the case of three control points, as follows.

Given a u in $0 \leq u \leq 1$ (see Figure 15.6(a)):

1. $a(u)$ interpolates between P_0 and P_1.

2. $b(u)$ between P_1 and P_2.

3. $d(u)$ between P_2 and P_3.

4. $e(u)$ between $a(u)$ and $b(u)$.

5. $f(u)$ between $b(u)$ and $d(u)$.

6. $c(u)$ between $e(u)$ and $f(u)$.

As u varies from 0 to 1, $c(u)$ describes the *cubic*, or *degree three*, or *fourth-order*, Bézier curve approximating the four control points P_0, P_1, P_2 and P_3.

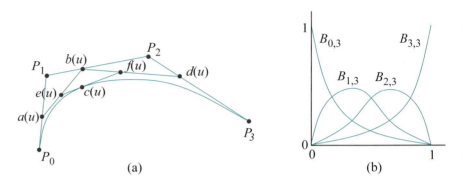

(a)   (b)

Figure 15.6: (a) Bézier curve of degree 3 (b) Bernstein polynomials of degree 3: $B_{0,3}(u) = (1-u)^3$, $B_{1,3}(u) = 3(1-u)^2 u$, $B_{2,3}(u) = 3(1-u)u^2$, $B_{3,3}(u) = u^3$.

Exercise 15.6. Prove that the parametric equation of the Bézier curve c approximating the four control points P_0, P_1, P_2 and P_3 is

$$c(u) = B_{0,3}(u)P_0 + B_{1,3}(u)P_1 + B_{2,3}(u)P_2 + B_{3,3}(u)P_3 \qquad (0 \le u \le 1)$$
$$(15.8)$$

where the Bernstein polynomials are

$$B_{0,3}(u) = (1-u)^3, \quad B_{1,3}(u) = 3(1-u)^2 u, \quad B_{2,3}(u) = 3(1-u)u^2, \quad B_{3,3}(u) = u^3$$

Figure 15.6(b) shows their graphs.

Exercise 15.7. If the ambient space is $\mathbb{R}^2$ write an equation analogous to (15.5) for the cubic Bézier curve.

Exercise 15.8. What is the equation of the cubic Bézier curve c with control points $[-2\ 2]^T$, $[0\ -3]^T$, $[3\ 4]^T$ and $[7\ 0]^T$? What are the points on c corresponding to the values 0, 0.5 and 1 of the parameter u?

Experiment 15.3. Run `bezierCurves.cpp` and choose order 4 to get a feel for cubic Bézier curves. Note again how the curve mimics its control polygon. **End**

Exercise 15.9. Make four observations for a fourth-order Bézier curve c, corresponding to the four made for Bézier curves of orders 2 and 3.

Exercise 15.10. What is the attraction of each of the control points P_0, P_1, P_2 and P_3 at $c(0.2)$, where c is their approximating Bézier curve?

Exercise 15.11. Write Equation (15.8) in a matrix form similar to (15.7).

Exercise 15.12. (**Programming**) Write a program `deCasteljau4.cpp`, in the style of `deCasteljau3.cpp`, to illustrate de Casteljau's method for four control points.

Remark 15.3. Cubic Bézier curves are the ones most commonly used in design applications as three is a sort of "Goldilocks" degree, high enough to allow the curve good flexibility, yet not too high as to be computationally cumbersome.

Here's an exercise to get you warmed up for the general case coming next.

Exercise 15.13. From only the cases $n = 1$, 2 and 3, that we have seen, it's clear how the *variable part* changes of the Bernstein polynomials from $B_{0,n}(u)$ to $B_{1,n}(u)$, ..., finally, to $B_{n,n}(u)$. In fact, the variable part of $B_{0,n}(u)$ is $(1-u)^n u^0$. (Of course, $u^0 = 1$.) Next, for $B_{1,n}(u)$, the power of $1-u$ decreases by one and that of u increases by one, so its variable part is $(1-u)^{n-1} u^1$. And so it continues, until the variable part of $B_{n,n}(u)$ is $(1-u)^0 u^n$.

How about the *constant coefficients* though? Let's see what they are.
For Bernstein polynomials of degree 1:   1   1
For Bernstein polynomials of degree 2:   1   2   1
For Bernstein polynomials of degree 3: 1   3   3   1
Do you see a pattern? (*Hints*: Pascal's triangle, binomial coefficients.) Can you write down now the parametric equation for a fifth-order Bézier curve, without going through a de Casteljau process?

15.1.4   General Bézier Curves

It should now be fairly clear how to generalize de Casteljau's method to construct the Bézier curve approximating an arbitrary number of control points. We'll show that the parametric equation for the Bézier curve c approximating $n + 1$ control points $P_0, P_1, \ldots, P_n$ is

$$c(u) = \sum_{i=0}^{n} B_{i,n}(u) P_i \qquad (0 \leq u \leq 1) \tag{15.9}$$

where $B_{i,n}(u)$, $0 \leq i \leq n$, called the ith Bernstein polynomial of degree n, is given by

$$B_{i,n}(u) = \binom{n}{i} (1-u)^{n-i} u^i \tag{15.10}$$

where $\binom{n}{i} = \frac{n!}{(n-i)!i!}$ is a binomial coefficient. The curve c is called a Bézier curve of *degree* n, or *order* $n + 1$.

We'll verify Equation (15.9) by induction based on the following recursive specification of de Casteljau's method.

Recursive de Casteljau

We'll start the recursive definition by specifying (again) that the Bézier curve approximating two control points P_0 and P_1 is the straight segment joining them, given by (repeating (15.3)):

$$c(u) = B_{0,1}(u)P_0 + B_{1,1}(u)P_1 \qquad (0 \leq u \leq 1) \tag{15.11}$$

Assume, then, that we can specify the Bézier curve approximating any n control points, for some given $n \geq 2$, and that, next, we are given $n+1$ control points $P_0, P_1, \ldots P_n$. Say the Bézier curve $c_0(u)$, $0 \leq u \leq 1$, approximates the first n of these $P_0, P_1, \ldots P_{n-1}$, and that $c_1(u)$, $0 \leq u \leq 1$, approximates the last n points $P_1, P_2, \ldots P_n$.

Recursive de Casteljau says, then, that the Bézier curve approximating all $n + 1$ control points $P_0, P_1, \ldots P_n$ is

$$c(u) = (1 - u)c_0(u) + uc_1(u) \qquad (0 \leq u \leq 1) \tag{15.12}$$

which is an "interpolation" between $c_0(u)$ and $c_1(u)$. The scheme is indicated in Figure 15.7.

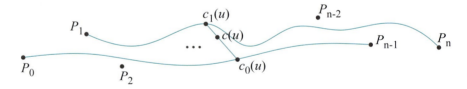

Figure 15.7: Recursive de Casteljau scheme: $c_0(u)$ approximates $P_0, P_1, \ldots, P_{n-1}$; $c_1(u)$ approximates $P_1, P_2, \ldots, P_n$; $c(u)$ interpolates between $c_0(u)$ and $c_1(u)$ to approximate $P_0, P_1, \ldots, P_n$.

Exercise 15.14. Verify that the recursive specification above yields the same equations as de Casteljau's method for three and four control points described earlier.

Note that in the case of three control points the earlier description matches exactly the recursive one above. However, in the case of four, we earlier did repeated linear interpolation between pairs of points on separate straight segments, while the recursive specification above would have us linearly interpolate just once between points on two separate quadratic Bézier curves.

Let's turn now to proving the general formula (15.9) by induction. Starting the induction is simply a matter of noting that (15.9) is identical to (15.11) when $n = 1$.

Suppose, inductively, that (15.9) is true with $n - 1$ in place of n, i.e., it is true for n control points. We'll prove it next for $n + 1$ control points, say, $P_0, P_1, \ldots P_n$. By the inductive hypothesis, the Bézier curve approximating the first n of these, $P_0, P_1, \ldots P_{n-1}$, is given by

$$c_0(u) = \sum_{i=0}^{n-1} B_{i,n-1}(u) P_i \quad (0 \leq u \leq 1)$$

and that approximating the last n points, $P_1, P_2, \ldots P_n$, by

$$c_1(u) = \sum_{i=0}^{n-1} B_{i,n-1}(u) P_{i+1} \quad (0 \leq u \leq 1)$$

The Bézier curve approximating all $n + 1$ control points $P_0, P_1, \ldots P_n$ by the recursive formula (15.12), therefore, is

$$
\begin{aligned}
c(u) \quad &= (1-u) c_0(u) + u c_1(u) \\
&= (1-u) \sum_{i=0}^{n-1} B_{i,n-1}(u) P_i + u \sum_{i=0}^{n-1} B_{i,n-1}(u) P_{i+1} \\
&= (1-u) \sum_{i=0}^{n-1} \binom{n-1}{i} (1-u)^{n-i-1} u^i P_i + \\
&\qquad u \sum_{i=0}^{n-1} \binom{n-1}{i} (1-u)^{n-i-1} u^i P_{i+1}
\end{aligned}
$$

(applying formula (15.10) for Bernstein polynomials of degree $n - 1$)

$$
\begin{aligned}
&= \quad (1-u) \sum_{i=0}^{n-1} \binom{n-1}{i} (1-u)^{n-i-1} u^i P_i + \\
&\qquad u \sum_{i=1}^{n} \binom{n-1}{i-1} (1-u)^{n-i} u^{i-1} P_i
\end{aligned}
$$

(changing the limits on the second summation by replacing i by $i - 1$)

$$
\begin{aligned}
&= (1-u)^n P_0 + \sum_{i=1}^{n-1} \left(\binom{n-1}{i} (1-u)^{n-i} u^i + \binom{n-1}{i-1} (1-u)^{n-i} u^i \right) P_i + \\
&\qquad u^n P_n
\end{aligned}
$$

(bringing together terms for $i = 1, \ldots, n - 1$ from the two summations)

$$= (1-u)^n\, P_0 + \sum_{i=1}^{n-1} \left(\binom{n}{i} (1-u)^{n-i} u^i \right) P_i + u^n\, P_n$$

(using the property of binomial coefficients that $\binom{n-1}{i} + \binom{n-1}{i-1} = \binom{n}{i}$)

$$= \sum_{i=0}^{n} B_{i,n}(u)\, P_i$$

This completes the inductive verification of Equation (15.9).

If the control points are $P_i = [x_i\ y_i\ z_i]^T$, $0 \le i \le n$, in real-world 3-space, then (15.9) can be written as

$$c(u) = \left[\sum_{i=0}^{n} B_{i,n}(u)x_i \quad \sum_{i=0}^{n} B_{i,n}(u)y_i \quad \sum_{i=0}^{n} B_{i,n}(u)z_i \right]^T \quad (0 \le u \le 1)$$

$$(15.13)$$

We collect facts about general Bézier curves in the following:

Proposition 15.1. *If c is the Bézier curve approximating the sequence of $n+1$ control points $P_0, P_1, \ldots, P_n$ – called a Bézier curve of order $n+1$, or degree n – then the following hold:*

(a) *c is polynomial of degree n in the parameter u. In particular, each coordinate value of c is polynomial of degree n in u.*

(b) *c is a weighted sum of the control points $P_0, P_1, \ldots, P_n$, where the weight of P_i, for $0 \le i \le n$, is its blending function $B_{i,n}(u)$.*

(c) *Every point of c is a convex combination of the control points $P_0, P_1, \ldots, P_n$; therefore, c lies inside the convex hull of $P_0, P_1, \ldots, P_n$.*

(d) *c interpolates the first and last control points, but not necessarily intermediate ones.*

(e) *(Affine Invariance) If the control points $P_0, P_1, \ldots, P_n$ belong to $\mathbb{R}^3$ and $g : \mathbb{R}^3 \to \mathbb{R}^3$ is an affine transformation, then the image curve $g(c)$ is the Bézier curve approximating the images $g(P_0), g(P_1), \ldots, g(P_n)$ of the control points.*

In other words, the transformed curve approximates the transformed control points.

(f) *(End Tangents) The tangent to c at P_0 lies along the straight line from P_0 to P_1 and the tangent to c at P_n lies along the straight line from P_{n-1} to P_n.*

Note: Further discussions of affine invariance and end tangents follow the proof.

Proof. Items (a) and (b) follow straightforwardly from Equation (15.9).

It's easily seen that Bernstein polynomials $B_{i,n}(u) = \binom{n}{i}(1-u)^{n-i}u^i$ all lie between 0 and 1, for each u in $0 \le u \le 1$. Further, for any u,

$$\sum_{i=0}^{n} B_{i,n}(u) = \sum_{i=0}^{n} \binom{n}{i}(1-u)^{n-i}u^i = (\,(1-u)+u\,)^n = 1 \qquad (15.14)$$

by the Binomial Theorem. It follows that the point $c(u) = \sum_{i=0}^{n} B_{i,n}(u)P_i$, for $0 \le u \le 1$, is indeed a convex combination of the points $P_0, P_1, \dots, P_n$, proving (c).

Incidentally, Equation (15.14) proves that the blending functions form a partition of unity over the parameter space $[0,1]$: a set of functions is said to a form a *partition of unity* over some domain if they are each non-negative and add up to 1 everywhere in that domain.

Item (d) is verified by checking that $c(0) = P_0$ and $c(1) = P_n$.

The proof of (e) exploits the fact that the blending functions form a partition of unity over the parameter space. Let the affine transformation $g : \mathbb{R}^3 \to \mathbb{R}^3$ be given by $g(P) = MP + D$, where M is a non-singular 3×3 matrix and D a 3-vector (the translational component). For any u in $0 \le u \le 1$, we then have

$$
\begin{aligned}
g(c(u)) &= Mc(u) + D \\[2mm]
&= M\Big(\sum_{i=0}^{n} B_{i,n}(u)P_i\Big) + D \\[2mm]
&= \sum_{i=0}^{n} B_{i,n}(u)(MP_i) + \Big(\sum_{i=0}^{n} B_{i,n}(u)\Big)D \\[2mm]
&\qquad \text{(invoking the partition-of-unity property } \textstyle\sum_{i=0}^{n} B_{i,n}(u) = 1) \\[2mm]
&= \sum_{i=0}^{n} B_{i,n}(u)(MP_i + D) \\[2mm]
&= \sum_{i=0}^{n} B_{i,n}(u)g(P_i)
\end{aligned}
$$

proving (e).

For (f), observe that the derivative

$$
\begin{aligned}
c'(u) &= \frac{\mathrm{d}}{\mathrm{d}u}(\,(1-u)^n P_0 + n(1-u)^{n-1}uP_1 + \dots) \\[2mm]
&= -n(1-u)^{n-1}P_0 + n(1-u)^{n-1}P_1 + \textit{ terms that} \\
&\quad \textit{each contain the factor } u \qquad\qquad\qquad (15.15)
\end{aligned}
$$

Therefore, the tangent vector at P_0, when $u = 0$, is

$$c'(0) = -nP_0 + nP_1 = n(P_1 - P_0)$$

as terms after the second of (15.15) all vanish. Evidently, then, the tangent at P_0 is in the direction toward P_1, proving the first part of (f). The second part follows symmetrically. $\qquad\square$

Remark 15.4. From (a) it follows that a Bézier curve is C^∞, or smooth (recall definitions from Section 10.1.6).

Experiment 15.4. Run `bezierCurves.cpp` and choose the higher orders. It's straightforward to enhance the code for orders even greater than 6. **End**

Affine Invariance of Bézier Curves

The affine invariance of Bézier curves given by Proposition 15.1(e) is extremely useful. Here's a simple example of what it means practically. Suppose a cubic Bézier curve c approximating the four control points

$$P_0 = [5\ 5]^T \qquad P_1 = [7\ 8]^T \qquad P_2 = [8\ 4]^T \qquad P_3 = [11\ 5]^T$$

is drawn as a 10-segment polyline l (Figure 15.8 left, where l is drawn at an offset to c).

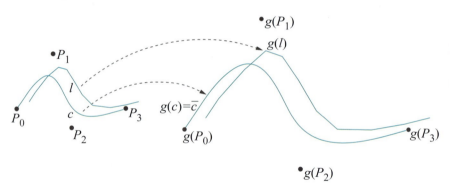

Figure 15.8: Left: A 10-segment fairly smooth-looking polyline l approximation (drawn at an offset) of the Bézier curve c with control points P_0, P_1, P_2 and P_3. Right: The magnification $g(l)$ (also at an offset) is not a good approximation of $g(c)$.

Next, suppose we want to magnify c twofold by scaling it a factor of 2 in each coordinate direction. For the purpose of drawing, if the scaling transformation, call it g, is applied simply to the polyline l, then the result $g(l)$ is likely too coarse an approximation of the magnified curve $g(c)$.

Affine invariance, however, says that $g(c)$ is the same as the cubic Bézier curve $\bar{c}$ approximating the four control points

$$g(P_0) = [10\ 10]^T \qquad g(P_1) = [14\ 16]^T \qquad g(P_2) = [16\ 8]^T \qquad g(P_3) = [22\ 10]^T$$

Therefore, one can "forget" the original polyline and, instead, approximate $\bar{c}$ at a resolution of one's choosing (e.g., with a 20-segment polyline).

Bottom Line: Affine invariance means that if a Bézier curve is affinely transformed to a new one, then the original control points transform to the new. Therefore, the only "data" required to generate the transformed Bézier curve are the transformed control points.

Exercise 15.15. The transformations

```
glScalef(1.0, 2.0, 2.0);
glTranslatef(2.0, 3.0, 0.0);
```

are applied to the cubic Bézier curve with control points

$$[2\ 1\ 1]^T \quad [3\ 3\ 2]^T \quad [-2\ 7\ -1]^T \quad [0\ 0\ 4]^T$$

Describe the resulting curve.

End Tangents and Joining Bézier Curves

Proposition 15.1(f) enables the user to smoothly join two Bézier curves c_0 (approximating control points $P_0, P_1, \ldots, P_n$) and c_1 (approximating control points $Q_0, Q_1, \ldots, Q_m$) by, for instance, making P_n coincide with Q_0 and arranging the three points $P_{n-1}, P_n(= Q_0)$ and Q_1 in that order on one straight line. See Figure 15.9. Another way to say this is that two Bézier curves meet smoothly if their control polygons meet smoothly.

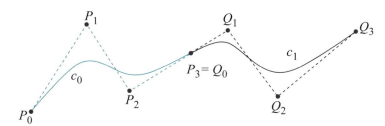

Figure 15.9: Two cubic Bézier curves (one blue, other black) meet smoothly at an endpoint.

Experiment 15.5. Run `bezierCurveTangent.cpp`. The second curve may be shaped by selecting a control point with the space bar and moving it with the arrow keys. See Figure 15.10. Visually verify Proposition 15.1(f). **End**

Exercise 15.16. (Programming) Show how to use Proposition 15.1(f) to arrange a sequence of control points, so that the approximating Bézier curve is a *smooth closed* loop. Illustrate with the help of `bezierCurves.cpp`.

Exercise 15.17. A Bézier loop is drawn approximating the six control points

$$[0\ 0\ 0]^T \quad [4\ 0\ 0]^T \quad [8\ -5\ 3]^T \quad [-1\ -5\ -3]^T \quad [x\ y\ z]^T \quad [0\ 0\ 0]^T$$

Figure 15.10: Screenshot of `bezierCurveTangent.-cpp`.

Suggest values for x, y and z for the second to last control point to make the loop smooth at $[0\ 0\ 0]^T$.

A curve made by joining Bézier curves end to end, but not necessarily smoothly, is called *piecewise Bézier*. See Figure 15.11.

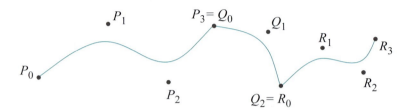

Figure 15.11: A piecewise Bézier curve consisting of three Bézier arcs.

Exercise 15.18. Can a piecewise Bézier curve be Bézier? For example, in Figure 15.11 one may ask if the union of the three Bézier curves is merely the ninth-order Bézier curve with control points $P_0, P_1, P_2, P_3 = Q_0, Q_1, Q_2 = R_0, R_1, R_2, R_3$. Consider, in particular, the case when the Bézier pieces happen all to join smoothly (which is not the case in Figure 15.11).

Exercise 15.19. The sequence $P_0, P_1, \ldots, P_n$ of the control points of a Bézier curve is obviously important. Jumbling them up will not give the same curve. How about if the sequence is reversed to $P_n, P_{n-1}, \ldots, P_0$?

Exercise 15.20. There is nothing special about the parameter space $[0, 1]$. Show how to change the parameter space of the Bézier curve, given by Equation (15.9), to $[u_1, u_2]$, where $u_1 < u_2$ may be arbitrary, without changing the curve's shape.

Exercise 15.21. Show that the blending function $B_{i,n}(u)$ of the ith control point P_i reaches its maximum at $u = \frac{i}{n}$, and at this point the value of $B_{i,n}(u)$ exceeds that of all the other blending functions. This means that the attraction of P_i is greatest on the point $c(\frac{i}{n})$ of the curve.

Polynomial Curves and Bézier Curves

As noted in Proposition 15.1(a), each coordinate value of a degree n Bézier curve is a polynomial of degree n in the parameter u. Recall from Section 10.1.4 that a polynomial curve (in $\mathbb{R}^3$) is of the form

$$b(u) = [f(u)\ g(u)\ h(u)]^T$$

where each coordinate value $f(u)$, $g(u)$ and $h(u)$ is polynomial in u. Bézier curves are, therefore, polynomial. How about the other way around? Are polynomial curves Bézier? It's nice to know, in fact, that all polynomial curves are Bézier. Precisely:

Proposition 15.2. *If*

$$b(u) = [f(u)\ g(u)\ h(u)]^T \qquad (0 \le u \le 1)$$

is a polynomial curve, each coordinate value being a polynomial of degree at most n, then one can find $n+1$ control points $P_0, P_1, \ldots P_n$, such that $b(u) = c(u)$, where c is the Bézier approximation of the P_i, $0 \le i \le n$. In other words, the Bézier approximation of these control points is the given polynomial curve.

Proof. The proof is beyond our scope here and the interested reader is referred to the text by Buss [21]. $\qquad\qquad\qquad\qquad\qquad\qquad\qquad\qquad\square$

Remark 15.5. It's certainly interesting that, despite the fact that they arise from the very special de Casteljau construction, the proposition says that the class of Bézier curves is just as general as the class of polynomial curves.

At this point let's pause a moment to appreciate the power and utility of Bézier curves, particularly in light of the preceding proposition. Suppose that a developers' group set out to design 1D primitives for a modeler. They might quite reasonably decide to support, in addition to straight lines and polylines, polynomial curves of degree 3, namely, of the form

$$p(u) = [f_3 u^3 + f_2 u^2 + f_1 u + f_0 \quad g_3 u^3 + g_2 u^2 + g_1 u + g_0 \quad h_3 u^3 + h_2 u^2 + h_1 u + h_0]^T$$

for $0 \le u \le 1$. That's 12 coefficient scalars $f_3, f_2, \ldots, h_0$ required to specify such a curve and a (very simple-minded) design decision would be to allow the user to edit the curve by changing each.

Contrast this with representing a polynomial curve of degree 3 as a cubic Bézier curve specified by four control points. The size of the representation is still 12 scalars – three coordinates per control point – and the preceding proposition says that we still get all 3D polynomial curves. Consider, though, how much more convenient it is to mold the curve by manipulating control points rather than coefficients!

In fact, is there at all an easy-to-understand relationship between the coefficients $f_3, f_2, \ldots, h_0$ and the shape of $p(u)$ as given above? Even for the simple plane paper graph of a curve, say, $y = 3x^3 - x^2 + 5x + 7$, do the four coefficients 3, -1, 5 and 7 themselves convey anything immediately meaningful about its shape?

15.2 Bézier Surfaces

From an understanding of Bézier curves it's a fairly intuitive next step to defining Bézier surfaces. Suppose we have an $(n+1) \times (m+1)$ *array* of control points

$$P_{i,j}, \quad \text{for } 0 \le i \le n, 0 \le j \le m$$

and wish to approximate these with a surface s. A construction of s via Bézier curves is as follows:

Think of the $(n+1) \times (m+1)$ array $P_{i,j}$ as $n+1$ different *sequences*, each of $m+1$ control points. In particular, the ith sequence, for $0 \le i \le n$, consists of $P_{i,0}, P_{i,0}, \dots, P_{i,m}$, these being the points along the ith row of the control points array. Construct the Bézier curve approximating each of these $n+1$ sequences to obtain $n+1$ different Bézier curves, each of order $m+1$. Say the Bézier curve approximating the ith sequence is c_i, $0 \le i \le n$. See Figure 15.12, where both n and m are 3.

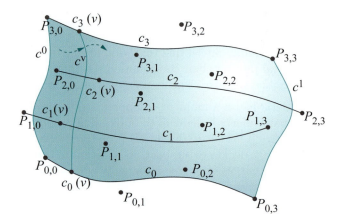

Figure 15.12: Constructing the Bézier surface approximating a 4×4 array of control points by sweeping a Bézier curve of order 4.

For each v in $0 \le v \le 1$, there are $n+1$ points, one on each curve c_i, corresponding to the parameter value v, namely, the sequence $c_0(v), c_1(v), \dots, c_n(v)$. Say, the Bézier curve c^v of order $n+1$ approximates these points. One such c^v is shown in the figure.

The union of all the Bézier curves c^v, for $0 \le v \le 1$, is the Bézier surface s approximating the control points array $P_{i,j}$, $0 \le i \le n$, $0 \le j \le m$. One can, as well, think of s as being *swept* by c^v, as v changes from 0 to 1.

The polyhedral surface composed of the quadrilateral faces $P_{i,j}P_{i+1,j}P_{i+1,j+1}P_{i,j+1}$, $0 \le i \le n-1$, $0 \le j \le m-1$, is called the *control polyhedron* of the Bézier surface specified by the control points $P_{i,j}$, $0 \le i \le n$, $0 \le j \le m$. As a Bézier curve mimics its control polygon, so a Bézier surface mimics its control polyhedron. See Figure 15.13.

Experiment 15.6. Run `sweepBezierSurface.cpp` to see an animation of the procedure. Press the left/right (or up/down) arrow keys to move the sweeping curve and the space bar to toggle between the two possible sweep directions. Figure 15.14 is a screenshot.

The 4×4 array of the Bézier surface's control points (drawn as small squares) consists of a blue, red, green and yellow row of four control points

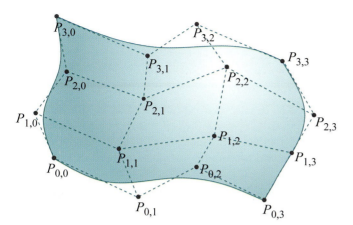

Figure 15.13: The Bézier surface approximating a 4×4 array of control points and its control polyhedron (dashed).

each. The four fixed Bézier curves of order 4 are drawn blue, red, green and yellow, respectively (the curves are in 3-space, which is a bit hard to make out because of the projection). The sweeping Bézier curve is black and its (moving) control points are drawn as larger squares. The currently swept part of the Bézier surface is the dark mesh. The current parameter value is shown at the top left. **End**

Determining the parametric equation of the Bézier surface s constructed as above is not difficult. The equation of c_i, the Bézier curve along the ith row of control points is

$$c_i(v) \;=\; \sum_{j=0}^{m} B_{j,m}(v) P_{i,j} \qquad (0 \le v \le 1)$$

for $0 \le i \le n$. Therefore, the equation of the Bézier curve c^v approximating the "column" control sequence $c_0(v), c_1(v), \ldots, c_n(v)$ is

$$
\begin{aligned}
c^v(u) &= \sum_{i=0}^{n} B_{i,n}(u) c_i(v) \\
&= \sum_{i=0}^{n} B_{i,n}(u) \left(\sum_{j=0}^{m} B_{j,m}(v) P_{i,j} \right) \\
&= \sum_{i=0}^{n} \sum_{j=0}^{m} B_{i,n}(u) B_{j,m}(v) P_{i,j} \qquad (0 \le u \le 1)
\end{aligned}
$$

Letting both u and v vary one obtains the following parametric equation for the Bézier surface s approximating the control points array $P_{i,j}, 0 \le i \le n$,

Figure 15.14: Screen-shot of `sweepBezier-Surface.cpp`.

$0 \leq j \leq m$:

$$s(u,v) = \sum_{i=0}^{n} \sum_{j=0}^{m} B_{i,n}(u) B_{j,m}(v) P_{i,j} \qquad (0 \leq u \leq 1,\ 0 \leq v \leq 1) \quad (15.16)$$

Exercise 15.22. If the control points of a Bézier surface are $P_{i,j} = [x_{i,j} \quad y_{i,j} \quad z_{i,j}]^T$, $0 \leq i \leq n$, $0 \leq j \leq m$, write a parametric equation for its x-, y- and z-values, analogous to (15.13) for Bézier curves. There will now, of course, be two parameter variables instead of the one for curves.

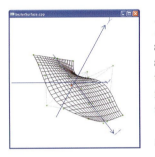

Experiment 15.7. Run `bezierSurface.cpp`, which allows the user to shape a Bézier surface by selecting and moving control points. Press the space and tab keys to select a control point. Use the left/right arrow keys to move the control point parallel to the x-axis, the up/down arrow keys to move it parallel to the y-axis and the page up/down keys to move it parallel to the z-axis.

Press 'x/X', 'y/Y' and 'z/Z' to turn the viewpoint. See Figure 15.15 for a screenshot. **End**

Figure 15.15: Screenshot of `bezierSurace.cpp`.

Exercise 15.23. If the procedure to construct the Bézier surface s via Bézier curves is "inverted" to first (a) construct $m + 1$ different Bézier curves, each of order $n + 1$, approximating a *column* of control points, and then (b) sweep the Bézier curve approximating the points corresponding to the same parameter value on each of these $m + 1$ curves, prove that the same surface s is obtained. In particular, derive the parametric form of the surface resulting from the inverted process and show it to be identical to Equation (15.16).

The program `sweepBezierSurface.cpp` of Experiment 15.6 allows the user to toggle between either process by pressing the space bar.

The following proposition is similar to Proposition 15.1 for Bézier curves:

Proposition 15.3. *If s is the Bézier surface approximating an $(n + 1) \times (m+1)$ array of control points $P_{i,j}$, $0 \leq i \leq n$, $0 \leq j \leq m$, then the following hold:*

(a) s is polynomial of degree n in one parameter variable u and polynomial of degree m in the other parameter variable v.

(b) s is a weighted sum of the control points $P_{i,j}$, $0 \leq i \leq n$, $0 \leq j \leq m$, where the weight of $P_{i,j}$ is the blending function $B_{i,n}(u)B_{j,m}(v)$ (a product of Bézier curve blending functions).

(c) Every point of s is a convex combination of the control points $P_{i,j}$, $0 \leq i \leq n$, $0 \leq j \leq m$, and, therefore, c lies inside the convex hull of the $P_{i,j}$.

(d) *s passes through the four corner control points $P_{0,0}$, $P_{n,0}$, $P_{0,m}$ and $P_{n,m}$, but not necessarily the others.*

(e) *(Affine Invariance) If the control points $P_{i,j}$, $0 \leq i \leq n$, $0 \leq j \leq m$, belong to $\mathbb{R}^3$ and $g : \mathbb{R}^3 \to \mathbb{R}^3$ is an affine transformation, then the image surface $g(s)$ is the Bézier surface approximating the images $g(P_{i,j})$, $0 \leq i \leq n$, $0 \leq j \leq m$, of the control points.*

In other words, the transformed surface approximates the transformed control points.

Proof. We begin by observing that the blending functions $B_{i,n}(u)B_{j,m}(v)$, $0 \leq i \leq n$, $0 \leq j \leq m$, form a partition of unity over the parameter space $[0,1] \times [0,1]$ because

$$\sum_{i=0}^{n} \sum_{j=0}^{m} B_{i,n}(u)B_{j,m}(v) = \sum_{i=0}^{n} B_{i,n}(u) \sum_{j=0}^{m} B_{j,m}(v) = 1 * 1 = 1$$

and leave the rest of the proof, which is similar to that of Proposition 15.1, to the reader. $\square$

Exercise 15.24. What kinds of curves are the u-parameter and v-parameter curves – recall these from Section 10.2.4 – on the Bézier surface

$$s(u,v) = \sum_{i=0}^{n} \sum_{j=0}^{m} B_{i,n}(u)B_{j,m}(v)P_{i,j} \ ?$$

Exercise 15.25. Recall the equation (10.19)

$$s(u,v) = (1-u)(1-v)\,p_1 + u(1-v)\,q_1 + (1-u)v\,p_2 + uv\,q_2, \quad u,v \in [0,1]$$

of a bilinear patch from Section 10.2.8. It's again a weighted sum of the "control" points p_1, p_2, q_1 and q_2. Do the blending functions form a partition of unity? Is a bilinear patch a Bézier surface?

Exercise 15.26. What condition would you impose on the control polyhedrons of two abutting Bézier surfaces, say, s defined by an $(n+1) \times (m+1)$ array of control points and s' defined by an $(n+1) \times (m'+1)$ array of control points – the number of rows is the same – so that they join smoothly?

Hint: A similar discussion for bicubic patches was in Section 10.3.2.

15.3 Summary, Notes and More Reading

This chapter was a fairly thorough introduction to the theory of the Bézier primitives. Our exploration was restricted, however, to the polynomial version, which itself is popularly used in design and, moreover, sets the stage for the rational primitives in a forthcoming chapter. Theory too has now caught up with practice: we learned to code polynomial Bézier curves and surfaces much earlier in Chapter 10.

There are a number of excellent books – Farin [42], Mortenson [90, 92], Rogers & Adams [113] and Vince [141] to name a few – which both complement the material here and take the reader beyond it. It is interesting to read in the first chapter of Farin's book an account by Bézier himself of the invention of the UNISURF CAD system that uses his primitives. In addition to those just mentioned, which are mostly math and modeling books, any CG book itself will likely have a section or two on Bézier theory and practice. The reader should have no trouble now in following discussions of Bézier primitives in even advanced CG texts, such as Akenine-Möller, Haines & Hoffman [1], Buss [21], Slater et al. [129] and Watt [142].

CHAPTER 16

B-Spline

O ur aim in this chapter is to master the theory underpinning B-spline primitives, the dominant class of primitives used in freeform design nowadays. As in the preceding chapter on Bézier theory, we'll restrict ourselves here to the polynomial version, reserving the more general rational class of NURBS (Non-Uniform Rational B-Spline) primitives for Chapter 18, as an application of projective spaces, which are the natural setting for these primitives.

Almost all 3D modelers support NURBS primitives – and so, of course, their polynomial subclass as well – in a WYSIWYG design environment. In such a setting, the user can get by merely pushing control points around, with little understanding of theory. OpenGL, on the other hand, provides an interface at a much lower level. In fact, there is almost a one-to-one correspondence between NURBS theory and OpenGL syntax. Consequently, some knowledge at least of the former is required in order to use the latter.

Unfortunately, as NURBS theory is more complex than Bézier, there really is no use-now-learn-later approach. This is the reason we did not introduce NURBS, or even its polynomial subclass, in the earlier chapter on drawing, as we did polynomial Bézier primitives. True, the lack of shortcuts and a fancy interface will be seen as drawbacks by those who care only about design and not so much about what is under the hood. On the other hand, OpenGL's minimalist setting is ideal for the purpose of grasping the underlying theory.

Our account of B-splines begins in Section 16.1 with an analysis of the weakness of Bézier primitives, motivating the progression to B-splines as a search, in fact, for better blending functions. The investigation of the B-spline primitives themselves begins with curves in Section 16.2, setting the stage with so-called knot vectors in anticipation of new blending functions that are polynomial in knot intervals. In subsections 16.2.1-16.2.3, we go from (uniform) first-order to quadratic B-spline curves, applying an intuitive "break-and-make" procedure to repeatedly increase the degree of the spline functions. The reader is asked to apply this procedure herself in 16.2.4 to fill in the details for cubic B-splines. A significant generalization is made in 16.2.5, not only by extending the theory to B-splines of arbitrary order, but by allowing the knot vector to be non-uniform as well. We'll see the utility of non-uniform knot vectors, particularly of repeated knots which empower the designer with the best of both worlds, Bézier and B-spline.

From B-spline curves to surfaces in Section 16.3 is exactly similar a process as from Bézier curves to surfaces. The topic of Section 16.4 is the OpenGL NURBS drawing primitives, though we use them in this chapter only to the extent of their polynomial functionality. Subsections 16.4.1 and 16.4.2 discuss drawing B-spline curves and surfaces, respectively. We describe how to light and texture a B-spline surface in 16.4.3. The useful technique of trimming a B-spline surface is described in 16.4.4. Section 16.5, with notes and suggestions for future reading, concludes the chapter.

16.1 Problems with Bézier Primitives: Motivating B-Splines

Bézier curves and surfaces are easy to use, especially in an interactive environment, and powerful enough to create complex designs. However, they suffer from two weaknesses:

1. *Lack of local control.*

   Observe that the blending function of each control point of a Bézier curve is non-zero over the entire open parameter interval $(0, 1)$; in other words, each has non-zero weight (attraction, pull, ...) at every point of the curve, except, possibly, the endpoints. For example, Figure 16.1(a) shows the blending functions of a cubic Bézier curve, which are, of course, the Bernstein polynomials of degree three.

   This makes modifying a Bézier curve difficult: moving any one control point alters the *entire* curve, not just near the control point. Albeit points on the curve far from the relocated control point move little because its weight is small at distant points, nevertheless, there is change. Moving control point P_1 in Figure 16.1(b), for example, from a reading of its blending function $B_{1,3}(u)$ in Figure 16.1(a) maximally affects the curve in the vicinity of $c(0.33)$, but all points on the curve, except for the endpoints, are altered to some extent.

   The situation for Bézier surfaces is similar, as each control point has non-zero weight at every point of the surface, except, possibly, the corners.

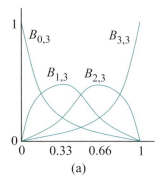

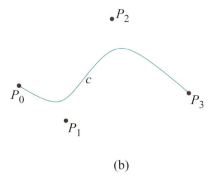

(a)   (b)

Figure 16.1: (a) Bernstein polynomials of degree 3:
$B_{0,3}(u) = (1-u)^3$, $B_{1,3}(u) = 3(1-u)^2 u$, $B_{2,3}(u) = 3(1-u)u^2$, $B_{3,3}(u) = u^3$ (b) A
cubic Bézier curve.

Typically, in designing a complex object with numerous control points a designer would prefer to be able to modify parts of the object independently, in other words, have local control, which in turn would necessitate restricting each control point to its own limited "region of influence". For example, in arranging Boris's smirk – see Figure 16.2 – the designer may want to leave his nose and eyes exactly as they are.

2. *The degree increases with the number of control points.*

The Bézier curve $c(u)$ approximating $n+1$ control points is polynomial of degree n in u. Evaluating a high-degree polynomial is expensive and repeated products lead to numerical instability. Complex curves, therefore, with multiple control points present a computational problem. And ditto for surfaces.

Figure 16.2: Mesh of Boris's head (courtesy of Sateesh Malla at www.sateeshmalla.com).

What to do about these problems? First, let's step back a bit to take the following abstract view of Bézier curves: a Bézier curve is the sum

$$c(u) = f_0(u)P_0 + f_1(u)P_1 + \ldots + f_n(u)P_n \qquad (0 \le u \le 1) \qquad (16.1)$$

of its control points P_i weighted by blending functions f_i which *happen to be* Bernstein polynomials. There's no reason they *have to be* Bernstein polynomials, provided that the resulting curve c – maybe no longer Bézier – does a satisfactory job of approximating the control points. The plan then is to try and find new blending functions which, hopefully, alleviate the Bézier difficulties.

Before proceeding, here's a bit of useful terminology: if a function f, defined on the interval domain $[a, b]$, is non-zero everywhere inside the subinterval $[a', b']$, excepting possibly its endpoints a' and b', and zero on the rest of $[a, b]$, then it is said to have *support* in $[a', b']$. Figure 16.3(a) depicts a function $f_i(u)$ defined on $[0, 1]$ with support in the subinterval $[a', b']$.

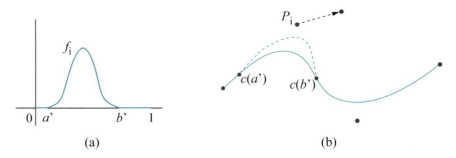

Figure 16.3: (a) Function f_i defined on $[0,1]$ has support in $[a', b']$ (b) Moving P_i, with associated blending function f_i, changes c only between $c(a')$ and $c(b')$.

Exercise 16.1. If the blending function f_i of control point P_i in expression (16.1) has support in the proper subinterval $[a', b']$ of the parameter interval $[0, 1]$, then show that moving P_i changes the arc of the approximating curve c only between $c(a')$ and $c(b')$. See Figure 16.3(b).

Exercise 16.2. Prove that the ith Bernstein polynomial of degree n for every i, $0 \le i \le n$, has support in the entire parameter interval $[0, 1]$ (keep in mind that the behavior of the polynomial outside of $[0, 1]$ is of no interest).

From the preceding two exercises, it seems, then, that the first problem with Bézier curves mentioned above arises because the blending function of every control point has support in the entire parameter interval $[0, 1]$. A solution, therefore, would be to find blending functions having support each in only part of that interval. Moreover, the second problem would be solved if the degree of the blending functions could be *decoupled* from the number of control points, so that increasing the latter did not necessarily raise the former. So now we have an idea of what we want, let's see what we can find.

Suppose, to begin with, that we ask for blending functions all quadratic, *no matter* the number of control points. The first thing to do then is find quadratics with limited support – whose graphs resemble that of f_i in Figure 16.3(a). Unfortunately, this is a hopeless task because there are none such: a quadratic is zero only at its at most two roots, not on any interval stretch like that between 0 and a', or b' and 1. But, look again at f_i. Except for the two straight zero parts at either end, the graph of f_i does resemble somewhat an upside-down parabola – see the graph of the parabola $f(u) = u^2$ in Figure 16.4(a).

Note: Curves drawn in this chapter are fairly accurate sketches, but not necessarily exact plots of their equations.

Here, then, is a drastic solution. Let's make a blending function f like f_i by assembling it from three parts – one quadratic (an upside-down parabola) and two straight zero – as follows:

$$f(u) = \begin{cases} 0, & -2 \le u \le -1 \\ -u^2 + 1, & -1 \le u \le 1 \\ 0, & 1 \le u \le 2 \end{cases}$$

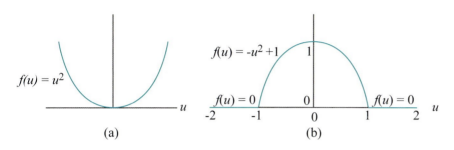

(a) (b)

Figure 16.4: (a) Parabola (b) Three-part function: one upside-down parabola and two straight.

There's no law that says that a formula has to be one line! So the specification of f is fine. Figure 16.4(b) shows its graph. As the two end parts are constant, they are actually "cheaper" than a quadratic!

We seem to be headed in the right direction. We have a blending function which is at most quadratic and which has support in $[-1, 1]$, just half of its whole domain $[-2, 2]$.

Note: If the reader is wondering about the new parameter interval $[-2, 2]$, keep in mind that there's nothing special about the parameter interval $[0, 1]$ we use most often, other than that it's convenient to write. Parameter intervals can be any $[a, b]$, with $a < b$. In the case above, $[-2, 2]$ helps avoid fractions.

The corners (C^1-discontinuities, to be precise) at $u = \pm 1$, where the straight parts of f meet the parabolic, are undesirable though, because discontinuities in the blending function will carry over to discontinuities in the approximating curve employing such a function. It'll be nice to be rid of them. How do we get a parabolic part to join a straight part without making a corner? Oddly enough, Figure 16.4(a) suggests an answer. Consider the part of the parabola $f(u) = u^2$ to the *right* of the y-axis and the (straight) part of the x-axis to the *left* of the y-axis: they meet smoothly at the origin! See Figure 16.5.

So here's the next draft. For $u \le 0$ and $u \ge 4$, define $f(u)$ to be 0, giving two long straight parts; define $f(u) = u^2$ between 0 and 1; and, $f(u) = (u - 4)^2$ between 3 and 4. See the blue curves in Figure 16.6. Particularly, $f(u) = u^2$ in $[0, 1]$ is part of the right wing of the parabola of Figure 16.4(a), while $f(u) = (u - 4)^2$ in $[3, 4]$ from the left wing of the same parabola (but shifted 4 units to the right). The two quadratics meet the

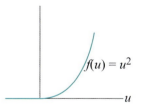

Figure 16.5: The right wing of the parabola $f(u) = u^2$ meeting the straight left half of the x-axis smoothly at the origin.

straight parts smoothly, so that's taken care of, but there's a piece missing in between (pretend you don't see the black curve!). Now, if we could only find a quadratic to sit smoothly atop the two side quadratics and cap the gap.

It turns out that a fairly intuitive choice works: drag $f(u) = u^2$ two units to the right, flip it upside down and then raise it two units. The equation is $f(u) = -(u - 2)^2 + 2$, giving the black curve in Figure 16.6. We leave verifications to the reader in the next two exercises.

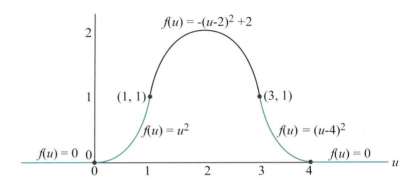

Figure 16.6: Five-part function: three parabolic and two straight parts. Joints are black points.

Exercise 16.3. Show that the curve $f(u) = -(u - 2)^2 + 2$ indeed meets $f(u) = u^2$ at $(1, 1)$ and $f(u) = (u - 4)^2$ at $(3, 1)$.

Exercise 16.4. Show that at each of the four *joints* $(0, 0)$, $(1, 1)$, $(3, 1)$ and $(4, 0)$ of the five-part function depicted in Figure 16.6 the tangent lines of the curves on either side are equal. Therefore, there is no C^1-discontinuity at a joint and the function is C^1-continuous everywhere.

Part answer: At $(1, 1)$, where $u = 1$, the tangent on the left is from $f(u) = u^2$ and on the right from $f(u) = -(u - 2)^2 + 2$. Now, $\frac{d}{du} u^2 = 2u$, which is 2 at $u = 1$, and $\frac{d}{du}(-(u - 2)^2 + 2) = -2(u - 2)$, which is also 2 at $u = 1$, so, indeed, the tangent lines on either side of the joint $(1, 1)$ are equal.

For the record, here's the 5-line formula specifying f:

$$f(u) = \begin{cases} 0, & u \leq 0 \\ u^2, & 0 \leq u \leq 1 \\ -(u - 2)^2 + 2, & 1 \leq u \leq 3 \\ (u - 4)^2, & 3 \leq u \leq 4 \\ 0, & 4 \leq u \end{cases} \tag{16.2}$$

f has support in $[0, 4]$ and, from the preceding exercise, is C^1-continuous throughout. Moreover, if its parameter interval is chosen to be an interval larger than $[0, 4]$, e.g., $[-2, 6]$, then we have indeed a C^1-continuous blending function with limited support.

The moral then is to look for blending functions among the class of piecewise polynomial functions: a function is *piecewise polynomial* if its domain can be split into subintervals in each of which it's polynomial. For example, f is composed of five polynomial pieces. From a computational point of view, evaluating a piecewise polynomial is not much harder than evaluating a polynomial. If one thinks in terms of C or C++ code, then there is simply an extra `if/else` ladder to determine the appropriate subinterval and corresponding polynomial.

The piecewise polynomials to be used as blending functions must be chosen carefully though. For example, looking back at Propositions 15.1 and 15.3 of the last chapter, it's desirable for the set of blending functions to form a partition of unity over the parameter space. Good things happen then: (a) points on the curve (or surface) are convex combinations of its control points, so the whole lies in the convex hull of its control points and (b) affine invariance.

Writing down all the properties we want, then, we put together a Wish List for blending functions. We ask that they

(a) be at least C^1-continuous piecewise polynomial,

(b) be of a low degree independent of the number of control points,

(c) each have support in only part of the parameter space, and,

(d) together form a partition of unity over the parameter space.

We're led to B-splines.

16.2 B-Spline Curves

Let's set the stage for the *B-spline blending functions* (or, as they are also called, *B-spline functions*, or *B-splines*, or *spline functions*) that we are going to define. Each will be piecewise polynomial, in other words, polynomial on subintervals. In anticipation, then, let's fix a particular parameter space and chop it up into subintervals. For convenience now, we choose $[0, r]$, and its r subintervals to be the equally sized

$$[0, 1], \quad [1, 2], \quad \ldots, \quad [r-1, r]$$

See Figure 16.7. The sequence

$$\{0, 1, \ldots, r\}$$

of successive interval endpoints is called the *knot vector* and the endpoints $0, 1, \ldots, r$ themselves, *knots*. Each subinterval $[i, i+1]$, for $0 \le i \le r - 1$, is a *knot interval*. We expect to define blending functions polynomial in each knot interval.

Figure 16.7: Parameter space $[0, r]$ with uniformly-spaced knots.

589

Remark 16.1. A knot vector as above with equally spaced knots is called a *uniform* knot vector. Later in this chapter we'll see non-uniform knot vectors as well.

Remark 16.2. The "B" in B-splines, the name given these functions by Schoenberg [122], a pioneer in their use, comes from "basis".

16.2.1 First-Order B-Splines

We'll start at the lowest level possible and define the *B-splines of degree 0* by means of constant functions. There are r B-splines of degree 0, each equal to 1 on one knot interval and 0 outside it. Precisely, the ith B-spline of degree 0, for $0 \le i \le r - 1$, denoted $N_{i,1}$, is defined as follows.
When $i = 0$:

$$N_{0,1}(u) = \begin{cases} 1, & 0 \le u \le 1 \\ 0, & \text{otherwise} \end{cases} \tag{16.3}$$

When $1 \le i \le r - 1$:

$$N_{i,1}(u) = \begin{cases} 1, & i < u \le i+1 \\ 0, & \text{otherwise} \end{cases} \tag{16.4}$$

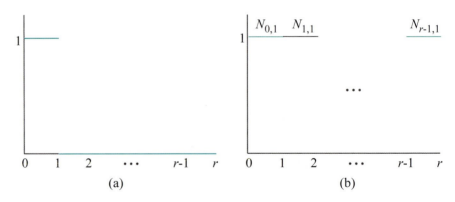

Figure 16.8: First-order B-splines: (a) $N_{0,1}$ (b) Non-zero parts of $N_{i,1}$, $0 \le i \le r - 1$, distinguished by alternate blue and black colors.

In other words, each $N_{i,1}$ is 1 on the knot interval $[i, i + 1]$, except, possibly, at the endpoints, and 0 outside it. Figure 16.8(a) shows the graph of $N_{0,1}$ over the entire parameter space $[0, r]$, while Figure 16.8(b) only the non-zero parts of the graphs of $N_{i,1}$, for $0 \le i \le r - 1$. The niggling technicality – see the first line of the two equations above – of having to define $N_{0,1}$ to be 1 on a closed interval, while the other B-splines of degree 0 are equal to 1 on a half-open interval, is unavoidable. For, we want the r B-splines of degree 0 to form together a partition of unity over $[0, r]$, so no two are allowed be 1 at the same point.

Experiment 16.1. Run `bSplines.cpp`, which shows the non-zero parts of the spline functions from first order to cubic over the uniformly spaced knot vector

$$[0, 1, 2, 3, 4, 5, 6, 7, 8]$$

Press the up/down arrow keys to choose the order. Figure 16.9 is a screenshot of the first order. The knot values can be changed as well, but there's no need to now. **End**

B-splines of degree 0 are commonly called *first-order B-splines*. If the knot vector is uniform, as above, they are called *uniform first-order B-splines*.

Interestingly, all items on the Wish List at the end of Section 16.1 are fulfilled by the first-order B-splines, except for continuity, where, in fact, they fail badly, because the $N_{i,1}$ are not even continuous (i.e., not even C^0-continuous). As we see next, unfortunately, this deficiency carries over to approximating curves made from first-order B-splines as well.

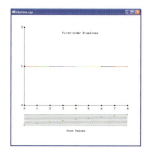

Figure 16.9: Screenshot of `bSplines.cpp` at first order.

First-Order B-Spline Curves

A *first-order B-spline approximation* of r control points $P_0, P_1, \ldots, P_{r-1}$ is called a first order B-spline curve. This is the curve c obtained from applying the first-order B-splines as blending functions to these control points, namely,

$$c(u) = \sum_{i=0}^{r-1} N_{i,1}(u) P_i \qquad (0 \leq u \leq r) \qquad (16.5)$$

What sort of a curve is c? Well, one would be hard pressed to call c a curve in the first place! Applying the definitions of $N_{i,1}$ from Equations (16.3)-(16.4) to Equation (16.5) above, one sees that c is *stationary* at P_0 for u from 0 to 1. When u crosses 1, c *jumps* to P_1, staying stationary again till u crosses 2, when c jumps to P_2 and so on. The graph of c is then just the collection of its own control points! See Figure 16.10. Obviously, if there are even two distinct control points then c is not C^0. Clearly, we'll have to move to higher orders of B-splines for satisfaction.

First-order B-Spline Properties

However, before leaving the first order, here are a few of their properties for future reference:

1. Each $N_{i,1}$ is piecewise polynomial, consisting of at most three pieces, each of which is constant.

2. $N_{i,1}$ has support in the single knot interval $[i, i + 1]$.

3. Each $N_{i,1}$ is not C^0 only at the endpoints of its supporting interval; elsewhere, it's C^∞. In other words, it's smooth – remember from Definition 10.7 that C^∞ is also called smooth – apart from its joints.

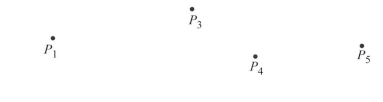

Figure 16.10: First-order B-spline approximation – the "curve" consists of its control points.

4. Together, the $N_{i,1}$ form a partition of unity over the parameter space $[0, r]$.

5. Except for $N_{0,1}$, the $N_{i,1}$ are translates of one another, i.e., the graph of one is a translate of that of another. This is a consequence of the knots being uniformly spaced.

6. A first-order B-spline approximation is, generally, not even C^0.

16.2.2   Linear B-Splines

The clear problem with first-order B-splines is that their polynomial degree 0 is too low, allowing them little flexibility in shape. Straight and horizontal is all they can be. Let's go one higher to degree 1. We'll do this in a particular way which will be easy to generalize down the road.

The trivial formula that

$$1 = u + (-u + 1) \tag{16.6}$$

allows one to "break" each B-spline $N_{i,1}$, of degree 0, into two functions $N_{i,1}^0$ and $N_{i,1}^1$ of degree 1. For example, $N_{0,1}$ breaks into $N_{0,1}^0$ and $N_{0,1}^1$, where

$$N_{0,1}^0(u) = \begin{cases} u, & 0 \leq u \leq 1 \\ 0, & \text{otherwise} \end{cases} \tag{16.7}$$

and

$$N_{0,1}^1(u) = \begin{cases} -u + 1, & 0 \leq u \leq 1 \\ 0, & \text{otherwise} \end{cases} \tag{16.8}$$

The two obviously add up to give back $N_{0,1}$, viz.,

$$N_{0,1}(u) = N_{0,1}^0(u) + N_{0,1}^1(u) \tag{16.9}$$

$N_{i,1}$, when $i > 0$, can likewise be broken into $N_{i,1}^0$ and $N_{i,1}^1$, where

$$N_{i,1}^0(u) = \begin{cases} u - i, & i < u \leq i+1 \\ 0, & \text{otherwise} \end{cases} \tag{16.10}$$

$$N_{i,1}^1(u) = \begin{cases} -u + i + 1, & i < u \leq i + 1 \\ 0, & \text{otherwise} \end{cases} \quad (16.11)$$

so that again

$$N_{i,1}(u) = N_{i,1}^0(u) + N_{i,1}^1(u) \quad (16.12)$$

Figure 16.11 shows the two parts of each first-order B-spline. For obvious reasons, we call the $N_{i,1}^0$'s "up" and the $N_{i,1}^1$'s "down". The up parts are all left or right translates of one another, as are the down parts, except that the technicality that their values at the left end of the knot interval $[0,1]$ are different from those at the left end of other knot intervals persists from first-order.

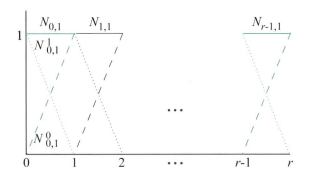

Figure 16.11: First-order B-splines each broken into an up part (dashed) $N_{i,1}^0$ and a down part (dotted) $N_{i,1}^1$. Successive $N_{i,1}$'s are distinguished by color.

Remark 16.3. This is important! For future reference, think of what we have just done as the following: each $N_{i,1}$ is broken into two functions over its support, one obtained from multiplying $N_{i,1}$ by a *straight-line function increasing from 0 to 1* from the left end of its support to the right, while the other from multiplying it by a *straight-line function decreasing from 1 to 0* over the same interval.

Equations (16.9) and (16.12) evidently guarantee that the $N_{i,1}^k$, for $k = 0, 1$ and $0 \leq i \leq r - 1$, together form a partition of unity because the $N_{i,1}$, $0 \leq i \leq r - 1$, do. But there are $2r$ of the functions $N_{i,1}^k$, which is twice as many as we need to blend r control points. Figure 16.11, in fact, suggests a way to pair them up nicely – join each up part to the following down part! Accordingly, define the *second order B-splines* (or *linear B-splines*), for $0 \leq i \leq r - 2$, as follows:

$$N_{i,2}(u) = N_{i,1}^0(u) + N_{i+1,1}^1(u) = \begin{cases} 0, & u \leq i \\ u - i, & i \leq u \leq i + 1 \\ -u + i + 2, & i + 1 \leq u \leq i + 2 \\ 0, & i + 2 \leq u \end{cases} \quad (16.13)$$

Figure 16.12 shows the non-zero parts of the linear B-splines $N_{i,2}$, $0 \leq i \leq r-2$, on the domain $[0,r]$. *See the magic*: pairing has removed all C^0-discontinuities! The linear B-splines are each continuous everywhere.

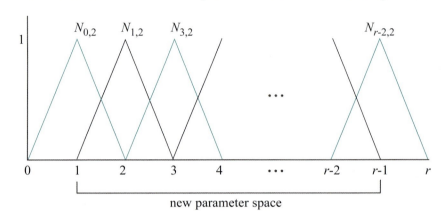

new parameter space

Figure 16.12: Non-zero parts of linear B-splines. Each is an inverted V. Successive ones are distinguished by color. The down part in the first knot interval and the up part in the last are discarded. The new (truncated) parameter space is $[1, r-1]$.

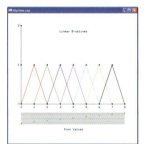

Figure 16.13: Screenshot of bSplines.cpp at second order.

Experiment 16.2. Run again bSplines.cpp and select the linear B-splines over the knot vector

$$[0, 1, 2, 3, 4, 5, 6, 7, 8]$$

Figure 16.13 is a screenshot.   **End**

Exercise 16.5. Verify that the multi-part formula above for $N_{i,2}(u)$ indeed follows from joining up and down parts (using the equations for $N_{i,1}^0$ and $N_{i,1}^1$ given earlier).

Remark 16.4. The technicality at the left endpoint of a supporting interval is now gone. The definition of $N_{i,2}$ is same for all i in $0 \leq i \leq r-2$.

Remark 16.5. Second order B-splines as defined above are often called *uniform linear B-splines* to emphasize the use of a uniform knot vector.

Note that the down part $N_{0,1}^1$ of $N_{0,1}$ and the up part $N_{r-1,1}^0$ of $N_{r-1,1}$ have no partners, so are discarded, which is why we have $r-1$ linear B-splines $N_{i,2}$, for $i = 0$ to $r-2$, versus the r first-order B-splines we started with. It's clear from Figure 16.12 that the parameter space must be truncated from $[0,r]$ to $[1, r-1]$ as well, for, otherwise, there's a problem with the partition-of-unity property in the two end knot intervals $[0,1]$ and $[r-1, r]$. Once this is done, though, we're in good shape or at least in significantly better shape than the first-order B-splines. All items in the Wish List at the end of Section 16.1 are now fulfilled except for C^1-continuity, but now the functions are at least C^0, if not quite C^1 (because of corners at the joints).

Linear B-Spline Curves

What sort of curve is the linear B-spline approximation c of $r - 1$ control points $P_0, P_1, \ldots, P_{r-2}$, which uses the linear B-splines as blending functions? It's defined as follows:

$$c(u) = \sum_{i=0}^{r-2} N_{i,2}(u)P_i \qquad (1 \leq u \leq r - 1) \qquad (16.14)$$

Exercise 16.6. Verify that the linear B-spline approximation c given by Equation (16.14) is the polygonal line through the control points in the sequence they are given. See Figure 16.14, where $r = 7$. This is certainly more respectable a curve than the first-order approximation.

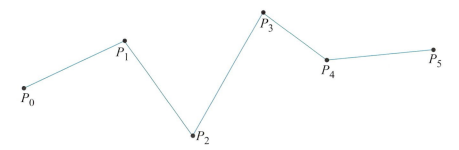

Figure 16.14: Linear B-spline approximation.

Terminology: A B-spline approximation of a sequence of control points is often called a *B-spline curve*, a *spline curve* or, simply, a *spline*. There is ambiguity sometimes, therefore, with the terminology for B-spline blending functions, but it'll be clear from the context if the term refers to a blending function or an approximating curve.

Linear B-Spline Properties

Here's a list of properties of linear B-splines similar to the one made earlier for first-order B-splines:

1. Each $N_{i,2}$ is piecewise polynomial, consisting of at most four pieces, each of which is linear, except for zero end pieces.

2. $N_{i,2}$ has support in $[i, i+2]$, the union of two consecutive knot intervals.

3. Each $N_{i,2}$ is C^0, but not C^1, at its joints. Apart from its joints it's smooth everywhere.

4. Together, the $N_{i,2}$ form a partition of unity over the parameter space $[1, r - 1]$.

5. The $N_{i,2}$ are translates of one another.

6. A linear B-spline approximation is C^0, but, generally, not C^1.

16.2.3 Quadratic B-Splines

Linear B-splines are certainly preferable to first-order ones, but we're still shy of C^1-continuity. If we could raise the degree of the polynomial pieces yet again, from 1 to 2, we might do better continuity-wise.

It turns out that the approach introduced in the last section of breaking first-order B-splines into up and down parts of one higher degree, and then pairing them up to make linear ones, generalizes. Consider first $N_{0,2}$, graphed in Figure 16.15(a). Recall Remark 16.3: to break $N_{0,2}$ into two, multiply it by a straight-line function increasing from 0 at the left end of its support to 1 at the right, as well as by the complementary function decreasing from 1 to 0. Since the supporting interval of $N_{0,2}$ is $[0, 2]$, the two straight-line functions called for are $u/2$ and $-u/2 + 1$, respectively, which are shown in Figure 16.15(a) as well.

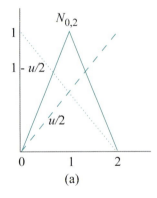
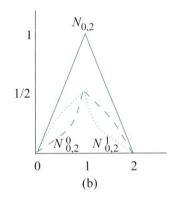

Figure 16.15: (a) The graphs of the two straight-line multiplying functions for $N_{0,2}$, one dashed and one dotted (b) The result of the multiplication: the up part $N_{0,2}^0$ (dashed) and the down part $N_{0,2}^1$ (dotted).

Accordingly, break $N_{0,2}$ as follows:

$$N_{0,2} = \frac{u}{2}\,N_{0,2} + \left(-\frac{u}{2} + 1\right) N_{0,2} \tag{16.15}$$

where the up part – it's not really increasing any more but we'll stick with the term – is

$$N_{0,2}^0(u) = \frac{u}{2}\,N_{0,2} = \begin{cases} 0, & u \le 0 \\ \frac{1}{2}u^2, & 0 \le u \le 1 \\ -\frac{1}{2}u^2 + u, & 1 \le u \le 2 \\ 0, & 2 \le u \end{cases} \tag{16.16}$$

and the down part

Section 16.2
B-Spline Curves

$$N_{0,2}^1(u) = (-\frac{u}{2} + 1) N_{0,2} = \begin{cases} 0, & u \leq 0 \\ -\frac{1}{2}u^2 + u, & 0 \leq u \leq 1 \\ \frac{1}{2}u^2 - 2u + 2, & 1 \leq u \leq 2 \\ 0, & 2 \leq u \end{cases} \quad (16.17)$$

The graphs of the two parts, resembling opposing shark fins, are shown in Figure 16.15(b).

Exercise 16.7. Verify the formulae for $N_{0,2}^0$ and $N_{0,2}^1$ by multiplying that for $N_{0,2}$ by $u/2$ and $-u/2 + 1$, respectively.

The other linear B-splines $N_{i,2}$, for $1 \leq i \leq r - 2$, can similarly be broken. Figure 16.16 shows the graphs of the up and down parts.

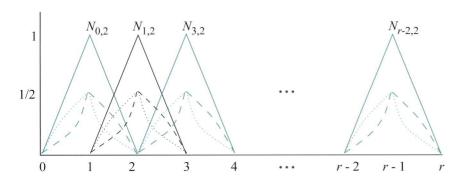

Figure 16.16: Linear B-splines each broken into an up (dashed) part $N_{i,2}^0$ and down (dotted) part $N_{i,2}^1$. Successive ones are distinguished by color.

Next, as in the first-order case, pair them up, adding each up part to the down part of the following linear B-spline. Non-zero pieces of the up and down parts did not overlap in the first-order case, so adding meant simply splicing graphs end to end. Now we do actually have to add on the overlaps.

And again magic! Two adjacent and opposing shark fins, one dashed and the other dotted, both with a sharp corner in the middle, add up to a smooth-looking floppy hat! See Figure 16.17. Precisely, the up part of one linear B-spline adds to the down part of the following one to make a *quadratic B-spline* (or, *third order B-spline*).

Figure 16.17 explains exactly what's happening. The graph of $N_{0,2}^0$ is blue dashed, while that of $N_{1,2}^1$ black dotted. The graph $N_{0,3}$ of their sum consists of the outer blue dashed arc on $[0,1]$, the outer black dotted arc on $[2,3]$ and the unbroken red arc on $[1,2]$ in the middle, the latter being the sum of the inner blue dashed and the inner black dotted. So it's in the middle interval $[1,2]$ that actual summing takes place. We'll see the summed equation itself momentarily.

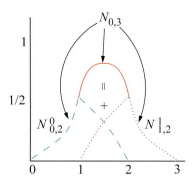

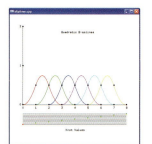

Figure 16.17: Adding $N_{0,2}^0$ and $N_{1,2}^1$ to make $N_{0,3}$. $N_{0,3}$ consists of three parts: on $[0,1]$ it's just $N_{0,2}^0$, on $[2,3]$ it's $N_{1,2}^1$, while in the middle, on $[1,2]$ it is the sum of $N_{0,2}^0$ and $N_{1,2}^1$.

Experiment 16.3. Run again `bSplines.cpp` and select the quadratic B-splines over the knot vector

$$[0,1,2,3,4,5,6,7,8]$$

Figure 16.18 is a screenshot. Note the joints indicated as black points. **End**

Figure 16.18: Screenshot of `bSplines.cpp` at third order.

$N_{0,3}$ is the first quadratic B-spline. Figure 16.19 depicts the sequence of quadratic B-splines $N_{i,3}$, $0 \le i \le r-3$, on the domain $[0,r]$.

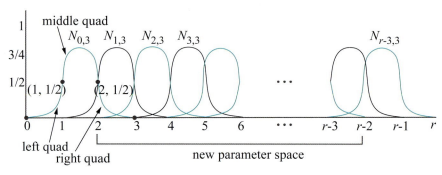

Figure 16.19: Non-zero parts of the quadratic B-splines; the four joints of the first one are indicated as points as well. Successive splines are distinguished by color.

Now for the equations of the quadratic B-splines. As they are evidently translates of one another, it's sufficient to write only that of the first one:

$$N_{0,3}(u) = N_{0,2}^0(u) + N_{1,2}^1(u) = \begin{cases} 0, & u \le 0 \\ \frac{1}{2}u^2, & 0 \le u \le 1 \\ \frac{3}{4} - (u - \frac{3}{2})^2, & 1 \le u \le 2 \\ \frac{1}{2}(-u + 3)^2, & 2 \le u \le 3 \\ 0, & 3 \le u \end{cases} \quad (16.18)$$

Exercise 16.8. Verify the preceding formula with the help of (16.16) and (16.17). Don't forget to shift the second equation one unit to the right for the formula for $N_{1,2}^1$.

Exercise 16.9. Use Equation (16.18) to determine the equation of $N_{1,3}(u)$ and, generally, $N_{i,3}(u)$.

Exercise 16.10. Verify that the first quadratic B-spline $N_{0,3}$ is C^1 *everywhere* by differentiating the functions on the RHS of (16.18) and comparing the tangents on either side at each joint (which is only where discontinuity might occur). The four joints of $N_{0,3}$, with x-values 0, 1, 2 and 3, are indicated in Figure 16.19.

Differentiating again, verify that $N_{0,3}$ is *not* C^2 at its joints.

As the quadratic B-splines are translates one of one another, it follows from the preceding exercise that they are all C^1 everywhere, though not C^2 at their joints.

Remark 16.6. Compare the 5-line formulas (16.2) and (16.18) to see that we've come now full circle back to almost the same piecewise quadratic blending function which we used to motivate B-splines in the first place!

As in the linear case, the parameter space must be truncated, this time to $[2, r-2]$, to ensure that the partition-of-unity property holds. The key to keep in mind is that partition-of-unity holds in those knot intervals on which there is defined a left, a middle and a right quadratic arc – from successive quadratic B-splines.

Pop the champagne bottles: we now officially have every item on the Wish List!

Quadratic B-Spline Curves

So what sort of curve is the quadratic B-spline approximation c of $r-2$ control points $P_0, P_1, \ldots, P_{r-3}$, defined by

$$c(u) = \sum_{i=0}^{r-3} N_{i,3}(u) P_i \qquad (2 \leq u \leq r-2) \qquad (16.19)$$

where the quadratic B-splines are used as blending functions?

First, and importantly, since the quadratic B-splines are all C^1, so is a quadratic B-spline approximation. We've gained at least respectable continuity then. However, as we ask the reader to show next, the property of interpolating the first and last control points has been lost (though not on our Wish List, this, nevertheless, is desirable).

Exercise 16.11. Prove that the quadratic spline curve c defined by Equation (16.19) begins at the midpoint of P_0 and P_1, ends at the midpoint of P_{r-4} and P_{r-3} and doesn't necessarily interpolate *any* of the control points. See Figure 16.20.

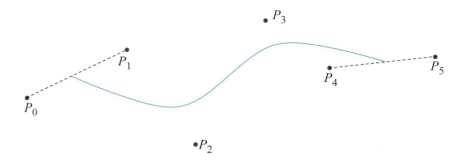

Figure 16.20: Quadratic B-spline approximation.

Darn, just when when we thought things were going our way, a potentially nasty bug rears its ugly head. Not to worry, as soon as we are able to loosen up the knot vector from being uniform, we'll be happily interpolating first and last control points again.

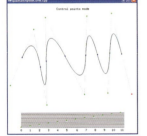

Figure 16.21: Screenshot of quadraticSpline-Curve.cpp.

Experimen**t** 16.4. Run `quadraticSplineCurve.cpp`, which shows the quadratic spline approximation of nine control points over a uniformly spaced vector of 12 knots. Figure 16.21 is a screenshot.

The control points are green. Press the space bar to select a control point – the selected one turns red – and the arrow keys to move it. The knots are the green points on the black bars at the bottom. At this stage there is no need to change their values. The blue points are the joints of the curve, i.e., images of the knots. Also drawn in light gray is the control polygon. **End**

Exerci**se** 16.12. What part of the quadratic spline curve c approximating the control points $P_0, P_1, \ldots, P_{r-3}$ is altered by moving only P_i? Your answer should be in terms of an arc of c between a particular pair of its joints. Verify using `quadraticSplineCurve.cpp`.

Quadratic B-Spline Properties

A list of properties for quadratic B-splines:

1. Each $N_{i,3}$ is piecewise polynomial, consisting of at most five pieces, each of which is quadratic, except for zero end pieces.

2. $N_{i,3}$ has support in $[i, i+3]$, the union of three consecutive knot intervals.

3. Each $N_{i,3}$ is C^1, but not C^2, at its joints. Apart from its joints it's smooth everywhere.

4. Together, the $N_{i,3}$ form a partition of unity over the parameter space $[2, r-2]$.

5. The $N_{i,3}$ are translates of one another.

6. A quadratic B-spline approximation is C^1, but, generally, not C^2.

When placing it in our Wish List, we expected to be rewarded for the partition-of-unity property by felicitous behavior of the B-spline approximating curves. The reader is asked to show next that indeed we are.

Exercise 16.13.

(a) Prove that the quadratic spline curve approximating a sequence of control points lies in the convex hull of the latter.

(b) Affine invariance: prove that an affine transformation of a quadratic spline curve is same as the quadratic spline curve approximating the transformed control points.

16.2.4 Cubic B-Splines

We're going to ask you to do most of the lifting in this section.

To start with, break the first quadratic B-spline $N_{0,3}$ into two parts: an "up" part obtained from multiplying it by a straight-line function increasing from 0 at the left end of its support to 1 at the right end and a "down" part from multiplying it by the complementary function decreasing from 1 to 0 over its support. Here's the equation showing the split:

$$N_{0,3} = \frac{u}{3} N_{0,3} + (-\frac{u}{3} + 1) N_{0,3} \qquad (16.20)$$

Exercise 16.14. Write equations for the up part

$$N_{0,3}^0(u) = \frac{u}{3} N_{0,3}$$

and the down part

$$N_{0,3}^1(u) = (-\frac{u}{3} + 1) N_{0,2}$$

in a manner analogous to Equations (16.16) and (16.17) for the quadratic B-splines. Both up and down parts are piecewise cubic.

Exercise 16.15. Verify by adding $N_{0,3}^0(u)$ and $N_{1,3}^1(u)$ that the equation of the first cubic B-spline is:

$$N_{0,4}(u) = \begin{cases} 0, & u \leq 0 \\ p(2 - u), & 0 \leq u \leq 1 \\ q(2 - u), & 1 \leq u \leq 2 \\ q(u - 2), & 2 \leq u \leq 3 \\ p(u - 2), & 3 \leq u \leq 4 \\ 0, & 4 \leq u \end{cases} \qquad (16.21)$$

where the functions p and q are given by:

$$p(u) = \frac{1}{6}(2 - u)^3$$

and

$$q(u) = \frac{1}{6}(3u^3 - 6u^2 + 4)$$

See Figure 16.22.

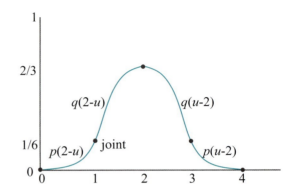

Figure 16.22: The first cubic B-spline function $N_{0,4}$.

Exercise 16.16. Verify that cubic B-splines are C^2, but not C^3, at their joints.

Exercise 16.17. Sketch the sequence of cubic B-splines $N_{i,4}$, for $0 \leq i \leq r - 4$, over $[0, r]$ similarly to Figure 16.19 for quadratic B-splines. What should be the new parameter range?

Experiment 16.5. Run **bSplines.cpp** and change the order to see a sequence of cubic B-splines. **End**

Cubic B-Spline Curves

The cubic spline curve c approximating $r - 3$ control points $P_0, P_1, \ldots, P_{r-4}$ is obtained as

$$c(u) = \sum_{i=0}^{r-4} N_{i,4}(u)P_i \qquad (3 \leq u \leq r - 3) \qquad (16.22)$$

Experiment 16.6. Run **cubicSplineCurve1.cpp**, which shows the cubic spline approximation of nine control points over a uniformly-spaced vector of 13 knots. The program is similar to **quadraticSplineCurve.cpp**. See Figure 16.23 for a screenshot.

Figure 16.23: Screenshot of **cubicSplineCurve1-.cpp**.

The control points are green. Press the space bar to select a control point – the selected one is colored red – then the arrow keys to move it. The knots are the green points on the black bars at the bottom. The blue points are the joints of the curve. The control polygon is a light gray. **End**

Cubic B-Spline Properties

A list of properties for cubic B-splines:

1. Each $N_{i,4}$ is piecewise polynomial, consisting of at most six pieces, each of which is cubic, except for zero end pieces.

2. $N_{i,4}$ has support in $[i, i+4]$, the union of four consecutive knot intervals.

3. Each $N_{i,4}$ is C^2, but not C^3, at its joints. Apart from its joints it's smooth everywhere.

4. Together, the $N_{i,4}$ form a partition of unity over the parameter space $[3, r-3]$.

5. The $N_{i,4}$ are translates of one another.

6. A cubic B-spline approximation is C^2, but, generally, not C^3.

Remark 16.7. Cubic B-splines are the most commonly used in design applications, because they offer the best trade-off between continuity and computational efficiency.

16.2.5 General B-Splines and Non-uniform Knot Vectors

It's probably evident now how to manufacture B-splines of arbitrary order over the uniform knot vector $\{0, 1, \ldots, r\}$. One would apply the *break-and-make* procedure to B-splines of each order to derive ones of one higher order. We formalize the derivation of B-splines of arbitrary order over $\{0, 1, \ldots, r\}$ recursively as follows:

Definition 16.1. The first-order B-splines $N_{i,1}$, $0 \leq i \leq r-1$, are as defined in Section 16.2.1.

Suppose, recursively, that the B-splines $N_{i,m-1}$, for $0 \leq i \leq r - m + 1$, have been defined for some order $m - 1 \geq 1$. Then define the ith B-spline $N_{i,m}$ of order m, for $0 \leq i \leq r - m$, by the equation:

$$N_{i,m}(u) = \left(\frac{u - i}{m - 1} \right) N_{i,m-1}(u) + \left(\frac{i + m - u}{m - 1} \right) N_{i+1,m-1}(u) \quad (16.23)$$

Equation (16.23) comes from a straightforward application of break-and-make. The summand

$$\left(\frac{u - i}{m - 1} \right) N_{i,m-1}(u)$$

is the up part of $N_{i,m-1}(u)$ obtained from multiplying it by the straight-line function $(u - i)/(m - 1)$ increasing from 0 at i, the left end of its support, to 1 at $i + m - 1$, the right end.

Likewise, the summand

$$\left(\frac{i+m-u}{m-1}\right) N_{i+1,m-1}(u)$$

is the down part of $N_{i+1,m-1}(u)$ obtained from multiplying it by the straight-line function $(i+m-u)/(m-1)$ decreasing from 1 to 0 from the left end $i+1$ to the right $i+m$ of its support.

Terminology: The *degree* of a B-spline is that of its polynomial pieces, while its *order* is its degree plus one.

Exercise 16.18. Make a six-point list of properties for uniform B-splines of the mth order like the ones earlier for uniform lower-order splines.

Before proceeding further, though, we'll loosen restrictions on the knot vector, which till now had been the uniform sequence

$$\{0, 1, \ldots, r\}$$

Keep in mind that the operative word is *uniform*, in particular, that knots are equally spaced and it does not matter that they are integers. For instance, if the knot vector were of the form

$$\{a, a + \delta, a + 2\delta, \ldots, a + r\delta\}$$

for some a, and some $\delta > 0$, e.g.,

$$\{1.3, 2.8, 4.3, \ldots, 1.3 + 1.5r\}$$

all calculations made so far would clearly go through again, though with different (and awkward) number values, and all properties of B-splines deduced previously would hold, too.

The restriction of uniformity is removed by allowing the knot vector to be any sequence of knots of the form

$$T = \{t_0, t_1, \ldots, t_r\}$$

where the t_i are *non-decreasing*, i.e.,

$$t_0 \leq t_1 \leq \ldots \leq t_r \tag{16.24}$$

Such knot vectors are called *non-uniform*. Yes, successive knots can even be equal and such so-called multiple knots have important applications, as we'll see.

Remark 16.8. The term non-uniform knot vector is a little unfortunate in that it actually means *not necessarily* uniform, because a uniform knot vector evidently satisfies (16.24) as well!

Hmm, do we start afresh working our way up from first-order splines, this time around over non-uniform knot vectors? Not at all. Pretty much all our earlier discussions go through again, including break-and-make. Without further ado then, here's the recursive definition of B-splines over non-uniform knot vectors.

Definition 16.2. Let

$$T = \{t_0, t_1, \ldots, t_r\} \tag{16.25}$$

be a non-uniform knot vector, where $r \geq 1$.

The (non-uniform) first-order B-spline functions $N_{i,1}$, for $0 \leq i \leq r - 1$, are defined as follows.

When $i = 0$:

$$N_{0,1}(u) = \begin{cases} 1, & t_0 \leq u \leq t_1 \\ 0, & \text{otherwise} \end{cases} \tag{16.26}$$

When $1 \leq i \leq r - 1$:

$$N_{i,1}(u) = \begin{cases} 1, & t_i < u \leq t_{i+1} \\ 0, & \text{otherwise} \end{cases} \tag{16.27}$$

The (non-uniform) mth order B-spline functions $N_{i,m}$, where the order m lies within $1 < m \leq r$, and the index i in $0 \leq i \leq r - m$, are recursively defined by:

$$N_{i,m}(u) = \left(\frac{u - t_i}{t_{i+m-1} - t_i} \right) N_{i,m-1}(u) + \left(\frac{t_{i+m} - u}{t_{i+m} - t_{i+1}} \right) N_{i+1,m-1}(u) \tag{16.28}$$

Note: The convention to follow in case the denominator of either of the two fractional terms is 0 – which may occur if there are equal knots – is the following: if the term is of the form $\frac{0}{0}$, then declare its value to be 1; if it is of the form $\frac{a}{0}$, where a is not 0, then declare its value to be 0.

This recursive formula (16.28), discovered by Cox, de Boor and Mansfield independently in 1972, known accordingly as the Cox-de Boor-Mansfield (CdM) formula or recurrence, was an important milestone in B-spline theory. However, it's fairly straightforward to understand given our own development so far:

Equations (16.26) and (16.27), respectively, replicate, with obvious changes, (16.3) and (16.4) for first-order B-splines over a uniform knot vector. Equation (16.28) follows (16.23). It formalizes break-and-make – the summands are the up and down parts, respectively, of two successive spline functions of one lower order. Figure 16.24 shows graphs of all four functions on the RHS of Equation (16.28).

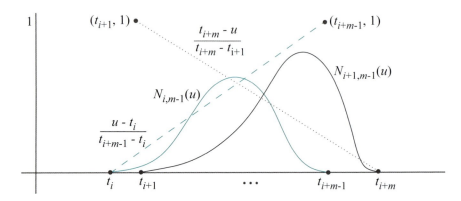

Figure 16.24: Graphs of the functions on the RHS of Equation (16.28): $N_{i,m-1}$ and $N_{i+1,m-1}$ and their respective linear multipliers $\frac{u-t_i}{t_{i+m-1}-t_i}$ and $\frac{t_{i+m}-u}{t_{i+m}-t_{i+1}}$.

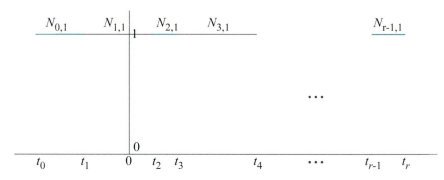

Figure 16.25: Non-zero parts of the first-order B-splines over a non-uniform knot vector.

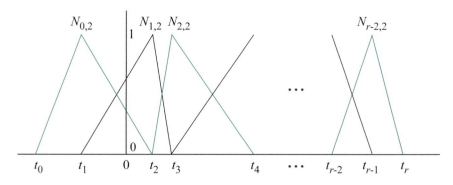

Figure 16.26: Non-zero parts of the linear B-splines over a non-uniform knot vector.

Figure 16.25 shows the graphs of the first-order B-splines over a non-uniform knot vector, while Figure 16.26 those of linear B-splines over the same knot vector.

The equations of the spline functions themselves are a little more complicated than in the case of integer knots for the simple reason that they now involve variables for knot values. For example, here's the equation, analogous to (16.18), for the first quadratic B-spline over a non-uniform knot vector:

$$
N_{0,3}(u) = \begin{cases}
0, & u \leq t_0 \\
\frac{u-t_0}{t_2-t_0}\frac{u-t_0}{t_1-t_0}, & t_0 \leq u \leq t_1 \\
\frac{u-t_0}{t_2-t_0}\frac{t_2-u}{t_2-t_1} + \frac{t_3-u}{t_3-t_1}\frac{u-t_1}{t_2-t_1}, & t_1 \leq u \leq t_2 \\
\frac{t_3-u}{t_3-t_1}\frac{t_3-u}{t_3-t_2}, & t_2 \leq u \leq t_3 \\
0, & t_3 \leq u
\end{cases} \tag{16.29}
$$

Not pretty, but B-spline computations are invariably done recursively, so a formula like this rarely needs to be written explicitly.

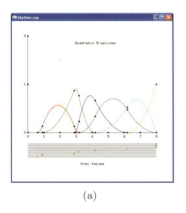
(a)

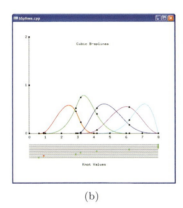
(b)

Figure 16.27: Screenshots of bSplines.cpp over a non-uniform knot vector with a triple knot at the right end: (a) Quadratic (b) Cubic.

Experiment 16.7. Run again bSplines.cpp. Change the knot values by selecting one with the space bar and then pressing the left/right arrow keys. Press delete to reset knot values. Note that the routine Bspline() implements the CdM formula (and its convention for 0 denominators).

In particular, observe the quadratic and cubic spline functions. Note how they lose their symmetry about a vertical axis through the center, and that no longer are they translates of one another.

Play around with making knot values equal – we'll soon be discussing the utility of multiple knots.

Figures 16.27(a) and (b) are screenshots of the quadratic and cubic functions, respectively, both over the same non-uniform knot vector with a triple knot at the right end. End

Example 16.1. Find the values of (a) $N_{3,3}(5)$ and (b) $N_{4,3}(5)$, if the knot vector is $\{0, 1, 2, 3, 4, 5, 5, 5, 6, 7, \ldots\}$, the non-negative integers, except that 5 has multiplicity three.

Answer: The successive knot values are

$$t_0 = 0, \ t_1 = 1, \ t_2 = 2, \ t_3 = 3, \ t_4 = 4, \ t_5 = 5, \ t_6 = 5, \ t_7 = 5, \ t_8 = 6, \ t_9 = 7, \ \ldots$$

(a) Instantiating the CdM formula (16.28):

$$N_{3,3}(u) \quad = \quad \frac{u - t_3}{t_5 - t_3} N_{3,2}(u) + \frac{t_6 - u}{t_6 - t_4} N_{4,2}(u)$$

Plugging in $u = 5$ and the given knot values:

$$N_{3,3}(5) \quad = \quad \frac{5 - 3}{5 - 3} N_{3,2}(5) + \frac{5 - 5}{5 - 4} N_{4,2}(5) \quad = \quad N_{3,2}(5) \qquad (16.30)$$

Using CdM again,

$$N_{3,2}(u) \quad = \quad \frac{u - t_3}{t_4 - t_3} N_{3,1}(u) + \frac{t_5 - u}{t_5 - t_4} N_{4,1}(u)$$

so that

$$\begin{aligned} N_{3,2}(5) \quad &= \quad \frac{5 - 3}{4 - 3} N_{3,1}(5) + \frac{5 - 5}{5 - 4} N_{4,1}(5) \\ &= \quad 2 * 0 + 0 * 1 \quad \text{(from Equations (16.26) and (16.27))} \\ &= \quad 0 \end{aligned}$$

Taking the above back to (16.30) we have

$$N_{3,3}(5) = 0$$

(b)

$$N_{4,3}(u) \quad = \quad \frac{u - t_4}{t_6 - t_4} N_{4,2}(u) + \frac{t_7 - u}{t_7 - t_5} N_{5,2}(u)$$

giving

$$\begin{aligned} N_{4,3}(5) \quad &= \quad \frac{5 - 4}{5 - 4} N_{4,2}(5) + \frac{5 - 5}{5 - 5} N_{5,2}(5) \\ &= \quad N_{4,2}(5) + \frac{0}{0} N_{5,2}(5) \\ &= \quad N_{4,2}(5) + N_{5,2}(5) \quad \text{(using convention } \tfrac{0}{0} = 1) \ (16.31) \end{aligned}$$

Using CdM again,

$$N_{4,2}(u) \quad = \quad \frac{u - t_4}{t_5 - t_4} N_{4,1}(u) + \frac{t_6 - u}{t_6 - t_5} N_{5,1}(u)$$

so that

$$N_{4,2}(5) \quad = \quad \frac{5-4}{5-4} \, N_{4,1}(5) \; + \; \frac{5-5}{5-5} \, N_{5,1}(5)$$

$$= \quad 1*1 \; + \; 0*0 \quad \text{(note by (16.27) that } N_{5,1} \text{ is zero everywhere)}$$

$$= \quad 1 \hspace{8cm} (16.32)$$

CdM again gives

$$N_{5,2}(u) \quad = \quad \frac{u-t_5}{t_6-t_5} \, N_{5,1}(u) \; + \; \frac{t_7-u}{t_7-t_6} \, N_{6,1}(u)$$

implying

$$N_{5,2}(5) \quad = \quad \frac{5-5}{5-4} \, N_{5,1}(5) \; + \; \frac{5-5}{5-5} \, N_{6,1}(5)$$

$$= \quad 1*0 \; + \; 1*0$$

$$= \quad 0 \hspace{8cm} (16.33)$$

Using (16.32) and (16.33) in (16.31) we have

$$N_{4,3}(5) = 1$$

Exercise 16.19. Find the values of $N_{5,3}(5)$ and $N_{6,3}(5)$ for the same knot vector as in the preceding example.

Exercise 16.20. Compute $N_{4,3}(7)$ again over the knot vector of the preceding example. You will have to invoke the convention that $\frac{a}{0} = 0$, if a is not 0.

General B-Spline Curves

The mth order B-spline approximation c of $r - m + 1$ control points $P_0, P_1, \ldots, P_{r-m}$ is the curve obtained by applying the mth order B-splines as blending functions. Its equation is:

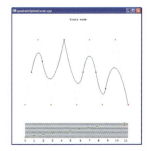

Figure 16.28: Screenshot of quadraticSpline-Curve.cpp with one double knot and one triple knot.

$$c(u) = \sum_{i=0}^{r-m} N_{i,m}(u) P_i \qquad (t_{m-1} \leq u \leq t_{r-m+1}) \hspace{2cm} (16.34)$$

Experiment 16.8. Run again `quadraticSplineCurve.cpp`. Press 'k' to enter knots mode and alter knot values using the left/right arrow keys and 'c' to return to control points mode. Press delete in either mode to reset.

Try to understand what happens if knots are repeated. Do you notice a loss of C^1-continuity when knots in the interior of the knot vector coincide? What if knots at the ends coincide? Figure 16.28 is a screenshot of `quadraticSplineCurve.cpp` with a double knot at 5 and a triple at the end at 11. **End**

Exercise 16.21. Can you find an arrangement of the knots for the quadratic spline curve to interpolate its first and last control points?

Exercise 16.22. Why does changing the value of only the first, or only the last knot, not affect the quadratic spline curve?

Experiment 16.9. Run again `cubicSplineCurve1.cpp`. Press 'k' to enter knots mode and alter knot values using the left/right arrow keys and 'c' to return to control points mode. Press delete in either mode to reset.   **End**

Exercise 16.23. Can you find an arrangement of the knots so that the cubic spline curve interpolates its first and last control points?

Exercise 16.24. What part of the mth order spline curve c approximating the control points $P_0, P_1, \ldots, P_{r-m}$ is altered by moving only P_i? Your answer should be in terms of an arc of c between a particular pair of its joints.

We collect information about mth order B-spline functions and their corresponding approximating spline curves in the following proposition.

Proposition 16.1. *Let*

$$T = \{t_0, t_1, \ldots, t_r\}$$

be a non-uniform knot vector, where $r \geq 1$.

The mth order B-spline functions $N_{i,m}$, for some order m lying within $1 \leq m \leq r$, and, where $0 \leq i \leq r - m$, satisfy the following properties:

(a) *Each $N_{i,m}$ is piecewise polynomial, consisting of at most $m + 2$ pieces, each of which is a degree $m - 1$ polynomial, except possibly for zero end pieces.*

(b) *$N_{i,m}$ has support in $[t_i, t_{i+m}]$, the union of m consecutive knot intervals.*

(c) *If the knots in T are distinct, each $N_{i,m}$ is C^{m-2}, but not C^{m-1}, at its joints. In this case, apart from its joints, each $N_{i,m}$ is smooth everywhere.*

(d) *The $N_{i,m}$ together form a partition of unity over the parameter space $[t_{m-1}, t_{r-m+1}]$.*

(e) *Every point of the mth order B-spline approximation c of $r - m + 1$ control points $P_0, P_1, \ldots, P_{r-m}$, defined by Equation (16.34), over the parameter space $[t_{m-1}, t_{r-m+1}]$, is a convex combination of the control points and lies inside their convex hull.*

(f) *(Affine Invariance) If $g : \mathbb{R}^3 \to \mathbb{R}^3$ is an affine transformation, and c is the mth order B-spline approximation of $r - m + 1$ control points $P_0, P_1, \ldots, P_{r-m}$ in $\mathbb{R}^3$, then the image curve $g(c)$ is the mth order B-spline approximation of the images $g(P_0), g(P_1), \ldots, g(P_{r-m})$ of the control points.*

(g) *If the knots in T are distinct, the mth order B-spline approximation c of $r-m+1$ control points $P_0, P_1, \ldots, P_{r-m}$ defined by Equation (16.34) is C^{m-2}, but, generally, not C^{m-1}.*

Proof. The proofs are a straightforward technical slog and we'll not write them out. $\square$

The following relation for a B-spline curve is useful to remember:

$$number\ of\ knots = number\ of\ control\ points + order \qquad (16.35)$$

Exercise 16.25. Deduce (16.35).
Hint: Count the number of knots and control points in (16.34).

Non-uniform Knot Vectors

So, of what use are non-uniform knot vectors?

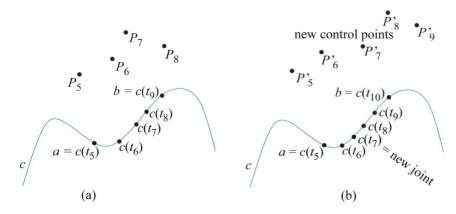

Figure 16.29: (a) Part of a cubic spline curve (b) With a new knot inserted.

One is to be able to control the influence that a control point has over an approximating curve. For example, consider the cubic spline curve c approximating control points $P_0, P_1, \ldots$ over the knot vector $\{t_0, t_1, \ldots\}$, as in Figure 16.29(a), which shows a few intermediate control points. Moving control point, say, P_5 alters only the arc of c between $a = c(t_5)$ and $b = c(t_9)$, as $N_{5,4}$ has support in $[t_5, t_9]$. Consequently, the closer or farther apart are the knots from t_5 to t_9, the more concentrated or diffuse the influence of P_5. This generalizes, of course, to all P_i, allowing the designer to vary the domain of influence of control points by rearranging knots.

Another practical consequence of non-uniform knot vectors is the technique of *knot insertion*, implemented in many commercial modelers, to allow the designer increasingly fine control over part of a spline curve. Clearly, the more knot images (joints, that is) there are in an arc of a curve, the more control points have influence over it and, therefore, the more finely it can be edited. Refer again to Figure 16.29(a). Currently, the shape of the arc between a and b is determined by the four control points P_5, P_6, P_7 and P_8. If one could insert a new knot, say, between t_6 and t_7 *without* changing the shape of the curve, there would then be five control points, instead of four, acting upon the same arc, affording the designer an added level of control.

Knots can, in fact, be inserted without changing either the shape of a spline curve or its degree, though, with a newly computed set of control points. See Figure 16.29(b), where a new knot has been inserted between t_6 and t_7, giving rise to a corresponding new joint. The joints have been re-labeled in sequence and a (hypothetical) new set of control points shown. We'll not go into the theory of knot insertion ourselves, referring the reader instead to more mathematical texts such as Buss [21], Farin [42] and Piegl & Tiller [107].

Multiple Knots

Coincident knots – *multiple knots* and *repeated knots* are the terms most commonly used – have a special application.

We'll motivate our discussion with a running example using the knot vector

$$T = \{t_0 = 0,\ t_1 = 1,\ t_2 = 2,\ t_3 = 3,\ t_4 = 3,\ t_5 = 4,\ t_6 = 5,\ t_7 = 6,\ \ldots\}$$

which has a double knot at $t_3 = t_4 = 3$. Generally, the *multiplicity* of a knot is the number of times it repeats.

The graphs of some of the B-spline functions over T are shown in Figure 16.30.

Exercise 16.26. Verify that the graphs of the first-order B-splines over T are correctly depicted in the top row of Figure 16.30 by applying the defining Equations (16.26) and (16.27). In particular, the first-order B-splines are all 1 on their supporting intervals, excluding possibly endpoints, and 0 elsewhere, *except* for $N_{3,1}$, which is 0 throughout.

Exercise 16.27. Derive the equations of the linear B-splines from the first-order ones – by plugging $m = 2$ into the recursive Equation (16.28) – to verify their graphs in the second row of Figure 16.30, as well as at the leftmost in the third. In particular, the linear B-splines over T are all C^0 and translates of one another, *except* for $N_{2,2}$ and $N_{3,2}$, neither of which is C^0.

Unfortunately, the artifact of vertical edges in the display when knots coincide makes it tricky to use `bSplines.cpp` to visually verify the linear B-spline graphs in Figure 16.30. However, there is no such issue with quadratic B-splines, so we ask the reader to do the following.

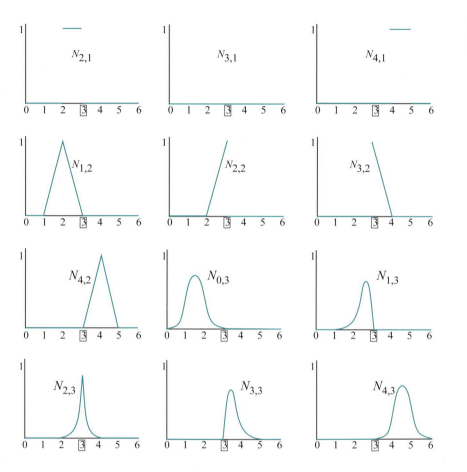

Figure 16.30: B-spline functions over the knot vector $T = \{0, 1, 2, 3, 3, 4, 5, \ldots\}$ with a double knot at 3 (distinguished inside a box).

Exercise 16.28. (Programming) Arrange the knots of `bSplines.cpp` to make their nine successive values 0, 1, 2, 3, 3, 4, 5, 6 and 7, which are the first few knots of T. Then verify visually the graphs of the five quadratic B-splines in Figure 16.30. In fact, all the quadratic B-splines over T are C^1 and translates of one another, *except* for $N_{1,3}$, $N_{2,3}$ and $N_{3,3}$, which are C^0 but not C^1.

Next, we investigate the behavior of the approximating B-spline curve in the presence of repeated knot values.

Exercise 16.29. Use Equation (16.34) and the graphs already drawn of the first-order and linear spline functions over T to verify that the first-order and linear spline curves approximating nine control points – arranged, alternately, in two horizontal rows – are correctly drawn in Figures 16.31(a) and (b), respectively.

In particular, the first-order approximation loses the control point P_3 (drawn hollow) altogether, while the linear approximation loses the segment P_2P_3 and, therefore, is no longer C^0.

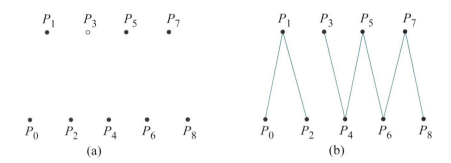

Figure 16.31: (a) First-order and (b) linear spline curves over the knot vector $T = \{0, 1, 2, 3, 3, 4, 5, 6, 7, \ldots\}$, approximating nine control points arranged alternately in two horizontal rows. The (hollow) control point P_3 is the only one missing from the first-order "curve", which consists of the remaining eight points. The second-order curve is the polyline $P_0P_1 \ldots P_8$ *minus* P_2P_3.

Experiment 16.10. Use the programs `quadraticSplineCurve.cpp` and `cubicSplineCurve1.cpp` to make the quadratic and cubic B-spline approximations over the knot vector $T = \{0, 1, 2, 3, 3, 4, 5, 6, 7, \ldots\}$ of nine control points placed as in Figure 16.31(a) (or (b)). See Figure 16.32(a) and (b) for screenshots of the quadratic and cubic curves, respectively.

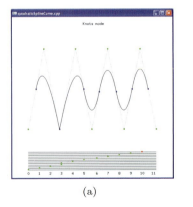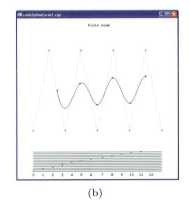

Figure 16.32: Screenshots of (a) `quadraticSplineCurve.cpp` and (b) `cubicSplineCurve1.cpp` over the knot vector $T = \{0, 1, 2, 3, 3, 4, 5, 6, 7, \ldots\}$ and approximating nine control points arranged in two horizontal rows.

The quadratic approximation loses C^1-continuity precisely at the control point P_2, which it now *interpolates* as the curve point $c(3)$. It's still C^0 everywhere.

It's not easy to discern visually, but the cubic spline drops from C^2 to C^1-continuous at $c(3)$.   **End**

Let's see next what happens with even higher multiplicity.

Experiment 16.11. Continuing with `cubicSplineCurve1.cpp` with control points as in the preceding experiment, press delete to reset and then make equal t_4, t_5 and t_6, creating a triple knot. Figure 16.33 is a screenshot of this configuration. Evidently, the control point P_3 is now interpolated at the cost of a drop in continuity there to mere C^0. Elsewhere, the curve is still C^2.   **End**

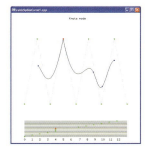

Figure 16.33: Screenshot of `cubicSplineCurve1.cpp` with a triple knot.

It seems, generally, that repeating a knot increases the influence of a particular control point, to the extent that if the repetition is sufficient then that control point itself is interpolated, though at the cost of continuity at the control point itself. This does not appear to be a particularly appealing trade-off unless a low-continuity artifact, e.g., a corner, is itself a design goal.

Let's examine more closely how the loss arises – evidently, because of the difference in the value of the derivative (of some order) of c on *either side* of a control point P. For example, the tangents to the arcs on either side of the interpolated control point P_2 of the quadratic spline curve in Figure 16.32(a) are different.

Consider now if P were an *endpoint* of c. Then continuity cannot be lost by derivatives differing on the two sides of P, for the simple reason that the curve is only to one side! And, yet, there is no reason why the influence of P cannot still be increased by repeating knots. We are on our way to recovering the property of interpolating end control points that was lost at first by quadratic spline curves.

Experiment 16.12. Make the first three and last three knots separately equal in `quadraticSplineCurve.cpp` (Figure 16.34(a)). Make the first four and last four knots separately equal in `cubicSplineCurve1.cpp` (Figure 16.34(b)). The first and last control points are interpolated in both. Do you notice any impairment in continuity? *No!*   **End**

Generally, if the first m and last m knots of an mth order spline curve are coincident, and there are no other multiple knots, then the curve interpolates its first and last control points without losing C^{m-2}-continuity anywhere. In fact, a knot vector which starts and ends with a multiplicity of m and whose intermediate knots are uniformly spaced is called a *standard knot vector*.

A standard knot vector for a quadratic spline with nine control points is

$$\{0, 0, 0, 1, 2, 3, 4, 5, 6, 7, 7, 7\}$$

The above is the canonical (and simplest) way to write standard knot vectors, though, for example

$$\{2.7, 2.7, 2.7, 3.5, 4.3, 5.1, 5.9, 6.7, 7.5, 8.3, 8.3, 8.3\}$$

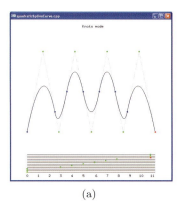
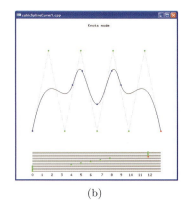

(a)   (b)

Figure 16.34: Screenshots of (a) `quadraticSplineCurve.cpp` and (b) `cubicSplineCurve1.cpp`, both with knots repeated at the end to interpolate the first and last control points.

would be equivalent.

The size of the standard knot vector is calculated from formula (16.35), viz.,

$$number\ of\ knots = number\ of\ control\ points + order$$

when given the two quantities on its RHS.

Exercise 16.30. Jot down a standard knot vector for a quadratic spline over 10 control points and for a cubic spline over 9 control points.

Exercise 16.31. Use the CdM formula to show that $N_{0,3}(t_2) = 1$ over the standard knot vector

$$T = \{0,\ 0,\ 0,\ 1,\ 2,\ \ldots,\ r-6,\ r-5,\ r-5,\ r-5\}$$

of size r for a quadratic spline. Use this to prove that the quadratic spline

$$c(u) = \sum_{i=0}^{r-3} N_{i,3}(u)P_i \qquad (t_2 = 0 \le u \le r-5 = t_{r-2})$$

approximating the $r - m + 1$ control points P_i, $0 \le i \le r - m$, over T indeed interpolates the first one, in particular, $c(t_2) = P_0$.

For the record here's a proposition:

Proposition 16.2. *A spline curve over a standard knot vector interpolates its first and last control points.*

Proof. The proof is a generalization of the preceding exercise to establish that the first control point is always interpolated. We'll leave the reader to do this by an induction. That the last control point is interpolated as well follows by symmetry. $\qquad \square$

The use of a standard knot vector for splines bequeaths yet another Bézier-like property – recall Proposition 15.1(f) – in addition to the interpolation of the end control points:

Proposition 16.3. *The tangent lines at the endpoints of a spline curve over a standard knot vector each pass through the adjacent control point.*

Proof. We'll not prove this in full generality, but only for quadratic splines in the next example. The general proof is not difficult, but tedious. □

Example 16.2. Prove that the tangent lines at the endpoints of a quadratic spline curve over a standard knot vector each pass through the adjacent control point.

Answer: We'll show that the tangent vector at the first control point passes through the second. The result at the other end follows by symmetry.

For quadratic splines, the standard knot vector is

$$T = \{0,\, 0,\, 0,\, 1,\, 2,\, \ldots\}$$

The quadratic spline curve approximating the control points $P_0, P_1, P_2, \ldots$ is

$$c(u) = N_{0,3}(u)P_0 + N_{1,3}(u)P_1 + N_{2,3}(u)P_2 + N_{3,3}(u)P_3 + \ldots$$

Now, the blending functions $N_{i,3}$, for $i \geq 3$, all vanish in $[t_2, t_3] = [0, 1]$. Consequently, in $[0, 1]$:

$$c(u) = N_{0,3}(u)P_0 + N_{1,3}(u)P_1 + N_{2,3}(u)P_2$$

Plugging the standard knot vector values into formula (16.29) for $N_{0,3}$ we get

$$N_{0,3}(u) = 1 - 2u + u^2, \quad u \in [0, 1]$$

One can use (16.29) to determine $N_{1,3}(u)$ as well by incrementing the subscripts on its RHS by 1. This gives

$$N_{1,3}(u) = 2u - \frac{3}{2}u^2, \quad u \in [0, 1]$$

Likewise, it's found that

$$N_{2,3}(u) = \frac{1}{2}u^2, \quad u \in [0, 1]$$

Therefore,

$$c(u) = \left(1 - 2u + u^2\right)P_0 + \left(2u - \frac{3}{2}u^2\right)P_1 + \left(\frac{1}{2}u^2\right)P_2, \quad u \in [0, 1]$$

Differentiating,

$$c'(u) = (-2 + 2u)\,P_0 + (2 - 3u)\,P_1 + u\,P_2, \quad u \in [0, 1]$$

Plugging in $u = 0$, one sees that

$$c'(0) = 2(P_1 - P_0)$$

which is indeed in the direction from P_0 to P_1.

We see it's for good reason, therefore, that standard knot vectors are most often used in B-spline design.

Exercise 16.32. Proposition 16.1(e) says that a spline curve is contained in the convex hull of (all) its control points. Prove the stronger statement that a spline curve of order m can be divided into successive stretches that each lie in the convex hull of only some m of its control points.

Bézier Curves and Spline Curves

It turns out that Bézier curves are special cases of spline curves:

Proposition 16.4. *The $(n+1)$th order Bézier curve approximating the $n+1$ control points*

$$P_0, P_1, \ldots, P_n$$

coincides with the $(n+1)$th order spline curve approximating the same control points over the particular standard knot vector

$$\{0, 0, \ldots, 0, 1, 1, \ldots, 1\}$$

consisting of $n+1$ 0's followed by $n+1$ 1's.

Proof. Again, in the following example, we'll restrict ourselves to establishing the quadratic case, leaving the general proof by induction to the mathematically inclined reader. □

Example 16.3. Show that the quadratic Bézier curve approximating the three control points P_0, P_1 and P_2 coincides with the quadratic spline curve approximating the same control points over the particular standard knot vector $\{0, 0, 0, 1, 1, 1\}$.

Answer: Recall from the previous chapter that the Bézier curve approximating P_0, P_1 and P_2 is

$$c_B(u) = (1-u)^2 P_0 + 2(1-u)u P_1 + u^2 P_2, \quad u \in [0, 1]$$

The quadratic spline approximating the same three points over the knot vector $T = \{t_0 = 0, \ t_1 = 0, \ t_2 = 0, \ t_3 = 1, \ t_4 = 1, \ t_5 = 1\}$ is

$$c_S(u) = N_{0,3}(u) P_0 + N_{1,3}(u) P_1 + N_{2,3}(u) P_2, \quad u \in [t_2, t_3] = [0, 1]$$

Therefore, we must show that spline blending functions of the preceding equation match the Bernstein polynomial blending functions of the one before it, over the knot interval $[0, 1]$. Refer to formula (16.29) for $N_{0,3}$. The fourth line on the RHS gives

$$
\begin{aligned}
N_{0,3}(u) &= \frac{t_3 - u}{t_3 - t_1} \frac{t_3 - u}{t_3 - t_2} \\
&= (1 - u)^2
\end{aligned}
$$

(after plugging in the knot values $t_0 = t_1 = t_2 = 0$ and $t_3 = 1$)

in $t_2 = 0 \le u \le 1 = t_3$, confirming a match with the first Bernstein polynomial.

We can use (16.29) for $N_{1,3}$ as well, making sure to increment the subscripts on the RHS by 1. This gives

$$
\begin{aligned}
N_{1,3}(u) &= \frac{u - t_1}{t_3 - t_1} \frac{t_3 - u}{t_3 - t_2} + \frac{t_4 - u}{t_4 - t_2} \frac{u - t_2}{t_3 - t_2} \\
&= 2(1 - u)u
\end{aligned}
$$

(after plugging in the knot values $t_0 = t_1 = t_2 = 0$ and $t_3 = 1$)

in $0 \le u \le 1$, matching the second Bernstein polynomial. We'll leave the reader to verify that $N_{2,3}(u) = u^2$, $u \in [0, 1]$, completing the answer.

In the opposite direction, the following is true because spline curves are piecewise polynomial (from the way they are constructed) and polynomial curves are Bézier (from Proposition 15.2).

Proposition 16.5. *A spline curve is piecewise Bézier.* □

Exercise 16.33. Why is it not possible that the preceding proposition can somehow be strengthened to say that spline curves are, in fact, Bézier, not just piecewise?
Hint: Bézier curves are smooth throughout.

16.3   B-Spline Surfaces

The construction of B-spline surfaces as a continuum of B-spline curves parallels exactly the construction of Bézier surfaces from Bézier curves described in Section 15.2. See Figure 16.35 for the following.

Suppose that we are given an $(n + 1) \times (n' + 1)$ array of control points

$$
P_{i,j}, \quad \text{for } 0 \le i \le n, \ 0 \le j \le n'
$$

and two spline orders m and m', and a knot vector

$$
T = \{t_0, t_1, \ldots, t_r\}, \quad \text{whose size satisfies } |T| = r + 1 = n + 1 + m
$$

(to ensure that *number of knots = number of control points + order*) and another knot vector

$$
T' = \{t_0, t_1, \ldots, t_{r'}'\}, \quad \text{whose size satisfies } |T'| = r' + 1 = n' + 1 + m'
$$

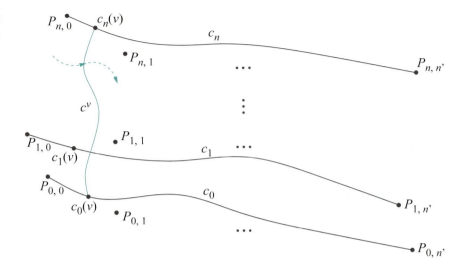

Figure 16.35: Constructing the B-spline surface approximating an array of control points by sweeping a B-spline curve. The B-spline curves depicted all interpolate both end control points, which need not always be the case in practice.

Think of the control points array as $n + 1$ different sequences, each of $n' + 1$ control points. In particular, the ith sequence, for $0 \leq i \leq n$, consists of $P_{i,0}, P_{i,1}, \ldots, P_{i,n'}$, lying along the ith row of the control points array. Construct the m'th order B-spline curve c_i, for $0 \leq i \leq n$, approximating the ith sequence, each using the knot vector T' over the parameter space $[t_{m'-1}, t_{r'-m'+1}]$.

For each v in $t_{m'-1} \leq v \leq t_{r'-m'+1}$, generate the mth order B-spline curve c^v approximating the control points sequence $c_0(v), c_1(v), \ldots, c_n(v)$, using the knot vector T over the parameter space $[t_{m-1}, t_{r-m+1}]$. The union of all these B-spline curves c^v, for $t_{m'-1} \leq v \leq t_{r'-m'+1}$, is the B-spline surface s approximating the control points array $P_{i,j}$, $0 \leq i \leq n$, $0 \leq j \leq m$. One may imagine s as being swept by c^v, as v varies from $t_{m'-1}$ to $t_{r'-m'+1}$.

Exercise 16.34. Prove that the parametric equation of the B-spline surface s constructed as above is

$$s(u, v) = \sum_{i=0}^{n} \sum_{j=0}^{n'} N_{i,m}^{T}(u) N_{j,m'}^{T'}(v) P_{i,j} \tag{16.36}$$

for $t_{m-1} \leq u \leq t_{r-m+1}$ and $t_{m'-1} \leq v \leq t_{r'-m'+1}$, and where $N_{i,m}^{T}$ (respectively, $N_{j,m'}^{T'}$) denotes the B-spline function $N_{i,m}$ over the knot vector T (respectively, $N_{j,m'}$ over T').

In other words, the surface is obtained from applying the blending function $N_{i,m}^{T}(u) N_{j,m'}^{T'}(v)$ to the control point $P_{i,j}$, over the parameter domain $t_{m-1} \leq u \leq t_{r-m+1}$, $t_{m'-1} \leq v \leq t_{r'-m'+1}$.

Hint: Mimic the proof of (15.16) for a Bézier surface.

Exercise 16.35. Formulate an analogue for B-spline surfaces of Proposition 16.1 for curves.

Remark 16.9. We've given thus far an account of NURBS (non-uniform rational B-spline) theory, *except* for the 'R', or rational, part. Instead of generally rational, our functions have all been polynomial. You could say that we have covered NUPBS, or simply NUBS, as the default for B-splines is polynomial! We'll put the 'R' into NURBS in Chapter 18 with the help of projective spaces.

16.4 Drawing B-Spline Curves and Surfaces

NURBS – the full-blown rational version of B-splines – curves and surfaces are implemented in the GLU library of OpenGL. Now that we have a fair amount of the theory, the GLU NURBS interface will turn out to be fairly simple to use, as the mapping between theory and syntax is almost one-to-one. We'll, of course, restrict ourselves to polynomial B-spline primitives for now, leaving the rational ones to a later chapter.

16.4.1 B-Spline Curves

We had already used OpenGL to draw polynomial B-spline curves in the programs `quadraticSplineCurve.cpp` and `cubicSplineCurve1.cpp` earlier this chapter, without caring then about the drawing syntax itself. Let's look at this now.

The command

```
gluNurbsCurve(*nurbsObject, knotCount, *knots, stride, *controlPoints,
              order, type)
```

defines a B-spline curve which is pointed by *nurbsObject*. The parameter *knotCount* is the number of knots in the knot vector – a one-dimensional array – pointed by *knots*. The parameter *order* is the order of the spline curve, *controlPoints* points to the one-dimensional array of control points, and *stride* is the number of floating point values between the start of the data set for one control point and that of the next in the control points array. The number of control points is not explicitly specified, but computed by OpenGL with the help of (16.35):

$$number\ of\ control\ points = number\ of\ knots - order$$

The parameter *type* is `GL_MAP1_VERTEX_3` or `GL_MAP1_VERTEX_4`, according as the spline curve is polynomial or rational.

A `gluNurbsCurve()` command must be bracketed between a `gluBeginCurve()`-`gluEndCurve()` pair of statements. The following statements from the drawing routine of `quadraticSplineCurve.cpp`, defining a quadratic B-spline curve approximating nine control points, should now be clear:

```
gluBeginCurve(nurbsObject);
gluNurbsCurve(nurbsObject, 12, knots, 3, ctrlpoints[0], 3,
              GL_MAP1_VERTEX_3);
gluEndCurve(nurbsObject);
```

Exercise 16.36. Refer to Section 10.3.1 for the syntax of the command `glMap1f()` defining a Bézier curve and compare it with that of `gluNurbsCurve()`.

There are certain initialization steps to be completed prior to a `gluNurbsCurve()` call. First, `gluNewNurbsRenderer()` returns the pointer to a NURBS object, which is passed to the subsequent `gluNurbsCurve()` call. Then optional `gluNurbsProperty()` calls control the quality of the rendering. They can activate as well a callback interface. There are several possible attributes for `gluNurbsProperty()` and we refer to the red book for details. Our own usage is kept to a simple minimum – the relevant statements from the `setup()` routine of `quadraticSplineCurve.cpp` are the following:

```
nurbsObject = gluNewNurbsRenderer();
gluNurbsProperty(nurbsObject, GLU_SAMPLING_METHOD, GLU_PATH_LENGTH);
gluNurbsProperty(nurbsObject, GLU_SAMPLING_TOLERANCE, 10.0);
```

The last two statements specify that the longest length of a line segment in a strip approximating a NURBS curve (or that of a quad edge, in the case of a mesh approximating a NURBS surface) is at most 10.0 pixels.

Experiment 16.13. Change the last parameter of the statement

```
gluNurbsProperty(nurbsObject, GLU_SAMPLING_TOLERANCE, 10.0);
```

in the initialization routine of `quadraticSplineCurve.cpp` from 10.0 to 100.0. The fall in resolution is noticeable. **End**

If you are wondering whether a B-spline curve can be drawn in a manner similar to that using `glMapGrid1f()` followed by `glEvalMesh1()` for a Bézier curve – sampling the curve uniformly through the parameter domain – the answer is yes. We don't use them ourselves but the needed calls are `gluNurbsProperty(`*nurbsObject*`, GLU_SAMPLING_METHOD, GLU_DOMAIN_DISTANCE)` and `gluNurbsProperty(`*nurbsObject*`, GLU_U_STEP, `*value*`)`. The reader is referred to the red book for implementation details.

Experiment 16.14. Run `cubicSplineCurve2.cpp`, which draws the cubic spline approximation of 30 movable control points, initially laid out on a circle, over a fixed standard knot vector. Press space and backspace to cycle through the control points and the arrow keys to move the selected control point. The delete key resets the control points. Figure 16.36 is a screenshot of the initial configuration.

The number of control points being much larger than the order, the user has good local control. **End**

Exercise 16.37. (Programming) Use `cubicSplineCurve2.cpp` to draw two smooth closed loops like those in Figure 16.37.

Figure 16.36: Screenshot of `cubicSplineCurve2-.cpp`.

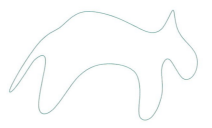

Figure 16.37: Use `cubicSplineCurve2.cpp` to draw a man and his cat.

16.4.2 B-Spline Surfaces

The OpenGL syntax for a B-spline surface is a straightforward extension of that for a B-spline curve. The `gluNurbsSurface()` command, which must be bracketed between a `gluBeginSurface()`-`gluEndSurface()` pair of statements, has the following form:

gluNurbsSurface(*nurbsObject, uknotCount, *uknots, vknotCount, *vknots, ustride, vstride, *controlPoints, uorder, vorder, type)

vknots points to the knot vector used with the control point row, in other words, to make the parameter curves c_i in the discussion in Section 16.3 of a B-spline curve sweeping a surface; *uknots* points to the knot vector used with the control point columns, i.e., to make the curves c^v in that discussion.

The parameter *vknotCount* is the number of knots in the vector pointed by *vknots*, *vorder* is the order of the B-spline curves c_i and *vstride* is the number of floating point values between the data set for one control point and the next in a row of the control points array. The parameters *uknotCount*, *uorder* and *ustride* represent similar values for the control point columns.

The parameter *type* is `GL_MAP2_VERTEX_3` or `GL_MAP2_VERTEX_4` for polynomial or rational surfaces, respectively; it can have other values as well to specify surface normals and texture coordinates.

Experiment 16.15. Run `bicubicSplineSurface.cpp`, which draws a spline surface approximation to a 15×10 array of control points, each movable in 3-space. The spline is cubic in both parameter directions and a standard knot vector is specified in each as well.

Press the space, backspace, tab and enter keys to select a control point. Move the selected control point using the arrow and page up and down keys. The delete key resets the control points. Press 'x/X', 'y/Y' and 'z/Z' to turn the surface. Figure 16.38 is a screenshot. **End**

Exercise 16.38. (Programming) Use `bicubicSplineSurface.cpp` to draw a hilly terrain and a boat.

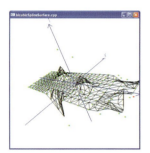

Figure 16.38: Screenshot of `bicubicSpline-Surface.cpp`.

16.4.3 Lighting and Texturing a B-Spline Surface

Lighting and texturing a B-spline surface is similar to doing likewise for a Bézier surface. Normals are required for lighting and the quickest way to create normals for a B-spline surface is to generate them automatically with a call, as for Bézier surfaces, to `glEnable(GL_AUTO_NORMAL)`.

And, again as for Bézier surfaces, determining texture coordinates for a B-spline surface requires, first, the creation of a "fake" B-spline surface in texture space on the same parameter rectangle as the real one – the reader should review if need be the discussion in Section 12.4 on specifying texture coordinates for a Bézier surfaces. OpenGL, subsequently, assigns as texture coordinates to the image on the real surface of a particular parameter point the image of that same point on the fake surface in texture space. Code will clarify.

Experiment 16.16. Run `bicubicSplineSurfaceLitTextured.cpp`, which textures the spline surface of `bicubicSplineSurface.cpp` with a red-white chessboard texture. Figure 16.39 is a screenshot. The surface is illuminated by a single positional light source whose location is indicated by a large black point. User interaction remains as in `bicubicSplineSurface.cpp`. Note that pressing the 'x'-'Z' keys turns only the surface, not the light source.

The bicubic B-spline surface, as well as the fake bilinear one in texture space, are created by the following statements in the drawing routine:

Figure 16.39: Screenshot of `bicubicSplineSurface-LitTextured.cpp`.

```
gluBeginSurface(nurbsObject);
gluNurbsSurface(nurbsObject, 19, uknots, 14, vknots,
        30, 3, controlPoints[0][0], 4, 4, GL_MAP2_VERTEX_3);
gluNurbsSurface(nurbsObject, 4, uTextureknots, 4, vTextureknots,
        4, 2, texturePoints[0][0], 2, 2, GL_MAP2_TEXTURE_COORD_2);
gluEndSurface(nurbsObject);
```

We'll leave the reader to parse in particular the third statement and verify that it creates a "pseudo-surface" – a 10×10 rectangle – in texture space on the same parameter domain $[0, 12] \times [0, 7]$ as the real one. **End**

Exercise 16.39. (Programming) Light and texture the B-spline surfaces you created for Exercise 16.38.

16.4.4 Trimmed B-Spline Surface

A powerful design tool is to *trim* (i.e., excise or remove) part of a B-spline surface. Here, first, is what happens theoretically.

Say the parametric specification of a surface s is given to be

$$x = f(u, v), \ y = g(u, v), \ z = h(u, v), \quad \text{where} \ (u, v) \in W = [u_1, u_2] \times [v_1, v_2]$$

The parametric equations map the rectangle W from uv-space onto the surface s in xyz-space. Moreover, a loop (closed curve) c on W maps to a loop c' on s. See Figure 16.40.

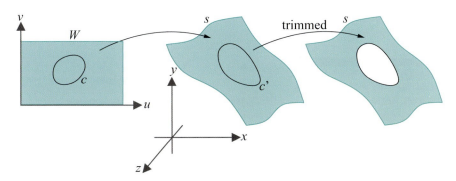

Figure 16.40: The loop c on the parameter space W is mapped to the loop c' on the surface s by the parametric equations for s. Then s is trimmed by c.

If the part of s inside, or outside, the loop c' is excised, then s is said to be trimmed by the loop c (probably, more accurate would be to say that it is trimmed by c', but the given usage is common). Figure 16.40 shows the inside trimmed. Loop c itself is called the *trimming loop*.

OpenGL allows B-spline surfaces to be trimmed. We use the program `trimmedBicubicBsplineSurface.cpp`, as a running example to explain OpenGL syntax for trimming.

Experiment 16.17. Run `trimmedBicubicBsplineSurface.cpp`, which shows the surface of `cubicBsplineSurface.cpp` trimmed by multiple loops. The code is modified from `bicubicBsplineSurface.cpp`, functionality remaining same. Figure 16.41(a) is a screenshot. **End**

All the code relevant to trimming is in the drawing routine:

```
gluBeginSurface(nurbsObject);
gluNurbsSurface(nurbsObject, 19, uknots, 14, vknots,
        30, 3, controlPoints[0][0], 4, 4, GL_MAP2_VERTEX_3);

gluBeginTrim(nurbsObject);
   gluPwlCurve(nurbsObject, 5, boundaryPoints[0], 2,
        GLU_MAP1_TRIM_2);
```

```
gluEndTrim(nurbsObject);

gluBeginTrim(nurbsObject);
    gluPwlCurve(nurbsObject, 11, circlePoints[0], 2,
                GLU_MAP1_TRIM_2);
gluEndTrim(nurbsObject);

gluBeginTrim(nurbsObject);
    gluNurbsCurve(nurbsObject, 10, curveKnots, 2, curvePoints[0], 4,
                  GLU_MAP1_TRIM_2);
gluEndTrim(nurbsObject);

gluEndSurface(nurbsObject);
```

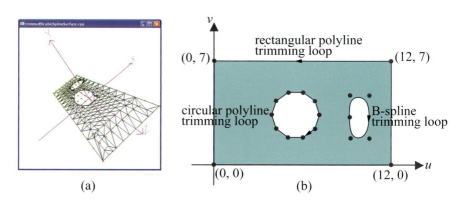

(a) (b)

Figure 16.41: (a) Screenshot of `trimmedBicubicBsplineSurface.cpp` (b) The three trimming loops – two polygonal and one B-spline.

Points to note:

1. Each trimming loop is defined within a `gluBeginTrim()`-`gluEndTrim()` pair of statements, which itself must lie within the `gluBeginSurface()`-`gluEndSurface()` pair. The trimming loop definitions are located after the `gluNurbsSurface()` definition.

2. Each trimming loop must be a closed curve in the parameter space.

3. There are two ways to define a trimming loop:

   (a) As a polygonal line loop defined by a

   $$glPwlCurve(*nurbsObject, \ pointsCount, \ *pointsArray, \\ stride, \ type)$$

   statement, where *pointsCount* is the number of vertices in an array of the form $\{v_0, v_1, \ldots, v_n\}$ pointed by *pointsArray* (it is required that $v_0 = v_n$).

There are two such polyline trimming loops in the program (see Figure 16.41(b)):

(i) The five vertices (first and last equal) of one are in the array `boundaryPoints`, describing the rectangular boundary of the parameter space itself, oriented *counter-clockwise*. We'll soon see why this particular bounding trimming loop is required.

(ii) The eleven vertices (again, first and last equal) of the other are in the array `circlePoints`, equally spaced along a circle, oriented *clockwise*.

(b) As a B-spline loop defined by

$$\text{gluNurbsCurve}(nurbsObject,\ knotCount,\ *knots,\ stride,$$
$$*controlPoints,\ order,\ type)$$

In the program there is a single such B-spline trimming loop, whose six control points (first and last equal) are in the array `curvePoints` oriented *clockwise* (Figure 16.41(b)).

4. The part outside a trimming loop oriented counter-clockwise is trimmed, while that inside a trimming loop oriented clockwise is trimmed.

Accordingly, the first trimming polyline loop of the program, which bounds the parameter space going counter-clockwise, trims off the *exterior* of the drawn surface, not trimming the surface itself per se. The other two trimming loops actually create holes in the surface.

Exercise 16.40. (Programming) Draw a terrain with a few extinct volcanoes (or smoking ones if you like particle systems).

16.5   Summary, Notes and More Reading

In this chapter we learned a fair amount of the theory underlying the extremely important and widely-used class of 3D design primitives – B-splines, both curves and surfaces. Emphasis was on motivating each new concept. We did *not* want to pull stuff out of a hat. A test if we were successful is for the reader to deduce some formula, e.g., (16.18) for the first quadratic B-spline $N_{0,3}$ over a uniform knot vector or the Cox-de Boor-Mansfield recurrence (16.28), using just pencil and paper, and not referring again to the text! This chapter prepares the reader, as well, for the rational version of the theory – NURBS – coming up in Chapter 18.

As for OpenGL, we learned not only how to draw B-spline curves and surfaces, but to illuminate, texture and trim the latter as well.

B-spline theory is extensive and there is a large literature. What we have covered in this chapter of the polynomial B-spline primitives, together with what is covered in Chapter 18 of NURBS, is ample for an applications programmer to function knowledgeably. However, the reader is well-advised to expand her knowledge, particularly, of such practical topics as "knot insertion", "degree elevation", etc. It's easy enough given the number of excellent books available – Bartels et al. [9], Farin [42], Mortenson [90], Piegl & Tiller [107] and Rogers & Adams [113] are a few that come to mind. The mathematically inclined reader, in particular, will find much to fascinate her in the more specialized nooks and crannies. Advanced 3D CG books, e.g., Akenine-Möller, Haines & Hoffman [1], Buss [21], Slater et al. [129] and Watt [142], should each have a thorough presentation of B-spline theory as well.

B-spline functions were first studied in the 1800s by the Russian mathematician Nicolai Lobachevsky. However, the modern theory began with Schoenberg's [121] application of spline functions to data smoothing and received particular impetus with the discovery in 1972 of the recursive formula (16.28) for B-spline functions by Cox [29], de Boor [32] and Mansfield. It has since seen explosive growth and B-spline (and NURBS) primitives are *de rigueur* in modern-day CG design.

CHAPTER 17

Hermite

O ur objective in this chapter is to learn a method of interpolating a set of control points, in other words, finding a curve (or surface) that passes through each. Bézier curves, as we know, mandatorily interpolate only their first and last control points, while Bézier surfaces only the four corner control points. B-spline curves and surfaces of quadratic and higher degree do not necessarily interpolate any of their control points. Nevertheless, we learned in Section 16.2.5 how to force a B-spline curve to interpolate a control point by raising the multiplicity of a knot. In fact, the so-called standard knot vector, with repeated end knots, is often used to ensure the interpolation of corner control points.

However, if a designer wishes to draw a curve or surface interpolating *all* its control points, then it's best to apply an intrinsically interpolating technique, rather than try to coax an approximating one like Bézier or B-spline into interpolating. A popular class of interpolating curves is that of the Hermite splines and this short chapter introduces this class, together with two special subclasses, that of the natural cubic splines and the cardinal splines. We discuss Hermite surface patches, as well, to interpolate 2D arrays of control points.

We begin with a discussion of general Hermite splines in Section 17.1. These curves, unfortunately, are guaranteed only to be piecewise smooth – they can have corners at control points. Moreover, the user is required to specify tangent vectors at all the control points. The subclass of natural cubic splines, the topic of Section 17.2, automatically determine these tangent vectors by imposing an additional C^2-continuity requirement. Cardinal splines, in Section 17.3, are based upon yet another scheme to automatically specify tangent vectors at control points.

We make a brief presentation of Hermite surfaces in Section 17.4 and conclude in Section 17.5.

17.1 Hermite Splines

A *Hermite spline*, also called a *cubic spline*, interpolating a sequence $P_0, P_1, \ldots, P_n$ of $n+1$ control points, is a *piecewise cubic* curve c passing through the control points. Each cubic arc of c joins successive pairs of control points, so that the entire spline comprises n cubic arcs joined end to end. Figure 17.1 shows a Hermite spline through four control points on a plane. There are corners at the middle two because the tangents of the cubics on either side don't agree.

Terminology: A *cubic arc* is a part of a cubic curve; e.g., an arc of the graph of $y = x^3$ is a cubic arc on the plane. Sometimes we'll loosen cubic to mean a polynomial of degree at most three, rather than exactly three.

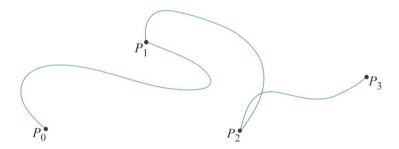

Figure 17.1: A (non-smooth) Hermite spline through four control points, composed of three cubic arcs.

$\mathcal{R}em\alpha rk$ 17.1. Hermite splines are named after the nineteenth-century French mathematician Charles Hermite.

$\mathcal{R}em\alpha rk$ 17.2. Curves of degree higher than three could be used to interpolate, or even lower, e.g., quadratic. However, three is a "Goldilocks" degree: a happy medium value, high enough to assure flexibility, and yet low enough to be computationally efficient.

Hermite interpolation for evident reasons is often called *cubic interpolation*.

We'll soon find a way to eliminate the corners in the interior and create a smooth Hermite spline through a given sequence of control points, but let's see first how to make a single cubic arc joining two arbitrary points P and Q.

Write the parametric equation of a general cubic curve c as

$$c(u) = A_3 u^3 + A_2 u^2 + A_1 u + A_0 \qquad (0 \leq u \leq 1) \qquad (17.1)$$

where each A_i, $0 \leq i \leq 3$, is a point – precisely, its vector of coordinates – in the ambient space. If you are wondering about polynomial coefficients which are vectors rather than scalars, then consider the following example.

Example 17.1. Suppose that we are interested in Hermite splines in the real world so that our ambient space is $\mathbb{R}^3$. Then the equation of a cubic curve is

$$c(u) = A_3 u^3 + A_2 u^2 + A_1 u + A_0 \quad (0 \le u \le 1)$$

where each A_i, $0 \le i \le 3$, is a point in 3-space.

To illustrate, say,

$$A_3 = [-1 \; 2 \; 0]^T, \quad A_2 = [3 \; 0 \; -2]^T, \quad A_1 = [4 \; 3 \; 4]^T, \quad \text{and} \quad A_0 = [0 \; 8 \; 7]^T,$$

Then

$$
\begin{aligned}
c(u) &= [-1 \; 2 \; 0]^T u^3 + [3 \; 0 \; -2]^T u^2 + [4 \; 3 \; 4]^T u + [0 \; 8 \; 7]^T \\
&= [-u^3 + 3u^2 + 4u \quad 2u^3 + 3u + 8 \quad -2u^2 + 4u + 7]^T
\end{aligned}
$$

over the interval $[0, 1]$. As one would expect, the cubic c in $\mathbb{R}^3$ is simply a scalar cubic in *each* of its three coordinates.

Example 17.2. Express in the form (17.1) the twisted cubic given parametrically by

$$x = t, \; y = t^2, \; z = t^3$$

Answer:

$$c(t) = [t \; t^2 \; t^3]^T = [0 \; 0 \; 1]^T t^3 + [0 \; 1 \; 0]^T t^2 + [1 \; 0 \; 0]^T t$$

Returning to the general form (17.1) of the cubic, rewrite it as a matrix equation:

$$c(u) = [u^3 \; u^2 \; u \; 1] \begin{bmatrix} A_3 \\ A_2 \\ A_1 \\ A_0 \end{bmatrix} \quad (0 \le u \le 1) \quad (17.2)$$

Note: The RHS is a product of a 1×4 matrix of scalars with a 4×1 matrix of vectors, but this is not a problem if we appropriately multiply a vector by a scalar while following the usual rules of matrix multiplication.

Differentiating (17.2) one obtains the derivative of c as

$$c'(u) = [3u^2 \; 2u \; 1 \; 0] \begin{bmatrix} A_3 \\ A_2 \\ A_1 \\ A_0 \end{bmatrix} \quad (0 \le u \le 1) \quad (17.3)$$

Substitute 0 and 1 for u in Equations (17.2) and (17.3) to find that

$$c(0) = A_0, \quad c(1) = A_3 + A_2 + A_1 + A_0, \quad c'(0) = A_1, \quad c'(1) = 3A_3 + 2A_2 + A_1 \quad (17.4)$$

It seems that if one could specify $c(0)$, $c(1)$, $c'(0)$ and $c'(1)$, then one would have four equations in the four unknowns A_0, A_1, A_2 and A_3, which should solve to find these coefficients and specify c (*alert*: that's four equations in four *vector* unknowns, so, e.g., if we are in 3-space, we'll have actually twelve equations in twelve *scalar* unknowns!). Since c goes from P to Q we know at least that $c(0) = P$ and $c(1) = Q$; as for the tangent vectors $c'(0)$ and $c'(1)$, we have freedom to specify them as please. Let's choose them to be two vectors denoted P' and Q', respectively. See Figure 17.2.

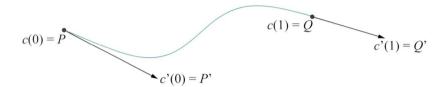

Figure 17.2: Four boundary constraints on a cubic curve c.

Accordingly, write (17.4) as

$$P = A_0, \quad Q = A_3 + A_2 + A_1 + A_0, \quad P' = A_1, \quad Q' = 3A_3 + 2A_2 + A_1 \tag{17.5}$$

which in matrix form is the equation

$$\begin{bmatrix} P \\ Q \\ P' \\ Q' \end{bmatrix} = \begin{bmatrix} 0 & 0 & 0 & 1 \\ 1 & 1 & 1 & 1 \\ 0 & 0 & 1 & 0 \\ 3 & 2 & 1 & 0 \end{bmatrix} \begin{bmatrix} A_3 \\ A_2 \\ A_1 \\ A_0 \end{bmatrix} \tag{17.6}$$

Solve this equation by inverting the coefficient matrix as follows

$$\begin{bmatrix} A_3 \\ A_2 \\ A_1 \\ A_0 \end{bmatrix} = \begin{bmatrix} 0 & 0 & 0 & 1 \\ 1 & 1 & 1 & 1 \\ 0 & 0 & 1 & 0 \\ 3 & 2 & 1 & 0 \end{bmatrix}^{-1} \begin{bmatrix} P \\ Q \\ P' \\ Q' \end{bmatrix}$$

$$= \begin{bmatrix} 2 & -2 & 1 & 1 \\ -3 & 3 & -2 & -1 \\ 0 & 0 & 1 & 0 \\ 1 & 0 & 0 & 0 \end{bmatrix} \begin{bmatrix} P \\ Q \\ P' \\ Q' \end{bmatrix} \tag{17.7}$$

to see that the four coefficients A_0, A_1, A_2 and A_3 can indeed be derived from the four boundary constraints P, Q, P' and Q'. The 4×4 matrix in the second line of the equation is called the *Hermite matrix* and denoted M_H, so (17.7) is written concisely as

$$[A_3 \; A_2 \; A_1 \; A_0]^T = M_H \, [P \; Q \; P' \; Q']^T \tag{17.8}$$

Finally, let's use (17.2) to write c's equation in terms of its boundary constraints:

$$
\begin{aligned}
c(u) &= [u^3\ u^2\ u\ 1]\,[A_3\ A_2\ A_1\ A_0]^T \\
&= [u^3\ u^2\ u\ 1]\,M_H\,[P\ Q\ P'\ Q']^T \\
&= (2u^3 - 3u^2 + 1)\,P + (-2u^3 + 3u^2)\,Q + \\
&\qquad (u^3 - 2u^2 + u)\,P' + (u^3 - u^2)\,Q' \qquad (17.9)
\end{aligned}
$$

in $0 \le u \le 1$, after performing the matrix multiplications in the second line. Therefore,

$$
c(u) = H_0(u)\,P + H_1(u)\,Q + H_2(u)\,P' + H_3(u)\,Q' \qquad (0 \le u \le 1) \quad (17.10)
$$

where the polynomials

$$
\begin{aligned}
H_0(u) &= 2u^3 - 3u^2 + 1, & H_1(u) &= -2u^3 + 3u^2, \\
H_2(u) &= u^3 - 2u^2 + u, & H_3(u) &= u^3 - u^2
\end{aligned}
$$

are called *Hermite blending polynomials*, which, of course, are blending functions, but obviously different from those used earlier in Bézier and B-spline theory (moreover, not just control points, but tangent vectors, too, enter the mix!). Their graphs are sketched in Figure 17.3. Certain symmetries are evident. Observe, as well, that $H_3(u)$ is non-positive in $0 \le u \le 1$, reaching a minimum value of nearly -0.15.

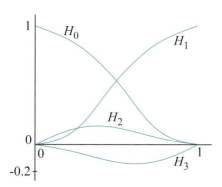

Figure 17.3: Hermite blending polynomials (not exact plots).

The curve $c(u)$ itself is called a *Hermite cubic*. Equation (17.10) is called the *geometric* form of the cubic because its expression is in terms of c's boundary constraints, while (17.1) is its *algebraic* form.

Remark 17.3. Readers familiar with programs such as Adobe Photoshop or Illustrator will recognize that the pen tool can be used to draw Hermite cubics by specifying endpoints and tangents there.

Exercise 17.1. Use calculus to determine the maximum value of $H_2(u)$ and the minimum value of $H_3(u)$ in the interval $[0, 1]$.

Exercise 17.2. Determine the symmetries among the Hermite blending polynomials. For example, that $H_0(u)$ and $H_1(u)$ are mirror images across the vertical line $u = \frac{1}{2}$ down the middle of the parameter interval $[0, 1]$ can be seen by substituting $(1 - u)$ for u in the equation of one to obtain that of the other.

Experiment 17.1. Run `hermiteCubic.cpp`, which implements Equation (17.10) to draw a Hermite cubic on a plane. Press space to select either a control point or tangent vector and the arrow keys to change it. Figure 17.4 is a screenshot. The actual cubic is simple to draw, but as you can see in the program we invested many lines of code to get the arrow heads right! **End**

Figure 17.4: Screenshot of `hermiteCubic.cpp`.

Exercise 17.3. What sort of curve is c if the two boundary constraints P' and Q' are both zero (i.e., if the two end velocities vanish)? Determine this from the geometric form of the Hermite cubic and verify in the program.

It's interesting to contrast (17.10) with the equation of the cubic Bézier curve (Equation (15.8)):

$$c(u) = B_{0,3}(u)P_0 + B_{1,3}(u)P_1 + B_{2,3}(u)P_2 + B_{3,3}(u)P_3 \qquad (0 \le u \le 1)$$

In the case of the Bézier curve, the control points are blended with weights equal to the Bernstein polynomials of degree 3; in the case of the Hermite cubic, the two end control points and their respective tangents are blended with weights equal to the Hermite blending polynomials, which are of degree 3 as well.

Remark 17.4. Since a Hermite cubic interpolates not only its two specified control points, but also the specified tangents there, it's said to be a first-order interpolation (versus a zeroth-order one which would interpolate merely control points).

Exercise 17.4. Prove the affine invariance of the cubic curve c given by Equation (17.10).

Note: Keeping in mind that an affine transformation is a linear transformation followed by a translation, we'll want its linear transformation part applied to all four boundary constraints P, Q, P' and Q', while the translation should apply only to P and Q.

Let's return to the original problem of joining successive pairs of the $n+1$ control points $P_0, P_1, \ldots, P_n$ by means of cubic arcs so that the resulting Hermite spline is smooth. A strategy that comes to mind from the discussion above is to ask the designer to specify, in addition to the $n+1$ control points, the tangent vectors $P'_0, P'_1, \ldots, P'_n$ at each, as indicated in Figure 17.5.

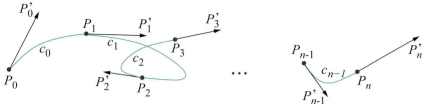

Figure 17.5: Specifying a Hermite spline by specifying the tangent vector at each control point.

Then, using (17.10) to manufacture each of the n successive Hermite cubic arcs c_i, $0 \leq i \leq n$, subject to the respective boundary constraints P_i, P_{i+1}, P_i' and P_{i+1}' yields a C^1-continuous Hermite spline, as the derivatives on either side of each internal control point agree by design.

However, asking the designer for $n+1$ tangent values, in addition to the control points themselves, may be a bit much. It would be nice to have an *automatic* way to deduce these tangent values from other constraints, *transparently* to the user. In fact, there is and we'll discuss next two popular types of Hermite splines arising from particular sets of constraints. These are the natural cubic and cardinal splines.

17.2 Natural Cubic Splines

A *natural cubic spline* is a Hermite spline that is C^2-continuous (i.e., its second derivative is continuous) *and* whose second derivative vanishes at its two end control points. It turns out, as we'll see, that these constraints are enough to uniquely determine the spline.

Assume that the $n+1$ control points through which a natural cubic spline passes are $P_0, P_1, \ldots, P_n$. Because of C^1-continuity (remember that C^2-continuity implies C^1-continuity) one assumes that the tangents at the control points are uniquely defined as well – say they are $P_0', P_1', \ldots, P_n'$, respectively. The values P_i', $0 \leq i \leq n$, are not user-specified; rather, they'll be computed.

Rewrite (17.9) as the equation of the cubic arc c_i from P_i to P_{i+1}:

$$c_i(u) = (2u^3 - 3u^2 + 1) P_i + (-2u^3 + 3u^2) P_{i+1} + (u^3 - 2u^2 + u) P_i' + (u^3 - u^2) P_{i+1}' \quad (0 \leq u \leq 1) \, (17.11)$$

Differentiating twice one finds the second derivative

$$c_i''(u) = (12u-6) P_i + (-12u+6) P_{i+1} + (6u-4) P_i' + (6u-2) P_{i+1}' \quad (0 \leq u \leq 1) \tag{17.12}$$

Observe now that the second-order constraints on a natural cubic spline can be written as the following system of $n+1$ equations:

$$c_0''(0) = 0, \quad c_{i-1}''(1) = c_i''(0) \text{ for } 1 \leq i \leq n-1, \quad c_{n-1}''(1) = 0 \tag{17.13}$$

(the middle equations say that the values of the second derivative on either side of each internal control point are equal, assuring C^2-continuity). Expand the constraint equations using (17.12):

$$
\begin{aligned}
-6P_0 + 6P_1 - 4P_0' - 2P_1' &= 0 \\
6P_{i-1} - 6P_i + 2P_{i-1}' + 4P_i' &= -6P_i + 6P_{i+1} - 4P_i' - 2P_{i+1}', 1 \le i \le n-1 \\
6P_{n-1} - 6P_n + 2P_{n-1}' + 4P_n' &= 0
\end{aligned}
$$

Simplifying and rearranging, we have the system

$$
\begin{aligned}
2P_0' + P_1' &= -3P_0 + 3P_1 \\
P_{i-1}' + 4P_i' + P_{i+1}' &= -3P_{i-1} + 3P_{i+1}, \qquad 1 \le i \le n-1 \\
P_{n-1}' + 2P_n' &= -3P_{n-1} + 3P_n
\end{aligned} \tag{17.14}
$$

of $n+1$ equations in $n+1$ unknowns, which can be solved for the P_i' in terms of the P_i. In fact, writing out the system (17.14) in matrix form one obtains

$$
\begin{bmatrix}
2 & 1 & 0 & 0 & 0 & 0 & \dots & 0 & 0 & 0 & 0 \\
1 & 4 & 1 & 0 & 0 & 0 & \dots & 0 & 0 & 0 & 0 \\
0 & 1 & 4 & 1 & 0 & 0 & \dots & 0 & 0 & 0 & 0 \\
0 & 0 & 1 & 4 & 1 & 0 & \dots & 0 & 0 & 0 & 0 \\
\dots & & & \dots & & & \dots & & & \dots & \\
0 & 0 & 0 & 0 & 0 & 0 & \dots & 0 & 1 & 4 & 1 \\
0 & 0 & 0 & 0 & 0 & 0 & \dots & 0 & 0 & 1 & 2
\end{bmatrix}
\begin{bmatrix}
P_0' \\ P_1' \\ P_2' \\ P_3' \\ \dots \\ P_{n-1}' \\ P_n'
\end{bmatrix}
=
\begin{bmatrix}
-3P_0 + 3P_1 \\ -3P_0 + 3P_2 \\ -3P_1 + 3P_3 \\ -3P_2 + 3P_4 \\ \dots \\ -3P_{n-2} + 3P_n \\ -3P_{n-1} + 3P_n
\end{bmatrix}
\tag{17.15}
$$

where the coefficient matrix is *tridiagonal* because it has non-zero entries only along the principal diagonal and its two neighbors. Tridiagonal matrices are particularly efficient to invert and, accordingly, equation systems with a tridiagonal coefficient matrix are efficiently solvable [140].

Finally, using the solved tangent values $P_0', P_1', \dots, P_n'$ and the geometric form (17.10) of the Hermite cubic, one determines the n Hermite cubic arcs between successive pairs from $P_0, P_1, \dots, P_n$. These arcs then join end to end to give the natural cubic spline through these $n+1$ control points.

Exercise 17.5. (Programming) Solve (17.15) by hand for only three control points P_0, P_1 and P_2. Write a program to draw a natural cubic spline through three control points, each of which can be moved on a plane.

Exercise 17.6. Investigate the local control (or lack thereof) of natural cubic splines. In particular, which of the cubic arcs of a natural cubic spline are affected by moving only one control point?

Hint: Playing with a natural cubic spline applet (there are many on the web) should suggest an answer.

17.3 Cardinal Splines

A *cardinal spline* is a C^1 Hermite spline whose tangent vector at each internal control point is determined by the location of its two adjacent control points in the following simple manner. Say the control points through which a cardinal spline passes are $P_0, P_1, \ldots, P_n$. The tangent vector at P_i, $1 \leq i \leq n-1$, is specified to be *parallel* to the vector from P_{i-1} to P_{i+1} by the equation

$$P_i' = \frac{1}{2}(1 - t_i)(P_{i+1} - P_{i-1}) \qquad (17.16)$$

See Figure 17.6.

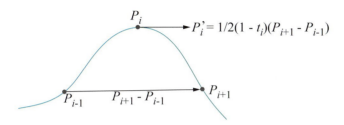

Figure 17.6: The tangent vector at an internal control point of a cardinal spline is parallel to the vector joining the adjacent control points – the tension parameter t_i is user-specified.

The constant of proportionality $\frac{1}{2}(1 - t_i)$ in (17.16) involves a designer-specified parameter t_i, called the *tension parameter*. The tension parameter is usually set between -1 and 1 at each internal control point. If it is set to 0 at *every* internal control point, one gets a popularly used special kind of cardinal spline called a *Catmull-Rom* spline. Specifically, the tangent vector at the internal control point P_i of a Catmull-Rom spline is

$$P_i' = \frac{1}{2}(P_{i+1} - P_{i-1}) \qquad (17.17)$$

Now, from (17.16), $1 \leq i \leq n-1$, one has only $n-1$ equations in the $n+1$ unknowns P_i', $0 \leq i \leq n$. Therefore, two more are required to uniquely solve for these unknowns and determine the cardinal spline through P_i, $0 \leq i \leq n$. Typically, as in the case of a natural cubic spline, these are obtained from requiring the second derivatives to vanish at the two end control points.

Exercise 17.7. Write a matrix equation analogous to (17.15) relating P_i' to P_i for a cardinal spline, assuming the additional constraints that the second derivatives vanish at the terminal control points. Is the coefficient matrix tridiagonal?

Exercise 17.8. What can you say of local control in cardinal splines? In other words, which of the cubic arcs of a cardinal spline are affected by moving a specific control point?

Exercise 17.9. Natural cubic splines are C^2 by definition. How about cardinal splines – are they C^2?

Hint: The answer is no in general and we ask the reader to try and come up with a counter-example. A Catmull-Rom spline through three control points which loses C^2-continuity in the middle is probably easiest.

17.4 Hermite Surface Patches

We'll give a brief introduction to the 2D version of Hermite curves, namely, Hermite surfaces. Analogously to (17.1), one can write the parametric equation of a *Hermite surface patch* (or *bicubic surface patch*) in algebraic form as

$$
\begin{aligned}
s(u, v) \;=\; & \sum_{i=0}^{3} \sum_{j=0}^{3} A_{i,j} u^i v^j \\
=\; & A_{3,3}\, u^3 v^3 + A_{3,2}\, u^3 v^2 + A_{3,1}\, u^3 v + A_{3,0}\, u^3 \\
& + A_{2,3}\, u^2 v^3 + A_{2,2}\, u^2 v^2 + A_{2,1}\, u^2 v + A_{2,0}\, u^2 \\
& + A_{1,3}\, uv^3 + A_{1,2}\, uv^2 + A_{1,1}\, uv + A_{1,0}\, u \\
& + A_{0,3}\, v^3 + A_{0,2}\, v^2 + A_{0,1}\, v + A_{0,0} \qquad (17.18)
\end{aligned}
$$

for $0 \le u, v \le 1$. The expression on the second line consists of 16 monomial summands, where A_{ij}, $0 \le i \le 3$, $0 \le j \le 3$, are points in the ambient space.

Going back to curves for a moment, observe that the geometric form (17.10), viz.,

$$
c(u) = H_0(u)\, P + H_1(u)\, Q + H_2(u)\, P' + H_3(u)\, Q'
$$

of the equation of a Hermite cubic is more useful than the algebraic (17.1), viz.,

$$
c(u) = A_3 u^3 + A_2 u^2 + A_1 u + A_0
$$

because it gives an equation in terms of *perceptible* boundary constraints, in particular, the endpoints P and Q and the tangent vectors P' and Q' there. Moreover, we were able to derive the algebraic form from the geometric because these four boundary constraints were sufficient to uniquely recover the four coefficients A_i, $0 \le i \le 3$, of the algebraic form.

So what would be a suitable set of boundary constraints for a geometric form of the equation of a Hermite patch? Clearly, one would want sixteen constraints leading to a unique determination of the 16 coefficients A_{ij}, $0 \le i \le 3$, $0 \le j \le 3$, on the RHS of (17.18).

Twelve choices are fairly clear. See Figure 17.7. They are the four corners $s(0,0), s(1,0), s(1,1), s(0,1)$ of the patch s (or, more precisely, the position vectors of these corners) and values of the partial derivatives with respect to u and v at each corner. The two partial derivatives at each corner are

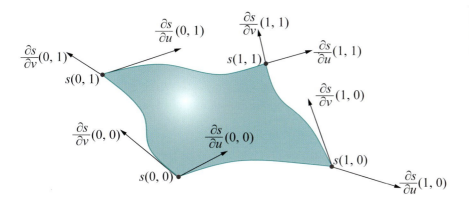

Figure 17.7: Twelve boundary constraints on a bicubic patch.

nothing but the tangent vectors to the two boundary curves meeting there. The four remaining boundary constraints are usually taken to be values of the second-order mixed partial derivatives at the corners, namely,

$$\frac{\partial^2 s}{\partial u \partial v}(0,0), \qquad \frac{\partial^2 s}{\partial u \partial v}(1,0), \qquad \frac{\partial^2 s}{\partial u \partial v}(0,1), \qquad \frac{\partial^2 s}{\partial u \partial v}(1,1)$$

These four are called *twist vectors* and have geometric significance as well, though not as straightforwardly as the first twelve.

We'll conclude our discussion by saying that it turns out that, indeed, the four corner position vectors, the eight tangent vectors at the corners and the four twist vectors together provide sixteen boundary constraints which are sufficient to uniquely specify a Hermite patch. We'll not go further into the derivation ourselves, but refer the interested reader to the chapter on Hermite surfaces in the book by Mortenson [90].

Lagrange Interpolation

At the conclusion of this chapter, we'll briefly describe a method of polynomial (in fact, *entirely* polynomial, not piecewise like Hermite) interpolation, called *Lagrange interpolation*, actually of more theoretical interest than practical value in design.

The *Lagrange polynomial* $f_{i,n}$, where n is a positive integer and i is an integer between 0 and n, is defined by the equation

$$f_{i,n}(u) = \prod_{0 \leq j \leq n, \, j \neq i} \frac{u - j}{i - j}$$

For example,

$$\begin{aligned} f_{2,4}(u) &= \frac{(u-0)(u-1)(u-3)(u-4)}{(2-0)(2-1)(2-3)(2-4)} \\ &= \frac{1}{4} u(u-1)(u-3)(u-4) \end{aligned}$$

Lagrange polynomials have the easily verified property that

$$f_{i,n}(u) = \begin{cases} 1, & u = i \\ 0, & u \in \{0, 1, \ldots, n\}, \ u \neq i \end{cases}$$

In other words, on the particular set of integers $\{0, 1, \ldots, n\}$, the Lagrange polynomial $f_{i,n}$ is 1 at exactly one point, namely i, and 0, elsewhere.

Exercise 17.10. Write the formula for $f_{0,4}(u)$ and check it for the above-mentioned property.

If, now, one uses the Lagrange polynomials as blending functions for $n + 1$ control points P_i, $0 \leq i \leq n$, obtaining the curve

$$c(u) = f_{0,n}(u)P_0 + f_{1,n}(u)P_1 + \ldots + f_{n,n}(u)P_n \qquad (0 \leq u \leq n)$$

then c, called a *Lagrange curve*, is a polynomial curve of degree n. It's seen easily from its definition that c interpolates all its control points; in particular, c is equal to P_i at the point i of the parameter domain $[0, n]$, for $0 \leq i \leq n$.

Exercise 17.11. Write the formula for the Lagrange curve interpolating the four control points

$$[0 \ -1 \ 3]^T \qquad [1 \ 2 \ -3]^T \qquad [5 \ -1 \ 4]^T \qquad [2 \ 0 \ 8]^T$$

Remark 17.5. Lagrange interpolation is rarely used in practice because it suffers from the Bézier-like problem that the degree of the interpolating curve grows with its number of control points. It lacks local control as well.

17.5 Summary, Notes and More Reading

After a couple of chapters on Bézier and B-spline approximation of control points, we learned in this chapter practical methods to interpolate. These will come in handy in design applications that do require interpolation and most 3D modelers, in fact, offer at least a flavor or two of Hermite interpolation, such as natural cubic and Catmull-Rom splines. It's true, though, in the majority of real-life applications that the only known constraints on a curve or surface are at its boundary, e.g., by the way a surface patch meets its neighbors, so the designer typically prefers using internal control points as attractors *a la* Bézier or B-spline, rather than having them tightly latched to an interpolating curve or surface.

For more about Hermite interpolation the reader should consult Farin [42] and Mortenson [90].

Part IX

The Projective Advantage

CHAPTER 18

Applications of Projective Spaces

P rojective spaces and transformations play an important role in computer graphics and our goal in this chapter is to study two crucial applications. The first is in the "shoot" part of shoot-and-print in the OpenGL pipeline and the next in developing rational versions of Bézier and B-spline theory.

It's best to come to this chapter with some familiarity with projective spaces. If you have this already, maybe from a college math course or from books on projective geometry such as Henle [67], Jennings [72] and Pedoe [104], you are set; if not, Appendix A, which is an introduction to projective spaces and transformations, has all you need. Appendix A has been written particularly for a CG audience, with connections constantly drawn to familiarly CG settings. In fact, you are strongly urged to flip through this appendix even if already acquainted with projective geometry.

However, we do realize there might be a significant readership as yet unfamiliar with projective spaces who, nevertheless, would like a view of their applications without necessarily going through all the math first. This chapter has been arranged to be accessible to such persons as far as possible. Before each part that invokes projective theory, the reader is alerted with a note containing the minimum information needed to make sense of it. Of course, understanding will not be 100%, but, hopefully, good enough for a first light on the applications. Familiarity at least with Section 5.2 on affine transformations, though, particularly the use of homogeneous coordinates, is assumed on everyone's part.

The first application of projective transformations in Section 18.1 is to accomplish the so-called projection transformation step in the synthetic-camera graphics pipeline – mapping the viewing volume to a box. This leads to a derivation of OpenGL's 4×4 projection matrices. A neat use of the OpenGL projection transformation that concludes the section is to draw perspective shadows cast by a local light source.

The second application is in Section 18.2 where we learn the rational versions of both Bézier and B-spline theory. This lengthy section begins with an extensive discussion of rational Bézier curves which, once assimilated, lends itself to fairly straightforward generalization, first to rational Bézier surfaces and then rational B-spline, or NURBS, primitives. Section 18.3 concludes the chapter.

18.1 OpenGL Projection Transformations

Way back in Section 2.2 we described OpenGL's rendering as conceptually a two-step process, shoot-and-print. Shooting consists of projecting – parallely in the case of a viewing box and perspectively in that of a viewing frustum – the scene onto the viewing face. Printing consists of scaling the viewing face to fit the OpenGL window. This account, though simplified, is not far from the actual implementation in the OpenGL graphics pipeline.

Scaling is straightforward, but projection is more difficult. The projection process itself is performed by OpenGL in two stages.

In the first stage, OpenGL transforms the viewing volume – a box defined by `glOrtho()` or a frustum by `glFrustum()` and `gluPerspective()` – into a *canonical viewing box*. The canonical viewing box is an axis-aligned cubical box centered at the origin with side lengths two. Figure 18.1 shows the canonical viewing box, as well as a generic viewing box and a generic viewing frustum. The transformation from the given viewing volume to the canonical viewing box is called the *projection transformation* of OpenGL.

We'll deduce equations for the projection transformation soon, but the crux of what it does geometrically is to take lines of sight to lines of sight, "straightening" them out in the process in the case of a frustum. See Figure 18.2 for a sectional view along the xz-plane. For example, the lines of sight l_1 and l_2, both in the box and frustum, are mapped by the projection transformation to the corresponding lines of sight l_1' and l_2' in the canonical viewing box. (The orientation of the lines of sight seems reversed by the transformation – we'll see why momentarily.) Moreover, the points p, q and r on both lines of sight l_1 are mapped to p', q' and r', respectively, on the line of sight l_1'. Rectangle X in the box and rectangle Y in the frustum are transformed to rectangle X' and the trapezoid Y', respectively, bold edge going to bold edge. The distortion from Y to Y' is precisely the foreshortening one would expect from a perspective view.

In the second stage of the two-stage projection process, OpenGL projects primitives in the canonical viewing box parallely onto its back face, the one lying on $z = -1$. It's because of this reversal of the direction of projection – a quirk of OpenGL as a projection to the front face would have worked just as well – that the orientation of the lines of sight is reversed.

Observe that projection in the canonical viewing box is exactly equivalent

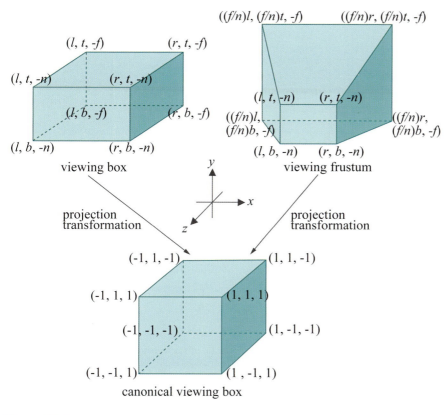

Figure 18.1: As part of the OpenGL rendering pipeline a `glOrtho(l, r, b, t, n, f)`-defined viewing box or `glFrustum(l, r, b, t, n, f)`-defined viewing frustum is transformed into the canonical viewing box by a projection transformation.

to that in the original viewing volume, precisely because the projection transformation preserves lines of sight. In Figure 18.2, for example, the point p, in both box and frustum, projects to the point r on the respective viewing face, while p' (the image of p by the projection transformation) projects to r' (the image of r by the projection transformation) in the canonical viewing box.

Of the two stages of the projection process, the second one is certainly computationally simpler, as it's a matter simply of tossing the z-values *after* they've been used in depth testing if need be. If depth testing is enabled, then z-values in the canonical box are used for this purpose, rather than the original ones from world space. It's valid to do so because if, say, the point q obscures the point p in the viewing volume prior to transformation, as in Figure 18.2, it does so after as well, the latter situation being detected by a simple comparison of the z-values of the transformed points in the box.

OpenGL accomplishes the projection transformation, from programmer-specified viewing volume to canonical viewing box, by means of a 4×4

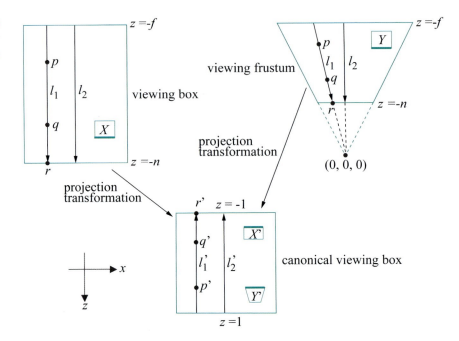

Figure 18.2: Sectional view along the xz-plane of the viewing volumes from Figure 18.1. Black arrows inside the viewing volumes are lines of sight: they are directed toward the $+z$ direction in the viewing box and frustum and toward the $-z$ direction in the canonical viewing box.

projection matrix whose nature depends on whether it is a box or frustum to be transformed into the canonical box. Our next objective is to derive the projection matrix in both cases.

Remark 18.1. To be fastidious we should now rephrase our earlier description of the print part of shoot-and-print to say that it scales the back face of the canonical box, rather than the front face of the viewing volume, to fit the OpenGL window.

18.1.1 Viewing Box to Canonical Viewing Box

The strategy to transform a `glOrtho()`-defined viewing box into the canonical box is straightforward: translate the viewing box so that its center coincides with that of the canonical one, then scale its sides so that they match those of the canonical box.

The center of the viewing box defined by a call to `glOrtho(l, r, b, t, n, f)` is at $[(r+l)/2 \quad (t+b)/2 \quad -(f+n)/2]^T$, while the center of the canonical box is at the origin $[0\ 0\ 0]^T$. Therefore, the displacement vector translating the first to the second is $[-(r+l)/2 \quad -(t+b)/2 \quad (f+n)/2]^T$.

The corresponding 4×4 translation matrix is

$$T(\,-(r+l)/2,\; -(t+b)/2,\; (f+n)/2\,)$$

(see Section 5.4 for a listing of affine transformation matrices in homogeneous form).

Since the viewing box is of size $(r-l) \times (t-b) \times (f-n)$, while the canonical box is of size $2 \times 2 \times 2$, the scaling transformation matching the sides of the former with those of the latter has the matrix

$$S(\,2/(r-l),\; 2/(t-b),\; 2/(f-n)\,)$$

Finally, to account for the reversal in direction of the lines of sight, the needed transformation is $(x, y, z) \mapsto (x, y, -z)$, whose matrix is

$$S(1, 1, -1)$$

Composing the preceding three transformations, one obtains the projection transformation, denoted $P(\texttt{glOrtho}(l,\ r,\ b,\ t,\ n,\ f))$, mapping the viewing box of the call $\texttt{glOrtho}(l,\ r,\ b,\ t,\ n,\ f)$ to the canonical viewing box. The projection matrix corresponding to $P(\texttt{glOrtho}(l,\ r,\ b,\ t,\ n,\ f))$, using eponymous notation, is

$$
\begin{aligned}
&P(\texttt{glOrtho}(l, r, b, t, n, f)) \\
=\; & S(1, 1, -1)\; S(\,2/(r-l),\; 2/(t-b),\; 2/(f-n)\,) \\
& T(\,-(l+r)/2,\; -(b+t)/2,\; (n+f)/2\,) \\
=\;&
\begin{bmatrix}
1 & 0 & 0 & 0 \\
0 & 1 & 0 & 0 \\
0 & 0 & -1 & 0 \\
0 & 0 & 0 & 1
\end{bmatrix}
\begin{bmatrix}
\frac{2}{r-l} & 0 & 0 & 0 \\
0 & \frac{2}{t-b} & 0 & 0 \\
0 & 0 & \frac{2}{f-n} & 0 \\
0 & 0 & 0 & 1
\end{bmatrix}
\begin{bmatrix}
1 & 0 & 0 & -\frac{r+l}{2} \\
0 & 1 & 0 & -\frac{t+b}{2} \\
0 & 0 & 1 & \frac{f+n}{2} \\
0 & 0 & 0 & 1
\end{bmatrix} \\
=\;&
\begin{bmatrix}
\frac{2}{r-l} & 0 & 0 & -\frac{r+l}{r-l} \\
0 & \frac{2}{t-b} & 0 & -\frac{t+b}{t-b} \\
0 & 0 & -\frac{2}{f-n} & -\frac{f+n}{f-n} \\
0 & 0 & 0 & 1
\end{bmatrix}
\qquad (18.1)
\end{aligned}
$$

As it is a composition of a translation and scalings, $P(\texttt{glOrtho}(l,\ r,\ b,\ t,\ n,\ f))$ is an affine transformation of $\mathbb{R}^3$.

Example 18.1. Determine how the point $[20\ 80\ 0]^T$ is transformed by the projection transformation corresponding to $\texttt{glOrtho}(0,\ 100,\ 0,\ 100,\ -1,\ 1)$.

Answer:

$$P(\text{glOrtho}(0, 100, 0, 100, -1, 1)) = \begin{bmatrix} \frac{2}{100} & 0 & 0 & -\frac{100}{100} \\ 0 & \frac{2}{100} & 0 & -\frac{100}{100} \\ 0 & 0 & -\frac{2}{2} & \frac{0}{2} \\ 0 & 0 & 0 & 1 \end{bmatrix}$$

$$= \begin{bmatrix} 0.02 & 0 & 0 & -1 \\ 0 & 0.02 & 0 & -1 \\ 0 & 0 & -1 & 0 \\ 0 & 0 & 0 & 1 \end{bmatrix}$$

Writing $[20\ 80\ 0]^T$ in homogeneous coordinates as $[20\ 80\ 0\ 1]^T$, one sees that it's transformed to the point

$$\begin{bmatrix} 0.02 & 0 & 0 & -1 \\ 0 & 0.02 & 0 & -1 \\ 0 & 0 & -1 & 0 \\ 0 & 0 & 0 & 1 \end{bmatrix} \begin{bmatrix} 20 \\ 80 \\ 0 \\ 1 \end{bmatrix} = \begin{bmatrix} -0.6 \\ 0.6 \\ 0 \\ 1 \end{bmatrix}$$

which is $[-0.6\ 0.6\ 0]^T$ in Cartesian coordinates.

Exercise 18.1. Determine how the following points are transformed by the projection transformation corresponding to glOrtho(-20, 20, 0, 50, -1, 1): (a) $[0\ 40\ 0]^T$ (b) $[50\ 20\ 0.5]^T$.

Note: Just as points inside the viewing box are transformed to points inside the canonical box, those outside, e.g., (b) of the exercise, are transformed to points outside the canonical box. The latter are clipped subsequently in the pipeline prior to rendering.

18.1.2 Viewing Frustum to Canonical Viewing Box

No affine transformation can map the viewing frustum defined by the call glFrustum(l, r, b, t, n, f) to the canonical viewing box, for the simple reason that this requires mapping intersecting lines (along edges of the frustum) to parallel ones (along edges of the box), while we know (see Proposition 5.1) that an affine transformation of $\mathbb{R}^3$ takes parallel straight lines to parallel straight lines and intersecting ones again to intersecting ones. We've run into a brick wall as far as affine transformations go. It's time to appeal to the projective.

Note to Readers Unfamiliar with Projective Geometry: Here's what you need to know for the rest of this particular section. Projective 3-space $\mathbb{P}^3$ consists of 4-tuples of the form $[x\ y\ z\ w]^T$, where these so-called homogeneous coordinates cannot all be zero. Two tuples represent the same point if one's a scalar multiple of the other, e.g., $[2\ 4\ 1\ -3]^T$ and $[4\ 8\ 2\ -6]^T$. Real 3-space $\mathbb{R}^3$ is embedded in $\mathbb{P}^3$ by mapping the point $[x\ y\ z]^T$ of $\mathbb{R}^3$ to $[x\ y\ z\ 1]^T$ of $\mathbb{P}^3$, e.g., $[2\ 4\ 1]^T$ maps to $[2\ 4\ 1\ 1]^T$.

Yet another thing to keep in mind is that, in addition to the points of $\mathbb{R}^3$ embedded into it as above, $\mathbb{P}^3$ has points corresponding to "directions" in $\mathbb{R}^3$. These points, which are called points at infinity, have a w-value of 0; e.g., $[0\ 0\ 1\ 0]^T$ is the point at infinity corresponding to the direction along the z-axis (both up and down directions along any line are regarded equal). Points with non-zero w-values, embedded from $\mathbb{R}^3$, are called regular points.

Finally, a projective transformation of $\mathbb{P}^3$ is defined by a non-singular 4×4 matrix and acts on tuples of $\mathbb{P}^3$ by multiplication from the left (similarly to how linear transformations of $\mathbb{R}^3$ act on 3-tuples).

You should jump now to the paragraph below containing Equation (18.2).

However, our experience with projective transformations – Example A.17 which illustrates a projective transformation of $\mathbb{P}^3$ mapping a trapezoid to a rectangle is particularly motivating – suggests applying one.

Projectively transforming $\mathbb{R}^3$ is analogous to projectively transforming $\mathbb{R}^2$. For the latter, we identified $\mathbb{R}^2$ with a "film" in $\mathbb{R}^3$, almost always the plane $z = 1$, to capture the transformation of 2D objects lifted to $\mathbb{P}^2$ (note that the allusion to films is developed in Appendix A). Likewise, to projectively transform $\mathbb{R}^3$, we'll identify it with the hyperplane $w = 1$ in four-dimensional $xyzw$-space $\mathbb{R}^4$ in order to capture the transformation of 3D objects lifted to $\mathbb{P}^3$.

In particular, for the current application we seek a projective transformation h^M of $\mathbb{P}^3$ which is captured on $\mathbb{R}^3$ as taking the viewing frustum specified by the call $\texttt{glFrustum}(l,\ r,\ b,\ t,\ n,\ f)$ to the canonical box – as depicted on the right of Figure 18.1. Suppose its defining matrix is

$$M = \begin{bmatrix} a_{11} & a_{12} & a_{13} & a_{14} \\ a_{21} & a_{22} & a_{23} & a_{24} \\ a_{31} & a_{32} & a_{33} & a_{34} \\ a_{41} & a_{42} & a_{43} & a_{44} \end{bmatrix} \qquad (18.2)$$

so that it maps the point $[x\ y\ z\ w]^T$ of $\mathbb{P}^3$ to $M[x\ y\ z\ w]^T$.

The four lines along the four sides of the frustum that meet at its apex, which is the regular point $[0\ 0\ 0\ 1]^T$, are mapped, respectively, to four lines along edges of the canonical box all parallel to the z-axis, meeting, therefore, at the point at infinity $[0\ 0\ 1\ 0]^T$. Accordingly, we ask that

$$h^M([0\ 0\ 0\ 1]^T) = [0\ 0\ 1\ 0]^T$$

giving the matrix equation

$$\begin{bmatrix} a_{11} & a_{12} & a_{13} & a_{14} \\ a_{21} & a_{22} & a_{23} & a_{24} \\ a_{31} & a_{32} & a_{33} & a_{34} \\ a_{41} & a_{42} & a_{43} & a_{44} \end{bmatrix} \begin{bmatrix} 0 \\ 0 \\ 0 \\ 1 \end{bmatrix} = \begin{bmatrix} 0 \\ 0 \\ d \\ 0 \end{bmatrix}$$

where d can be any non-zero scalar because homogeneous coordinates

$[0\ 0\ 1\ 0]^T$ and $[0\ 0\ d\ 0]^T$ represent the same point, implying that

$$a_{14} = 0, \quad a_{24} = 0, \quad a_{34} = d, \quad a_{44} = 0$$

It turns out that choosing $d = -\frac{2fn}{f-n}$ simplifies manipulations down the road so we'll write

$$M = \begin{bmatrix} a_{11} & a_{12} & a_{13} & 0 \\ a_{21} & a_{22} & a_{23} & 0 \\ a_{31} & a_{32} & a_{33} & -\frac{2fn}{f-n} \\ a_{41} & a_{42} & a_{43} & 0 \end{bmatrix}$$

The mappings

$$\begin{aligned}
h^M([l\ b\ -n\ 1]^T) &= [-1\ -1\ -1\ 1]^T \\
h^M([r\ b\ -n\ 1]^T) &= [1\ -1\ -1\ 1]^T \\
h^M([l\ t\ -n\ 1]^T) &= [-1\ 1\ -1\ 1]^T \\
h^M([r\ t\ -n\ 1]^T) &= [1\ 1\ -1\ 1]^T
\end{aligned}$$

from the mapping of the four vertices at the front of the frustum to the corresponding ones at the back of the canonical box give the four matrix equations

$$\begin{bmatrix} a_{11} & a_{12} & a_{13} & 0 \\ a_{21} & a_{22} & a_{23} & 0 \\ a_{31} & a_{32} & a_{33} & -\frac{2fn}{f-n} \\ a_{41} & a_{42} & a_{43} & 0 \end{bmatrix} \begin{bmatrix} l \\ b \\ -n \\ 1 \end{bmatrix} = \begin{bmatrix} -c_1 \\ -c_1 \\ -c_1 \\ c_1 \end{bmatrix}$$

$$\begin{bmatrix} a_{11} & a_{12} & a_{13} & 0 \\ a_{21} & a_{22} & a_{23} & 0 \\ a_{31} & a_{32} & a_{33} & -\frac{2fn}{f-n} \\ a_{41} & a_{42} & a_{43} & 0 \end{bmatrix} \begin{bmatrix} r \\ b \\ -n \\ 1 \end{bmatrix} = \begin{bmatrix} c_2 \\ -c_2 \\ -c_2 \\ c_2 \end{bmatrix}$$

$$\begin{bmatrix} a_{11} & a_{12} & a_{13} & 0 \\ a_{21} & a_{22} & a_{23} & 0 \\ a_{31} & a_{32} & a_{33} & -\frac{2fn}{f-n} \\ a_{41} & a_{42} & a_{43} & 0 \end{bmatrix} \begin{bmatrix} l \\ t \\ -n \\ 1 \end{bmatrix} = \begin{bmatrix} -c_3 \\ c_3 \\ -c_3 \\ c_3 \end{bmatrix}$$

$$\begin{bmatrix} a_{11} & a_{12} & a_{13} & 0 \\ a_{21} & a_{22} & a_{23} & 0 \\ a_{31} & a_{32} & a_{33} & -\frac{2fn}{f-n} \\ a_{41} & a_{42} & a_{43} & 0 \end{bmatrix} \begin{bmatrix} r \\ t \\ -n \\ 1 \end{bmatrix} = \begin{bmatrix} c_4 \\ c_4 \\ -c_4 \\ c_4 \end{bmatrix}$$

(c_i, $1 \leq i \leq 4$, are non-zero scalars) leading to 16 equations simultaneously in 16 unknowns:

$$l\,a_{11} + b\,a_{12} - n\,a_{13} = -c_1$$
$$l\,a_{21} + b\,a_{22} - n\,a_{23} = -c_1$$
$$l\,a_{31} + b\,a_{32} - n\,a_{33} - \frac{2fn}{f-n} = -c_1$$
$$l\,a_{41} + b\,a_{42} - n\,a_{23} = c_1$$
$$\cdots = \cdots$$
$$r\,a_{41} + t\,a_{42} - n\,a_{43} = c_4$$

These can be solved – not difficult, but tedious – to find

$$a_{11} = \frac{2n}{r-l}, \quad a_{12} = 0, \quad a_{13} = \frac{r+l}{r-l}, \quad a_{21} = 0, \quad a_{22} = \frac{2n}{t-b},$$
$$a_{23} = \frac{t+b}{t-b}, \quad a_{31} = 0, \quad a_{32} = 0, \quad a_{33} = -\frac{f+n}{f-n}, \quad a_{41} = 0,$$
$$a_{42} = 0, \quad a_{43} = -1, \quad c_1 = n, \quad c_2 = n, \quad c_3 = n, \quad c_4 = n.$$

It follows that the projection transformation mapping the viewing frustum of the call `glFrustum(`l, r, b, t, n, f`)` to the canonical viewing box is given by the matrix

$$P(\texttt{glFrustum}(l,r,b,t,n,f)) = \begin{bmatrix} \frac{2n}{r-l} & 0 & \frac{r+l}{r-l} & 0 \\ 0 & \frac{2n}{t-b} & \frac{t+b}{t-b} & 0 \\ 0 & 0 & -\frac{f+n}{f-n} & -\frac{2fn}{f-n} \\ 0 & 0 & -1 & 0 \end{bmatrix} \quad (18.3)$$

That $a_{43} \neq 0$ confirms that $P(\texttt{glFrustum}(l$, r, b, t, n, $f))$ is not affine, as a 4×4 projective transformation matrix is affine if and only if its last row is all 0 except for the last element.

Exercise 18.2. Characterize those regular points that $P(\texttt{glFrustum}(l$, r, b, t, n, $f))$ maps to points at infinity.

At this time we ask the reader to open the red book to Appendix F, where OpenGL's 4×4 projection matrices are given and compare their values to those in Equations (18.1) and (18.3) above. Seeing these together with its 4×4 matrices for translation, rotation and scaling, listed in Appendix F as well, and derived by us in Section 5.4, the reader may tend to agree that OpenGL "lives" in projective 3-space.

Example 18.2. Determine how the point $[0\ 0\ -10]^T$ is transformed by the projection transformation corresponding to `glFrustum(-5, 5, -5, 5, 5, 25)`.

Answer:

$$P(\texttt{glFrustum}(-5,5,-5,5,5,25)) = \begin{bmatrix} \frac{10}{10} & 0 & \frac{0}{10} & 0 \\ 0 & \frac{10}{10} & \frac{0}{10} & 0 \\ 0 & 0 & -\frac{30}{20} & -\frac{250}{20} \\ 0 & 0 & -1 & 0 \end{bmatrix}$$

$$= \begin{bmatrix} 1 & 0 & 0 & 0 \\ 0 & 1 & 0 & 0 \\ 0 & 0 & -1.5 & -12.5 \\ 0 & 0 & -1 & 0 \end{bmatrix}$$

Writing $[0\ 0\ -10]^T$ in homogeneous coordinates as $[0\ 0\ -10\ 1]^T$, one sees that it's transformed to the point

$$\begin{bmatrix} 1 & 0 & 0 & 0 \\ 0 & 1 & 0 & 0 \\ 0 & 0 & -1.5 & -12.5 \\ 0 & 0 & -1 & 0 \end{bmatrix} \begin{bmatrix} 0 \\ 0 \\ -10 \\ 1 \end{bmatrix} = \begin{bmatrix} 0 \\ 0 \\ 2.5 \\ 10 \end{bmatrix}$$

which is $[0\ 0\ 0.25]^T$ in Cartesian coordinates.

The last step, where the homogeneous coordinates are divided by the w-value – in this case $[0\ 0\ 2.5\ 10]^T$ by 10 – to project the transformed point back into xyz-space ($w = 1$) is often called *perspective division*, especially when performed as a part of the graphics pipeline.

Exercise 18.3. Determine how the following points are transformed by the projection transformation corresponding to `glFrustum(-10, 10, -10, 10, 1, 10)`: (a) $[1\ 1\ -2]^T$ (b) $[10\ 20\ -1]^T$ (c) $[5\ 5\ 0]^T$.

If any of them is mapped to a point at infinity – whose w-value is 0 – simply identify it as such. Obviously, you'll not be able to complete the projection transformation for such points, as they will not pass perspective division. We'll discuss in Section 19.1 of the next chapter how they are, in fact, handled in the pipeline.

Exercise 18.4. Write an equation similar to (18.3) for the projection matrix corresponding to the GLU call `gluPerspective(`*fovy, aspect, n, f*`)`.

Remark 18.2. Do keep in mind the terminological distinction that a *projective transformation* is one of projective space, while a *projection transformation* is a particular transformation in the graphics pipeline, which is implemented *by means of* a projective transformation, if the viewing volume is a frustum.

18.1.3 Projection Matrix Stack

We know now how the projection matrix corresponding to a programmer-specified projection command – `glOrtho()`, `glFrustum()` or `gluPerspective()` – is computed by OpenGL. How, then, is this matrix stored and applied in the graphics pipeline? The answer is in a manner exactly similar to modelview matrices.

As the current modelview matrix is at the top of the modelview matrix stack, so the *current projection matrix* is the topmost of the *projection matrix stack*. Again, as for modelview statements, a projection statement is applied by multiplying the current projection matrix on the right by the matrix corresponding to that statement. Moreover, the projection matrix stack can be pushed and popped, and the current projection matrix accessed and manipulated, just as the modelview matrix stack. Refer to Section 5.4.6 for commands to access the current modelview matrix.

Seeing is believing.

Experiment 18.1. Run `manipulateProjectionMatrix.cpp`, a simple modification of `manipulateModelviewMatrix.cpp` of Chapter 5. Figure 18.3 is a screenshot, though the output to the OpenGL window is of little interest. Of interest, though, are the new statements in the `resize()` routine that output the current projection matrix just before and after the call `glFrustum(-5.0, 5.0, -5.0, 5.0, 5.0, 100.0)`.

Compare the second matrix output to the command window with $P($`glFrustum(-5.0, 5.0, -5.0, 5.0, 5.0, 100.0)`$)$ computed with the help of Equation (18.3).   End

Figure 18.3: Screenshot of `manipulateProjection-Matrix.cpp`.

Exercise 18.5. (Programming) Continue with the preceding experiment by replacing the projection statement

    glFrustum(-5.0, 5.0, -5.0, 5.0, 5.0, 100.0)

with

    glOrtho(-10.0, 10.0, -10.0, 10.0, 0.0, 20.0)

Compare the second matrix output to the command window with

    P(glOrtho(-10.0, 10.0, -10.0, 10.0, 0.0, 20.0))

as computed using Equation (18.3).

Remark 18.3. It's unlikely that you'll ever need to access the projection matrix stack in a typical program, other than in the mandatory definition of the viewing volume, which would be exactly one of `glOrtho()` or `glFrustum()` or `gluPerspective()`.

18.1.4 Perspective Shadows

We'll make an interesting application of the OpenGL projection transformation to draw *perspective shadows*, cast by a *local* light source from which rays are not parallel.

Experiment 18.2. Run `ballAndTorusPerspectivelyShadowed.cpp`, a program which adds to `ballAndTorusShadowed.cpp` (Experiment 4.34) a back wall, lying along the $z = -35$ plane, and shadows of the ball and torus cast on it by a light source at the origin. Press space to start the ball traveling around the torus and the up and down arrow keys to change its speed. Figure 18.4 is a screenshot. **End**

It's not hard to understand how the shadows on the back wall are drawn by the program, if one imagines them to be the B/W picture of the ball and torus taken by a point camera at the origin with its film along the plane of the wall. The strategy, accordingly, is to mimic OpenGL's procedure to compute such a picture, in particular, to transform the frustum into the canonical box and then parallely project to its back face.

The last block of statements in the `setup()` routine, namely,

```
glMatrixMode(GL_PROJECTION);
glPushMatrix();
glLoadIdentity();
glFrustum(-1.0, 1.0, -1.0, 1.0, 35.0, 100.0);
glGetFloatv(GL_PROJECTION_MATRIX, matrixData);
glPopMatrix();
glMatrixMode(GL_MODELVIEW); // Exit the projection matrix stack.
```

generates the particular projection matrix M corresponding to the frustum F specified by

```
glFrustum(-1.0, 1.0, -1.0, 1.0, 35.0, 100.0)
```

and stores it in the `matrixData` array. The viewing plane of this frustum is along the back wall $z = -35$, while its viewing face, a 2×2 square, matches in size and shape the back face of the canonical box. This means that projection onto the viewing face of the frustum F is *equivalent* to transformation by M into the canonical box followed by projection to the latter's back face.

Note: If the viewing face of F differed in dimensions from the back face of the canonical box, shadows would be distorted by scaling.

The block of statements

```
glPushMatrix();
glTranslatef(0.0, 0.0, -35.0); // Translate the shadow to the
                               // back wall.
glScalef(1.0, 1.0, 0.0); // Collapse z-values to make a 2D shadow.
glMultMatrixf(matrixData); // Projective transformation to the
                           // canonical box.
drawFlyingBallAndTorus(1); // Draw black ball and torus.
glPopMatrix();
```

in the drawing routine draws the blackened ball and torus, transforms them into the canonical box by multiplying by M, collapses z-values in order

Figure 18.4: Screenshot of ballAndTorus-Perspectively-Shadowed.cpp.

to obtain their projections onto the xy-plane and, finally, translates the projections 35 units in the $-z$-direction, placing them as shadows on the back wall.

Note: The ball and torus actually lie outside the frustum F and transformation by M maps them outside the canonical box as well. Nevertheless, their images are still distorted in the right manner so that projection to the back plane of the box gives authentic perspective shadows.

Exercise 18.6. (Programming) Change the parameters of the projection statement in the `setup()` routine of `ballAndTorusPerspectively-Shadowed.cpp` from

```
glFrustum(-1.0, 1.0, -1.0, 1.0, 35.0, 100.0)
```

to

```
glFrustum(-5.0, 5.0, -5.0, 5.0, 35.0, 100.0)
```

Explain what you see. Change the statement again to

```
glFrustum(-1.0, 1.0, -1.0, 1.0, 10.0, 100.0)
```

and then to

```
glFrustum(-1.0, 1.0, -1.0, 1.0, 50.0, 100.0)
```

Again, explain what you see in either case.

Exercise 18.7. (Programming) Draw the perspective shadow of a ball, thrown in a parabolic arc, cast on the floor by a local overhead light source.

18.2 Rational Bézier and NURBS Curves and Surfaces

Our second application of projective geometry is to set the stage for rational Bézier primitives, as well as to put the 'R' – 'R' stands for rational, of course – into NURBS. In Chapters 15 and 16 we investigated the polynomial versions of Bézier and NURBS theory, respectively.

We'll begin with rational Bézier curves, as conceptually they are the simplest and notationally least cumbersome. Once we have rational Bézier curves under our belts, extending our understanding to rational Bézier surfaces and then to NURBS curves and surfaces will not be difficult.

18.2.1 Rational Bézier Curves Basics

Recall Equation (15.13) of a Bézier curve in $\mathbb{R}^3$ specified by $n+1$ control points $P_i = [x_i \ y_i \ z_i]^T$, $0 \le i \le n$:

$$C(u) = \left[\sum_{i=0}^{n} B_{i,n}(u)x_i \quad \sum_{i=0}^{n} B_{i,n}(u)y_i \quad \sum_{i=0}^{n} B_{i,n}(u)z_i \right]^T \quad (0 \le u \le 1)$$

$$(18.4)$$

Note to Readers Unfamiliar with Projective Geometry: Here's what you need to know for most of this particular section. Projective 2-space $\mathbb{P}^2$ consists of 3-tuples of the form $[x \ y \ z]^T$, where these so-called homogeneous coordinates cannot all be zero. Two tuples represent the same point if one's a scalar multiple of the other, e.g., $[0 \ -1 \ 2]^T$ and $[0 \ -4 \ 8]^T$.

Real 2-space $\mathbb{R}^2$ is embedded in $\mathbb{P}^2$ by mapping the point $[x \ y]^T$ of $\mathbb{R}^2$ to $[x \ y \ 1]^T$ of $\mathbb{P}^2$, e.g., $[2 \ 4]^T$ maps to $[2 \ 4 \ 1]^T$. Conversely, a point $[x \ y \ z]^T$ of $\mathbb{P}^2$, with $z \ne 0$, is an image by this embedding of the point $[x/z \ y/z]^T$ of $\mathbb{R}^2$.

Getting back to the equation at the top of the section, what if none of the P_i has coordinates all zero, so that one can imagine each to be a projective point with homogeneous coordinates $[x_i \ y_i \ z_i]^T$, rather than the real point $[x_i \ y_i \ z_i]^T$? Certainly, then, Equation (18.4) defines a point $C(u)$ in $\mathbb{P}^2$ for every u in $0 \le u \le 1$, as long as all its three components are not simultaneously zero either. In this case, one could call C *the* projective Bézier curve over the projective control points $P_i, 0 \le i \le n$, *provided* that it doesn't depend on the choice of the P_i's homogeneous coordinates, for, otherwise, (18.4) would give different curves $C(u)$ for different choices and not be a proper definition at all.

Let's see if C is, in fact, independent of the choice of homogeneous coordinates for its control points. Accordingly, write $P_i = [w_i x_i \ w_i y_i \ w_i z_i]^T$, where $w_i \ne 0$, for $0 \le i \le n$, and plug into (18.4):

$$D(u) = \left[\sum_{i=0}^{n} B_{i,n}(u)w_i x_i \quad \sum_{i=0}^{n} B_{i,n}(u)w_i y_i \quad \sum_{i=0}^{n} B_{i,n}(u)w_i z_i \right]^T \quad (0 \le u \le 1)$$

Is $D(u) = C(u)$? Not necessarily! Playing a bit with the equation it's clear that there's no way to "pull the w_i's out of the square brackets" and write

$$\left[\sum_{i=0}^{n} B_{i,n}(u)w_i x_i \quad \sum_{i=0}^{n} B_{i,n}(u)w_i y_i \quad \sum_{i=0}^{n} B_{i,n}(u)w_i z_i \right]^T$$

$$= w \left[\sum_{i=0}^{n} B_{i,n}(u)x_i \quad \sum_{i=0}^{n} B_{i,n}(u)y_i \quad \sum_{i=0}^{n} B_{i,n}(u)z_i \right]^T$$

for some one scalar w, unless all the w_i's happen to be equal to w.

Ouch! Major road block? Quit and start over? Nah, we'll just make a feature of the bug! Choosing different homogeneous coordinates for the projective control points gives different projective Bézier curves? Well then, more choice for the designer!

Let's start with control points all on the real plane as, at the end of the day, we'll be modeling in real space, not projective. However, first, identify $\mathbb{R}^2$ with the plane $z = 1$ in $\mathbb{R}^3$, i.e., $[x \ y]^T \in \mathbb{R}^2$ with $[x \ y \ 1]^T \in \mathbb{R}^3$.

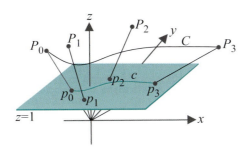

Figure 18.5: Four real control points $p_i = [x_i \ y_i \ 1]^T$, with weights w_i, are lifted to the projective control points $P_i = [w_i x_i \ w_i y_i \ w_i]^T$, $0 \leq i \leq 3$. The (black) polynomial projective Bézier curve C projects to the (blue) rational real Bézier curve c.

Choose $n + 1$ control points $p_i = [x_i \ y_i \ 1]^T$, $0 \leq i \leq n$, in $\mathbb{R}^2$, as well as $n + 1$ non-zero scalars w_i, $0 \leq i \leq n$. Lift each p_i to the projective point $P_i = [w_i x_i \ w_i y_i \ w_i]^T$ in $\mathbb{P}^2$, expressed using *these particular* homogeneous coordinates. See Figure 18.5. The scalar w_i is called the *weight* of the control point p_i. The projective *polynomial* Bézier curve C specified by the control points $P_i = [w_i x_i \ w_i y_i \ w_i]^T$ is

$$C(u) = \left[\sum_{i=0}^{n} B_{i,n}(u) w_i x_i \quad \sum_{i=0}^{n} B_{i,n}(u) w_i y_i \quad \sum_{i=0}^{n} B_{i,n}(u) w_i \right]^T \quad (0 \leq u \leq 1)$$
$$(18.5)$$

To return to $\mathbb{R}^2$, divide C throughout by its z-coordinate to get the plane curve

$$c(u) = \left[\frac{\sum_{i=0}^{n} B_{i,n}(u) w_i x_i}{\sum_{i=0}^{n} B_{i,n}(u) w_i} \quad \frac{\sum_{i=0}^{n} B_{i,n}(u) w_i y_i}{\sum_{i=0}^{n} B_{i,n}(u) w_i} \quad 1 \right]^T \quad (0 \leq u \leq 1)$$
$$(18.6)$$

assuming that $\sum_{i=0}^{n} B_{i,n}(u) w_i \neq 0$ in $0 \leq u \leq 1$, so there's never division by zero. Rewriting (18.6) as a proper equation in $\mathbb{R}^2$ by dropping the z-value 1, we have

$$c(u) = \left[\frac{\sum_{i=0}^{n} B_{i,n}(u) w_i x_i}{\sum_{i=0}^{n} B_{i,n}(u) w_i} \quad \frac{\sum_{i=0}^{n} B_{i,n}(u) w_i y_i}{\sum_{i=0}^{n} B_{i,n}(u) w_i} \right]^T \quad (0 \leq u \leq 1) \quad (18.7)$$

which is said to be the *rational Bézier curve* in $\mathbb{R}^2$ approximating the control points $p_i = [x_i \ y_i]^T$, with respective weights w_i, $0 \leq i \leq n$.

If the weights w_i, $0 \leq i \leq n$, are all positive, then it's guaranteed that the denominator $\sum_{i=0}^{n} B_{i,n}(u)w_i$ in (18.7) is positive, and so non-zero, in $0 \leq u \leq 1$. Consequently, this condition on the weights is, typically, assumed as a design constraint. We'll make a tacit assumption ourselves of positive weights henceforth.

It's a bit hard to make out from (18.7) exactly what's going on. Let's consider a particular case with few control points, say $n = 2$, for three control points. Write out the Bernstein polynomials in (18.7) to obtain the following equation for the quadratic rational Bézier curve on three control points $[x_0 \; y_0]^T$, $[x_1 \; y_1]^T$ and $[x_2 \; y_2]^T$, with weights w_0, w_1 and w_2, respectively:

$$c(u) = \left[\frac{w_0 x_0 (1-u)^2 + 2w_1 x_1 (1-u)u + w_2 x_2 u^2}{w_0 (1-u)^2 + 2w_1 (1-u)u + w_2 u^2} \right. $$
$$\left. \frac{w_0 y_0 (1-u)^2 + 2w_1 y_1 (1-u)u + w_2 y_2 u^2}{w_0 (1-u)^2 + 2w_1 (1-u)u + w_2 u^2} \right]^T \quad (18.8)$$

in $0 \leq u \leq 1$.

Compare with (15.5), which is

$$c(u) = [x_0(1-u)^2 + 2x_1(1-u)u + x_2 u^2 \quad y_0(1-u)^2 + 2y_1(1-u)u + y_2 u^2]^T$$

in $0 \leq u \leq 1$, the equation of the quadratic polynomial Bézier curve on the same three control points. Observe, first, that both the x- and y-values on the RHS of (18.8) are rational functions, i.e., *ratios* of two polynomials, particularly of two quadratics in this case. The values on the RHS of the equation for the polynomial curve, on the other hand, are simply quadratic polynomials. The following three exercises shed further light on the quadratic rational Bézier curve.

Exercise 18.8. Putting $u = 0$ and 1 in Equation (18.8), show that, whatever the assignment of weights, a quadratic rational Bézier curve always interpolates both its first and last control points.

Exercise 18.9. Show that if the weights of its three control points are equal, then a quadratic rational Bézier curve coincides with the quadratic polynomial Bézier curve specified by the same control points.

Evidently, then, at least for three control points, a polynomial Bézier curve is simply a special case of a rational one.

Exercise 18.10. The quadratic polynomial Bézier curve is a weighted sum of its control points. One can rewrite (15.5) above as follows to see this.

$$c(u)$$
$$= [x_0(1-u)^2 + 2x_1(1-u)u + x_2 u^2 \quad y_0(1-u)^2 + 2y_1(1-u)u + y_2 u^2]^T$$
$$= (1-u)^2 [x_0 \; y_0]^T + 2(1-u)u [x_1 \; y_1]^T + u^2 [x_2 \; y_2]^T$$

in $0 \leq u \leq 1$, where the weights in the second line, namely, $(1-u)^2$, $2(1-u)u$ and u^2, are the so-called blending functions of the control points.

As we know, these particular blending functions, called degree 2 Bernstein polynomials, form a partition of unity.

How about the quadratic rational Bézier curve of (18.8)? Write it similarly as a sum

$$c(u) = (\ldots) \, [x_0 \ y_0]^T + (\ldots) \, [x_1 \ y_1]^T + (\ldots) \, [x_2 \ y_2]^T$$

weights being rational blending functions, rather than polynomial. Do these new blending functions still form a partition of unity?

Going from quadratic rational to cubic rational with four weighted control points is straightforward, as we ask the reader to show next.

Exercise 18.11. Write an equation analogous to (18.8) for a cubic rational Bézier curve.

Experiment 18.3. Run `rationalBezierCurve1.cpp`, which draws the cubic rational Bézier curve specified by four control points on the plane at *fixed* locations, but with *changeable* weights.

The control points on the plane (light gray triangular mesh) are all red, except for the currently selected one, which is black. Press space to cycle through the control points. The control point weights are shown at the upper-left, that of the currently selected one being changed by pressing the up/down arrow keys. The rational Bézier curve on the plane is red as well. Figure 18.6 is a screenshot.

Drawn in green are all the lifted control points, except for that of the currently selected control point, which is black. The projective polynomial Bézier curve approximating the lifted control points is green too. The lifted control points are a larger size as well.

Note: The lifted control points and the projective Bézier curve are primitives in $\mathbb{P}^2$, of course, but represented in $\mathbb{R}^3$ using their homogeneous coordinates.

Also drawn is a cone of several gray lines through the projective Bézier curve which intersects the plane in its projection, the rational Bézier curve.

Observe that increasing the weight of a control point pulls the (red rational Bézier) curve toward it, while reducing it has the opposite effect. Moreover, the end control points are always interpolated regardless of assigned weights. It's sometimes hard to discern the very gradual change in the shape of the curve as one varies the weights. A trick is to press delete for the curve to spring back to its original configuration, at which moment the difference should be clear.

It seems, then, that the control point weights are an additional set of "dials" at the designer's disposal for use to edit the curve.

The code of `rationalBezierCurve1.cpp` is instructive as well, as we'll see in the next section on drawing. **End**

Figure 18.6: Screenshot of `rationalBezier-Curve1.cpp`.

We've been studying rational Bézier curves on the plane for really no other reason than that, though we started with the 3D equation (18.4), we soon projected it down to 2D. Deriving the equation for a rational Bézier curve in 3-space is not hard and, in fact, almost a repeat of the 2D process, as we ask the reader to show next.

Exercise 18.12. Identify $\mathbb{R}^3$ with the hyperplane $w = 1$ in $xyzw$-space $\mathbb{R}^4$, just as we identify $\mathbb{R}^2$ with the plane $z = 1$ in xyz-space $\mathbb{R}^3$. Suppose $p_i = [x_i \ y_i \ z_i \ 1]^T$, $0 \leq i \leq n$, are $n + 1$ control points in $\mathbb{R}^3$ with assigned weights w_i, $0 \leq i \leq n$.

Lift each p_i to the projective point $P_i = [w_i x_i \ \ w_i y_i \ \ w_i z_i \ \ w_i]^T$ in $\mathbb{P}^3$, expressed using those particular homogeneous coordinates. Reasoning as earlier in the 2D case, show that the equation of the rational Bézier curve in $\mathbb{R}^3$ approximating the control points $p_i = [x_i \ y_i \ z_i]^T$, with respective weights w_i, $0 \leq i \leq n$, is

$$c(u) = \left[\frac{\sum_{i=0}^{n} B_{i,n}(u) w_i x_i}{\sum_{i=0}^{n} B_{i,n}(u) w_i} \quad \frac{\sum_{i=0}^{n} B_{i,n}(u) w_i y_i}{\sum_{i=0}^{n} B_{i,n}(u) w_i} \quad \frac{\sum_{i=0}^{n} B_{i,n}(u) w_i z_i}{\sum_{i=0}^{n} B_{i,n}(u) w_i} \right]^T$$
(18.9)

for $0 \leq u \leq 1$, which, of course, is the analogue of the 2D equation (18.7) that we have already derived.

18.2.2 Drawing Rational Bézier Curves

OpenGL can draw rational Bézier curves in 3-space. To draw the curve with control points $p_i = [x_i \ y_i \ z_i]^T$ and weights w_i, $0 \leq i \leq n$, the command is

```
glMap1f(GL_MAP1_VERTEX_4,  t1,  t2,  stride,  order,  *controlPoints)
```

where *controlPoints* points to the $(n + 1) \times 4$ array

$$\{\{w_0 x_0 \ w_0 y_0 \ w_0 z_0 \ w_0\}, \{w_1 x_1 \ w_1 y_1 \ w_1 z_1 \ w_1\}, \ldots, \{w_n x_n \ w_n y_n \ w_n z_n \ w_n\}\}$$
(18.10)

and other parameters have the same meaning as for the command

```
glMap1f(GL_MAP1_VERTEX_3,  t1,  t2,  stride,  order,  *controlPoints)
```

with which we are familiar from drawing polynomial Bézier curves in Section 10.3.1.

Returning to `rationalBezierCurve1.cpp`, let's see if OpenGL's commands to draw a rational Bézier curve in 3-space have been correctly invoked. Say the four planar control points of the program are $[x_i \ y_i]^T$, $0 \leq i \leq 3$, represented in homogeneous coordinates by $[x_i \ y_i \ 1]^T$, the values of the latter being stored in the array `controlPoints`. Their respective weights w_i are stored in the array `weights`.

Note: We are using variable names for convenience, of course. The actual values in the program, as you can see, are $[7.0 \ 2.0]^T$ for $[x_1 \ y_1]^T$, 1.5 for w_1 and so on.

The array `ControlPointsLifted` is filled by the routine `compute-ControlPointsLifted()` with the lifted coordinate values $[w_i x_i \ w_i y_i \ w_i]^T$, $0 \le i \le 3$. The green Bézier curve is the 3D polynomial Bézier approximation of the lifted points drawn using `glMap1f(GL_MAP1_VERTEX_3, ...)`.

The array `controlPointsHomogeneous` is filled by `computeControl-PointsHomogeneous()` with the values $[w_i x_i \ w_i y_i \ w_i \ w_i]^T$. From our understanding of the syntax of `glMap1f(GL_MAP1_VERTEX_4, ...)` – compare, in particular, array `controlPointsHomogeneous` with array (18.10) above – the red rational Bézier curve approximates the control points $[x_i \ y_i \ 1]^T$ in $\mathbb{R}^3$ with weights w_i, $0 \le i \le 3$. By (18.9) the equation of the latter is seen to be

$$c(u) = \left[\frac{\sum_{i=0}^n B_{i,n}(u) w_i x_i}{\sum_{i=0}^n B_{i,n}(u) w_i} \quad \frac{\sum_{i=0}^n B_{i,n}(u) w_i y_i}{\sum_{i=0}^n B_{i,n}(u) w_i} \quad 1 \right]^T \quad (0 \le u \le 1)$$

which is precisely the 2D rational Bézier approximation of the control points $[x_i \ y_i]^T$, $0 \le i \le 3$, drawn on the plane $z = 1$.

We have verified, therefore, that the green and red curves of `rational-BezierCurve1.cpp` are indeed the particular Bézier approximations claimed in Experiment 18.3.

So what do rational Bézier curves have that the polynomial curves do not? Let's see...

18.2.3 Rational Bézier Curves and Conic Sections

Experiment 18.4. Run `rationalBezierCurve2.cpp`, which draws a red quadratic rational Bézier curve on the plane specified by the three control points $[1,0]^T$, $[1,1]^T$ and $[0,1]^T$. See Figure 18.7. Also drawn is the unit circle centered at the origin. Press the up/down arrow keys to change the weight of the middle control point $[1,1]^T$. The weights of the two end control points are fixed at 1.

Decrease the weight of the control point $[1,1]^T$ from its initial value of 1.5. It seems that at some value between 0.70 and 0.71 the curve lies exactly along a quarter of the circle (the screenshot of Figure 18.7 is at 1.13). This is no accident, as the following exercise shows. **End**

Figure 18.7: Screenshot of `rationalBezier-Curve2.cpp` with the weight of the middle control point 1.13.

Exercise 18.13. Plug the values

$$[x_0 \ y_0]^T = [1,0]^T, \qquad [x_1 \ y_1]^T = [1,1]^T, \qquad [x_2 \ y_2]^T = [0,1]^T$$

of the control points of the preceding experiment, together with the weights

$$w_0 = 1, \qquad w_1 = 1/\sqrt{2}, \qquad w_2 = 1$$

into Equation (18.8). Show, then, that the rational functions $x(u)$ giving the x-value and $y(u)$ the y-value, satisfy $x(u)^2 + y(u)^2 = 1$.

One sees from the preceding exercise that the quadratic rational Bézier curve specified by the control points $[1, 0]^T$ with weight 1, $[1, 1]^T$ with weight $1/\sqrt{2}$ ($\simeq 0.7071$) and $[0, 1]^T$ with weight 1 is indeed a quarter of a circle. It follows that any whole circle can be obtained by joining end to end at most four quadratic rational Bézier curves. In fact, this generalizes to a very close relationship between quadratic rational Bézier curves and conic sections:

Proposition 18.1. *Any bounded arc of a conic section can be obtained by joining end to end a finite number of quadratic rational Bézier curves.*

In the other direction, any quadratic rational Bézier curve is an arc of a conic section. □

The proof is beyond our scope here. We refer the interested reader to the text by Buss [21].

Remark 18.4. The qualifier "bounded" in the proposition is necessary simply because a rational Bézier curve is bounded by definition, so that no unbounded arc of a conic section (e.g., an entire parabola or wing of a hyperbola) can be assembled from a finite number of rational Bézier curves.

Remark 18.5. If the reader is wondering how a quadratic rational Bézier curve which happens to be a straight line segment, e.g., if its three control points are collinear, can be an arc of a conic section, keep in mind that straight lines are, in fact, degenerate conic sections (refer Exercise 10.21).

Now, not even a circle, the simplest of conic sections, can be constructed from polynomial Bézier curves, because no non-trivial arc of a circle has a polynomial parametrization, as we saw in Example 10.8. This is important! Using rational Bézier curves, though not polynomial ones, one can draw conic sections, including circles, ellipses, parabolas and hyperbolas, all curves which arise naturally in diverse applications.

Score one for the rationals!

Remark 18.6. The original Utah Teapot, discussed toward the end of Section 10.3.2, composed of bicubic polynomial Bézier patches, is not – and can never be – perfectly round! To make it so, it has to be redesigned with the help of rational patches.

18.2.4 Properties of Rational Bézier Curves

We ask the reader to establish some properties of rational Bézier curves in general. Observe first that the x and y components of the rational Bézier curve c given by Equation (18.7), written below again,

$$c(u) = \left[\frac{\sum_{i=0}^{n} B_{i,n}(u) w_i x_i}{\sum_{i=0}^{n} B_{i,n}(u) w_i} \quad \frac{\sum_{i=0}^{n} B_{i,n}(u) w_i y_i}{\sum_{i=0}^{n} B_{i,n}(u) w_i} \right]^T \quad (0 \leq u \leq 1)$$

are both ratios of polynomials of degree n in u. Accordingly, c is said to be the rational Bézier curve of *degree n*, or *order $n+1$*, the latter being the number of control points.

The earlier Exercises 18.8 and 18.9 both generalize to rational Bézier curves of arbitrary order, as we see next.

Exercise 18.14. Prove that a rational Bézier curve (of arbitrary order) always interpolates both its first and last control points, no matter what the assignment of weights.

Exercise 18.15. Prove that a rational Bézier curve (of arbitrary order) whose control points have all equal weights coincides with the polynomial Bézier curve specified by the same control points.

Therefore, generally, a polynomial Bézier curve is simply a special case of a rational one.

Let's massage (18.7) into a form which will afford us a familiar way of understanding rational Bézier curves:

$$c(u) = \left[\frac{\sum_{i=0}^{n} B_{i,n}(u) w_i x_i}{\sum_{i=0}^{n} B_{i,n}(u) w_i} \quad \frac{\sum_{i=0}^{n} B_{i,n}(u) w_i y_i}{\sum_{i=0}^{n} B_{i,n}(u) w_i} \right]^T$$

$$= \left[\sum_{i=0}^{n} \frac{B_{i,n}(u) w_i}{\sum_{i=0}^{n} B_{i,n}(u) w_i} x_i \quad \sum_{i=0}^{n} \frac{B_{i,n}(u) w_i}{\sum_{i=0}^{n} B_{i,n}(u) w_i} y_i \right]^T$$

$$= \sum_{i=0}^{n} \frac{B_{i,n}(u) w_i}{\sum_{i=0}^{n} B_{i,n}(u) w_i} p_i \qquad (18.11)$$

in $0 \le u \le 1$, where the control point $p_i = [x_i \ y_i]^T$, $0 \le i \le n$.

One sees from Equation (18.11) that a rational Bézier curve is a weighted sum of its control points, as is a polynomial Bézier curve, but using a different set of blending functions as weights Instead of the Bernstein polynomial $B_{i,n}(u)$, the rational function

$$\frac{B_{i,n}(u) w_i}{\sum_{i=0}^{n} B_{i,n}(u) w_i}$$

blends control point p_i.

Exercise 18.16. Verify that the blending functions of a rational Bézier curve form a partition of unity. Therefore, a rational Bézier curve is (a) constrained to lie in the convex hull of its control points, and (b) affinely invariant.
Hint: See the proof of Proposition 15.1.

Exercise 18.17. Prove that if the weight w_i of one particular control point p_i is increased in Equation (18.11), then the value

$$\frac{B_{i,n}(u) w_i}{\sum_{i=0}^{n} B_{i,n}(u) w_i}$$

of its blending function increases everywhere in the open interval $0 < u < 1$, while that of every other control point decreases.

This explains the phenomenon observed in Experiment 18.3, that increasing a control point's weight attracts the curve to it.

Figure 18.8: Screenshot of `rationalBezier-Curve3.cpp`.

Experiment 18.5. Run `rationalBezierCurve3.cpp`, which shows a rational Bézier curve on the plane specified by six control points. See Figure 18.8 for a screenshot. A control point is selected by pressing the space key, moved with the arrow keys and its weight changed by the page up/down keys. Pressing delete resets.   **End**

From a design point of view then a control point's weight is a dial to turn up or down its attraction on the curve. It adds a level of control to edit a rational Bézier curve beyond what is available for a polynomial one.

That's score two for the rationals!

18.2.5   Rational Bézier Curves and Projective Invariance

Note to Readers Unfamiliar with Projective Geometry: This section investigates how a projective transformation transforms a Bézier curve. It begins, though, with the effect of so-called snapshot transformations, a subclass of the projective, defined in Appendix A. Informally, a snapshot transformation is the change induced in how an object is seen by altering the alignment of a point camera. Unfortunately, just this may not be enough to follow the entire discussion in this section, but to go farther it seems unavoidable to refer to Appendix A. Our suggestion to the reader not inclined to peruse that appendix is to simply read once Proposition 18.2, which describes how a rational Bézier curve changes through projective transformation, and take it for granted.

What happens when a snapshot transformation (snapshot transformations, a special subclass of the projective, are introduced in Section A.5 of Appendix A) is applied to a Bézier curve, either polynomial or rational? Let's try and repeat Experiment A.1, where we ran the program `turnFilm1.cpp` to compare snapshots of parallel power lines taken with the film along the $z = 1$ and $x = 1$ plane, respectively, but, with power lines now replaced by a polynomial Bézier curve drawn on the $z = 1$ plane.

See Figure 18.9, where the control points p_0, p_1 and p_2 lie on the $z = 1$ plane, and the (red solid) quadratic polynomial Bézier curve c approximates them. The points p_0, p_1 and p_2 and the curve c are projected along lines toward the origin to the points p_0', p_1' and p_3' and the (red dashed) curve c' on the plane $x = 1$. Therefore, c' is the snapshot transformation of c.

Is c' the polynomial Bézier curve approximating p_0', p_1' and p_3'? No! That happens to be a different (green dashed) curve $\bar{c}$. Coding is believing. . .

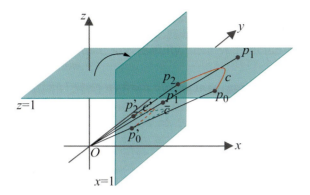

Figure 18.9: The (red solid) quadratic polynomial Bézier curve c on the $z = 1$ plane is specified by the control points p_0, p_1 and p_2. The points p_0, p_1 and p_2 and the curve c project to the points p_0', p_1' and p_3' and the (red dashed) curve c', respectively, on the plane $x = 1$. The (green dashed) curve $\bar{c}$, different from c', is the polynomial Bézier approximation of p_0', p_1' and p_3'.

Experiment 18.6. Run `turnFilm2.cpp`, which animates the snapshot transformation of a polynomial Bézier curve described above. Three control points and their red dashed approximating polynomial Bézier curve are initially drawn on the $z = 1$ plane. See Figure 18.10(a). The locations of the control points, and so of their approximating curve as well, are *fixed* in world space. However, they will *appear* to move as the film rotates.

Initially, the film lies along the $z = 1$ plane. Pressing the right arrow key rotates it toward the $x = 1$ plane, while pressing the left arrow key rotates it back. The film itself, of course, is never seen. As the film changes position, so do the control points and the red dashed curve, these being the *projections* (snapshot transformations, particularly) onto the current film of the control points and their approximating curve (all fixed, as said, in world space). Also drawn on the film is a green dashed curve, which is the polynomial Bézier curve approximating the current projections of the control points.

Note: The control points and their approximating curve, all fixed on the $z = 1$ plane, and corresponding to the control points p_0, p_1 and p_2 and the solid red curve in Figure 18.9, are *not* drawn by the program – only their snapshot transformations on the turning film.

Initially, when the plane of the film coincides with that on which the control points are drawn, viz., $z = 1$, the projection onto the film of the polynomial Bézier curve approximating the control points (the red dashed curve) coincides with the polynomial Bézier curve approximating the projected control points (the green dashed curve). This is to be expected because the control points coincide with their projections. However, as the film turns away from the $z = 1$ plane, the red and green dashed curves begin to separate. Their final configuration, when the film lies along $x = 1$, is shown in Figure 18.10(b).

There is more functionality to the program that we'll discuss momentarily.

End

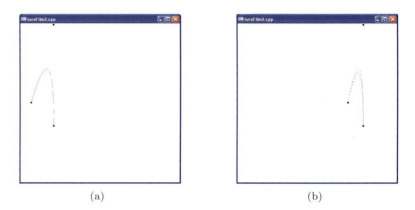

(a)   (b)

Figure 18.10: Screenshots of `turnFilm2.cpp`: (a) Initial configuration (b) Final.

So, if the snapshot transformation c' of the approximating polynomial Bézier curve (the red dashed curve of `turnFilm2.cpp`) is not the polynomial Bézier curve $\bar{c}$ approximating the transformed control points (the green dashed curve), then what is it?

It's not hard to deduce the answer by comparing the earlier Figure 18.5 with Figure 18.9. Imagine the plane $z = 1$ of the former figure replaced by $x = 1$ of the latter. Accordingly, points p_0, p_1 and p_2 of Figure 18.9 are liftings of their respective projections p'_0, p'_1 and p'_2 on $x = 1$, the weights associated with the latter three being the respective x-*values* of the first three.

Conclusion: the snapshot transformation of the polynomial Bézier curve on the control points p_0, p_1 and p_2, from the plane $z = 1$ to $x = 1$, is not the polynomial Bézier curve approximating the projected control points p'_0, p'_1 and p'_2, but, rather, the rational Bézier curve approximating them, with the weight of p'_i equal to the x-value of p_i, $0 \le i \le 2$.

What about snapshot transforming an arbitrary rational Bézier curve, rather than a polynomial one? Exactly the same principle applies. The result is a rational Bézier curve approximating the transformed control points, with *new* weights.

Figure 18.11 explains how the new weights are calculated in the simple case of transforming from $z = 1$ to $x = 1$. If the control point p on the $z = 1$ plane is $[x \ y \ 1]^T$ with weight w, then its lifted control point P in $\mathbb{R}^3$ is $[wx \ wy \ w]^T$. The projection of p, as that of P, on $x = 1$ is the point $p' = [1 \ y/x \ 1/x]^T$. Therefore, the weight of p' so that its lifting coincides with P is wx.

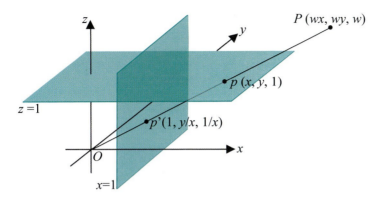

Figure 18.11: Both the control point p on $z = 1$, with weight w, and its lifting P project to the point p' on $x = 1$.

Suppose, now, that c is the rational Bézier curve approximating $n + 1$ control points $[x_i \ y_i \ 1]^T$ on the plane $z = 1$, with weights w_i, $0 \le i \le n$. It follows that the snapshot transformation of c from $z = 1$ to $x = 1$ is the rational Bézier curve on the transformed control points $p'_i = [1 \ y_i/x_i \ 1/x_i]^T$, with weights $w_i x_i$, $0 \le i \le n$.

Example 18.3. Compute the snapshot transformation of the rational Bézier curve c on the $z = 1$ plane with control points $[1 \ -1 \ 1]^T$, $[2 \ 1 \ 1]^T$ and $[4 \ 3 \ 1]^T$, and respective weights 0.5, 2.0 and 1.0, onto the $x = 1$ plane.

Answer: From the preceding discussion the transformation of c onto the $x = 1$ plane is the rational Bézier curve with control points $[1 \ -1 \ 1]^T$, $[1 \ 0.5 \ 0.5]^T$ and $[1 \ 0.75 \ 0.25]^T$, and respective weights 0.5, 4 and 4.

Example 18.4. A polynomial Bézier curve c is drawn in 3-space with control points at $[2 \ 2 \ 5]^T$, $[3 \ 1 \ 4]^T$ and $[0 \ 4 \ 2]^T$. What is its projection on the $z = 1$ plane?

Answer: The projection of c on $z = 1$ is the rational Bézier curve with control points at $[0.4 \ 0.4 \ 1]^T$, $[0.75 \ 0.25 \ 1]^T$ and $[0 \ 2 \ 1]^T$, and respective weights 5, 4 and 2.

Exercise 18.18. Compute the snapshot transformation of the rational Bézier curve c on the $z = 1$ plane with control points at $[2 \ 2 \ 1]^T$, $[1 \ 4 \ 1]^T$ and $[5 \ 1 \ 1]^T$, and respective weights 1.0, 4.0 and 0.5, onto the $y = 1$ plane.

Exercise 18.19. A polynomial Bézier curve c is drawn in 3-space with control points at $[4 \ 1 \ 5]^T$, $[2 \ 2 \ 3]^T$ and $[1 \ 2 \ 2]^T$. What is its projection on the $z = 1$ plane?

Experiment 18.7. Fire up `turnFilm2.cpp` once again. Pressing space at any time draws, instead of the green dashed curve, a blue dashed *rational* Bézier curve approximating the projected control points on the current plane of the film. The control point weights of the blue dashed curve are computed according to the strategy just described. Voilà! The blue dashed rational curve and the red dashed projection are inseparable. **End**

Exercise 18.20. (Programming) Verify that `turnFilm2.cpp` does as just claimed. In particular, check that the weights of the projected control points used to draw the blue dashed curve are correctly calculated as the new weights following a snapshot transformation.

Hint: The code is a little tricky as the projection of the control points on the turning film are computed "by hand", via the routine `computeProjectedControlPoints()`. What this routine does, in fact, is simulate the rotation of the film clockwise about the y-axis by computing the projection of the control points on the plane $z = 1$, after rotating the control points *counter-clockwise* about the y-axis (but leaving the film fixed). For this reason, the first viewing transformation, which is used to turn the film, is not applied to the projected control points, but rather a second one keeping the camera pointed at the $z = 1$ plane.

The routine `computeWeightedProjectedControlPoints()` assigns the new weights to the projected control points that then are used to draw the blue dashed curve.

Let's pause a moment to take stock. A snapshot transformation of a polynomial Bézier curve may not even be a polynomial Bézier curve. However, that of a rational Bézier curve is not only a rational Bézier curve, but the control points of the transformed curve are transformations of the original control points. Moreover, their new weights can be computed from values of the original weights and original control points. We call this property the *snapshot invariance* of rational Bézier curves.

In fact, rational Bézier curves are *projectively invariant*:

Proposition 18.2. *Applying a projective transformation of $\mathbb{P}^2$ to a rational Bézier curve in $\mathbb{R}^2$ gives another rational Bézier curve in $\mathbb{R}^2$ whose control points are the transformations of the original control points, and with altered weights which can be computed from the values of the original weights and original control points.*

Proof. Again, do keep in mind that a projective transformation acts on a curve in $\mathbb{R}^2$ by transforming its lifting (which belongs to $\mathbb{P}^2$). The proof of the proposition, though not difficult, involves a fair amount of algebraic manipulation which we'll not get into. The mathematically inclined reader should try to prove it for herself. Otherwise, refer to Piegl and Tiller [107]. □

Projective invariance versus only affine. Make that 3-0 in favor of the rationals then!

In Exercise 18.16 we deduced the affine invariance of rational Bézier curves as a consequence of the partition-of-unity property of their blending functions. It's also a consequence of the preceding proposition because affine transformations are a subclass of the projective.

Exercise 18.21. Prove that an affine transformation of a rational Bézier curve in $\mathbb{R}^2$ does not alter its weights.

Exercise 18.22. Find the flaw in the following argument:
Rational Bézier curves are projectively invariant. Polynomial Bézier curves are special cases of rational Bézier curves with weights all equal. Therefore, polynomial Bézier curves are projectively invariant as well.

Recall that a projective transformation can map a regular point to a point at infinity (and vice versa) with respect to a particular embedding of $\mathbb{R}^2$ in $\mathbb{P}^2$. In fact, one may even contemplate control points of a Bézier curve at infinity! Here's an interesting application to obtain a very familiar curve as a rational Bézier curve with one control point indeed at infinity:

Exercise 18.23. Embed $\mathbb{R}^2$ in $\mathbb{P}^2$ as the plane $z = 1$. Prove that the polynomial Bézier curve in $\mathbb{P}^2$ with control points $[1\ 0\ 1]^T$, $[0\ 1\ 0]^T$ and $[-1\ 0\ 1]^T$ projects to the upper-half of the unit circle centered at the origin of $\mathbb{R}^2$. Observe that the middle control point is at infinity with respect to $z = 1$.

18.2.6 Rational Bézier Curves in the Real World

Except for Exercise 18.12, our discussion thus far in this section has been exclusively of rational Bézier curves on the plane. Extension to curves in 3-space, however, is straightforward. For example, Equation (18.9)

$$c(u) = \left[\frac{\sum_{i=0}^n B_{i,n}(u)w_i x_i}{\sum_{i=0}^n B_{i,n}(u)w_i} \quad \frac{\sum_{i=0}^n B_{i,n}(u)w_i y_i}{\sum_{i=0}^n B_{i,n}(u)w_i} \quad \frac{\sum_{i=0}^n B_{i,n}(u)w_i z_i}{\sum_{i=0}^n B_{i,n}(u)w_i} \right]^T$$

$0 \le u \le 1$, of a rational Bézier curve in $\mathbb{R}^3$ approximating the control points $[x_i\ y_i\ z_i]^T$, with respective weights w_i, $0 \le i \le n$, which the reader was asked to deduce in Exercise 18.12, adds the expected z-component to its 2D counterpart (18.7)

$$c(u) = \left[\frac{\sum_{i=0}^n B_{i,n}(u)w_i x_i}{\sum_{i=0}^n B_{i,n}(u)w_i} \quad \frac{\sum_{i=0}^n B_{i,n}(u)w_i y_i}{\sum_{i=0}^n B_{i,n}(u)w_i} \right]^T \quad (0 \le u \le 1)$$

Exercise 18.24. Show that the projection of a rational 3D Bézier curve on any plane is a rational 2D Bézier curve.

Exercise 18.25. (Programming) Write a 3D version of `rational-BezierCurve3.cpp` of Experiment 18.5 with control points which can be moved in 3-space, and with changeable weights. Add functionality to rotate the viewpoint.

18.2.7 Rational Bézier Surfaces

With the spadework for rationalization already mostly done, the next step up to rational Bézier surfaces is not going to be much more than a matter of jotting down equations one by one with an eye still on the curves.

Recall from Section 15.2 the equation

$$s(u,v) = \sum_{i=0}^{n} \sum_{j=0}^{m} B_{i,n}(u) B_{j,m}(v) p_{i,j} \qquad (0 \leq u \leq 1,\ 0 \leq v \leq 1) \qquad (18.12)$$

of a polynomial Bézier surface in 3-space with control points $p_{i,j}$, for $0 \leq i \leq n$ and $0 \leq j \leq m$, and the process of "sweeping by a Bézier curve" by which it was derived.

Following a similar process, one can write the equation of a *rational Bézier surface* specified by control points $p_{i,j}$ with respective weights $w_{i,j}$, $0 \leq i \leq n$ and $0 \leq j \leq m$:

$$s(u,v) = \sum_{i=0}^{n} \sum_{j=0}^{m} \frac{B_{i,n}(u) B_{j,m}(v)\, w_i}{\sum_{i=0}^{n} \sum_{j=0}^{m} B_{i,n}(u) B_{j,m}(v)\, w_i} p_{i,j} \qquad (0 \leq u \leq 1,\ 0 \leq v \leq 1)$$

$$(18.13)$$

From (18.12) to (18.13) the change is simply in the blending functions, now rational, rather than polynomial. We ask the reader next to determine equations for a rational Bézier surface in forms analogous to those that we have already deduced for curves.

Exercise 18.26. Find equations for rational Bézier surfaces in $\mathbb{R}^3$ analogous to Equations (18.4)-(18.8) for rational Bézier curves.

It should come as no surprise to the reader that rational Bézier surfaces are projectively invariant and, therefore, affine and snapshot invariant as well. Moreover, they can represent exactly parts of paraboloids, ellipsoids and hyperboloids, and other quadric surfaces. From a designer's perspective, a control point's weight, as in the case of a rational Bézier curve, is an additional dial to turn up or down its attractive pull on the surface.

All the advantages of rational Bézier curves over polynomial propagate, therefore, to rational Bézier surfaces.

Drawing Rational Bézier Surfaces

Expectedly, the main change in drawing polynomial versus rational Bézier surfaces, as we saw in Section 18.2.2 going from polynomial to rational Bézier curves, is replacing "VERTEX_3" with "VERTEX_4" to include the extra weight parameter w, in addition to x, y and z, per control point.

Experiment 18.8. Run rationalBezierSurface.cpp, based on bezier-Surface.cpp, which draws a rational Bézier surface with the functionality that the location and weight of each control point can be changed. Press the space and tab keys to select a control point. Use the arrow and page up/down keys to translate the selected control point. Press '</>' to change its weight. Press delete to reset. The 'x/X', 'y/Y' and 'z/Z' keys turn the viewpoint. Figure 18.12 is a screenshot.

Mark the use of `glMap2f(GL_MAP2_VERTEX_4, ...)`, as also of `glEnable(GL_MAP2_VERTEX_4)`. The 2's in the syntax are for a surface. **End**

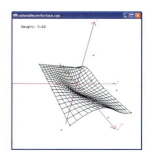

18.2.8 The 'R' in NURBS

With all the groundwork laid in rational Bézier theory, putting the 'R' now into NURBS (Non-Uniform Rational B-Splines) is going to be rather anti-climactic.

Recall Equation (16.34) of the mth order B-spline curve c approximating $r - m + 1$ control points $p_0, p_1, \ldots, p_{r-m}$ over the knot vector $\{t_0, t_1, \ldots, t_r\}$:

$$c(u) = \sum_{i=0}^{r-m} N_{i,m}(u) p_i \qquad (t_{m-1} \leq u \leq t_{r-m+1}) \tag{18.14}$$

where the blending function of the ith control point is the mth order B-spline function $N_{i,m}$, $0 \leq i \leq r - m$.

Following a development *exactly parallel* to that for rational Bézier curves, one can write for a NURBS curve an equation analogous to (18.11), which expresses a rational Bézier curve as a weighted sum of its control points, the weights being rational blending functions. In fact, the equation for a NURBS curve approximating $r - m + 1$ control points p_i, with weights w_i, $0 \leq i \leq r - m$, over the knot vector $\{t_0, t_1, \ldots, t_r\}$, is

$$c(u) = \sum_{i=0}^{r-m} \frac{N_{i,m}(u) w_i}{\sum_{i=0}^{r-m} N_{i,m}(u) w_i} p_i \qquad (t_{m-1} \leq u \leq t_{r-m+1}) \tag{18.15}$$

where, of course, the blending functions are now ratios of terms composed of B-splines.

We'll leave finding the equation of a NURBS surface to the reader in the following exercise.

Exercise 18.27. Recall Equation (16.36) of a B-spline surface approximating an $(n + 1) \times (n' + 1)$ array of control points over a pair of non-uniform knot vectors. Rewrite it for a NURBS surface, taking into account control point weights. What is the blending function of the control point $p_{i,j}$, $0 \leq i \leq n$, $0 \leq j \leq n'$?

Exercise 18.28. Prove that NURBS curves and surfaces are affinely invariant. (In fact, they are projectively invariant.)
Hint: Think partition of unity.

Drawing NURBS Curves and Surfaces

The reader may wish to review Section 16.4 where the GLU NURBS interface is explained and used to draw polynomial B-spline curves and surfaces. With the practicalities of the transition from drawing polynomial Bézier primitives to rational ones already learned from earlier in this chapter, those for drawing rational NURBS primitives are straightforward and left to the reader. The following two exercises ask her to apply the NURBS interface to draw a rational curve and a rational surface, respectively.

Exercise 18.29. (Programming) Modify `cubicSplineCurve1.cpp` of Experiment 16.6 to draw a cubic NURBS curve so that the weight of the selected control point can be changed, in addition to all the original functionality. You must use the call `gluNurbsCurve`(GL_MAP1_VERTEX_4, . . .).

Exercise 18.30. (Programming) Modify `bicubicSplineSurface.cpp` of Experiment 16.15 to draw a NURBS surface.

18.3   Summary, Notes and More Reading

In this chapter we studied two important applications to CG of projective spaces: (a) the projection transformation to convert a viewing volume into the canonical box in the synthetic-camera pipeline, and (b) rational Bézier and B-spline theory. The first demonstrates the practical importance of projective geometry in the CG rendering pipeline. The second is important for a deeper understanding of design because rational primitives, in particular NURBS, are the de facto standard in CAD.

There are several excellent sources for the reader to follow up on both rational Bézier and NURBS primitives. A few are the books by Buss [21], Farin [42, 43], Mortenson [90], Piegl & Tiller [107] and Rogers [111].

Part X

The Time is Pipe

CHAPTER 19

Fixed-Functionality Pipelines

A t the end of Chapter 4 about moving and shaping objects and manipulating the OpenGL camera, we said that it was like having gotten our driver's license. It's time now to look at the engine under the hood to understand the whole process, from ignition to motion. So in this chapter we are going to study graphics rendering pipelines – processes that transform a user-defined scene into an image on a raster display.

The particular topic of this chapter, though, is fixed-functionality pipelines where, once the programmer has specified the scene, she has little further say in the rendering process. In this category falls, first, the first-generation synthetic-camera pipeline, which, in fact, our OpenGL programs thus far have all invoked. The basic ray tracing pipeline, based on a global illumination model – versus a local in the case of the synthetic camera – is fixed-functionality as well, as is radiosity, another global illumination model often implemented in tandem with ray tracing.

BSP (Binary Space Partitioning) tree algorithms used to be a staple of early fixed-functionality pipelines. They are distinguished by the geometric method that they apply to incorporate hidden surface removal into the rendering process. Although rarely implemented in pipelines any more – having been dislodged by the ubiquitous z-buffer which is simpler and faster for the purpose of hidden surface removal – BSP trees are important enough a template for space-partitioning for a student of CG to be familiar with.

We begin in Section 19.1 with the fixed-functionality synthetic-camera pipeline. This is the pipeline that the first generation of OpenGL (versions $1.x$) implements and the one used so far in this book. Our description of this particular pipeline began, in fact, with the shoot-and-print analogy of Chapter 2. We'll put all the pieces together now to get a fairly complete idea of its implementation.

Section 19.2 introduces ray tracing, the most popular global illumination model and its rendering pipeline. As its name suggests, ray tracing is based upon following individual light rays through a scene. It is a near photorealistic way of rendering, but computationally so expensive as to be almost never used in real-time applications such as games. For off-line applications, though, e.g., movies, where computational resources and time are not major constraints, ray tracing is far more authentic an alternative to synthetic-camera-based rendering.

Radiosity, another global lighting model and the topic of Section 19.3, is frequently implemented together with ray tracing, as the two have complementary models of light transport.

BSP trees are the topic of Section 19.4 and we conclude in Section 19.5.

19.1 Synthetic-Camera Pipeline

We're in a position now to put together, with pieces learned so far in this book, a complete synthetic-camera rendering pipeline, though without a lot of the bells and whistles that practical implementations come with. Let's begin with a review of our understanding of how a piece of OpenGL code turns into a picture on the monitor.

Following the programmer's definition of a scene, the first thing to happen to it is modelview transformation as described, particularly, in Sections 4.1-4.2. Section 5.4 explains how to compute the matrix corresponding to any given modelview transformation – the transformation being applied to a vertex by multiplying it by the transformation's matrix from the left. Next, the scene is "captured on film" by applying the shoot-and-print paradigm of Section 2.2: primitives are projected to the front face of the viewing box or frustum (shoot) and then scaled to fit the OpenGL window (print).

In Section 18.1 we saw that the shoot process itself is implemented in two stages. First comes a projection transformation mapping the viewing volume to the canonical viewing box, which itself consists of multiplying the vertices in homogeneous $xyzw$-coordinates by the projection matrix, and then, possibly, a perspective division step to divide out the w-value. The next stage is a parallel projection to the canonical box's back plane of the parts of primitives *inside* it, because only these are rendered to the screen.

Implicit in the second stage, then, is the clipping of primitives to within the canonical box. This can be accomplished for 1D and 2D primitives, respectively, with use of the Cohen-Sutherland line clipper from Section 14.1 (particularly, its extension to 3-space suggested in Exercise 14.5) and the Sutherland-Hodgeman polygon clipper from Section 14.2 (Exercise 14.11 suggests the 3-space version). Note that clipping 0D primitives, i.e., points, is a trivial matter of tossing those whose coordinates place them outside the canonical box.

The last print step, where the back face of the canonical box is scaled to the OpenGL window, involves choosing and coloring a set of pixels in the latter for each primitive on the former, which, of course, is rasterization. Again, 0D primitives, or points, are easily rasterized, while 1D and 2D primitives can be processed with the use, respectively, of the line and polygon rasterizers from Sections 14.3 and 14.4. Just prior to rasterization, depth testing may be invoked to decide which part, if any, of a primitive is obscured by others, in which case this part is not allowed to colorize its corresponding pixels.

That's it. This is enough to give us a skeletal code-to-image pipeline. Texture, lighting and other capabilities can be worked in later. Let's keep it simple to start with.

19.1.1 Pipeline: Preliminary Version

Time now for specifics. Let's follow a vertex, given in homogeneous coordinates, down a simple pipeline that follows the strategy outlined above:

Synthetic-camera Rendering Pipeline (Preliminary Version)

1. $[x \ y \ z \ 1]^T \quad \longrightarrow \quad [x^M \ y^M \ z^M \ 1]^T$
   *Modelview transformation =
   multiplication by the modelview matrix.*

2. $\quad\quad\quad\quad \longrightarrow \quad [x^{PM} \ y^{PM} \ z^{PM} \ w^{PM}]^T$
   Multiplication by the projection matrix.

3. $\quad\quad\quad\quad \longrightarrow \quad \left[\frac{x^{PM}}{w^{PM}} \ \frac{y^{PM}}{w^{PM}} \ \frac{z^{PM}}{w^{PM}}\right]^T$
   Perspective division.

4. $\quad\quad\quad\quad \longrightarrow \quad \left[\frac{x^{PM}}{w^{PM}} \ \frac{y^{PM}}{w^{PM}} \ \frac{z^{PM}}{w^{PM}}\right]^T$
   Clipping to the canonical box.

5. $\quad\quad\quad\quad \longrightarrow \quad \left[\frac{x^{PM}}{w^{PM}} \ \frac{y^{PM}}{w^{PM}}\right]^T$
   *Projection to the back of the canonical box
   (z-values possibly retained for depth testing).*

6. $\quad\quad\quad\quad \longrightarrow \quad [i \ j]^T$
   Rasterization.

Notes:

(i) Superscripts indicate the transforming matrix, e.g., $[x^M \ y^M \ z^M \ 1]^T = M[x \ y \ z \ 1]^T$. Notation: M = modelview, P = projection and PM their product.

(ii) Multiplication by the projection matrix P (Stage 2) + perspective division (Stage 3) = projection transformation of Section 18.1, which transforms the viewing volume into the canonical viewing box.

(iii) Perspective division is a non-operation in the case of a glOrtho()-defined viewing box as w^{PM} are all 1.

(iv) All the x-, y-, z- and w-values, with or without superscripts, are floating points up to and including Stage 5. It's only at the final Stage 6 that the vertex "jumps" from $\mathbb{R}^2$ (the back face of the canonical box) to a discrete $m \times n$ raster (in the frame buffer), in other words, from being an $[x \ y]^T$ *floating point* tuple to an $[i \ j]^T$ *integer* tuple.

Figure 19.1 is a pictorial view. The pipeline does make sense. However, there are two significant technicalities to deal with before it can be put it into production. First is the problem of handling zero w-values in Stage 3; the next is the issue of so-called perspective correction that needs to be taken into account when colorizing a primitive in the raster in Stage 6. The next two subsections discuss these two technicalities, respectively. *Both are fairly mathematical, so if you are not so inclined, then skip to Section 19.1.4 and take the revised pipeline there for granted.*

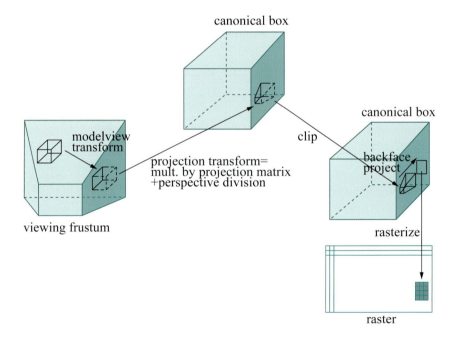

Figure 19.1: Synthetic-camera rendering pipeline (the dashed part of the drawn box is outside the viewing frustum).

19.1.2   Perspective Division by Zero

You may need to review Section 18.1 as our discussion here is a follow-on of the account in that section of the projection transformation.
Perspective division in Stage 3 of the pipeline could involve division by zero. Prima facie, however, this appears to be not much of a problem because the canonical box, into which the viewing volume is transformed by multiplication by the projection matrix, consists only of regular points (with respect to $w = 1$). Therefore, a point that's mapped to a point at infinity – with w-value 0 and, hence, outside the canonical box – never belonged to the viewing volume in the first place. So it seems all we have to do is add a filter before the perspective division stage to simply eject points with w-value equal to 0. Unfortunately, the problem is a bit more complicated, as the following experiment indicates.

$\mathbf{Experiment}$ **19.1.** Replace the box `glutWireCube(5.0)` of `box.cpp` with the line segment

```
glBegin(GL_LINES);
    glVertex3f(1.0, 0.0, -10.0);
    glVertex3f(1.0, 0.0, 0.0);
glEnd();
```

and delete the `glTranslatef(0.0, 0.0, -15.0)` statement. You see a short segment, the clipped part of the defined line segment, whose first endpoint $[1\ 0\ -10]^T$ is inside the viewing frustum defined by the program's projection statement `glFrustum(-5.0, 5.0, -5.0, 5.0, 5.0, 100.0)`, while the second $[1\ 0\ 0]^T$ is outside. Figure 19.2 is a screenshot.

Figure 19.2: Screenshot of Experiment 19.1.

Here's what's interesting though – the second endpoint is mapped to a point at infinity by multiplication by OpenGL's projection matrix! This is easy to verify. Simply take the dot product of $[0\ 0\ -1\ 0]$, which is the last row of the projection matrix corresponding to `glFrustum(-5.0, 5.0, -5.0, 5.0, 5.0, 100.0)` as given by Equation (18.3), and $[1\ 0\ 0\ 1]$, the homogeneous coordinates of the second endpoint, to find that the endpoint's transformed w-value is 0 (the other coordinate values are irrelevant). $\mathbf{End}$

The conclusion from the experiment is that even though vertices that map to infinity don't belong in the viewing volume, they may be corners of primitives that partially do. So we just can't toss them – we'll have to make sure that the primitives that they belong to are handed off correctly to the clipper. This requires a little care. It's convenient at this time to climb a dimension down to 2D to visualize the right strategy. Example A.17 of Appendix A is perfect for this purpose.

Note: If you have not yet read Appendix A on projective spaces and transformations, then simply take the following transformation for granted. However, to understand how it was derived, you will need to refer to the appendix.

The projective transformation h^M of $\mathbb{P}^2$ given by

$$M = \begin{bmatrix} -1/2 & 0 & 0 \\ 0 & -3/2 & 1 \\ 0 & -1/2 & 0 \end{bmatrix}$$

transforms the trapezoid q on the plane $z = 1$, aka $\mathbb{R}^2$, with vertices at

$$[-1\ 1]^T,\ [1\ 1]^T,\ [2\ 2]^T \text{ and } [-2\ 2]^T$$

to the rectangle q' with vertices at

$$[-1\ 1]^T,\ [1\ 1]^T,\ [1\ 2]^T \text{ and } [-1\ 2]^T$$

precisely the 2D analogue of transforming a frustum to a box – instead of a frustum we now have a trapezoid and instead of a box a rectangle. See Figure 19.3.

Note: Keep in mind that h^M maps the plane point $[x\ y]^T$ to the one on the plane $z = 1$ with homogeneous coordinates $M[x\ y\ 1]^T$. For example, to determine $h^M([-1\ 1]^T)$, we compute $M[-1\ 1\ 1]^T = [1/2\ -1/2\ -1/2]^T$. Dividing the latter through by, and then dropping its z-value, we see that $h^M([-1\ 1]^T) = [-1\ 1]^T$.

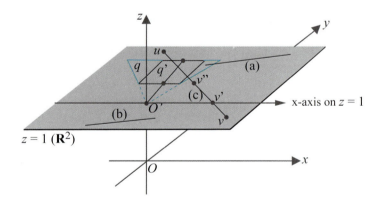

Figure 19.3: The synthetic camera in Flatland: the point camera is at O', the "viewing trapezoid" is q, the "canonical rectangle" q'.

Let's see how to deal with a line segment primitive, say uv, on $\mathbb{R}^2$, subject to transformation by h^M, and, subsequently, clipping to the "canonical" rectangle q'. Exercise A.28 of Appendix A tells us the nature of $h^M(uv)$. It is either a segment (if no point of uv maps to a point at infinity) or two semi-infinite segments (if an interior point maps to infinity) or one semi-infinite segment (if one endpoint maps to infinity) or empty (if both endpoints map to infinity).

It's checked easily that points of the plane mapped by h^M to points at infinity are precisely those on the x-axis. Here, then, is how to clip $h^M(uv)$ to the canonical rectangle – assumed available is a Cohen-Sutherland line clipper for this rectangle with the additional ability to clip semi-infinite segments *a la* Exercise 14.4. The three cases that can arise are listed below and for each an example segment correspondingly labeled is seen in Figure 19.3.

(a) Both u and v are above the x-axis (i.e., with positive y-values):

Pass the transformed segment $h^M(u)h^M(v)$ to the clipper. Note that the transformed segment itself is not drawn in the figure.

(b) Both u and v are below or on the x-axis:

Pre-clip uv altogether as it doesn't intersect the viewing trapezoid.

(c) One endpoint, say u, is above the x-axis and the other endpoint v on or below:

Determine the point v' of intersection of uv with the x-axis. Pass the image $h^M(uv')$, which is a semi-infinite segment, to the clipper (even if v is below the x-axis, the image of $v'v$, another semi-infinite segment, cannot intersect the trapezoid and need not be transmitted).

Note: If the clipper has been extended to handle semi-infinite segments in the manner suggested in Exercise 14.4, then it will need as input the finite endpoint of $h^M(uv')$, as well as the direction in which it is infinite. The finite endpoint is, of course, $h^M(u)$, while the direction it is infinite is toward $h^M(v'')$, where v'' is any point between u and v', e.g., the midpoint. Mind that $h^M(v')$, being a point at infinity, cannot decide the direction itself.

Exercise 19.1. Determine what is transmitted to the extended clipper in the 2D scenario above in the following cases.

(1) $u = [0\ 2]^T$ and $v = [3\ 1]^T$

(2) $u = [0\ -2]^T$ and $v = [3\ 0]^T$

(3) $u = [0\ 2]^T$ and $v = [3\ -1]^T$

Part answer:

(3) Here, u is above and v below the x-axis, so we are in case (c) above. The point where uv intersects the x-axis is $v' = [2\ 0]^T$. Take v'' to be the midpoint $[1\ 1]^T$ of uv'.

Therefore, passed to the extended clipper is a semi-infinite segment which has a finite end at the point on $z = 1$ with homogeneous coordinates

$$h^M(u) = \begin{bmatrix} -1/2 & 0 & 0 \\ 0 & -3/2 & 1 \\ 0 & -1/2 & 0 \end{bmatrix} [0\ 2\ 1]^T = [0\ -2\ -2]^T$$

and is infinite toward the point on $z = 1$ with homogeneous coordinates

$$h^M(v'') = \begin{bmatrix} -1/2 & 0 & 0 \\ 0 & -3/2 & 1 \\ 0 & -1/2 & 0 \end{bmatrix} [1\ 1\ 1]^T = [-\frac{1}{2}\ -\frac{1}{2}\ -\frac{1}{2}]^T$$

One sees, therefore, that the segment passed to the clipper has a finite end at $[0\ 1]^T$ and is infinite toward $[1\ 1]^T$.

We're going to leave it at this, hoping the reader is convinced that the approach just described to handle vertices, otherwise leading to perspective division by zero, can be implemented, even in 3D, by appropriately enhancing Stage 3 of the six-stage synthetic-camera rendering pipeline.

19.1.3 Rasterization with Perspective Correction

Rasterization is more than a matter of plugging in Bresenham's rasterizer for lines and the scan-based rasterizer for polygons (Sections 14.3 and 14.4, respectively). The reason is that both these rasterizing algorithms choose the pixels comprising a primitive but say nothing about how to color them.

However, coloring the pixels of a rasterized primitive seems merely a question of linearly interpolating the values specified at its vertices through its interior. It really ought to be since we made such a fuss in Chapter 7 about how nice are points, line segments and triangles – the fundamental primitives of OpenGL – because values at their vertices can, in fact, be unambiguously interpolated through their interiors! Well, it is, pretty much...

Figure 19.4: Screenshot of perspective-Correction.cpp.

Experiment 19.2. Run `perspectiveCorrection.cpp`. You see a thick straight line segment which starts at a red vertex at its left and ends at a green one at its right. Also seen is a big point just above the line, which can be slid along it by pressing the left/right arrow keys. The point's color can be changed, as well, between red and green by pressing the up/down arrow keys. Figure 19.4 is a screenshot.

The color-tuple of the segment's left vertex, as you can verify in the code, is $(1.0, 0.0, 0.0)$, a pure red, while that of the right is $(0.0, 1.0, 0.0)$, a pure green. As expected by interpolation, therefore, there is a color transition from red at the left end of the segment to green at its right.

The number at the topmost right of the display indicates the fraction of the way the big movable point is from the left vertex of the segment to the right. The number below it indicates the fraction of the "way" its color is from red to green – precisely, if the value is u then the color of the point is $(1 - u, u, 0)$.

Initially, the point is at the left and a pure red; in other words, it is 0 distance from the left end, and its color 0 distance from red. Change both values to 0.5 – the color of the point does *not* match that of the segment below it any more. It seems, therefore, that the midpoint of the line is not colored $(0.5, 0.5, 0.0)$, which is the color of the point. Shouldn't it be so, though, by linear interpolation, as it is half-way between two end vertices colored $(1.0, 0.0, 0.0)$ and $(0.0, 1.0, 0.0)$, respectively? **End**

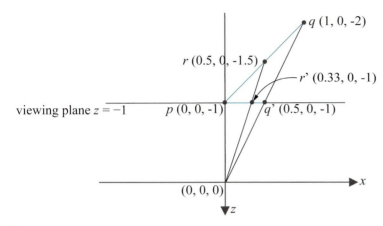

Figure 19.5: The line segment drawn in `perspectiveCorrection.cpp` is pq and its projection on the viewing face pq'.

The apparent conundrum of the preceding experiment is not hard to resolve. Figure 19.5, an xz-section of world space, shows what's happening. The line segment drawn in `perspectiveCorrection.cpp` is from $p = [0\ 0\ -1]^T$ to $q = [1\ 0\ -2]^T$, as specified in the `drawScene()` routine. The midpoint of pq is $r = [0.5\ 0\ -1.5]^T$. Moreover, the perspective projections of p, q and r on the viewing plane $z = -1$ are p itself, q' and r', respectively. The coordinates shown in the figure of q' and r' can be easily verified by properties of similar triangles.

One sees, then, that r', the projection of the midpoint of the segment pq, is *not* the midpoint of the projected segment pq', but rather approximately $0.66\ (= 0.33/0.5)$ of the way from its left end p. With this in mind, return to the program to set the color fraction of the movable point to 0.5 and its distance fraction to 0.66 – now you do see a match! If perspective projections preserved convex combinations, like linear transformations, then midpoints would map to midpoints, but, unfortunately, as we have just found out, they do not.

The conclusion, then, is that the colors at its endpoints *are* linearly interpolated along the user-specified line segment, which is a virtual object in world space; however, as perspective projection does not preserve convex combination, colors of the projected endpoints are not linearly interpolated through the segment drawn on screen.

We understand the issue now, so let's square it with our rasterization procedure by incorporating an additional *perspective correction* factor.

A point $p = [p_x\ p_y\ p_z]^T$ in the viewing frustum is mapped by projection transformation to the point $p' = [p'_x\ p'_y\ p'_z]^T$ in the canonical viewing box; parallel projection to the back face of the box then maps p' to $\bar{p} = [p'_x\ p'_y]^T$. See Figure 19.6(a).

Moreover, a point $tp + (1-t)q$ on the segment joining two points p and q in the viewing frustum maps to a point $up' + (1-u)q'$ on the segment joining their respective images p' and q', though, as we understand now, not necessarily does $u = t$. See Figure 19.6(b).

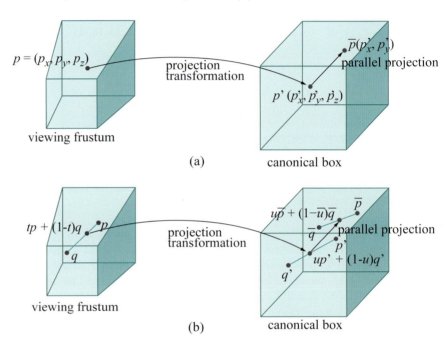

Figure 19.6: (a) A point is mapped by the projection transformation from a viewing frustum to the canonical viewing box, followed by parallel projection to the latter's back face. (b) Likewise for a line segment: the projection transformation does not preserve convex combinations, but parallel projection does.

We want to find the function $u \to t$ that gives the pre-image $tp+(1-t)q$ of $up'+(1-u)q'$. This will serve our purpose of perspective correction, for we'll color the point $u\bar{p} + (1-u)\bar{q}$ – identifying the box's back face with the raster – with the color values $tC(\bar{p}) + (1-t)C(\bar{q})$, instead of $uC(\bar{p}) + (1-u)C(\bar{q})$ as in the case of uncorrected interpolation. $C(\bar{p}) = C(p)$ and $C(\bar{q}) = C(q)$ are, of course, the programmer-specified colors at p and q, respectively. Observe that correction is required only for the projection transformation, not the parallel projection to the back face of the canonical box, as the latter is a linear map preserving convex combinations.

Finding the function $u \to t$ is a matter of some calculation. Write $p = [p_x\ p_y\ p_z]^T$ in homogeneous coordinates as $[p\ 1]^T = [p_x\ p_y\ p_z\ 1]^T$ and q

as $[q\ 1]^T = [q_x\ q_y\ q_z\ 1]^T$. Let P be the projection matrix. Denote the results of multiplying $[p\ 1]^T$ and $[q\ 1]^T$ by P as follows:

$$P[p\ 1]^T = [p_x^*\ p_y^*\ p_z^*\ -p_z]^T \quad \text{and} \quad P[q\ 1]^T = [q_x^*\ q_y^*\ q_z^*\ -q_z]^T \quad (19.1)$$

where the starred symbols are variables to be determined, while the two w-values on the RHS's follow because the last row of the projection matrix P is always $[0\ 0\ -1\ 0]$ (see Equation (18.3)). Applying perspective division, denoted D, next, we have

$$DP[p\ 1]^T = [-\frac{p_x^*}{p_z}\ -\frac{p_y^*}{p_z}\ -\frac{p_z^*}{p_z}\ 1]^T = [p_x'\ p_y'\ p_z'\ 1]^T = [p'\ 1]^T$$

and

$$DP[q\ 1]^T = [-\frac{q_x^*}{q_z}\ -\frac{q_y^*}{q_z}\ -\frac{q_z^*}{q_z}\ 1]^T = [q_x'\ q_y'\ q_z'\ 1]^T = [q'\ 1]^T$$

where the second equality in both equations above follows because DP, in fact, is the projection transformation mapping p to p' and q to q'. The preceding two equations imply that

$$p_x^* = -p_z p_x',\ p_y^* = -p_z p_y',\ p_z^* = -p_z p_z',\ q_x^* = -q_z q_x',\ q_y^* = -q_z q_y',\ q_z^* = -q_z q_z' \tag{19.2}$$

Consider, next, an interpolated point $tp + (1-t)q$ between p and q. Multiplying it by P:

$$\begin{aligned}
&P\,[tp + (1-t)q\quad 1]^T\\
&= P\,(t\,[p\ 1]^T\ +\ (1-t)\,[q\ 1]^T\,)\\
&= t\,(P\,[p\ 1]^T)\ +\ (1-t)\,(P\,[q\ 1]^T)\\
&= t[p_x^*\ p_y^*\ p_z^*\ -p_z]^T\ +\ (1-t)[q_x^*\ q_y^*\ q_z^*\ -q_z]^T \quad \text{(applying (19.1))}\\
&= [tp_x^* + (1-t)q_x^*\ \ tp_y^* + (1-t)q_y^*\ \ tp_z^* + (1-t)q_z^*\ \ -tp_z - (1-t)q_z]^T
\end{aligned}$$

Applying D by dividing through by the w-value:

$$\begin{aligned}
&DP\,[tp + (1-t)q\quad 1]^T\\
&= \left[-\frac{tp_x^* + (1-t)q_x^*}{tp_z + (1-t)q_z}\ \ -\frac{tp_y^* + (1-t)q_y^*}{tp_z + (1-t)q_z}\ \ -\frac{tp_z^* + (1-t)q_z^*}{tp_z + (1-t)q_z}\ \ 1\right]^T\\
&= \left[\frac{tp_z p_x' + (1-t)q_z q_x'}{tp_z + (1-t)q_z}\ \ \frac{tp_z p_y' + (1-t)q_z q_y'}{tp_z + (1-t)q_z}\ \ \frac{tp_z p_z' + (1-t)q_z q_z'}{tp_z + (1-t)q_z}\ \ 1\right]^T\\
&\qquad \text{(using (19.2))}\\
&= \frac{tp_z}{tp_z + (1-t)q_z}\,[p_x'\ p_y'\ p_z'\ 1]^T\ +\ \frac{(1-t)q_z}{tp_z + (1-t)q_z}\,[q_x'\ q_y'\ q_z'\ 1]^T\\
&= \frac{tp_z}{tp_z + (1-t)q_z}\,[p'\ 1]^T\ +\ \frac{(1-t)q_z}{tp_z + (1-t)q_z}\,[q'\ 1]^T\\
&= u\,[p'\ 1]^T\ +\ (1-u)\,[q'\ 1]^T
\end{aligned}$$

where

$$u = \frac{tp_z}{tp_z + (1-t)q_z}$$

Inverting the preceding relationship gives the desired function $u \to t$:

$$
\begin{aligned}
t &= \frac{uq_z}{(1-u)p_z + uq_z} \\
&= \frac{q_z}{(\frac{1}{u}-1)p_z + q_z}, \text{ if } u > 0 ; \quad 0, \text{ if } u = 0
\end{aligned}
\qquad (19.3)
$$

Whew! But we're all set now. Here, then, is how to apply perspective correction when coloring pixels along a line segment. Say the rasterization $R(S)$ of a line segment S joining the points $p = [p_x\, p_y\, p_z]^T$ and $q = [q_x\, q_y\, q_z]^T$ in the viewing frustum consists of $N + 1$ pixels in the raster, as depicted in Figure 19.7. The end pixel of $R(S)$ corresponding to p is (i_1, j_1) and that to q is (i_2, j_2). Precisely, (i_1, j_1) is obtained from mapping p to $\bar{p}$ on the back face of the canonical box by projection transformation and parallel projection, followed by mapping $\bar{p}$ to a point on the raster by the scaling transformation matching the back face of the canonical box to the raster and, finally, followed by a rounding to integer coordinates. Likewise, (i_2, j_2) is obtained from $\bar{q}$. Suppose, as well, that $R(S)$ makes an angle of at most $45°$ with the positive i-axis – other dispositions of $R(S)$ can be handled by symmetry – so that $i_2 = i_1 + N$, which means that each successive pixel of $R(S)$ from left to right has one higher i-value.

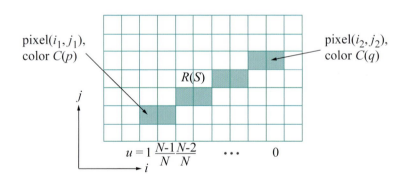

Figure 19.7: The rasterization $R(S)$ of a line segment S consists of $N + 1$ pixels, each corresponding to a particular u-value (a few u-values are shown vertically below the corresponding pixel).

Each of the $N + 1$ pixels of $R(S)$, counting from the left, corresponds successively to a point $u\bar{p} + (1-u)\bar{q}$ of $\overline{pq}$, where $u = 1, \frac{N-1}{N}, \frac{N-2}{N}, \ldots, 0$. The first few u-values are indicated at the bottom of a pixel's column in the figure.

The color tuples $C(p)$ and $C(q)$ of the two end pixels are, of course, the programmer-specified colors of the corresponding end vertices. It remains to color the in-between pixels. The pixel next to the leftmost corresponds to $u = \frac{N-1}{N}$, therefore, in turn, by Equation (19.3), to

$$t = \frac{q_z}{(\frac{1}{\frac{N-1}{N}} - 1)p_z + q_z} = \frac{q_z}{\frac{p_z}{N-1} + q_z}$$

In other words, that pixel corresponds to the point $tp + (1-t)q$ of S, where t is given by the preceding equation. Consequently, the color to apply is $tC(p) + (1-t)C(q)$. Likewise, the color to apply to the next pixel is $tC(p) + (1-t)C(q)$, after updating t to

$$t = \frac{q_z}{\frac{2p_z}{N-2} + q_z}$$

by setting $u = \frac{N-2}{N}$ in (19.3). The procedure of decrementing u by $\frac{1}{N}$, updating t and applying the interpolated colors $tC(p) + (1-t)C(q)$ to the next pixel is repeated until the pixel just before the rightmost is colored, which, of course, completes the coloring of $R(S)$. This procedure can be integrated into Bresenham's line rasterizer: simultaneously picking the pixels along a line segment *and* coloring them with perspective correction.

We'll leave the reader to convince herself that, going from 1D to 2D, a similar perspective correction can be incorporated into triangle and polygon rasterization.

Remark 19.1. Not only color values, but other numerical data defined per vertex, e.g., normals, can be interpolated across rasterized primitives with perspective correction as well – generally, such interpolation is called *perspectively correct*.

19.1.4 Revised Pipeline

Enhancements to the preliminary version of the synthetic-camera rendering pipeline are needed then to Stage 3 to handle perspective division by zero and Stage 6 to incorporate perspective correction into rasterization. Once these are done we have all the pieces in place to go into "production". For the record, the enhanced version (with additions in bold) is shown below. Figure 19.8 after it is a schematic overview of the steps.

This is a complete synthetic-camera pipeline in that it will transform a user-specified scene correctly into a picture on the monitor. However, it is skeletal. The OpenGL pipeline, as we'll see next, adds several features.

Synthetic-camera Rendering Pipeline

1. $[x\ y\ z\ 1]^T$ $\longrightarrow$ $[x^M\ y^M\ z^M\ 1]^T$
    Modelview transformation =
    multiplication by the modelview matrix.

2. $\longrightarrow$ $[x^{PM}\ y^{PM}\ z^{PM}\ w^{PM}]^T$
    Multiplication by the projection matrix.

3. $\longrightarrow$ $\left[\frac{x^{PM}}{w^{PM}}\ \frac{y^{PM}}{w^{PM}}\ \frac{z^{PM}}{w^{PM}}\right]^T$
    Perspective division **with mechanism**
    to handle zero w-values.

4. $\longrightarrow$ $\left[\frac{x^{PM}}{w^{PM}}\ \frac{y^{PM}}{w^{PM}}\ \frac{z^{PM}}{w^{PM}}\right]^T$
    Clipping to the canonical box.

5. $\longrightarrow$ $\left[\frac{x^{PM}}{w^{PM}}\ \frac{y^{PM}}{w^{PM}}\right]^T$
    Projection to the back of the canonical box
    (z-values possibly retained for depth testing).

6. $\longrightarrow$ $[i\ j]^T$
    Rasterization **with perspective correction.**

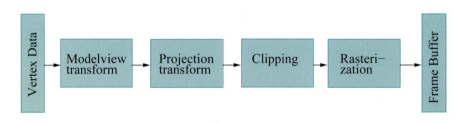

Figure 19.8: Synthetic-camera rendering pipeline scheme (complete but minimal).

19.1.5   OpenGL Pipeline

The OpenGL 1.x rendering pipeline, while keeping the gist of the minimal synthetic-camera pipeline described above, enhances it with several new capabilities to make it significantly more powerful. The additions are indicated in a darker shade in Figure 19.9.

The first addition is texturing, where vertex data (vertex and texture coordinates, particularly) and a set of controlling parameters (filters, environment settings, etc.) are used to combine the texture images into the

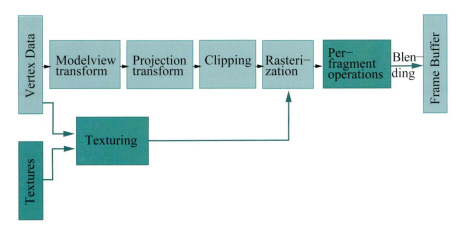

Figure 19.9: OpenGL rendering pipeline scheme (main features). Additions to the minimal synthetic-camera pipeline are darkly shaded.

raster. We learned the fundamentals of texturing ourselves in Chapter 12.

Next, instead of simply copying the raster into the frame buffer, the user is allowed to define per-fragment operations to determine which fragments – a raster pixel with color data and, possibly, a z-value is called a fragment – from the raster do get copied into the frame buffer. The per-fragment operations allowed in OpenGL consist of four tests which occur in the following order:

1. Scissor test

2. Alpha test

3. Stencil test

4. Depth test

If a fragment fails an early test then it is eliminated immediately and does not proceed to subsequent tests. We are already familiar with the stencil test (Section 13.6) and depth test (from as far back as Section 2.8). The scissor test is just a special case of the stencil test, where stencil tags are used to mask a rectangular region of the OpenGL window. It has been made a separate test because it can be optimized in hardware, unlike the general stencil test. The alpha test allows the user to accept or reject a fragment depending upon its alpha value.

Finally, blending is marked on the arrow leading into the frame buffer because, as we know from Section 13.1, it requires data from both the incoming fragments and their respective currently resident destination pixels.

If the reader is wondering that lighting is missing from the pipeline of Figure 19.9, then note that lighting computations are done along the top path, starting from vertex data, which includes normal values as well. It's simply to avoid clutter in the figure that specific lighting calculation stages have been omitted, as have some other processing stages, such as fogging and antialiasing.

Remark 19.2. The Shading Language in second-generation OpenGL (2.0 and higher) significantly impacts the pipeline of Figure 19.9 by making programmable parts of it that are currently of fixed functionality. In particular, so-called vertex and fragment shaders allow the programmer to dictate to a great extent how vertices and fragments are processed. This permits greater flexibility and, therefore, creativity, in modeling scenes than with the fixed-functionality pipeline. We'll be studying the Shading Language ourselves in the last chapter.

19.1.6   1D Primitive Example

Let's chase a 1D primitive down the synthetic-camera rendering pipeline of Section 19.1.4, which is *the* OpenGL pipeline minus bells and whistles.

Example 19.1. The projection statement in the reshape routine of a program is

```
glFrustum(-5, 5, -5, 5, 5, 25)
```

The only primitive definition and only modelview transformation in the drawing routine are

```
glTranslatef(0, 5, 0);
glBegin(GL_LINES);
   glColor3f(1, 0, 0); glVertex3f(0, 0, -10);
   glColor3f(0, 1, 0); glVertex3f(25, 5, -20);
glEnd();
```

All other statements in the program are routine.

Apply the synthetic-camera rendering pipeline to determine the rasterization of the line segment drawn by the program in a 100×100 raster. Determine as well the z-values corresponding to the segment's pixels.

Answer: The segment endpoints in homogeneous coordinates are

$$p = [0 \ 0 \ -10 \ 1]^T \quad \text{and} \quad q = [25 \ 5 \ -20 \ 1]^T$$

The matrix corresponding to the translation is (from Equation (5.28))

$$M = \begin{bmatrix} 1 & 0 & 0 & 0 \\ 0 & 1 & 0 & 5 \\ 0 & 0 & 1 & 0 \\ 0 & 0 & 0 & 1 \end{bmatrix}$$

The matrix corresponding to the projection statement is (from Equation (18.3))

$$P = \begin{bmatrix} 1 & 0 & 0 & 0 \\ 0 & 1 & 0 & 0 \\ 0 & 0 & -1.5 & -12.5 \\ 0 & 0 & -1 & 0 \end{bmatrix}$$

Apply the modelview transformation first, multiplying both endpoints by M:

$$Mp = [0\ 5\ -10\ 1]^T \quad \text{and} \quad Mq = [25\ 10\ -20\ 1]^T$$

At this point we note that both z-values are negative, so we are in a situation analogous to case (a) at the end of the discussion in Section 19.1.2 and can proceed down the pipeline without worrying about invoking enhancements to handle zero w-values in Stage 3. Accordingly, multiplying by P next:

$$PMp = P[0\ 5\ -10\ 1]^T = [0\ 5\ 2.5\ 10]^T$$

and

$$PMq = P[25\ 10\ -20\ 1]^T = [25\ 10\ 17.5\ 20]^T$$

Perspective division, then, gives the Cartesian coordinates of the transformed endpoints as follows:

$$[0/10\ 5/10\ 2.5/10]^T = [0\ 0.5\ 0.25]^T$$

and

$$[25/20\ 10/20\ 17.5/20]^T = [1.25\ 0.5\ 0.875]^T$$

As the first point lies in the canonical box, while the second outside of only the $x = 1$ plane, clipping involves a single intersection computation – that of the transformed segment with the $x = 1$ plane. We'll leave the reader to verify by means of elementary geometry that the intersection, in fact, is $[1\ 0.5\ 0.75]^T$, so that the endpoints of the clipped segment are

$$[0\ 0.5\ 0.25]^T \quad \text{and} \quad [1\ 0.5\ 0.75]^T$$

We must determine the color tuple to assign the new second endpoint. It's checked that the new endpoint as a convex combination of the old ones is:

$$[1\ 0.5\ 0.75]^T = u[0\ 0.5\ 0.25]^T + (1-u)[1.25\ 0.5\ 0.875]^T$$

where $u = 0.2$. Therefore, by Equation (19.3), it corresponds to the point $tp + (1-t)q$ on pq, where

$$t = \frac{0.875}{(\frac{1}{1/0.2} - 1)0.25 + 0.875} = 0.467$$

Accordingly, the color tuple assigned the new endpoint is

$$0.467(1,0,0) + 0.533(0,1,0) = (0.467, 0.533, 0)$$

Projecting the first endpoint of the clipped segment to the back face of the canonical box gives then the point $[0\ 0.5]^T$, with color value $(1,0,0)$ and z-value 0.25 (retained possibly for depth testing). Likewise, the second endpoint projects to $[1\ 0.5]^T$ with color value $(0.467, 0.533, 0)$ and z-value

0.75.

Time to leap from world space to screen space!

Generally, if the raster is $m \times n$ and pixel centers have integer coordinates (i, j), where $0 \leq i \leq m - 1$ and $0 \leq j \leq n - 1$, then the area of the raster is an axis-aligned rectangle, whose lower-left corner is $(-0.5, -0.5)$ and upper-right $(m - 0.5, n - 0.5)$. See Figure 19.10. The back face of the canonical box, on the other hand, can be imagined a 2×2 square with corner coordinates $x = \pm 1$ and $y = \pm 1$ on the xy-plane (now that projection is done, we can "forget" z-values, except when they are needed for depth testing). Accordingly, functions that scale the back face onto the raster – we're doing the print part of shoot-and-print now – are:

$$x \rightarrow \frac{x + 1}{2} m - 0.5 \quad \text{and} \quad y \rightarrow \frac{y + 1}{2} n - 0.5$$

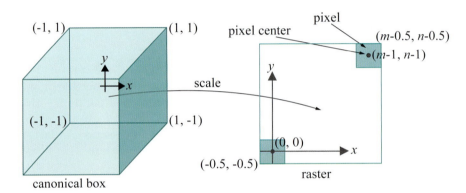

Figure 19.10: Scaling from the 2×2 back face of the canonical box, located on an xy-plane, to the $m \times n$ area of the raster.

Applying these functions to the projected endpoints on the back face of the canonical box, with $m = n = 100$, we get:

$$[0 \ 0.5]^T \rightarrow [49.5 \ 74.5]^T \quad \text{and} \quad [1 \ 0.5]^T \rightarrow [99.5 \ 74.5]^T$$

Rounding, one has the endpoint pixels on the raster as $(49, 74)$ and $(99, 74)$, respectively. Data associated with these pixels are the color values $(1, 0, 0)$ and $(0.467, 0.533, 0)$ and z-values 0.25 and 0.75, respectively.

As the rasterized segment is horizontal, choosing pixels along it is trivial – $(49, 74), (50, 74), \ldots, (99, 74)$ – obtaining a raster line segment of length $N = 50$ (containing 51 pixels).

It remains to color the in-between pixels using perspective correction, as well as assign their z-values. The u-value corresponding to pixel $(50, 74)$, second from left, is $1 - \frac{1}{50} = 0.98$, and t-value, therefore (applying (19.3)):

$$\frac{0.75}{(\frac{1}{0.98} - 1)0.25 + 0.75} = 0.993$$

Accordingly, its color tuple is

$$0.993(1, 0, 0) + 0.007(0.467, 0.533, 0) = (0.996, 0.004, 0)$$

and z-value

$$0.98 * 0.25 + 0.02 * 0.75 = 0.26$$

using u itself, rather than t, to interpolate.

Important: z-values *need not* be perspectively corrected as their values in the canonical box, following projection transformation, are valid.

We'll leave the reader to calculate the color and z-values of a few more pixels or, better still, write a routine to generate them all.

19.1.7 Exercising the Pipeline

Exercise 19.2. Redo the preceding example with only the part in the drawing routine changed to

```
glRotatef(45, 0, 0, 1);
glBegin(GL_LINES);
    glColor3f(1, 1, 1); glVertex3f(5, 0, -10);
    glColor3f(0, 0, 0); glVertex3f(10, 10, -5);
glEnd();
```

Exercise 19.3. Repeat the previous exercise with the drawing routine changed again to

```
glRotatef(90, 0, 1, 0);
glBegin(GL_LINES);
    glColor3f(1, 0, 0); glVertex3f(1, 0, -1);
    glColor3f(1, 0, 0); glVertex3f(-4, 4, 0);
glEnd();
```

and the projection statement, as well, to

```
glOrtho(-5, 5, -5, 5, 5, -5)
```

Exercise 19.4. Repeat the previous exercise with the drawing routine changed once more to

```
glTranslatef(1, 1, 1);
glBegin(GL_LINES);
    glColor3f(1, 1, 1); glVertex3f(5, 0, -10);
    glColor3f(0, 0, 0); glVertex3f(10, 10, 5);
glEnd();
```

and the projection statement back to

```
glFrustum(-5, 5, -5, 5, 5, 25)
```

Note: You can roughly check your result for each of the preceding exercises by comparing it with the output of a minimal OpenGL program containing the given statements.

Exercise 19.5. (Programming) This is a substantial programming project: implement the synthetic-camera pipeline to render (only) 0D and 1D primitives (drawn in 3-space, of course). Use the OpenGL window to simulate the raster as in DDA.cpp.

19.2   Ray Tracing Pipeline

The ray tracing pipeline is an "alternate" to the synthetic-camera graphics pipeline. The reason for the quotes is that the ray tracing approach is very different from that of the synthetic-camera-based approach and rarely does a programmer have the option of simply exchanging one for the other. Why this is the case will be apparent once we understand how ray tracing works.

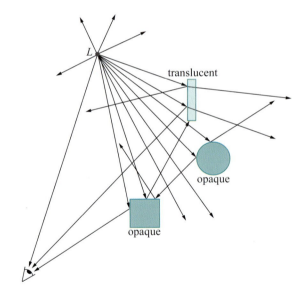

Figure 19.11: Tracing rays *from* a light source *L* – only few reach the eye.

The idea behind ray tracing is straightforward: to follow light rays from each source as they interact with the scene – reflecting off opaque objects one to another, and both reflecting off and refracting through translucent ones, in the process casting shadows and creating reflections – till they finally reach the eye. However, implementing this idea as just stated is not a particularly well-advised endeavor, as (a) there is an infinite continuum of light rays emanating from each source, and (b) even after somehow discretizing them to a finite number, only a fraction thereof reach the viewer. See Figure 19.11 for an idea of the situation.

Ray tracing, instead, implements the plan "backwards". Rays are traced *from* the eye, one through each pixel, so that no computation is expended on rays which are ultimately invisible. Each ray is followed through the pixel and into and around the scene, possibly bouncing off opaque objects and passing through translucent ones, for a *finite* amount of time, till a determination is made of its color. Of course, an implementation has to "cut off" each ray after a finite number of steps and determine the color it has picked up through its interactions with objects till then, otherwise the ray will continue traveling indefinitely.

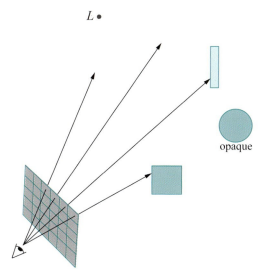

Figure 19.12: Tracing rays *from* the eye, one through each pixel. Rays are "stopped" when they strike an object.

See Figure 19.12 for a very simple scene. The screen is virtual – akin to the front face of an OpenGL viewing box or frustum. In this particular figure, rays either go off to infinity (there are two such drawn) or stop upon hitting the surface of an object (there are two such as well). We don't (as yet) follow rays beyond their first intersection with an object's surface.

This, in fact, suggests a simple first implementation of ray tracing: color pixels, rays through which go off to infinity without collision, the background color; assign every other pixel the color of the first point of intersection of the ray through it with an object's surface. This particular color is determined from Phong's lighting model – see Chapter 11, in particular the lighting equation (11.12).

Here's pseudo-code:

Ray Tracer Version 1: Non-recursive local

```
void topLevelRoutineCallsTheRayTracer()
{
   positionEye = position of eye in world space;
   for (each pixel P of the virtual screen)
   {
      d = unit vector from positionEye toward the center point of P;
      color of P = rayTracer(positionEye, d);
   }
}


Color rayTracer(p, d)
{
   if (ray from point p in the direction d does not intersect the
       surface of any object in the scene) return backgroundColor;
   else
   {
      q = first point of intersection with an object's surface;
      computedColor = color computed at q using Phong's lighting model;
      return computedColor;
   }
}
```

Notes:

1. The *base* case of Version 1, when the ray is stopped at an intersection with an object, uses Phong for color calculation. It is typical, in fact, of ray tracers to invoke a local lighting model at the base case.

2. Intersection detection, implicit in the code, is the most computationally intensive part of the ray tracer. We'll not go into intersection computation in our account of ray tracing, but focus instead on color calculations.

Interestingly, the ray tracer version above renders the same image as a synthetic-camera pipeline – *a la* OpenGL – implementing Phong's lighting model with depth testing. The only difference is that depth testing via the z-buffer has been replaced by ray tracing to determine visible surfaces (one surface obscuring another if it blocks rays from reaching the other).

19.2.1 Going Global: Shadows

The next step up is shadow computation. This is simple to do. If a ray through a pixel intersects a surface, then send a *feeler ray* from the point of intersection toward each light source. If the feeler ray hits an object before reaching a light source, then the point of intersection is in the shadow of the struck object and not illuminated directly by the source. Recall

in this connection that, according to Phong's model, only the diffuse and
specular components of light reflected off a surface depend upon direct, i.e.,
straight-line, illumination from the light source, while the ambient does not.

See Figure 19.13. Point p_1 is in the shadow of the ball S_1 cast by light
from L_1 – because the feeler ray from it toward L_1 is cut off by S_1 – so it
reflects only the ambient component of light from that particular source. On
the other hand, p_1 is directly illuminated by L_2, so reflects all components
of light from that source. Point p_2 is in the shadows of S_1, of which it is
a point itself, cast both by L_1 and L_2. Point p_3 is illuminated directly by
both sources.

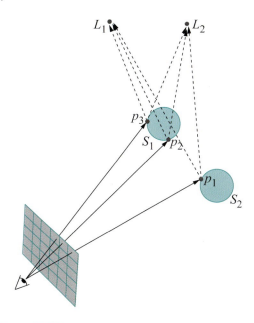

Figure 19.13: Shadow computation: feeler rays are dashed.

Here's pseudo-code for a shadow-computing ray tracer (a top-level routine
the same as that of the first version is not repeated):

Ray Tracer Version 2: Non-recursive global, with shadows

```
void topLevelRoutineCallsTheRayTracer(); // See Version 1.

Color rayTracer(p, d)
{
    if (ray from point p in the direction d does not intersect the
        surface of any object) return backgroundColor;
    else
    {
        q = first point of intersection;
```

```
computedColor = black; // Color values all set to zero.
for (each light source L)
{
    // Object not shadowed.
    if (feeler ray from q toward L does not intersect the
        surface of any object before reaching L)
            computedColor  += color computed at q due to light from
                              L, using Phong's lighting model;

    // Object shadowed.
    else computedColor += ambient component of color computed at
                          q due to light from L using Phong's
                          lighting model;
}
return computedColor;
}
}
```

A hugely significant development in Version 2 is that the lighting model has now gone *global*: object-object light interaction comes into play in computing shadows. A local lighting model, on the other hand, does not take into account other objects when coloring a particular one. Recollect how we had to compute and insert the shadows ourselves in the programs ballAndTorusShadowed.cpp and ballAndTorusPerspectively-Shadowed.cpp. Now this global version of ray tracing not only gives us shadows, but ones as authentic as those drawn by Mother Nature, in particular, the laws of light.

19.2.2   Going Even More Global: Recursive Reflection and Transmission

We are ready now for the full blast of ray tracing power. So far, we've stopped at the first intersection of a ray from the eye with a surface. In reality, rays from a light source can bounce from object to object, or even pass through them, several times before reaching the viewer, giving rise to such phenomena as reflection and translucence. To model this in keeping with ray tracing's backward approach of following rays from the eye into the scene, one must allow a ray to continue even after it hits an object. The physics of light suggests that a ray striking an object is partially reflected off its surface and partially transmitted through it, depending on the characteristics of the material, as well as the color of the light. For example, an opaque object transmits almost zero light and reflects the remainder according to its surface finish and color, while a translucent one transmits most.

Accordingly, we'll enhance Ray Tracer Version 2 such that each ray from the eye that strikes a surface spawns two additional rays: a *reflected ray* in the direction of perfect reflection and a *transmitted ray* passing through the surface, possibly with its direction altered by refraction. The two spawned rays are treated *exactly* as the incoming ray and may each spawn additional rays themselves upon intersection with a surface. If you are thinking recursion, then that's exactly where we're headed.

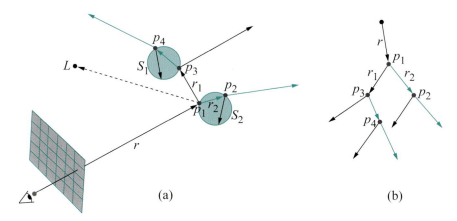

Figure 19.14: (a) Reflection and transmission: reflected rays are black, transmitted blue. One dashed feeler ray is drawn. (b) Ray tree (not all edges are labeled).

As an example, Figure 19.14(a) follows a single ray r from the eye through a few intersections with two translucent balls. The resulting binary *ray tree* data structure is shown in Figure 19.14(b). Observe that the transmitted rays are refracted by the material of the balls. The color computed at a point now has three components – one computed locally, one returned by the reflected ray, and one by the transmitted ray – as given by the following equation:

$$computedColor = color_{local} + coef_{refl}\ color_{refl} + coef_{tran}\ color_{tran}$$

For example, at point p_1 of the figure, $color_{local}$ is computed using Phong (exactly as in Version 2, with help of feeler rays to find "visible" light sources – the feeler ray from p_1 to L_1 is shown in the figure); $color_{refl}$ is the value returned recursively by the reflected ray r_1, attenuated by a material-dependent multiplicative factor $coef_{refl}$, which specifies the fraction of the incoming ray r that is reflected; $color_{tran}$ is likewise returned recursively by the transmitted ray r_2 and attenuated by $coef_{tran}$.

Pseudo-code is below. The new top-level routine passes a non-negative integer depth parameter `maxDepth` to the ray tracer to cut off recursion after a finite number of levels.

Ray Tracer Version 3: Recursive global, with shadows, reflection and transmission

```
void topLevelRoutineCallsTheRayTracer()
{
```

```
positionEye = position of eye in world space;
for (each pixel P of the virtual screen)
{
    d = unit vector from positionEye toward the center point of P;
    color of P = rayTracer(positionEye, d, maxDepth);
}
}

Color rayTracer(p, d, depth)
{
    if (ray from point p in the direction d does not intersect the
        surface of any object) return backgroundColor;
    else
    {
        q = first point of intersection;
        computedColor = black; // Color values all set to zero.

      // Local component, copy of Version 2 calculations.
        for (each light source L)
        {
            // Object not shadowed.
            if (feeler ray from q toward L does not intersect the
                surface of any object before reaching L)
                    computedColor  += color computed at q, due to light from
                                        L, using Phong's lighting model;

            // Object shadowed.
            else computedColor += ambient component of color computed at
                                    q, due to light from L, using Phong's
                                    lighting model;
        }

        // Global component.
        if (depth > 0)
        {
            d1 = unit vector from q in direction of perfect reflection;
            d2 = unit vector from q in direction of transmission;

            // Reflected component added in recursively.
            computedColor += coefRefl * rayTracer(q, d1, depth-1)

            // Transmitted component added in recursively.
            computedColor += coefTran * rayTracer(q, d2, depth-1)
        }
        return computedColor;
    }
}
```

Notes:

1. Determining where an incident ray strikes an object and spawns a reflected and a transmitted ray obviously requires intersection computation. Subsequent calculation of the direction of the reflected and transmitted rays requires computation of the normal to the surface at the point of incidence as well:

   (a) The direction of the reflected ray is given by the laws of reflection, which say that both the incident and reflected rays make the same angle with the normal to the surface, and that all three lie on the same plane. See Figure 19.15, where the equation for reflection is $A = B$.

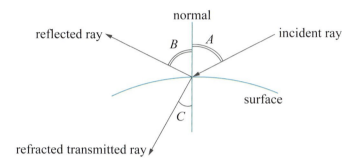

Figure 19.15: Calculating the direction of the reflected and transmitted rays: $A =$ angle of incidence, $B =$ angle of reflection, $C =$ angle of refraction.

   (b) Routines to compute the direction of transmission can be simple or as fancy as the need for realism dictates.

   For instance, refraction is often taken into account with the help of Snell's law, which says that the ratio of the sine of the angle of incidence to the sine of the angle of refraction is equal to the ratio of the speed of light in the medium of the incident ray to that in the medium of the refracted ray; moreover, the incident ray, refracted ray and normal to the surface all lie on the same plane.

   The ratio of the speed of light in two different media is the inverse ratio of the refractive indices of the media. Therefore, in Figure 19.15, one can write the equation for refraction as $\frac{\sin A}{\sin C} = \frac{\eta_2}{\eta_1}$, where η_1 is the refractive index of the medium on the side of the incident ray and η_2 that on the side of the refracted ray.

2. The direction of the reflected ray is taken to be that of perfect mirror-like reflection. This models well the transport of specular light but not that of diffuse. For the latter is needed *multiple* reflected rays

– remember that diffuse light is scattered in all directions by the lit object – which would make the ray tracing process computationally overwhelming.

This inability of ray tracing to realistically model diffuse illumination is a weakness often overcome by combining it with radiosity, another global lighting model which is specially designed to track the dispersion of diffuse light. We'll discuss radiosity in the next section.

Remark 19.3. It's interesting to observe that a ray tracer does not ask for a small set of simple drawing primitives, e.g., points, line segments and triangles, as needed for efficient implementation of the synthetic-camera model. As long as their intersection with a given ray can be computed and the normal at a given point determined, arbitrary curved surfaces may be rendered.

Here's code to show off how realistic ray traced rendering can be.

Experiment 19.3. We're going to use POV-Ray (Persistence of Vision Ray Tracer), a freely downloadable ray tracer from `povray.org` [109]. Download and install POV-Ray. The executable is about 10 MB and there are Linux, Mac OS and Windows versions. It comes packaged with a nicely written tutorial and a reference manual. However, if for some reason you don't want to install POV-Ray, we have a compiled and rendered image file for you to simply open and compare with OpenGL's rendering.

If you have successfully installed POV-Ray, then open `sphereIn-BoxPOV.pov` from that program; if not, use any editor.

The code itself is fairly self-explanatory. It's written in POV-Ray's scene description language (SDL), which, unlike OpenGL, is *not* a library meant to be called from a C++ program – the SDL is stand-alone. We've obviously tried to follow the settings in our OpenGL program `sphereInBox1.cpp` as far as possible. The camera and a white light source are placed identically as in `sphereInBox1.cpp`. The red box, as in `sphereInBox1.cpp`, is an axis-aligned cube of side lengths two centered at the origin. It comprises six polygonal faces, each originally drawn as a square with vertices at $(-1, -1)$, $(1, -1)$, $(1, 1)$ and $(-1, 1)$ on the xy-plane, and then appropriately rotated and translated. The top face is opened to an angle of $60°$. Finally drawn is a green sphere of radius one. The material finishes are minimally complex, just enough to obtain reflection and a specular highlight on the sphere.

If you have installed POV-Ray, then press the Run button at the top; otherwise, open the output image file `sphereInBoxPOV.jpg` in our `Code` folder. Figure 19.16(a) is a screenshot. Impressive, is it not, especially if you compare with the output in Figure 19.16(b) of `sphereInBox1.cpp`? The inside of the box, with the interplay of light evident in shadows and reflections, is far more realistic in the ray-traced picture. *End*

So what gives? If ray tracing is so much more realistic than the synthetic-camera-based OpenGL pipeline, then why bother with the latter (or, for

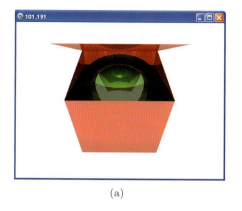 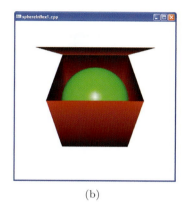

(a)   (b)

Figure 19.16: Ray tracing versus OpenGL: screenshot of (a) `sphereInBoxPOV.pov` (b) `sphereInBox1.cpp`.

that matter, write fat books about it)?! If you noticed how long it took to render the output of `sphereInBoxPOV.pov` – a few seconds at least on a decent desktop – then you have the answer. Ray tracing is *very very* computationally intensive. Intersection computations don't come cheap and they have to be done for every ray at every level in every ray tree generated, and there's one ray tree for each of maybe a million pixels. To even open the lid of the simple box of `sphereInBoxPOV.pov` in *real-time*, in the manner of `sphereInBox1.cpp`, is impossible for any modern-day desktop. Interactive animation, therefore, of remotely complex scenes (read games) is likely to remain beyond the reach of ray traced rendering for a while now.

On the other hand, still-life and movies, where either there is either no animation or it is all done off-line, are perfect applications for ray tracing. Computer animation in Hollywood is almost exclusively ray traced, individual frames of complex and highly realistic animated sequences sometimes taking hours each to render on special-purpose hardware (often clusters of computers called *render farms*). Incidentally, POV-Ray, too, has the capability to sequence an animation from individually generated frames (refer to their tutorial).

Remark 19.4. The holy grail of ray tracing research is, in fact, real-time ray-traced rendering.

A somewhat amusing, though fairly authentic, comparison of the synthetic-camera pipeline with ray tracing is to say that the former is "object-oriented", while the latter "screen-oriented". See Figure 19.17: on the left, objects (primitives) are dropped into the synthetic-camera pipeline to emerge rasterized, while on the right, pixels are dropped into the ray tracing pipeline to emerge colorized.

In case you enjoyed the little of POV-Ray that we showed and want to try your own hand at ray tracing, here are a couple of exercises.

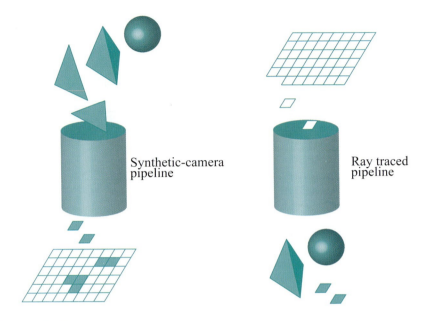

Figure 19.17: The (object-oriented) synthetic-camera pipeline versus the (screen-oriented) ray traced pipeline.

Exercise 19.6. (Programming) Use POV-Ray to generate a ray traced "combination" of `ballAndTorusReflected.cpp` and `ballAndTorus-PerspectivelyShadowed.cpp` with both shadows and reflections. It will be a single still shot, of course. Make sure that both the ball and torus, in addition to the floor and wall, are highly reflective, so that, in fact, they all reflect each other. You may need additional light sources to liven the scene.

Exercise 19.7. (Programming) Animate the opening of the lid of the box of `sphereInBoxPOV.pov` by generating a sequence of stills – one for every degree the lid turns would mimic `sphereInBox1.cpp`.

19.3 Radiosity

19.3.1 Introduction

Radiosity is a global lighting model which uses principles of heat transfer to track the dispersion of diffuse light around a scene.

It is quite often that a significant component of the light illuminating a scene is, in fact, multiply reflected diffuse light. For an example, consider a living room scene like the one depicted in Figure 19.18, populated with non-shiny furniture and lit by early morning rays. In such a setting there is little specular transport of light (i.e., by mirror-like reflection). Instead,

Figure 19.18: Living room lit mostly with diffuse light (courtesy www.freshome.com).

in addition to the ambient component, which is fairly constant throughout, there tends to be mostly diffuse activity. For example, light from the floor and walls reflect diffusely onto the shelves and furniture fabric. Even parts of the environment obscured from direct lighting, such as the floor between the sofas, are not in a well-defined shadow, but mildly illuminated by light reflecting off adjacent objects.

Such multiple diffuse reflections are not modeled by ray tracing at all. If you see again the design of Ray Tracer Version 3 in the previous section, a ray upon intersecting an object spawns one ray in the direction of transmission and another in the direction of perfect reflection. The direction of transmission is, of course, that of light passing through the object, while the direction of perfect reflection is that of the specular component of incident light.

Radiosity complements ray tracing by modeling diffuse illumination. Together, they can deliver highly realistic rendering, ray tracing emphasizing the shadows and highlights, and radiosity recording the softer lights.

The radiosity algorithm that we'll describe begins by dividing the scene into some number n of small flat, typically polygonal, *patches*, $P_i, 1 \leq i \leq n$, e.g., see Figure 19.19. A triangulated scene is, of course, automatically patchified. However, even then, one may want to refine certain triangles, or combine others to coarsen the given triangulation in response to two opposing forces in patchification: the smaller and more numerous the patches the more authentic is the lighting calculation; on the other hand, the time complexity of the radiosity algorithm, which is $O(n^2)$, increases rapidly with the number of patches. The best strategy is an adaptive one where a region over which light intensities are expected to vary rapidly is finely patchified, while one of steadier light levels more coarsely.

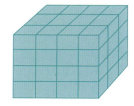

Figure 19.19: Patchified box.

Remark 19.5. Patchifying to compute radiosity is a *finite element* method.

The *brightness* or *radiosity* of a patch is the light energy per unit time per unit area leaving the patch, measured, typically, in a unit such as $joules/(second \times meter^2)$ (equivalent to $watts/meter^2$). The brightness varies with the frequency of the light in a manner that determines the perceived color of the patch; e.g., a red patch emits the greatest intensity at the red end of the visible spectrum. However, for simplicity's sake, we'll develop the theory assuming the brightness of a patch P_i as a single scalar value B_i, while a real implementation will have three different scalar values corresponding to the RGB brightnesses.

19.3.2 Basic Theory

Our starting point is the following equation which, in fact, holds for each i, $1 \le i \le n$,

$$A_i B_i = A_i E_i + R_i \sum_{j=1}^{n} F_{ji} A_j B_j \qquad (19.4)$$

where A_i is the area of patch P_i, B_i its brightness, E_i its emission rate, R_i the reflective scaling factor and, finally, F_{ji}, the so-called *form factor* between patches P_j and P_i.

The equation simply states that the amount of light energy leaving a patch P_i, which is the *area* × *brightness* term on the LHS, is equal to (a) the amount it emits as a source *plus* (b) the amount it reflects of incoming light, the two additive terms, respectively, on the RHS.

The value of (a) as the product of the patch's area and emission rate is clear. For (b), note first that the form factor F_{ji} denotes the fraction of the total light energy leaving patch P_j that reaches P_i. Therefore, $F_{ji}(A_j B_j)$ is, in fact, the amount of light leaving P_j for P_i; multiplied by P_i's reflective scaling factor R_i, this gives the amount of light from P_j actually reflected from P_i. Accordingly, the value of (b) is the summation of $R_i F_{ji}(A_j B_j)$ over all patches P_j, which is precisely the second term on the RHS of the equation above.

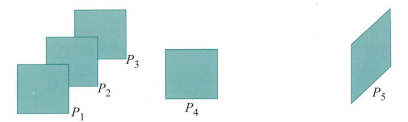

Figure 19.20: Form factor between patches depends on their respective orientation, the distance between them and if there is occlusion by other patches.

Shortly, we'll be computing form factors mathematically but it's easy to understand intuitively that F_{ji} depends on the orientation of P_j and P_i relative to each other, their distance and, further, if there is occlusion by intermediate patches between the two. For example, in Figure 19.20, the form factor between patches P_1 and P_2 and between P_2 and P_3 is high, because the two pairs are side by side and parallel, while that between P_1 and P_3 low because P_2 is between them. The form factor between P_4 and any one of P_1, P_2 and P_3 is low because of unfavorable orientation. The form factor between P_5 and any one of P_1, P_2 and P_3 is low as well because of distance.

Form factors will be seen to satisfy the *reciprocity equation*:

$$F_{ij}A_i = F_{ji}A_j \qquad (19.5)$$

Assuming this reciprocity for now, rewrite Equation (19.4) as

$$A_iB_i = A_iE_i + R_i \sum_{j=1}^{n} F_{ij}A_iB_j \qquad (19.6)$$

Dividing out A_i, one gets the *radiosity equations*:

$$B_i = E_i + R_i \sum_{j=1}^{n} F_{ij}B_j, \quad 1 \le i \le n \qquad (19.7)$$

which is a set of simultaneous linear equations in the brightnesses B_i, the latter being the only unknowns, provided we already have at hand the emissivities E_i and reflectivities R_i from a knowledge of material properties, and provided we can compute, as well, the form factors F_{ij} from the patch geometry.

Rearranging the radiosity equations as

$$(1 - R_iF_{ii})B_i - \sum_{1 \le j \le n,\ j \ne i} R_iF_{ij}B_j = E_i, \quad 1 \le i \le n$$

one can write them in matrix form as

$$\begin{bmatrix} 1 - R_1F_{11} & -R_1F_{12} & \cdots & -R_1F_{1n} \\ -R_2F_{21} & 1 - R_2F_{22} & \cdots & -R_2F_{2n} \\ \vdots & \vdots & \cdots & \vdots \\ -R_nF_{n1} & -R_nF_{n2} & \cdots & 1 - R_nF_{nn} \end{bmatrix} \begin{bmatrix} B_1 \\ B_2 \\ \vdots \\ B_n \end{bmatrix} = \begin{bmatrix} E_1 \\ E_2 \\ \vdots \\ E_n \end{bmatrix} \qquad (19.8)$$

Denoting

$$\mathbf{B} = \begin{bmatrix} B_1 \\ B_2 \\ \vdots \\ B_n \end{bmatrix}, \quad \mathbf{E} = \begin{bmatrix} E_1 \\ E_2 \\ \vdots \\ E_n \end{bmatrix}, \quad \text{and} \quad Q = \begin{bmatrix} R_1F_{11} & R_1F_{12} & \cdots & R_1F_{1n} \\ R_2F_{21} & R_2F_{22} & \cdots & R_2F_{2n} \\ \vdots & \vdots & \cdots & \vdots \\ R_nF_{n1} & R_nF_{n2} & \cdots & R_nF_{nn} \end{bmatrix}$$

a succinct matrix form of the radiosity equations is obtained from (19.8):

$$(I_n - Q)\mathbf{B} = \mathbf{E} \qquad (19.9)$$

where, of course, I_n is the $n \times n$ identity matrix.

Therefore, once we know how to do the following two tasks efficiently, we'll be in a position to practically implement the theory developed thus far:

(a) Compute form factors.

(b) Solve the radiosity equation (19.9) to determine patch brightnesses.

We discuss the two in the next sections.

19.3.3 Computing Form Factors

Consider two patches P_i and P_j. Even though the respective normal directions, n_i and n_j, remain constant over the patches, assumed flat, the amount of light from a point on P_i reaching a point on P_j, e.g., p_i and p_j in Figure 19.21, varies as the points vary, depending on the distance between them and the angle the line joining them makes with n_i and n_j, respectively. Therefore, one must integrate over the two patches – points being represented by infinitesimal areas – in order to determine the total light reaching P_j from P_i. In the figure, small triangles indicate the infinitesimal areas dp_i and dp_j at p_i and p_j, respectively.

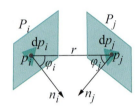

Figure 19.21: Computing form factors.

In fact, if patches are presumed to be Lambertian, i.e., they reflect light uniformly in all directions from every surface point, then it can be proved that the form factor F_{ij}, the fraction of the total light emanating from F_i that reaches F_j, is given by:

$$F_{ij} = \frac{1}{A_i} \int\limits_{p_i \in P_i} \int\limits_{p_j \in P_j} v_{ij} \frac{\cos \phi_i \cos \phi_j}{\pi r^2} \, dp_j \, dp_i \qquad (19.10)$$

where, A_i is the area of P_i, ϕ_i and ϕ_j are the angles between the segment $p_i p_j$ and the normals n_i and n_j, respectively, r is the length of $p_i p_j$, and v_{ij} is a Boolean which is 1 if p_j is visible from p_i and 0 otherwise.

Exercise 19.8. It is easy now to deduce the reciprocity equation (19.5) from the formula for a form factor. Do so.

Except for the simplest cases, the double integral in Formula (19.10) is impossible to compute exactly. The *hemicube method*, however, is a clever approximation algorithm developed by Cohen and Greenberg [24], which takes advantage of fast hardware-based z-buffers.

Write formula (19.10) as

$$F_{ij} = \frac{1}{A_i} \int\limits_{p_i \in P_i} \left(\int\limits_{p_j \in P_j} v_{ij} \frac{\cos \phi_i \cos \phi_j}{\pi r^2} \, dp_j \right) dp_i \qquad (19.11)$$

The inner integral can be imagined to be the form factor between p_i – or, more precisely, an infinitesimal patch dp_i containing p_i as in Figure 19.21 – and P_j, while F_{ij} itself is the average of these form factors over points of P_i.

The first assumption of the hemicube method is that the form factor between p_i and P_j does not vary significantly as p_i varies over P_i, which is justified if the distance between P_i and P_j is large in comparison to their respective sizes. In such a case, the average F_{ij} can be approximated by a single value, say that of the form factor at a fixed point p_i located centrally in P_i; precisely,

$$F_{ij} = \int_{p_j \in P_j} v_{ij} \frac{\cos \phi_i \cos \phi_j}{\pi r^2} \, dp_j \qquad (19.12)$$

obtained from assuming the inner integral of (19.11) to be constant over P_i.

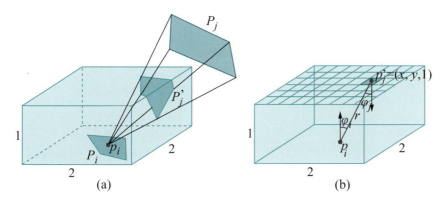

Figure 19.22: (a) Projecting a patch onto a hemicube (b) Computing the delta form factor.

The next assumption is that this p_i-P_j form factor itself can be approximated by replacing P_j with its projection P'_j on an (imaginary) hemicube – half a cube – with its base lying on the plane of P_i and centered at p_i. Figure 19.22(a) shows such a hemicube, being half of a cube of side lengths 2. The justification for this assumption is as follows.

As P_i is Lambertian, light from each of its points emanates uniformly in all directions, which means that the light from p_i uniformly illuminates a hemisphere with its base along P_i and center at p_i. So the projection of P_j onto such a hemisphere would be an "ideal" replacement for P_j. However, for the sake of computational advantages, which will be perceived momentarily, the hemi*sphere* is replaced with a hemi*cube* centered at p_i and of dimensions indicated in Figure 19.22. So we get the following approximation from (19.12) by replacing the patch P_j with its projection P'_j on the replacement hemicube:

$$F_{ij} = \int_{p'_j \in P'_j} v_{ij} \frac{\cos \phi_i \cos \phi_j}{\pi r^2} \, dp'_j \qquad (19.13)$$

The hemicube algorithm, next, discretizes the computation of the preceding integral by dividing the hemicube into a grid of squares, called (suggestively, as we shall see) pixels, and treating each as an infinitesimal area dp'_j. Figure 19.22(b) shows a division into pixels of the top face. This process effectively replaces the integral with a finite sum.

It's in the evaluation of this sum that the beauty of the hemicube method lies. Here's what it asks: for each of the five faces of the hemicube – top and four sides – render the scene with the eye at p_i and the front face of the viewing frustum coinciding with that hemicube face. Presto! Screen pixels now correspond to pixels on the hemicube face so that occlusion – the pesky v_{ij} in the integral – is automatically taken care of by means of the z-buffer.

To determine all v_{ij}, then, color code each patch – with, typically, 2^{24} colors to choose from,' there should be plenty to assign a unique one to each patch – and render the scene with depth testing to find the screen pixels of a given color, which determines the projection of the patch of that color on a hemicube face. For example, the projection P'_j of patch P_j in Figure 19.22(a) has a part on the top and one on the side of the hemicube. If P_j were coded, say, red, then the red pixels, when the scene is rendered with the hemicube top as the viewing face, comprise the part of the top not occluded (in the figure there happens to be no occlusion of P_j at all).

Consider, next, a single pixel belonging to patch P'_j, lying on the top face and centered at the point $p'_j = (x, y, 1)$, e.g., the darker one in Figure 19.22(b). We have $r = \sqrt{x^2 + y^2 + 1}$, $\phi_i = \phi_j$ and $\cos \phi_i = \cos \phi_j = 1/r$. Moreover, the area of the pixel is $\frac{4}{wh}$, where the screen size is w pixels $\times$ h pixels. Therefore, the contribution of the top face of the hemicube to the integral (19.13) is approximated by the sum

$$\sum_{\substack{\text{pixel on top face} \\ \text{of color code of } P_j}} \frac{(1/r)(1/r)}{\pi r^2} \frac{4}{wh} = \frac{4}{\pi wh} \sum_{\substack{\text{pixel on top face} \\ \text{of color code of } P_j}} \frac{1}{(x^2 + y^2 + 1)^2}$$

(19.14)

Exercise 19.9. Write sums analogous to (19.14) for the contributions of each of the four side faces of the hemicube to the integral (19.13).

The implementation of the hemicube algorithm should now be clear. First, color code patches. Next, for each patch, and for each of the five faces of the hemicube centered at the middle of the patch, render the scene using that particular face as the viewing face and then process the resulting screen by tallying the contribution of each pixel according to its color. The contribution of a pixel to the form factor between p_i and the patch of the pixel's color, e.g.,

$$\frac{4}{\pi(x^2 + y^2 + 1)^2 wh}$$

for a pixel on top of the hemicube, is called a *delta form factor*. Accordingly, the computation of each form factor is reduced to the process of incrementing it from zero, by a delta form factor at each step, as the screen is swept row by row, pixel by pixel, for each of the five renderings.

19.3.4 Solving the Radiosity Equation to Determine Patch Brightnesses

The final piece before we can practically implement the radiosity method is an efficient algorithm to solve the radiosity equation (copied from (19.9))

$$(I_n - Q)\mathbf{B} = \mathbf{E}$$

to determine the *patch brightness vector* $\mathbf{B}$, where the matrix

$$I_n - Q = \begin{bmatrix} 1 - R_1 F_{11} & R_1 F_{12} & \cdots & R_1 F_{1n} \\ R_2 F_{21} & 1 - R_2 F_{22} & \cdots & R_2 F_{2n} \\ \vdots & \vdots & \cdots & \vdots \\ R_n F_{n1} & R_n F_{n2} & \cdots & 1 - R_n F_{nn} \end{bmatrix}$$

Trying to solve the preceding equation by writing $\mathbf{B} = (I_n - Q)^{-1}\mathbf{E}$ and straightforwardly inverting $I_n - Q$ would be prohibitively expensive as the computation involved is $O(n^3)$, where the number n of patches is typically in the thousands. However, certain properties of the matrix $I_n - Q$, derived from properties of the form factors, lead to an efficient method to approximate its inverse.

First,

$$F_{ii} = 0, \quad 1 \le i \le n$$

because patches, being flat, cannot self-reflect. Moreover, we can assume as well that

$$\sum_{j=1}^{n} F_{ij} = 1, \quad 1 \le i \le n$$

which means all light leaving any given patch strikes other patches, by closing off the environment with black (i.e., non-reflective) patches. These properties of the form factors, together with that each reflectivity R_i is at most 1, imply that the principal diagonal of $I_n - Q$ consists of all 1's and that the sum of non-diagonal entries in any row of $I_n - Q$ is at most 1. One can then prove that

$$(I_n - Q)^{-1} = I_n + Q + Q^2 + \ldots$$

where the series on the right converges (which might remind the reader of the power series expansion $(1 - x)^{-1} = 1 + x + x^2 + \ldots$, which converges if $|x| < 1$). Therefore,

$$\mathbf{B} = (I_n - Q)^{-1}\mathbf{E} = \mathbf{E} + Q\mathbf{E} + Q^2\mathbf{E} + \ldots$$

allowing the patch brightness vector $\mathbf{B}$ to be approximated to arbitrary accuracy by adding sufficiently many terms of the series on the right. Care needs still to be taken, as a simple-minded computation of the term $Q^k\mathbf{E}$ by repeatedly multiplying by Q is nearly $O(n^3)$ again. A simple observation, however, helps cut the cost.

Denote successive partial sums of the series $\mathbf{E} + Q\mathbf{E} + Q^2\mathbf{E} + \ldots$ by $\mathbf{B}_0, \mathbf{B}_1, \ldots$. In other words, $\mathbf{B}_0 = \mathbf{E}$, $\mathbf{B}_1 = \mathbf{E} + Q\mathbf{E}$, $\mathbf{B}_2 = \mathbf{E} + Q\mathbf{E} + Q^2\mathbf{E}$ and so on. Then we have the recurrence

$$\mathbf{B}_{k+1} = \mathbf{E} + Q\mathbf{B}_k, \quad k \geq 0$$

so that each successive term of the sequence $\mathbf{B}_0, \mathbf{B}_1, \mathbf{B}_2, \ldots$ of partial sums converging to $\mathbf{B}$ can be computed from the previous by a matrix-vector product and a vector-vector addition, the two operations together being of $O(n^2)$ complexity. In fact, if the matrix Q is sparse, likely if the form factor between patches at a distance greater than some threshold value are set to 0, then the complexity may be closer to linear.

Exercise 19.10. Consider the operator Ψ that acts on n-vectors by

$$\Psi(X) = \mathbf{E} + QX$$

Prove that the solution $\mathbf{B}$ of the radiosity equation is a *fixed point* of Ψ, i.e., a vector X such that $\Psi(X) = X$.

The preceding exercise leads to *Jacobi's iterative method* to approximate the fixed point $\mathbf{B}$ of Ψ as follows. Choose arbitrarily a start vector X'. Then repeatedly apply Ψ to X' to obtain a sequence $X', \Psi(X'), \Psi(\Psi(X')), \ldots$. Properties of the radiosity equations guarantee that this sequence converges to the unique fixed point of the operator Ψ. We'll not discuss the theory underlying Jacobi's method any further ourselves, but the interested reader is referred to Hageman & Young [62].

Exercise 19.11. Prove that Jacobi's iterative method to approximately determine the solution $\mathbf{B}$ of the radiosity equation, using a *zero* start vector, is precisely equivalent to the power series method of approximating $\mathbf{B}$.

Remark 19.6. A useful physical insight into the sequence $\mathbf{B}_0, \mathbf{B}_1, \mathbf{B}_2, \ldots$ of partial sums converging to $\mathbf{B}$ is that the first term represents only emitted light, the second emitted light together with diffuse light coming to the eye after a single reflection, the third emitted light together with diffuse light after at most two reflections and so on.

In practical terms, this means that using the sequence $\mathbf{B}_0, \mathbf{B}_1, \mathbf{B}_2, \ldots$ as brightness vectors illuminates the scene in an increasingly authentic manner, which leads to the cost-saving technique of executing each iteration to compute $\mathbf{B}_i$ only on demand, called *progressive refinement*.

19.3.5 Implementing Radiosity

Figure 19.23 shows the four steps of the radiosity algorithm. The first three are *view-independent* and may be pre-computed for a scene whose geometry does not change. The last rendering step, of course, depends on the location of the viewer. To reduce aliasing artifacts at patch borders, instead of rendering each patch with its computed brightness, each *vertex* is assigned a brightness computed from its adjacent patches, e.g., a weighted average. Subsequently, vertex colors are interpolated through each patch via Gouraud shading.

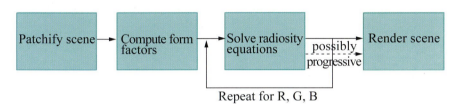

Figure 19.23: The radiosity algorithm.

Remark 19.7. As the first three steps of the radiosity algorithm are view-independent, while the last, even though dependent on the viewer's location, is not particularly computationally intensive, radiosity can be efficiently incorporated into a real-time walk-through of a static scene, e.g., a building interior.

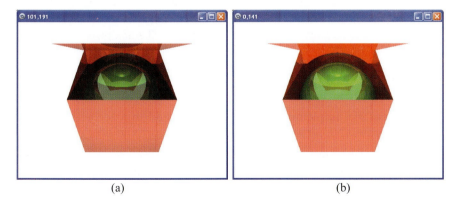

Figure 19.24: Without and with radiosity: screenshot of `sphereInBoxPOV.pov` with (a) radiosity disabled (b) radiosity enabled.

Experiment 19.4. Run again `sphereInBoxPOV.pov`. Then run again after uncommenting the line

```
global_settings{radiosity{}}
```

at the top to enable radiosity computation with default settings. The difference is significant, is it not?

Figure 19.24(a) is the output without radiosity (or see it separately as the image file `sphereInBoxPOV.jpg` in our `Code` folder), while Figure 19.24(b) is the output with radiosity (`sphereInBoxPOVWithRadiosity.jpg` in `Code`). There clearly is much more light going around inside the box in the latter rendering.   **End**

Exercise 19.12. If the lighting in a scene changes, then which steps of the radiosity algorithm need to be redone? How about if the geometry changes, e.g., with a ball looping in and out of a torus?

19.4 BSP Trees

The BSP (Binary Space Partitioning) tree algorithm for hidden surface removal, invented by Fuchs, Kedem and Naylor [47], is an implementation of the so-called *painter's method*. The painter's method draws objects from back to front toward the eye, each new object possibly drawn over earlier ones, which automatically engenders hidden surface removal. See Figure 19.25.

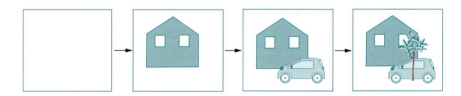

Figure 19.25: The painter's method draws the scene "toward" the eye: first the house, then the car and, finally, the tree.

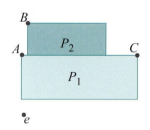

Figure 19.26: Vertex A of P_1 is closer to eye e than vertex B of P_2, while B is closer than C of P_1.

How does one implement the painter's method in a scene specified as a set of polygons? One would obviously like to sort the polygons relative to the eye e, so that those farther away appear earlier in the sorted order than nearer ones, and then draw them in this order. The first problem that one runs into along this line of thinking is determining which of two given polygons, say P_1 and P_2, is "farther from" e. Measuring the distance from e to vertices of P_1 and P_2 is not a workable approach, as there may be one vertex of P_1 that is closer to e than some vertex of P_2, while another vertex of P_1 is farther from e than one of P_2, e.g., as in Figure 19.26.

The figure itself suggests a possible solution. If one imagines P_1 and P_2 as vertical rectangles lying on parallel planes p_1 and p_2, respectively, then evidently P_2 lies *behind* p_1 in that P_2 lies on the side of p_1 opposite the eye (i.e., in the half-space of p_1 not containing the eye). Clearly, then, P_2 should be taken to be farther than P_1 and drawn before P_1, because it cannot obscure P_1. Unfortunately, for arbitrary polygons P_1 and P_2, it may not be the case that the plane of either separates the other from the eye, as, e.g., in Figure 19.27, where the plane of each intersects the other.

Suppose, for the moment, that a scene comprising a set S of polygons does, in fact, contain at least one, say P, such that every polygon of S, other than P, lies entirely on one side or the other of the plane p containing P (i.e., each polygon of S, other than P, lies entirely in one or other half-space of p). In this case, we say that P *splits* S. Suppose, as well, that the eye e does not lie on p but to one side as well. See Figure 19.28, where the polygon P splits the set S into two, S' and S'', and the eye e lies on the same side of p as S''. Clearly, then, a legitimate drawing order has the polygons of S' first (drawn in some appropriate order among themselves), then P and, finally, the polygons of S'' (again in some appropriate order).

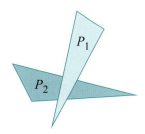

Figure 19.27: Each of P_1 and P_2 intersects the plane containing the other – the planes themselves are not drawn.

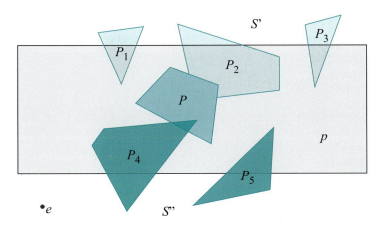

Figure 19.28: The polygon P splits the set of six polygons. P's plane p is drawn a light gray. The three to one side of p, viz., P_1, P_2 and P_3, form the subset S', while the two, P_4 and P_5, on the other form S''. The eye e lies on the same side of p as S''.

If the equation of the plane p is $ax + by + cz + d = 0$, then the sign of the quantity $ax + by + cz + d$, positive or negative, determines the side of p that a point $[x\ y\ z]^T$ lies. For the sake of compact notation we'll denote by h a function whose sign determines the side of p that a primitive lies (assuming, of course, that the primitive does lie wholly to one side or the other). For example, $h(e)$ and $h(P_1)$ have opposite signs in Figure 19.28.

Below, then, is a first cut at an algorithm to draw the polygons in S on the assumption that a splitting polygon P of S can indeed be found. Moreover, the recursive **draw()** calls are based on the premise that every time a subset of S is partitioned by a splitting polygon into two smaller subsets, a splitting polygon can *again* be found in either (provided the size of the subset is greater than 1; otherwise, of course, there is no need to partition further). In other words, we suppose that we can keep partitioning by means of splitting polygons until we are down to singletons.

Drawing Algorithm Version 1

```
void draw(S, e) // Input S, a set of polygons, and e, eye location.
{
   if (|S| == 0) ; // Nothing to do.
   else if (|S| == 1)
   {
      P = the one member of S;
      draw(P);
   }
   else
   {
      P = a splitting polygon of S; // Can be found by assumption.
      S'  = polygons Q of S - P such that h(Q) > 0;
      S'' = polygons Q of S - P such that h(Q) < 0;
      if ( h(e) > 0 )
      {
         draw(S'', e);
         draw(P);
         draw(S', e):
      }
      else // h(e) < 0
      {
         draw(S', e);
         draw(P);
         draw(S'', e):
      }
   }
}
```

In the scenario of the preceding algorithm, the splitting polygons can be naturally arranged in the form of a binary tree, called a *BSP tree*: the children (if they exist) of each splitting polygon P being the splitting polygons of the two subsets that result from partitioning by P. A BSP tree node corresponding to a particular splitting polygon stores its data via a pointer to the polygon itself in the list of scene primitives, in addition to the obligatory left and right child pointers. Figure 19.29 shows an example of a set S of seven polygons lying on parallel planes and a corresponding BSP tree (only one pointer has been drawn from a tree node to a polygon).

Observe that the BSP tree of a given set of polygons need not be unique, but depends on the splitting polygon selected from each subset (in the situation of Figure 19.29 any member of a subset of S splits it and we have been intentionally arbitrary in making choices). Moreover, the BSP tree, once constructed, is a static data structure – provided the scene doesn't change – independent of the eye. It can, therefore, be pre-computed. It's size is linear in the number of polygons.

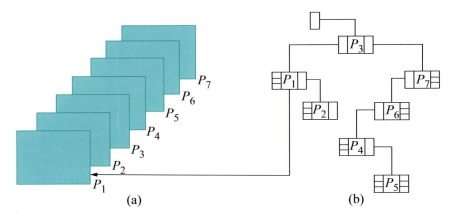

Figure 19.29: (a) Seven polygons on parallel planes (b) A corresponding BSP tree (only one polygon pointer drawn to avoid clutter).

Here is recursive pseudo-code to construct a BSP tree for an input set S of polygons, keeping the earlier assumption that splitting polygons can be repeatedly found:

BSP Tree Construction Algorithm Version 1

```
pointer makeBSPTree(S) // Input S, a set of polygons.
{
    allocate empty BSP tree node  root;

    if ( |S| == 1 )
    {
        P = the one member of S;
        root.value = P;
        root.left_child = NULL;
        root.right_child = NULL;
    }
    else
    {
        P = a splitting polygon in S; // Can be found by assumption.
        root.value = P;
        S'   = polygons Q of S - P such that h(Q) > 0;
        S''  = polygons Q of S - P such that h(Q) < 0;
        if ( |S'| == 0 ) root.left_child = NULL;
        else root.left_child = makeBSPTree(S');
        if ( |S''| == 0 ) root.right_child = NULL;
        else root.right_child = makeBSPTree(S'');
    }
```

```
    return pointer to root;
}
```

Returning to the question of drawing, we ask the reader in the next exercise to devise an algorithm to draw a scene by choosing splitting polygons given a BSP tree.

Exercise 19.13. Given input a pointer T to the root of a BSP tree constructed for the set S of polygons comprising a scene – assume that we were indeed able to split successive subsets until the tree was complete – and the location of the eye e, write a routine `draw(T, e)` to draw the scene by modifying Drawing Algorithm Version 1. For future reference, call your routine the BSP Tree Based Drawing Algorithm.

Exercise 19.14. What is the complexity of the BSP Tree Based Drawing Algorithm – does it depend on the height of the BSP tree or its number of nodes?

Hint: The drawing algorithm traverses the BSP tree, visiting each node once.

Exercise 19.15. In OpenGL, for instance, not only is the location of the camera given, but the direction in which it's pointed and its up direction as well (think `gluLookAt()`). Where would these come into play in a drawing routine which uses a BSP tree?

Hint: They are needed at the time of drawing a polygon (e.g., `draw(P)` in Version 1 of the drawing algorithm) and *not* when traversing the BSP tree. The reason is that if, say, the plane of a polygon P_1 separates the eye e from another polygon P_2, then P_2 cannot obscure P_1 no matter how a camera at e is pointed or its top aligned, implying that P_2 can be drawn ahead of P_1 regardless of these parameters.

We come now to the final piece of the puzzle. How do we assure the existence of a splitting polygon in each subset as we successively partition them? *Answer*: We cannot. Instead, we'll resort to some amount of violence, if need be, to members of the subset in order to be able to extract a splitting polygon. The idea is actually simple. Given a set S of polygons, choose one (arbitrarily), say P. If another polygon Q of S intersects the plane p of P, then subdivide Q into subpolygons that do not intersect p (mind that touching is not considered intersecting in the realm of BSP trees). We'll call this the process of *subdivision of Q by P*.

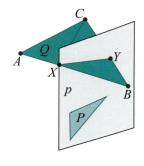

Figure 19.30: Subdividing Q by P.

For example, $Q = ABC$ in Figure 19.30, which intersects the plane p of P in the straight segment XY, can be subdivided into the quadrilateral $AXYC$ and the triangle YXB. The subdivision of Q by P need not be unique as long as the outcome is a set of polygons that don't intersect p. For example, if drawing primitives are triangles, as in OpenGL, the subdivision of Q in the figure into the three triangles AXC, CXY and YXB might be preferable. If Q does not intersect p, then its subdivision by P is, of course, a null process.

Accordingly, if every polygon of the set S, other than P, is subdivided by P, then the result is a *refinement* of S, in that each polygon in this collection is a subpolygon of some polygon of S. Evidently, from the way it was made, this refinement is split by P. From the point of view of drawing the originally specified scene, there is no loss in refining, because the union of the polygons in the refinement is exactly the same as that of those in the original set. It's just that the number of polygons has increased. Care must be taken, however, that the new polygons are consistent with the ones from which they are derived with respect to orientation, normal direction and so forth.

We obtain now our final version of the BSP tree construction algorithm, which makes no assumption on the polygon set S, by a simple modification of the first version to include refinement.

BSP Tree Construction Algorithm Version 2

```
pointer makeBSPTree(S) // Input S, a set of polygons.
{
   allocate empty BSP tree node  root;

   if ( |S| == 1 )
   {
      P = the one member of S;
      root.value = P;
      root.left_child = NULL;
      root.right_child = NULL;
   }
   else
   {
      P = a random polygon in S;
      root.value = P;
      S    = refinement of S obtained by subdividing by P each
               polygon of S other than P;
      S'   = polygons Q of S - P such that h(Q) > 0;
      S''  = polygons Q of S - P such that h(Q) < 0;
      if ( |S'| == 0 ) root.left_child = NULL;
      else root.left_child = makeBSPTree(S');
      if ( |S''| == 0 ) root.right_child = NULL;
      else root.right_child = makeBSPTree(S'');
   }

   return pointer to root;
}
```

This algorithm, together with the BSP Tree Based Drawing Algorithm, answer to Exercise 19.13, completes a routine for drawing with hidden surface removal.

Exercise 19.16. Construct a BSP tree for the "cycle" of three triangles

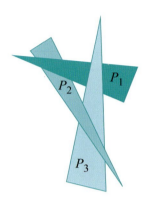

disposed as in Figure 19.31, each intersecting the plane of the other two, subdividing as necessary.

Remark 19.8. An implementation of BSP trees would have to take into account additional practical issues that we have not discussed, e.g., what to do if the eye e lies *on* the plane of a polygon in the tree (we have been tacitly assuming that it always lies to a side) and how to prevent *slivers* from arising when subdividing a polygon (e.g., if B is very close to p in Figure 19.30, though still on the side opposite to that of A and C, then the triangle YXB is undesirably long and narrow).

Remark 19.9. The complexity of the BSP tree returned by the second version of the construction algorithm depends on the choice of P used to subdivide the other members of S. However, the complexity of determining an optimal sequence of subdividing polygons is generally not worth the improvement in the tree itself. A random choice of the subdividing polygon seems to work best in practice.

Remark 19.10. BSP trees were used extensively for visibility determination, particularly in flight simulators, in the 60's and 70's (flight simulators were the killer app for high-end real-time rendering systems at that time). However, through the 1980's, BSP trees continued to be superseded by the simple-minded z-buffer by virtue of the latter's sheer speed, derived from being implemented on dedicated hardware (the first graphics cards were released in 1981).

However, the space-partitioning technique that BSP trees implement has numerous other applications: the frustum culling technique we studied in Chapter 6 is an example.

19.5   Summary, Notes and More Reading

In this chapter we went into particularly gory detail about the synthetic-camera pipeline that OpenGL implements, the fixed-functionality variant in particular. The reader should now be in a position to even implement a barebones version of her own. The synthetic-camera pipeline is based on a local illumination model. We were introduced as well to two global models, that of ray tracing and radiosity, and saw how much more realism they afford than the synthetic camera, though at hugely more computation cost. We also learned about BSP trees, a classical technique which integrates hidden surface removal into the rendering process.

The seminal work on ray tracing was by Appel [4] and Whitted [143], and on radiosity by Goral [54]. The book by Jim Blinn [16], a CG pioneer, has several insightful articles, written in his particularly entertaining style, on various pipeline-related topics. Segal-Akeley [123] is a must-read high-level overview of the OpenGL pipeline written by two members of the original design team and, of course, the red book itself is a canonical source. Good textbooks to pick up for more advanced reading about ray tracing and radiosity are those by Akenine-Möller, Haines & Hoffman [1], Buss [21] and Watt [142]. Cohen & Wallace [25] and Sillion & Puech [126] are especially about radiosity.

CHAPTER 20

Programmable Pipelines

rogrammers mutiny! We're going to throw off our shackles and take over the engine room!

The OpenGL Shading Language (abbreviated as GLSL, also called GLslang) was developed originally by the OpenGL ARB (Architecture Review Board) to give programmers direct control over parts of the graphics processing pipeline. Included in the core of OpenGL 2.0, released in 2004, it represented the first major upgrade of OpenGL since its creation in 1992. The way the GLSL operates is to allow the programmer to write programs, called shaders, to supplant parts of the graphics pipeline formerly of fixed-functionality.

Historically, the GLSL evolved as a response to the increasing capabilities of graphics cards (also called graphics processing units, or GPUs) and the need to expose these capabilities to the application programmer. Before the standardization of the GLSL, a programmer had to write code in hardware-specific assembly language to access individual GPU features – a difficult and inefficient task at best. Just as high-level programming languages like C evolved from assembly in order to hide low-level and hardware-specific calls from the developer and give her a structured and stable environment, so did the GLSL.

As a language, the GLSL itself is based on C, so coding will not be a problem for us. In addition to most of C's functionality, the GLSL necessarily has dedicated functions and variables for the shaders to interact with each other, as well as with the main OpenGL program to which they are attached (and, thereby, with the rest of the pipeline).

The goal of this chapter is not an extensive coverage of the GLSL, but a thorough introduction. We begin in Section 20.1 with an overview of the programmable pipeline, learn how to attach shaders to an OpenGL program and run over the data types of the GLSL. Section 20.2 describes how a program communicates with its shaders, shipping them data from the fixed-functionality parts of the pipeline, and how the shaders communicate between themselves. The communication interface between these entities is implemented by means of specially qualified variables. It is precisely because shaders are programmable and at the same time have access to almost all hitherto opaque data in the pipeline that the GLSL is so powerful.

Per-pixel lighting – versus the per-vertex lighting of the first-generation fixed pipeline – is a popular practical application of the GLSL that we'll see in Section 20.3. In Section 20.4 we learn how to import and manipulate textures in the programmable pipeline. We conclude in Section 20.5.

20.1   GLSL Basics

Shaders are the programs written by the user to replace fixed-functionality. One each of two kinds of shaders can be attached to an OpenGL program – *vertex shaders* and *fragment shaders*. Both kinds are written in the same C-like GLSL but target different parts of the graphics pipeline. The vertex shader operates on incoming vertex data (including coordinates, normal values, colors, etc.) before handing over its output for clipping and then rasterization. The fragment shader, on the other hand, operates on fragments in the raster before passing them on to the bottom stage of the classical pipeline, which remains intact, to perform its own per-fragment operations such as the scissor, alpha, stencil and depth tests, as well as blending. Figure 20.1 is a simplified diagram. Observe that texturing is within the purview of the fragment shader, which is justified technically as it is a process of combining texels with fragments in the raster.

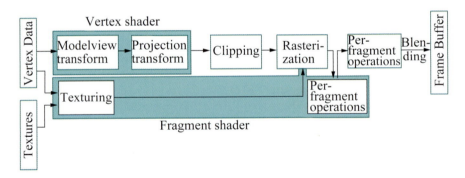

Figure 20.1: GLSL pipeline simplified: shaded regions indicate parts of the classical pipeline now programmable.

A vertex shader operates on each vertex coming down the pipeline, while a fragment shader on each fragment in the raster. Once attached, shaders do have minimum responsibilities that they have to discharge and that cannot be left to fixed parts of the pipeline. The vertex shader, for example, must output at least the vertex's coordinates in world space. It cannot, say, choose to operate only on a vertex's normal values, leaving coordinate computation to fixed-functionality (though, of course, it is free to just mimic the fixed-functionality computation of vertex coordinates and not do anything more). Likewise, the fragment shader must at least assign each fragment a color or decide to discard it from the pipeline altogether.

20.1.1 Attaching Shaders

Time now to go see what we have been talking about. In other words, let's code. Our guinea pig will be our trusty workhorse from way back when – `square.cpp`.

$\mathbb{E}$xperiment 20.1. Fire up `redSquare.cpp` in the folder `Code/GLSL/Red-Square`.

Note: For how to set up the environment to run GLSL programs see Appendix B. Each of our GLSL programs is in a similarly named folder in the `GLSL` subdirectory of `Code`, with two accompanying shader files. Make sure, when running a GLSL program, to keep it in the same directory as its two shader files.

Now, `redSquare.cpp` is *exactly* `square.cpp` with the *barest* minimum amount of code added to be able to attach a vertex shader, called `passThrough.vs`, and a fragment shader, called `red.fs`. The output is a red square in the OpenGL window, as in Figure 20.2(a). $\mathbb{E}$nd

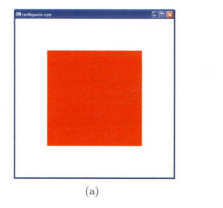
(a)

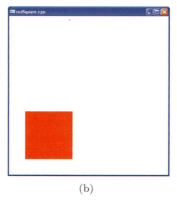
(b)

Figure 20.2: Screenshots: (a) `redSquare.cpp` (b) Experiment 20.2.

Before we get to the shaders, let's see what, in fact, has been added to the program itself to "GLSLify" it. Comments starting with "NEW" indicate new lines of code.

Linking in the OpenGL Extension Wrangler Library (GLEW), as we have done with directives at the top of the program, is advisable for the run-time support of OpenGL extensions that GLEW provides.

The integer globals `programHandle`, `vertexShaderHandle` and `fragment-ShaderHandle` are to store references to program, vertex shader and fragment shader objects, respectively. The routine `readShader()` is a generic routine to read external text files, and not GLSL-specific.

The `setShaders()` routine is the critical one that initializes the two shaders. Let's go through it line by line. The commands

```
char* vertexShader = readShader(vertexShaderFile);
char* fragmentShader = readShader(fragmentShaderFile);
```

read the two shader source files, which are simply text files, and store them internally as character strings. Next,

```
programHandle = glCreateProgram();
vertexShaderHandle = glCreateShader(GL_VERTEX_SHADER);
fragmentShaderHandle = glCreateShader(GL_FRAGMENT_SHADER);
```

create an empty program object and two empty shader objects, returning references to them. The pair of statements

```
glShaderSource(vertexShaderHandle, 1,
                      (const char**) &vertexShader, NULL);
glShaderSource(fragmentShaderHandle, 1,
                      (const char**) &fragmentShader, NULL);
```

attaches the vertex and fragment shader sources to the respective shader objects. The two statement pairs

```
glCompileShader(vertexShaderHandle);
glCompileShader(fragmentShaderHandle);

glAttachShader(programHandle, vertexShaderHandle);
glAttachShader(programHandle, fragmentShaderHandle);
```

compile the vertex and fragment shader source code strings and attach the resulting shader objects to the program object. Finally,

```
glLinkProgram(programHandle);
glUseProgram(programHandle);
```

links the program and creates an executable which is installed into the current rendering state.

The `main()` routine actually sets the programmable pipeline into motion by initializing GLEW and invoking the shaders. That's it. That is all the additional overhead to attaching shaders – essentially, the `setShaders()` routine, which is the same across all our GLSL programs.

Minimal Shaders

Now let's see what the shaders associated with redSquare.cpp do. Apparently not much. The vertex shader is passThrough.vs.

Note: Our vertex shaders all have the extension .vs, while fragment shaders the extension .fs, but this is not a standard, and they are treated simply as text files by the programming environment.

The only operational statement

gl_Position = gl_ProjectionMatrix * gl_ModelViewMatrix * gl_Vertex

in passThrough.vs is fairly intuitive. As each vertex comes down the pipeline, its coordinate vector is multiplied by the modelview and projection matrices and the result returned in a position vector. This is precisely what the fixed pipeline does and, as this particular shader does no more, it is called a *pass-through* vertex shader.

Note: The naming convention is gl_*state* for a variable containing the program state set by gl*state**, e.g., gl_Vertex for the variable set by glVertex*().

As indicated by comments in passThrough.vs, the *derived* matrix gl_ModelViewProjectionMatrix is equal to the product

gl_ProjectionMatrix * gl_ModelViewMatrix

and can replace it. An even more optimized option, if it is desired only to multiply a vertex's coordinates by the projection and modelview matrices, is to use the special command

gl_Position = ftransform()

which guarantees the same result as fixed-functionality.

The fragment shader red.fs is equally simple-minded, setting all fragment colors to red with the single statement

gl_FragColor = vec4(1.0, 0.0, 0.0, 1.0)

20.1.2 Data Types

It seems, even from the very concise shaders passThrough.vs and red.fs attached to redSquare.cpp, that the GLSL has vector and matrix data types (e.g., vec4) to manipulate state variables such gl_Vertex, gl_Position, gl_ProjectionMatrix and such. This is correct.

To begin with, the GLSL has the classical C types

      float   int   bool

to declare a single floating point, integer and Boolean quantity, respectively. In addition, the GLSL can declare 2-, 3- and 4-component floating point vectors, respectively, with

| vec2 | vec3 | vec4 |

Likewise, integer vectors are declared with

| ivec2 | ivec3 | ivec4 |

and Boolean vectors with

| bvec2 | bvec3 | bvec4 |

Matrices, which are always floating point and square, of sizes 2×2, 3×3 or 4×4 are declared, respectively, with

| mat2 | mat3 | mat4 |

The final set of special non-C data types, called *samplers*, are handles to access textures:

| sampler1D | sampler2D | sampler3D |
| samplerCube | sampler1DShadow | sampler2DShadow |

The first three declare handles to 1D, 2D and 3D textures, respectively; the fourth a handle to a cube-mapped texture; the fifth and sixth handles to 1D and 2D depth textures, respectively, with comparison.

Finally, the complex data types structures (`struct`) and arrays (`[ ]`), functioning just as in C, are available as well.

20.1.3 Swizzling

Components of a vector can be selected and even rearranged and duplicated using the *swizzle* operator ".". Vector components are accessed using the following three sets of names:

| x, y, z, w | r, g, b, a | s, t, p, q |

For example, x, r and s each denotes the first component; y, g and t each the second component; and so on. The sets cannot be mixed though. The following snippet illustrates how the swizzle operator behaves on the right-hand side – read the comments:

```
vec4 pos1 = vec4(1.0, 2.0, 3.0, 4.0);
vec4 pos2 = pos1.yxzw; // Rearrangement: pos2 = (2.0, 1.0, 3.0, 4.0)
vec4 pos3 = pos1.rrba; // Duplication: pos3 = (1.0, 1.0, 3.0, 4.0)
vec4 pos4 = vec4(pos1.xyz, 5.0); // pos4 = (1.0, 2.0, 3.0, 5.0).
vec2 pos5 = pos1.xy; // pos5 = (1.0, 2.0).
vec4 pos6 = pos1.xgga; // Illegal: mixing names from different sets.
```

On the left-hand side, though, the swizzle operator does not accept repeated components:

```
vec4 pos1 = vec4(1.0, 2.0, 3.0, 4.0);
pos1.xy = vec2(5.0, 6.0); // pos1 = (5.0, 6.0, 3.0, 4.0).
pos1.yx = vec2(5.0, 6.0); // pos1 = (6.0, 5.0, 3.0, 4.0).
pos1.xx = vec2(5.0, 6.0); // Illegal - x is repeated.
```

Here's a simple application of swizzling in a vertex shader.

Experiment 20.2. Replace the vertex shader code for `redSquare.cpp` with

```
void main()
{
   vec4 scaledPos =  vec4(0.5 * gl_Vertex.xy, 0.0, 1.0);
   gl_Position = gl_ModelViewProjectionMatrix * scaledPos;
}
```

As expected, the xy-values of the square's vertices are both halved. See Figure 20.2(b) for a screenshot. **End**

Exercise 20.1. (Programming) In the preceding experiment, if the first line of the new shader is simply

```
vec4 scaledPos =  0.5 * gl_Vertex;
```

instead of

```
vec4 scaledPos =  vec4(0.5 * gl_Vertex.xy, 0.0, 1.0);
```

what happens? Why?

20.2 Communication

Even the simple shaders attached to `redSquare.cpp` are evidently doing some amount of rudimentary communication with the program. The vertex shader `passThrough.vs` accesses the program's state – particularly, the current modelview and projection matrices through the variables `gl_ModelViewMatrix` and `gl_ProjectionMatrix`, respectively – in addition, of course, to the incoming vertex's coordinates contained in the variable `gl_Vertex`. It outputs the vertex's transformed coordinates into the variable `gl_Position`. The fragment shader `red.fs`, too, outputs to a variable `gl_FragColor` defining the current fragment's color.

20.2.1 Overview

A very simple overview of the GLSL's communication scheme is as follows.

The program has so-called *built-in* variables – all prefixed with "gl_" – that the shaders can access. Built-in variables are of two kinds: read-only (like `gl_Vertex`, `gl_ModelViewMatrix`, etc.) for the shaders to access incoming data and the program's state, and writable (like `gl_Position`, gl_FragColor, etc.) for the shaders to output the result of their computation for use by the program (in other words, back to the pipeline).

Shaders, too, have built-in variables for communication between one another. In addition to the built-in variables, the programmer can define variables herself, called *user-defined* obviously, for the program to communicate with the shaders, as well as for inter-shader communication.

Before we write programs with our own user-defined variables, here's another with only built-in variables, but with somewhat more going on than redSquare.cpp.

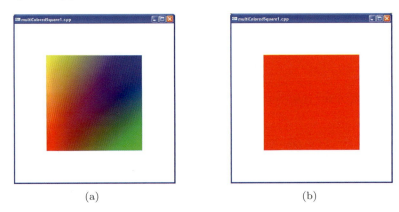

Figure 20.3: Screenshots of multiColoredSquare1.cpp: (a) Front (b) Back.

Experiment **20.3.** Run multiColoredSquare1.cpp. The program itself is a copy of redSquare.cpp, except for a different color at each square vertex *and* enabling of two-sided coloring with a call to glEnable(GL_-VERTEX_PROGRAM_TWO_SIDE) in the setup routine. The output initially is a multi-colored square (Figure 20.3(a)).

The vertex shader simpleColorizer.vs writes out both a front and a back color to the built-in variables gl_FrontColor and gl_BackColor, respectively:

```
gl_FrontColor = gl_Color;
gl_BackColor = vec4(1.0, 0.0, 0.0, 1.0);
```

It reads the front color from the user-defined colors, which it accesses through the built-in state variable gl_Color, while the back color is a fixed red.

The fragment shader passThrough.fs, on the other hand, simply sets

```
gl_FragColor = gl_Color;
```

Now, the way the GLSL works, the fragment shader *does not* receive its gl_Color values from the program; rather they are computed by interpolation from either the gl_FrontColor or gl_BackColor values specified in the vertex shader, depending on the visible face. The use of the same name gl_Color to represent actually different variables in the two shaders – the vertex shader using gl_Color to access the program, while the fragment shader to access its sibling – can be a source of confusion. One needs to keep the context in mind when using this variable. As the current fragment shader does no more than assign colors interpolated from the vertex shader, it is called a *pass-through* fragment shader.

As the square itself is oriented counter-clockwise and, therefore, front-facing, the fragment shader computes its gl_Color values by interpolation from gl_FrontColor, which in turn tracks the vertex color values as specified in the program. Consequently, a multi-colored square is drawn.

A fun way to reverse the square's orientation next is with a bit of swizzling. Accordingly, replace the vertex shader code with

```
void main()
{
   gl_FrontColor = gl_Color;
   gl_BackColor = vec4(1.0, 0.0, 0.0, 1.0);

   vec4 transposePos = gl_Vertex.yxzw; // Interchanges x and y
   // coordinate values, reversing the order of the vertices.

   gl_Position = gl_ModelViewProjectionMatrix * transposePos;
}
```

to see now a back-facing red square (Figure 20.3(b)).   **End**

20.2.2   Specifying Interface Variables with Qualifiers

We see next how variables at the interfaces between the program and the two shaders, as well as at that between the shaders themselves, are specified. Particularly, these variables will be of the GLSL types described earlier in Section 20.1.2, but each with an additional *qualifier* from the following:

> attribute   uniform   varying

attribute *qualifier*:   Attribute-qualified variables (or attribute variables or, simply, attributes) are used by the program to communicate per-vertex data to the vertex shader. An attribute is read by the vertex shader with each incoming vertex; therefore, it can be updated by the program for each vertex as well. Built-in attributes include gl_Vertex, gl_Color and gl_Normal, among others.

Vertex shaders can only read attributes, not write them. Attributes cannot be accessed by a fragment shader at all.

For efficiency of implementation, attribute types are restricted to floating point scalars, vectors and matrices.

`uniform` *qualifier*:   Uniform-qualified variables (or uniform variables or, simply, uniforms) are used by the program to communicate per-primitive data to either shader or to both together. A uniform is not read and cannot be updated within a `gl_Begin(`*primitive*`)`-`glEnd()` pair in the program, only outside. Therefore, its value remains constant through a primitive and it cannot be used to send per-vertex values. Built-in uniforms include `gl_ModelViewMatrix`, `gl_ProjectionMatrix` and `gl_ModelViewProjectionMatrix`, among others.

Shaders can only read uniforms, not write them.

`varying` *qualifier*:   Varying-qualified variables (or varying variables or, simply, varyings) are used by the vertex shader to communicate data to the fragment shader. There is, however, an important particularity of varyings: the vertex shader writes them per-vertex; subsequently, however, this data is *interpolated in a perspectively-correct manner* across each primitive's fragments before being read by the fragment shader. For this reason varyings are often called *interpolators*, which actually is a more appropriate term. Built-in varyings include `gl_Color`, `gl_FrontColor` and `gl_BackColor`, among others.

Fragment shaders can only read varyings, not write them.

Remark 20.1. As we learned from Experiment 20.3, the vertex shader writes to the built-in varyings `gl_FrontColor` and `gl_BackColor`, while the fragment shader reads the built-in varying `gl_Color`, which is interpolated from one of the first two, depending on the primitive's visible face. Moreover, `gl_Color` is the name of an attribute as well, which is accessed by the vertex shader to read the program's color state.

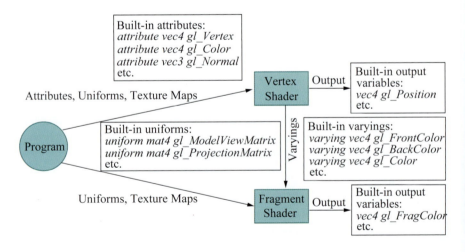

Figure 20.4: GLSL's communication scheme and a few popular built-in variables.

Figure 20.4 illustrates the qualifier definitions in the GLSL's communication scheme and shows a few of the most commonly used built-in variables. Texture maps are communicated by the program to the shaders as well and we'll see their use soon.

Remark 20.2. In addition to the three above, the GLSL defines four other type qualifiers: `const` (just like its C namesake for compile-time constants), `in`, `out` and `inout` (the latter three qualify formal function parameters).

Let's get our feet wet with the easiest of the three interface qualifiers to apply to a user-defined variable: varying.

Experiment 20.4. Run `multiColoredSquare2.cpp`. The code is exactly as `redSquare.cpp`, except this time the output is a multi-colored square because of the new shaders. Figure 20.5 is a screenshot. **End**

Figure 20.5: Screenshot of multiColored-Square2.cpp.

What we have done in `multiColoredSquare2.cpp` is use the shaders to color the square's vertices in a somewhat offbeat manner: by converting the position of each into a color by dividing its xyz-values by 100 to get RGB's within 0 to 1. For example, the vertex at $(80.0, 20.0, 0.0)$ gets the color $(0.8, 0.2, 0.0)$, a reddish hue; likewise, the vertex at $(20.0, 80.0, 0.0)$ gets the color $(0.2, 0.8, 0.0)$, which is more green.

Since the fragment shader `positionToColor.fs` cannot read `gl_Vertex`, a varying is used to pass the needed data to it from the vertex shader `positionToColor.vs`. In particular, the global declaration

```
varying vec3 vertexColor;
```

in both shaders links them through the varying `vertexColor`. Next

```
vertexColor = 0.01 * gl_Vertex.xyz;
```

in the vertex shader writes `vertexColor` with one-hundredth of the vertex xyz-values, while

```
gl_FragColor = vec4(vertexColor, 1.0);
```

in the fragment shader turns `vertexColor` into a legitimate color vector by tacking a 1 on for the alpha value.

Let's code a couple of uniforms next.

Experiment 20.5. Run `wavyCylinder1.cpp`. This program, based on `cylinder.cpp`, draws a cylinder with a wavy surface, allowing the user to control the number of waves, as well as change its color from red to green. Press the up/down arrow keys to change the waviness, the left/right arrow keys to change the color and 'x'-'Z' keys to turn the cylinder. Figure 20.6 shows the cylinder initially. **End**

Figure 20.6: Screenshot of wavyCylinder1.cpp.

The waviness and color capabilities of `wavyCylinder1.cpp` are added into `cylinder.cpp` via the vertex and fragment shaders, respectively. The vertex shader receives a parameter value from the program through the uniform waveParamShader to control the number of waves, while the fragment shader receives a parameter value from the program through the uniform colorParamShader to determine the cylinder's color. Let's understand the shaders themselves before deciphering how the uniforms are linked to the program.

The vertex shader waves1.vs first reads the cylinder's vertex coordinates gl_Vertex into the local vector variable v. Subsequently, the statements

```
v.x *= 1 + 0.1 * sin(waveParamShader * PI * v.z);
v.y *= 1 + 0.1 * sin(waveParamShader * PI * v.z);
```

scale the cross-section of the cylinder, which is parallel to the xy-plane, by the factor

$$1 + 0.1 \sin(\texttt{waveParamShader} * \pi * \texttt{v.z})$$

This scaling factor follows a sine function along the z-axis, which is the axis of the cylinder. The greater the parameter waveParamShader, the more rapidly does the scaling factor vary with z-value v.z. Finally,

```
gl_Position = gl_ModelViewProjectionMatrix * v;
```

outputs the position of the cylinder vertex.

The fragment shader colorInterpolate.fs is simple. The statement

```
gl_FragColor = vec4(1.0 - colorParamShader, colorParamShader,
                                    0.0, 1.0);
```

uses the parameter colorParamShader to interpolate the fragment color between red and green.

Let's see now how the uniforms are actually linked to the program. The global declaration

```
uniform float waveParamShader = 2.0;
```

in the vertex shader declares the uniform waveParamShader. It will be linked to a counterpart variable waveParamProgram, which has been declared globally in the program. The two statements

```
parameterLocation = glGetUniformLocation(programHandle,
                                    "waveParamShader");
glUniform1f(parameterLocation, waveParamProgram);
```

in the specialKeyInput() routine – where waveParamProgram is manipulated – complete the linking: the first statement obtains the location of the uniform waveParamShader, while the second binds it to the program variable waveParamProgram.

Likewise, the fragment shader uniform colorParamShader is linked to its program counterpart colorParamProgram.

Coding an attribute is next.

Experiment 20.6. Run `wavyCylinder2.cpp`. The output and controls are exactly as for `wavyCylinder1.cpp`. The difference between the two is in the mechanism by which the cross-section of the cylinder is scaled, which we discuss next. **End**

The scaling factor applied to the cylinder in `wavyCylinder2.cpp`, the same as in `wavyCylinder1.cpp` and parametrized by the program variable `waveParamProgram` as well, is computed by the program itself and shipped to its vertex shader `waves2.vs` via an attribute, rather than being computed in the vertex shader as for `wavyCylinder1.cpp`. The triangle strip definition inside the `for` loop

```
for(j = 0; j < q; j++)
{
    glBegin(GL_TRIANGLE_STRIP);
    for(i = 0; i <= p; i++)
    {
        scaleFactorProgram = 1 + 0.1 * sin(waveParamProgram * PI *
                                           h(i,j+1));
        glVertexAttrib1f(attributeIndex, scaleFactorProgram);
        glArrayElement( (j+1)*(p+1) + i );

        scaleFactorProgram = 1 + 0.1 * sin(waveParamProgram * PI *
                                           h(i,j));
        glVertexAttrib1f(attributeIndex, scaleFactorProgram);
        glArrayElement( j*(p+1) + i );
    }
    glEnd();
}
```

in the `drawScene()` routine computes the scaling factor `scaleFactorProgram` per vertex, using a formula parametrized by `waveParamProgram`, which yields the same value as the corresponding formula in the vertex shader `waves1.vs` of `wavyCylinder1.cpp`.

The method of linking the program variable `scaleFactorProgram` to its counterpart vertex shader attribute `scaleFactorShader` is a little different from that of linking to a uniform. The command

```
glBindAttribLocation(programHandle, attributeIndex,
                     "scaleFactorShader");
```

just before the `for` loop in `drawScene()`, associates the vertex shader attribute `scaleFactorShader` with the index `attributeIndex` (current value 1). Subsequently, each command

```
glVertexAttrib1f(attributeIndex, scaleFactorProgram);
```

in the triangle strip definition sets the value of `scaleFactorShader`, already associated with `attributeIndex`, to the current value of `scaleFactorProgram`.

The fragment shader `colorInterpolate.fs` of `wavyCylinder2.cpp` is copied over from `wavyCylinder1.cpp`.

Exercise 20.2. (Programming) Rewrite `throwBall.cpp` with the help of shaders – in particular, replace the call to translate the ball in the program with statements to change its coordinates in the vertex shader.

20.3 Per-Pixel Lighting

We come now to an application which definitely takes the GLSL beyond first-generation OpenGL: *per-pixel* lighting. The context is bump mapping, which we first encountered as a special effect in Section 13.7. The idea of bump mapping is to give an illusion of detail on a surface by perturbing its normals so that light reflects off it as though it were actually detailed. We applied this idea in Experiment 13.12 of Section 13.7 to make a plane appear corrugated in the program `bumpMapping.cpp`. We remarked then that bump mapping is particularly effective with the per-pixel lighting of Phong's shading model, where normal values are interpolated across primitives, rather than being fixed at vertices. We'll demonstrate now that indeed this is the case.

First, though, we'll warm up by replicating, via a vertex shader, *per-vertex* lighting – which consists of Phong lighting at each vertex, followed by Gouraud shading to interpolate colors through the primitives. Per-vertex lighting is implemented in the traditional fixed OpenGL pipeline and is what we have seen in all our lit programs to date.

Experiment 20.7. Run `bumpMappingPerVertexLighting.cpp`, which is code-wise almost exactly `bumpMapping.cpp`, but with a couple of shaders attached. Interaction is the same as well: press space to toggle between bump mapping on and off. Figures 20.7(a) and (b) are screenshots of `bumpMapping.cpp` and `bumpMappingPerVertexLighting.cpp`, respectively, doing bump mapping. Yes, they are exactly the same and we'll see momentarily why! **End**

The big difference between `bumpMapping.cpp` and `bumpMappingPerVertexLighting.cpp` is that is all the lighting calculations for the latter happen in its vertex shader `perVertexLightingSimple.vs`. These calculations, however, replicate exactly those of the first-generation pipeline by implementing Phong's lighting model to determine the color intensities at each vertex, and, moreover, with environment parameters set as for `bumpMapping.cpp`. Subsequently, the fragment shader `passThrough.fs` applies Gouraud shading by interpolating the vertex intensities through each triangle, again as in the first-generation pipeline. This is the reason, then, for the identical output from the two programs. We'll soon examine the vertex shader `perVertexLightingSimple.vs` in detail.

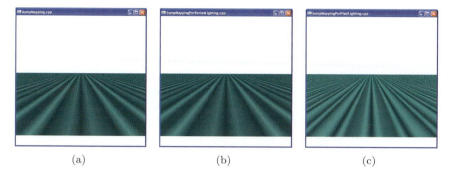

(a)   (b)   (c)

Figure 20.7: Screenshots of (a) `bumpMapping.cpp` (b) `bumpMappingPerVertex-Lighting.cpp` (c) `bumpMappingPerPixelLighting.cpp`.

Before examining the vertex shader `perVertexLightingSimple.vs`, where all the action is, let's first see a general definition of those built-in uniforms through which it accesses the current OpenGL state, particularly for material and lighting properties.

A Few of GLSL's Built-in Uniforms

Below is a code-like listing of those built-in uniforms that come in handy in lighting applications. It is extracted from the orange book [115], where you will find a listing of all GLSL built-ins.

```
// Matrix State
uniform mat3 gl_NormalMatrix;    // Derived

// Material State
struct gl_MaterialParameters
{
    vec4 emission;     // Ecm
    vec4 ambient;      // Acm
    vec4 diffuse;      // Dcm
    vec4 specular;     // Scm
    float shininess;   // Srm
};

uniform gl_MaterialParameters gl_FrontMaterial;
uniform gl_MaterialParameters gl_BackMaterial;

// Light State
struct gl_LightSourceParameters
{
    vec4 ambient;                  // Acli
    vec4 diffuse;                  // Dcli
    vec4 specular;                 // Scli
    vec4 position;                 // Ppli
```

```
    vec4 halfVector;              // Derived: Hi
    vec3 spotDirection;           // Sdli
    float spotExponent;           // Srli
    float spotCutoff;             // Crli
                                  // (range: [0.0, 90.0], 180.0)
    float spotCosCutoff;          // Derived: cos(Crli)
                                  // (range: [1.0, 0.0], -1.0)
    float constantAttenuation;    // K0
    float linearAttenuation;      // K1
    float quadraticAttenuation;   // K2
};
```

```
uniform gl_LightSourceParameters gl_LightSource[gl_MaxLights];
```

```
struct gl_LightModelParameters
{
    vec4 ambient;   // Acs
};
```

```
uniform gl_LightModelParameters gl_LightModel;
```

Now let's get to the vertex shader `perVertexLightingSimple.vs`. The first statement

```
normal = normalize(gl_NormalMatrix * gl_Normal)
```

multiplies the current normal, stored in the attribute `gl_Normal`, by the uniform `gl_NormalMatrix`, which is derived from the modelview matrix as the correct transformation for normals (see Section 11.11.5 for the derivation of the normal matrix from the modelview). Then it normalizes the result with the help of the built-in function `normalize()`.

Next

```
lightDirection = normalize(gl_LightSource[0].position.xyz)
```

extracts the normalized light direction as a 3-vector from the light's position vector. The position's w value, which is 0 because the given light source is directional, is ignored. Keep in mind, as well, that the light direction is the same for all vertices, again because the light source is directional.

The statement

```
halfVector = normalize(gl_LightSource[0].halfVector.xyz);
```

returns the normalized halfway vector between light and eye directions. The `halfVector` field of `gl_LightSource[0]` is a derived one storing the bisector between `gl_LightSource[0].position` and the vector $[0, 0, 1, 0]$. Therefore, it is correct to use the value of the preceding equation as the halfway vector *only if* the viewpoint is infinite (an infinite viewpoint is where the eye is assumed in the positive z-direction – see Section 11.4 for a discussion of local versus infinite viewpoints). The program, in fact, asserts an infinite viewpoint.

The next few statements

```
emission = gl_FrontMaterial.emission;
globalAmbient =  gl_LightModel.ambient * gl_FrontMaterial.ambient;
ambient = gl_LightSource[0].ambient * gl_FrontMaterial.ambient;
diffuse = max(dot(normal, lightDirection), 0.0)
            * (gl_LightSource[0].diffuse * gl_FrontMaterial.diffuse);
specular = pow( max(dot(normal, halfVector), 0.0),
                            gl_FrontMaterial.shininess )
            * (gl_LightSource[0].specular * gl_FrontMaterial.specular);
gl_FrontColor =  emission + globalAmbient + ambient + diffuse +
                specular;
```

need little narration. If you refer to the first lighting equation (11.10) in Section 11.2.4, then the statements above are a word for word implementation (assuming only one light source, of course). Note that, as there is neither a spotlight nor distance attenuation to take into account, implementing the first equation (11.10) effectively implements the full OpenGL lighting equation (11.12).

The pass-through fragment shader completes the job of Gouraud shading the interiors of primitives by applying interpolated color values. Note that there is no user-defined communication – it's all through built-in variables.

Experiment 20.8. If you are skeptical that we have actually replicated fixed-functionality lighting calculations in the vertex shader `perVertex-LightingSimple.vs` and wondering if we are still somehow sneaking the output from fixed-functionality, then replace the `gl_FrontColor` specification in that shader with

```
gl_FrontColor = vec4(1.0, 0.0, 0.0, 1.0);
```

Figure 20.8 is a screenshot. There is no doubt, is there, that it's the vertex shader that's in charge of color calculation?!   **End**

With an understanding of how to code per-vertex lighting in shaders, the upgrade next to per-pixel is not hard.

Figure 20.8: Screenshot of Experiment 20.8.

Experiment 20.9. Run `bumpMappingPerPixelLighting.cpp`. The program itself is identical to `bumpMappingPerVertexLighting.cpp` – the difference is in the shaders, which now implement Phong shading, or per-pixel lighting as it is called. Again, press space to toggle between bump mapping on and off. Figure 20.7(c) is a screenshot.   **End**

Let's first look at the vertex shader `perPixelLightingSimple.vs` of `bumpMappingPerPixelLighting.cpp`. The statements

```
normal = normalize(gl_NormalMatrix * gl_Normal);
lightDirection = normalize(gl_LightSource[0].position.xyz);
halfVector = normalize(gl_LightSource[0].halfVector.xyz);
```

are copied over from the per-vertex shader `perPixelLightingSimple.vs`, but now the left-side variables are all three global varyings, rather than local variables, in order to transmit their values to the fragment shader for interpolation across triangles.

The variables `emission`, `globalAmbient` and `ambient`, as well, are transmitted as varyings to the fragment shader after their values have been set by the statements

```
emission = gl_FrontMaterial.emission;
globalAmbient =  gl_LightModel.ambient * gl_FrontMaterial.ambient;
ambient = gl_LightSource[0].ambient * gl_FrontMaterial.ambient;
```

These three light components will then simply be interpolated across interior pixels. Nothing further need be done, as these components are the same as in the case of per-vertex lighting, even in the interior of triangles.

The two statements next pass the light's diffuse and specular color intensities, respectively scaled by the material's corresponding color values, as varyings to the fragment shader for interpolation, *and further attenuation per pixel*, as we shall see:

```
diffuse = gl_LightSource[0].diffuse * gl_FrontMaterial.diffuse;
specular = gl_LightSource[0].specular * gl_FrontMaterial.specular;
```

Next let's see what the fragment shader `perPixelLightingSimple.fs` does. First,

```
normalPerPixel = normalize(normal);
```

normalizes the received interpolated normal values. There is no need to do likewise for the received interpolated light direction and halfway vector values because the two are constant over all vertices, given that the program has a single directional light source and an infinite viewpoint; therefore, their interpolated values are constant, as well, and of unit length already.

The next two statements, respectively, attenuate the interpolated diffuse and specular values received from the vertex shader with a factor computed exactly as in the case of per-vertex lighting, *except* now the interpolated normal value is used at each pixel:

```
diffusePerPixel = max(dot(normalPerPixel, lightDirection), 0.0) *
                  diffuse;
specularPerPixel = pow( max(dot(normalPerPixel, halfVector), 0.0),
                   gl_FrontMaterial.shininess ) * specular;
```

Finally,

```
gl_FragColor =  emission + globalAmbient + ambient +
                diffusePerPixel + specularPerPixel;
```

accumulates all the color components and applies them per fragment.

The superiority of per-pixel lighting is apparent upon comparing Figures 20.7(a) and (b) with Figure 20.7(c): the corrugations are far more sharply defined in the latter.

Exercise 20.3. (Programming) Rewrite `litCylinder.cpp`, using the GLSL to move lighting calculation over to the shaders. Mind that both front and back faces of the cylinder are visible. Write both per-vertex and per-pixel lit versions.

Exercise 20.4. (Programming) Write a per-pixel lit version of `lightAndMaterial1.cpp`. Make sure to take into account distance attenuation and that both lights are positional.

20.4 Textures

Textures can not only be imported into the programmable pipeline without difficulty, but manipulated in there to great effect. We'll see a simple example next in a program which interpolates between two textures.

Experiment 20.10. Run `interpolateTextures.cpp`, which allows the user to interpolate between (or, blend, if you like) two textures painted on a square. Figure 20.9 shows screenshots of the start, a part way and end configurations. **End**

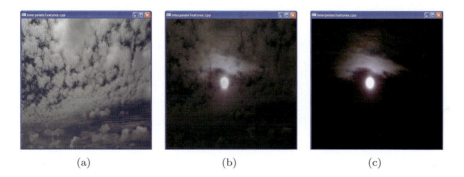

(a)	(b)	(c)

Figure 20.9: Screenshots of `interpolateTextures.cpp`: (a) Start (b) Part way (c) End.

The program itself simply loads and activates two textures, assigns texture coordinates to the vertices of the square and, most importantly, links to the shaders, passing to them the texture maps and coordinates. Specifically, the statements

```
glActiveTexture(GL_TEXTURE0);
glBindTexture(GL_TEXTURE_2D, texture[0]);
parameterLocation = glGetUniformLocation(programHandle, "skyTexture");
glUniform1i(parameterLocation, 0);
```

links the sampler `skyTexture`, declared as a uniform in the fragment shader `textureInterpolator.fs`, with the sky image. A similar set of statements links the sampler `nightskyTexture`, also declared in the fragment shader, with the night sky image.

The assignments

```
gl_TexCoord[0] = gl_MultiTexCoord0;
gl_TexCoord[1] = gl_MultiTexCoord0;
```

in the vertex shader `textureSimple.vs` simply access the texture coordinates to use with the two textures – which happen to be the same in this case – and place them in the built-in varyings on the left for interpolation and transmission to the fragment shader. Finally, the equation

```
gl_FragColor = (1.0 - paramShader) * texture2D(skyTexture,
               gl_TexCoord[0].st) +
               paramShader * texture2D(nightSkyTexture,
               gl_TexCoord[1].st);
```

in the fragment shader, parametrized by the uniform `paramShader`, interpolates between the two textures.

Exercise 20.5. (Programming) Rewrite `litTexturedCylinder.cpp` with the help of shaders. In particular, you will have to implement the `GL_MODULATE` option to combine light with texture.

20.5   Summary, Notes and More Reading

This chapter gave an introduction to the programmable pipeline and, particularly, OpenGL's Shading Language, the GLSL, which is used to do the programming. The canonical source for all things OpenGL, including the GLSL is, of course, the OpenGL site [99]. The standard text reference for the GLSL is by Rost & Licea-Kane [115], known as the orange book.

Particularly exciting is that, with OpenGL ES 2.0, shaders have gone mobile with a vengeance: anything that can be done in a shader has been removed from fixed-functionality and *has to be done* in a shader! OpenGL ES 2.0 is a "lean, mean, shadin' machine" as the OpenGL site calls it. The mobile shading language, GLSL ES, itself is very similar to the desktop version of this chapter, so the reader should be able to begin coding shaders for small devices without trouble.

Another popular high-level shading language is Cg (or, C for Graphics) developed by Nvidia and Microsoft, which, in fact, is nearly identical to Microsoft's own proprietary HLSL (High Level Shading Language). The particular advantage of Cg for Windows developers is that it allows simultaneous development of shaders for DirectX and OpenGL. However, the GLSL, being part of OpenGL 2.0, enjoys wider vendor and platform support.

Projective Spaces and Transformations

P rojective geometry is at the heart of computer graphics whichever view you take of it, practical or theoretical. The various transformations of real 3-space we learned to use for the purpose of animation in Chapter 4 and studied mathematically in Chapter 5 are, in fact, most naturally viewed as transformations of projective 3-space, following a so-called lifting of the scene from real to projective space. A consequence is that representing these transformations as projective is more efficient from a computational point of view, a fact that OpenGL takes constant advantage of in its design. Capturing the scene after a perspective projection on film – "shooting" as we imagine the OpenGL point camera to do – involves a projective transformation as well.

In fact, it's not an exaggeration to say that projective geometry is the mathematical foundation of modern-day CG, and that API's such as OpenGL "live" in projective 3-space. Unfortunately, though, because projective geometry works its magic deep inside the graphics pipeline, its importance often is not realized.

There are several books out there which discuss projective geometry – Coxeter [30], Henle [67], Jennings [72], Pedoe [104] and Samuel [117] come to mind – from mainly a geometer's point of view, as well as a few, such as Baer [6] and Kadison & Kromann [73], which take an algebraic standpoint. All these books, however, seem written primarily for a student of mathematics. There seems none yet dedicated to answering the computer scientist's (almost certainly a CG person) question of projective geometry, "What can you do for me?"

This appendix is a small attempt to fill this gap in the literature and introduce projective spaces and transformations from a CG point of view.

Projective spaces generalize real space. They are not difficult to understand, but geometric primitives, such as lines and planes, behave somewhat differently in a projective space than a real one. By applying a camera-view analogy from the outset, we try to convey a physical-based intuition for basic concepts, establishing at the same time connection with CG.

This appendix is long and the mathematics often admittedly abstract, but the payback for persevering through it comes in the form of a wealth of applications, including the projection transformation in the graphics pipeline, as well as the rational Bézier and all-important NURBS primitives, which are all topics of Chapter 18 on applications of projective spaces.

Logically, this appendix could as well have been a chapter of the book, just prior to Chapter 18. However, we decided against upsetting the fairly easy gradient of the book from the first chapter to the last with the insertion of a mathematical "hill". In fact, Chapter 18 on applications has been written so that the reader reluctant to take on the venture into projective theory can still make her way through it with minimal loss. This is not in any way to diminish the importance of the material in this appendix, but merely recognition of the reality that there are numbers of people out there who would make fine CG professionals, but care little for abstract mathematics.

We begin in Section A.1 by invoking a camera's point of view to motivate the definition of the projective plane. The geometry of this plane, including its surprising point-line duality and coordinatization by means of the homogeneous coordinate system, is the topic of Sections A.2 and A.3. In Section A.4 we study the structure of the projective plane and learn that the real plane can be embedded in the projective, which in turn yields a classification of projective points into regular ones and those at infinity.

A particularly intuitive kind of projective transformation, the so-called snapshot transformation, comes next in Section A.5. Section A.6 covers a few applications of homogeneous polynomial equations, including an algebraic insight into the projective plane's point-line duality, and an algebraic method to compute the outcome of a snapshot transformation. Following a brief discussion of projective spaces of arbitrary dimension in Section A.7, we move on to projective transformations.

Projective transformations are first defined algebraically in Section A.8 and then understood geometrically in A.9. In Section A.10 we relate projective, snapshot and affine transformations, and see that projective transformations are more powerful than either of the other two. The process of determining the projective transformation to accomplish some particular mapping – often beyond the reach of an affine transformation – is the topic of Section A.11.

A.1 Motivation and Definition of the Projective Plane

Consider a viewer taking pictures with a point camera with a film in front of it. Light rays from objects in the scene travel toward the camera and their intersections with the film render the scene. See Figure A.1. Captured on film is the (perspective) projection of the objects. In the case of OpenGL, this is precisely the situation when the user defines a viewing frustum: the point camera is at the apex of the frustum, while the film lies along its front face.

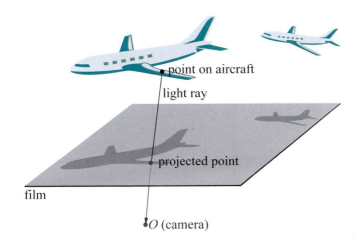

point on aircraft

light ray

projected point

film

O (camera)

Figure A.1: Perceiving objects with a point camera and a plane film.

Clearly, points in the scene that lie on the same (straight) line through the camera cannot be distinguished by the viewer. In fact, all objects, e.g., points and line segments, that lie on one line l through the camera cannot be distinguished by the viewer. They all project to and are perceived as a single point on the film. See Figure A.2(a). Assume for the moment that the film is two-sided and that objects behind project onto it as well (depicted is one such point). For now, ignore as well that lines through the camera parallel to the film, e.g., l', do not intersect it at all. This is owing to the alignment of the film, which can always be changed.

So one can say that the viewer perceives *every* line through the camera as a point. What then does he perceive as a line? The likely answer is a plane. Indeed, any plane through the camera intersects the film in a line, though, again, the film may have to be re-aligned so as not to be parallel to the plane. See Figure A.2(b).

Lines are points, planes are lines, ... Let's take a moment to formalize, as a new space, the world as it is perceived through a point camera at the origin. Recall that a *radial* primitive is one which passes through the origin.

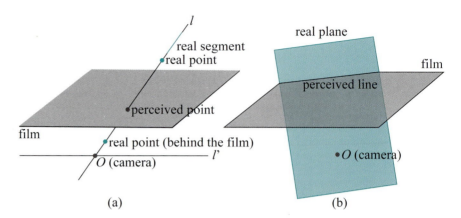

Figure A.2: Perceiving points, lines and planes by projection.

Definition A.1. A radial line in 3-space $\mathbb{R}^3$ is called a *projective point*. The set of all projective points lying on any one radial plane in $\mathbb{R}^3$ is called a *projective line*. (See Figure A.3.)

The set of all projective points is called 2-dimensional *projective space* and denoted $\mathbb{P}^2$. $\mathbb{P}^2$ is also called the *projective plane*.

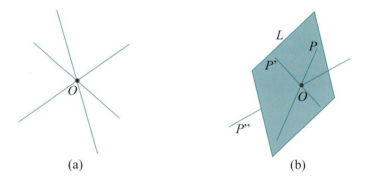

Figure A.3: (a) Projective points are radial lines (b) A projective line consists of all projective points on a radial plane: projective points P and P' belong to the projective line L, while P'' does not. *Keep the distinction in mind that, though we have labeled the plane L, the projective line L actually consists of all the projective points, e.g., P and P', that lie on this plane, and is different from the plane itself.*

$\mathcal{R}em\alpha rk$ A.1. We are taking a significant step up in abstraction in leaving $\mathbb{R}^2$ for $\mathbb{P}^2$. The real plane $\mathbb{R}^2$ is easy to visualize as, well, a real plane. e.g., a table top or a sheet of paper. Not so the projective plane. There is no real object to which it corresponds nicely.

Things such as a line, which is a set of points in one space, being just a point of another may seem a bit strange as well. It's mostly a matter of getting used to though – like learning a foreign language. As with a new language, some words translate literally, but some don't simply because the concept isn't familiar (what's sandstorm in Eskimoan?).

It's recommended that the reader stick close to the real-based definitions at first. A thought process like "Hmm, the projective point P belongs to the projective line L. Well then, that means the real line which is P sits inside the real plane which is L" may seem cumbersome at first, but projective primitives will seem less and less strange as we go along.

The term "projective" arose because objects on the projective plane are perceived by projection onto a real one, which for us is the film. Observe that in Figure A.3(b) we denote by L both a radial plane (a primitive in $\mathbb{R}^3$), as well as the projective line (a primitive in $\mathbb{P}^2$) consisting of projective points that lie on that plane. There should be no cause for ambiguity as it'll be clear from the context which we mean.

Terminology: We'll generally use lower case letters to denote primitives in $\mathbb{R}^2$ and upper case for those in $\mathbb{P}^2$.

Remark A.2. The dimension of $\mathbb{P}^2$ is two (as indicated by the superscript). This is because, while points in $\mathbb{R}^3$ have three "degrees of freedom", radial lines in $\mathbb{R}^3$ have only two. We'll elaborate on the dimension of the projective plane in Section A.7.

A.2 Geometry on the Projective Plane and Point-Line Duality

We have, then, on the projective plane $\mathbb{P}^2$ projective points and projective lines, just as on the real plane $\mathbb{R}^2$ we have real points and lines. It's interesting to compare the relationship between points and lines in the two spaces.

Recall the following two facts from Euclidean geometry (geometry in real space is called Euclidean):

(a) There is a unique line containing two distinct points in $\mathbb{R}^2$.

(b) Two distinct lines in $\mathbb{R}^2$ intersect in a unique point, *except* if they are parallel, in which case they do not intersect at all.

What is the situation in projective geometry)?

Two distinct projective points P and P' correspond to two distinct radial lines in $\mathbb{R}^3$ and there is a unique radial plane L in $\mathbb{R}^3$ containing the latter two. See Figure A.4(a).

It follows that:

(A) There is a unique projective line containing two distinct projective points in $\mathbb{P}^2$.

How about two distinct projective lines? The corresponding two distinct radial planes, say, L and L' in $\mathbb{R}^3$, intersect in a unique radial line corresponding to some projective point P. See Figure A.4(b). We have:

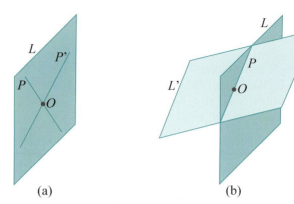

Figure A.4: (a) Radial lines corresponding to projective points P and P' are contained in a unique radial plane corresponding to the projective line L (b) Radial planes corresponding to projective lines L and L' intersect in a unique radial line corresponding to the projective point P.

(B) Two distinct projective lines in $\mathbb{P}^2$ intersect in a unique projective point.

No exceptions! There's no such thing as parallelism in $\mathbb{P}^2$! Any two different lines always intersect in a point. Two points–one line, two lines–one point, *always*: $\mathbb{P}^2$ has better so-called *point-line duality* than $\mathbb{R}^2$. We'll have more to say about the point-line duality of $\mathbb{P}^2$ as we go along.

Exercise A.1. Consider three distinct projective lines L, L' and L''. We know that their pairwise intersections are three projective points, say, P, P' and P''. Give examples where (a) all three points are identical and (b) all three are distinct. Can only two of them be distinct? If all three are distinct can they be collinear, i.e., lie on one projective line?

A.3 Homogeneous Coordinates

We want to *coordinatize* $\mathbb{P}^2$, if possible, in a manner similar to that of $\mathbb{R}^2$ by Cartesian coordinates. This is important for the purpose of geometric calculations. For example, Cartesian coordinates on the real plane allow us to make a statement such as "The equation of the line through the $[-2 \ -5]^T$ and $[1 \ 1]^T$ is $y - 2x + 1 = 0$, which is satisfied as well by $[0 \ -1]^T$, so that all three points are collinear."

So how does one coordinatize $\mathbb{P}^2$? As follows:

Definition A.2. The *homogeneous coordinates* of a projective point can be the Cartesian coordinates of any real point on it, other than the origin. (Yes, homogeneous coordinates are not unique, and a projective point has many different homogeneous coordinates. This may seem strange at first but read on...)

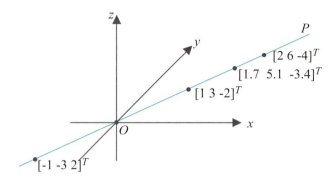

Figure A.5: The coordinates of any point on P, except the origin, can be used as its homogeneous coordinates – four possibilities are shown.

Example A.1. The projective point P corresponding to the radial line through $[1 \; 3 \; -2]^T$ has, as shown in Figure A.5, among others, homogeneous coordinates $[1 \; 3 \; -2]^T$, $[2 \; 6 \; -4]^T$, $[-1 \; -3 \; 2]^T$ and $[1.7 \; 5.1 \; -3.4]^T$. In fact, any tuple of the form $[c \; 3c \; -2c]^T$, where $c \neq 0$, can serve as homogeneous coordinates for P.

Terminology: To avoid clutter in diagrams, we'll often write homogeneous coordinates $[x \; y \; z]^T$ as (x, y, z).

That a projective point can have infinitely many different homogeneous coordinates may seem odd, but it's not really a problem because two distinct projective points cannot share the same homogeneous coordinates. This is because two distinct radial lines do not share any point other than the origin. In other words, even though projective points have non-unique homogeneous coordinates, there is no risk of ambiguity. As an analogy, think of a roomful of people, each having multiple nicknames, but no two having a nickname in common – there is no danger of confusion then. As a non-zero tuple $[x \; y \; z]^T$ gives homogeneous coordinates of a unique projective point, we'll often refer to *the* projective point $[x \; y \; z]^T$ or write, say, the projective point $P = [x \; y \; z]^T$.

If you are wondering if $\mathbb{P}^2$ can at all be coordinatized in a unique manner, as is $\mathbb{R}^2$ by Cartesian coordinates, the answer is that there is no "natural" way to do this. Don't take our word for it, but give it some thought and you'll see the pitfalls. For example, a likely approach is to choose the coordinates of *one* real point from the radial line corresponding to each projective point. But then one has to come up with a *well-defined* way of choosing such a point, in other words, an algorithm that given input a radial line uniquely outputs a point from it. Try and devise such an algorithm!

Remark A.3. An important difference between the Cartesian and homogeneous coordinate systems is the lack of an origin in the latter. No matter how one sets up a Cartesian coordinate system in $\mathbb{R}^3$, i.e., no matter how one sets up the coordinate axes, the origin $(0, 0, \ldots, 0)$ is always distinguished as a special point. This is not the case for the homogeneous coordinate system in $\mathbb{P}^2$ – no projective point is special. It is truly homogeneous!

Example A.2. Find homogeneous coordinates of the projective point P of intersection of the projective lines L and L', corresponding, respectively, to the radial planes $2x + 2y - z = 0$ and $x - y + z = 0$.

Answer: Solving the simultaneous equations

$$
\begin{aligned}
2x + 2y - z &= 0 \\
x - y + z &= 0
\end{aligned}
$$

one finds that points on their intersecting line are of the form

$$y = -3x, \quad z = -4x$$

Therefore, homogeneous coordinates of P are (arbitrarily choosing $x = 1$)

$$[1 \ -3 \ -4]^T$$

Exercise A.2. Find homogeneous coordinates of the projective point P of intersection of the projective lines L and L' corresponding, respectively, to the radial planes $-x - y + z = 0$ and $3x + 2y = 0$.

Exercise A.3. Find the equation of the radial plane in $\mathbb{R}^3$ corresponding to the projective line L which intersects the two projective points $P = [1 \ 2 \ 3]^T$ and $P' = [2 \ -1 \ 0]^T$.

A.4 Structure of the Projective Plane

We're going to try and understand the structure of $\mathbb{P}^2$ by relating it to that of $\mathbb{R}^2$. In fact, we'll start off by using the homogeneous coordinate system of $\mathbb{P}^2$ to *embed* $\mathbb{R}^2$ inside it.

A.4.1 Embedding the Real Plane in the Projective Plane

Associate a point $p = [x \ y]^T$ of $\mathbb{R}^2$ with the projective point $\phi(p) = [x \ y \ 1]^T$. The easiest way to picture this association is to first identify $\mathbb{R}^2$ with the $z = 1$ plane; i.e., $[x \ y]^T$ of $\mathbb{R}^2$ is identified with $[x \ y \ 1]^T$ of $z = 1$. See Figure A.6. Following which, the association $p \mapsto \phi(p)$ is simply each real point with the radial line through it, in particular, the real point $[x \ y \ 1]^T$ (Cartesian coordinates) with the projective point $[x \ y \ 1]^T$ (homogeneous coordinates).

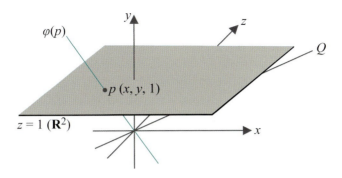

Figure A.6: Real point p on the plane $z = 1$ is associated with the projective point $\phi(p)$. Projective point Q, lying on the plane $z = 0$, is not associated with any real point.

The association $p \mapsto \phi(p)$ is clearly one-to-one as distinct points of $z = 1$ give rise to distinct radial lines through them. It's not onto as points of $\mathbb{P}^2$ that lie *on* the plane $z = 0$ or, equivalently, that are parallel to $z = 1$, do not intersect $z = 1$ and, therefore, are not associated with any point of $\mathbb{R}^2$ (e.g., Q in the figure). Precisely, points of $\mathbb{P}^2$ with homogeneous coordinates of the form $[x \; y \; 0]^T$ are not associated with any point of $\mathbb{R}^2$.

$\mathbb{R}^2$, therefore, is embedded by ϕ as the *proper* subset of $\mathbb{P}^2$ consisting of radial lines intersecting $z = 1$. We're at the point now where we can try to understand how we ended up trading parallelism in $\mathbb{R}^2$ for perfect point-line duality in $\mathbb{P}^2$.

A.4.2   A Thought Experiment

Here's a thought experiment. Two parallel lines l and l' lie on $\mathbb{R}^2$, aka the plane $z = 1$ in $\mathbb{R}^3$, a distance of d apart. Points p and p' on l and l', respectively, start a distance d apart and begin to travel at the same speed and in the same direction on their individual lines. See Figure A.7. Evidently, they remain d apart no matter how far they go. Well, of course, as l and l' are parallel!

Consider next what happens to the projective points $\phi(p)$ and $\phi(p')$ associated with p and p', respectively. See again Figure A.7 to convince yourself that both $\phi(p)$ and $\phi(p')$ draw closer and closer to the particular radial line l'' on the plane $z = 0$ that is parallel to l and l'. As it lies on $z = 0$, l'' corresponds to a projective point P'' not associated with any real; in fact, P'''s homogeneous coordinates are of the form $[x \; y \; 0]^T$.

Observe that the projective point $\phi(p)$ itself travels along a projective line L – the one whose radial plane contains l. We'll call L the projective line corresponding to l. Likewise, the projective point $\phi(p')$ travels along the projective line L' corresponding to l'. Moreover, L and L' intersect in P''. See Figure A.8.

Let's take stock of the situation so far. The parallel lines l and l' on the

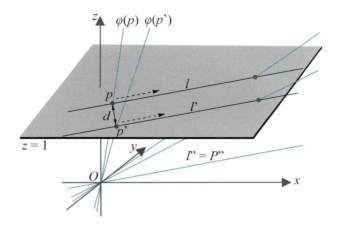

Figure A.7: The real points p and p' travel along parallel lines l and l'. Associated projective points $\phi(p)$ and $\phi(p')$ travel with p and p'.

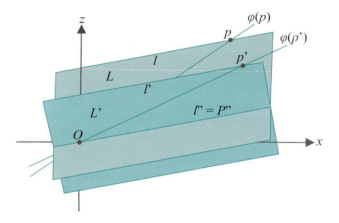

Figure A.8: $\phi(p)$ travels along L and $\phi(p')$ along L'. L and L' meet at P''.

real plane never meet, but the projective lines L and L' corresponding to them in $\mathbb{P}^2$ meet in P''. Moreover, every point of L or L', *except* for P'', is associated by ϕ to a point of l or l', respectively. We can say then that the projective line L equals its real counterpart l *plus* the extra point P''; L', likewise, is its real counterpart l' *plus* P''. And it's at this point P'', beyond the reals, that the two projective lines meet, while their real counterparts never do.

Example A.3. What if both points p and p', and together with them $\phi(p)$ and $\phi(p')$, travel along their respective lines in directions opposite to those indicated in Figure A.7? What if only one reversed its direction?

Answer: If both p and p' reversed directions, then again they would travel forever exactly d apart. If only one of the two reversed its direction, then, of course, the distance between them would continuously increase.

However, in either case, $\phi(p)$ and $\phi(p')$ draw closer, again both to P''. It seems that, whatever the sense of travel is of $\phi(p)$ and $\phi(p')$ along their respective projective lines L and L', they approach that one point of intersection of these two lines. Two points traveling in opposite directions along a real line ultimately grow farther and farther apart. A projective line, on the other hand, apparently behaves more like a circle.

A.4.3   Regular Points and Points at Infinity

Recall *equivalence relations* and *equivalence classes* from undergrad discrete math. In particular, recall that the lines of $\mathbb{R}^2$ can be split into equivalence classes by the equivalence relation of being parallel. Consider any equivalence class $\mathbf{l}$ of parallel lines of $\mathbb{R}^2$, the latter being identified with the plane $z = 1$ in $\mathbb{R}^3$ as before. There is a unique radial line l on the plane $z = 0$ parallel to the members of $\mathbf{l}$. See Figure A.9.

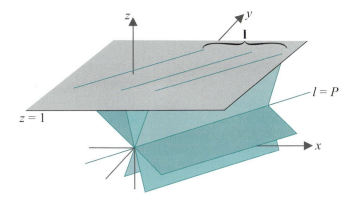

Figure A.9: The line l (= projective point P) is parallel to lines in $\mathbf{l}$. P is said to be the point at infinity along the equivalence class $\mathbf{l}$ of parallel lines.

Denote the projective point corresponding to l by P. Projective lines corresponding to lines in $\mathbf{l}$ all meet at P, because their radial planes each contain l. The point P, which is not associated with any real point by ϕ, as it lies on $z = 0$, is called the *point at infinity* along $\mathbf{l}$ or, simply, the point at infinity along any one of the lines in $\mathbf{l}$. Conversely, any radial line l on the plane $z = 0$ is the point at infinity along the equivalence class of lines in $\mathbb{R}^2$ parallel to it. In other words, the correspondences

$$\text{equivalence class of parallel lines in } \mathbb{R}^2 \quad \leftrightarrow \quad \text{radial line on } z = 0$$
$$\leftrightarrow \quad \text{point at infinity of } \mathbb{P}^2$$

are both one-to-one. Note that points at infinity of $\mathbb{P}^2$ are precisely those with homogeneous coordinates of the form $[x \ y \ 0]^T$.

Returning to the thought experiment of Section A.4.2, one can imagine points at infinity plugging the "holes" along the "border" of $\mathbb{R}^2$ through which parallel lines "run off" without meeting, which explains why every pair of lines on the projective plane meets.

Projective points which are not points at infinity are called *regular points*. Regular points have homogeneous coordinates of the form $[x\ y\ z]^T$, where z is *not* zero. Moreover, regular points intersect $z = 1$, so are associated each by ϕ^{-1} with a point of $\mathbb{R}^2$ (remember ϕ takes a real point of $\mathbb{R}^2$, represented by the plane $z = 1$, to the projective point whose corresponding radial line passes through that point). Accordingly, one can write:

$$\mathbb{P}^2 = \mathbb{R}^2 \cup \{\text{points at infinity}\} = \{\text{regular points}\} \cup \{\text{points at infinity}\}$$

The union of all points at infinity, called the *line at infinity*, is the projective line whose radial plane is $z = 0$. Therefore, one can as well write:

$$\mathbb{P}^2 = \mathbb{R}^2 \cup \text{line at infinity} = \{\text{regular points}\} \cup \text{line at infinity}$$

Our embedding ϕ of $\mathbb{R}^2$ as a subset of $\mathbb{P}^2$ depends on the plane $z = 1$, particularly because we identify $z = 1$ with $\mathbb{R}^2$ and subsequently associate each point of $\mathbb{R}^2$ with the radial line in $\mathbb{R}^3$ through it. Is there anything special about the plane $z = 1$? Not at all. It just seemed convenient. In fact, we could have used any *any* non-radial plane p.

Exercise **A.4.** Why does p have to be non-radial?

Example **A.4.** Instead of $z = 1$, identify $\mathbb{R}^2$ with the plane $x = 2$ in $\mathbb{R}^3$. Accordingly, embed $\mathbb{R}^2$ into $\mathbb{P}^2$ by associating $[x\ y]^T$ with the radial line through $[2\ x\ y]^T$. Which now are the regular points and which the points at infinity of $\mathbb{P}^2$?

Answer: The regular points of $\mathbb{P}^2$ are the radial lines in $\mathbb{R}^3$ that intersect the plane $x = 2$. These are precisely the radial lines that do not lie on the plane $x = 0$. The points at infinity are the radial lines that do lie on the plane $x = 0$. Equivalently, regular points have homogeneous coordinates of the form $[x\ y\ z]^T$, where $x \neq 0$, while points at infinity have homogeneous coordinates of the form $[0\ y\ z]^T$.

Exercise **A.5.** Identify $\mathbb{R}^2$ with the plane $x + y + z = 1$ in $\mathbb{R}^3$, embedding it into $\mathbb{P}^2$ by associating $[x\ y]^T$ with the radial line through $[x\ y\ 1 - x - y]^T$. Which now are the regular points and which the points at infinity of $\mathbb{P}^2$?

It may seem strange at first that the separation of $\mathbb{P}^2$ into regular points and points at infinity depends on the particular embedding of $\mathbb{R}^2$ in $\mathbb{P}^2$. However, this situation becomes clearer after a bit of thought. It's related, as a matter of fact, to the discussion at the beginning of the chapter, where we motivated projective spaces by observing that lines through a point camera are perceived as points on the plane film. Even though all lines through the camera do not intersect the film, we argued this to be merely an artifact of the alignment of the film, the latter being changeable. Therefore, we concluded that all radial lines should be taken as points in projective space.

We come now full circle back to this initially motivating scenario. Embedding $\mathbb{R}^2$ in $\mathbb{P}^2$ corresponds exactly to choosing an alignment of the film – the film is a copy of $\mathbb{R}^2$ and each point on it associated with the light ray (= radial line in $\mathbb{R}^3$ = point of $\mathbb{P}^2$) through that point to the camera. Light rays toward the camera that intersect the film are regular points of $\mathbb{P}^2$ and visible, while those that do not are points at infinity and invisible. Moreover, the line at infinity corresponds to the plane through the camera parallel to the film. And, of course, we are at perfect liberty to align the film, i.e., embed $\mathbb{R}^2$ in $\mathbb{P}^2$, as we like, different choices leading to different sets of visible and invisible light rays.

A.5 Snapshot Transformations

Here's another interesting thought experiment.

Example A.5. A point camera is at the origin with two power lines passing over it, both parallel to the x-axis. One lies along the line $y = 2, z = 2$ (i.e., the intersection of the planes $y = 2$ and $z = 2$) and the other along the line $y = -2, z = 2$.

Take "snapshots" of the power lines with the film aligned along (a) the plane $z = 1$ and (b) the plane $x = 1$. Sketch and compare the two snapshots.

Answer: This is one you might want to try yourself before reading on!

See Figure A.10. Figure A.10(a) shows the snapshot (or, projection) of the power lines on the plane $z = 1$. They are two *parallel* lines $y = 1$ and $y = -1$. This is not hard to understand: by simple geometry, the line $y = 2, z = 2$ projects toward the origin (the camera) to the line $y = 1$ on the plane $z = 1$; likewise, $y = -2, z = 2$ projects to $y = -1$ on $z = 1$.

Figure A.10(b) shows the snapshot on the plane $x = 1$. It is the two *intersecting* lines $z = y$ and $z = -y$ making an X-shape. This requires explanation. The top of the X, above its center, is formed from intersections with the film of light rays through points on the power lines with x-value greater than zero, while the bottom from rays through points with x-value less than zero. The rays from the points on either power line with x-value equal to zero do not strike the film.

The point $[1\ 0\ 0]^T$ at the middle of the X is included in the snapshot, though no ray from either power line passes through it, because it's the intersection with the film of the "limit" of the rays from points on either power line as they run off to infinity. It's convenient to imagine the limits of visible rays as being visible as well and we ask the reader to accept this. In geometric drawing parlance $[1\ 0\ 0]^T$ is the *vanishing point* of the power lines – it's where they *seem* to meet on the film $x = 1$.

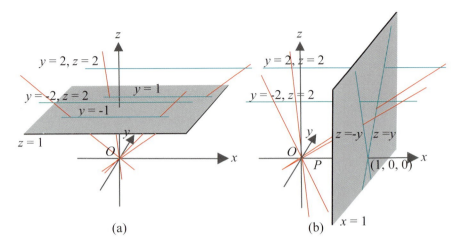

Figure A.10: Power lines $y = \pm 2, z = 2$ projected onto the planes (a) $z = 1$ and (b) $x = 1$. Red lines depict light rays. The x-axis corresponds to the projective point P.

Contemplate now the situation from the point of view of projective geometry. The projective lines corresponding to the two power lines meet at the projective point P corresponding to the x-axis, as the radial planes through the power lines intersect in the x-axis. Now, P is a point at infinity with respect to the plane $z = 1$ (because the x-axis doesn't intersect this plane), while it's a regular point with respect to the plane $x = 1$ (because the x-axis intersects this plane at $[1\ 0\ 0]^T$). In terms of shooting pictures, then, the camera with its film along $z = 1$ cannot see where the two power lines meet, so they appear parallel. However, with its film along $x = 1$ the camera sees them meet at $[1\ 0\ 0]^T$.

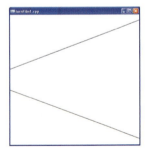

Figure A.11: Screenshot of `turnFilm1.cpp`.

Ex**periment A.1.** Run `turnFilm1.cpp`, which animates the setting of the preceding exercise by means of a viewing transformation. Initially, the film lies along the $z = 1$ plane. Pressing the right arrow key rotates it toward the $x = 1$ plane, while pressing the left one reverses the rotation. Figure A.11 is a screenshot midway. You cannot, of course, see the film, only the view of the lines as captured on it.

The reason that the lower part of the X-shaped image of the power lines cannot be seen is that OpenGL film doesn't capture rays hitting it from behind, as the viewing plane is a clipping plane too. Moreover, if the lines seem to actually meet to make a V after the film turns a certain finite amount, that's because they are very long and your monitor has limited resolution!

This program itself is simple with the one statement of interest being `gluLookAt()`, which we ask the reader to examine next. **End**

Exerci**se A.6. (P**rogramming**)** Verify that the `gluLookAt()` statement of `turnFilm1.cpp` indeed simulates the film's rotation as claimed between the $z = 1$ and $x = 1$ planes.

Example A.6. Refer to Figure A.10(b). Suppose two power lines *actually* lie along the two intersecting lines $z = y$ and $z = -y$ on the plane $x = 1$, which is the snapshot on the plane $x = 1$ of the power lines of the preceding example. What would *their* snapshot look like on the films $z = 1$ and $x = 1$?

Answer: Exactly as in the preceding Example A.5, as depicted in Figures A.10(a) and (b)! It's not possible to distinguish between these two pairs of power lines – the pair in Example A.5 being "really" parallel and the current one "really" intersecting – with a point camera at the origin.

A somewhat whimsical take on all this is to imagine a Matrix-like world where one can never know reality. Perception is limited to whatever is captured on film. Therefore, one agent's intersecting power lines are just as real as the other's parallel ones...

It's useful to think of one snapshot of Example A.5 or A.6 as a *transformation* of the other. Keep in mind that if a snapshot appears as the two parallel lines $y = \pm 1$ on the film $z = 1$, then it always appears as the two intersecting lines $z = \pm y$ on the film $x = 1$, *regardless* of what the "real" objects are.

Convince yourself of this by mentally tilting one of the power lines in Figure A.10(a) on the radial plane (not drawn) through it, so that its projection on the $z = 1$ plane does not change. The power line's projection on the $x = 1$ plane remains unchanged, as well, because the set of light rays from it through the camera doesn't change. For this reason, it makes sense to talk of transforming one snapshot to another, without any reference to the real scene. We'll informally call such transformations *snapshot transformations*.

Remark A.4. Snapshot transformations as described are not really transformations in the mathematical sense, as they don't map some space to itself but, rather, one plane (film) to another. A rigorous formulation is possible, though likely not worth the effort, as we'll see soon that snapshot transformations are subsumed within the class of projective transformations, which we'll be studying in depth. Nevertheless, the notion of a snapshot transformation is geometrically intuitive and useful.

Here are more for you to ponder.

Exercise A.7. In each case below you are told what the snapshot looks like on the film $z = 1$, aka $\mathbb{R}^2$, and asked what is captured on the film $x = 1$. The $z = 1$ shots are drawn in Figure A.12, each labeled the same as the item to which it corresponds. You don't have to find equations for your answer for $x = 1$. Just a sketch or a verbal description is enough.

Answers are in italics. Figure A.13 justifies the answer to (h).

(a) Two lines that intersect at the origin, neither being the y-axis (on $\mathbb{R}^2$). *Two parallel lines.* Why the caveat? What happens if one is the y-axis?

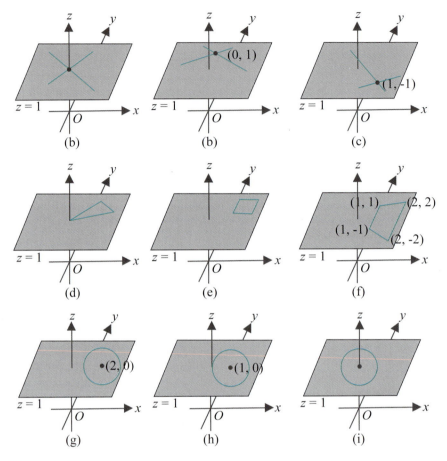

Figure A.12: Transform these snapshots on the plane $z = 1$ to the plane $x = 1$. Some points on the plane $z = 1$ are shown with their xy coordinates. Labels correspond to items of Exercise A.7.

(b) Two lines that intersect at the point $[0 \ 1]^T$, neither being the y-axis.
   Two parallel lines.

(c) Two lines that intersect at the point $[1 \ -1]^T$.
   Two intersecting lines.

(d) A triangle in the upper-right quadrant with one vertex at the origin but, otherwise, not touching the axes.
   An infinitely long U-shape with straight sides.

(e) A square in the upper-right quadrant not touching any of the axes.
   A quadrilateral with no two parallel sides.

(f) A trapezoid symmetric about the x-axis with vertices at $[1\ 1]^T$, $[1\ -1]^T$, $[2\ -2]^T$ and $[2\ 2]^T$.
  A rectangle.

(g) A unit radius circle centered at $[2\ 0]^T$.
  An ellipse.

(h) A unit radius circle centered at $[1\ 0]^T$.
  A parabola – see Figure A.13.

(i) A unit radius circle centered at the origin.
  A hyperbola.

Remark A.5. Exercise A.7(f) seems innocuous enough, but it is very important. It's generalization to 3D will help convert viewing frustums to rectangular boxes in the graphics pipeline.

Exercise A.8. Refer to the geometric construction of conic sections in Section 10.1.5 as plane sections of a double cone, and show that any non-degenerate conic section can be snapshot transformed to another such.

Exercise A.9. (Programming) Write code similar to `turnFilm1.cpp` to animate the snapshot transformation of Exercise A.7(h). Again, you'll see only part of the parabola because OpenGL cannot see behind its film.

It's not hard to see that none of the snapshot transformations of Exercise A.7, except for (c) and (g), can be accomplished using OpenGL modeling transformations. This is because they are not affine – recall from Section 5.4.5 that OpenGL implements only affine transformations.

Remark A.6. We just said that most of the snapshot transformations of Exercise A.7 are not affine and yet seem to be suggesting with the preceding Exercise A.9 that they may be implemented by means of an OpenGL viewing transformation. We know, however, that the latter is equivalent to a sequence of modeling transformations and, therefore, affine.

The apparent conundrum is not hard to resolve. The result of the viewing transformation of, e.g., `turnFilm1.cpp`, is indeed a snapshot transformation in terms of what is *seen on the screen*. In other words, the transformation from the OpenGL window prior to applying the viewing transformation to that after is a snapshot transformation. However, the viewing transformation serves only to change the scene to one which OpenGL *projects* onto the window as the new one. A snapshot transformation, therefore, is more than a viewing transformation – it's a viewing transformation *plus* a projection.

Exercise A.10. By considering how to turn the film, i.e., viewing plane, show that implementing a snapshot transformation in OpenGL is equivalent to:

(a) setting the *centerx*, *centery*, *centerz*, *upx*, *upy* and *upz* parameters of the viewing transformation

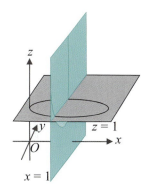

Figure A.13: Answer to Exercise A.7(h).

gluLookAt(0, 0, 0, *centerx*, *centery*, *centerz*, *upx*, *upy*, *upz*)

and

(b) setting the *near* parameter of the perspective projection call

glFrustum(*left*, *right*, *bottom*, *top*, *near*, *far*)

where the other five parameters can be kept fixed at some initially chosen values.

A.6 Homogeneous Polynomial Equations

The only application we've made so far of homogeneous coordinates is to embed $\mathbb{R}^2$ in $\mathbb{P}^2$. We haven't used them yet to write equations of curves on the projective plane. Let's now try to do this.

We'll start with the simplest curve on the projective plane, in fact, a projective line. We want an equation – as for straight lines in real geometry – that will say if a projective point belongs to a projective line. For example, an equation such as $2x + y - 1 = 0$ for a straight line on the real plane gives the condition for a real point $[x\ y]^T$ to lie on that line.

Now, a projective point is a radial line and a projective line a radial plane. Moreover, a radial line lies on a radial plane if and only if any point of it, other than the origin, lies on that plane (the origin always does). See Figure A.14.

Therefore, a projective point $P = [x\ y\ z]^T$ belongs to a projective line L, whose radial plane has the equation $ax + by + cz = 0$, if and only if the real point $[x\ y\ z]^T$ lies on the real plane $ax + by + cz = 0$. It follows that the equation of L is identical to that of its radial plane:

$$ax + by + cz = 0 \tag{A.1}$$

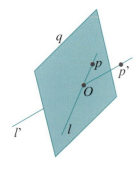

Figure A.14: Point p of radial line l lies on radial plane q, implying that l lies on q; point p' of l' doesn't lie on q, implying that no point of l', other than the origin, lies on q.

Accordingly, a projective point $P = [x\ y\ z]^T$ belongs to L if it satisfies (A.1). Does it matter if we choose some other homogeneous coordinates for P? No, because

$$a(kx) + b(ky) + c(kz) = k(ax + by + cz) = 0$$

so any homogeneous coordinates $[kx\ ky\ kz]^T$ for P satisfy Equation (A.1).

Exercise A.11. Prove that if the projective line L is specified by the equation

$$ax + by + cz = 0$$

then it is specified by any equation of the form

$$(ma)x + (mb)y + (mc)z = 0$$

where $m \neq 0$, as well.

Exercise A.12. What is the equation of the projective line through the projective points $[2\ 1\ -1]^T$ and $[3\ 4\ 2]^T$?

Answer: Suppose that the line is L with equation

$$ax + by + cz = 0$$

Since $[2\ 1\ -1]^T$ and $[3\ 4\ 2]^T$ lie on L they must satisfy its equation, giving

$$\begin{aligned}2a + b - c &= 0 \\ 3a + 4b + 2c &= 0\end{aligned}$$

Any solution to these simultaneous equations, not all zero, then determines L. As there are more variables then equations, let's set one of them, say c, arbitrarily to 1, to get the equations

$$\begin{aligned}2a + b - 1 &= 0 \\ 3a + 4b + 2 &= 0\end{aligned}$$

These solve to give $a = 1.2$ and $b = -1.4$. The equation of the projective line L is, therefore,

$$1.2x - 1.4y + z = 0$$

Exercise A.13. What is the projective point of intersection of the projective lines $3x + 2y - 4z = 0$ and $x - y + z = 0$?

Exercise A.14. When are three projective points $[x\ y\ z]^T$, $[x'\ y'\ z']^T$ and $[x''\ y''\ z'']^T$ collinear, i.e., when do they belong to the same projective line? Find a simple condition involving a determinant.

A.6.1   More About Point-Line Duality

In Section A.4.2 we tried to understand the point-line duality of the projective plane from a geometric point of view. We'll examine the phenomenon now from an algebraic standpoint.

The correspondence from the set of projective points to the set of projective lines given by

projective point $[a\ b\ c]^T$ $\mapsto$ projective line $ax + by + cz = 0$ (A.2)

is well-defined as, whatever homogeneous coordinates we choose for a projective point, the image is the same projective line (by Exercise A.11). Moreover, the correspondence is easily seen to be one-to-one and onto.

Definition A.3. The projective line $ax + by + cz = 0$ is said to be the *dual* of the projective point $[a\ b\ c]^T$ and vice versa.

Exercise A.15. Prove that a projective point P belongs to a projective line L if and only if the dual of L belongs to the dual of P.

The preceding exercise implies that if some statement about the incidence of projective points and lines is true, then so is the dual statement, obtained by replacing "point" with "line" and "line" with "point".

Exercise A.16. What is the dual of the following statement? "There is a unique projective line incident to two distinct projective points."

From this last exercise one sees, then, the point-line duality of the projective plane as a consequence of the one-to-one correspondence (A.2) between projective points and lines. We ask the reader to contemplate if there exists a similar correspondence between real points and lines.

A.6.2 Lifting an Algebraic Curve from the Real to the Projective Plane

Let's see next projective curves more complex than a line. Consider, then, the curve Q' in $\mathbb{P}^2$ consisting of the projective points intersecting the parabola q

$$y - x^2 = 0 \tag{A.3}$$

on $\mathbb{R}^2$ (the plane $z = 1$). See Figure A.15.

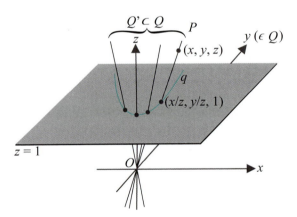

Figure A.15: Lifting a parabola drawn on the real plane $z = 1$ to the projective plane.

The intersection of the projective point $P = [x \ y \ z]^T$ with the plane $z = 1$ is the real point $[x/z \ y/z \ 1]^T$, assuming $z \neq 0$, for, otherwise, there is no intersection. Now, $[x/z \ y/z \ 1]^T$ satisfies the equation of the parabola q if

$$y/z - (x/z)^2 = 0 \quad \Longrightarrow \quad yz - x^2 = 0$$

Accordingly, the curve consisting of projective points $[x \; y \; z]^T$ which satisfy

$$yz - x^2 = 0 \tag{A.4}$$

is called the *lifting Q* of q from the real to the projective plane. Q is sometimes simply called the lifting of q and also the *projectivization* of q. In this particular case, as a lifting of a parabola, Q is a *parabolic projective curve*.

The camera analogy is that Q is that set of rays seen, by intersection with the film $z = 1$, as q. However, Q is actually one point *bigger* than Q', the set of projective points intersecting the parabola q on $\mathbb{R}^2$, as it includes the projective point $[0 \; 1 \; 0]^T$, the y-axis of 3-space, which satisfies (A.4), but does not intersect q. We can justify the inclusion of this extra point, with the help of the proviso from Section A.5 that a limit of visible rays is visible, as follows.

From its equation $y - x^2 = 0$, a point of q is of the form $[x \; x^2 \; 1]^T$, for any x. Therefore, the homogeneous coordinates of a projective point intersecting q are $[x \; x^2 \; 1]^T$, for any x, as well. Rewriting these coordinates as $[\frac{1}{x} \; 1 \; \frac{1}{x^2}]^T$ we see that its limit as $x \to \infty$ is indeed $[0 \; 1 \; 0]^T$. More intuitively, *a la* the thought experiment of Section A.4.2, as a point p travels off along either wing of the parabola, the projective point $\phi(p)$, corresponding to the line through p, approaches $[0 \; 1 \; 0]^T$, the projective point corresponding to the y-axis.

Definition A.4. A *homogeneous polynomial* is one whose terms each have the same degree, the degree of a term being the sum of the powers of the variables in the term. This common degree is called the degree of the homogeneous polynomial.

An equation with a homogeneous polynomial on the left and 0 on the right is called a homogeneous polynomial equation.

The equations $ax + by + cz = 0$ of a projective line and $yz - x^2 = 0$ of a parabolic projective curve are homogeneous polynomial equations of degree one and two, respectively. That they are both homogeneous is no accident, as we'll soon see.

Exercise A.17. Suppose that $p(x_1, x_2, \ldots, x_n)$ is a homogeneous polynomial in n variables. Then, if $[x_1 \; x_2 \; \ldots \; x_n]^T$ satisfies the equation $p(x_1, x_2, \ldots, x_n) = 0$, so does $[cx_1 \; cx_2 \; \ldots \; cx_n]^T$, for any scalar c. *Hint:* Show, first, that, if $p(x_1, x_2, \ldots, x_n)$ is homogeneous of degree r, then

$$p(cx_1, cx_2, \ldots, cx_n) = c^r p(x_1, x_2, \ldots, x_n)$$

For example, for the homogeneous polynomial $yz - x^2$ of degree 2,

$$(cy)(cz) - (cx)^2 = c^2(yz - x^2)$$

So, in this case, if (x, y, z) satisfies $yz - x^2 = 0$, then so does (cx, cy, cz), because $(cy)(cz) - (cx)^2 = 0$ as well, by the equation just above.

The preceding exercise implies that a homogeneous polynomial equation of the form $p(x, y, z) = 0$ is legitimately an equation in $\mathbb{P}^2$, because a point of $\mathbb{P}^2$ can be tested if it satisfies $p(x, y, z) = 0$, independently of the homogeneous coordinates used to represent it. Here are some definitions.

Definition A.5. An *algebraic curve* on the real plane consists of points satisfying an equation of the form

$$p(x, y) = 0$$

where p is a polynomial in the two variables x and y. The degree of the curve is the highest degree of a term belonging to $p(x, y)$.

A *projective algebraic curve* on the projective plane consists of points satisfying an equation of the form

$$p(x, y, z) = 0$$

where p is a homogeneous polynomial in the three variables x, y and z. The degree of the curve is the degree of $p(x, y, z)$.

Familiar algebraic curves of degree one include straight lines, e.g., $2x + y - 3 = 0$, and of degree two conic sections, e.g., the hyperbola $xy - 1 = 0$.

At the start of this section we lifted a parabola from the real to the projective plane. Here are a couple more examples of lifting.

Example A.7. Lift the straight line

$$ax + by + c = 0$$

drawn on the plane $z = 1$, to $\mathbb{P}^2$.

Answer: The projective point $[x \; y \; z]^T$ intersects the plane $z = 1$ at the real point $[x/z \; y/z \; 1]^T$ (assuming $z \neq 0$), which belongs to the given straight line if, replacing x by x/z and y by y/z in the latter's equation:

$$ax/z + by/z + c = 0$$

Multiplying throughout by z, one gets the homogeneous polynomial equation of degree 1

$$ax + by + cz = 0$$

confirming that the lifting of a straight line on $z = 1$ is the projective line corresponding to it. A projective line is, of course, a projective algebraic curve of degree 1.

Example A.8. Lift the algebraic curve of degree 3

$$x^3 + 3x^2y + y^2 + x + 2 = 0$$

drawn on the plane $z = 1$, to $\mathbb{P}^2$.

Answer: The projective point $[x \ y \ z]^T$ intersects the plane $z = 1$ at the real point $[x/z \ y/z \ 1]^T$ (assuming $z \neq 0$). Accordingly, replace x by x/z and y by y/z in the given polynomial equation:

$$(x/z)^3 + 3(x/z)^2(y/z) + (y/z)^2 + x/z + 2 = 0$$
$$\implies \quad x^3/z^3 + 3x^2y/z^3 + y^2/z^2 + x/z + 2 = 0$$
$$\implies \quad x^3 + 3x^2y + y^2z + xz^2 + 2z^3 = 0$$

defining the lifted curve, a projective algebraic curve of degree 3.

Exercise A.18. Lift the algebraic curve of degree 5

$$xy^4 - 2x^2y^2 + 3xy^2 + y^3 - xy + 2 = 0$$

drawn on the plane $z = 1$, to $\mathbb{P}^2$.

It should be fairly clear at this point that the lifting of an algebraic curve $p(x, y) = 0$ is a projective algebraic curve $\bar{p}(x, y, z) = 0$ of the same degree. We leave a formal proof to the reader in the following exercise.

Exercise A.19. Show that the lifting of an algebraic curve $p(x, y) = 0$ of degree r is a projective algebraic curve $\bar{p}(x, y, z) = 0$ of degree r.

Definition A.6. The process of going from the equation of an algebraic curve on the real plane to the homogeneous polynomial equation of its lifting is called *homogenization*.

It's worth keeping mind that the process of homogenization depends on the particular plane on which the algebraic equation holds. For example, in Examples A.7 and A.8 and Exercise A.18 the plane was $z = 1$. This need not always be the case as we see in the next example.

Example A.9. Homogenize the polynomial equation

$$y^2 + z^2 + z = 0$$

drawn on the plane $x = 2$.

Answer: The projective point $[x \ y \ z]^T$ intersects the plane $x = 2$ at the real point $[2 \ 2y/x \ 2z/x]^T$ (assuming $x \neq 0$, and multiplying $[x \ y \ z]^T$ by $2/x$). Accordingly, replace y by $2y/x$ and z by $2z/x$ in the given polynomial equation:

$$(2y/x)^2 + (2z/x)^2 + 2z/x = 0 \quad \implies \quad 4y^2/x^2 + 4z^2/x^2 + 2z/x = 0$$

Multiplying throughout by x^2 one gets the homogenized polynomial equation

$$4y^2 + 4z^2 + 2xz = 0$$

Not surprisingly, giving the algebraic equation on different real planes corresponds, simply, to specifying the algebraic curve as seen by the viewer on differently aligned films. The lifting itself, of course, is the set of rays intersecting the film in the given curve, which does not change.

Exercise A.20. Homogenize the polynomial equation

$$3x^4 + 2x^2y + 2y^3 + 2x^2 + xy + x = 0$$

drawn on the plane $z = 1$.

Exercise A.21. Homogenize the polynomial equation

$$x^3 + 2xz - z^4$$

drawn on the plane $y = 2$.

Remark A.7. It's possible to define the homogenization of a polynomial in an abstract manner independent of reference to a particular plane. See Jennings [72].

One sees, then, that the algebraic analogue of lifting an algebraic curve from the real to the projective plane is homogenization. The reverse process of projecting a (projective algebraic) curve onto a real plane consists of taking the section of the projective points composing the curve with the given plane. Algebraically, this means simultaneously solving the equation of the curve and that of the plane – a process not surprisingly called *de-homogenization*.

Example A.10. Project the curve

$$yz - x^2 = 0$$

in $\mathbb{P}^2$ onto the real plane $z = 1$.

Answer: De-homogenize the equation of the curve by simultaneously solving

$$
\begin{aligned}
yz - x^2 &= 0 \\
z &= 1
\end{aligned}
$$

to get

$$y - x^2 = 0$$

which is the equation of a parabola.

Exercise A.22. Project the curve of the preceding example onto the real plane $x = 1$.

Exercise A.23. Project the curve

$$4y^2 + 4z^2 + 2xz = 0$$

in $\mathbb{P}^2$ onto the real plane $y = -2$.

A.6.3 Snapshot Transformations Algebraically

It should make sense now that the snapshot transformation of an algebraic curve c from one real plane p to another p' can be determined by (a) first homogenizing the equation of c to lift it to the projective plane and then (b) de-homogenizing to project it back onto p'.

Example A.11. Let's solve the snapshot transformation problem of Exercise A.7(h) algebraically. The equation of the unit circle, centered at $[1 \ 0]^T$ on the $z = 1$ plane, is

$$x^2 + y^2 - 2x = 0$$

Homogenizing, one gets

$$x^2 + y^2 - 2xz = 0$$

To project onto the plane $x = 1$, de-homogenize by simultaneously solving

$$
\begin{aligned}
x^2 + y^2 - 2xz &= 0 \\
x &= 1
\end{aligned}
$$

to get

$$y^2 - 2z + 1 = 0 \quad \Longrightarrow \quad z = \frac{1}{2}y^2 + \frac{1}{2}$$

which indeed agrees with the sketch of a parabola in Figure A.13.

Exercise A.24. Solve Exercises A.7(g) and (i) algebraically.

A.7 The Dimension of the Projective Plane and Its Generalization to Higher Dimensions

Note: The next few paragraphs about $\mathbb{P}^2$ as a surface require recollecting some of the material from Section 10.2.12 on surface theory. If the reader is not inclined to do so, then she can safely skip ahead to Definition A.7. It won't affect her understanding of anything that follows.

Why do we say that the projective plane is a projective space of dimension 2? Because, as we'll see momentarily, $\mathbb{P}^2$ is a surface. In fact, it's a regular C^∞ surface, *except* that it is not a subset of $\mathbb{R}^3$: it's not possible to embed $\mathbb{P}^2$ in $\mathbb{R}^3$. One must go at least one dimension higher to $\mathbb{R}^4$.

Ignoring for now the question of the space in which it's embedded, it's not hard to find a coordinate patch containing any given point $P \in \mathbb{P}^2$. Suppose, for the moment, that P intersects the point p on the plane $z = 1$ (our favorite copy of $\mathbb{R}^2$). See Figure A.16. Let W be a closed rectangle

Section A.7
THE DIMENSION OF THE
PROJECTIVE PLANE
AND ITS
GENERALIZATION TO
HIGHER DIMENSIONS

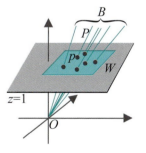

Figure A.16: The coordinate patch B containing P in $\mathbb{P}^2$ is in one-to-one correspondence with the rectangle W containing p in $\mathbb{R}^2$ (a few points in W and their corresponding projective points are shown).

containing p and B be the set of projective points intersecting W. The function

$$\text{point} \quad \mapsto \quad \text{the radial line through it}$$

from W to B is a one-to-one correspondence that makes B a coordinate patch.

And what if P doesn't intersect $z = 1$, i.e., if P is a point at infinity with respect to $z = 1$? Remember, there's nothing special about $z = 1$! Simply choose another non-radial plane with respect to which P is regular.

The reader has guessed by now that there exist projective spaces of various dimensions. True.

Definition A.7. A radial line in $\mathbb{R}^{n+1}$ is said to be an *n-dimensional projective point*. The set of all n-dimensional projective points is *n-dimensional projective space*, denoted $\mathbb{P}^n$.

$\mathbb{P}^0$, not very interestingly, is a one-point space as there is only one line, radial or otherwise, in $\mathbb{R}^1$. We'll try to convince the reader next, without being mathematically precise, that $\mathbb{P}^1$ is a circle.

Let U be the upper-half of a circle centered at the origin of $\mathbb{R}^2$. Associate with each radial line in $\mathbb{R}^2$ its intersection(s) with U. See Figure A.17, where, e.g., the radial line P is associated with the point p. Each radial line in $\mathbb{R}^2$ is then associated with a unique point of U, except for the x-axis, which we denote Q; Q intersects U in two points q_1 and q_2. And, the other way around, every point of U is associated with a unique radial line, except only for q_1 and q_2, which are associated with the same one Q. It follows, then, that the set $\mathbb{P}^1$ of all radial lines in $\mathbb{R}^2$ is in one-to-one correspondence with the space obtained by "identifying" the two endpoints q_1 and q_2 of U as one. But this latter space is clearly a circle (imagine U as a length of string whose ends are brought together).

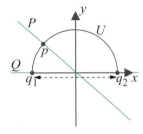

Figure A.17: Identifying $\mathbb{P}^1$ with a circle.

One can set up homogeneous coordinates for an arbitrary $\mathbb{P}^n$ in a manner similar to what we did for $\mathbb{P}^2$. For example, the homogeneous coordinates of a point $P \in \mathbb{P}^3$ are the coordinates of any point, other than the origin, on the radial line in $\mathbb{R}^4$ to which it corresponds. So the homogeneous coordinates of the point in $\mathbb{P}^3$ corresponding to the radial line through $[x \ y \ z \ w]^T$, where x, y, z and w are not all zero, is any tuple of the form $[cx \ cy \ cz \ cw]^T$, where $c \neq 0$.

It's hard to visualize $\mathbb{P}^3$ and higher-dimensional projective spaces for the same reason that it's hard to visualize $\mathbb{R}^4$ and higher-dimensional real spaces. The trick is to develop one's intuition in $\mathbb{P}^2$, as many of its properties do generalize.

A.8 Projective Transformations Defined

That the homogeneous coordinates of a point $P \in \mathbb{P}^2$ are of the form $[x \ y \ z]^T$ suggests defining transformations of $\mathbb{P}^2$ by mimicking the definition of a linear transformation of real 3-space. In particular, if

$$M = \begin{bmatrix} a_{11} & a_{12} & a_{13} \\ a_{21} & a_{22} & a_{23} \\ a_{31} & a_{32} & a_{33} \end{bmatrix}$$

is a 3×3 matrix, then tentatively define a transformation of $\mathbb{P}^2$ by

$$[x \ y \ z]^T \mapsto M[x \ y \ z]^T \tag{A.5}$$

This definition has the virtue at least of being unambiguous because

$$[cx \ cy \ cz]^T \mapsto M[cx \ cy \ cz]^T = c(M[x \ y \ z]^T)$$

which represents the same point as $M[x \ y \ z]^T$, implying that the choice of any homogeneous coordinates for P gives the same image by the transformation.

The potential glitch to consider before putting (A.5) into production is if it maps a non-zero tuple to a zero tuple, for then it maps the homogeneous coordinates of a point $P \in \mathbb{P}^2$ to a value not even belonging to $\mathbb{P}^2$. However, we know from basic linear algebra that there is a non-zero tuple $[x \ y \ z]^T$ such that

$$M[x \ y \ z]^T = [0 \ 0 \ 0]^T$$

if and only if M is a singular matrix; otherwise, M maps non-zero tuples to non-zero tuples. We conclude that defining a transformation of $\mathbb{P}^2$ by (A.5) is indeed valid provided M is non-singular. Ergo:

Definition A.8. If M is a non-singular 3×3 matrix, then the transformation

$$[x \ y \ z]^T \mapsto M[x \ y \ z]^T$$

denoted h^M, is called a *projective transformation* of the projective plane. The transformation f^M of $\mathbb{R}^3$ – the linear transformation defined by M – is called a *related* linear transformation.

A simple relation between h^M and f^M is the following: if the radial line corresponding to a point P of $\mathbb{P}^2$ is l, then that corresponding to $h^M(P)$ is $f^M(l)$, the image of l by f^M.

Exercise A.25. Prove that if M is a non-singular 3×3 matrix and c is a scalar such that $c \neq 0$, then M and cM define the same projective transformation of $\mathbb{P}^2$, i.e., $h^M = h^{cM}$.

Remark A.8. The preceding exercise implies that actually there is not a unique linear transformation related to a projective transformation h^M, because f^{cM} is related to $h^{cM} = h^M$, for any non-zero c. However, when we do have a specific M that we are using to define h^M, then we'll often speak of *the* related linear transformation f^M.

Exercise A.26. Prove that a projective transformation h^M of $\mathbb{P}^2$ takes projective lines to projective lines.

Hint: The related (non-singular) linear transformation f^M takes radial planes in $\mathbb{R}^3$ to radial planes in $\mathbb{R}^3$.

Exercise A.27. Prove that the composition $h^M \circ h^N$ of two projective transformations of $\mathbb{P}^2$ is equal to the projective transformation h^{MN}.

A.9 Projective Transformations Geometrically

Our definition of projective transformations was purely algebraic. We would like to picture, if possible, how they transform primitives in $\mathbb{P}^2$. Now, projective primitives are "seen" by projection onto the real plane – by capture on a point camera's film as we've been putting it. Let's find out, then, what a projective transformation looks like through a point camera.

Here's what we plan to do. Start with a primitive s, on the plane $z = 1$, our favorite copy of $\mathbb{R}^2$, as the designated film. Suppose that the given projective transformation is h^M. Then we'll transform the lifting S of s by h^M to $h^M(S)$. Finally, we'll project $h^M(S)$ back to $z = 1$ to obtain a new primitive s'. It's precisely the change from s to s' which is seen as the transformation h^M by a point camera at the origin. For example, in Figure A.18, a boxy car is changed (fancifully) into a sleek convertible.

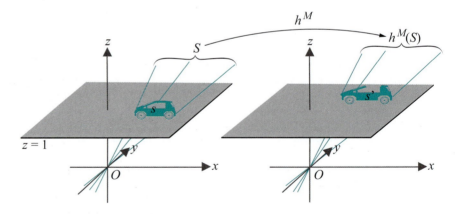

Figure A.18: Projective transformation of a car (purely conceptual!).

Back to reality, let's begin with a simple example. Consider a straight segment s joining two points p and q on $z = 1$. Given a projective transformation h^M, we want to determine s'. The lifting S of s, which is the set of all radial lines intersecting s, is not hard to visualize: it forms an "infinite double triangle" which lies on the radial plane containing s and the origin. See Figure A.19(a). The radial lines through p and q are denoted P and Q, respectively.

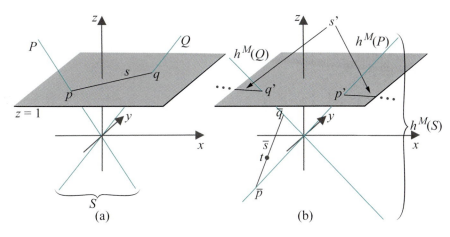

Figure A.19: (a) A segment s on $\mathbb{R}^2$ and its lifting S (b) f^M transforms s to $\bar{s}$ and S to $h^M(S)$, while s' is the intersection of $h^M(S)$ with $z = 1$.

The related linear transformation f^M transforms s to a segment $\bar{s} = \overline{\bar{p}\,\bar{q}}$, where $f^M(p) = \bar{p}$ and $f^M(q) = \bar{q}$. See Figure A.19(b). Note that $\bar{s}$ can be anywhere in 3-space, depending on f^M, unlike s and s', which are both on $z = 1$.

Moreover, each radial line in S, the lifting of s, is transformed by f^M to a radial line in $h^M(S)$. Each radial line in $h^M(S)$, of course, intersects $\bar{s}$. A diagram depicting a particular disposition of $\bar{s}$, where it intersects the xy-plane in a single point t, is shown in Figure A.19(b).

The transformed primitive s' is the intersection of the radial lines in $h^M(S)$ with $z = 1$. At this time we ask the reader to complete the following exercise to find out for herself what it looks like, depending on the situation of $\bar{s}$.

Exercise A.28. Show that exactly one of (a)-(c) is true:

(a) $\bar{s}$ does not intersect the xy-plane, equivalently, every radial line in $h^M(S)$ is a regular point with respect to $z = 1$.

In this case, s' is the segment between the points p' and q' where $h^M(P)$ and $h^M(Q)$, respectively, intersect $z = 1$ (remember that P and Q are the radial lines through p and q, the endpoints of s, respectively). Sketch this case.

(b) $\bar{s}$ intersects the xy-plane at one point, equivalently, exactly one radial line in $h^M(S)$ is a point at infinity with respect to $z = 1$. Now, there are two subcases:

(b1) If the intersection point, call it t, is in the interior of $\bar{s}$, then s' consists of the *entire* infinite straight line through p' and q', where $h^M(P)$ and $h^M(Q)$, respectively, intersect $z = 1$, *minus* the finite open segment between p' and q'. This situation is sketched in Figure A.19(b).

(b2) If the intersection is an endpoint of $\bar{s}$, say $\bar{p}$, then s' is a straight line infinite in one direction and with an endpoint at q', where $h^M(Q)$ intersects $z = 1$, in the other. Sketch this case.

(c) $\bar{s}$ lies on the xy-plane, equivalently, every radial line in $h^M(S)$ is a point at infinity with respect to $z = 1$.

In this case, s' is empty.

The answer to the preceding exercise is not tidy, but in most practical situations it will be case (a), the most benign of the three, that applies.

So we know now what we set out to find: what the projective transformation of the lifting of a segment looks like on film. Generally, for any primitive s on the plane, if s' is the "film-capture" of the transformation by h^M of the lifting of s, we'll call s' the *projective transformation* of s by h^M and denote it $h^M(s)$ – giving thus a geometric counterpart of the algebraic definition of a projective transformation in Section A.8. Although h^M is well-defined, it is not a transformation of $\mathbb{R}^2$ in general because $h^M(p)$ may not even exist for a point $p \in \mathbb{R}^2$, particularly if p's corresponding projective point is taken by h^M to a point at infinity (which has no film-capture).

In our usage, therefore, h^M can represent either a transformation of projective space (as defined in Section A.8) or one of real primitives (as just defined above). There is no danger of ambiguity as the nature of the argument in $h^M(*)$ will make clear how it's being used.

ExampIe A.12. The segment s joins $p = [1 \ -1]^T$ and $q = [-2 \ -2]^T$ on the plane $z = 1$, the latter identified with $\mathbb{R}^2$. The projective transformation $h^M : \mathbb{P}^2 \to \mathbb{P}^2$ is specified by

$$M = \begin{bmatrix} 0 & 0 & -1 \\ 0 & 1 & 0 \\ 1 & 0 & 0 \end{bmatrix}$$

which is the matrix corresponding to a rotation f^M of $\mathbb{R}^3$ by $90°$ about the y-axis, clockwise when seen from the positive side of the y-axis. Determine $h^M(s)$.

Answer: f^M transforms s to the segment $\bar{s} = \bar{p}\bar{q}$, where $\bar{p}$ and $\bar{q}$ are the images by f^M of p and q, respectively. Multiplying p and q, written as points of $z = 1$, on the left by M we get:

$$\bar{p} = M[1 \ -1 \ 1]^T = [-1 \ -1 \ 1]^T$$

and

$$\bar{q} = M[-2 \ -2 \ 1]^T = [-1 \ -2 \ -2]^T$$

As the z-values of $\bar{p}$ and $\bar{q}$ are of different signs, an interior point of $\bar{s}$ lies on the xy-plane. Therefore, we are in case (b1) of Exercise A.28 above.

Let P and Q denote the radial lines through p and q, respectively. The radial line $h^M(P)$ through $\bar{p}$ meets $z = 1$ at $h^M(p) = [-1 \ -1 \ 1]^T$, which is $\bar{p}$ itself. The radial line $h^M(Q)$ through $\bar{q}$ meets $z = 1$ at $h^M(q) = [\frac{1}{2} \ 1 \ 1]^T$ (multiplying the coordinate tuple of $\bar{q}$ by $-\frac{1}{2}$ to make its z-value equal to 1).

Applying Exercise A.28 case (b1), $h^M(s)$ is the entire straight line through the points $[-1 \ -1]^T$ and $[\frac{1}{2} \ 1]^T$ minus the finite open segment joining $[-1 \ -1]^T$ to $[\frac{1}{2} \ 1]^T$.

Example A.13. The rectangle r lies on the plane $z = 1$, the latter identified with $\mathbb{R}^2$. Its vertices are $p_1 = [0.5 \ 1]^T$, $p_2 = [0.5 \ -1]^T$, $p_3 = [1 \ -1]^T$ and $p_4 = [1 \ 1]^T$. See Figure A.20. Determine $h^M(r)$, where h^M is the same projective transformation as in the preceding example.

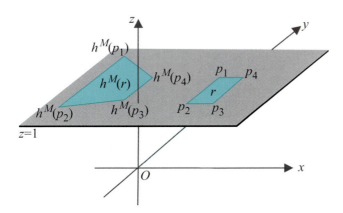

Figure A.20: Rectangle r is transformed to the trapezoid $h^M(r)$.

Answer: f^M transforms r to the rectangle $\bar{r}$ with vertices $\bar{p}_i = f^M(p_i)$, $1 \le i \le 4$. Multiplying each p_i, written as points of $z = 1$, on the left by M we get:

$$\begin{aligned}
\bar{p}_1 &= M[0.5 \ 1 \ 1]^T &= [-1 \ 1 \ 0.5]^T \\
\bar{p}_2 &= M[0.5 \ -1 \ 1]^T &= [-1 \ -1 \ 0.5]^T \\
\bar{p}_3 &= M[1 \ -1 \ 1]^T &= [-1 \ -1 \ 1]^T \\
\bar{p}_4 &= M[1 \ 1 \ 1]^T &= [-1 \ 1 \ 1]^T
\end{aligned}$$

As the z-value of every $\bar{p}_i$, $1 \le i \le 4$, is greater than 0, none of the edges of $\bar{r}$ intersects the xy-plane. According to case (a) of Exercise A.28 then, $h^M(r)$ is the quadrilateral with vertices at the points $h^M(p_i)$, where the radial lines through $\overline{p_i}$, $1 \le i \le 4$, intersect $z = 1$. See Figure A.20. Multiply the coordinate tuple of each $\overline{p_i}$ by a scalar to make its z-value equal to 1, to find that

$$
\begin{aligned}
h^M(p_1) &= [-2\ 2\ 1]^T \\
h^M(p_2) &= [-2\ -2\ 1]^T \\
h^M(p_3) &= [-1\ -1\ 1]^T \\
h^M(p_4) &= [-1\ 1\ 1]^T
\end{aligned}
$$

One sees, therefore, that $h^M(r)$ has vertices at $[-2\ 2]^T$, $[-2\ -2]^T$, $[-1\ -1]^T$ and $[-1\ 1]^T$, which makes it a trapezoid.

It's interesting to note that no affine transformation of $\mathbb{R}^2$ can map a rectangle to a trapezoid: because affine transformations preserve parallelism (see Proposition 5.1), at best they can transform a rectangle to a parallelogram.

Clearly, with the help of Exercise A.28 we can determine the projective transformation of any shape specified by straight edges. More general shapes are curved and curves specified by equations. Let's see, for example, how a parabola is projectively transformed.

Example A.14. Determine how the parabola $y - x^2 = 0$ on $z = 1$, the latter identified with $\mathbb{R}^2$, is mapped by the same projective transformation h^M as in the previous example.

Answer: The point $[x\ y]^T$ on $z = 1$, which has coordinates $[x\ y\ 1]^T$ in $\mathbb{R}^3$, is transformed by f^M to the point $[\overline{x}\ \overline{y}\ \overline{z}]^T$, where

$$
\begin{bmatrix} \overline{x} \\ \overline{y} \\ \overline{z} \end{bmatrix} = \begin{bmatrix} 0 & 0 & -1 \\ 0 & 1 & 0 \\ 1 & 0 & 0 \end{bmatrix} \begin{bmatrix} x \\ y \\ 1 \end{bmatrix} = \begin{bmatrix} -1 \\ y \\ x \end{bmatrix}
$$

which gives

$$
\overline{x} = -1, \qquad \overline{y} = y, \qquad \overline{z} = x
$$

The image $[x'\ y']^T$ of $[x\ y]^T$ by h^M, then, is the point $[\overline{x}/\overline{z}\ \overline{y}/\overline{z}]^T$, where the radial line through $[\overline{x}\ \overline{y}\ \overline{z}]^T$ intersects $z = 1$. Therefore:

$$
x' = \overline{x}/\overline{z} = -1/x \quad \Longrightarrow \quad x = -1/x'
$$

and

$$
y' = \overline{y}/\overline{z} = y/x \quad \Longrightarrow \quad y = y'x = -y'/x' \text{ (using } x = -1/x' \text{ from above)}
$$

Plugging these expressions for x and y into the equation of the parabola $y - x^2 = 0$, we have the equation

$$
-y'/x' - 1/x'^2 = 0, \text{ equivalently,} \quad x'y' + 1 = 0
$$

of the transformed curve, which describes a hyperbola.

Here's another rather interesting example.

Example A.15. Determine how points of $\mathbb{R}^2$, identified with $z = 1$, are transformed by the projective transformation h^M of $\mathbb{P}^2$ specified by

$$M = \begin{bmatrix} 1 & 0 & 7 \\ 0 & 1 & 0 \\ 0 & 0 & 1 \end{bmatrix}$$

Answer: The point $[x \ y]^T$ on $z = 1$, which has coordinates $[x \ y \ 1]^T$ in $\mathbb{R}^3$, is transformed by f^M to the point $[\overline{x} \ \overline{y} \ \overline{z}]^T$, where

$$\begin{bmatrix} \overline{x} \\ \overline{y} \\ \overline{z} \end{bmatrix} = \begin{bmatrix} 1 & 0 & 7 \\ 0 & 1 & 0 \\ 0 & 0 & 1 \end{bmatrix} \begin{bmatrix} x \\ y \\ 1 \end{bmatrix} = \begin{bmatrix} x + 7 \\ y \\ 1 \end{bmatrix}$$

giving

$$\overline{x} = x + 7, \quad \overline{y} = y, \quad \overline{z} = 1$$

The image $[x' \ y']^T$ of $[x \ y]^T$ by h^M, then, is the point $[\overline{x}/\overline{z} \ \overline{y}/\overline{z} \ 1]^T$, where the radial line through $[\overline{x} \ \overline{y} \ \overline{z}]^T$ intersects $z = 1$. Therefore,

$$x' = \overline{x}/\overline{z} = x + 7 \quad \text{and} \quad y' = \overline{y}/\overline{z} = y$$

which is nothing but a *translation* by 7 units in the x-direction.

A projection transformation has just done something beyond the reach of linear transformations, for a linear transformation cannot translate. Translations are in the domain of affine transformations. In fact, in Example A.13, we saw a projective transformation convert a rectangle into a trapezoid, something beyond even affine transformations. For transformations inspired by and defined by matrix-vector multiplication, just like linear transformations, projective transformations certainly seem to carry plenty of firepower. It turns out that this makes them particularly worthy allies in the advancement of computer graphics.

Incidentally, we did not pull the matrix M above out of a hat: it is the transformation matrix of a 3D shear whose plane is the xy-plane and line the x-axis (recall 3D shears from Section 5.4.8).

Exercise A.29. Find a projective transformation to translate points of $\mathbb{R}^2$ 3 units in the x-direction and 2 in the y-direction, i.e., whose displacement vector is $[3 \ 2]^T$.
Hint: Think another shear.

Exercise A.30. Determine how the segment s on $\mathbb{R}^2$, the latter identified with the plane $z = 1$, joining $p = [2 \ -2]^T$ and $q = [-2 \ 1]^T$, is mapped by the projective transformation h^M of $\mathbb{P}^2$ specified by

$$M = \begin{bmatrix} 0 & 1 & 1 \\ 1 & 0 & 1 \\ 1 & 1 & 0 \end{bmatrix}$$

Exercise A.31. Determine how the hyperbola $xy = 1$ on $z = 1$, the latter identified with $\mathbb{R}^2$, is mapped by the same projective transformation h^M as in the previous exercise.

Part answer: The problem is not hard but there is a fair amount of manipulation.

The point $[x \; y]^T$ on $z = 1$, which has coordinates $[x \; y \; 1]^T$ in $\mathbb{R}^3$, is transformed by f^M to the point $[\overline{x} \; \overline{y} \; \overline{z}]^T$, where

$$
\begin{bmatrix} \overline{x} \\ \overline{y} \\ \overline{z} \end{bmatrix} = \begin{bmatrix} 0 & 1 & 1 \\ 1 & 0 & 1 \\ 1 & 1 & 0 \end{bmatrix} \begin{bmatrix} x \\ y \\ 1 \end{bmatrix}
$$

Let's flip this equation over with the help of an inverse matrix:

$$
\begin{bmatrix} x \\ y \\ 1 \end{bmatrix} = \begin{bmatrix} 0 & 1 & 1 \\ 1 & 0 & 1 \\ 1 & 1 & 0 \end{bmatrix}^{-1} \begin{bmatrix} \overline{x} \\ \overline{y} \\ \overline{z} \end{bmatrix} = \frac{1}{2} \begin{bmatrix} -1 & 1 & 1 \\ 1 & -1 & 1 \\ 1 & 1 & -1 \end{bmatrix} \begin{bmatrix} \overline{x} \\ \overline{y} \\ \overline{z} \end{bmatrix}
$$

which gives

$$
x = \frac{1}{2}(-\overline{x} + \overline{y} + \overline{z}) \qquad y = \frac{1}{2}(\overline{x} - \overline{y} + \overline{z}) \qquad 1 = \frac{1}{2}(\overline{x} + \overline{y} - \overline{z})
$$

Plugging these expressions into the equation of the hyperbola $xy = 1 = 1^2$ we get:

$$
\frac{1}{4}(-\overline{x} + \overline{y} + \overline{z})(\overline{x} - \overline{y} + \overline{z}) = \frac{1}{4}(\overline{x} + \overline{y} - \overline{z})^2
$$

Now, the image $[x' \; y']^T$ of $[x \; y]^T$ by h^M is the point $[\overline{x}/\overline{z} \; \; \overline{y}/\overline{z}]^T$, where the radial line through $[\overline{x} \; \overline{y} \; \overline{z}]^T$ intersects $z = 1$. We ask the reader to complete the exercise by dividing the preceding equation by $\overline{z}^2$ throughout to obtain an equation relating x' and y', and identifying the corresponding curve.

Exercise A.32. Determine how the straight line $x + y + 1 = 0$ on $z = 1$ is mapped by the same projective transformation h^M as in the previous exercise.

Exercise A.33. We saw in Example 5.4 that affine transformations preserve convex combinations and barycentric coordinates. Show that projective transformations in general do not.

Remark A.9. Projective transformations of $\mathbb{P}^2$ can be thought of as a powerful class of *pseudo-transformations* of $\mathbb{R}^2$ – pseudo because a projective transformation may map a regular point to a point at infinity, in which case the corresponding point of $\mathbb{R}^2$ has no valid image. If one is careful, however, to restrict its domain to a region of $\mathbb{R}^2$ where it *is* valid throughout, one may be able to exploit the ability of a projective transformation to do more than an affine one.

A.10 Relating Projective, Snapshot and Affine Transformations

We'll explore in this section the inter-relationships between projective, snapshot and affine transformations.

A.10.1 Snapshot Transformations via Projective Transformations

Snapshot transformations, being transformations of an object seen through a point camera as the film changes alignment, are geometrically intuitive. They are, in fact, a kind of projective transformation, as we'll now see.

Consider again Example A.13 for motivation. We saw that the rectangle r on the plane $z = 1$ (aka $\mathbb{R}^2$) with vertices at $p_1 = [0.5 \ 1]^T$, $p_2 = [0.5 \ -1]^T$, $p_3 = [1 \ -1]^T$ and $p_4 = [1 \ 1]^T$ is mapped to the trapezoid $r' = h^M(r)$ with vertices at $p_1' = [-2 \ 2]^T$, $p_2' = [-2 \ -2]^T$, $p_3' = [-1 \ -1]^T$ and $p_4' = [-1 \ 1]^T$, by the projective transformation h^M specified by

$$
M = \begin{bmatrix} 0 & 0 & -1 \\ 0 & 1 & 0 \\ 1 & 0 & 0 \end{bmatrix}
$$

See Figure A.21(a).

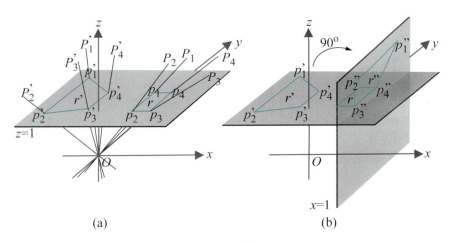

(a) (b)

Figure A.21: (a) Projective transformation h^M maps rectangle r to trapezoid $r' = h^M(r)$ (b) r' is the "same" as r'', the picture of r captured on a film along $x = 1$.

We observed, as well, that the related linear transformation f^M is a rotation of $\mathbb{R}^3$ by $90°$ about the y-axis, which is clockwise when seen from the positive side of the y-axis.

Denote the radial line through p_i by P_i, $1 \leq i \leq 4$, and their respective images $h^M(P_i)$ by P_i'. Now rotate the radial lines P_i', *as well as* the plane $z = 1$, an angle of 90° about the y-axis, this time counter-clockwise when seen from the positive side of the y-axis, in order to undo the effect of f^M; in other words, apply $f^{M^{-1}}$. We see the following:

(a) The radial line P_i', of course, rotates back onto (its pre-image) the radial line P_i, $1 \leq i \leq 4$.

(b) The plane $z = 1$ is taken by the rotation onto the plane $x = 1$.

(c) The trapezoid r', as a consequence of (a) and (b), rotates onto a trapezoid r'' with vertices at the intersections p_i'' of P_i with $x = 1$, for $1 \leq i \leq 4$. See Figure A.21(b) (note that the edge of r that happens to lie on the intersection of the planes $z = 1$ and $x = 1$ is shared with r'').

But r'' is precisely the snapshot transformation of r from the film along $z = 1$ to the one along $x = 1$! Here's what is happening. The image r' is obtained by applying the rotation f^M to the radials P_i and intersecting them with the plane $z = 1$, while r'' is obtained from r' by applying the reverse rotation $f^{M^{-1}}$, which takes the radials back to the where they were, and, at the same time, changes the intersecting plane from $z = 1$ to $x = 1$. Therefore, the transformation from r to r'' comes from a change in the plane (= film) intersecting the radials, which is precisely a snapshot transformation.

One sees, therefore, that, generally, a snapshot transformation in which the film is re-aligned by a rotation f about a radial axis is equivalent to a projective transformation whose related linear transformation is f^{-1}, in that the images are identical, though situated differently in space (precisely, the two images differ by a rigid transformation of $\mathbb{R}^3$). But, how about snapshot transformations where the new alignment of the film cannot be obtained from the original by mere rotation? To answer this question, we ask the reader, first, to prove the following, which says that an arbitrary snapshot transformation can be composed from two very simple ones.

Exercise A.34. Prove that any plane p in $\mathbb{R}^3$ can be aligned with any other p' by a translation parallel to itself followed by a rotation about a radial axis.

Therefore, any snapshot transformation is the composition of two: first, a snapshot transformation from one film to a parallely translated one and then another, where one film is obtained from the other by a rotation about a radial axis.

Hint: See Figure A.22.

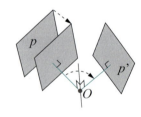

Figure A.22: Aligning plane p with p' by a parallel displacement, so that their respective distances from the origin are equal, followed by a rotation.

We have already seen how a snapshot transformation from one film to a rotated one is equivalent to a projective transformation. A snapshot transformation to a parallely translated one is also equivalent to a projective transformation, as the next exercise asks the reader to show.

Exercise A.35. Suppose that two parallel non-radial planes p and p' in $\mathbb{R}^3$ are at a distance of c and c' from the origin, respectively. Then the snapshot transformation from p to p' is equivalent to the projective transformation h^M, where

$$M = \begin{bmatrix} \frac{c'}{c} & 0 & 0 \\ 0 & \frac{c'}{c} & 0 \\ 0 & 0 & \frac{c'}{c} \end{bmatrix} = \frac{c'}{c} I$$

(i.e., a projective transformation whose related linear transformation is a *uniform* scaling of $\mathbb{R}^3$ by a factor of $\frac{c'}{c}$ in all directions).

Hint: See Figure A.23.

Putting the pieces together we have the following proposition:

Proposition A.1. *A snapshot transformation k from a non-radial plane p in $\mathbb{R}^3$ to another p' is equivalent to a projective transformation h^M of $\mathbb{P}^2$, in the sense that the images of primitives by k and h^M are identical modulo a rigid transformation of $\mathbb{R}^3$.*

In particular, k is equivalent to the projective transformation h^M which is the composition of a projective transformation h^{dI}, whose related linear transformation is a uniform scaling, with a projective transformation h^N, whose related linear transformation is a rotation of $\mathbb{R}^3$ about a radial axis.

In other words, k is equivalent to h^{dN}, where d is a scalar and N is the matrix of a rotation of $\mathbb{R}^3$ about a radial axis. □

Exercise A.36. Determine the projective transformation equivalent to the snapshot transformation from the plane $z = 1$ to the plane $x = 2$.

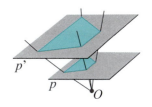

Figure A.23: A snapshot transformation to a parallel plane is equivalent to a scaling by a constant factor in all directions.

A.10.2   Affine Transformations via Projective Transformations

We begin by asking if there exist projective transformations that respect regular points, i.e., map regular points to regular points. Such a transformation could then be *entirely* captured on film because it takes no point of the film out of it, as would happen, say, if a regular point were mapped to one at infinity. Looking back at Remark A.9, one could then say that such a projective transformation is no longer "pseudo", but a true transformation of $\mathbb{R}^2$.

So suppose the film lies along the plane (surprise) $z = 1$. What condition must a projective transformation h^M, where

$$M = \begin{bmatrix} a_{11} & a_{12} & a_{13} \\ a_{21} & a_{22} & a_{23} \\ a_{31} & a_{32} & a_{33} \end{bmatrix}$$

satisfy in order to transform each point regular with respect to $z = 1$ to another such? Homogeneous coordinates of regular points are of the form

$[x \ y \ 1]^T$. Now,

$$h^M([x \ y \ 1]^T) = M[x \ y \ 1]^T = \begin{bmatrix} a_{11}x + a_{12}y + a_{13} \\ a_{21}x + a_{22}y + a_{23} \\ a_{31}x + a_{32}y + a_{33} \end{bmatrix}$$

For this image point to be regular we must have

$$a_{31}x + a_{32}y + a_{33} \neq 0$$

However, if either one of a_{31} and a_{32} is non-zero, or if a_{33} is zero, then it's possible to find values of x and y such that $a_{31}x + a_{32}y + a_{33} = 0$. The conclusion then is that for h^M to transform all regular points to regular points, one must have both a_{31} and a_{32} equal to zero and a_{33} non-zero. Therefore, M must be of the form

$$\begin{bmatrix} a_{11} & a_{12} & a_{13} \\ a_{21} & a_{22} & a_{23} \\ 0 & 0 & a_{33} \end{bmatrix}$$

with $a_{33} \neq 0$. By Exercise A.25, M can be multiplied by $1/a_{33}$ to still represent the same projective transformation, so one can assume $a_{33} = 1$, implying that the form of M is

$$\begin{bmatrix} a_{11} & a_{12} & a_{13} \\ a_{21} & a_{22} & a_{23} \\ 0 & 0 & 1 \end{bmatrix}$$

In this case, h^M transforms $[x \ y \ 1]^T$ to

$$\begin{bmatrix} a_{11} & a_{12} & a_{13} \\ a_{21} & a_{22} & a_{23} \\ 0 & 0 & 1 \end{bmatrix} \begin{bmatrix} x \\ y \\ 1 \end{bmatrix} = \begin{bmatrix} a_{11}x + a_{12}y + a_{13} \\ a_{21}x + a_{22}y + a_{23} \\ 1 \end{bmatrix}$$

Tossing the last coordinate, it transforms $[x \ y]^T \in \mathbb{R}^2$ to

$$\begin{bmatrix} a_{11}x + a_{12}y + a_{13} \\ a_{21}x + a_{22}y + a_{23} \end{bmatrix} = \begin{bmatrix} a_{11} & a_{12} \\ a_{21} & a_{22} \end{bmatrix} \begin{bmatrix} x \\ y \end{bmatrix} + \begin{bmatrix} a_{13} \\ a_{23} \end{bmatrix}$$

which is precisely an affine transformation!

We conclude that a projective transformation of $\mathbb{P}^2$ that respects regular points gives nothing but an affine transformation of $\mathbb{R}^2$. Conversely, it's not hard to see that any affine transformation of $\mathbb{R}^2$ can be obtained as a projective transformation preserving regular points. We record these facts in the following proposition.

Proposition A.2. *An affine transformation of $\mathbb{R}^2$ is equivalent to a projective transformation of $\mathbb{P}^2$, in particular, one that respects regular points.*

Conversely, a projective transformation of $\mathbb{P}^2$ that respects regular points is equivalent to an affine transformation of $\mathbb{R}^2$.

Evidently, the constraint to respect regular points is a burden on projective transformations. It dumbs them down to affine and all the excitement of parallel lines turning into intersecting ones, rectangles into trapezoids and circles into hyperbolas is lost!

However, one does see now a good reason for the use of homogeneous coordinates of real points in computing affine transformations. When first we did this in Section 5.2.3, it seemed merely a neat maneuver to obtain an affine transformation as a single matrix-vector multiplication. The bigger picture is that affine transformations are a subclass of the projective. Therefore, as the latter are obtained (by definition) from matrix-vector multiplication, so can the former, provided we relocate to projective space, in other words, use homogeneous coordinates.

A Roundup of the Three Kinds of Transformations

Snapshot and affine transformations are subclasses of the projective, as we have just seen. How about the relationship between these two subclasses themselves? Are snapshot transformations affine or affine transformations snapshot?

At the start of Section A.10.1 we saw a projective transformation, equivalent, in fact, to a snapshot transformation, map a rectangle to a trapezoid. This is not possible for an affine transformation to do, as it is obliged to preserve parallelism (Proposition 5.1). Therefore, snapshot transformations are certainly not all affine.

A shear on the plane, an affine transformation, can map a rectangle to a non-rectangular parallelogram. We leave the reader to convince herself that this is not possible for a snapshot transformation. So not all affine transformations are snapshot.

We see then that neither of the two subclasses, snapshot and affine, of projective transformations contains the other. However, what transformations, if any, do they have in common? We ask the reader herself to characterize the transformations at the intersection of affine and snapshot in the next exercise.

Exercise A.37. Prove that projective transformations which are both affine and snapshot are precisely those whose related linear transformation is a uniform scaling.

Figure A.24: Venn diagram of transformation classes of $\mathbb{R}^2$.

The final important question on the relationship between the three classes is if the union of snapshot and affine covers projective transformations or if the latter is strictly bigger. In Exercise A.40 in the next section we'll see an example of a projective transformation neither snapshot nor affine. Therefore, indeed, the class of projective transformations is strictly bigger than the union of snapshot and affine. Figure A.24 summarizes the relationship between the three classes.

A.11 Designing a Projective Transformation to Specification

We know from elementary linear algebra that a linear transformation is specified by specifying its values on a basis. Here's a like-minded claim for projective transformations of $\mathbb{P}^2$.

Proposition A.3. *If two sets $\{P_1, P_2, P_3, P_4\}$ and $\{Q_1, Q_2, Q_3, Q_4\}$ of four points each from $\mathbb{P}^2$ are such that no three in any one set are collinear, then there is a unique projective transformation of $\mathbb{P}^2$ that maps P_i to Q_i, for $1 \leq i \leq 4$.*

Proof. Choose non-zero vectors p_1, p_2, p_3 and p_4 from $\mathbb{R}^3$ lying on P_1, P_2, P_3 and P_4, respectively, and non-zero vectors q_1, q_2, q_3 and q_4 lying on Q_1, Q_2, Q_3 and Q_4, respectively.

Since P_1, P_2 and P_3 do not lie on one projective line, p_1, p_2 and p_3 do not lie on one radial plane. The latter three form, therefore, a basis of $\mathbb{R}^3$. Likewise, q_1, q_2 and q_3 form a basis of $\mathbb{R}^3$ as well.

Let c_1, c_2 and c_3 be arbitrary scalars, all three non-zero, whose values will be determined. As q_1, q_2 and q_3 form a basis of $\mathbb{R}^3$, so do $c_1 q_1$, $c_2 q_2$ and $c_3 q_3$. Therefore, there is a unique non-singular linear transformation $f^M : \mathbb{R}^3 \to \mathbb{R}^3$ such that

$$f^M(p_i) = c_i q_i, \text{ for } 1 \leq i \leq 3$$

which, then, is related to a projective transformation $h^M : \mathbb{P}^2 \to \mathbb{P}^2$, such that

$$h^M(P_i) = Q_i, \text{ for } 1 \leq i \leq 3$$

It remains to make $h^M(P_4) = Q_4$.

As p_1, p_2 and p_3 form a basis of $\mathbb{R}^3$, there exist unique scalars α, β and γ such that

$$p_4 = \alpha p_1 + \beta p_2 + \gamma p_3$$

Now, α, β and γ are all three non-zero, for, otherwise, p_4 lies on the same radial plane as two of p_1, p_2 and p_3, which implies that P_4 lies on the same projective line as two of P_1, P_2 and P_3, contradicting an initial hypothesis. Likewise, there exist unique non-zero scalars, λ, μ and ν such that

$$q_4 = \lambda q_1 + \mu q_2 + \nu q_3$$

For

$$h^M(P_4) = Q_4$$

to hold, then, one requires a scalar $c_4 \neq 0$ such that

$$
\begin{aligned}
f^M(p_4) &= c_4 q_4 \\
&= c_4(\lambda q_1 + \mu q_2 + \nu q_3) \\
&= \lambda c_4 q_1 + \mu c_4 q_2 + \nu c_4 q_3
\end{aligned}
\tag{A.6}
$$

However,

$$\begin{aligned}
f^M(p_4) &= f^M(\alpha p_1 + \beta p_2 + \gamma p_3) \\
&= \alpha f^M(p_1) + \beta f^M(p_2) + \gamma f^M(p_3) \\
&= \alpha c_1 q_1 + \beta c_2 q_2 + \gamma c_3 q_3
\end{aligned} \tag{A.7}$$

Combining (A.6) and (A.7) one has

$$\alpha c_1 q_1 + \beta c_2 q_2 + \gamma c_3 q_3 = \lambda c_4 q_1 + \mu c_4 q_2 + \nu c_4 q_3$$

As q_1, q_2 and q_3 is a basis of $\mathbb{R}^3$, it follows that

$$\alpha c_1 = \lambda c_4, \quad \beta c_2 = \mu c_4, \quad \gamma c_3 = \nu c_4$$

giving

$$c_1 = (\lambda/\alpha)c_4, \quad c_2 = (\mu/\beta)c_4, \quad c_3 = (\nu/\gamma)c_4$$

determining c_1, c_2, c_3 and c_4 uniquely, up to a constant of proportionality, thereby completing the proof. $\square$

The following corollary, which is a straightforward application of the proposition, is particularly important.

Corollary A.1. *Any non-degenerate quadrilateral, i.e., one with no three collinear vertices, in $\mathbb{R}^2$ can be projectively transformed to any other such.* $\square$

More than just theoretically, the proposition is important in that it suggests how to go about finding projective transformations specified at only a few points.

Example A.16. Determine the projective transformation h^M of $\mathbb{P}^2$ mapping the projective points

$$P_1 = [1\ 0\ 0]^T, \ P_2 = [0\ 1\ 0]^T, \ P_3 = [0\ 0\ 1]^T \text{ and } P_4 = [1\ 1\ 1]^T$$

to the respective images

$$Q_1 = [2\ 1\ 3]^T, \ Q_2 = [-1\ -1\ 1]^T, \ Q_3 = [0\ 1\ 1]^T \text{ and } Q_4 = [0\ 0\ 6]^T$$

Answer: Choose (not particularly imaginatively)

$$p_1 = [1\ 0\ 0]^T, \ p_2 = [0\ 1\ 0]^T, \ p_3 = [0\ 0\ 1]^T \text{ and } p_4 = [1\ 1\ 1]^T$$

in $\mathbb{R}^3$ lying on P_i, $1 \leq i \leq 4$, and

$$q_1 = [2\ 1\ 3]^T, \ q_2 = [-1\ -1\ 1]^T, \ q_3 = [0\ 1\ 1]^T \text{ and } q_4 = [0\ 0\ 6]^T$$

lying on Q_i, $1 \leq i \leq 4$.

The linear transformation $f^M : \mathbb{R}^3 \to \mathbb{R}^3$ such that $f^M(p_i) = c_i q_i$, for $1 \le i \le 3$, where c_1, c_2 and c_3 are non-zero scalars, is easily verified to be given by

$$M = \begin{bmatrix} 2c_1 & -c_2 & 0 \\ c_1 & -c_2 & c_3 \\ 3c_1 & c_2 & c_3 \end{bmatrix}$$

One can verify as well that

$$p_4 = p_1 + p_2 + p_3$$

and

$$q_4 = q_1 + 2q_2 + q_3$$

Therefore,

$$f^M(p_4) = f^M(p_1 + p_2 + p_3) = f^M(p_1) + f^M(p_2) + f^M(p_3) = c_1 q_1 + c_2 q_2 + c_3 q_3$$

Accordingly, if $f^M(p_4) = c_4 q_4$, for some $c_4 \ne 0$, then

$$c_1 q_1 + c_2 q_2 + c_3 q_3 = c_4(q_1 + 2q_2 + q_3) = c_4 q_1 + 2c_4 q_2 + c_4 q_3$$

which implies that

$$c_1 = c_4, \quad c_2 = 2c_4, \quad c_3 = c_4$$

Setting $c_4 = 1$, one has: $c_1 = 1$, $c_2 = 2$, $c_3 = 1$, $c_4 = 1$. One concludes that the required projective transformation h^M is given by

$$M = \begin{bmatrix} 2 & -2 & 0 \\ 1 & -2 & 1 \\ 3 & 2 & 1 \end{bmatrix}$$

The following example will help in an application of projective transformations in the graphics pipeline.

Example A.17. Determine the projective transformation h^M of $\mathbb{P}^2$ that transforms the trapezoid q on the plane $z = 1$ (aka $\mathbb{R}^2$) with vertices at

$$p_1 = [-1 \ 1]^T, \ p_2 = [1 \ 1]^T, \ p_3 = [2 \ 2]^T \text{ and } p_4 = [-2 \ 2]^T$$

to the rectangle q' on the same plane with vertices at

$$p_1' = [-1 \ 1]^T, \ p_2' = [1 \ 1]^T, \ p_3' = [1 \ 2]^T \text{ and } p_4' = [-1 \ 2]^T$$

See Figure A.25.

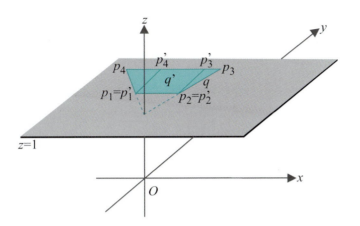

Figure A.25: Transforming the trapezoid q on $z = 1$ to the rectangle (bold) q'.

Answer: Suppose that the required projective transformation h^M is defined by the matrix

$$M = \begin{bmatrix} a_{11} & a_{12} & a_{13} \\ a_{21} & a_{22} & a_{23} \\ a_{31} & a_{32} & a_{33} \end{bmatrix}$$

We have to determine the a_{ij} up to a non-zero multiplicative constant.

The two sides $p_1 p_4$ and $p_2 p_3$ of the trapezoid q meet at the regular point (with respect to $z = 1$) $[0\ 0\ 1]^T$, while the corresponding sides $p_1' p_4'$ and $p_2' p_3'$ of the rectangle q' are parallel and meet at the point at infinity $[0\ 1\ 0]^T$. The transformation must, therefore, map $[0\ 0\ 1]^T$ to $[0\ 1\ 0]^T$, yielding our first equation

$$h^M([0\ 0\ 1]^T) = [0\ 1\ 0]^T$$

(the RHS could be $c[0\ 1\ 0]^T$ for any non-zero scalar c, but there's no loss in assuming that $c = 1$) which translates to the matrix equation

$$\begin{bmatrix} a_{11} & a_{12} & a_{13} \\ a_{21} & a_{22} & a_{23} \\ a_{31} & a_{32} & a_{33} \end{bmatrix} \begin{bmatrix} 0 \\ 0 \\ 1 \end{bmatrix} = \begin{bmatrix} 0 \\ 1 \\ 0 \end{bmatrix}$$

giving

$$a_{13} = 0, \quad a_{23} = 1, \quad a_{33} = 0$$

So we write

$$M = \begin{bmatrix} a_{11} & a_{12} & 0 \\ a_{21} & a_{22} & 1 \\ a_{31} & a_{32} & 0 \end{bmatrix}$$

That we have $h^M(p_1) = p_1'$ and $h^M(p_2) = p_2'$ gives two more matrix equations

$$\begin{bmatrix} a_{11} & a_{12} & 0 \\ a_{21} & a_{22} & 1 \\ a_{31} & a_{32} & 0 \end{bmatrix} \begin{bmatrix} -1 \\ 1 \\ 1 \end{bmatrix} = \begin{bmatrix} -c \\ c \\ c \end{bmatrix} \quad \text{and} \quad \begin{bmatrix} a_{11} & a_{12} & 0 \\ a_{21} & a_{22} & 1 \\ a_{31} & a_{32} & 0 \end{bmatrix} \begin{bmatrix} 1 \\ 1 \\ 1 \end{bmatrix} = \begin{bmatrix} d \\ d \\ d \end{bmatrix}$$

where c and d are arbitrary non-zero scalars, yielding the six equations

$$\begin{aligned}
-a_{11} + a_{12} &= -c \\
-a_{21} + a_{22} + 1 &= c \\
-a_{31} + a_{32} &= c \\
a_{11} + a_{12} &= d \\
a_{21} + a_{22} + 1 &= d \\
a_{31} + a_{32} &= d
\end{aligned}$$ (A.8)

Subtracting the first equation from the fourth, adding the second and fifth, and adding the third and sixth, one gets

$$a_{11} = \frac{c+d}{2}, \qquad a_{22} = \frac{c+d}{2} - 1, \qquad a_{32} = \frac{c+d}{2}$$

implying that

$$a_{22} = a_{11} - 1 \quad \text{and} \quad a_{32} = a_{11}$$

Likewise, adding the first and fourth equations, subtracting the second from the fifth, and subtracting the third from the sixth, one gets

$$a_{12} = a_{21} = a_{31}$$

We can now write

$$M = \begin{bmatrix} a_{11} & a_{12} & 0 \\ a_{12} & a_{11} - 1 & 1 \\ a_{12} & a_{11} & 0 \end{bmatrix}$$

That $h^M(p_3) = p_3'$ and $h_M(p_4) = p_4'$ give another two matrix equations

$$\begin{bmatrix} a_{11} & a_{12} & 0 \\ a_{12} & a_{11} - 1 & 1 \\ a_{12} & a_{11} & 0 \end{bmatrix} \begin{bmatrix} 2 \\ 2 \\ 1 \end{bmatrix} = \begin{bmatrix} e \\ 2e \\ e \end{bmatrix} \quad \text{and}$$

$$\begin{bmatrix} a_{11} & a_{12} & 0 \\ a_{12} & a_{11} - 1 & 1 \\ a_{12} & a_{11} & 0 \end{bmatrix} \begin{bmatrix} -2 \\ 2 \\ 1 \end{bmatrix} = \begin{bmatrix} -f \\ 2f \\ f \end{bmatrix}$$

where e and f are arbitrary non-zero scalars. Again one obtains six equations, as in (A.8), which can be solved to find that

$$a_{11} = -1/2 \quad \text{and} \quad a_{12} = 0$$

We have, finally, that

$$M = \begin{bmatrix} -1/2 & 0 & 0 \\ 0 & -3/2 & 1 \\ 0 & -1/2 & 0 \end{bmatrix}$$

(or, a non-zero scalar multiple of the matrix on the RHS).

Exercise A.38. The projective transformation h^M of the preceding example mapped, by design, the regular point $[0\ 0\ 1]^T$ to the point at infinity $[0\ 1\ 0]^T$. What other regular points, if any, does it map to a point at infinity?

Exercise A.39. Determine the projective transformation h^M of $\mathbb{P}^2$ that transforms the rectangle q on the plane $z = 1$ with vertices at

$$p_1 = [0.5\ 1]^T,\ p_2 = [0.5\ -1]^T,\ p_3 = [1\ -1]^T \text{ and } p_4 = [1\ 1]^T$$

to the trapezoid q' on $z = 1$ with vertices at

$$p_1' = [-2\ 2]^T,\ p_2' = [-2\ -2]^T,\ p_3' = [-1\ -1]^T \text{ and } p_4' = [-1\ 1]^T$$

(see Example A.13 earlier for the solution).

Exercise A.40. Prove that there exist projective transformations that are neither affine nor snapshot.

Suggested approach: Corollary A.1 implies that a square can be projectively transformed to any non-degenerate quadrilateral. Non-degenerate quadrilaterals q' that can be obtained from a square q by a snapshot transformation are the intersections of a non-radial plane with the "cone" C through q (see Figure A.26). Those that can be obtained by an affine transformation, on the other hand, are parallelograms.

Therefore, if one can find a non-degenerate quadrilateral q'' which is neither a parallelogram, nor the intersection of C with a plane, then one shows that there exists a projective transformation neither affine nor snapshot.

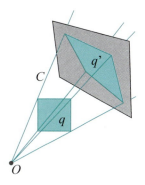

Figure A.26: The square q is mapped to the quadrilateral q' by a snapshot transformation.

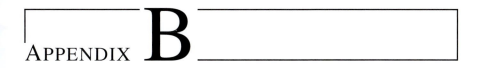

APPENDIX B

Installing OpenGL and Running Code

(by Chansophea Chuon)

W e explain here how to install OpenGL and run the book programs, and create and execute your own, on Windows, Mac OS and Ubuntu Linux platforms.

B.1 Microsoft Windows XP and Higher

Download and install Microsoft Visual C++ 2010 Express edition from `http://www.microsoft.com/express/`. After Visual C++ has been successfully installed, do the following.

- Install GLUT:

  1. Download and unzip the file `glut-3.7.6-bin.zip` from `http://www.xmission.com/~nate/glut.html`.

     (a) Copy `glut32.dll` to `C:\Windows\System32`.
     (b) Copy `glut32.lib` to `C:\Program Files\Microsoft Visual Studio 10.0\VC\lib`
     (c) Copy `glut.h` to `C:\Program Files\Microsoft Visual Studio 10.0\VC\include\GL`. Note that you may have to create the `GL` directory.

- Install GLEW (if your graphics card supports at least OpenGL 1.5):

789

1. Download `glext.h` from `http://www.opengl.org/registry/api/glext.h` to `C:\Program Files\Microsoft Visual Studio 10.0\VC\include\GL`.

2. Download and unzip the file `glew-1.5.4-win32.zip` from `http://glew.sourceforge.net/`.

   (a) Copy `glew32.dll` from the `bin` directory to `C:\Windows\System32`.

   (b) Copy `glew32.lib` and `glew32s.lib` from the `lib` directory to `C:\Program Files\Microsoft Visual Studio 10.0\VC\lib`

   (c) Copy `glew.h`, `glxew.h` and `wglew.h` from the `include\GL` directory to `C:\Program Files\Microsoft Visual Studio 10.0\VC\include\GL`.

Now you are ready to run the book programs and write and run your own. Follow the steps in the next pages.

- Open Visual C++ 2010 from the Start Menu to bring up the welcome screen. See Figure B.1.

Figure B.1: Visual Studio 2010 Express welcome screen.

- Create a new project by going to File → New → Project. See Figure B.2.

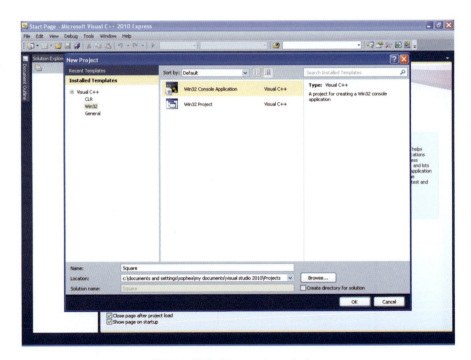

Figure B.2: New project window.

- Select Win32 from the Installed Templates panel and then Win32 Console Application from the next panel. Name your project and select the folder where you want to save it. Uncheck the box which says "Create directory for solution". Click OK to bring up a wizard welcome window. See Figure B.3.

Figure B.3: Win32 application wizard welcome window.

- Click Application Settings for the settings dialog box. See Figure B.4.

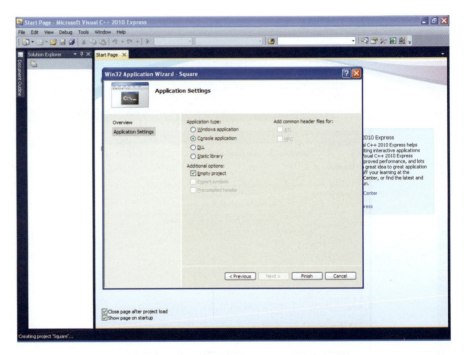

Figure B.4: Win32 application settings dialog box.

- Uncheck the Precompiled header box, check the Empty project box and choose Console application. Click Finish to see a new project window. See Figure B.5

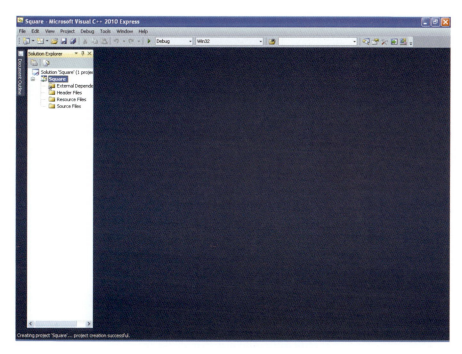

Figure B.5: New project window.

- Right click on Source Files and choose Add → New Item to bring up a dialog box. See Figure B.6.

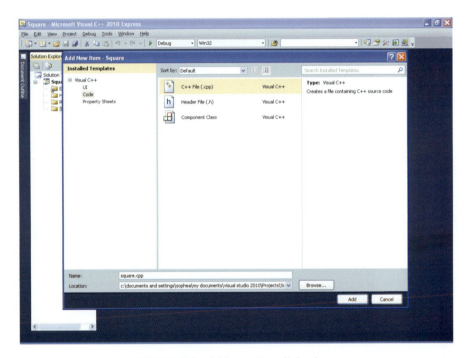

Figure B.6: Add new item dialog box.

- Select Code from the Installed Templates panel and C++ File(.cpp) from the next panel. Name your file and click Add to see an empty code panel in the project window titled with your chosen name. See Figure B.7.

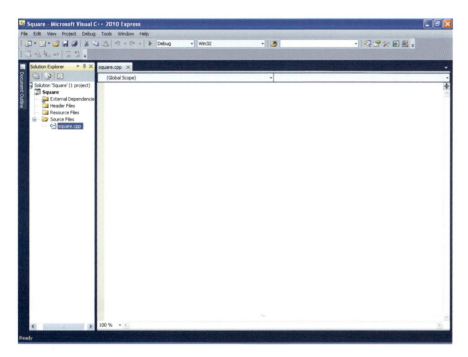

Figure B.7: Project window with empty code panel.

- Copy any of our book programs into or write your own in the code panel. See Figure B.8.

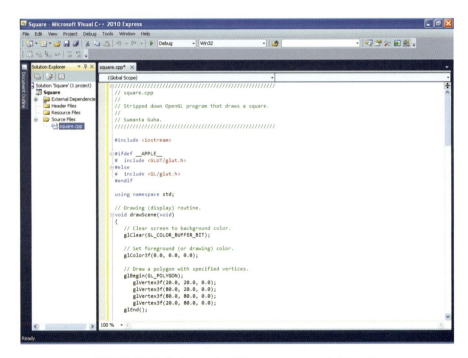

Figure B.8: Code panel with `square.cpp` copied into it.

- Save and build your project by going to Debug → Build Solution. Then execute the program with Debug → Start Debugging. If the program has been built successfully, then you should see no error in the output window.

Congratulations! Output from our `square.cpp` program is shown in Figure B.9.

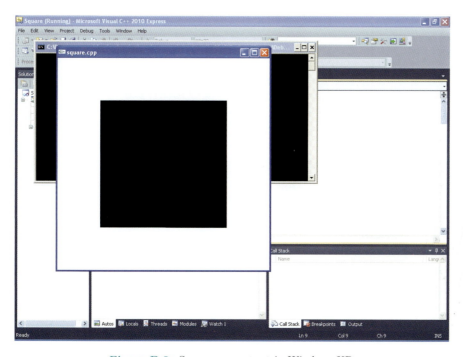

Figure B.9: Square.cpp output in Windows XP.

B.2   Mac OS X Snow Leopard

The steps to installing and running OpenGL on the Mac OS X are few because OpenGL is integrated into the operating system:

- Install GLEW by going to `http://glew.darwinports.com/` and following instructions there.

- Open a book program or write your own using your preferred editor. Let's assume the program file is `square.cpp`.

- Open the terminal window, change to the directory where the source code is located and compile the program with the command `g++ square.cpp -o square -framework Cocoa -framework OpenGL -framework GLUT`.

- Run the program by entering the command `./square`. Figure B.10 shows the output of our `square.cpp` program.

  Note: For a GLSL program the command is `g++ glslProgram.cpp -o glslProgram -framework Cocoa -framework OpenGL -framework GLUT -I/opt/local/include -L/opt/local/lib -lGLEW`

Figure B.10: Square.cpp output in Mac OS X Snow Leopard.

B.3 Ubuntu Linux

Here are the steps to setting up and running OpenGL on the Ubuntu Linux distribution:

- Go to System → Administration → Synaptic Package Manager and check the boxes to install g++, libglut3-dev and libglew-dev. Synaptic Package Manager will ask you to install dependent packages if necessary for the C++, GLUT and GLEW libraries.

- Open a book program or write your own using your preferred editor. Let's assume the program file is square.cpp.

- Open the terminal window, change to the directory where the source code is located and compile the program with the command gcc square.cpp -o square -I/usr/include -L/usr/lib -lglut -lGL -lGLU -lX11.

- Run the program by entering the command ./square. Figure B.11 shows the output of square.cpp program.

  Note: For a GLSL program the command is gcc glslProgram.cpp -o glslProgram -I/usr/include -L/usr/lib -lglut -lGLEW -lGL -lGLU -lX11.

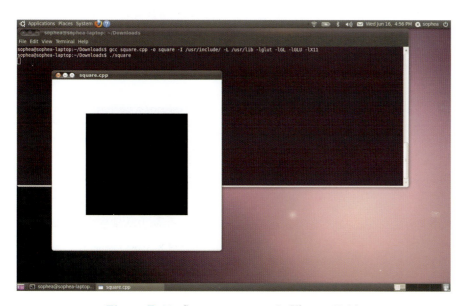

Figure B.11: Square.cpp output in Ubuntu 10.04.

APPENDIX **C**

Math Self-Test

This self-test is designed to help you assess your math readiness. The essentials in order to study computer graphics are coordinate geometry, trigonometry, linear algebra and calculus, all at elementary level.

Try to answer all 30 questions. Time is not an issue. And feel free to dust off old math books you may have stashed away in some corner, stroll over to the school or public library, or, even, peek into the internet. The principle, of course, is that each and all of these activities will be allowed throughout your career as a student and practitioner of CG (except, maybe, when you are actually in an exam). Having just the right formula or solution method pop off the top of your head is fantastic, but it's fine as well, given a problem, that you know how to *go about* solving it.

Give yourself 4 points for each correct answer (solutions follow in the next section). If you score at least 100, come on in, the water's fine.* If you're between 80 and 100 then the questions you missed tell where the rust is and, as long as you are willing to put in the extra work, you should be okay. If less than 80 then you need to sit down with yourself and be perfectly honest: is it simply rust that will come off or things that I've just never had in school but trust myself to be able to pick up or is this the kind of stuff that makes me want to curl up into a fetal position?

A word about math and CG, especially to those who did not fare well in the test. If you are motivated to study CG then picking up the math on the way isn't just possible, it can be a lot of fun. Its application to CG will bring to life stuff that caused your eyes to roll in high school. "The middle of the spacecraft is light because of the interpolated color values from the ends of the long triangle," or "I'm going to calculate a matrix to

---

*There's more math you'll learn while studying CG (some from this book itself) than is covered in the test. Doing well here simply means you're unlikely to have serious problems.

skew the evil character's head." are a lot different from "Groan, that's 12 different theorems and a chapter-load of trig formulas I have to cram for the mid-term."

If you are interested, there are several books out there dedicated to teaching the math needed for CG. A few that come to mind are Dunn [36], Lengyel [81], Mortenson [92] and Vince [141].

Use the following if you need to (some are approximations): $\sin 30° = \cos 60° = 0.5$, $\sin 45° = \cos 45° = 0.707$, $\sin 60° = \cos 30° = 0.866$, $\pi = 3.141$, $\sqrt{2} = 1.414$, $\sqrt{3} = 1.732$.

The first seven questions refer to Figure C.1.

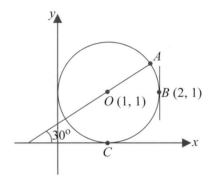

Figure C.1: Circle of unit radius.

1. What is the equation of the circle?

2. What are the coordinates of the point C?

3. What is the equation of the tangent to the circle at B?

4. What is the length of the short arc of the circle from A to B?

5. What are the coordinates of the point A?

6. If the circle is moved (without turning) so that its center lies at $(-3, -4)$, where then does the point B lie?

7. Suppose another circle is drawn with center at A and passing through O. The two circles intersect in two points. What angles do their tangents make at these points?

8. If a straight line on a plane passes through the points $(3, 1)$ and $(5, 2)$, which, if any, of the following two points does it pass through as well: $(9, 4)$ and $(12, 6)$?

9. What are the coordinates of the midpoint of the straight line segment joining the points $(3, 5)$ and $(4, 7)$?

10. At what point do the straight lines $3x + 4y - 6 = 0$ and $4x + 7y - 8 = 0$ intersect?

11. What is the equation of the straight line through the point $(3, 0)$ that is parallel to the straight line $3x - 4y - 6 = 0$?

12. What is the equation of the straight line through the point $(3, 0)$ that is perpendicular to the straight line $3x - 4y - 6 = 0$?

13. What are the coordinates of the point that is the reflection across the line $y = x$ of the point $(3, 1)$?

14. What is the length of the straight line segment on the plane joining the origin $(0, 0)$ to the point $(3, 4)$? In 3-space (xyz-space) what is the length of the straight line segment joining the points $(1, 2, 3)$ and $(4, 6, 8)$?

15. Determine the value of $\sin 75°$ using only the trigonometric values given at the top of the test (in other words, don't use your calculator to do anything other than arithmetic operations).

16. What is the dot product (or, scalar product, same thing) of the two vectors u and v in 3-space, where u starts at the origin and ends at $(\frac{1}{\sqrt{2}}, \frac{1}{\sqrt{2}}, 0)$ and v starts at the origin and ends at $(\frac{1}{\sqrt{2}}, \frac{1}{\sqrt{2}}, \sqrt{3})$, i.e., $u = \frac{1}{\sqrt{2}}\mathbf{i} + \frac{1}{\sqrt{2}}\mathbf{j}$ and $v = \frac{1}{\sqrt{2}}\mathbf{i} + \frac{1}{\sqrt{2}}\mathbf{j} + \sqrt{3}\mathbf{k}$.

Use the dot product to calculate the angle between u and v.

17. Determine a vector that is perpendicular to *both* the vectors u and v of the preceding question.

18. For the block in Figure C.2, what are the coordinates of the corner point F?

19. For the block again, what is the angle CDE?

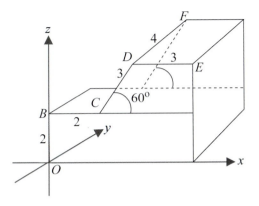

Figure C.2: Solid block (some edges are labeled with their length).

20. For the unit sphere (i.e., of radius 1) centered at the origin, depicted in Figure C.3, the equator ($0°$ latitude) is the great circle cut by the xy-plane, while $0°$ longitude is that half of the great circle cut by the xz-plane where x-values are non-negative.

What are the xyz coordinates of the point P whose latitude and longitude are both $45°$?

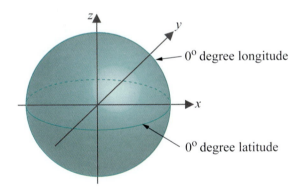

Figure C.3: Unit sphere.

21. Multiply two matrices:

$$\begin{bmatrix} 2 & 4 \\ 3 & 1 \end{bmatrix} \times \begin{bmatrix} 1 & 1 & 0 \\ 2 & 1 & -1 \end{bmatrix}$$

22. Calculate the value of the following two determinants:

$$\begin{vmatrix} 2 & 4 \\ 3 & 1 \end{vmatrix} \quad \text{and} \quad \begin{vmatrix} -1 & 2 & -3 \\ 0 & 5 & -2 \\ 0 & 3 & 3 \end{vmatrix}$$

23. Calculate the inverse of the following matrix:

$$\begin{bmatrix} 4 & 7 \\ 2 & 4 \end{bmatrix}$$

24. If the Dow Jones Industrial Average were a straight-line (or, linear, same thing) function of time (it isn't) and if its value on January 1, 2007 is 12,000 and on January 1, 2009 it's 13,500, what is the value on January 1, 2010?

25. Are the following vectors linearly independent?

$$[2\ 3\ 0]^T \quad [3\ 7\ -1]^T \quad [1\ -6\ 3]^T$$

26. Determine the linear transformation of $\mathbb{R}^3$ that maps the standard basis vectors

$$[1\ 0\ 0]^T \quad [0\ 1\ 0]^T \quad [0\ 0\ 1]^T$$

to the respective vectors

$$[-1\ -1\ 1]^T \quad [-2\ 3\ 2]^T \quad [-3\ 1\ -2]^T$$

27. What is the equation of the normal to the parabola

$$y = 2x^2 + 3$$

at the point $(2, 11)$?

28. If x and y are related by the equation

$$xy + x + y = 1$$

find a formula for $\frac{dy}{dx}$.

29. The formula for the height at time t of a projectile shot vertically upward from the ground with initial velocity u is

$$h = ut - \frac{1}{2}gt^2$$

assuming only the action of gravitational acceleration g (ignoring wind resistance and other factors).

What is the velocity of the projectile at time t? What is the maximum height reached by the projectile?

30. At what points do the curves $y = \sin x$ and $y = \cos x$ meet for values of x between 0 and 2π? What angles do they make at these points?

Math Self-Test Solutions

Score 4 points for each correct answer. For an assessment of your total, read the part before the start of the test.

Use the following if you need to (some are approximations): $\sin 30° = \cos 60° = 0.5$, $\sin 45° = \cos 45° = 0.707$, $\sin 60° = \cos 30° = 0.866$, $\pi = 3.141$, $\sqrt{2} = 1.414$, $\sqrt{3} = 1.732$.

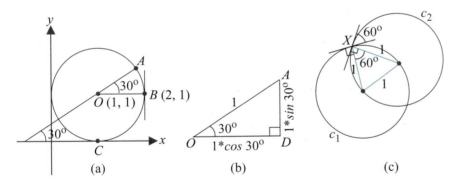

| | | |
| (a) | (b) | (c) |

Figure D.1: (a) Circle of unit radius (b) Right-angled triangle (c) Two circles of unit radius.

The first seven questions refer to Figure D.1(a).

1. What is the equation of the circle?

    Answer: Generally, the equation of a circle on the xy-plane centered at (a, b) and of radius r is

    $$(x - a)^2 + (y - b)^2 = r^2$$

    So the equation of the drawn circle is

    $$(x-1)^2 + (y-1)^2 = 1^2 \quad \text{which evaluates to} \quad x^2 + y^2 - 2x - 2y + 1 = 0$$

2. What are the coordinates of the point C?

   Answer: $(1, 0)$.

3. What is the equation of the tangent to the circle at B?

   Answer: $x = 2$.

4. What is the length of the short arc of the circle from A to B?

   Answer: The angle subtended at the center by this arc is $\angle AOB = 30°$ (Figure D.1(a)).

   Therefore, the length of the arc is $\frac{30}{360}$ of the circumference $= \frac{30}{360} * 2\pi * 1 = \frac{\pi}{6} = 3.141/6 = 0.5235$.

5. What are the coordinates of the point A?

   Answer: Suppose the horizontal line through O and the vertical line through A intersect at D (Figure D.1(b)). The hypotenuse OA of the right-angled triangle ODA is of length 1 (= radius of the circle). Therefore, the length of $OD = 1 * \cos 30° = 0.866$ and the length of $AD = 1 * \sin 30° = 0.5$.

   These lengths are the displacements of A in the x and y direction, respectively, from O. Therefore, the coordinates of A are $(1, 1) + (0.866, 0.5) = (1.866, 1.5)$.

6. If the circle is moved (without turning) so that its center lies at $(-3, -4)$, where then does the point B lie?

   Answer: Since the center is originally at $(1, 1)$ the translation that moves it to $(-3, -4)$ consists of a displacement of -4 in the x-direction and -5 in the y direction. The same translation applies to B, so B moves $(2, 1) + (-4, -5) = (-2, -4)$.

7. Suppose another circle is drawn with center at A and passing through O. The two circles intersect in two points. What angles do their tangents make at these points?

   Answer: The two circles c_1 and c_2 intersect at the points X and Y (Figure D.1(c)). The triangle OAX is equilateral as all its sides are of length 1, the radius of either circle. All its angles, therefore, are $60°$.

   Now, the tangents to c_1 and c_2 at X are perpendicular, respectively, to OX and AX. Therefore, the angle between them is the same as that between OX and AX, which is $60°$.

   Symmetrically, the tangents to the two circles at Y intersect at $60°$ as well.

8. If a straight line on a plane passes through the points $(3, 1)$ and $(5, 2)$, which, if any, of the following two points does it pass through as well: $(9, 4)$ and $(12, 6)$?

Answer: Suppose the equation of the straight line is $y = mx + c$ (any non-vertical straight line has an equation of this form, called slope-intercept form). Since it passes through $(3, 1)$ and $(5, 2)$, we have the two equations

$$1 = 3m + c$$
$$2 = 5m + c$$

Solving simultaneously, we get $m = \frac{1}{2}$ and $c = -\frac{1}{2}$, yielding the equation $y = \frac{1}{2}x - \frac{1}{2}$ for the line. Of the two points $(9, 4)$ and $(12, 6)$, only $(9, 4)$ satisfies this equation and so lies on it.

9. What are the coordinates of the midpoint of the straight line segment joining the points $(3, 5)$ and $(4, 7)$?

Answer: The coordinates of the midpoint are $(\frac{3+4}{2}, \frac{5+7}{2}) = (3.5, 6)$.

10. At what point do the straight lines $3x + 4y - 6 = 0$ and $4x + 7y - 8 = 0$ intersect?

Answer: Simultaneously solving the two equations we get the intersection as $(2, 0)$.

11. What is the equation of the straight line through the point $(3, 0)$ that is parallel to the straight line $3x - 4y - 6 = 0$?

Answer: Any line parallel to $3x - 4y - 6 = 0$ may be written as $3x - 4y - c = 0$, where c can be an arbitrary number. If such a line passes through $(3, 0)$ then this point's coordinates must satisfy $3x - 4y - c = 0$.

In other words, $3 * 3 - 4 * 0 - c = 0 \implies c = 9$.

Therefore, the required equation is $3x - 4y - 9 = 0$.

12. What is the equation of the straight line through the point $(3, 0)$ that is perpendicular to the straight line $3x - 4y - 6 = 0$?

Answer: Rewrite the given straight line's equation in slope-intercept form: $y = \frac{3}{4}x - \frac{3}{2}$. Its gradient, therefore, is $\frac{3}{4}$. Now the gradient of a straight line that is perpendicular to one of gradient m is $-\frac{1}{m}$.

Therefore, the gradient of the straight line perpendicular to the given one is $-\frac{4}{3}$ and its equation is of the form $y = -\frac{4}{3}x + c$. Since it passes through $(3, 0)$ we have $0 = -\frac{4}{3} * 3 + c \implies c = 4$.

Therefore, the required equation is $y = -\frac{4}{3}x + 4$ or $3y + 4x - 12 = 0$.

13. What are the coordinates of the point that is the reflection across the line $y = x$ of the point $(3, 1)$?

   Answer: Reflecting a point across the line $y = x$ interchanges its x and y coordinates. So the reflection of $(3, 1)$ is $(1, 3)$.

14. What is the length of the straight line segment on the plane joining the origin $(0, 0)$ to the point $(3, 4)$? In 3-space (xyz-space) what is the length of the straight line segment joining the points $(1, 2, 3)$ and $(4, 6, 8)$?

   Answer: The (Euclidean) distance between two points (x_1, y_1) and (x_2, y_2) on the plane is $\sqrt{(x_2 - x_1)^2 + (y_2 - y_1)^2}$, while that between two points (x_1, y_1, z_1) and (x_2, y_2, z_2) in 3-space is (similarly) $\sqrt{(x_2 - x_1)^2 + (y_2 - y_1)^2 + (z_2 - z_1)^2}$.

   Therefore, the length of the straight line segment joining $(0, 0)$ and $(3, 4)$ is $\sqrt{(3 - 0)^2 + (4 - 0)^2} = \sqrt{25} = 5$ and that joining $(1, 2, 3)$ and $(4, 6, 8)$ is

   $$\sqrt{(4 - 1)^2 + (6 - 2)^2 + (8 - 3)^2}$$
   $$= \sqrt{50} = \sqrt{25 * 2} = \sqrt{25} * \sqrt{2} = 5 * 1.414 = 7.07$$

   using only the value of $\sqrt{2}$ given above.

15. Determine the value of $\sin 75°$ using only the trigonometric values given at the top of the test (in other words, don't use your calculator to do anything other than arithmetic operations).

   Answer: Use the formula

   $$\sin(A + B) = \sin A \cos B + \cos A \sin B$$

   to write

   $$\begin{aligned} \sin 75° &= \sin(45° + 30°) \\ &= \sin 45° \cos 30° + \cos 45° \sin 30° \\ &= 0.707 * 0.866 + 0.707 * 0.5 \\ &= 0.966 \end{aligned}$$

16. What is the dot product (or, scalar product, same thing) of the two vectors u and v in 3-space, where u starts at the origin and ends at $(\frac{1}{\sqrt{2}}, \frac{1}{\sqrt{2}}, 0)$ and v starts at the origin and ends at $(\frac{1}{\sqrt{2}}, \frac{1}{\sqrt{2}}, \sqrt{3})$, i.e., $u = \frac{1}{\sqrt{2}}\mathbf{i} + \frac{1}{\sqrt{2}}\mathbf{j}$ and $v = \frac{1}{\sqrt{2}}\mathbf{i} + \frac{1}{\sqrt{2}}\mathbf{j} + \sqrt{3}\mathbf{k}$.

   Use the dot product to calculate the angle between u and v.

   Answer:

$u \cdot v = (\frac{1}{\sqrt{2}}, \frac{1}{\sqrt{2}}, 0) \cdot (\frac{1}{\sqrt{2}}, \frac{1}{\sqrt{2}}, \sqrt{3}) = \frac{1}{\sqrt{2}} * \frac{1}{\sqrt{2}} + \frac{1}{\sqrt{2}} * \frac{1}{\sqrt{2}} + 0 * \frac{1}{\sqrt{3}} = 1$

Moreover,

$|u| = \sqrt{(\frac{1}{\sqrt{2}})^2 + (\frac{1}{\sqrt{2}})^2 + 0^2} = 1$ and $|v| = \sqrt{(\frac{1}{\sqrt{2}})^2 + (\frac{1}{\sqrt{2}})^2 + (\sqrt{3})^2} = 2.$

Now, $u \cdot v = |u||v| \cos \theta$, where θ is the angle between u and v. Therefore, $\cos \theta = \frac{u \cdot v}{|u||v|} = \frac{1}{1*2} = \frac{1}{2}$, which means $\theta = 60°$.

17. Determine a vector that is perpendicular to *both* the vectors u and v of the preceding question.

Answer: The cross-product of two (non-zero and non-collinear) vectors is perpendicular to both of them. Now, $u \times v = (\frac{1}{\sqrt{2}}\mathbf{i} + \frac{1}{\sqrt{2}}\mathbf{j}) \times (\frac{1}{\sqrt{2}}\mathbf{i} + \frac{1}{\sqrt{2}}\mathbf{j} + \sqrt{3}\mathbf{k}) =$

$$\begin{vmatrix} \mathbf{i} & \frac{1}{\sqrt{2}} & \frac{1}{\sqrt{2}} \\ \mathbf{j} & \frac{1}{\sqrt{2}} & \frac{1}{\sqrt{2}} \\ \mathbf{k} & 0 & \sqrt{3} \end{vmatrix} = \frac{\sqrt{3}}{\sqrt{2}}\mathbf{i} - \frac{\sqrt{3}}{\sqrt{2}}\mathbf{j}$$

Therefore, the vector that starts at the origin and ends at $(\frac{\sqrt{3}}{\sqrt{2}}, -\frac{\sqrt{3}}{\sqrt{2}}, 0)$ is perpendicular to both u and v (the answer is not unique).

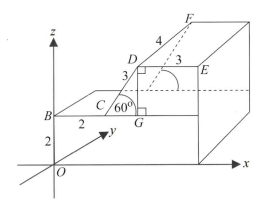

Figure D.2: Solid block (some edges are labeled with their length).

18. For the block in Figure D.2, what are the coordinates of the corner point F?

Answer: Drop the perpendicular DG from D to the straight line through B and C (Figure D.2).

The x-coordinate of F is $|BC| + |CG| = 2 + 3\cos 60° = 2 + 3 * 0.5 = 3.5$.

The y-coordinate of F is $|DF| = 4$.

The z-coordinate of F is $|AB| + |GD| = 2 + 3\sin 60° = 2 + 3*0.866 = 4.598$.

Therefore, $F = (3.5, 4, 4.598)$.

19. For the block again, what is the angle CDE?

    Answer: $\angle CDE = \angle CDG + \angle GDE = 30° + 90° = 120°$.

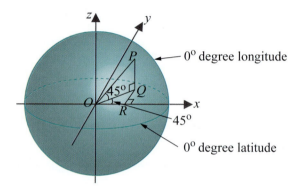

Figure D.3: Unit sphere.

20. For the unit sphere (i.e., of radius 1) centered at the origin, depicted in Figure D.3, the equator ($0°$ latitude) is the great circle cut by the xy-plane, while $0°$ longitude is that half of the great circle cut by the xz-plane where x-values are non-negative.

    What are the xyz coordinates of the point P whose latitude and longitude are both $45°$?

    Answer: Drop the perpendicular from P to Q on the xy-plane and then the perpendicular from Q to R on the x-axis (Figure D.3).

    Now, $|OP| = 1$, so that $|PQ| = 1 * \sin 45° = \frac{1}{\sqrt{2}}$ and $|OQ| = 1 * \cos 45° = \frac{1}{\sqrt{2}}$.

    Moreover, $QR = |OQ|\sin 45° = \frac{1}{\sqrt{2}} * \frac{1}{\sqrt{2}} = \frac{1}{2}$ and $OR = |OQ|\cos 45° = \frac{1}{\sqrt{2}} * \frac{1}{\sqrt{2}} = \frac{1}{2}$.

    Now, the x, y and z coordinates of P are $|OR|$, $|QR|$ and $|PQ|$, respectively. Therefore, $P = (\frac{1}{2}, \frac{1}{2}, \frac{1}{\sqrt{2}})$.

21. Multiply two matrices:

$$\begin{bmatrix} 2 & 4 \\ 3 & 1 \end{bmatrix} \times \begin{bmatrix} 1 & 1 & 0 \\ 2 & 1 & -1 \end{bmatrix}$$

Answer:

$$\begin{bmatrix} 2 & 4 \\ 3 & 1 \end{bmatrix} * \begin{bmatrix} 1 & 1 & 0 \\ 2 & 1 & -1 \end{bmatrix} = \begin{bmatrix} 10 & 6 & -4 \\ 5 & 4 & -1 \end{bmatrix}.$$

22. Calculate the value of the following two determinants:

$$\begin{vmatrix} 2 & 4 \\ 3 & 1 \end{vmatrix} \quad \text{and} \quad \begin{vmatrix} -1 & 2 & -3 \\ 0 & 5 & -2 \\ 0 & 3 & 3 \end{vmatrix}$$

Answer:

$$\begin{vmatrix} 2 & 4 \\ 3 & 1 \end{vmatrix} = 2*1 - 4*3 = -10$$

$$\begin{vmatrix} -1 & 2 & -3 \\ 0 & 5 & -2 \\ 0 & 3 & 3 \end{vmatrix} = -1 * \begin{vmatrix} 5 & -2 \\ 3 & 3 \end{vmatrix} - 0 * \begin{vmatrix} 2 & -3 \\ 3 & 3 \end{vmatrix} + 0 * \begin{vmatrix} 2 & -3 \\ 5 & -2 \end{vmatrix}$$

$$= -(5*3 - (-2)*3) = -21$$

23. Calculate the inverse of the following matrix:

$$\begin{bmatrix} 4 & 7 \\ 2 & 4 \end{bmatrix}$$

Answer: To obtain the inverse we have to replace each element by its cofactor, take the transpose and, finally, divide by the determinant of the original matrix.

Replacing each element by its cofactor we get the matrix

$$\begin{bmatrix} 4 & -2 \\ -7 & 4 \end{bmatrix}$$

Taking the transpose next gives

$$\begin{bmatrix} 4 & -7 \\ -2 & 4 \end{bmatrix}$$

Finally, dividing by the determinant $4*4 - 2*7 = 2$ of the original matrix, we have its inverse

$$\begin{bmatrix} 2 & -3.5 \\ -1 & 2 \end{bmatrix}$$

24. If the Dow Jones Industrial Average were a straight-line (or, linear, same thing) function of time (it isn't) and if its value on January 1, 2007 is 12,000 and on January 1, 2009 it's 13,500, what is the value on January 1, 2010?

    Answer: As a linear function of time then the DJIA grows 1500 points in two years, or 750 per year, which takes it to 14,250 on January 1, 2010.

25. Are the following vectors linearly independent?

$$[2\ 3\ 0]^T \qquad [3\ 7\ -1]^T \qquad [1\ -6\ 3]^T$$

    Answer: The vectors are linearly independent if the only solution to the equation

$$c_1[2\ 3\ 0]^T + c_2[3\ 7\ -1]^T + c_3[1\ -6\ 3]^T = [0\ 0\ 0]^T$$

    is $c_1 = c_2 = c_3 = 0$.

    The vector equation above is equivalent to the following set of three simultaneous equations – one from each position in the vectors – in c_1, c_2 and c_3:

$$\begin{aligned} 2c_1 + 3c_2 + c_3 &= 0 \\ 3c_1 + 7c_2 - 6c_3 &= 0 \\ -c_2 + 3c_3 &= 0 \end{aligned}$$

    Solving we find solutions not all 0, e.g., $c_1 = 5$, $c_2 = -3$ and $c_3 = -1$, proving that the given set of vectors is not linearly independent.

26. Determine the linear transformation of $\mathbb{R}^3$ that maps the standard basis vectors

$$[1\ 0\ 0]^T \qquad [0\ 1\ 0]^T \qquad [0\ 0\ 1]^T$$

    to the respective vectors

$$[-1\ -1\ 1]^T \qquad [-2\ 3\ 2]^T \qquad [-3\ 1\ -2]^T$$

    Answer: The required linear transformation is defined by the matrix whose columns are, respectively, the images of the successive basis vectors. In particular, then, its matrix is

$$\begin{bmatrix} -1 & -2 & -3 \\ -1 & 3 & 1 \\ 1 & 2 & -2 \end{bmatrix}$$

27. What is the equation of the normal to the parabola

$$y = 2x^2 + 3$$

at the point $(2, 11)$?

Answer: $\frac{dy}{dx} = 4x$, so at the point $(2, 11)$ the gradient of the tangent is 4*2=8. The gradient of the normal, therefore, is $-\frac{1}{8}$. Its equation, accordingly, is

$$\frac{y - 11}{x - 2} = -\frac{1}{8} \quad \text{or} \quad x + 8y - 90 = 0$$

28. If x and y are related by the equation

$$xy + x + y = 1$$

find a formula for $\frac{dy}{dx}$.

Answer: Differentiating the equation throughout with respect to x:

$$(\frac{d}{dx}x)y + x(\frac{d}{dx}y) + \frac{d}{dx}x + \frac{d}{dx}y = \frac{d}{dx}1$$

which simplifies to

$$y + x\frac{dy}{dx} + 1 + \frac{dy}{dx} = 0$$

giving

$$\frac{dy}{dx} = -\frac{y + 1}{x + 1}$$

29. The formula for the height at time t of a projectile shot vertically upward from the ground with initial velocity u is

$$h = ut - \frac{1}{2}gt^2$$

assuming only the action of gravitational acceleration g (ignoring wind resistance and other factors).

What is the velocity of the projectile at time t? What is the maximum height reached by the projectile?

Answer: Its velocity at time t is

$$\frac{dh}{dt} = u - gt$$

When the projectile attains maximum height, its velocity is 0. Therefore, $u - gt = 0$, implying that $t = \frac{u}{g}$. Therefore, the maximum height reached by the projectile is obtained by substituting $t = \frac{u}{g}$ in the formula for its height, which gives the maximum height as

$$u\frac{u}{g} - \frac{1}{2}g\left(\frac{u}{g}\right)^2 = \frac{u^2}{g} - \frac{u^2}{2g} = \frac{u^2}{2g}$$

30. At what points do the curves $y = \sin x$ and $y = \cos x$ meet for values of x between 0 and 2π? What angles do they make at these points?

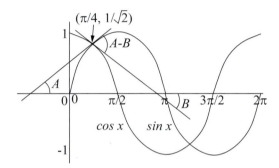

Figure D.4: Graphs of $\sin x$ and $\cos x$ (*sketch, not exact*).

Answer: When the curves (see Figure D.4) meet, their y-values are equal, so $\sin x = \cos x$, giving $\tan x = \frac{\sin x}{\cos x} = 1$. The two values in the interval $[0, 2\pi]$ where $\tan x = 1$ are $\frac{\pi}{4}$ and $\frac{5\pi}{4}$. Therefore, the two points at which the curves meet are $\left(\frac{\pi}{4}, \frac{1}{\sqrt{2}}\right)$ and $\left(\frac{5\pi}{4}, -\frac{1}{\sqrt{2}}\right)$ ($\cos \frac{\pi}{4} = \sin \frac{\pi}{4} = \frac{1}{\sqrt{2}}$ and $\cos \frac{5\pi}{4} = \sin \frac{5\pi}{4} = -\frac{1}{\sqrt{2}}$).

For the first curve $\frac{dy}{dx} = \cos x$, while for the second $\frac{dy}{dx} = -\sin x$. At the point of intersection $\left(\frac{\pi}{4}, \frac{1}{\sqrt{2}}\right)$, therefore, the gradients of the two curves are $\cos \frac{\pi}{4} = \frac{1}{\sqrt{2}}$ and $-\sin \frac{\pi}{4} = -\frac{1}{\sqrt{2}}$, respectively.

If the tangent lines at $\left(\frac{\pi}{4}, \frac{1}{\sqrt{2}}\right)$ make angles A and B, respectively, with the x-axis, then the gradients are precisely the tan of these angles. Therefore, $\tan A = \frac{1}{\sqrt{2}}$ and $\tan B = -\frac{1}{\sqrt{2}}$ (B is a negative angle). The angle between the curves – which by definition is the angle between their tangents – at $\left(\frac{\pi}{4}, \frac{1}{\sqrt{2}}\right)$ is $A - B$. Now,

$$\tan(A - B) = \frac{\tan A - \tan B}{1 + \tan A \tan B} = \frac{\frac{1}{\sqrt{2}} + \frac{1}{\sqrt{2}}}{1 - \frac{1}{\sqrt{2}}\frac{1}{\sqrt{2}}} = 2\sqrt{2}$$

which means $A - B = \tan^{-1} 2\sqrt{2} = \tan^{-1} 2.828 = 1.231$ radians or $70.526°$ (approximately). The angle between the curves at the other point of intersection is the same by symmetry.

Bibliography

[1] T. Akenine-Möller, E. Haines, N. Hoffman, *Real-Time Rendering*, 3rd Edn., A K Peters, 2008.

[2] E. Angel, *Interactive Computer Graphics: A Top-Down Approach Using OpenGL*, 5th Edn., Addison Wesley, 2008.

[3] E. Angel, *OpenGL; A Primer*, 3rd Edn., Addison Wesley, 2007.

[4] A. Appel, Some techniques for the machine rendering of solids, *Proceedings of the Spring Joint Computer Conference*, 1968, 37-45.

[5] F. Ayres, E. Mendelson, *Schaum's Outlines: Calculus*, 5th Edn., McGraw-Hill, 2008.

[6] R. Baer, *Linear Algebra and Projective Geometry*, Dover Publications, 2005.

[7] T. Banchoff, J. Wermer, *Linear Algebra through Geometry*, 2nd Edn., Springer-Verlag, 1993.

[8] M. F. Barnsley, *Fractals Everywhere*, 2nd Edn., Morgan Kaufmann, 2000.

[9] R. H. Bartels, J. C. Beatty, B. A. Barsky, *An Introduction to Splines for Use in Computer Graphics and Geometric Modeling*, Morgan Kaufmann, 1995.

[10] M. de Berg, M. van Kreveld, M. Overmars, O. Schwarzkopf, *Computational Geometry: Algorithms and Applications*, Springer-Verlag, 2009.

[11] R. S. Berns, *Billmeyer and Saltzman's Principles of Color Technology*, 3rd Edn., Wiley-Interscience, 2000.

[12] P. E. Bézier, How Renault uses numerical control for car body design and tooling, *Society of Automotive Engineers' Congress*, SAE paper 680010, Detroit, 1968.

[13] P. E. Bézier, Mathematical and practical possibilities of UNISURF, in *Computer Aided Geometric Design: Proceedings of a Conference Held at the University of Utah*, R. Barnhill and R. Riesenfeld, editors, Academic Press, 1974, 127-152.

[14] J. F. Blinn, Models of light reflection for computer synthesized pictures, *Computer Graphics (Proceedings SIGGRAPH 1977)* **11** (1977), 192-198.

[15] J. F. Blinn, Simulation of wrinkled surfaces, *Computer Graphics (Proceedings SIGGRAPH 1978)* **12** (1978), 286-292.

[16] J. F. Blinn, *Jim Blinn's Corner: A Trip Down the Graphics Pipeline*, Morgan Kaufmann, 1996.

[17] J. F. Blinn, *Jim Blinn's Corner: Dirty Pixels*, Morgan Kaufmann, 1998.

[18] J. Blinn, M. Newell, Texture and reflection in computer generated images, *Communications of the ACM*, **19** (1976), 456-547.

[19] D. Bourg, *Physics for Game Developers*, O'Reilly, 2001.

[20] J. E. Bresenham, Algorithm for computer control of digital plotter, *IBM Systems Journal* **4** (1965), 25-30.

[21] S. R. Buss, *3-D Computer Graphics: A Mathematical Introduction with OpenGL*, Cambridge University Press, 2003.

[22] E. Catmull, R. Rom, A class of local interpolating splines, in *Computer Aided Geometric Design: Proceedings of a Conference Held at the University of Utah*, R. Barnhill and R. Riesenfeld, editors, Academic Press, 1974, 317-326.

[23] C. Chuon, S. Guha, Volume Cost Based Mesh Simplification, *Proceedings 6th International Conference on Computer Graphics, Imaging and Visualization (CGIV 09)* (2009), 164-169.

[24] M. F. Cohen, D. P. Greenberg, The hemi-cube: a radiosity solution for complex environments, *Computer Graphics (Proceedings SIGGRAPH 1985)* **19** (1985), 31-40.

[25] M. F. Cohen, J. R. Wallace, *Radiosity and Realistic Image Synthesis*, Morgan Kaufmann, 1993.

[26] Comparison of OpenGL and Direct3D (in Wikipedia), `http://en.wikipedia.org/wiki/Direct3D\_vs.\_OpenGL`.

[27] Computational Geometry Algorithms Library, `http://www.cgal.org`.

[28] R. L. Cook, K. E. Torrance, A reflectance model for computer graphics, *ACM Transaction on Graphics* **1** (1982), 7-24.

[29] M. G. Cox, The numerical evaluation of B-splines, *Journal of the Institute of Mathematics and its Applications* **10** (1972), 134-149.

[30] H. S. M. Coxeter, *Projective Geometry*, Springer, 2nd Edn., 2003.

[31] F. C. Crow, The origins of the teapot, *IEEE Computer Graphics and Applications* **7** (1987), 8-19.

[32] C. de Boor, On calculating with B-splines, *Journal of Approximation Theory* **6** (1972), 50-62.

[33] P. de Casteljau, Outillages méthodes calcul, Technical report, A. Citroen, Paris, 1959.

[34] P. de Casteljau, Courbes et surfaces à poles, Technical report, A. Citroen, Paris, 1963.

[35] M. P. Do Carmo, *Differential Geometry of Curves and Surfaces*, Prentice Hall, 1976.

[36] F. Dunn, I. Parberry, *3D Math Primer for Graphics and Game Development*, Jones and Bartlett Publishers, 2002.

[37] D. H. Eberly, *3D Game Engine Design: A Practical Approach to Real-Time Computer Graphics*, 2nd Edn., Morgan Kaufmann, 2006.

[38] D. H. Eberly, *Game Physics*, 2nd Edn., Morgan Kaufmann, 2010.

[39] H. Edelsbrunner, *Geometry and Topology for Mesh Generation*, Cambridge University Press, 2006.

[40] C. Ericson, *Real-Time Collision Detection*, Morgan Kaufmann, 2005.

[41] K. Falconer, *Fractal Geometry: Mathematical Foundations and Applications*, 2nd Edn., John Wiley & Sons, 2003.

[42] G. Farin, *Curves and Surfaces for CAGD: A Practical Guide*, 5th Edn., Morgan Kaufmann, 2001.

[43] G. Farin, *NURBS: from Projective Geometry to Practical Use*, 2nd Edn., A K Peters, 1999.

[44] J. D. Foley, A. van Dam, S. K. Feiner, J. F. Hughes, *Computer Graphics: Principles and Practice*, 2nd Edn., Addison Wesley, 1995.

[45] J. D. Foley, A. van Dam, S. K. Feiner, J. F. Hughes, R. L. Phillips, *Introduction to Computer Graphics*, Addison Wesley, 1993.

[46] S. H. Friedberg, A. J. Insel, L. E. Spence, *Linear Algebra*, 4th Edn., Prentice Hall, 2002.

[47] H. Fuchs, Z. M. Kedem, B. F. Naylor, On visible surface generation by a priori tree structures, *Computer Graphics (Proceedings SIGGRAPH 1980)* **14** (1980), 124-133.

[48] D. C. Giancoli, *Physics for Scientists and Engineers with Modern Physics*, 4th Edn., Prentice Hall, 2008.

[49] GIMP, `http://www.gimp.org`.

[50] A. S. Glassner, *Andrew Glassner's Notebook: Recreational Computer Graphics*, Morgan Kaufmann, 1999.

[51] A. S. Glassner, *Andrew Glassner's Other Notebook: Further Recreations in Computer Graphics*, A K Peters, 2002.

[52] GLUI, `http://www.cs.unc.edu/~rademach/glui`.

[53] GLUT, `http://www.opengl.org/resources/libraries/glut.html`.

[54] C. M. Goral, K. E. Torrance, D. P. Greenberg, B. Battaile, Modeling the interaction of light between diffuse surfaces, *Computer Graphics (Proceedings SIGGRAPH 1984)* **18** (1984), 213-222.

[55] H. Gouraud, Continuous shading of curved surfaces, *IEEE Transactions on Computers* **20** (1971), 623-629.

[56] S. Govil-Pai, *Principles of Computer Graphics: Theory and Practice Using OpenGL and Maya*, Springer, 2005.

[57] W. H. Greub, *Linear Algebra*, 4th Edn., Springer, 2004.

[58] S. Guha, Joint separation of geometric clusters and the extreme irregularities of regular polyhedra, *International Journal of Computational Geometry and Applications* **15** (2005), 491-510.

[59] A. J. Hanson, *Visualizing Quaternions*, Morgan Kaufmann, 2006.

[60] P. Haeberli, K. Akeley, The accumulation buffer: hardware support for high-quality rendering, *Computer Graphics (Proceedings 17th Annual Conference on Computer Graphics and Interactive Techniques)* **24** (1990), 309-318.

[61] P. Haeberli, M. Segal, Texture mapping as a fundamental drawing primitive, *Proceedings Fourth Eurographics Workshop on Rendering* (1993), 259-266 (on-line version at `http://www.sgi.com/misc/grafica/texmap`).

[62] L. A. Hageman, D. M. Young, *Applied Iterative Methods*, Dover Publications, 2004.

[63] X. D. He, K. E. Torrance, F. X. Sillion, D. P. Greenberg, A comprehensive physical model for light reflection, *Computer Graphics (Proceedings SIGGRAPH 1991)* **25** (1991), 175-186.

[64] X. D. He, P. O. Heynen, R. L. Phillips, K. E. Torrance, D. H. Salesin, D. P. Greenberg, A fast and accurate light reflection model, *Computer Graphics (Proceedings SIGGRAPH 1992)* **26** (1992), 253-254.

[65] D. Hearn, M. P. Baker, *Computer Graphics with OpenGL*, 3rd Edn., Prentice Hall, 2004.

[66] P. S. Heckbert, Survey of texture mapping, *IEEE Computer Graphics and Applications* **6** (1986), 56-67.

[67] M. Henle, *Modern Geometries: Non-Euclidean, Projective, and Discrete Geometry*, 2nd Edn., Prentice Hall, 2001.

[68] F. S. Hill, Jr., S. M. Kelley, *Computer Graphics Using OpenGL*, 3rd Edn., Prentice Hall, 2006.

[69] K. M. Hoffman, R. Kunze, *Linear Algebra*, 2nd Edn., Prentice Hall, 1971.

[70] International Meshing Roundtable, `http://www.imr.sandia.gov`.

[71] R. Jackson, L. MacDonald, K. Freeman, *Computer Generated Colour: A Practical Guide to Presentation and Display*, John Wiley & Sons, 1994.

[72] G. A. Jennings, *Modern Geometry with Applications*, Springer, 1997.

[73] L. Kadison, M. T. Kromann, *Projective Geometry and Modern Algebra*, Birkhuser Boston, 1996.

[74] M. J. Kilgard, Improving shadows and reflections via the stencil buffer, `http://developer.nvidia.com/object/Stencil\_Buffer\_Tutorial.html`.

[75] Khronos Group OpenGL ES, `http://www.opengles.org`.

[76] B. Kolman, D. R. Hill, *Introductory Linear Algebra: An Applied First Course*, 8th Edn., Prentice Hall, 2004.

[77] E. Kreyszig, *Differential Geometry*, Dover Publications, 1991.

[78] J. B. Kuipers, *Quaternions and Rotation Sequences: A Primer with Applications to Orbits, Aerospace and Virtual Reality*, Princeton University Press, 2002.

[79] A. Kumar, V. Kwatra, B. Singh, S. Kapoor, Dynamic Binary Space Partitioning for Hidden Surface Removal, *Proceedings Indian Conference on Computer Vision, Graphics and Image Processing (ICVGIP 1998)*, 1998.

[80] D. C. Lay, *Linear Algebra and Its Applications*, 3rd Edn., Addison Wesley, 2005.

[81] E. Lengyel, *Mathematics for 3D Game Programming & Computer Graphics*, 2nd Edn., Charles River Media, 2004.

[82] Y. Liang, B. Barsky, A new concept and method for line clipping, *ACM Transactions on Graphics* **3** (1984), 1-22.

[83] Lighthouse 3D, `http://www.lighthouse3d.com/opengl`.

[84] M. M. Lipschutz, *Schaum's Outlines: Differential Geometry*, McGraw-Hill, 1969.

[85] D. Luebke, M. Reddy, J. D. Cohen, A. Varshney, B. Watson, R. Huebner, *Level of Detail for 3D Graphics*, Morgan Kaufmann, 2002.

[86] F. D. Luna, *Introduction to 3D Game Programming with DirectX 10*, Jones and Bartlett Publishers, 2008.

[87] B. Mandelbrot, *The Fractal Geometry of Nature*, W. H. Freeman, 1983.

[88] T. McReynolds, D. Blythe, *Advanced Graphics Programming Using OpenGL*, Morgan Kaufmann, 2005.

[89] Mesa 3D, `http://www.mesa3d.org`.

[90] M. E. Mortenson, *Geometric Modeling*, 3rd Edn., Industrial Press, 2006.

[91] M. E. Mortenson, *Geometric Transformations for 3D Modeling*, 2nd Edn., Industrial Press, 2007.

[92] M. E. Mortenson, *Mathematics for Computer Graphics Applications*, 2nd Edn., Industrial Press, 1999.

[93] T. K. Mukherjee, *private communication*, 2008.

[94] J. Munkres, *Elements of Algebraic Topology*, Westview Press, 1996.

[95] J. Munkres, *Topology*, 2nd Edn., Prentice Hall, 2000.

[96] Nate Robins, `http://www.xmission.com/~nate/tutors.html`.

[97] S. K. Nayar, M. Oren, Generalization of the Lambertian model and implications for machine vision, *International Journal of Computer Vision* **14** (1995), 227-251.

[98] NeHe Productions, `http://nehe.gamedev.net`.

[99] OpenGL, `http://www.opengl.org`.

[100] OpenGL Architecture Review Board, *OpenGL Programming Guide*, 4th Edn., Addison Wesley, 2004.

[101] OpenGL Architecture Review Board, *OpenGL Reference Manual*, 4th Edn., Addison Wesley, 2004.

[102] B. O'Neill, *Elementary Differential Geometry*, 2nd Edn., Academic Press, 2006.

[103] J. O'Rourke, *Computational Geometry in C*, Cambridge University Press, 2001.

[104] D. Pedoe, *Geometry: A Comprehensive Course*, Dover Publications, 1988.

[105] B. T. Phong, Illumination for computer generated pictures, *Communications of the ACM* **18** (1975), 311-317.

[106] Physics in Graphics, `http://physicsingraphics.endofinternet.org`.

[107] L. Piegl, W. Tiller, *The NURBS Book*, 2nd Edn., Springer, 1996.

[108] P. Poulin, A. Fournier, A model for anisotropic reflection, *Computer Graphics (Proceedings SIGGRAPH 1990)* **24** (1990), 273-282.

[109] POV-Ray, `http://www.povray.org`.

[110] A. Pressley, *Elementary Differential Geometry*, 2nd Edn., Springer, 2010.

[111] D. F. Rogers, *An Introduction to NURBS: With Historical Perspective*, Morgan Kaufmann, 2000.

[112] D. F. Rogers, *Procedural Elements for Computer Graphics*, 2nd Edn., McGraw-Hill, 1997.

[113] D. F. Rogers, J. A. Adams, *Mathematical Elements for Computer Graphics*, 2nd Edn., McGraw-Hill, 1989.

[114] S. Roman, *Advanced Linear Algebra*, 3rd Edn., Springer, 2007.

[115] R. J. Rost, B. Licea-Kane, *OpenGL Shading Language*, 3rd Edn., Addison Wesley, 2009.

[116] H. Samet, *Foundations of Multidimensional and Metric Data Structures*, Morgan Kaufmann, 2006.

[117] P. Samuel, *Projective Geometry*, Springer-Verlag, 1988.

[118] H. M. Schey, *Div, Grad, Curl, and All That: An Informal Text on Vector Calculus*, 4th Edn., W. W. Norton & Co., 2005.

[119] H. Schildt, *STL Programming from the Ground Up*, Osborne/McGraw-Hill, 1998.

[120] C. Schlick, An inexpensive BRDF model for physically-based rendering, *Computer Graphics Forum* **13** (1994), 233-246.

[121] I. Schoenberg, Contributions to the problem of approximation of equidistant data by analytic functions, *Quarterly of Applied Mathematics* **4** (1946), 45-99.

[122] I. Schoenberg, On spline functions, in *Inequalities*, O. Sisha, editor, Academic Press, 1967, 249-274.

[123] M. Segal, K. Akeley, The design of the OpenGL graphics interface, 1994 (on-line version at `http://www.sun.com/software/graphics/opengl/OpenGLdesign.pdf`).

[124] P. Shirley, S. Marschner, *Fundamentals of Computer Graphics*, 3rd Edn., A K Peters, 2000.

[125] ACM SIGGRAPH, `http://www.siggraph.org`.

[126] F. X. Sillion, C. Puech, *Radiosity and Global Illumination*, Morgan Kaufmann, 1994.

[127] G. F. Simmons, *Calculus With Analytic Geometry*, 2nd Edn., McGraw-Hill, 1996.

[128] I. M. Singer, J. A. Thorpe, *Lecture Notes on Elementary Topology and Geometry*, Springer-Verlag, 1976.

[129] M. Slater, A. Steed, Y. Chrysanthou, *Computer Graphics and Virtual Environments: From Realism to Real-Time*, Addison Wesley, 2001.

[130] M. R. Spiegel, *Schaum's Outlines: Vector Analysis and an Introduction to Tensor Analysis*, McGraw-Hill, 1968.

[131] Stewart Calculus, `http://www.stewartcalculus.com`.

[132] G. Strang, *Introduction to Linear Algebra*, 4th Edn., Wellesley-Cambridge Press, 2003.

[133] I. E. Sutherland, *Sketchpad: A Man-Machine Graphical Communication System*, MIT Thesis, 1963.

[134] I. E. Sutherland, G. W. Hodgeman, Reentrant polygon clipping, *Communications of the ACM* **17** (1974), 32-42.

[135] A. Thorn, *DirectX 9 Graphics: The Definitive Guide to Direct3D*, Jones and Bartlett Publishers, 2005.

[136] Trolltech, `http://www.trolltech.com`.

[137] UNC GAMMA Research Group, `http://www.cs.unc.edu/~geom`.

[138] G. van den Bergen, *Collision Detection in Interactive 3D Environments*, Morgan Kaufmann, 2003.

[139] J. M. van Verth, L. M. Bishop, *Essential Mathematics for Games and Interactive Applications: A Programmer's Guide*, Morgan Kaufmann, 2004.

[140] W. T. Vetterling, W. H. Press, S. A. Teukolsky, B. P. Flannery, *Numerical Recipes in C++: The Art of Scientific Computing*, 2nd Edn., Cambridge University Press, 2002.

[141] J. Vince, *Mathematics for Computer Graphics*, 3rd Edn., Springer, 2010.

[142] A. Watt, *3D Computer Graphics*, 3rd Edn., Addison Wesley, 1999.

[143] J. T. Whitted, An improved illumination model for shaded display, *Communications of the ACM* **23** (1980), 343-349.

[144] L. Williams, Pyramidal parametrics, *Computer Graphics (Proceedings SIGGRAPH 1983)* **17** (1983), 1-11.

BIBLIOGRAPHY

[145] R. C. Wrede, M. Spiegel, *Schaum's Outlines: Advanced Calculus*, 3rd Edn., McGraw-Hill, 2010.

[146] R. S. Wright, Jr., B. Lipchak, N. Haemel, *OpenGL Superbible*, 4th Edn., Addison-Wesley, 2007.

[147] G. Wyszecki, W. S. Stiles, *Color Science: Concepts and Methods, Quantitative Data and Formulae*, 2nd Edn., Wiley-Interscience, 2000.

[148] Z. Xiang, R. Plastock, *Schaum's Outlines: Computer Graphics*, 2nd Edn., McGraw-Hill, 2000.

[149] I. M. Yaglom, *Geometric Transformations I*, Mathematical Association of America, 1962.

[150] I. M. Yaglom, *Geometric Transformations II*, Mathematical Association of America, 1968.

[151] I. M. Yaglom, *Geometric Transformations III*, Mathematical Association of America, 1973.

Subject Index

world
    coordinate system, 30, 32
    space, 30

X
XYZ color model, *see* CIE

Y
yaw, 248

Z
z-buffer, *see* buffer, depth
zoom, 157, 478

Command Index

Program Index (page of first experiment)